The commentary text in this handbook is written to assist users in understanding and applying the provisions of NFPA 472 and NFPA 473. The commentary explains the reasoning behind the Standard's requirements and provides numerous tables, photographs, and illustrations. The commentary text is printed in brown to easily distinguish it from the text of NFPA 472 and NFPA 473. Note that the commentary is not part of NFPA 472 and NFPA 473 and therefore is not enforceable.

EX, photograph shows exposed patients at a hazardous materials incident. (Courtesy of Fairfax County Fire and Rescue Department)

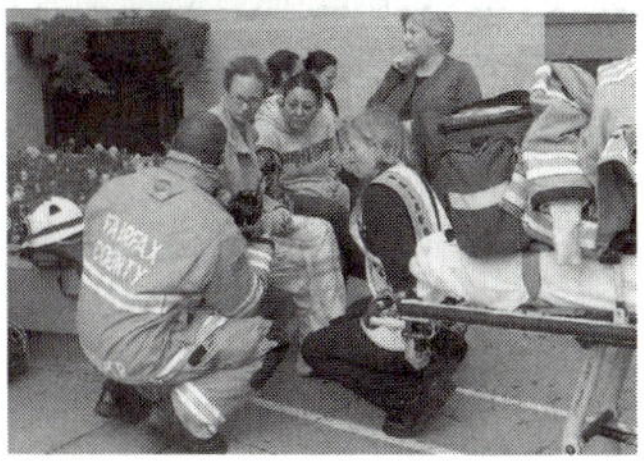

to guarding against cross contamination and treatment of hazardous materials exposures. This will aid in ensuring that the most appropriate receiving facility is chosen to ensure the best care for the patient and safety of attending personnel.

4.3.2.3 The BLS level responder shall describe the BLS protocols and SOPs at hazardous materials/WMD incidents as developed by the AHJ and the prescribed role of medical control and poison control centers, as follows:

The local emergency response plan, protocols, and procedures must define the actions that the BLS level responder should follow when confronted with a hazardous material–related mass causality incident. The action plan should provide direction for times when normal communications are disrupted. An alternative means of contacting medical control or poison control centers is required.

(1) During mass casualty incidents
(2) Where exposures have occurred
(3) In the event of disrupted radio communications

4.3.2.4 The BLS level responder shall identify the formal and informal mutual aid resources (hospital- and nonhospital-based) for the field management of multicasualty incidents, as follows:

(1) Mass-casualty trailers with medical supplies
(2) Mass-decedent capabilities
(3) Regional decontamination units
(4) Replenishment of medical supplies during long-term incidents
(5) Rehabilitation units for the EMS responders
(6) Replacement transport units for vehicles lost to mechanical trouble, collision, theft, and contamination

4.3.2.5 The BLS level responder shall identify the special hazards associated with inbound and outbound air transportation of patients exposed to hazardous materials/WMD.

ination of the
l possibly the
its might off-
Additionally,
gement of air

Commentary art is set within brown lines and labeled "Exhibit." The caption is printed in brown. The commentary exhibits, including both drawings and photographs, provide detailed views of NFPA 472 and NFPA 473 concepts and are numbered sequentially throughout each chapter.

EXHIBIT I.12.5 This is one of two body bolsters usually found on a tank car. (Source: Union Pacific Railroad)

EXHIBIT I.12.6 A head shield is commonly found on tank cars. (Source: Union Pacific Railroad)

Some tank cars are equipped with a series of continuous parallel pipes or coils mounted internally (inside the tank) or externally (outside the tank). Steam, water, or hot oil from an external source is run through these coils to heat thick or solidified materials (i.e., asphalts, fused solids, heavy fuel oils, phenol, sulfur, metallic sodium, or petroleum waxes) to make them flow more easily when loading and unloading. Exterior heater coils are shown in Exhibit I.12.7.

EXHIBIT I.12.7 This is an example of heater coils mounted on the exterior of a tank car. (Source: Union Pacific Railroad)

(d) Jacket

The jacket is an outer covering, which is typically 11 gauge (1/8 in.) steel, used to hold both insulation and jacketed thermal protection in place as well as to protect the insulation or thermal protection from the weather. Wooden blocks or metal brackets hold the jacket away from the tank.

(e) Lining and cladding

The interiors of some tank cars are lined or clad with materials to protect the tank from the corrosive or reactive effects of the contents or to maintain the purity of the contents. A lining

Hazardous Materials/Weapons of Mass Destruction Response Handbook

FIFTH EDITION

Edited by

David G. Trebisacci, CIH, CSP

With the complete text of the 2008 editions of NFPA® 472, *Standard for Competence of Responders to Hazardous Materials/Weapons of Mass Destruction Incidents,* and NFPA® 473, *Standard for Competence of EMS Responders Responding to Hazardous Materials/Weapons of Mass Destruction Incidents*

NFPA®

National Fire Protection Association®
Quincy, Massachusetts

Product Manager: Debra Rose
Developmental Editor: Khela Thorne
Project Editor: Michael Gammell
Permissions Editor: Josiane Domenici
Copy Editor: Marla Marek

Composition: Modern Graphics, Inc.
Art Coordinator: Cheryl Langway
Cover Designer: McCusker Communications
Manufacturing Manager: Ellen Glisker
Printer: Courier/Westford

Notice Concerning Liability: Publication of this handbook is for the purpose of circulating information and opinion among those concerned for fire and electrical safety and related subjects. While every effort has been made to achieve a work of high quality, neither the NFPA® nor the contributors to this handbook guarantee the accuracy or completeness of or assume any liability in connection with the information and opinions contained in this handbook. The NFPA and the contributors shall in no event be liable for any personal injury, property, or other damages of any nature whatsoever, whether special, indirect, consequential, or compensatory, directly or indirectly resulting from the publication, use of, or reliance upon this handbook.

This handbook is published with the understanding that the NFPA and the contributors to this handbook are supplying information and opinion but are not attempting to render engineering or other professional services. If such services are required, the assistance of an appropriate professional should be sought.

NFPA codes and standards are made available for use subject to Important Notices and Legal Disclaimers, which appear at the end of this handbook and can also be viewed at *www.nfpa.org/disclaimers.*

Notice Concerning Code Interpretations: This fifth edition of *Hazardous Materials/Weapons of Mass Destruction Response Handbook* is based on the 2008 editions of NFPA 472, *Standard for Competence of Responders to Hazardous Materials/Weapons of Mass Destruction Incidents,* and NFPA 473, *Standard for Competence of EMS Responders Responding to Hazardous Materials/Weapons of Mass Destruction Incidents.* All NFPA codes, standards, recommended practices, and guides are developed in accordance with the published procedures of the NFPA by technical committees comprised of volunteers drawn from a broad array of relevant interests. The handbook contains the complete text of NFPA 472 and NFPA 473 and any applicable Formal Interpretations issued by the Association. These documents are accompanied by explanatory commentary and other supplementary materials.

The commentary and supplementary materials in this handbook are not a part of the codes and do not constitute Formal Interpretations of the NFPA (which can be obtained only through requests processed by the responsible technical committees in accordance with the published procedures of the NFPA). The commentary and supplementary materials, therefore, solely reflect the personal opinions of the editor or other contributors and do not necessarily represent the official position of the NFPA or its technical committees.

NFPA No.: 472HB08
ISBN-10: 0-87765-752-1
ISBN-13: 978-0-87765-752-1
Library of Congress Control No.: 2008922489

Printed in the United States of America
08 09 10 11 12 5 4 3 2 1

In Memoriam

The NFPA Technical Committee on Hazardous Materials Response Personnel dedicates the 2008 edition of the *Hazardous Materials/Weapons of Mass Destruction Response Handbook* to Chief John M. Eversole, who passed away on May 20, 2007.

John Eversole was an early member of the NFPA Technical Committee on Hazardous Materials Response Personnel and served for 20 years, including 9 years as Committee Chairman. Continuing the important work of preceding committee chairs Warren Isman and Peter McMahon, John always made sure it would "work in the street" and fought hard to keep emergency responders safe. He strived to make a difference, and he achieved his goal in a big way.

John was a 32-year veteran of the Chicago Fire Department and founded that department's Hazardous Materials Incident Team in 1985. Born in St. Louis, John's love for the fire service began as a founding member of the Lewis College Fire Department. His initial training was conducted at the Chicago Fire Academy, and John was instantly hooked. He joined the Chicago Fire Department on February 16, 1969, and was assigned to Truck 26 on the city's busy west side.

John's career path included assignments at Squad 2, Engine 95 and Truck 36, as well as an assignment to the Fire Academy as an instructor. He was appointed as the Hazardous Materials Team Coordinator in 1987. John earned the civil service ranks of Lieutenant, Captain, and Battalion Chief.

In 2000, John was appointed Chief of Special Functions in charge of Hazardous Materials, Technical Rescue, Specialty Apparatus, Air/Sea Rescue, and the Office of Fire Investigation. John was also involved with the formation of the Deep Tunnel Team and the Collapse Rescue Team.

John helped train the Lake County (IL) Hazardous Materials Team and was instrumental in developing the first hazardous materials training program for the Illinois Fire Service Institute. He also taught hazmat classes for the National Fire Academy, the International Association of Fire Chiefs, and numerous other organizations.

John retired from active duty in 2001, but he remained extremely active. He continued to serve as the Chairman of both the NFPA and the IAFC Hazardous Materials Committees, as well as a member of the Inter-Agency Board. John was a frequent guest speaker and a respected hazardous materials subject matter expert who testified numerous times before both Congress and the Executive Branch.

John Eversole was a tireless advocate for the fire service and he will be truly missed. But most importantly, he was a good husband, father, and friend.

Contents

Preface

As work began on the 2008 edition of the *Hazardous Materials/Weapons of Mass Destruction Response Handbook*, the emerging terrorism threat of weapons of mass destruction (WMD) and the use of hazardous materials in criminal activities were significantly changing the traditional philosophies of hazardous materials emergency response. The subsequent development of tactical and operational procedures to meet the demands posed by these threats blurred the classic distinction between offensive and defensive response operations that have long been the cornerstone of both the National Fire Protection Association's NFPA 472, *Standard for Competence of Responders to Hazardous Materials/Weapons of Mass Destruction Incidents,* and the Occupational Safety and Health Administration's 29 CFR 1910.120 (q), Hazardous Waste Operations and Emergency Response (HAZWOPER).

Consequently, the purpose of this edition of the handbook is to renew the emphasis on basic hazmat and emergency medical response procedures, while introducing new material to reflect a "changed world" reality.

In preparing the 2008 edition of NFPA 472, the Technical Committee on Hazardous Materials Response Personnel worked closely with a number of organizations, including the American Society for Testing and Materials (ASTM) E54 Committee on Homeland Security Applications — Emergency Preparedness, Training, and Procedures; the Interagency Board for Equipment Standardization and Interoperability (IAB); the Federal Bureau of Investigation's Hazmat Unit; the U.S. Capitol Police; the National Association of Bomb Squad Commanders; the National Sheriffs Association; and many other emergency medical services, emergency management, and law enforcement agencies.

The technical committee established working groups to review NFPA 472 and determine how the standard could better meet traditional hazardous materials response issues and at the same time take into account the challenge of responding to terrorism and the criminal use of hazardous materials. These task groups were successful at making NFPA 472 and NFPA 473, *Standard for Competence of EMS Responders Responding to Hazardous Materials/Weapons of Mass Destruction Incidents,* more responsive to the needs and concerns of organizations well beyond the fire service. The working groups established several operational philosophies:

1. Emergency response operations to a terrorism or criminal scenario using hazardous materials are based on the fundamental concepts of hazardous materials response. In other words, responders cannot safely and effectively respond to a terrorism or criminal scenario involving hazardous materials/WMD if they do not first understand hazardous materials response.
2. NFPA 472 should apply to all emergency responders who would respond to the emergency phase of a hazardous materials/WMD incident, regardless of the individual's response discipline.
3. Emergency responders should be trained to perform their expected tasks, regardless of their discipline and organizational affiliation. Given the real-world demands of limited time and resources, training should focus on an individual's expected duties and tasks.

4. Personnel not directly involved in providing on-scene emergency response services (for example, hospital first-receivers) should not be covered in NFPA 472.
5. Competencies for emergency medical services personnel should remain in NFPA 473.

Part I of this handbook is the text of NFPA 472 and accompanying commentary. Included in this part are sections specifying competencies for awareness level personnel, operations level responders, hazardous materials technicians, incident commanders, hazardous materials officers, hazardous materials safety officers, specialist employees, and specialty technicians.

Part II of this handbook is the text and accompanying commentary of NFPA 473. This part includes the competencies for Basic Life Support (BLS) and Advanced Life Support (ALS) responders.

Part III provides seven supplements to this handbook that contain information on lessons learned from terrorist attacks, operating safely in hazardous environments, fire fighter fatalities, propane tank explosions, selection of chemical-protective ensembles using NFPA standards, response levels, and incident mitigation.

One major change in the 2008 edition of the handbook was the withdrawal of NFPA 471, *Recommended Practice for Responding to Hazardous Materials Incidents,* by the NFPA Standards Council in June 2007. The technical committee retained the information that was deemed current and applicable and added it to the appropriate sections of NFPA 472 and NFPA 473 and to the supplements of this handbook. Other changes to this edition of the handbook include the following updates to Part I, which covers NFPA 472:

1. *Awareness level personnel.* The term *responders* has been dropped from the definition of awareness level and replaced with *awareness level personnel.* The technical committee views these individuals as those who, in the course of their normal duties, might be first on-scene yet might not be emergency responders.

2. *Operations level responders.* An individual who is tasked to respond to the scene of a hazardous materials/WMD incident during the emergency phase is viewed as an operations level responder. This level includes fire, rescue, law enforcement, emergency medical services, private industry, and other allied professionals. Competencies for operations level responders have been divided into two categories:

- *Core competencies* (Chapter 5). These competencies are required of all emergency responders at this level. This chapter is essentially the competencies from the 2002 edition of NFPA 472, Chapter 5, minus the product control and personal protective clothing competencies.
- *Mission-specific competencies* (Chapter 6). These competencies are optional and are provided so that the authority having jurisdiction (AHJ) can match the expected tasks and duties of its personnel with the competencies required to perform those tasks. Mission-specific competencies are available for operations level responders who are assigned to perform the following tasks:
    i. Use personal protective equipment, as provided by the AHJ
    ii. Perform technical decontamination
    iii. Perform mass decontamination
    iv. Perform product control
    v. Perform air monitoring and sampling
    vi. Perform victim rescue and recovery operations
    vii. Preserve evidence and perform sampling
    viii. Respond to illicit laboratory incidents

Operations level mission-specific competencies are to be performed under the guidance of a hazardous materials technician, allied professional, or standard operating procedure. The competencies for personnel previously trained to the Operations Level of the 2002 edition of NFPA 472 can now be referenced as follows:

i. Core Competencies (Chapter 5)
ii. Personal Protective Equipment (Section 6.2)
iii. Product Control (Section 6.5)

3. *Hazardous materials technician.* Although the definition of a *hazardous materials technician* has been modified to reflect the usage of a risk-based response process and the definition of *hazardous materials response team* has been changed to specifically reference the performance of technician-level skills, there are no other major changes. Given that hazardous materials response teams are a typed resource under the National Incident Management System (NIMS) and to ensure consistency in operational capabilities, the technical committee felt strongly that the concept of "mission-specific" could not be applied to the hazardous materials technician level.

4. *Specialist employee.* Although there are no competency changes, the title has been changed from *private sector specialist employee* to *specialist employee* for consistency with the 29 CFR 1910.120(q) terminology and usage of the term in the field.

5. *Hazardous materials officer.* Although there are no significant competency changes, the definition has been modified to reflect that in some response organizations this individual can function as an adviser to the incident commander or as a technical specialist.

6. *Competencies for hazardous materials technician with a radioactive material specialty.* These new competencies apply to responders already trained to the hazardous materials technician level and were developed by a working group representing the U.S. Department of Energy and state and local radiation emergency responders. The technical committee decided to place these non-mandatory competencies in an annex for informational purposes at this time.

7. *Competencies for operations level responders assigned agent-specific responsibilities.* These agent-specific competencies are for responders who are already trained to NFPA 472, Chapter 5 — Core Competencies for Operations Level Responders — and Section 6.2 — Personal Protective Equipment. Agent-specific competencies have been provided for chemical, biological, and radiological agents. The technical committee decided to place these non-mandatory competencies in the annexes for informational purposes at this time.

Acknowledgments

I would like to express my appreciation to those who have preceded us in the quest for the development of safe standards for the hazardous materials responder, including Martin Henry and Gary Tokle, early staff liaisons to the technical committee and former assistant vice presidents of NFPA's Public Fire Protection Division. Many thanks also go to the subsequent staff liaisons who continued the revisions of NFPA 471, NFPA 472, and NFPA 473, including Chuck Smeby, Don Leblanc, and Jerry Laughlin, for their contributions to revisions of NFPA 471, NFPA 472, and NFPA 473. All of these individuals worked tirelessly to develop standards that offer improvements to operational methods and the professional competencies of both hazardous materials and emergency medical services responders.

I would also like to express my gratitude to the present and past members of the Technical Committee on Hazardous Materials Response Personnel — to those individuals who have eagerly contributed their expertise, those who took the time to submit proposals and comments for the committee's consideration, and the many others who unselfishly contributed written material or artwork and reviewed various portions of this handbook.

I also owe a debt of gratitude to the technical committee task groups, especially the following task group leaders for their commentary and review: Tom Clawson, Manny Ehrlich, Rem Gaade, Joe Gorman, Daryl Louder, Tony Mussorfiti, Greg Noll, Steve Patrick, Bob Royall, Glen Rudner, Rob Schnepp, Charlie Wright, and Wayne Yoder.

I would like to thank the many members of the NFPA technical and editorial staff, including Bruce Teele, who reviewed several sections of the commentary for consistency with other NFPA standards; Debra Rose, Product Manager; Khela Thorne, Developmental Editor; Michael Gammell, Project Editor; Josiane Domenici, Permissions Editor; and Marla Marek, Copy Editor. Their help has been invaluable.

Finally, I would like to thank my wife, Palmalee, and my children Tatiana and Christian, who understand the importance of this handbook and patiently supported me through its revision.

— David G. Trebisacci

About the Editor

David Trebisacci is the Program Manager of the National Board on Fire Service Professional Qualifications (Pro Board), which is a national certification program for fire fighters and emergency responders. He also provides advisory support to National Fire Protection Association (NFPA) technical committees that are responsible for standards related to emergency management and business continuity (NFPA 1600, *Standard on Disaster/Emergency Management and Business Continuity Programs*), hazardous materials response, and pre-incident planning (NFPA 1620, *Recommended Practice for Pre-Incident Planning*).

Trebisacci was also the Executive Secretary of the Marine Chemist Qualification Board, which is charged with the certification and oversight of the Marine Chemist Program. He also provided Occupational Safety and Health Administration (OSHA) standards training to the U.S. Coast Guard and maritime industry groups, and advisory support to several technical committees related to explosives, fire safety on vessels and in the shipyard, storage tanks, shipbuilding, and marine terminals.

Prior to his experience at NFPA, Trebisacci was employed in the maritime industry as a chemist, with the responsibility of providing technical support to shipyard fire brigades, testing confined spaces, and maintaining shipyard chemical quality control. His experience in industry included ensuring workplace compliance with OSHA, U.S. Department of Transportation, U.S. Environmental Protection Agency, U.S. Navy, and U.S. Coast Guard regulations and standards. Trebisacci holds a bachelor's degree in chemistry, and Certified Safety Professional (CSP) and Certified Industrial Hygienist (CIH) certifications.

PART I

NFPA® 472, *Standard for Competence of Responders to Hazardous Materials/Weapons of Mass Destruction Incidents*, 2008 Edition, with Commentary

Part I of this handbook presents the full text of NFPA 472, *Standard for Competence of Responders to Hazardous Materials/Weapons of Mass Destruction Incidents*, and explanatory commentary to guide the reader through the code. NFPA 472 identifies the levels of competence required of responders to incidents involving hazardous materials and/or weapons of mass destruction for awareness level personnel, operations level responders, hazardous materials technicians, incident commanders, hazardous materials officers, hazardous materials safety officers, and other specialist employees.

The 2008 edition of NFPA 472 is designed to address traditional hazardous materials response issues and the emerging issues presented by terrorism and the criminal use of hazardous materials.

An asterisk (*) following a code paragraph number indicates that advisory annex material pertaining to that paragraph appears in Annex A. Paragraphs that begin with the letter A are extracted from Annex A of the code. Although printed in black ink, this nonmandatory material is purely explanatory in nature. For ease of use, this handbook places Annex A material immediately after the code paragraph to which it refers.

In addition to code text and annexes, Part I includes explanatory commentary that provides the history and other background information for specific paragraphs in the code.

The text, figures, and tables of NFPA 472 appear in black. The commentary and its exhibits and tables are printed in brown.

Administration

CHAPTER 1

Chapter 1 provides the administrative text and requirements for the 2008 edition of NFPA 472, *Standard for Professional Competence of Responders to Hazardous Materials/Weapons of Mass Destruction Incidents*. This chapter covers the scope, purpose, and application of the standard.

NFPA 472 applies to all personnel whose normal duties may require them to be first on the scene. Responders include those from fire service, emergency medical service, law enforcement, and other public sector agencies as well as responders from private industry, including industrial fire brigades and workers engaged in both transportation and fixed hazardous materials handling and disposal operations.

For those organizations that must comply with either the Occupational Health and Safety Administration (OSHA) or the Environmental Protection Agency (EPA) hazardous materials emergency response regulations [1], Commentary Table I.1.1 compares these federal regulations to the specific competency levels outlined in NFPA 472.

COMMENTARY TABLE I.1.1 NFPA and OSHA Comparison

NFPA 472, 2008 Edition	*OSHA 1910.120*
Chapter 4, Awareness Level Personnel	1910.120(q)(6)(i), First Responder Awareness Level
Chapter 5, Operations Level Responders	1910.120(q)(6)(ii), First Responder Operations Level
Chapter 7, Hazardous Materials Technician	1910.120(q)(6)(iii), Hazardous Materials Technician
Chapter 8, Incident Commander	1910.120(q)(6)(v), On Scene Incident Commander
Chapter 9, Specialist Employee	1910.120(q)(5), Specialist Employee
Chapter 10, Hazardous Materials Officer	No OSHA equivalent
Chapter 11, Hazardous Materials Safety Officer	No OSHA equivalent
Chapter 12, Technician with a Tank Car Specialty	No OSHA equivalent
Chapter 13, Technician with a Cargo Tank Specialty	No OSHA equivalent
Chapter 14, Technician with an Intermodal Tank Specialty	No OSHA equivalent

1.1 Scope

1.1.1* This standard shall identify the minimum levels of competence required by responders to emergencies involving hazardous materials/weapons of mass destruction (WMD).

The key to achieving a given level of competence required in 1.1.1 is training and testing. Because NFPA 472 is a performance-based standard, training hour requirements are not given. Instead, NFPA 472 deals with the objectives and abilities that a responder must attain in order to be considered competent. The performance-based approach was chosen by committee members because hazardous materials emergency response is a constantly changing field of knowledge, skills, and control methods, based on continually developing technology, chemicals, and needs. The knowledge and skills required for competency change as technology develops.

Competencies can identify the minimum job performance requirements for specific hazardous materials responder levels. NFPA 472 can be used for designing training and evaluation programs, certifying responders, measuring and critiquing on-the-job performance, defining hiring practices, and setting organizational policies, procedures, and goals.

Requirements for the technician with a specialty designation have been included since the 1997 edition of NFPA 472 and now include the following four specialty areas of expertise:

1. Tank cars (Chapter 12)
2. Cargo tanks (Chapter 13)
3. Intermodal tanks (Chapter 14)
4. Marine tank vessels (Chapter 15)

The inclusion of these specialty levels does not limit the technician level in the performance of those tasks that have always been assigned at that level. Instead, the Technical Committee on Hazardous Materials Response Personnel has added specific language to explain that the capabilities and operations performed by the technician level can still be performed by technicians in hazardous materials incidents involving tank cars, cargo tanks, intermodal tanks, and marine tank vessels.

Since the 1997 edition, NFPA 472 has also included two other levels of competencies:

1. Hazardous materials officer (Chapter 10)
2. Hazardous materials safety officer (Chapter 11)

The hazardous materials officer could be the "officer-in-charge" of the hazardous materials team, the hazardous materials group supervisor, or a hazardous materials technical specialist. The hazardous materials officer can be a part of the entry team or can closely supervise the operation from the warm zone. At some larger incidents, more than one hazardous materials officer could be needed for each separate team working at the incident.

A response team can be composed of members at several different levels. For example, a team might have personnel at the operations level and at the technician level, or it might be made up of personnel at specialist employee A and B levels. The levels of competencies in NFPA 472 apply to individuals rather than to response teams.

A.1.1.1 Outside the United States, hazardous materials might be called dangerous goods *(see Annex H)*. Weapons of mass destruction (WMD) are known by many different abbreviations and acronyms, including CBRNE (chemical, biological, radiological, nuclear, explosive), B-NICE (biological, nuclear, incendiary, chemical, explosive), COBRA (chemical, ordinance, biological, radiological agents), and NBC (nuclear, biological, chemical).

1.1.2 This standard shall apply to any individual or member of any organization who responds to hazardous materials/WMD incidents.

1.1.3 This standard shall cover the competencies for awareness level personnel, operations level responders, hazardous materials technicians, incident commanders, hazardous materials officers, hazardous materials safety officers, and other specialist employees.

1.2 Purpose

1.2.1 The purpose of this standard shall be to specify minimum competencies required for those who respond to hazardous materials/WMD incidents and necessary for a risk-based response to these incidents.

1.2.2 The competencies contained herein shall help reduce the numbers of accidents, injuries, and illnesses during response to hazardous materials/WMD incidents and shall help prevent exposure to hazardous materials/WMD, thus reducing the possibility of fatalities, illness, and disabilities to emergency response personnel.

Hazardous materials are an integral part of the industrial society in which we live, and incidents involving such materials are inevitable. Many responders might experience a complex hazardous materials incident only once in their careers. However, those dedicated to providing emergency services must be prepared to manage such an incident effectively and safely. Those involved in manufacturing, using, storing, and transporting hazardous materials should also be trained to safely undertake initial protective actions when an unplanned release occurs and to assist emergency responders who are called upon to minimize and control the hazard.

Protecting the responders and ensuring their safety at the scene of an incident is one of the principal purposes specified in 1.2.2 of NFPA 472. Adhering to the provisions of NFPA 472 at each level of competency should provide a high degree of safety, despite the hazards encountered. The successful mitigation of a hazardous materials incident depends on proper organization and planning. The technical committee has used a standard system to organize the approach and mitigation of a hazardous materials incident, and that system is represented in Exhibit I.1.1.

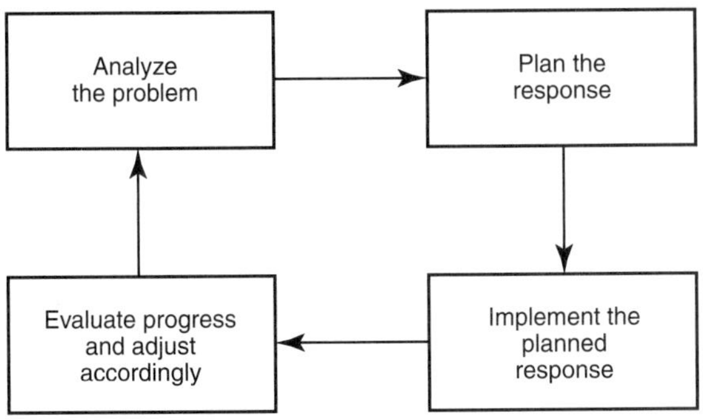

EXHIBIT I.1.1 *This diagram shows the duties of initial response personnel that are associated with emergencies involving hazardous materials.*

Each duty involves a series of tasks and steps that must be considered and resolved by decisions and actions. These duties, when supported by the response community, are the framework for an appropriate, risk-based response to hazardous materials incidents. Reasoned decisions based on this risk-based approach minimize harm resulting from a hazardous materials incident while reducing the risk to the responders.

1.3 Application

It shall not be the intent of this standard to restrict any jurisdiction from exceeding these minimum requirements.

NFPA 472 is not intended to be the sole description of ultimate performance regarding hazardous materials incident response, which is stated in Section 1.3. The authority having jurisdiction (AHJ) should not take the contents of the standard as sufficient for all possible incidents. Rather, the included requirements resulted from a consensus process that specified these are the minimum requirements. Therefore, an appropriate action for the AHJ would be to consider the standard and decide to add to the requirements as needed in order to address local conditions or specific situations.

REFERENCES CITED IN COMMENTARY

1. Title 29, Code of Federal Regulations, Part 1910.120, "Hazardous Waste Operations and Emergency Response (HAZWOPER)." U.S. Government Printing Office, Washington, DC 20402.

Referenced Publications

CHAPTER 2

This chapter lists the publications that are referenced within the mandatory chapters of NFPA 472. These mandatory referenced publications are needed for effective use of and compliance with NFPA 472. The requirements contained within these references constitute part of the requirements of NFPA 472. Annex H lists nonmandatory publications that are referenced within the nonmandatory annexes of NFPA 472.

2.1 General

The documents or portions thereof listed in this chapter are referenced within this standard and shall be considered part of the requirements of this document.

2.2 NFPA Publications

National Fire Protection Association, 1 Batterymarch Park, Quincy, MA 02169-7471.

NFPA 704, *Standard System for the Identification of the Hazards of Materials for Emergency Response,* 2007 edition.

2.3 Other Publications

2.3.1 U.S. Government Publications.

U.S. Government Printing Office, Superintendent of Documents, Washington, DC 20402.

Emergency Response Guidebook, U.S. Department of Transportation, 2004 edition.
Title 18, U.S. Code, Section 2332a, "Use of Weapons of Mass Destruction."
Title 29, Code of Federal Regulations, Part 1910.12.

2.3.2 Other Publications.

Merriam-Webster's Collegiate Dictionary, 11th edition, Merriam-Webster, Inc., Springfield, MA, 2003.

2.4 References for Extracts in Mandatory Sections (Reserved)

Definitions

CHAPTER 3

The definitions presented in Chapter 3 either are NFPA primary definitions or have been established by the Technical Committee on Hazardous Materials Response Personnel for specific use in the 2008 edition of NFPA 472. Many new definitions have been added to this edition, including risk-based response, weapons of mass destruction, agent-specific competencies, core competencies, and mission-specific competencies. Operations level responder definitions are also located here in Chapter 3 but will mainly apply to the competencies required in Chapters 5 and 6. All the terms defined in Chapter 3 will assist the responder to understand the competencies as they are presented throughout the document. The committee has made every effort to present definitions that are commonly used terms and widely understood by emergency responders. The phrase *hazardous materials/weapons of mass destruction* is truncated to the acronym *HM/WMD* or the term *hazmat/WMD* in various places throughout the commentary of this handbook.

3.1 General

The definitions contained in this chapter shall apply to the terms used in this standard. Where terms are not defined in this chapter or within another chapter, they shall be defined using their ordinarily accepted meanings within the context in which they are used. *Merriam-Webster's Collegiate Dictionary,* 11th edition, shall be the source for the ordinarily accepted meaning.

3.2 NFPA Official Definitions

3.2.1* Approved. Acceptable to the authority having jurisdiction.

A.3.2.1 Approved. The National Fire Protection Association does not approve, inspect, or certify any installations, procedures, equipment, or materials; nor does it approve or evaluate testing laboratories. In determining the acceptability of installations, procedures, equipment, or materials, the AHJ may base acceptance on compliance with NFPA or other appropriate standards. In the absence of such standards, said authority may require evidence of proper installation, procedure, or use. The AHJ may also refer to the listings or labeling practices of an organization that is concerned with product evaluations and is thus in a position to determine compliance with appropriate standards for the current production of listed items.

3.2.2* Authority Having Jurisdiction (AHJ). An organization, office, or individual responsible for enforcing the requirements of a code or standard, or for approving equipment, materials, an installation, or a procedure.

In a document dealing with the very broad concept of hazardous materials response, authorities having jurisdiction (AHJ) include officials at all levels of government, from federal to local authorities.

A.3.2.2 Authority Having Jurisdiction (AHJ). The phrase "authority having jurisdiction," or its acronym AHJ, is used in NFPA documents in a broad manner, since jurisdictions and approval agencies vary, as do their responsibilities. Where public safety is primary, the authority having jurisdiction may be a federal, state, local, or other regional department or individual such as a police chief, sheriff, fire chief; fire marshal; chief of a fire prevention bureau, labor department, or health department; building official; electrical inspector; or others having statutory authority. For insurance purposes, an insurance inspection department, rating bureau, or other insurance company representative may be the authority having jurisdiction. In many circumstances, the property owner or his or her designated agent assumes the role of the authority having jurisdiction; at government installations, the commanding officer or departmental official may be the authority having jurisdiction.

3.2.3* Listed. Equipment, materials, or services included in a list published by an organization that is acceptable to the authority having jurisdiction and concerned with evaluation of products or services, that maintains periodic inspection of production of listed equipment or materials or periodic evaluation of services, and whose listing states that either the equipment, material, or service meets appropriate designated standards or has been tested and found suitable for a specified purpose.

A.3.2.3 Listed. The means for identifying listed equipment may vary for each organization concerned with product evaluation; some organizations do not recognize equipment as listed unless it is also labeled. The authority having jurisdiction should utilize the system employed by the listing organization to identify a listed product.

3.2.4 Shall. Indicates a mandatory requirement.

3.2.5 Should. Indicates a recommendation or that which is advised but not required.

The term *should* is not used in the main body of an NFPA standard, but it can be found in the annexes, which contain explanatory or recommended information.

3.2.6 Standard. A document, the main text of which contains only mandatory provisions using the word "shall" to indicate requirements and which is in a form generally suitable for mandatory reference by another standard or code or for adoption into law. Nonmandatory provisions shall be located in an appendix or annex, footnote, or fine-print note and are not to be considered a part of the requirements of a standard.

3.3 General Definitions

3.3.1* Allied Professional. That person who possesses the knowledge, skills, and technical competence to provide assistance in the selection, implementation, and evaluation of mission-specific tasks at a hazardous materials weapons of mass destruction (WMD) incident.

Allied professionals might also be referred to as subject matter experts (SME) in a mission-specific area.

A.3.3.1 Allied Professional. Examples are a Certified Industrial Hygienist (CIH), Certified Safety Professional (CSP), Certified Health Physicist (CHP), Certified Hazardous Materials Manager (CHMM), and similar credentialed or competent individuals as determined by the AHJ.

3.3.2 Analyze. The process of identifying a hazardous materials/weapons of mass destruction (WMD) problem and determining likely behavior and harm within the training and capabilities of the emergency responder.

3.3.3 Area of Specialization.

3.3.3.1 Individual Area of Specialization. The qualifications or functions of a specific job(s) associated with chemicals and/or containers used within an organization.

*3.3.3.2 **Organization's Area of Specialization.*** Any chemicals or containers used by the specialist employee's employer.

3.3.4 Awareness Level Personnel. (29 CFR 1910.12: First Responder at the Awareness Level) Personnel who, in the course of their normal duties, could encounter an emergency involving hazardous materials/weapons of mass destruction (WMD) and who are expected to recognize the presence of the hazardous materials/weapons of mass destruction (WMD), protect themselves, call for trained personnel, and secure the scene. *(See Annex H).*

Awareness level personnel include truck drivers; train crew members; municipal, industrial, or chemical plant workers; and others whose duties require them to work in facilities where hazmat/WMD are manufactured, transported, stored, used, or could be otherwise released. Awareness level personnel are not expected to take any action that would require a great deal of training and experience. Rather, their actions are basic, defensive, and limited. Awareness level personnel are those likely to witness or discover a hazmat/WMD release during their normal job activities and who would be expected, as part of their responsibilities, to activate the emergency notification system.

The term *responder* as it relates to the awareness level was dropped from the 2008 edition of NFPA 472. The technical committee defines personnel who are tasked to respond as part of the emergency response to a hazmat/WMD incident as those individuals who are trained to the operational core competencies of Chapter 5.

3.3.5 CANUTEC. The Canadian Transport Emergency Center, operated by Transport Canada, which provides emergency response information and assistance on a 24-hour basis for responders to hazardous materials/weapons of mass destruction (WMD) incidents.

3.3.6 CHEMTREC. The Chemical Transportation Emergency Response Center, a public service of the American Chemistry Council, which provides emergency response information and assistance on a 24-hour basis for responders to hazardous materials/weapons of mass destruction (WMD) incidents.

3.3.7 Competence. Possessing knowledge, skills, and judgment needed to perform indicated objectives.

Knowledge and skills can be measured, but judgment is not as easily evaluated, and judgment and decision-making skills can vary substantially with the circumstances. Nonetheless, training and experience can effectively improve a responder's emergency decision-making process. The various competencies outlined in this performance-based standard focus on the skills necessary rather than the hours of training needed to achieve those skills. Necessary training hours can vary from individual to individual and jurisdiction to jurisdiction, depending on prior individual training and experience and on the requirements of the AHJ. In addition to training, experience and practice through drills, tabletop exercises, and other hands-on practice and training exercises is invaluable.

3.3.8* Confined Space. An area large enough and so configured that a member can bodily enter and perform assigned work but which has limited or restricted means for entry and exit and is not designed for continuous human occupancy.

The U.S. Occupational Safety and Health Administration (OSHA) has promulgated regulations relating to worker safety in confined spaces, as defined earlier. These regulations introduced a system of working in such spaces that includes hazard recognition and risk assessment; testing, evaluating, and monitoring; and permits for entry, work, and rescue. See www.osha.gov for more information.

A.3.3.8 Confined Space. Additionally, a confined space is further defined as having one or more of the following characteristics:

(1) The area contains or has the potential to contain a hazardous atmosphere, including an oxygen-deficient atmosphere.
(2) The area contains a material with the potential to engulf a member.
(3) The area has an internal configuration such that a member could be trapped by inwardly converging walls or a floor that slopes downward and tapers to a small cross section.
(4) The area contains any other recognized serious hazard.

3.3.9 Confinement. Those procedures taken to keep a material, once released, in a defined or local area.

Confinement procedures and tactics, such as spill control, are primarily defensive in nature. These tactics usually expose responders to a lower level of risk.

3.3.10 Container. A receptacle used for storing or transporting material of any kind.

Although some regulations and codes define a container by placing size limitations on its capacity, this handbook does not. A container in this case would be anything designed or intended to hold a hazardous material.

3.3.11 Containment. The actions taken to keep a material in its container (e.g., stop a release of the material or reduce the amount being released).

Containment often involves plugging or patching a container to stop a leak. Committing personnel to this type of operation must be carefully considered and must take into account the level of training they have received. Many, if not most, containment activities can be considered to be "offensive" in nature, involve a higher risk of exposure, and typically require training to the technician or the specialist employee A level (see Exhibit I.3.1).

EXHIBIT I.3.1 *A technician or private sector specialist employee with Level A training can enter the hot zone for assessment and initial rescue.*

3.3.12 Contaminant. A hazardous material, or the hazardous component of a weapon of mass destruction (WMD), that physically remains on or in people, animals, the environment, or equipment, thereby creating a continuing risk of direct injury or a risk of exposure.

Inherent in the definition of contaminant is the concept that the offending material is present where it does not belong and that it is somehow toxic or harmful to persons, animals, or the environment. Contaminants could be present in any form—solid, gas, liquid, or vapor. Dealing with each hazard type requires a different set of skills and operations.

3.3.13 Contamination. The process of transferring a hazardous material, or the hazardous component of a weapon of mass destruction (WMD), from its source to people, animals, the environment, or equipment, that can act as a carrier.

The term *contamination* is one of the most important considerations for emergency responders from a health and safety standpoint. The importance of determining whether personnel or equipment have been contaminated cannot be stressed too strongly. Because personnel are often unaware that contamination has occurred, procedures must be established ahead of time to ensure proper monitoring and decontamination procedures are in place and followed at each incident scene.

3.3.13.1 ***Cross Contamination.*** The process by which a contaminant is carried out of the hot zone and contaminates people, animals, the environment, or equipment.

3.3.14 Control. The procedures, techniques, and methods used in the mitigation of hazardous material/weapons of mass destruction (WMD) incidents, including containment, extinguishment, and confinement.

The term *control* can be used interchangeably with the word *mitigation*. Every measure taken to control a hazardous materials incident is part of the mitigation process. Limiting the degree of contamination by whatever means available is also part of the control or mitigation process.

3.3.15* Control Zones. The areas at hazardous materials/weapons of mass destruction incidents within an established/a controlled perimeter that are designated based upon safety and the degree of hazard.

The choice of basic terms related to control zones such as *hot*, *warm*, and *cold* is based on the fact that the words are simple and easily understood and that they clearly suggest the nature of the situation one would expect to encounter in any area with such a designation. (See also the definitions of cold zone, hot zone, and warm zone.) The relationship between these zones at the incident site is shown in Exhibit I.3.2.

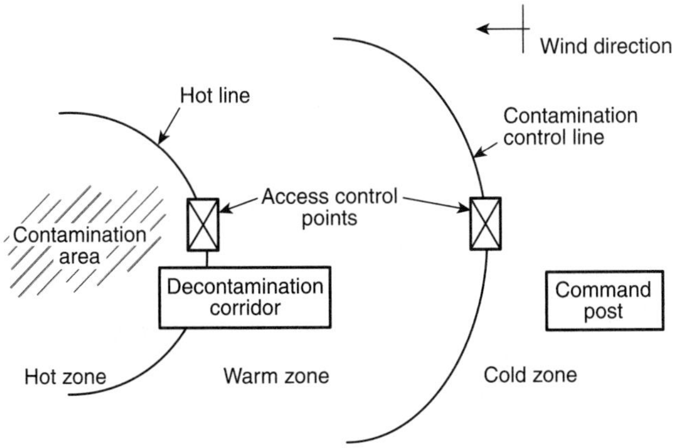

EXHIBIT I.3.2 *Control zones are shown in relation to the incident site.*

A.3.3.15 Control Zones. Law enforcement agencies might utilize different terminology for site control, for example, *inner and outer perimeters* as opposed to *hot or cold zones*. The operations level responder should be familiar with the terminology and procedures used by the AHJ and coordinate on-scene site control operations with law enforcement.

Many terms are used to describe these control zones; however, for the purposes of this standard, these zones are defined as the hot, warm, and cold zones.

3.3.15.1 Cold Zone. The control zone of hazardous materials/weapons of mass destruction incidents that contains the incident command post and such other support functions as are deemed necessary to control the incident.

The outer boundary of the cold zone is referred to as the isolation perimeter or the outer perimeter. Only emergency responders or incident support personnel can operate inside this perimeter, and in most instances the public does not have access to the cold zone.

3.3.15.2 Decontamination Corridor. The area usually located within the warm zone where decontamination is performed.

3.3.15.3 Hot Zone. The control zone immediately surrounding hazardous materials/weapons of mass destruction (WMD) incidents, which extends far enough to prevent adverse effects of hazards to personnel outside the zone.

The hot zone is the area where the hazmat/WMD release has occurred or could take place. The hot zone may also be referred to as the inner perimeter by law enforcement personnel. It is the area where there is a high potential for exposure or contamination to the materials involved, and where personal protective clothing and equipment are required based upon the hazards present.

The boundary between the hot zone and warm zone should be indicated by some physical means, such as barrier tape, barricades, or some other means of marking.

3.3.15.4* Warm Zone. The control zone at hazardous materials/weapons of mass destruction (WMD) incidents where personnel and equipment decontamination and hot zone support takes place.

Two functions of the warm zone are to contain the contaminants and prevent a contaminant's spread to the cold zone and beyond (see Exhibit I.3.3). To some extent, the warm zone serves as a buffer between the hot zone and cold zone. The level of contamination in the warm zone should decrease the closer one gets to the cold zone, not only because of the decontamination function, but also because of the increasing space.

__EXHIBIT I.3.3__ When the hazmat team is ready to exit the hot zone, they must pass through the warm zone and decontamination area to wash their chemical-protective clothing (CPC) free from any potentially dangerous material. (Courtesy of Hildebrand and Noll Technical Resources, LLC)

The warm zone includes control points for the decontamination corridor. Support activities can include staging of back-up personnel and equipment, staging of evidence, and personnel and equipment decontamination. Additionally, portions of this area might be used as safe refuge for initial patient evacuation and triage.

A.3.3.15.4 Warm Zone. The warm zone includes control points for the decontamination corridor, thus helping to reduce the spread of contamination. This support may include staging of backup personnel and equipment, staging of evidence, and personnel and equipment decontamination. Additionally, portions of this area may be used as a safe refuge for initial patient evacuation and triage.

3.3.16 Coordination. The process used to get people, who could represent different agencies, to work together integrally and harmoniously in a common action or effort.

Determining the specific individual and agency in charge of a hazmat/WMD incident can sometimes be difficult in the aftermath of a hazmat/WMD incident, depending upon its scope and nature. For example, after a major incident many responders, in an effort to be of assistance, will report to the scene and might be inclined to begin rescue or recovery operations independently of other operations that could be in progress. Command might also be transferred as an incident develops.

3.3.17* Decontamination. The physical and/or chemical process of reducing and preventing the spread of contaminants from people, animals, the environment, or equipment involved at hazardous materials/weapons of mass destruction (WMD) incidents.

A.3.3.17 Decontamination. There are two types of decontamination (commonly known as "decon") performed by emergency responders: gross and technical.

Gross decontamination is performed on the following:

(1) Entry team members before their technical decontamination
(2) Victims during emergency decontamination
(3) Persons requiring mass decontamination

Technical decontamination is performed on entry team members. Decontamination sometimes performed on victims in a hospital setting is generally referred to as *definitive decontamination*, but is not covered in this standard.

The types of decontamination (except *definitive decontamination*) are further defined in A.3.3.17.1 through A.3.3.17.4.

3.3.17.1 **Emergency Decontamination.*** The physical process of immediately reducing contamination of individuals in potentially life-threatening situations with or without the formal establishment of a decontamination corridor.

A.3.3.17.1 Emergency Decontamination. This process can be as simple as removal of outer or all garments from the individual to washing down with water from a fire hose or emergency safety shower. The sole purpose is to quickly separate as much of the contaminant as possible from the individual to minimize exposure and injury.

3.3.17.2 **Gross Decontamination.*** The phase of the decontamination process during which the amount of surface contaminants is significantly reduced.

A.3.3.17.2 Gross Decontamination. Victims of a hazardous material release that is potentially life threatening due to continued exposure from contamination are initially put through a gross decontamination, which will significantly reduce the amount of additional exposure. This is usually accomplished by mechanical removal of the contaminant or initial rinsing from handheld hose lines, emergency showers, or other nearby sources of water. Responders operating in a contaminated zone in personal protective equipment (PPE) are put through gross decontamination, which will make it safer for them to remove the PPE without exposure and for members assisting them.

3.3.17.3 **Mass Decontamination.*** The physical process of reducing or removing surface contaminants from large numbers of victims in potentially life-threatening situations in the fastest time possible.

A.3.3.17.3 Mass Decontamination. Mass decontamination is initiated where the number of victims and time constraints do not allow the establishment of an in-depth decontamination process. Mass decontamination is a gross decontamination process utilizing large volumes of low-pressure water to reduce the level of contamination. A soap-and-water solution or universal decontamination solution would be more effective; however, availability of such solutions in sufficient quantities cannot always be ensured.

Extensive research into mass decontamination operations at terrorist incidents involving hazardous materials and chemical warfare agents has been conducted by the U.S. Army's Research, Development, and Engineering Command (RDECOM), and the resulting guidelines and documents are available on the Internet *(see K.1.2.5)*.

Mass decontamination should be established quickly to reduce the harm being done to the victims by the contaminants. Initial operations will likely be through handheld hose lines or master streams supplied from fire apparatus while a more formal process is being set up. Examples of mass decontamination methods are the ladder pipe decontamination system and the emergency decontamination corridor system, both of which are described in RDECOM's guidelines.

3.3.17.4 **Technical Decontamination.*** The planned and systematic process of reducing contamination to a level that is as low as reasonably achievable (ALARA).

A.3.3.17.4 Technical Decontamination. Technical decontamination is the process subsequent to gross decontamination designed to remove contaminants from responders, their equipment, and victims. It is intended to minimize the spread of contamination and ensure responder safety. Technical decontamination is normally established in support of emergency responder entry operations at a hazardous materials incident, with the scope and level of technical decontamination based on the type and properties of the contaminants involved. In non life-threatening contamination incidents, technical decontamination can also be used on victims of the initial release. Examples of technical decontamination methods are the following:

(1) Absorption
(2) Adsorption
(3) Chemical degradation
(4) Dilution
(5) Disinfecting
(6) Evaporation
(7) Isolation and disposal
(8) Neutralization
(9) Solidification
(10) Sterilization
(11) Vacuuming
(12) Washing

The specific decontamination procedure to be used at an incident is typically selected by a hazardous materials technician *(see 7.3.4)* and is subject to the approval of the incident commander.

3.3.18 Degradation. (1) A chemical action involving the molecular breakdown of a protective clothing material or equipment due to contact with a chemical. (2) The molecular breakdown of the spilled or released material to render it less hazardous during control operations.

3.3.19* Demonstrate. To show by actual performance.

A.3.3.19 Demonstrate. This performance can be supplemented by simulation, explanation, illustration, or a combination of these.

3.3.20 Describe. To explain verbally or in writing using standard terms recognized by the hazardous materials/weapons of mass destruction (WMD) response community.

3.3.21 Emergency Response Guidebook (ERG). A reference book, written in plain language, to guide emergency responders in their initial actions at the incident scene.

The *Emergency Response Guidebook* (ERG) [1] was developed by the U.S. Department of Transportation, Transport Canada, and the Secretariat of Communications and Transportation of Mexico (SCT) for use by fire fighters, police, and other emergency services personnel. The guidebook's primary purpose is to assist emergency personnel in identifying the specific or generic materials involved in an incident and to protect themselves and the public during the initial response to an incident. The ERG is periodically updated to reflect new products and technology.

The ERG is free to emergency responders, and commercially available through government printing office bookstores. See Exhibits I.3.4 and I.3.5.

3.3.22 Endangered Area. The actual or potential area of exposure associated with the release of a hazardous material/weapon of mass destruction (WMD).

The size of an endangered area is a key element in determining the magnitude of a hazardous materials incident. Obviously, a hazardous material that poses a threat to a 500 ft^2 (47 m^2) area is much easier to manage than one that affects one square mile. The Table of Initial Isolation and Protective Action Distances in the ERG provides guidance in determining the size of the endangered area based on the contaminants present and the type of container involved.

3.3.23 Evaluate. The process of assessing or judging the effectiveness of a response operation or course of action within the training and capabilities of the emergency responder.

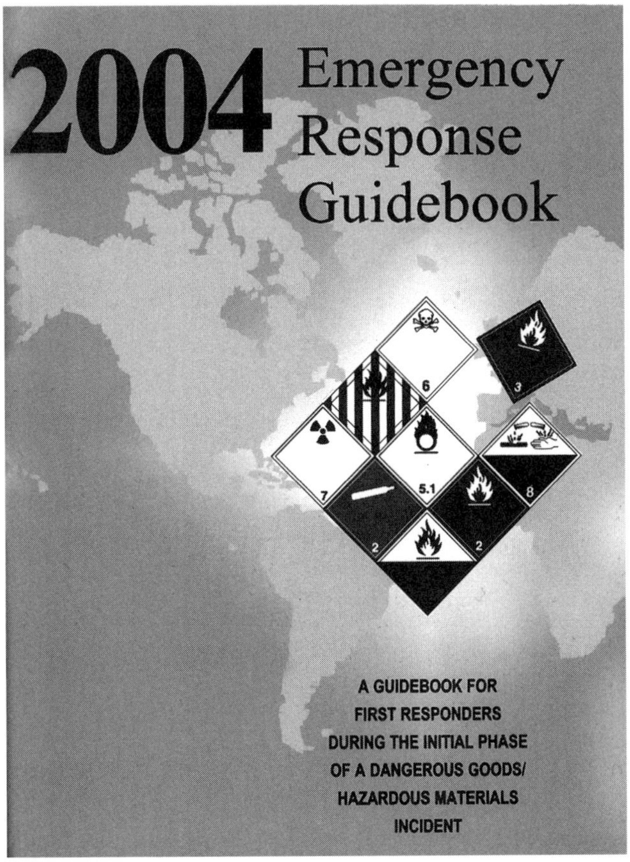

EXHIBIT I.3.4 *The Emergency Response Guidebook was developed by Canadian, Mexican, and U.S. authorities for use during the initial response to an incident.*

ID No.	Guide No.	Name of Material
1579	153	4-Chloro-o-toluidine hydrochloride
1579	153	4-Chloro-o-toluidine hydrochloride, solid
1580	154	Chloropicrin
1581	123	Chloropicrin and Methyl bromide mixture
1581	123	Methyl bromide and Chloropicrin mixture
1582	119	Chloropicrin and Methyl chloride mixture
1582	119	Methyl chloride and Chloropicrin mixture
1583	154	Chloropicrin mixture, n.o.s.
1585	151	Copper acetoarsenite
1586	151	Copper arsenite
1587	151	Copper cyanide
1588	157	Cyanides, inorganic, n.o.s.
1588	157	Cyanides, inorganic, solid, n.o.s.
1589	125	CK
1589	125	Cyanogen chloride, inhibited
1589	125	Cyanogen chloride, stabilized
1590	153	Dichloroanilines
1590	153	Dichloroanilines, liquid
1590	153	Dichloroanilines, solid
1591	152	o-Dichlorobenzene
1593	160	Dichloromethane
1593	160	Methylene chloride
1594	152	Diethyl sulfate
1594	152	Diethyl sulphate
1595	156	Dimethyl sulfate
1595	156	Dimethyl sulphate
1596	153	Dinitroanilines
1597	152	Dinitrobenzenes

Page 38

ID No.	Guide No.	Name of Material
1597	152	Dinitrobenzenes, liquid
1597	152	Dinitrobenzenes, solid
1598	153	Dinitro-o-cresol
1599	153	Dinitrophenol, solution
1600	152	Dinitrotoluenes, molten
1601	151	Disinfectant, solid, poisonous, n.o.s.
1601	151	Disinfectant, solid, toxic, n.o.s.
1601	151	Disinfectants, solid, n.o.s. (poisonous)
1602	151	Dye, liquid, poisonous, n.o.s.
1602	151	Dye, liquid, toxic, n.o.s.
1602	151	Dye intermediate, liquid, poisonous, n.o.s.
1602	151	Dye intermediate, liquid, toxic, n.o.s.
1603	155	Ethyl bromoacetate
1604	132	Ethylenediamine
1605	154	Ethylene dibromide
1606	151	Ferric arsenate
1607	151	Ferric arsenite
1608	151	Ferrous arsenate
1610	159	Halogenated irritating liquid, n.o.s.
1611	151	Hexaethyl tetraphosphate
1611	151	Hexaethyl tetraphosphate, liquid
1611	151	Hexaethyl tetraphosphate, solid
1612	123	Hexaethyl tetraphosphate and compressed gas mixture
1613	154	Hydrocyanic acid, aqueous solution, with less than 5% Hydrogen cyanide
1613	154	Hydrocyanic acid, aqueous solution, with not more than 20% Hydrogen cyanide

ID No.	Guide No.	Name of Material
1613	154	Hydrogen cyanide, aqueous solution, with not more than 20% Hydrogen cyanide
1614	152	Hydrogen cyanide, anhydrous, stabilized (absorbed)
1614	152	Hydrogen cyanide, stabilized (absorbed)
1616	151	Lead acetate
1617	151	Lead arsenates
1618	151	Lead arsenites
1620	151	Lead cyanide
1621	151	London purple
1622	151	Magnesium arsenate
1623	151	Mercuric arsenate
1624	154	Mercuric chloride
1625	141	Mercuric nitrate
1626	157	Mercuric potassium cyanide
1627	141	Mercurous nitrate
1629	151	Mercury acetate
1630	151	Mercury ammonium chloride
1631	154	Mercury benzoate
1634	154	Mercuric bromide
1634	154	Mercurous bromide
1634	154	Mercury bromides
1636	154	Mercuric cyanide
1636	154	Mercury cyanide
1637	151	Mercury gluconate
1638	151	Mercury iodide
1639	151	Mercury nucleate
1640	151	Mercury oleate
1641	151	Mercury oxide
1642	151	Mercuric oxycyanide
1642	151	Mercury oxycyanide, desensitized

ID No.	Guide No.	Name of Material
1643	151	Mercury potassium iodide
1644	151	Mercury salicylate
1645	151	Mercuric sulfate
1645	151	Mercuric sulphate
1645	151	Mercury sulfate
1645	151	Mercury sulphate
1646	151	Mercury thiocyanate
1647	151	Ethylene dibromide and Methyl bromide mixture, liquid
1647	151	Methyl bromide and Ethylene dibromide mixture, liquid
1648	127	Acetonitrile
1648	127	Methyl cyanide
1649	131	Motor fuel anti-knock mixture
1649	131	Tetraethyl lead, liquid
1650	153	beta-Naphthylamine
1650	153	beta-Naphthylamine, solid
1650	153	Naphthylamine (beta)
1650	153	Naphthylamine (beta), solid
1651	153	Naphthylthiourea
1652	153	Naphthylurea
1653	151	Nickel cyanide
1654	151	Nicotine
1655	151	Nicotine compound, solid, n.o.s.
1655	151	Nicotine preparation, solid, n.o.s.
1656	151	Nicotine hydrochloride
1656	151	Nicotine hydrochloride, liquid
1656	151	Nicotine hydrochloride, solid
1656	151	Nicotine hydrochloride, solution
1657	151	Nicotine salicylate
1658	151	Nicotine sulfate, solid
1658	151	Nicotine sulfate, solution
1658	151	Nicotine sulphate, solid

Page 39

EXHIBIT I.3.5 *This sample page from the 2004 Emergency Response Guidebook reflects new products and technology*

3.3.24 Example. An illustration of a problem serving to show the application of a rule, principle, or method (e.g., past incidents, simulated incidents, parameters, pictures, and diagrams).

3.3.25* Exposure. The process by which people, animals, the environment, and equipment are subjected to or come in contact with a hazardous material/weapon of mass destruction (WMD).

An exposure is quickly assumed to be an external one, but attention must be paid to the special hazards of internal exposures. For example, an internal radiation exposure from ingesting radioactive material can be more damaging to the body than an external exposure.

A.3.3.25 Exposure. The magnitude of exposure is dependent primarily on the duration of exposure and the concentration of the hazardous material. This term is also used to describe a person, animal, the environment, or a piece of equipment. The exposure can be external, internal, or both.

3.3.26* Fissile Material. Material whose atoms are capable of nuclear fission (capable of being split).

A.3.3.26 Fissile Material. Department of Transportation (DOT) regulations define fissile material as plutonium-239, plutoniun-242, uranium-233, uranium-235, or any combination of these radionuclides. This material is usually transported with additional shipping controls that limit the quantity of material in any one shipment. Packaging used for fissile material is designed and tested to prevent a fission reaction from occurring during normal transport conditions as well as hypothetical accident conditions.

3.3.27 Hazard/Hazardous. Capable of posing an unreasonable risk to health, safety, or the environment; capable of causing harm.

3.3.28* Hazardous Material. A substance (either matter — solid, liquid, or gas — or energy) that when released is capable of creating harm to people, the environment, and property, including weapons of mass destruction (WMD) as defined in 18 U.S. Code, Section 2332a, as well as any other criminal use of hazardous materials, such as illicit labs, environmental crimes, or industrial sabotage.

A.3.3.28 Hazardous Material. The following are explanations of several CBRN-related terms:

(1) *CBRN.* An abbreviation for chemicals, biological agents, and radiological particulate hazards.
(2) *CBRN terrorism agents.* Chemicals, biological agents, and radiological particulates that could be released as the result of a terrorist attack. Chemical terrorism agents include solid, liquid, and gaseous chemical warfare agents and toxic industrial chemicals. Chemical warfare agents include, but are not limited to, GB (Sarin), GD (Soman), HD (sulfur mustard), VX, and specific toxic industrial chemicals. Many toxic industrial chemicals (e.g., chlorine and ammonia) are identified as potential chemical terrorism agents because of their availability and the degree of injury they could inflict. Biological agents are bacteria, viruses, or the toxins derived from biological material.
(3) *Chemical terrorism agents.* Liquid, solid, gaseous, and vapor chemical warfare agents and toxic industrial chemicals used to inflict lethal or incapacitating casualties, generally on a civilian population as a result of a terrorist attack.
(4) *Biological terrorism agents.* Liquid or particulate agents that can consist of a biologically derived toxin or pathogen to inflict lethal or incapacitating casualties.
(5) *Radiological particulate terrorism agents.* Particles that emit ionizing radiation in excess of normal background levels used to inflict lethal or incapacitating casualties, generally on a civilian population, as the result of a terrorist attack.
(6) *Toxic industrial chemicals.* Highly toxic solid, liquid, or gaseous chemicals, that have been identified as mass casualty threats that could be used to inflict casualties, generally on a civilian population, during a terrorist attack.

3.3.29* Hazardous Materials Branch/Group. The function within an overall incident management system that deals with the mitigation and control of the hazardous materials/weapons of mass destruction (WMD) portion of an incident.

The hazardous materials branch provides an organizational structure that allows the necessary supervision and control of the operations that are essential at a hazardous materials incident. These operations include the tactical objectives carried out in the hot zone, the control of site access, and decontamination. Depending upon the size and scope of the incident, the hazardous materials branch can be referred to as the hazard group. See NFPA 1561, *Standard on Emergency Services Incident Management System* [2], for additional information related to incident management systems.

A.3.3.29 Hazardous Materials Branch/Group. This function is directed by a hazardous materials officer and deals principally with the technical aspects of the incident.

3.3.30* Hazardous Materials Officer. (NIMS: Hazardous Materials Branch Director/Group Supervisor.) The person who is responsible for directing and coordinating all operations involving hazardous materials/weapons of mass destruction (WMD) as assigned by the incident commander.

A.3.3.30 Hazardous Materials Officer. This individual might also serve as a technical specialist for incidents that involve hazardous materials/WMD.

The person designated as the hazardous materials officer is responsible for managing the hazardous materials response team under the overall direction of the incident commander. Rescue operations in the control zones also come under the direction of the hazardous materials officer. This position is also called the hazardous materials group supervisor.

3.3.31* Hazardous Materials Response Team (HMRT). An organized group of trained response personnel operating under an emergency response plan and applicable standard operating procedures who perform hazardous material technician level skills at hazardous materials/weapons of mass destruction (WMD) incidents.

This definition has been revised in the 2008 edition to reflect that hazardous materials response teams have the personnel, equipment, and staffing to deliver technician level skills at HM/WMD incidents.

The hazardous materials response team (HMRT) can be made up of various personnel. In some cases, all of the responders might come from the same agency. In others, responders might be drawn from various agencies and/or disciplines. However, all members of a hazardous materials response team must operate under one set of operating procedures, have an organizational structure, and train together on a regular basis as shown in Exhibit I.3.6.

Personnel on the HMRT do not need to be trained to the same level. The important factor, however, is that the duties and responsibilities of each member be commensurate with their level of training. For example, if the incident calls for operations to control the release and for work in specialized chemical-protective clothing, these functions should only be attempted by certified hazardous materials technicians or specialist employees A. See Section 9.4 of NFPA 472 for competence requirements for a specialist employee A.

A.3.3.31 Hazardous Materials Response Team (HMRT). The team members respond to releases or potential releases of hazardous materials/WMD for the purpose of control or stabilization of the incident.

EXHIBIT I.3.6 *Safety is ensured by frequent practices and drills. In this illustration, personnel train for a hazmat mass casualty incident. (Courtesy of Biddeford, ME Hazmat)*

3.3.32* Hazardous Materials Safety Officer. (NIMS: Assistant Safety Officer — Hazardous Material.) The person who works within an incident management system (IMS) (specifically, the hazardous materials branch/group) to ensure that recognized hazardous materials/WMD safe practices are followed at hazardous materials/weapons of mass destruction (WMD) incidents.

A.3.3.32 Hazardous Materials Safety Officer. The hazardous materials safety officer will be called on to provide technical advice or assistance regarding safety issues to the hazardous materials officer and incident safety officer at a hazardous materials/WMD incident.

3.3.33* Hazardous Materials Technician. Person who responds to hazardous materials/weapons of mass destruction (WMD) incidents using a risk-based response process by which they analyze a problem involving hazardous materials/weapons of mass destruction (WMD), select applicable decontamination procedures, and control a release using specialized protective clothing and control equipment.

The role of the hazardous materials technician is significantly different from the roles of awareness level personnel and operations level responders. At those levels, individuals are likely to be the first on the scene or those whose normal duties include initial response to emergencies that involve hazardous materials. Accordingly, their actions are limited because their primary duty is to reduce or eliminate access to the immediate area of the incident and to minimize the potential harm of a release of hazardous materials. The hazardous materials technician, on the other hand, assumes a more aggressive role in controlling an incident. The hazardous materials technician needs to approach the point of release to plug, patch, or otherwise stop a hazardous substance spill.

A.3.3.33 Hazardous Materials Technician. These persons might have additional competencies that are specific to their response mission, expected tasks, and equipment and training as determined by the AHJ.

3.3.33.1* Hazardous Materials Technician with a Cargo Tank Specialty. Person who provides technical support pertaining to cargo tanks, provides oversight for product removal and movement of damaged cargo tanks, and acts as a liaison between the hazardous materials technician and other outside resources.

A.3.3.33.1 Hazardous Materials Technician with a Cargo Tank Specialty. The hazardous materials technicians are expected to use specialized chemical-protective clothing and specialized control equipment.

3.3.33.2 Hazardous Materials Technician with a Marine Tank Vessel Specialty. Person who provides technical support pertaining to marine tank vessels, provides oversight for product removal and movement of damaged marine tank vessels, and acts as a liaison between the hazardous materials technician and other outside resources.

3.3.33.3* Hazardous Materials Technician with an Intermodal Tank Specialty. Person who provides technical support pertaining to intermodal tanks, provides oversight for product removal and movement of damaged intermodal tanks, and acts as a liaison between the hazardous materials technician and other outside resources.

A.3.3.33.3 Hazardous Materials Technician with an Intermodal Tank Specialty. See A.3.3.33.1.

3.3.33.4* Hazardous Materials Technician with a Tank Car Specialty. Person who provides technical support pertaining to tank cars, provides oversight for product removal and movement of damaged tank cars, and acts as a liaison between the hazardous materials technician and other outside resources.

A.3.3.33.4 Hazardous Materials Technician with a Tank Car Specialty. See A.3.3.33.1.

3.3.34 Identify. To select or indicate verbally or in writing using standard terms to establish the fact of an item being the same as the one described.

3.3.35 Incident. An emergency involving the release or potential release of hazardous materials/weapons of mass destruction (WMD).

3.3.36* Incident Commander (IC). The individual responsible for all incident activities, including the development of strategies and tactics and the ordering and the release of resources.

A.3.3.36 Incident Commander (IC). This position is equivalent to the on-scene incident commander as defined in OSHA 1910.120(8), Hazardous Waste Operations and Emergency Response. The IC has overall authority and responsibility for conducting incident operations and is responsible for the management of all incident operations at the incident site.

The National Incident Management System establishes the following functions as the primary responsibilities of the incident commander (IC):

- Have clear authority and know agency policy
- Ensure incident safety
- Establish the incident command post (ICP)
- Set priorities, determine incident objectives and strategies to be followed
- Establish the incident command system (ICS) organization needed to manage the incident
- Approve the incident action plan (IAP)
- Coordinate command and general staff activities
- Approve resource requests and use of volunteers and auxiliary personnel
- Order demobilization as needed
- Ensure after-action reports are completed

The IC is responsible for the overall control of operations. Where multiple agencies are likely to be involved, which is generally the case when hazardous materials are involved, the roles and responsibilities of the various agencies should be clarified ahead of time. The actual command can change from one person to another as the incident develops and becomes more complex.

The important factor from an incident management standpoint is that a designated person is in charge and that the person is clearly identified in the local emergency response plan. The IC works from the strategic level and develops the overall response objectives. The IC, however, should not personally become involved in carrying out tactical operations. Broadly speaking, the IC is responsible for the safety of response personnel and the public, controlling the incident, and minimizing harm to the environment and property.

Even though the IC is responsible for directing and coordinating the response, some management functions might have to be delegated to others. The IC might establish a hazardous materials group or hazardous materials branch to manage activities in the warm and hot control zones. These would be managed by a hazardous materials group supervisor or materials branch director, who reports to the IC or to operations.

A number of different agencies might be responsible for controlling and cleaning up the more complex hazardous materials incidents. These agencies would be local, county, state, or federal agencies, and some incidents would involve more than one agency from each jurisdiction. A vital step during pre-incident planning is to identify these agencies. The plans must establish protocols that spell out which agency is to be the lead agency during an incident. Some incident management systems establish a unified command for multiagency or multijurisdictional incidents. Each participating agency maintains its authority, responsibility, or accountability. Collectively, these lead persons manage the incident. (See Exhibit I.3.7.)

3.3.37 Incident Command System. A management system designed to enable effective and efficient on-scene incident management by integrating a combination of facilities, equipment,

EXHIBIT I.3.7 *When a unified command has been established, representatives from each agency work together to establish a single set of objectives for the entire incident.*

personnel, procedures, and communications operating within a common organizational structure.

3.3.38* Incident Management System (IMS). A plan that defines the roles and responsibilities to be assumed by personnel and the operating procedures to be used in the management and direction of emergency operations to include the incident command system, multi-agency coordination system, training, and management of resources.

An effective incident management system is one of the most important aspects of operating safely during an emergency. NFPA 1561 provides excellent information for developing an effective incident management system and clearly addresses the components that must be included for personnel to operate at an incident in an organized and safe manner.

A.3.3.38 Incident Management System (IMS). The IMS provides a consistent approach for all levels of government, private sector, and volunteer organizations to work effectively and efficiently together to prepare for, respond to, and recover from domestic incidents, regardless of cause, size, or complexity. An IMS provides for interoperability and compatibility among all capability levels of government, the private sector, and volunteer organizations. The IMS includes a core set of concepts, principles, terminology, and technologies covering the incident command system, multiagency coordination systems, training, and identification and management of resources.

3.3.39 Match. To provide with a counterpart.

3.3.40* Material Safety Data Sheet (MSDS). A form, provided by manufacturers and compounders (blenders) of chemicals, containing information about chemical composition, physical and chemical properties, health and safety hazards, emergency response, and waste disposal of the material.

Exhibit I.3.8 shows a sample Material Safety Data Sheet (MSDS). Note that this MSDS is provided only as an illustration; its content should not be relied on as it may not be the most recent information that is available.

A.3.3.40 Material Safety Data Sheet (MSDS). Under the Global Harmonization System, the MSDS is known as an SDS (Safety Data Sheet) and contains more detailed information.

3.3.41 Monitoring Equipment. Instruments and devices used to identify and quantify contaminants.

Monitoring equipment is extremely useful for the following reasons:

1. To determine the presence and nature of hazardous material
2. To help determine the personal protective equipment that can be used safely
3. To help establish control zones
4. To identify or classify unknowns
5. To determine the level of contamination or decontamination

EXHIBIT I.8 *A Material Safety Data Sheet contains information about the material that can be useful to incident responders. (Courtesy of Praxair Technology, Inc.)*

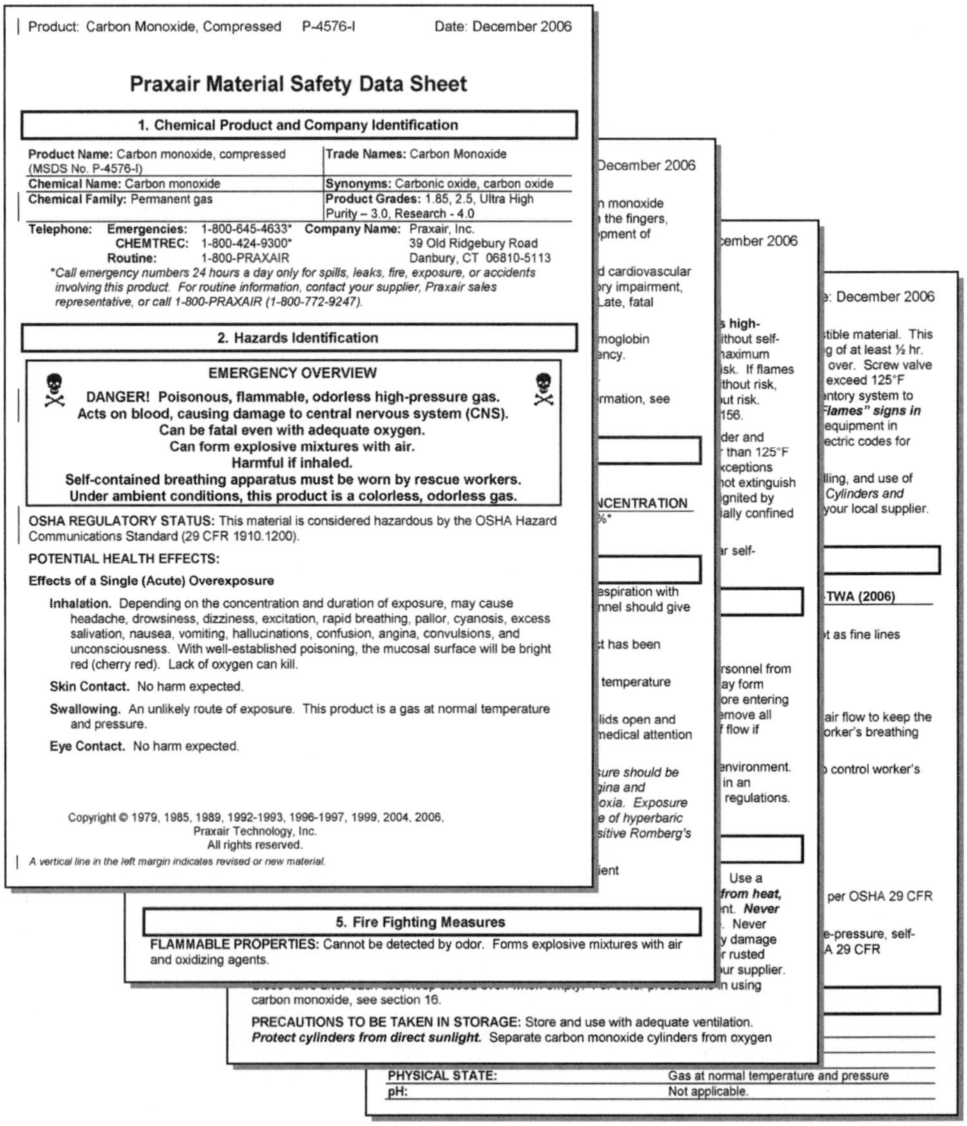

Product: Carbon Monoxide, Compressed P-4576-I Date: December 2006

Praxair Material Safety Data Sheet

1. Chemical Product and Company Identification

Product Name: Carbon monoxide, compressed (MSDS No. P-4576-I)	**Trade Names:** Carbon Monoxide
Chemical Name: Carbon monoxide	**Synonyms:** Carbonic oxide, carbon oxide
Chemical Family: Permanent gas	**Product Grades:** 1.85, 2.5, Ultra High Purity – 3.0, Research - 4.0

Telephone: **Emergencies:** 1-800-645-4633* **Company Name:** Praxair, Inc.
CHEMTREC: 1-800-424-9300* 39 Old Ridgebury Road
Routine: 1-800-PRAXAIR Danbury, CT 06810-5113

**Call emergency numbers 24 hours a day only for spills, leaks, fire, exposure, or accidents involving this product. For routine information, contact your supplier, Praxair sales representative, or call 1-800-PRAXAIR (1-800-772-9247).*

2. Hazards Identification

EMERGENCY OVERVIEW

DANGER! Poisonous, flammable, odorless high-pressure gas.
Acts on blood, causing damage to central nervous system (CNS).
Can be fatal even with adequate oxygen.
Can form explosive mixtures with air.
Harmful if inhaled.
Self-contained breathing apparatus must be worn by rescue workers.
Under ambient conditions, this product is a colorless, odorless gas.

OSHA REGULATORY STATUS: This material is considered hazardous by the OSHA Hazard Communications Standard (29 CFR 1910.1200).

POTENTIAL HEALTH EFFECTS:

Effects of a Single (Acute) Overexposure

Inhalation. Depending on the concentration and duration of exposure, may cause headache, drowsiness, dizziness, excitation, rapid breathing, pallor, cyanosis, excess salivation, nausea, vomiting, hallucinations, confusion, angina, convulsions, and unconsciousness. With well-established poisoning, the mucosal surface will be bright red (cherry red). Lack of oxygen can kill.

Skin Contact. No harm expected.

Swallowing. An unlikely route of exposure. This product is a gas at normal temperature and pressure.

Eye Contact. No harm expected.

A vertical line in the left margin indicates revised or new material.

5. Fire Fighting Measures

FLAMMABLE PROPERTIES: Cannot be detected by odor. Forms explosive mixtures with air and oxidizing agents.

carbon monoxide, see section 16.

PRECAUTIONS TO BE TAKEN IN STORAGE: Store and use with adequate ventilation. *Protect cylinders from direct sunlight.* Separate carbon monoxide cylinders from oxygen

PHYSICAL STATE:	Gas at normal temperature and pressure
pH:	Not applicable.

3.3.42 Objective. A goal that is achieved through the attainment of a skill, knowledge, or both, that can be observed or measured.

3.3.43* Packaging. Any container that holds a material (hazardous or nonhazardous).

A.3.3.43 Packaging. Packaging for hazardous materials includes bulk and nonbulk packaging.

> ***3.3.43.1* Bulk Packaging.*** Any packaging, including transport vehicles, having a liquid capacity of more than 119 gal (450 L), a solids capacity of more than 882 lb (400 kg), or a compressed gas water capacity of more than 1001 lb (454 kg).

A.3.3.43.1 Bulk Packaging. Bulk packaging can be either placed on or in a transport vehicle or vessel or constructed as an integral part of the transport vehicle.

> ***3.3.43.2 Nonbulk Packaging.*** Any packaging having a liquid capacity of 119 gal (450 L) or less, a solids capacity of 882 lb (400 kg) or less, or a compressed gas water capacity of 1001 lb (454 kg) or less.
>
> ***3.3.43.3* Radioactive Materials Packaging.*** Any packaging for radioactive materials including excepted packaging, industrial packaging, Type A, Type B, and Type C packaging.

The strength and reliability of radioactive materials packaging are especially important in transportation because of the possibility of devastating incidents and the potential harm that could result from released radiation. At least four groups in the United States promulgate rules governing the transport of radioactive material, including the U.S. Department of Transportation (DOT), U.S. Nuclear Regulatory Commission, U.S. Department of Energy, and the U.S. Postal Service. In addition, each state has regulations governing this type of transport. The DOT regulations are generally more detailed in 49 CFR 173, Subpart I (Class 7 Radioactive Materials) [3]. The Canadian Nuclear Safety Commission (CNSC) is the regulatory authority in Canada.

Regulations cannot eliminate all accidents in transportation, so the emphasis is on ensuring safety in routine handling situations for minimally hazardous material and ensuring integrity under all circumstances for highly dangerous materials.

These goals of containment and safety focus on the package and its ability in the following three areas:

1. Contain the material (prevent leaks)
2. Prevent unusual occurrences (such as criticality)
3. Reduce external radiation to safe levels through shielding

Concentrations of radioactive materials in Type A packages must not exceed the limits established in 49 CFR 173.431 [4]. The contents are non-life endangering amounts of radioactive materials. Type A packages are designed to survive normal transportation handling and minor accidents. The outer package might be cardboard box, wood crate, or metal drum. The shape and the material is not regulated, but Type A packages must pass certain performance tests. Shippers must have documentation showing test results for the container design. The packages are identified with the words *Type A* on the package.

Type B packages are stronger than Type A and strong enough to survive a severe accident and to transport material with radioactivity levels higher than those allowed in Type A containers. Limits on radioactivity in a Type B package are provided in 49 CFR 173.431.

The package must be a metal drum or large shielded transport container. Type B packages must also pass certain performance tests that are more rigorous than those for Type A packages. The Nuclear Regulatory Commission issues a certificate of compliance for Type B packages. Type B packages are identified with the words *Type B* on the package and shipping papers. Examples of Type A and Type B packaging are shown in Exhibit I.3.9.

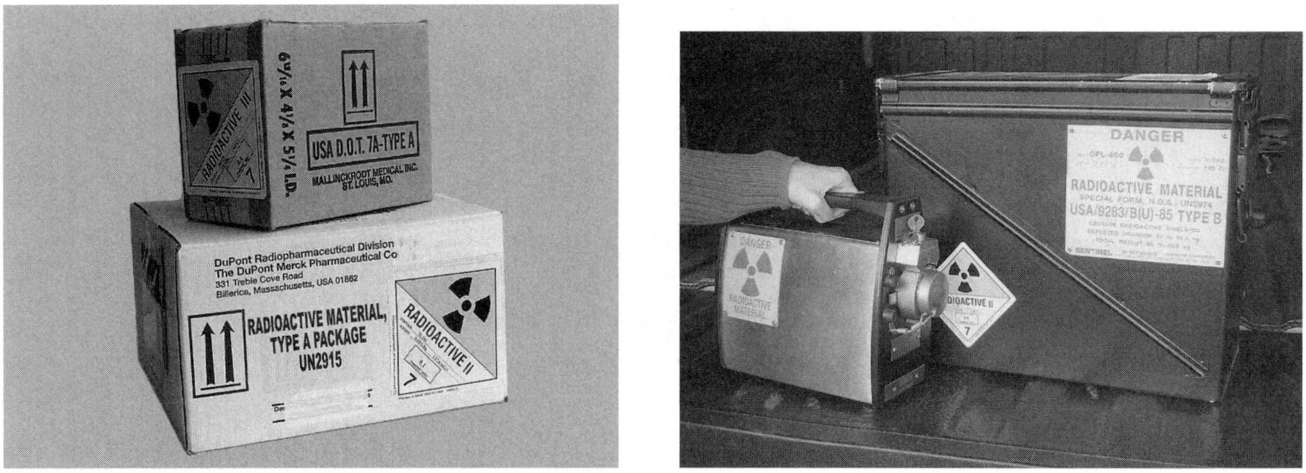

EXHIBIT I.3.9 This exhibit shows examples of radioactive materials packaging Type A and Type B: (a) Radiopharmaceuticals are commonly transported in Type A packages like the ones pictured here, and (b) Type B packages can range from small hand-held radiography cameras like the one shown here to heavily shielded steel casks that weigh well over 100 tons.

A.3.3.43.3 Radioactive Materials Packaging. Excepted packaging is packaging used to transport materials with extremely low levels of radioactivity that meet only general design requirements for any hazardous material. Excepted packaging ranges from a product's fiberboard box to a sturdy wooden or steel crate, and typical shipments include limited quantities of materials, instruments, and articles such as smoke detectors. Excepted packaging will contain non-life-endangering amounts of radioactive material.

Industrial packaging is packaging used to transport materials that present limited hazard to the public and environment. Examples of these materials are contaminated equipment and radioactive waste solidified in materials such as concrete. This packaging is grouped into three categories (IP-I, IP-2, IP-3), based on the strength of packaging. Industrial packaging will contain non-life-endangering amounts of radioactive material.

Type A packaging is used to transport radioactive materials with concentrations of radioactivity not exceeding the limits established in 49, CFR, Part 173.431. Typically, Type A packaging has an inner containment vessel made of glass, plastic, or metal and packing material made of polyethylene, rubber, or vermiculite. Examples of materials shipped in Type A packaging include radiopharmaceuticals and low-level radioactive waste. Type A packaging will contain non-life-endangering amounts of radioactive material.

Type B packaging is used to transport radioactive materials with radioactivity levels higher than those allowed in Type A packaging, such as spent fuel and high-level radioactive waste. Limits on activity contained in a Type B packaging are provided in Title 49, CFR 173.431. Type B packaging ranges from small drums [208 L (55 gal)], to heavily shielded steel casks that sometimes weigh more than 100 metric tons (98 tons). Type B packaging can contain potentially life-endangering amounts of radioactive material.

Type C packaging is used for consignments, transported by aircraft, of high-activity radioactive materials that have not been certified as "low dispersible radioactive material" (including plutonium). They are designed to withstand severe accident conditions associated with air transport without loss of containment or significant increase in external radiation levels. The Type C packaging performance requirements are significantly more stringent than those for Type B packaging. Type C packaging is not authorized for domestic use but can be authorized for international shipments of these high-activity radioactive material consignments. Regulations require that both Type B and Type C packaging be marked with a trefoil symbol to ensure that the package can be positively identified as carrying radioactive material. The trefoil symbol must be resistant to the effects of both fire and water so that it will be likely to survive a severe accident and serve as a warning to emergency responders.

The performance requirements for Type C packaging include those applicable to Type B packaging with enhancements on some tests that are significantly more stringent than those for Type B packaging. For example, a 321.8 km/hr (200 mph) impact onto an unyielding target is required instead of the 9.1 m (30 ft) drop test required of a Type B packaging; a 60-minute fire test is required instead of the 30-minute test for Type B packaging; and a puncture/tearing test is required. These stringent tests are expected to result in packaging designs that will survive more severe aircraft accidents than Type B packaging designs.

3.3.44 Penetration. The movement of a material through a suit's closures, such as zippers, buttonholes, seams, flaps, or other design features of chemical-protective clothing, and through punctures, cuts, and tears.

Responders to hazardous materials incidents must ensure that the protection provided by chemical protective clothing (CPC) is not compromised by openings in the suit's material. See Exhibit I.3.10.

3.3.45 Permeation. A chemical action involving the movement of chemicals, on a molecular level, through intact material.

As shown in Exhibit I.3.11, different fabrics have different levels of resistance to chemical permeation, and all fabrics absorb chemicals over a period of time. Guidelines for manufacturer permeation testing and certification of CPC are provided in NFPA 1991, *Standard on Vapor-Protective Ensembles for Hazardous Materials Emergencies* [5], and NFPA 1992, *Standard on Liquid Splash-Protective Ensembles and Clothing for Hazardous Materials Emergencies* [6]. When purchasing CPC, an important step is to verify that the garments meet the appropriate standard and are certified as meeting that standard. In any event, the user should exercise extreme caution that the garments selected are appropriate for the hazardous material at the incident.

Proper care of CPC, including appropriate decontamination and storage procedures, should be emphasized in all training programs and standard operating procedures. Proper care, or the lack of it, can significantly affect the protection afforded by CPC.

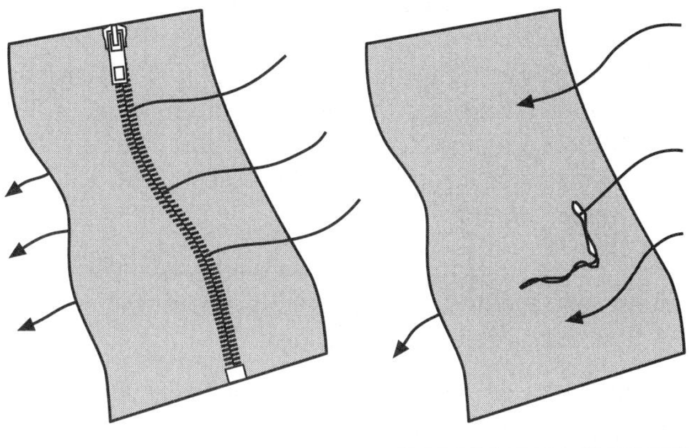

EXHIBIT I.3.10 *Penetration involves the movement of a material through a suit's closures, such as zippers, buttonholes, and seams, or through rips, tears, or flaws in the fabric.*

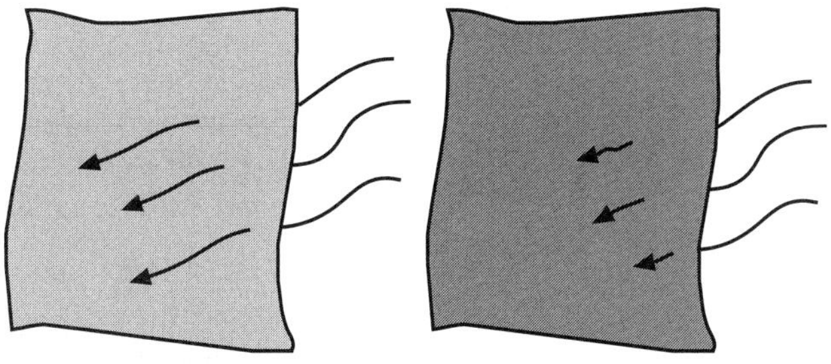

EXHIBIT I.3.11 *Permeation involves chemical movement, on a molecular level, through intact material, and different fabrics permit different levels of permeation.*

3.3.46* Personal Protective Equipment. The equipment provided to shield or isolate a person from the chemical, physical, and thermal hazards that can be encountered at hazardous materials/weapons of mass destruction (WMD) incidents.

Personal protective equipment is designed to protect the individual responder against anticipated or expected hazards. Responders must be adequately trained to use, care for, and select appropriate personal protective equipment.

A.3.3.46 Personal Protective Equipment. Personal protective equipment includes both personal protective clothing and respiratory protection. Adequate personal protective equipment should protect the respiratory system, skin, eyes, face, hands, feet, head, body, and hearing.

3.3.47 Plan.

3.3.47.1* Emergency Response Plan. A plan developed by the authority having jurisdiction, with the cooperation of all participating agencies and organizations, that details specific actions to be performed by all personnel who are expected to respond during an emergency.

A.3.3.47.1 Emergency Response Plan. Emergency response plans can be developed at organizational, agency, local, state, and federal levels.

Whether developed for a community, county, state, or industrial facility, an emergency response plan must identify the hazards that are present within given sites in a community or region and must establish written procedures for handling such incidents. Regarding hazardous materials response, the Code of Federal Regulations [7] is very clear. In addition to emergency responders and governmental authorities, employers whose workers could potentially be exposed to hazardous materials while on the job are required to develop emergency response plans. To avoid duplication, private sector employers can use the appropriate plan developed by local or state authorities as part of their own emergency response plan.

3.3.47.2* Incident Action Plan. An oral or written plan approved by the incident commander containing general objectives reflecting the overall strategy for managing an incident.

A.3.3.47.2 Incident Action Plan. It can include the identification of operational resources and assignments. It can also include attachments that provide direction and important information for management of the incident during one or more operational periods.

3.3.47.3 Site Safety and Control Plan. A site safety and control plan should be completed and approved by the hazardous materials officer, the hazardous materials safety officer, and the incident commander for inclusion in the incident action plan. The plan must be briefed to personnel operating within the hot zone by the hazardous materials safety officer or the hazardous materials officer prior to entry mission initiation. The initial site safety and control plan for the first operational period can be written or oral. The plan should be documented as soon as resources allow.

3.3.48* Planned Response. The incident action plan, with the site safety and control plan, consistent with the emergency response plan and/or standard operating procedures for a specific hazardous material/weapon of mass destruction (WMD) incident.

Appropriate control actions are identified based on the magnitude of the problem. This process considers available resources and training levels of responders and then determines the direction the response effort must take to influence the events and to favorably change the outcome. The material in 3.3.48 also describes the various areas of concern that should be included in any complete incident plan.

A.3.3.48 Planned Response. The following site safety plan considerations are from the EPA's *Standard Operating Safety Guides:*

(1) Site description
(2) Entry objectives
(3) On-site organization
(4) On-site control
(5) Hazard evaluations
(6) Personal protective equipment
(7) On-site work plans
(8) Communication procedures
(9) Decontamination procedures
(10) Site safety and health plan

3.3.49 Predict. The process of estimating or forecasting the future behavior of a hazardous materials/weapons of mass destruction (WMD) container and/or its contents within the training and capabilities of the emergency responder.

3.3.50* Protective Clothing. Equipment designed to protect the wearer from heat and/or from hazardous materials, or from the hazardous component of a weapon of mass destruction contacting the skin or eyes.

A.3.3.50 Protective Clothing. Protective clothing is divided into three types:

(1) Structural fire-fighting protective clothing
(2) High temperature–protective clothing
(3) Chemical-protective clothing
    (a) Liquid splash–protective clothing
    (b) Vapor-protective clothing

3.3.50.1* Chemical-Protective Clothing. Items made from chemical-resistive materials, such as clothing, hood, boots, and gloves, that are designed and configured to protect the wearer's torso, head, arms, legs, hands, and feet from hazardous materials.

A.3.3.50.1 Chemical-Protective Clothing. Chemical-protective clothing (garments) can be constructed as a single- or multipiece garment. The garment can completely enclose the wearer either by itself or in combination with the wearer's respiratory protection, attached or detachable hood, gloves, and boots.

Chemical-protective clothing (CPC) allows responders to work for a specified length of time in or near an area contaminated by hazardous materials by isolating their bodies from chemical hazards. Exhibit I.3.12 shows responders donning CPC ensembles. The type of CPC ensemble needed varies depending on the type of hazardous materials and conditions present. Each CPC ensemble and ensemble elements that are certified as compliant with the NFPA standard, either NFPA 1991 for vapor-protective ensembles or NFPA 1992 for liquid splash-protective ensembles, comes with a list of specific chemicals that the ensemble has been tested for and certified as providing protection from these chemicals. Vapor protective ensembles are also known as Level A ensembles, and liquid splash-protective ensembles could also be Level B or Level C ensembles.

3.3.50.2* High Temperature–Protective Clothing. Protective clothing designed to protect the wearer for short-term high temperature exposures.

A.3.3.50.2 High Temperature–Protective Clothing. This type of clothing is usually of limited use in dealing with chemical commodities.

Specialized protective ensembles, as defined in 3.3.50.2, are designed to provide brief protection against high radiant heat exposure to temperatures as high as 2000°F (1093°C).

EXHIBIT I.3.12 *County hazmat technicians don CPC ensembles to respond to an incident. (Courtesy of the Seneca County Hazmat Team)*

Requirements for the design, performance, testing, and certification of proximity fire-fighting ensembles are provided in NFPA 1971, *Standard on Protective Ensembles for Structural Fire Fighting and Proximity Fire Fighting* [8]. These protective ensembles consist of a one- or two-piece upper and lower torso garments, helmet, shroud, gloves, and footwear that provide the radiant reflective protection for the wearer.

Fire entry suits are another type of specialized high-temperature ensemble designed to protect the wearer against abnormally high temperatures during direct entry into the flames. Fire entry suits are designed and constructed for specific expected exposures, temperatures, and duration of the anticipated operations that will be conducted.

3.3.50.3* Liquid Splash–Protective Clothing. The garment portion of a chemical-protective clothing ensemble that is designed and configured to protect the wearer against chemical liquid splashes but not against chemical vapors or gases.

A.3.3.50.3 Liquid Splash–Protective Clothing. This type of protective clothing is a component of EPA Level B chemical protection. Liquid splash–protective clothing should meet the requirements of NFPA 1992, *Standard on Liquid Splash-Protective Ensembles and Clothing for Hazardous Materials Emergencies.*

3.3.50.4* Structural Fire-Fighting Protective Clothing. The fire resistant protective clothing normally worn by fire fighters during structural fire-fighting operations, which includes a helmet, coat, pants, boots, gloves, PASS device, and a fire resistant hood to cover parts of the head and neck not protected by the helmet and respirator facepiece.

A.3.3.50.4 Structural Fire-Fighting Protective Clothing. Structural fire-fighting protective clothing provides limited protection from heat but might not provide adequate protection from the harmful gases, vapors, liquids, or dusts that are encountered during hazardous materials/WMD incidents.

Structural fire-fighting protective ensembles certified as compliant with NFPA 1971 consist of upper and lower torso single or two-piece garments (coat and trousers, or single piece coverall style), helmet, hood, gloves, and footwear. A personal alert safety system (PASS) is worn either attached to the upper or lower torso garment or as an integrated part of the self-contained breathing apparatus (SCBA). When properly worn, this ensemble provides protection from the hazards normally encountered in structural fire-fighting operations. In most cases, this clothing does not offer adequate protection from hazardous materials. However, NFPA 1971 contains additional optional requirements for chemical, biological, radiological, and nuclear (CBRN) protection for the total ensemble including specific CBRN respiratory protection, for protection for the wearer without having to add or subtract any parts of the ensemble, and without compromising the fire-fighting protection afforded by the ensemble. This additional CBRN protection is intended for protection of fire department first responders who could be involved at a CBRN terrorism incident prior to the arrival of additional specialized responders with more highly protective ensembles for protracted operations.

3.3.50.5* Vapor-Protective Clothing. The garment portion of a chemical-protective clothing ensemble that is designed and configured to protect the wearer against chemical vapors or gases.

A.3.3.50.5 Vapor-Protective Clothing. This type of protective clothing is a component of EPA Level A chemical protection. Vapor-protective clothing should meet the requirements of NFPA 1991, *Standard on Vapor-Protective Ensembles for Hazardous Materials Emergencies.*

Totally encapsulating vapor-protective ensembles (Level A ensembles) must be worn when the greatest level of skin, respiratory, and eye protection is needed. This protection should be worn when the hazard cannot be determined or is identified as immediately dangerous to life and health.

3.3.51 Qualified. Having knowledge of the installation, construction, or operation of apparatus and the hazards involved.

3.3.52* Respiratory Protection. Equipment designed to protect the wearer from the inhalation of contaminants.

A.3.3.52 Respiratory Protection. Respiratory protection is divided into three types:

(1) Positive pressure self-contained breathing apparatus
(2) Positive pressure air-line respirators
(3) Air-purifying respirators

Because inhalation of toxins is one of the principal causes of serious injury to responders, respiratory protection is of the utmost importance. Guidance on selection of appropriate respiratory protection is found in Chapter 5 of NFPA 472.

Positive pressure SCBA is the type of breathing apparatus fire fighters normally use. The wearer's mobility is not restricted except for the physical limitation of the SCBA's size, and SCBA provides the highest level of respiratory protection based on laboratory testing. The most commonly used units are National Institute for Occupational Safety and Health (NIOSH) approved for rated times at 30 to 60 minutes of protection. However, closed-circuit SCBA are available that provide longer periods of use. NFPA 1981, *Standard on Open-Circuit Self-Contained Breathing Apparatus (SCBA) for Emergency Services* [9], provides design, performance, testing, and certification requirements for SCBA. The actual time protection provided by SCBA is substantially reduced for persons not at their best level of physical fitness and for those performing exhausting and demanding work. On-scene medical evaluation and careful timekeeping help to prevent reliance on SCBA beyond the optimal time limits established for each worker and their unit.

A major change in the 2007 edition of NFPA 1981 was the mandatory requirement for all emergency services SCBA to also be NIOSH certified as CBRN SCBA in accordance with the NIOSH *Statement of Standard for NIOSH CBRN SCBA Testing* [10]. This requirement provides respiratory protection from CBRN terrorism agents (specified chemicals, biological agents, and radiological particulate) that could be released as a result of a terrorism attack.

While major metropolitan areas might be more likely targets of a terrorist event, emergency responders from smaller communities and rural areas could be called upon to respond to urban areas where the emergency services require assistance. Terrorists themselves might reside in small or rural communities while they await their opportunity to strike. They might have the chemical, biological, or nuclear material in their possession, making the possibility of exposure greater for small communities and rural areas. Terrorist attacks aside, CBRN-certified SCBA offers greater protection for the emergency services for a very minimal cost.

CBRN protection offers verification of enhanced protection for emergency responders that is not otherwise available. Without CBRN protection evaluation, no SCBA components are tested for permeation, penetration, corrosion resistance, or other detrimental effects from exposure to toxic industrial chemicals during hazardous materials incidents and hazardous chemical warfare atmospheres. NIOSH benchmark testing of non-CBRN hardened SCBA against CBRN agents demonstrated that CBRN agents and toxic industrial chemicals could cause catastrophic failures within minutes of exposure.

The test challenge agents for CBRN protection were selected by NIOSH based on a comprehensive review of available technical data and consultations with other government agencies, such as the U.S. Department of Defense, the U.S. Department of Justice, the U.S. Department of Energy, etc. Various chemical data lists were analyzed, including lists from the Environmental Protection Agency (EPA), Agency for Toxic Substances and Disease Registry (ATSDR), NFPA 1994, *Standard on Protective Ensembles for First Responders to CBRN Terrorism Incidents* [11], *U.S. Army Center for Health Promotion and Preventative Medicine (USACHPPM) Technical Guide 24* [12], and classified sources. This review established a total of 151 toxic industrial chemicals (TICs) and chemical warfare agents (CWAs) as potential candidates for challenge agents. The candidate agents were evaluated for permeation (molecularly diffusing through material) and penetration (seeping through interfacing components) characteristics as part of a review of their physical properties. This evaluation concluded that Sarin (GB) and Sulfur Mustard (HD) could be selected as the two representative agents for the penetration/permeation test for the complete listing of 151 CWAs and TICs due to their physical properties and molecular structure. NIOSH is unaware of any data that indicate the CBRN-certified SCBA provide less protection against TICs than their industrial counterparts.

The evaluation for CBRN protection provides verification and assurance that the component and material combinations in the approved SCBA configurations provide high resistance to permeation and penetration of hazardous atmospheres of toxic industrial chemicals and materials into the breathing air. This is of importance to all responders subject to extreme exposures to any hazardous industrial chemicals and materials.

Positive pressure air-line respirators give the user an unlimited supply of air from a remote source and are lighter than SCBA. However, these respirators restrict the travel distance of the user to a maximum of 300 ft (91 m) and must be equipped with an escape breathing air cylinder. In addition, the path of travel must be kept clear of obstructions while the user is negotiating the incident scene. Nonetheless, these respirators are an ideal method of supplying air to the emergency worker under certain conditions.

Air purifying respirators (APRs) and powered air purifying respirators (PAPRs) use a filter or a sorbent to remove airborne contaminants. The respirators should not be used by the initial responders at an incident involving hazardous materials. APRs and PAPRs should be used only when the hazard and the concentration of the hazardous material are known and the ambient atmosphere at the scene is verified to contain more than 19.5 percent oxygen.

3.3.53* Response. That portion of incident management in which personnel are involved in controlling hazardous materials/weapons of mass destruction (WMD) incidents.

A.3.3.53 Response. The activities in the response portion of a hazardous materials/WMD incident include analyzing the incident, planning the response, implementing the planned response, evaluating progress, and terminating the emergency phase of the incident.

3.3.54 Risk-Based Response Process. Systematic process by which responders analyze a problem involving hazardous materials/weapons of mass destruction (WMD), assess the hazards, evaluate the potential consequences, and determine appropriate response actions based upon facts, science, and the circumstances of the incident.

This definition of risk-based response process is new in the 2008 edition of NFPA 472. Analysis of hazardous materials/WMD incidents where emergency responders have been injured or killed (as well as "close calls") has shown that risk evaluation is a critical element for a safe and effective response.

3.3.55 Safely. To perform the assigned tasks without injury to self or others, to the environment, or to property.

From the inception of NFPA 472, the committee's intent has been to outline the *minimum* competencies responders need to increase their level of safety. Throughout the document, the technical committee has stressed conducting operations safely.

3.3.56 Scenario. A sequence or synopsis of actual or imagined events used in the field or classroom to provide information necessary to meet student competencies; can be based upon threat assessment.

3.3.57 SETIQ. The Emergency Transportation System for the Chemical Industry in Mexico.

SETIQ is the Mexican equivalent to CHEMTREC.

3.3.58 Specialist Employees.

> ***3.3.58.1* Specialist Employee A.*** That person who is specifically trained to handle incidents involving chemicals or containers for chemicals used in the organization's area of specialization.

A.3.3.58.1 Specialist Employee A. Consistent with the organization's emergency response plan and/or standard operating procedures, the specialist employee A is able to analyze an incident involving chemicals within the organization's area of specialization, plan a response to that incident, implement the planned response within the capabilities of the resources available, and evaluate the progress of the planned response. Specialist employees are those persons who, in the course of their regular job duties, work with or are trained in the hazards of specific chemicals or containers within their organization's area of specialization. In response to emergencies involving hazardous materials/WMD in their organization's area of specialization, they could be called on to provide technical advice or assistance to the incident commander relative to specific chemicals or containers for chemicals. Specialist employees should receive training or demonstrate competency in their area of specialization annually. Specialist employees also should receive additional training to meet applicable DOT, OSHA, EPA, and other appropriate state, local, or provincial occupational health and safety regulatory requirements. Specialist employees respond to hazardous materials/WMD incidents under differing circumstances. They respond to incidents within their facility, inside and outside their assigned work area, and outside their facility. Persons responding away from the facility or within the facility outside their assigned work area respond as members of a hazardous materials response team or as specialist employees as outlined in this definition and in Chapter 9. When responding to incidents away from their assigned work area, specialist employees should be permitted to perform only at the response level at which they have been trained.

Persons responding to a hazardous materials/WMD incident within their work area are not required to be trained to the levels specified by this chapter. Persons within their work area who have informed the incident management structure of an emergency as defined in the emergency response plan who have adequate personal protective equipment and adequate training in the procedures they are to perform and who have employed the buddy system can take limited action in the danger area (e.g., turning a valve) before the emergency response team arrives. The limited action taken should be addressed in the emergency response plan.

Once the emergency response team arrives, these persons should be restricted to the actions that their training level allows and should operate under the incident command structure.

3.3.58.2 Specialist Employee B.* That person who, in the course of his or her regular job duties, works with or is trained in the hazards of specific chemicals or containers within the individual's area of specialization.

Specialist employees B are expected by their organization to respond *only* within their field of expertise. The specialists might have expertise in a number of fields, including the following:

- Product specialties
- Industrial hygiene/toxicological specialties
- Environmental waste/remediation specialties
- Monitoring specialties
- Container specialties
- Material handling, such as loaders and unloaders
- Maintenance specialties

Specialist employees B only need to meet those competencies required in their area of specialization.

A.3.3.58.2 Specialist Employee B. Because of the employee's education, training, or work experience, the specialist employee B can be called on to respond to incidents involving specific chemicals or containers. The specialist employee B can be used to gather and record information, provide technical advice, and provide technical assistance (including work within the hot zone) at the incident consistent with the organization's emergency response plan and/or standard operating procedures and the emergency response plan. See 3.3.47.1.

3.3.58.3 Specialist Employee C.* That person who responds to emergencies involving chemicals and/or containers within the organization's area of specialization.

A.3.3.58.3 Specialist Employee C. Consistent with the organization's emergency response plan and/or standard operating procedures, the specialist employee C can be called on to gather and record information, provide technical advice, and/or arrange for technical assistance. A specialist employee C does not enter the hot or warm zone at an emergency. See 3.3.15.

3.3.59 Stabilization. The point in an incident when the adverse behavior of the hazardous material, or the hazardous component of a weapon of mass destruction (WMD), is controlled.

The term *stabilization* does not mean that the incident is over but rather that the hazardous conditions are not likely to escalate or intensify further. Stabilization could also mean that the cleanup operations following stabilization are prolonged and that specialized types of equipment and expertise are needed before the operations are completed.

3.3.60* Termination. That portion of incident management after the cessation of tactical operations in which personnel are involved in documenting safety procedures, site operations, hazards faced, and lessons learned from the incident.

A.3.3.60 Termination. Termination is divided into three phases: debriefing the incident, post incident analysis, and critiquing the incident.

3.3.61* UN/NA Identification Number. The four-digit number assigned to a hazardous material/weapon of mass destruction (WMD), which is used to identify and cross-reference products in the transportation mode.

A.3.3.61 UN/NA Identification Number. United Nations (UN) numbers are four-digit numbers used in international commerce and transportation to identify hazardous chemicals or classes of hazardous materials. These numbers generally range between 0000 and 3500 and usually are preceded by the letters "UN" (e.g., "UN1005") to avoid confusion with number codes.

North American (NA) numbers are identical to UN numbers. If a material does not have a UN number, it may be assigned an NA number. These usually are preceded by "NA" followed by a four-digit number starting with 8 or 9.

3.3.62* Weapon of Mass Destruction (WMD). (1) Any destructive device, such as any explosive, incendiary, or poison gas bomb, grenade, rocket having a propellant charge of more than four ounces, missile having an explosive or incendiary charge of more than one quarter ounce (7 grams), mine, or device similar to the above; (2) any weapon involving toxic or poisonous chemicals; (3) any weapon involving a disease organism; or (4) any weapon that is designed to release radiation or radioactivity at a level dangerous to human life.

A.3.3.62 Weapon of Mass Destruction (WMD). The source of this definition is 18 USC 2332a.

3.4 Operations Level Responders Definitions

If an individual is tasked to respond to the scene of a hazardous materials/WMD incident during the emergency phase, that individual is considered to be an operations level responder. This would include fire, rescue, law enforcement, emergency medical services, private industry, and other allied professionals. Competencies for operations level responders have been broken into the following two categories:

1. *Chapter 5 – Core Competencies.* These competencies are required of all emergency responders at this level. This chapter is essentially the Chapter 5 competencies from the 2002 edition of the code, minus the product control and personal protective clothing competencies.
2. *Chapter 6 – Mission-Specific Competencies.* These competencies are optional and are provided so that the AHJ can match the expected tasks and duties of its personnel with the required competencies to perform those tasks.

Mission-specific competencies are available for Operations Level Responders who are assigned to perform the following tasks:

- Use personal protective equipment, as provided by the AHJ
- Perform technical decontamination
- Perform mass decontamination
- Perform product control
- Perform air monitoring and sampling
- Perform victim rescue and recovery operations
- Evidence preservation and sampling
- Respond to illicit laboratory incidents

Operations level mission-specific tasks must be performed under the guidance of a hazardous materials technician, allied professional, or standard operating procedure. The competencies for personnel previously trained to the operations level (2002 edition) are now referenced as follows:

- Chapter 5 – Core Competencies
- Chapter 6.2 – Mission-specific Competencies: Personal Protective Equipment
- Chapter 6.6 – Mission-specific Competencies: Product Control

See the NFPA 472 Operations Level Responder Matrix (Table A.5.1.1.1) for examples on the application and use of the operations level core and mission-specific competencies.

3.4.1 Agent-Specific Competencies. The knowledge, skills, and judgment needed by operations level responders who have completed the operations level competencies and who are designated by the authority having jurisdiction to respond to releases or potential releases of a specific group of WMD agents.

3.4.2 Core Competencies. The knowledge, skills, and judgment needed by operations level responders who respond to releases or potential releases of hazardous materials/weapons of mass destruction (WMD).

3.4.3 Mission-Specific Competencies. The knowledge, skills, and judgment needed by operations level responders who have completed the operations level competencies and who are designated by the authority having jurisdiction to perform mission specific tasks, such as decontamination, victim/hostage rescue and recovery, evidence preservation, and sampling.

3.4.4* Operations Level Responders. Persons who respond to hazardous materials/weapons of mass destruction (WMD) incidents for the purpose of implementing or supporting actions to protect nearby persons, the environment, or property from the effects of the release.

A.3.4.4 Operations Level Responders. The source of this definition is 29 CFR 1910.120. These responders can have additional competencies that are specific to their response mission, expected tasks, and equipment and training as determined by the AHJ.

3.4.5 Operations Level Responders Assigned to Perform Air Monitoring and Sampling. Persons, competent at the operations level, who are assigned to implement air monitoring and sampling operations at hazardous materials/weapons of mass destruction (WMD) incidents.

3.4.6 Operations Level Responders Assigned to Perform Evidence Preservation and Sampling. Persons, competent at the operations level, who are assigned to preserve forensic evidence, take samples, and/or seize evidence at hazardous materials/weapons of mass destruction (WMD) incidents involving potential violations of criminal statutes or governmental regulations.

3.4.7 Operations Level Responders Assigned to Perform Mass Decontamination During Hazardous Materials/Weapons of Mass Destruction (WMD) Incidents. Persons, competent at the operations level, who are assigned to implement mass decontamination operations at hazardous materials/weapons of mass destruction (WMD) incidents.

3.4.8 Operations Level Responders Assigned to Perform Product Control. Persons, competent at the operations level, who are assigned to implement product control measures at hazardous materials/weapons of mass destruction (WMD) incidents.

3.4.9 Operations Level Responders Assigned to Perform Technical Decontamination During Hazardous Materials/Weapons of Mass Destruction (WMD) Incidents. Persons, competent at the operations level, who are assigned to implement technical decontamination operations at hazardous materials/weapons of mass destruction (WMD) incidents.

3.4.10 Operations Level Responders Assigned to Perform Victim Rescue/Recovery During Hazardous Materials/Weapons of Mass Destruction (WMD) Incidents. Persons, competent at the operations level, who are assigned to rescue and/or recover exposed and contaminated victims at hazardous materials/weapons of mass destruction (WMD) incidents.

3.4.11 Operations Level Responders Assigned to Respond to Illicit Laboratory Incidents. Persons, competent at the operations level, who, at hazardous materials/weapons of mass destruction (WMD) incidents involving potential violations of criminal statutes specific to the illegal manufacture of methamphetamines, other drugs, or weapons of mass destruction (WMD), are assigned to secure the scene, identify the laboratory/process, and preserve evidence.

3.4.12 Operations Level Responders Assigned Responsibilities for Biological Response. Persons, competent at the operations level, who, at hazardous materials/weapons of mass destruction (WMD) incidents involving biological materials, are assigned to support the hazardous materials technician and other personnel, provide strategic and tactical recommendations to the on-scene incident commander, serve in a technical specialist capacity to provide technical oversight for operations, and act as a liaison between the hazardous materials technician, response personnel, and other outside resources regarding biological issues.

3.4.13 Operations Level Responders Assigned Responsibilities for Chemical Response. Persons, competent at the operations level, who, at hazardous materials/weapons of mass destruction (WMD) incidents involving chemical materials, are assigned to support the hazardous materials technician and other personnel, provide strategic and tactical recommendations to the on-scene incident commander, serve in a technical specialist capacity to provide technical oversight for operations, and act as a liaison between the hazardous material technician, response personnel, and other outside resources regarding chemical issues.

3.4.14 Operations Level Responders Assigned Responsibilities for Radioactive Material Response. Persons, competent at the operations level, who, at hazardous materials/weapons of mass destruction (WMD) incidents involving radioactive materials, are assigned to support the hazardous materials technician and other personnel, provide strategic and tactical recommendations to the on-scene incident commander, serve in a technical specialist capacity to provide technical oversight for operations, and act as a liaison between the hazardous material technician, response personnel, and other outside resources regarding radioactive material issues.

3.4.15 Operations Level Responders Assigned to Use Personal Protective Equipment During Hazardous Materials/Weapons of Mass Destruction (WMD) Incidents. Persons, competent at the operations level, who are assigned to use of personal protective equipment at hazardous materials/weapons of mass destruction (WMD) incidents.

REFERENCES CITED IN COMMENTARY

1. *Emergency Response Guidebook* (ERG). Office of Hazardous Materials Initiatives and Training, PHH-50, Pipeline and Hazardous Materials Safety Administration, U.S. Department of Transportation, 1200 New Jersey Avenue, SE East Building, 2nd Floor, Washington, DC 20590, http://hazmat.dot/gydebook.
2. NFPA 1561, *Standard on Emergency Services Incident Management System*, National Fire Protection Association, Quincy, MA, 2008.
3. Title 49, Code of Federal Regulations, Part 173, Subpart I (Class 7 Radioactive Materials). U.S. Department of Transportation, 1200 New Jersey Ave, SE, Washington, DC 20590.
4. Title 49, Code of Federal Regulations, Part 173.431. U.S. Department of Transportation, 1200 New Jersey Ave, SE, Washington, DC 20590.
5. NFPA 1991, *Standard on Vapor-Protective Ensembles for Hazardous Materials Emergencies,* National Fire Protection Association, Quincy, MA, 2005.
6. NFPA 1992, *Standard on Liquid Splash-Protective Ensembles and Clothing for Hazardous Materials Emergencies*, National Fire Protection Association, Quincy, MA, 2005
7. Title 29, Code of Federal Regulations, Part 1910.120, U.S. Government Printing Office, Washington, DC.
8. NFPA 1971, *Standard on Protective Ensembles for Structural Fire Fighting and Proximity Fire Fighting*, National Fire Protection Association, Quincy, MA, 2007.
9. NFPA 1981, *Standard on Open-Circuit Self-Contained Breathing Apparatus (SCBA) for Emergency Services,* National Fire Protection Association, Quincy, MA 2007.
10. NIOSH Statement of Standard for NIOSH CBRN SCBA Testing, NIOSH, 395 E Street, S.W. Suite 9200, Patriots Plaza Building, Washington, DC 20201.

11. NFPA 1994, *Standard on Protective Ensembles for First Responders to CBRN Terrorism Incidents*, National Fire Protection Association, Quincy, MA, 2008.
12. *U.S. Army Center for Health Promotion and Preventative Medicine (USACHPPM) Technical Guide 244.* United States Army Center for Health Promotion & Preventive Medicine, 5158 Blackhawk Road, Aberdeen Proving Ground, MD 21010-5403.

Competencies for Awareness Level Personnel

CHAPTER 4

The term *awareness level personnel* as discussed in this chapter includes anyone who in the course of their normal duties could encounter hazardous materials or weapons of mass destruction. Awareness level personnel are trained to recognize these kinds of materials, protect themselves from the materials, inform others (including trained response personnel), and secure the area.

4.1 General

4.1.1 Introduction.

4.1.1.1 Awareness level personnel shall be persons who, in the course of their normal duties, could encounter an emergency involving hazardous materials/weapons of mass destruction (WMD) and who are expected to recognize the presence of the hazardous materials/WMD, protect themselves, call for trained personnel, and secure the area.

The committee adopted the term *awareness level personnel* in the 2008 edition to reflect the new terminology for personnel who have a potential of encountering hazardous materials/ WMD as part of their regular duties. These personnel include public works employees, maintenance workers, and others who can see an event occur and can follow the actions prescribed in this chapter.

4.1.1.2 Awareness level personnel shall be trained to meet all competencies of this chapter.

4.1.1.3 Awareness level personnel shall receive additional training to meet applicable governmental occupational health and safety regulations.

Regardless of the entity they represent, awareness level personnel should have the required skills to accomplish the activities at a hazardous materials/WMD incident that are critical to safely and effectively performing the duties described in this chapter.

4.1.2 Goal.

4.1.2.1 The goal of the competencies at the awareness level shall be to provide personnel already on the scene of a hazardous materials/WMD incident with the knowledge and skills to perform the tasks in 4.1.2.2 safely and effectively.

4.1.2.2 When already on the scene of a hazardous materials/WMD incident, the awareness level personnel shall be able to perform the following tasks:

(1) Analyze the incident to determine both the hazardous material/WMD present and the basic hazard and response information for each hazardous material/WMD agent by completing the following tasks:

   (a) Detect the presence of hazardous materials/WMD.

When detecting or confirming the presence of hazardous materials, personnel must take appropriate safety precautions. Personnel at this level do not have any specialized protective clothing or equipment and must therefore exercise caution when confirming the presence of hazardous materials or WMD at an incident.

At this level, personnel must also be made aware of the potential of further activity taking place if the incident has been caused intentionally by individuals seeking to cause harm for any purpose.

(b) Survey a hazardous materials/WMD incident from a safe location to identify the name, UN/NA identification number, type of placard, or other distinctive marking applied for the hazardous materials/WMD involved.

(c) Collect hazard information from the current edition of the DOT *Emergency Response Guidebook.*

The first step in analyzing the incident, as required in 4.1.2.2(1), is to determine whether hazardous materials/WMD are present. A few typical indicators that awareness level personnel should immediately recognize include the following:

- Operators or witnesses who report the presence of a hazard or a breach of a container
- Individuals who exhibit signs of exposure
- Occupancy type (such as paint supply stores, gas stations, warehouses, industrial and manufacturing plants, etc.)
- Placards, labels, and markings
- Type of containers involved
- Written resources (such as shipping papers indicating the presence of hazardous materials)
- Fume exhaust stacks on building roofs and the presence of fires or explosions

Hazardous materials can also be indicated on pre-incident surveys, pre-plans, and response plans maintained by local emergency response committees (LEPCs). NFPA 472 clearly indicates that training in recognizing and identifying a hazardous materials/WMD incident for awareness level personnel is required and that the employer, whether in the public or private sector, is responsible for seeing that employees likely to be on the scene receive such training.

(2) Implement actions consistent with the emergency response plan, the standard operating procedures, and the current edition of the DOT *Emergency Response Guidebook* by completing the following tasks:

(a) Initiate protective actions.

(b) Initiate the notification process.

The actions required in 4.1.2.2(2) must be taken with appropriate safety precautions. Personnel at this level do not have any specialized protective clothing or equipment and must, therefore, exercise extreme caution when confirming the presence of hazardous materials/WMD. These personnel also must be trained in the appropriate community response procedures and know how to initiate them. Personnel should be familiar with their own organization's response plan and with their role. In many cases, these actions consist of notifying the local emergency responders, such as the fire department, and securing the immediate area to prevent exposure to others.

4.2 Competencies — Analyzing the Incident

4.2.1* Detecting the Presence of Hazardous Materials/WMD.

Given examples of various situations, awareness level personnel shall identify those situations where hazardous materials/WMD are present and shall meet the following requirements:

A.4.2.1 The AHJ should identify local situations where hazardous materials/WMD might be encountered. This can include areas where hazardous materials are transported, local industries and facilities where hazardous materials are used or stored, and locations where illicit laboratories might be likely.

(1)* Identify the definitions of both *hazardous material* (or *dangerous goods*, in Canada) and *WMD*.

A.4.2.1(1) See Annex I.

(2) Identify the UN/DOT hazard classes and divisions of hazardous materials/WMD and identify common examples of materials in each hazard class or division.

Because proper identification of hazardous materials/WMD is extremely important, the actions of personnel trained to the awareness level can be critical to a successful emergency response. By being able to identify the hazard classes as is required in 4.2.1(2), these personnel have a better understanding of potential problems that might arise (see Annex J of NFPA 472).

(3)* Identify the primary hazards associated with each UN/DOT hazard class and division.

A.4.2.1(3) See Annex J.

By accurately identifying the type of hazardous materials/WMD present and the primary hazards the materials involve as required in 4.2.1(3), awareness level personnel can begin to take the correct protective actions early, if it is safe to do so. In addition, awareness level personnel can give this information to the hazardous materials/WMD response team members, who will then understand the type of incident they are responding to and be able to request any specialized equipment or additional resources needed. For example, a responder would initiate considerably different actions dealing with a Class 1, Division 1.1 material (mass detonating explosives) than dealing with a Class 2, Division 2.2 (nonflammable gases) material. See Commentary Table I.4.1 for a full listing of U.S. Department of Transportation (DOT) hazard classes.

(4) Identify the difference between hazardous materials/WMD incidents and other emergencies.

The adverse consequences of exposure to a hazardous material can be far-reaching and severe. In a large-scale emergency, responders of all levels can be at risk simply because a complex incident, such as the train wreck shown in Exhibit I.4.1, involves so many factors, including hazardous material exposure. Hazardous materials/WMD emergencies stand apart from other types of emergencies because they present such a large potential for doing great harm and because personnel must be specifically trained and equipped to deal with them properly. NFPA 472 also requires that personnel take care to avoid worsening the situation.

Incidents involving WMDs might have additional differences including the following:

- **Intent:** An act of terrorism involving a WMD is different from normal emergencies in that it is intended to cause damage, inflict harm, and kill.
- **Severity and Complexity:** WMD incidents can involve large numbers of casualties or unusual materials (such as radioactive materials) with which awareness level personnel might have little practical experience.
- **Crime Scene Management:** Terrorist attacks are crimes, and preservation of evidence becomes an extremely important consideration during a response to a terrorist attack.
- **Incident Command:** Most terrorist incidents require some form of unified command. Law enforcement will have jurisdiction over all incidents involving terrorism.
- **Secondary Devices/Attacks and Armed Resistance:** Attacks designed to incapacitate emergency responders include the following:
  - Secondary events intended to kill, incapacitate, or delay emergency responders
  - Armed resistance and assault
  - Use of weapons
  - Booby traps

COMMENTARY TABLE I.4.1 International Hazard Classes and Divisions

Classes and Divisions	*Examples of Materials (by Hazard Class or Division)*	*General Hazard Properties (Not All-Inclusive)*
Class 1—Explosives and blasting agents		
Division 1.1—explosives with mass explosion hazard	Dynamite, TNT, black powder	Explosive; exposure to heat, shock, or contamination could result in thermal and mechanical hazards
Division 1.2—explosives with projection hazard	Projectiles with bursting charges	
Division 1.3—explosives with fire, minor blast, or minor projection hazard	Propellant explosives, rocket motors, special fireworks	
Division 1.4—explosive devices with minor explosion hazard	Common fireworks	
Division 1.5—very insensitive explosives	Ammonium nitrate–fuel oil mix, blasting agent	
Division 1.6—extremely insensitive explosives		
Class 2—Gases		
Division 2.1—flammable gases	Propane, butadiene, acetylene, methyl chloride	Under pressure; container may rupture violently (fire and nonfire); may be flammable, poisonous, corrosive, asphyxiant, and/or thermally unstable
Division 2.2—nonflammable, nonpoisonous (nontoxic) gas	Carbon dioxide, anhydrous ammonia	
Division 2.3—poisonous (toxic) gas by inhalation	Arsine, phosgene, chlorine, methyl bromide	
Class 3—Flammable liquids		
(flashpoint less than 141°F [60°C])	Acetone, amyl acetate, gasoline, methyl alcohol	Flammable; container may rupture violently from heat/fire; may be corrosive, toxic, and/or thermally unstable
Class 4—Flammable solids and reactive liquids and solids		
Division 4.1—flammable solids	Nitrocellulose, matches	Flammable, some spontaneously; may be water reactive, toxic, and/or corrosive; may be extremely difficult to extinguish
Division 4.2—spontaneously combustible materials	Phosphorus, aluminum alkyls, charcoal	
Division 4.3—dangerous when wet materials	Calcium carbide, potassium	
Class 5—Oxidizers and organic peroxides		
Division 5.1—oxidizers	Ammonium nitrate fertilizer	Supplies oxygen to support combustion; sensitive to heat, shock, friction, and/or contamination
Division 5.2—organic peroxides	Dibenzoyl peroxide	
Class 6—Toxic (poisonous) materials		
Division 6.1—poisonous (toxic) material	Aniline, arsenic, tear gas, carbon tetrachloride	Toxic by inhalation, ingestion, and skin/eye contact; may be flammable
Division 6.2—infectious substances	Anthrax, botulism, tetanus	
Class 7—Radioactive materials	Cobalt, uranium hexafluoride	May cause burns and biologic effects; may be in form of energy or matter
Class 8—Corrosive materials	Hydrochloric acid, sulfuric acid, sodium hydroxide	Disintegration of contacted tissues; may be fuming and/or water reactive
Class 9—Miscellaneous hazardous materials	Adipic acid, molten sulfur, dry ice, PCBs	

Source: NFPA *Fire Protection Handbook*, 20th edition, Table 13.8.1.

EXHIBIT I.4.1 *At this California incident, fires from a train wreck ignited chemicals transported on board and one person died. (Source: Photo by William Wilson Lewis III/AP Worldwide)*

(5) Identify typical occupancies and locations in the community where hazardous materials/WMD are manufactured, transported, stored, used, or disposed of.

Individuals involved with hazardous materials/WMD response planning at the awareness level are required in 4.2.1(5) to know where in a given community or industrial or manufacturing location that hazardous materials/WMD are most likely to be found. This knowledge helps personnel select appropriate emergency actions that can be planned and coordinated as necessary with responders from the public and private sectors. Local planning for response should be based, in part, on the knowledge that particular hazardous materials/WMD might be present at certain fixed sites within a specified area or community. For example, an area fire chief should know where the auto body painting businesses are located in the community and the chemicals and solvents that these sites are likely to contain. Similarly, the plant manager of a pharmaceutical company should be aware of what chemicals and chemical compounds are on the site at any given time. The manager should be able to advise personnel what chemicals are present, where and how they are stored, and what, if any, built-in precautions have been taken.

(6) Identify typical container shapes that can indicate the presence of hazardous materials/WMD.

The configuration of some containers is so unusual that it signals the presence of some hazardous materials/WMD. Containers that provide clues include those used for radioactive materials, pressurized products, cryogenics, and corrosives. Exhibits I.4.2 through I.4.8 show some typical container shapes that should alert awareness level personnel to the presence of hazardous materials. Examples of container shapes that should clue awareness level personnel to the possible presence of hazardous materials can be found in the front of the current edition of the *Emergency Response Guidebook* (ERG) [1].

(7) Identify facility and transportation markings and colors that indicate hazardous materials/WMD, including the following:

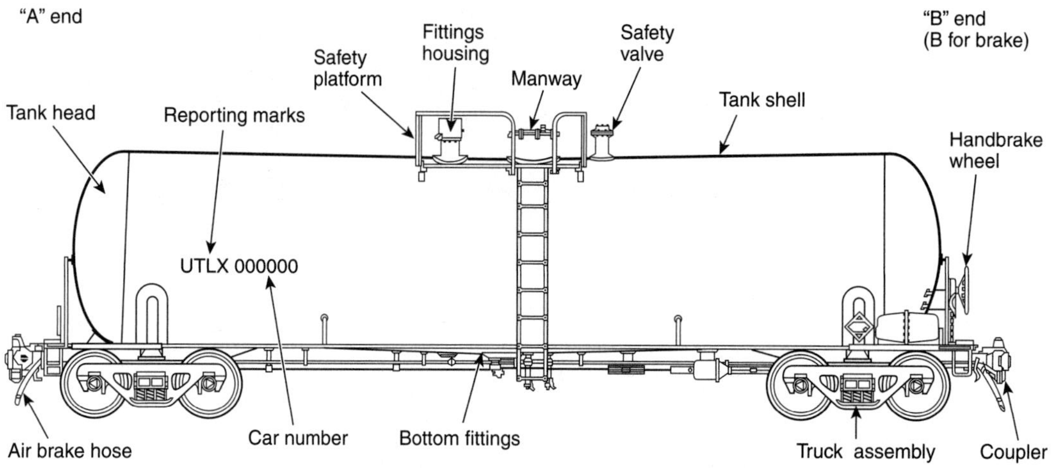

EXHIBIT I.4.2 *Some of the basic features of a tank car tank are identified in this illustration.*

EXHIBIT I.4.3 *A typical older-style nonpressure tank car is shown here with an expansion dome.*

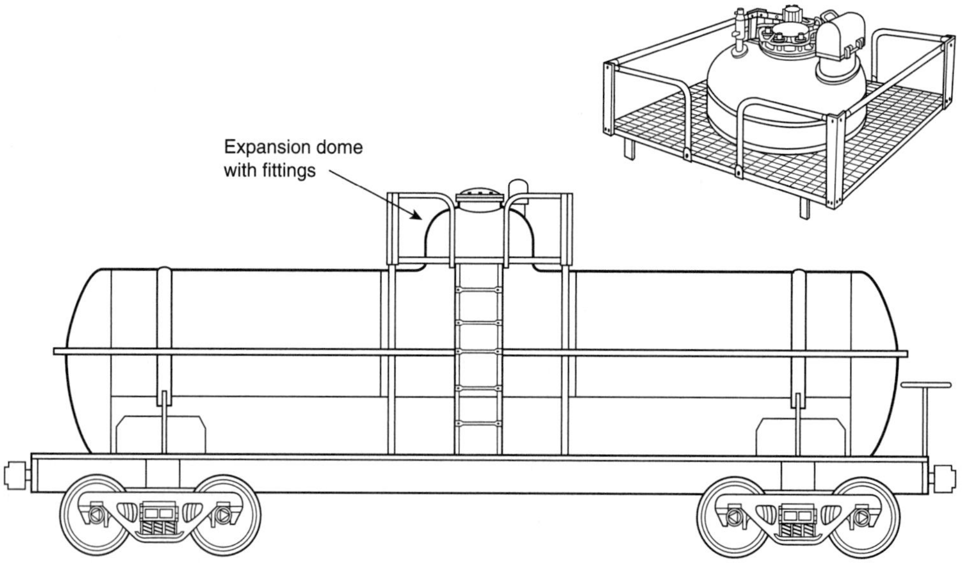

(a) Transportation markings, including UN/NA identification number marks, marine pollutant mark, elevated temperature (HOT) mark, commodity marking, and inhalation hazard mark

Several numbers used on placards are shown in Exhibit I.4.9. Exhibit I.4.9(a) and (c) illustrate two different ways of showing the four-digit product identification number on a placard. Exhibit I.4.9(b) illustrates the use of the class or division number and the product identification number on the same placard. For example, the class or division number might be displayed at the bottom of the placard, as the number 6 is shown in Exhibit I.4.9(b). Personnel should also be familiar with the intermodal container requirements in the *European Agreement Concerning the International Carriage of Dangerous Goods by Road* (ADR) [2] and the *International Regulations Concerning the Carriage of Dangerous Goods by Rail* (RID) [3] markings. The colors of the placards also indicate the hazard class. For example, yellow is

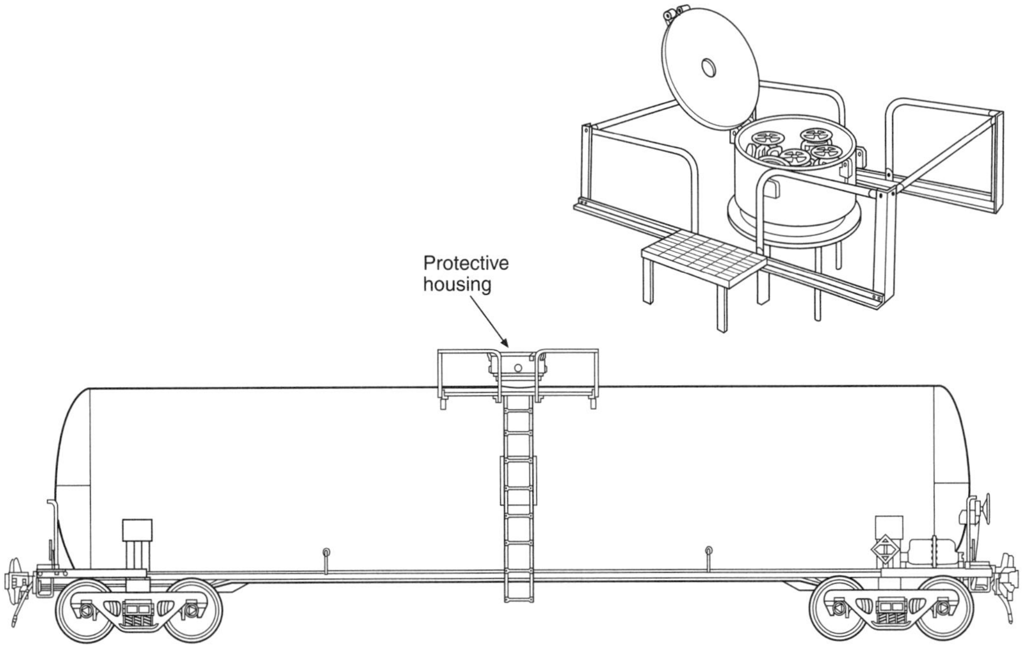

EXHIBIT I.4.4 *A typical pressure tank car is shown here with protective housing (inset).*

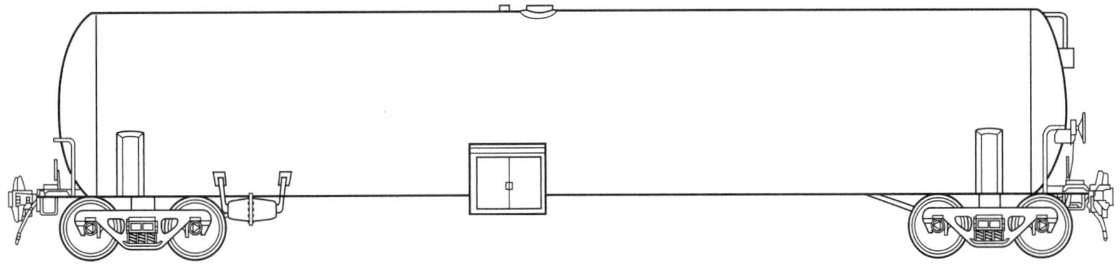

EXHIBIT I.4.5 *A typical cryogenic liquid tank car is shown in this illustration.*

EXHIBIT I.4.6 *Various modes of transportation can be used for tank containers.*

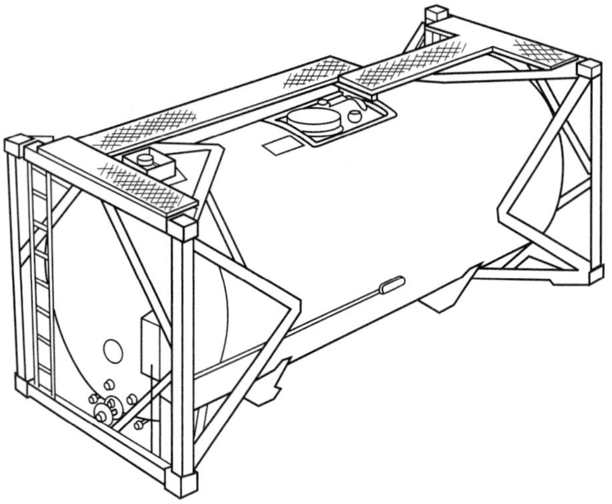

EXHIBIT I.4.7 *The diagram illustrates a nonpressure tank container.*

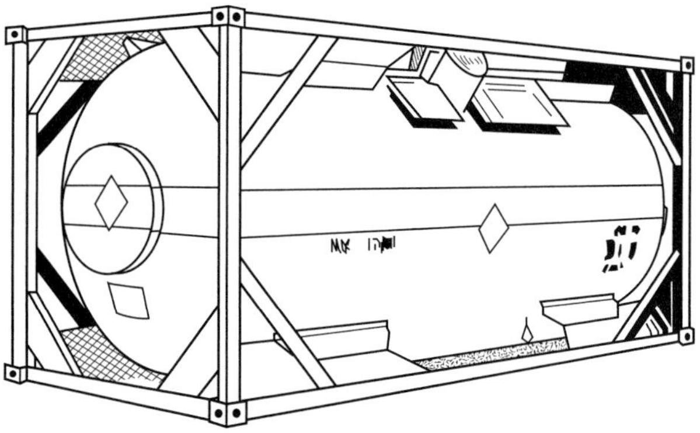

EXHIBIT I.4.8 *The diagram depicts a typical pressure tank container.*

(a) (b) (c)

EXHIBIT I.4.9 *These placards show how numbers are displayed: (a) and (c) show two ways to depict the four-digit product identification number on a placard and (b) shows the class or division number on the placard as well.*

used for Class 5, Oxidizing Substances, and black and white is used to denote Class 8, Corrosives.

(b) NFPA 704, *Standard System for the Identification of the Hazards of Materials for Emergency Response*, markings

See 4.2.1(8).

(c)* Military hazardous materials/WMD markings

A.4.2.1(7)(c) The responder should understand the standard military fire hazard and chemical hazard markings.

The military uses different types of markings for shipments on military facilities and knowledge of this labeling is required by 4.2.1(7)(c). The military marking system establishes the following four hazard classes:

1. Class 1 Hazard: Materials that present a mass detonation hazard.
2. Class 2 Hazard: Materials that present an explosive with fragmentation hazard.
3. Class 3 Hazard: Materials with a mass fire hazard.
4. Class 4 Hazard: Materials that present a moderate fire hazard.

In addition, four special warnings are indicated separately as follows:

- Chemical hazard
  - Highly toxic
  - Harassing agents
  - White phosphorus munitions
- Apply no water
- Wear protective breathing apparatus
- Special hazard communication markings

These warnings can be found in some facilities, and they identify hazardous materials/WMD. The U.S. military uses both special hazard symbols and detonation hazard symbols, which are shown in Exhibit I.4.10.

(d) Special hazard communication markings for each hazard class

(e) Pipeline markings

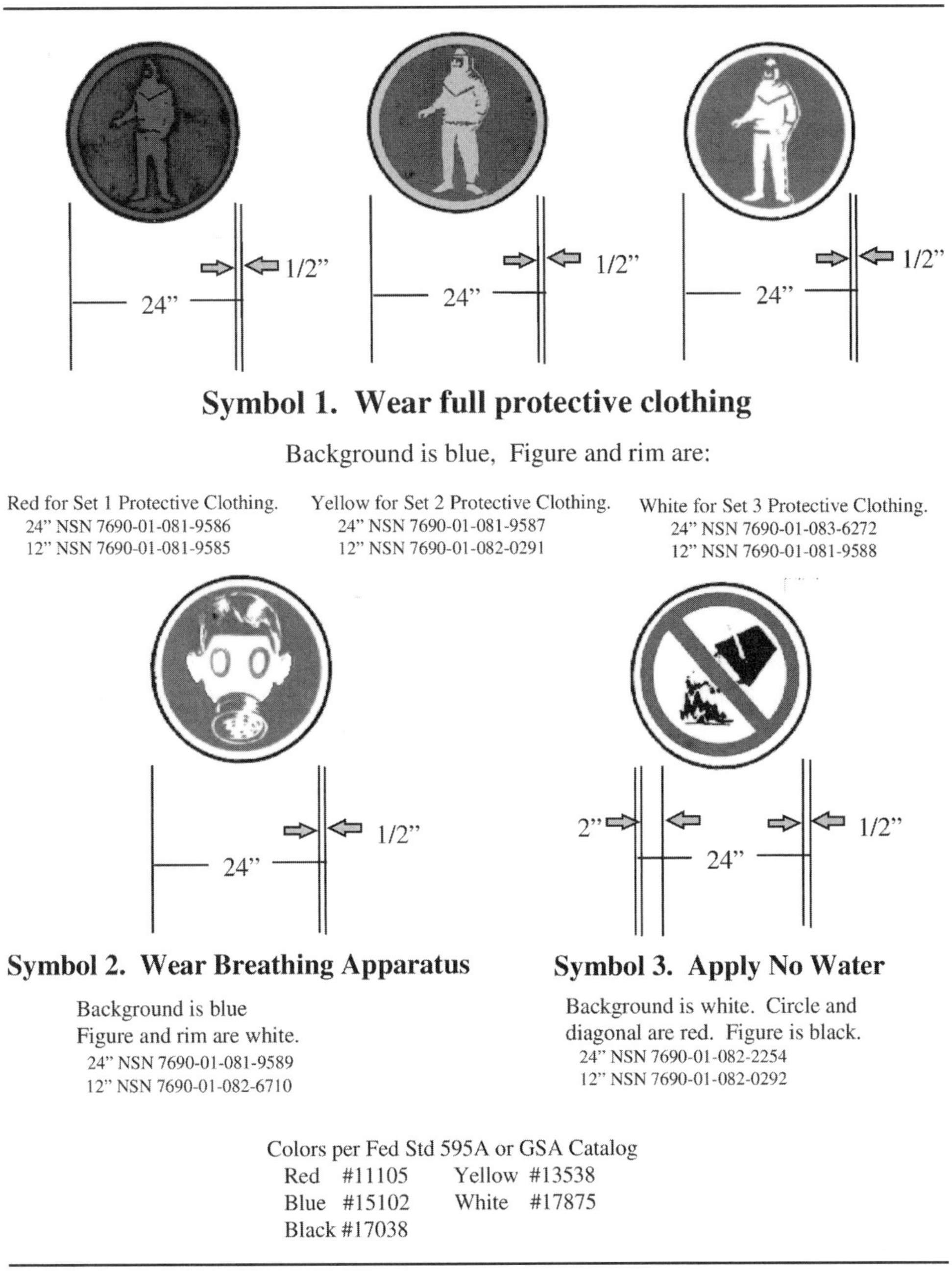

EXHIBIT I.4.10 *U.S. Military Special Hazard Symbols. (Source: U.S. Air Force Manual 91-201, Explosives Safety Standards, Figure 2.2, October 2001)*

Pipeline markings are usually metal signs or plastic poles placed adjacent to and above a hazardous materials pipeline. The markings contain information about the ownership of the pipeline, the product being carried, and a 24-hour emergency contact number.

(f) Container markings

Often, markings on a container provide some indication as to the type of product it holds. These markings may include product names such as chlorine.

(8) Given an NFPA 704 marking, describe the significance of the colors, numbers, and special symbols.

NFPA's *Fire Protection Guide to Hazardous Materials* contains a great deal of information about hazardous chemicals, their fire hazard properties, and hazardous chemical reactions [4]. In addition, the guide includes NFPA 704, *Standard System for the Identification of the Hazards of Materials for Emergency Response* [5]. The NFPA 704 marking system is based on the "704 diamond," which visually presents information on the following three principal categories of hazard as well as the degree of severity of each hazard:

1. Health (Blue)
2. Flammability (Red)
3. Instability (Yellow)

In addition, the fourth part of the diamond at the six o'clock position is reserved for indicating special hazards. (See Exhibit I.4.11.) These special hazards are as follows:

1. Reactivity with water
2. Oxidizing ability (see Annex J of NFPA 472)

The NFPA 704 diamond symbol is intended to provide immediate general information to awareness level personnel (and emergency responders), who are required in 4.2.1(8) to be able to interpret the placard.

The five degrees of hazard, in descending order, have the following general meanings to fire fighters.

- **Degree of Hazard 4:** Fire is too dangerous to approach with standard fire-fighting equipment and procedures. Withdraw and obtain expert advice on how to handle fire.
- **Degree of Hazard 3:** Fire can be fought using methods intended for extremely hazardous situations, such as remote-control monitors or personal protective equipment that prevents all bodily contact.
- **Degree of Hazard 2:** Fire can be fought with standard procedures, but hazards are present that can be handled safely only with certain special equipment or procedures.

EXHIBIT I.4.11 This quick reference explains how the NFPA 704 identification system is used.

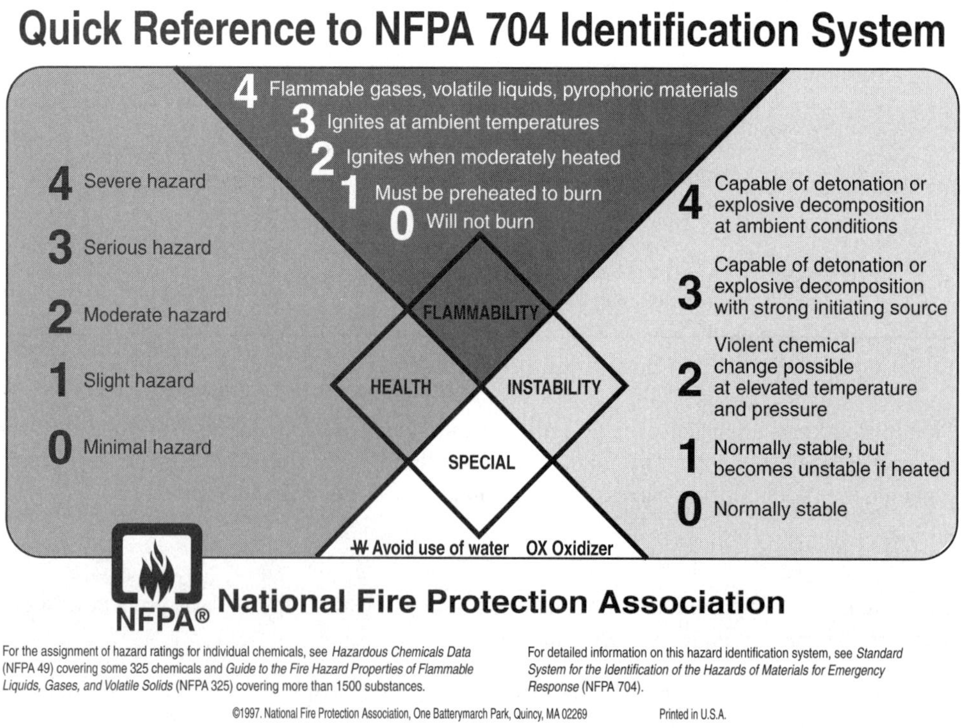

- **Degree of Hazard 1:** Nuisance hazards are present that require some care, but standard fire-fighting procedures can be used.
- **Degree of Hazard 0:** No special hazards are present; therefore, no special measures are needed.

NFPA 704 describes the hazard categories and the security levels that the various numbers indicate for each hazard. The following, which is adapted from NFPA 704, summarizes the hazard information and recommends protective actions.

The numbers from 0 through 4 are placed in the three upper squares of the diamond to show the degree of hazard present for each of the three hazard categories. The 0 indicates the lowest degree of hazard (but not NO hazard), and the 4 indicates the highest. The fourth square, at the bottom, is used for special information. Two symbols for this bottom space are recognized by NFPA 704. Either symbol, or both, can appear in this square, or it may be blank.

The recognized symbols are as follows:

- A letter W with a bar through it (W) indicates that a material is unusually reactive with water. This symbol does not mean that water should not be used, since some forms of water, such as fog or spray, can be used in some cases. What it does mean is that water can cause a hazard, so it must be used very cautiously pending proper information.
- The letters OX indicate an oxidizer.

Although not recognized by NFPA 704, some users insert the letters ALK for alkaline materials, and ACID for acidic materials, in this square.

Health Hazards. In general, the health hazard in fire fighting is that of a single exposure, the duration of which can vary from a few seconds up to an hour. The physical exertion demanded in fire fighting or other emergencies can be expected to intensify the effects of any exposure. In assigning degrees of danger, local conditions must be considered. The following explanation is based on the use of protective equipment normally worn by fire fighters.

- **Degree of Hazard 4:** Materials that under emergency conditions can be lethal. A few whiffs of the vapor could cause death, or the vapor or liquid could be fatal on penetrating the fire fighter's normal full-protective clothing. The normal full-protective clothing and breathing apparatus available to the average fire department does not provide adequate protection against inhalation or skin contact with these materials.
- **Degree of Hazard 3:** Materials that under emergency conditions can cause serious or permanent injury. Full-protective clothing, self-contained breathing apparatus (SCBA), gloves, boots, and bands around legs, arms, and waist should be provided. No skin surfaces should be exposed.
- **Degree of Hazard 2:** Materials that under emergency conditions can cause temporary incapacitation or residual injury. Full-face mask self-contained breathing apparatus that provides eye protection should be provided.
- **Degree of Hazard 1:** Materials that under emergency conditions can cause significant irritation. Wearing SCBA might be desirable.
- **Degree of Hazard 0:** Materials that, under emergency conditions, would offer no health hazard beyond that of ordinary combustible materials.

Flammability Hazards. Susceptibility to burning is the basis for assigning degrees within this category (see Exhibit I.4.11). The method of attacking the fire is influenced by this susceptibility factor.

- **Degree of Hazard 4:** Flammable gases, flammable cryogenic materials, very volatile flammable liquids, Class IA liquids, or materials that ignite spontaneously in air (pyrophoric). If possible, shut on/off flow and keep cooling water streams on exposed tanks or containers. Withdrawal might be necessary.
- **Degree of Hazard 3:** Materials that can be ignited under almost all normal temperature conditions (Class IB). Water might be ineffective because of the low flash point.

- **Degree of Hazard 2:** Materials that must be moderately heated or exposed to relatively high ambient temperatures before ignition occurs (Class II and Class IIIA liquids). Water spray can be used to extinguish the fire because the material can be cooled below its flash point.
- **Degree of Hazard 1:** Materials that must be preheated before ignition can occur (Class IIIB liquids). Water can cause frothing if it gets below the surface of the liquid and turns to steam. If this occurs, water fog gently applied to the surface causes a frothing that extinguishes the fire.
- **Degree of Hazard 0:** Materials that do not burn.

Instability Hazards. The assignment of relative degrees of hazard in the reactivity category is based on the susceptibility of materials to release energy either by themselves or in combination with other materials (see Exhibit I.4.11). Fire exposure was one of the factors considered along with conditions of shock and pressure.

- **Degree of Hazard 4:** Materials that are readily capable of detonation or explosive decomposition at normal temperatures and pressures. Includes materials that are sensitive to localized thermal or mechanical shock. If they are involved in a massive fire, vacate the area.
- **Degree of Hazard 3:** Materials that are capable of detonation, explosive decomposition, or explosive reaction, but require a strong initiating source, or that must be heated under confinement before initiation. Includes materials that are sensitive to thermal or mechanical shock at elevated temperatures and pressures, or that can react explosively with water without requiring heat or confinement. Fire fighting should be conducted from behind explosion-resistant locations.
- **Degree of Hazard 2:** Materials that readily undergo a violent chemical change at elevated temperatures and pressures. Includes materials that react violently with water or that can form potentially explosive mixtures with water. Use portable monitors, hose holders, or straight hose streams from a distance to cool the tanks and the material in them. Use caution.
- **Degree of Hazard 1:** Materials that are normally stable but can become unstable at elevated temperatures and pressures or that react vigorously but not violently with water. Includes materials that change or decompose on exposure to air, light, or moisture. Use normal precautions as in approaching any fire.
- **Degree of Hazard 0:** Materials that are normally stable and thus do not present any reactivity hazard to fire fighters.

Special Information. When W appears at the bottom in the fourth space (see Exhibit I.4.11):

- **Degree of Hazard 4:** W is not used with reactivity hazard 4.
- **Degree of Hazard 3:** In addition to the three hazards, these materials can react explosively with water. Explosion protection is essential if water is to be used.
- **Degree of Hazard 2:** In addition to the three hazards, these materials can react with water or form potentially explosive mixtures with water.
- **Degree of Hazard 1:** In addition to the three hazards, these materials can react vigorously but not violently with water.
- **Degree of Hazard 0:** W is not used with reactivity hazard 0.

Methods of Presentation. Several methods for presentation of the ratings are shown in Exhibit I.4.11. A basic requirement is that numbers be spaced as though they were in the diamond outline. Chapter 6 of NFPA 704 presents a recommended layout and sizes for the symbol, a distance-legibility table, and several examples using the symbol.

Assigning Degrees of Hazard. Numbers (degrees of hazard) for use in the diamond are assigned on the basis of the worst hazard expected in the area, whether it is from hazards of the original material or of its combustion or breakdown products. The effects of local conditions must be considered. For instance, a drum of carbon tetrachloride sitting in a well-ventilated storage shed presents a different hazard than a drum sitting in an unventilated basement.

Advantages of the NFPA 704 System. The NFPA 704 system can warn against hazards under fire conditions of materials that other information systems class as nonhazardous. For example, edible tallow produces toxic and irritating combustion products. The tallow would be given a "2" degree of health hazard, indicating the need for air-supplied respiratory equipment.

NFPA 704 markings also can warn against overall fire hazards in an area. On the door of a laboratory or storage room, the system can warn of the worst hazards likely in a fire situation. Such information is useful both in preplanning and in actual fires.

The NFPA 704 system also can be used without a supplementary manual. Because of its simplicity, the general meanings of the numbers are easily understood and the whole symbol is read and interpreted quickly on the spot, in poor light, and at a distance.

Limitations of the NFPA 704 System. The NFPA 704 system supplies only minimum information on the hazards themselves. Because the system informs on protective measures, the same number can be used for various types of hazards so that, for instance, a health hazard "3" means "serious hazard" without saying whether the hazard is corrosive to the skin or toxic by absorption through the skin. Thus, the symbol is most useful to trained or informed persons.

(9) Identify U.S. and Canadian placards and labels that indicate hazardous materials/WMD.
(10) Identify the following basic information on material safety data sheets (MSDS) and shipping papers for hazardous materials:

In 4.2.1(10), awareness level personnel are required to identify the following information:

- Manufacturer's name and location
- Name and family of the chemical
- Hazardous ingredients
- Physical data
- Fire and explosion hazard data
- Health hazard data
- Spill or leak procedures
- Special protection information
- Special precautions needed when dealing with the material

While these are the sections required in an MSDS by mandatory OSHA standards, OSHA encourages supplemental information be provided as well. ANSI standard Z400.1, *Hazardous Industrial Chemicals — Material Safety Data Sheets — Preparation* [6], specifies 16 different sections for an MSDS. OSHA now recommends that the ANSI format be used, although this format is not presently required. See Commentary Table I.4.2 for the minimum information needed on a safety data sheet (SDS).

The Global Harmonization System (GHS) for Hazard Classification and Communication sets forth recommendations for minimum information to be provided on a SDS, the GHS equivalent of an MSDS, which is shown in Exhibit I.4.12. These sheets are being used worldwide. SDSs include essentially the same information as recommended by ANSI for an MSDS, with some minor differences. The minimum information required for an SDS is shown in Exhibit I.4.12.

(a) Identify where to find MSDS.

Employers are required by 4.2.1(10)(a) to maintain a MSDS for every hazardous chemical used at their facilities. This requirement is part of the Hazard Communication Standard Title 29 CFR Part 1910.1200 [7]. MSDSs are also available from manufacturers, suppliers, and shippers; and they are sometimes attached to non-bulk containers and shipping papers.

COMMENTARY TABLE I.4.2 *Safety Data Sheet Minimum Requirements*

Globally Harmonized System (GHS) Requirements	
Category	***Information***
1. Identification of the substance or mixture and of the supplier	• GHS product identifier. • Other means of identification. • Recommended use of the chemical and restrictions on use. • Supplier's details (including name, address, phone number, etc.). • Emergency phone number.
2. Hazards identification	• GHS classification of the substance/mixture and any national or regional information. • GHS label elements, including precautionary statements. (Hazard symbols may be provided as a graphical reproduction of the symbols in black and white or the name of the symbol, e.g., flame, skull and crossbones.) • Other hazards which do not result in classification (e.g., dust explosion hazard) or are not covered by the GHS.
3. Composition/information on ingredients	*Substance* • Chemical identity. • Common name, synonyms, etc. • Chemical abstract service (CAS) number, European Communities (EC) number, etc. • Impurities and stabilizing additives which are themselves classified and which contribute to the classification of the substance. *Mixture* • The chemical identity and concentration or concentration ranges of all ingredients which are hazardous within the meaning of the GHS and are present above their cutoff levels. Note: For information on ingredients, the competent authority rules for Confidential Business Information (CBI) take priority over the rules for product identification.
4. First aid measures	• Description of necessary measures, subdivided according to the different routes of exposure, i.e., inhalation, skin and eye contact, and ingestion. • Most important symptoms/effects, acute and delayed. • Indication of immediate medical attention and special treatment needed, if necessary.
5. Firefighting measures	• Suitable (and unsuitable) extinguishing media. • Specific hazards arising from the chemical (e.g., nature of any hazardous combustion products). • Special protective equipment and precautions for firefighters.
6. Accidental release measures	• Personal precautions, protective equipment and emergency procedures. • Environmental precautions. • Methods and materials for containment and cleaning up.
7. Handling and storage	• Precautions for safe handling. • Conditions for safe storage, including any incompatibilities.
8. Exposure controls/personal protection	• Control parameters, e.g., occupational exposure limit values or biological limit values. • Appropriate engineering controls. • Individual protection measures, such as personal protective equipment.
9. Physical and chemical properties	• Appearance (physical state, color, etc.). • Odor. • Odor threshold. • pH. • melting point/freezing point.

COMMENTARY TABLE I.4.2 *Continued*

Globally Harmonized System (GHS) Requirements	
Category	***Information***
	• initial boiling point and boiling range. • flash point. • evaporation rate. • flammability (solid, gas). • upper/lower flammability or explosive limits. • vapor pressure. • vapor density. • relative density. • solubility(ies). • partition coefficient: n-octanol/water. • autoignition temperature. • decomposition temperature.
10. Stability and reactivity	• Chemical stability. • Possibility of hazardous reactions. • Conditions to avoid (e.g., static discharge, shock or vibration). • Incompatible materials. • Hazardous decomposition products.
11. Toxicological information	Concise but complete and comprehensible description of the various toxicological (health) effects and the available data used to identify those effects, including: • Information on the likely routes of exposure (inhalation, ingestion, skin and eye contact); • Symptoms related to the physical, chemical and toxicological characteristics; • Delayed and immediate effects and also chronic effects from short- and long-term exposure; • Numerical measures of toxicity (such as acute toxicity estimates).
12. Ecological information	• Ecotoxicity (aquatic and terrestrial, where available). • Persistence and degradability. • Bioaccumulative potential. • Mobility in soil. • Other adverse effects.
13. Disposal considerations	• Description of waste residues and information on their safe handling and methods of disposal, including the disposal of any contaminated packaging.
14. Transport information	• United Nations (UN) Number. • UN Proper shipping name. • Transport Hazard class(es). • Packing group, if applicable. • Marine pollutant (Yes/No). • Special precautions which a user needs to be aware of or needs to comply with in connection with transport or conveyance either within or outside their premises.
15. Regulatory information	• Safety, health and environmental regulations specific for the product in question.
16. Other information including information on preparation and revision of the SDS	

Source: Adapted from Occupational Safety and Health Administration, "A Guide to *The Globally Harmonized System of Classification and Labeling of Chemicals* (GHS)," Figure 4.14, www.osha.gov/dsg/hazcom/ghs.html#4.0.

Hazardous Material Data Sheet	Containment System ID.

Material Name ______________________ DOT ID № ______ STCC № ________

Synonyms ______________________

Hazard Class ______________________

NFPA 704 Marking: Health ______ Flammability ______ Instability ______ Special ______

Physical Properties

Form	Color	Odor	Chemical Formula	Molecular Wt.
☐ Solid ☐ Liquid ☐ Gas				

Chemical Properties

Actual Temp.	Boiling Point	Melting Point	Vapor Pressure	Expansion Ratio	Specific Gravity	Vapor Density	Soluble?	Degree of Solubility
							☐ Yes	

Physical Hazards

Flammable (heat/fire) ☐ Yes
Cryogenic (cold) ☐ Yes
Oxidizer (supports combustion) ☐ Yes
Explosive ☐ Yes
Reactive ☐ Yes
To What? ______

Actual Temperature	Flash Point	Ignition Temperature
Actual Concentration	**Flammable Range**	**Toxic Products of Combustion**

Health Hazards

Acute Health Hazards:
Poisonous ☐ Yes
Corrosive ☐ Yes
To What? ______
Asphyxiation ☐ Yes
Etiologic ☐ Yes
Radiation ☐ Yes
Type: alpha beta gamma

Chronic Health Hazards:
Carcinogen ☐ Yes
Mutagen ☐ Yes
Teratogen ☐ Yes
Aquatic Hazard ☐ Yes

Actual Concentration	Non-Life-Threatening Exposure Limits		
	TLV-TWA(PEL)	TLV-C	TLV-STEL
Odor Threshold	**Life-Threatening Exposure Limits**		
	IDLH	LC_{50}	LD_{50}

Route of Entry ☒ = YES		Toxicity Rating	Notes
Inhalation	☐	1 2 3 4 5 6	
Dermal	☐	1 2 3 4 5 6	
Ingestion	☐	1 2 3 4 5 6	

Response Information

Evacuation Distances ______________________

First Aid ______________________

____

Personal Protective Equipment ______________________

Decontamination ______________________

Extinguishing Agents ______________________

Neutralizing Agents ______________________

EXHIBIT I.4.12 *A hazardous material data sheet is used for recording information obtained during the task of collecting and interpreting hazard and response information.*

(b) Identify major sections of an MSDS.

The major categories on an MSDS that indicate the presence of hazardous materials include hazardous ingredients, fire and explosion hazard data, health hazard data, reactivity data, spill or leak procedures, special protection information, and special precautions.

(c) Identify the entries on shipping papers that indicate the presence of hazardous materials.

Shipping papers can contain several pieces of information that indicate the presence of a hazardous material. The areas required in 4.2.1(10)(c) to be identified include the material's proper shipping name (which can be referenced in the blue-bordered pages of the ERG), the material's hazard class or division, ID number (which can be referenced in the ERG's yellow-bordered pages), the packing group (an indication of what type packaging is required for some products), total quantity, and emergency response telephone number.

Shipping documents are stored in the cab of a motor vehicle, the possession of a train crew member, a holder on the bridge of a vessel, or in the possession of an aircraft pilot. Exhibit I.4.13 illustrates the key elements of shipping papers.

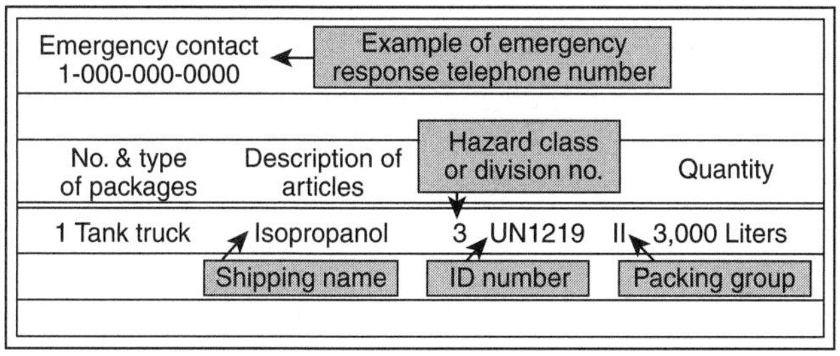

EXHIBIT I.4.13 *The key elements of a shipping document provide crucial information about the material being transported. (Source: Emergency Response Guidebook, 2004)*

(d) Match the name of the shipping papers found in transportation (air, highway, rail, and water) with the mode of transportation.
(e) Identify the person responsible for having the shipping papers in each mode of transportation.

Shipping papers provide important information about what is being transported. Once the mode of transport is identified, awareness level personnel can determine what the shipping paper is called, where it is located in or on the particular vehicle or vessel, and the responsible person who might have additional information or knowledge of the material. Commentary Table I.4.3 offers a listing by transit route of the likely location of shipping papers and the person responsible.

COMMENTARY TABLE I.4.3 ***A List of the Shipping Papers by Route of Transportation, Title of Shipping Paper, Location, and Responsible Person***

Mode of Transportation	***Title of Shipping Paper***	***Location of Shipping Papers***	***Responsible Person***
Highway	Bill of lading or freight bill	Cab of vehicle	Driver
Rail	Train list or train consist and/or waybill	With member of train crew (conductor or engineer)	Conductor
Water	Dangerous cargo manifest	Wheelhouse or pipelike container or barge	Captain or master
Air	Air bill with shipper's certification for restricted articles	Cockpit (may also be found attached to the outside of packages)	Pilot

Source: Adapted from *Fire Protection Handbook*, 20th edition, Table 13.8.3.

(f) Identify where the shipping papers are found in each mode of transportation.

See commentary for 4.2.1(10)(c) and Commentary Table I.4.3.

(g) Identify where the papers can be found in an emergency in each mode of transportation.

See the commentary for 4.2.1(10)(c) and Commentary Table I.4.3.

(11)*Identify examples of clues (other than occupancy/ location, container shape, markings/color, placards/ labels, MSDS, and shipping papers) the sight, sound, and odor of which indicate hazardous materials/WMD.

A.4.2.1(11) These clues include odors, gas leaks, fire or vapor cloud, visible corrosive actions or chemical reactions, pooled liquids, hissing of pressure releases, condensation lines on pressure tanks, injured victims, or casualties.

(12) Describe the limitations of using the senses in determining the presence or absence of hazardous materials/WMD.

One problem in using senses to evaluate the presence of hazardous materials/WMD is that if awareness level personnel are close enough to use them, he or she may already have been endangered. In addition, many gases are odorless, tasteless, and colorless. In the case of hydrogen sulfide, exposure to this toxic gas can deaden the olfactory senses, resulting in an inability to smell it. Individuals might therefore incorrectly assume the hazard has gone away.

(13)*Identify at least four types of locations that could be targets for criminal or terrorist activity using hazardous materials/WMD.

A.4.2.1(13) The following are examples of potential criminal or terrorist targets:

(1) Public assembly areas
(2) Public buildings
(3) Mass transit systems
(4) Places with high economic impact
(5) Telecommunications facilities
(6) Places with historical or symbolic significance
(7) Military installations
(8) Airports
(9) Industrial facilities

The word *terrorism* is defined in Title 18, Chapter 113, Section 2331 of the U.S. Code of Federal Regulations in the following ways [8]:

(1) The term *international terrorism* means activities that

(A) Involve violent acts or acts dangerous to human life that are a violation of the criminal laws of the United States or of any State, or that would be a criminal violation if committed within the jurisdiction of the United States or of any State;

(B) Appear to be intended

(i) To intimidate or coerce a civilian population;
(ii) To influence the policy of a government by intimidation or coercion; or
(iii) To affect the conduct of a government by mass destruction, assassination, or kidnapping; and

(C) Occur primarily outside the territorial jurisdiction of the United States, or transcend national boundaries in terms of the means by which they are accomplished, the persons they appear intended to intimidate or coerce, or the locale in which their perpetrators operate or seek asylum;

(5) The term *domestic terrorism* means activities that

(A) Involve acts dangerous to human life that are a violation of the criminal laws of the United States or of any State;

(B) Appear to be intended

(i) To intimidate or coerce a civilian population;
(ii) To influence the policy of a government by intimidation or coercion, or
(iii) To affect the conduct of a government by mass destruction, assassination, or kidnapping; and

(C) Occur primarily within the territorial jurisdiction of the United States.

Terrorists are arrested and convicted under existing criminal statutes, which is why criminal and terrorist activities are grouped together in A.4.2.1(13). Terrorist activities are currently receiving more attention than criminal activities because of the demonstrated willingness of terrorists to go to more extensive measures than domestic criminals to achieve their objectives.

Regarding the example of places of historical interest or symbolic significance, different locations could be targeted for domestic or international terrorists, and planning must consider all of the possibilities. For example, a domestic terrorist might target a women's health clinic where abortions are performed, but an international terrorist might target an historical site like the Washington Monument. Whether domestic or international, any terrorist incident could involve hazardous materials/WMD.

Public assembly targets might be selected by terrorists because of the opportunity to harm or intimidate large numbers of people in a single incident. Enclosed structures with a high density of people can be attacked in a dramatic way with a bomb, or the supply of food and water can be contaminated so that the harm is not immediately known.

Public buildings present a target of perhaps high occupant density but low occupant awareness of hazards, exits, and places of shelter.

A mass transportation system provides a confined location where large numbers gather, especially at intersection points where different systems meet or where different subway lines cross.

Places with high economic impact would include the gold depository at Fort Knox, but a domestic terrorist might target an area's largest employer.

Terrorists might want to disrupt telecommunications facilities because they are important as a means both for maintaining normal lifestyles and also for responding to a terrorist incident in progress and in the recovery time afterward.

Military installations might be targeted by domestic or international terrorists. These installations are normally high-risk targets for attack, but when a terrorist finds a weakness, attacking a military installation can have a high psychological impact.

Airports can be a target in the same way as other facilities where large numbers of people gather but also can be an entry point for hijacking a plane, as occurred on September 11, 2001.

Targeting an industrial facility can have a double terrorism impact. An incident involving a significant industrial facility can cause serious impact, not only economically in the long term from lost production or laid-off workers, but also immediately from intentional release of hazardous materials/WMD into the environment to injure workers, residents, and emergency personnel.

(14)*Describe the difference between a chemical and a biological incident.

A.4.2.1(14) A chemical incident is characterized by a rapid onset of medical symptoms (minutes to hours) and can have observed signatures such as colored residue, dead foliage, pungent odor, and dead insect and animal life. With biological incidents, the onset of symptoms usually requires days to weeks, and there are typically no characteristic signatures because biological agents are usually odorless and colorless. The area affected can be greater due to the migration of infected individuals because of the delayed onset of symptoms. An infected person could transmit the disease to another person.

(15)*Identify at least four indicators of possible criminal or terrorist activity involving chemical agents.

A.4.2.1(15) The following are examples of indicators of possible criminal or terrorist activity involving chemical agents:

(1) The presence of hazardous materials/WMD or laboratory equipment that is not relevant to the occupancy
(2) Intentional release of hazardous materials/WMD
(3) Unexplained patterns of sudden onset of similar, nontraumatic illnesses or deaths (patterns that might be geographic, by employer, or associated with agent dissemination methods)
(4) Unexplained odors or tastes that are out of character with the surroundings
(5) Multiple individuals exhibiting unexplained signs of skin, eye, or airway irritation
(6) Unexplained bomb- or munitions-like material, especially if it contains a liquid
(7) Unexplained vapor clouds, mists, and plumes
(8) Multiple individuals exhibiting unexplained health problems such as nausea, vomiting, twitching, tightness in chest, sweating, pinpoint pupils (miosis), runny nose (rhinorrhea), disorientation, difficulty breathing, convulsions, or death
(9) Trees, shrubs, bushes, food crops, and/or lawns that are dead, discolored, abnormal in appearance, or withered (not due to a current drought and not just a patch of dead weeds)
(10) Surfaces exhibiting oily droplets/films and unexplained oily film on water surfaces
(11) An abnormal number of sick or dead birds, animals, or fish
(12) Unusual security, locks, bars on windows, covered windows, or barbed wire

(16)*Identify at least four indicators of possible criminal or terrorist activity involving biological agents.

A.4.2.1(16) The following are examples of indicators of possible criminal or terrorist activity involving biological agents:

(1) Unusual number of sick or dying people or animals (any number of symptoms; time before symptoms are observed dependent on the agent used but usually days to weeks)
(2) Healthcare facilities reporting multiple casualties with similar signs or symptoms
(3) Unscheduled or unusual spray being disseminated, especially if outdoors during period of darkness
(4) Abandoned spray devices (devices with no distinct odors)

(17) Identify at least four indicators of possible criminal or terrorist activity involving radiological agents.
(18) Identify at least four indicators of possible criminal or terrorist activity involving illicit laboratories (clandestine laboratories, weapons lab, ricin lab).

Exhibit I.4.14 shows a hazmat team member responding to an illicit laboratory incident.

(19) Identify at least four indicators of possible criminal or terrorist activity involving explosives.

Explosive/incendiary attack indicators include the following:

- Warning or threat of an attack or received intelligence
- Reports of an explosion
- Explosion

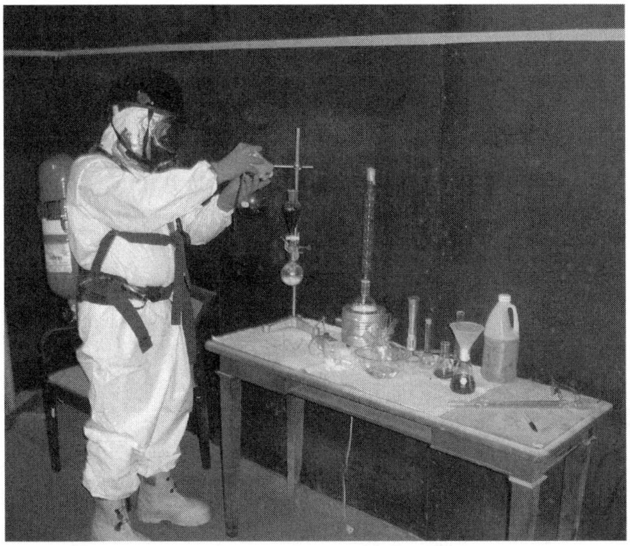

EXHIBIT I.4.14 *This is an example of an illicit laboratory.*

- Accelerant odors
- Multiple fires or explosions
- Incendiary device or bomb components
- Unexpectedly heavy burning or high temperatures
- Unusually fast burning fires
- Unusually colored smoke or flames
- Presence of propane or other flammable gas cylinders in unusual locations
- Unattended packages/backpacks/objects left in high traffic/public areas
- Fragmentation damage/injury
- Damage that exceeds that usually seen during gas explosions including shattered reinforced concrete or bent structural steel
- Crater(s)
- Scattering of small metal objects such as nuts, bolts, nails used as shrapnel

Exhibit I.4.15 shows a hazmat team investigating an explosive/incendiary attack.

(20)*Identify at least four indicators of secondary devices.

A.4.2.1(20) An evaluation of the scene for secondary devices would include the following safety steps:

(1) Evaluate the scene for likely areas where secondary devices might be placed.
(2) Visually scan operating areas for a secondary device.
(3) Avoid touching or moving anything that might conceal an explosive device.
(4) Designate and enforce scene control zones.
(5) Evacuate victims, other responders, and nonessential personnel as quickly and as safely as possible.

Awareness level personnel should pay attention to anything that arouses their curiosity and attracts their attention, including the following:

- Containers with unknown liquids or materials.
- Unusual devices or containers with electronic components such as wires, circuit boards, cellular phones, antennas and other items attached or exposed.
- Devices containing quantities of fuses, fireworks, match heads, black powder, smokeless powder, incendiary materials or other unusual materials.

EXHIBIT I.4.15 Responders are investigating possible criminal or terrorist activity involving explosives.

- Materials attached to or surrounding an item such as nails, bolts, drill bits, marbles, etc., that could be used for shrapnel.
- Ordinance such as blasting caps, detcord, military explosives, commercial explosives, grenades, etc.
- Any combination of the previously described items.

4.2.2 Surveying Hazardous Materials/WMD Incidents.

Given examples of hazardous materials/WMD incidents, awareness level personnel shall, from a safe location, identify the hazardous material(s)/WMD involved in each situation by name, UN/NA identification number, or type placard applied and shall meet the following requirements:

(1) Identify difficulties encountered in determining the specific names of hazardous materials/WMD at facilities and in transportation.

Even if awareness level personnel know placarding and labeling systems, and are also familiar with other methods of identifying the presence of hazardous materials/WMD, they can still have difficulty determining which materials are involved in a specific incident, which is addressed in 4.2.2(1). Awareness level personnel might not be able to get close enough to make accurate identification and, in some cases, the labels or placards might be missing; the placards or labels might list only the class or division, not the specific product identifier; shipments might contain mixed loads of hazardous materials/WMD and require only the "dangerous" placard; the shipper could err in the labeling and placarding; or the shipping papers could be inaccessible.

(2) Identify sources for obtaining the names of, UN/NA identification numbers for, or types of placard associated with hazardous materials/WMD in transportation.

One of the best ways to identify products or types of placards, as is required in 4.2.2(2), is to use the current edition of ERG. The shipping papers, if available, should also contain both the four-digit identification number and the proper shipping name of the material.

(3) Identify sources for obtaining the names of hazardous materials/WMD at a facility.

At fixed facilities, the sources required to be known in accordance with 4.2.2(3) are the names of hazardous materials/WMD found on the MSDS and in the emergency planning documents, and signs or other markings on storage containers.

4.2.3* Collecting Hazard Information.

Given the identity of various hazardous materials/WMD (name, UN/NA identification number, or type placard), awareness level personnel shall identify the fire, explosion, and health hazard information for each material by using the current edition of the DOT *Emergency Response Guidebook* and shall meet the following requirements:

(1)*Identify the three methods for determining the guidebook page for a hazardous material/WMD.

A.4.2.3(1) Three methods for determining the appropriate guidebook page include the following:

(1) Using the numerical index for UN/NA identification numbers
(2) Using the alphabetical index for chemical names
(3) Using the Table of Placards and Initial Response Guides

To find this information in the ERG, awareness level personnel must accomplish one of the following tasks:

- Identify the material by finding the four-digit UN/NA identification number, which is explained in A.3.3.61, on a placard or orange panel, in a shipping paper, or on the package, then locate the number in the yellow-bordered pages and determine the appropriate guide page.
- Locate the name of the material in the shipping papers or on the placard or package, then locate the material in the alphabetical listing of products on the blue-bordered pages and determine the appropriate guide page.
- Locate a matching placard in the table of placards and consult the three-digit guide number found next to the look-alike placard, if a name or identification number is not available but awareness level personnel can see the placard. If the material is an explosive, awareness level personnel should consult one of the four guides listed inside the front cover of the ERG.

(2) Identify the two general types of hazards found on each guidebook page.

Each guide page of the ERG contains information on the fire and explosion hazard, and on the health hazard of the specific hazardous material or material class. In accordance with 4.2.3(2), awareness level personnel must be able to identify the two general types of hazards.

A.4.2.3 It is the intent of this standard that the awareness level personnel be taught the noted competency to a specific task level. This task level is required to have knowledge of the contents of the current edition of the DOT *Emergency Response Guidebook* or other reference material provided.

Awareness level personnel should be familiar with the information provided in those documents so they can use it to assist with accurate notification of an incident and take protective actions.

If other sources of response information, including the MSDS, are provided to the hazardous materials/WMD responder at the awareness level in lieu of the current edition of the DOT *Emergency Response Guidebook*, the responder should identify hazard information similar to that found in the current edition of the DOT *Emergency Response Guidebook*.

4.3* Competencies — Planning the Response

(Reserved)

Section 4.3 is reserved because the committee felt that awareness level personnel would not be involved in planning for an emergency response but would apply standard operating procedures established by the organization or found in the local emergency response plan. At this level, the responsibilities of awareness level personnel are to identify a hazardous material, notify the authorities, and isolate the material if possible.

A.4.3 No competencies are currently required at this level.

4.4 Competencies — Implementing the Planned Response

4.4.1* Initiating Protective Actions.

Given examples of hazardous materials/WMD incidents, the emergency response plan, the standard operating procedures, and the current edition of the DOT *Emergency Response Guidebook,* awareness level personnel shall be able to identify the actions to be taken to protect themselves and others and to control access to the scene and shall meet the following requirements:

The competencies in 4.4.1 are designed to ensure that awareness level personnel can implement the appropriate protective actions based on the information they have acquired while analyzing the incident. Emergency response plans should establish the methods and procedures that facility owners and operators, as well as local emergency and medical response personnel, are to follow. Personnel at all levels must understand their role and its importance. If a response is to be handled effectively, awareness level personnel must accurately assess the situation and initiate the appropriate measures.

A.4.4.1 Jurisdictions that have not developed an emergency response plan can refer to the National Response Team document NRT-1, *Hazardous Materials Emergency Planning Guide.*

The National Response Team, composed of 16 federal agencies having major responsibilities in environmental, transportation, emergency management, worker safety, and public health areas, is the national body responsible for coordinating federal planning, preparedness, and response actions related to oil discharges and hazardous substance releases.

Under the Superfund Amendments and Reauthorization Act of 1986, the NRT is responsible for publishing guidance documents for the preparation and implementation of hazardous substance emergency plans.

Addressed in A.4.4.1, NRT member agencies are the Environmental Protection Agency (chair), Department of Transportation (U.S. Coast Guard) (vice chair), Department of Commerce (National Oceanic and Atmospheric Administration), Department of the Interior, Department of Agriculture, Department of Defense, Department of State, Department of Justice, Department of Transportation (Research and Special Programs Administration), Department of Health and Human Services, Federal Emergency Management Agency, Department of Energy, Department of Labor, Nuclear Regulatory Commission, General Services Administration, and Department of the Treasury.

National Response Team document *Hazardous Materials Emergency Planning Guide* (NRT-1) was first published and distributed in March 1987 and discussed the elements required to develop an effective hazardous materials/WMD emergency response plan and addressed the planning process [9]. The NRT released a 2001 update to NRT-1 to address outdated information and to include guidance on integrating local emergency response plans prepared and updated by LEPCs with planning requirements in recent legislation.

(1) Identify the location of both the emergency response plan and/or standard operating procedures.
(2) Identify the role of the awareness level personnel during hazardous materials/WMD incidents.

In order to fulfill the competency as required in 4.4.1(2), the definition of personnel at the awareness level is identified in Chapter 3. The definition emphasizes the need to act defensively in the name of safety and to call for trained personnel. The role is further identified in the goal statement for awareness level personnel.

(3) Identify the following basic precautions to be taken to protect themselves and others in hazardous materials/WMD incidents:

At the awareness level, personnel are required in 4.4.1(3) to take protective actions to isolate the hazard and to evacuate threatened persons from the immediate area. If evacuation is not possible, the personnel are to provide in-place protection until additional resources become available. In some cases, in-place protection is all that is required. The *Emergency Response Guidebook* recommends that persons protected in place be warned to stay far away from windows with a direct line of sight of the scene because windows can explode during a fire or explosion and shower glass or metal fragments.

(a) Identify the precautions necessary when providing emergency medical care to victims of hazardous materials/WMD incidents.

Personnel at the awareness level do not need to overly worry about providing emergency medical care to victims of hazardous materials/WMD incidents. Suffice it to say that hazards exist for both victim and responder. The victim could be contaminated, and decontamination procedures must be considered. In addition, many awareness level personnel might not be wearing respiratory protection or any other personal protective clothing that would protect them from the more severe hazards. The number of times both victim and responder can be exposed is limited only by the specific circumstances of the incident. In accordance with 4.4.1(3)(a), personnel must understand potential problems so that they do not become victims while attempting to rescue someone else.

(b) Identify typical ignition sources found at the scene of hazardous materials/WMD incidents.

In accordance with 4.4.1(3)(b), personnel must recognize sources of ignition, which include open flames; smoking materials; cutting and welding operations; heated surfaces; frictional heat; radiant heat; static, electrical, and mechanical sparks; and spontaneous ignition, such as occurs during heat-producing chemical reactions or is produced by pyrophoric materials.

(c)* Identify the ways hazardous materials/WMD are harmful to people, the environment, and property.

A.4.4.1(3)(c) These include thermal, mechanical, poisonous, corrosive, asphyxiating, radiological, and etiologic. They can also include psychological harm.

The term *etiologic* used in A.4.4.1(3)(c) refers to the set of factors that contributes to the cause of a disease.

(d)* Identify the general routes of entry for human exposure to hazardous materials/WMD.

Each hazard class has the possibility of having multiple routes of entry into the body, but the more common ones can be described as follows:

- Contact: The process in which a corrosive material (Class 8) damages skin or body tissue through touching. Acids and alkalis can cause severe burns. If the skin is broken or an open wound is present, another entry route exists. Explosions (Class 1) would have a serious contact consequence.

- Absorption: The process in which one substance penetrates the inner structure of another. Hydrogen cyanide (Class 2), for example, can be absorbed through the skin with fatal results.
- Inhalation: Breathing the substance, which can cause severe damage. Examples of damaging substances include chlorine and ammonia (Class 2), poisonous materials and infectious substances (Class 6), and some radioactive materials (Class 7).
- Ingestion: The introduction of a hazardous material into the body through the mouth. Toxic substances can be present in drinking water and in food. Examples of hazardous materials/WMD are poisonous materials and infectious substances (Class 6). Radioactive materials (Class 7) can do even more harm when ingested.

Obviously, proper personal protective equipment is important, and personnel must fully understand both the potential hazards and the appropriate safeguards.

A.4.4.1(3)(d) General routes of entry for human exposure are contact, absorption, inhalation, and ingestion. Absorption includes entry through the eyes and through punctures.

Absorption includes entry through the skin, eyes, or membranes. Inhalation includes breathing the material. Ingestion involves taking the material in through the mouth. Injection includes entry through a wound or cut.

(4)* Given examples of hazardous materials/WMD and the identity of each hazardous material/WMD (name, UN/NA identification number, or type placard), identify the following response information:

(a) Emergency action (fire, spill, or leak and first aid)
(b) Personal protective equipment necessary
(c) Initial isolation and protective action distances

At the awareness level, personnel are generally expected to take protective actions to isolate the hazard and to evacuate or direct threatened persons out of the hot zone without ever entering the hot zone themselves. In some cases, sheltering in-place protection is all that is required. The ERG recommends that persons sheltered in-place be warned to stay far away from windows with a direct line of sight of the scene because windows can explode during a fire or explosion and shower them with glass or metal fragments.

A.4.4.1(4) If other sources of response information, including the MSDS, are provided to the hazardous materials/WMD responder at the awareness level in lieu of the current edition of the DOT *Emergency Response Guidebook*, the responder should identify response information similar to that found in the current edition of the DOT *Emergency Response Guidebook*.

The importance that personnel demonstrate their proficiency in using the various emergency response guides and other sources of information they might be required to consult is expressed in A.4.4.1(4). The ERG provides general information about the potential hazards of a number of products and outlines the emergency procedures to be used in handling incidents involving these products. Personnel should be given several exercises that allow them to demonstrate their skills at locating and interpreting the appropriate information both from the ERG and other documents.

(5) Given the name of a hazardous material, identify the recommended personal protective equipment from the following list:

(a) Street clothing and work uniforms
(b) Structural fire-fighting protective clothing
(c) Positive pressure self-contained breathing apparatus
(d) Chemical-protective clothing and equipment

In the orange-bordered pages of the ERG, under the heading of Public Safety, each guide has personal protective equipment recommendations that personnel are required in 4.4.1(5)(d) to recognize. This section also includes equipment that is not recommended. For example, the guide for oxidizers warns that structural fire fighters' protective clothing only provides limited protection.

(6) Identify the definitions for each of the following protective actions:

Protective actions are those steps taken to preserve the health and safety of emergency personnel and the public during an incident involving the release of hazardous materials/WMD.

(a) Isolation of the hazard area and denial of entry

Everyone not directly involved in the emergency response operations should be kept away from the affected area, and unprotected emergency personnel should not be allowed within the isolation area.

(b) Evacuation

Evacuation is the movement of everyone from a threatened area to a safer place. To perform an evacuation, enough time must be available to warn people, to get them ready to go, and to leave the area. Evacuation is likely to be the best protective action if enough time is available. Personnel should begin evacuating people who are nearby and those who are outdoors in direct view of the scene. Evacuees should be sent upwind by a specific route to a definite place far enough away from the contaminated area that they do not have to be moved again if the wind shifts. As additional help is acquired, the area should be expanded to be evacuated downwind and crosswind at least to the extent recommended in Commentary Table I.4.4. Even after people move to the distances recommended, they are not completely safe from harm and may have to be decontaminated. The evacuation distances were adjusted for many of the materials in the 2004 edition of the ERG.

(c)* Sheltering in-place

A.4.4.1(6)(c) "In-place protection," "sheltering in-place," and "protection in-place" all mean the same thing.

Sheltering in-place protection is used when an evacuation cannot be performed or when evacuating the public would put them at greater risk than directing them to stay. When using in-place protection, people are directed to go quickly inside a building and remain there until the danger has passed. The people inside the building should be told to close all doors and windows and to shut off all ventilating, heating, and cooling systems.

In-place protection might not be the best option if explosive vapors are present, if a long time is needed to clear the area of gas, or if the building cannot be tightly closed. Vehicles are not as effective as buildings for sheltering in-place protection but can offer some protection for a short period if the vehicle windows are closed and the ventilating system is shut off.

(7) Identify the size and shape of recommended initial isolation and protective action zones.

The ERG provides initial isolation zones and protective action distances for vapors from hazardous materials/WMD that can produce poisonous effects. The shapes of those areas are shown in Exhibit I.4.16. Examples of these distances are provided in Commentary Table I.4.4 and the Table of Initial Isolation and Protective Action Distances on the green-bordered pages in the ERG.

(8) Describe the difference between small and large spills as found in the Table of Initial Isolation and Protective Action Distances in the DOT *Emergency Response Guidebook*.

Awareness level personnel are required to know the ERG definition of a small spill and large spill and to note the time of day that the incident has occurred. In determining the isolation

COMMENTARY TABLE I.4.4 *Sample of Table of Initial Isolation and Protective Action Distances*

ID		Small Spills (from a small package or small leak from a large package)						Large Spills (from a large package or from many small packages)					
		First Isolate in all Directions		Then Protect persons Downwind during —				First Isolate in all Directions		Then Protect persons Downwind during —			
				Day		Night				Day		Night	
No.	Name of Material	Meters	(Feet)	Kilometers	(Miles)	Kilometers	(Miles)	Meters	(Feet)	Kilometers	(Miles)	Kilometers	(Miles)
1005 1005 1005 1005 1005	Ammonia, anhydrous Ammonia, anhydrous, liquefied Ammonia, solution, with more than 50% Ammonia Anhydrous ammonia Anhydrous ammonia, liquefied	30 m	(100 ft)	0.1 km	(0.1 mi)	0.1 km	(0.1 mi)	60 m	(200 ft)	0.6 km	(0.4 mi)	2.2 km	(1.4 mi)
1008 1008	Boron trifluoride Boron trifluoride, compressed	30 m	(100 ft)	0.1 km	(0.1 mi)	0.6 km	(0.4 mi)	180 m	(600 ft)	1.8 km	(1.1 mi)	4.8 km	(3.0 mi)
1016 1016	Carbon monoxide Carbon monoxide, compressed	30 m	(100 ft)	0.1 km	(0.1 mi)	0.1 km	(0.1 mi)	90 m	(300 ft)	0.7 km	(0.4 mi)	2.4 km	(1.5 mi)
1017	Chlorine	30 m	(100 ft)	0.2 km	(0.2 mi)	1.2 km	(0.8 mi)	240 m	(800 ft)	2.4 km	(1.5 mi)	7.4 km	(4.6 mi)
1023 1023	Coal gas Coal gas, compressed	30 m	(100 ft)	0.2 km	(0.1 mi)	0.2 km	(0.1 mi)	60 m	(200 ft)	0.4 km	(0.2 mi)	0.5 km	(0.3 mi)
1026 1026 1026	Cyanogen Cyanogen, liquefied Cyanogen gas	30 m	(100 ft)	0.2 km	(0.2 mi)	1.2 km	(0.8 mi)	120 m	(400 ft)	1.1 km	(0.7 mi)	4.3 km	(2.7 mi)
1040 1040	Ethylene oxide Ethylene oxide with Nitrogen	30 m	(100 ft)	0.1 km	(0.1 mi)	0.2 km	(0.1 mi)	90 m	(300 ft)	0.8 km	(0.5 mi)	2.4 km	(1.5 mi)
1045 1045	Fluorine Fluorine, compressed	30 m	(100 ft)	0.2 km	(0.1 mi)	0.5 km	(0.3 mi)	90 m	(300 ft)	0.8 km	(0.5 mi)	3.5 km	(2.2 mi)
1048	Hydrogen bromide, anhydrous	30 m	(100 ft)	0.1 km	(0.1 mi)	0.5 km	(0.3 mi)	180 m	(600 ft)	1.8 km	(1.1 mi)	5.7 km	(3.6 mi)
1050	Hydrogen chloride, anhydrous	30 m	(100 ft)	0.1 km	(0.1 mi)	0.4 km	(0.3 mi)	360 m	(1200 ft)	3.6 km	(2.2 mi)	10.4 km	(6.5 mi)
1051	AC (when used as a weapon)	60 m	(200 ft)	0.2 km	(0.1 mi)	0.5 km	(0.3 mi)	500 m	(1500 ft)	1.7 km	(1.0 mi)	3.9 km	(2.4 mi)

Source: *Emergency Response Guidebook*, 2004, Table of Initial Isolation and Protective Action Distances, p. 302.

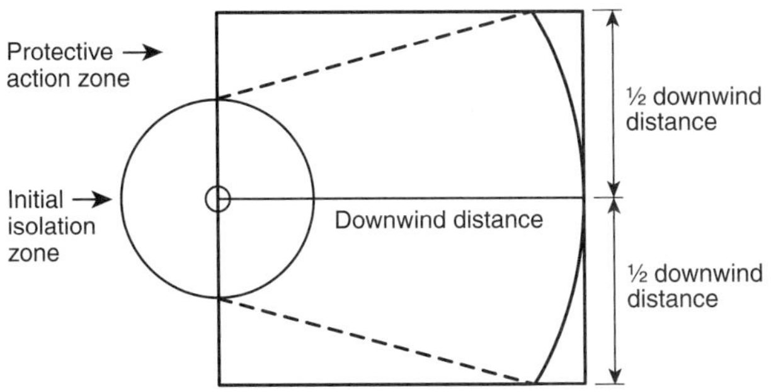

EXHIBIT I.4.16 *The shapes of the initial isolation zone and the protective action zone are identified here.*

and protective action distances, the ERG assumed that the maximum pool size for a small spill that formed a liquid pool was 48 ft (15 m) in diameter. A large spill pool was assumed to be a maximum of 60 ft (18 m) in diameter. The distances were calculated following a 30-minute period from the start of the release. This reason is why the ERG cautions that the distances are valid only for the 30-minute period following a spill.

(9) Identify the circumstances under which the following distances are used at a hazardous materials /WMD incidents:

(a) Table of Initial Isolation and Protective Action Distances

These distances are used only for products whose vapors present an inhalation hazard, where the release does not involve a fire, and where no more than 30 minutes have elapsed between the spill and the response.

(b) Isolation distances in the numbered guides

The isolation distances in the ERG are to be used when a hazardous material or its container is exposed to fire or when the product's vapors have potentially poisonous effects.

(10) Describe the difference between the isolation distances on the orange-bordered guidebook pages and the protective action distances on the green-bordered ERG *(Emergency Response Guidebook)* pages.

The orange-bordered pages in the emergency action section of the ERG provide isolation distances for selected materials that are involved in a fire. The distances found in this table are the recommended downwind protective action distances for materials whose poisonous vapors present inhalation hazards.

For example, ethylene oxide has an isolation distance (as found in the orange-bordered pages) of 1 mi (1600 m) when involved in a fire. When ethylene oxide is not involved in a fire, this same material has a recommended isolation distance of 100 ft (30 m) for a small spill and 300 ft (90 m) for a large spill, according to the 2004 edition of the ERG.

The competency in 4.4.1(10) points out how important it is that personnel understand how to use the ERG or similar guidebooks.

(11) Identify the techniques used to isolate the hazard area and deny entry to unauthorized persons at hazardous materials/WMD incidents.

At this level, personnel do not have many resources available to them. However, several steps can be taken to isolate a hazard area. A vehicle could be used to block a road or driveway, or a rope or some other type of barricade could be placed across the entrance to the area to block access. Awareness level personnel can also notify law enforcement officials to begin diverting traffic from the scene. In a fixed facility, personnel can close a door or gate, use the public address system to announce the problem to the facility's occupants or notify security.

(12)*Identify at least four specific actions necessary when an incident is suspected to involve criminal or terrorist activity.

A.4.4.1(12) The following are examples of actions that might be taken:

(1) Take the appropriate actions to protect yourself and other personnel.
(2) Communicate the suspicion during the notification process.
(3) Isolate potentially exposed people or animals.
(4) Document the initial observation
(5) Be alert for booby traps and explosive devices.

Awareness level personnel are urged to approach with caution whenever an incident is suspected to involve criminal or terrorist activity. The possibility exists that biological or chemical weapons could have contaminated an area, and secondary devices could be present.

4.4.2 Initiating the Notification Process.

Given scenarios involving hazardous materials/WMD incidents, awareness level personnel shall identify the initial notifications to be made and how to make them, consistent with the emergency response plan and/or standard operating procedures.

Awareness level personnel must be familiar with the notification process they must follow to begin an effective response to a hazardous materials/WMD incident. This might only involve notifying the local fire or police department. In some fixed facilities, internal notification procedures can be used to initiate the response of private sector specialists, the plant fire brigade, or security personnel. Whatever the procedures, the proper notification process must be immediately set in motion.

4.5* Competencies — Evaluating Progress

(Reserved)

A.4.5 No competencies are currently required at this level.

4.6* Competencies — Terminating the Incident

(Reserved)

A.4.6 No competencies are currently required at this level.

REFERENCES CITED IN COMMENTARY

1. *Emergency Response Guidebook,* U.S. Department of Transportation, Washington, DC, 2004.
2. *European Agreement Concerning the International Carriage of Dangerous Goods by Road* (ADR), United Nations, New York, NY, 2007.
3. *International Regulations Concerning the Carriage of Dangerous Goods by Rail* (RID), International Organization for international Carriage by Rail (OTIF), United Nations Economic Commission for Europe, Geneva, Switzerland.
4. Spencer, A., and Colonna, G., eds., *Fire Protection Guide to Hazardous Materials,* 13th ed., National Fire Protection Association, Quincy, MA, 2002.

5. NFPA 704, *Standard System for the Identification of the Hazards of Materials for Emergency Response*, National Fire Protection Association, Quincy, MA, 2007.
6. ANSI Standard Z400.1, *Hazardous Industrial Chemicals — Material Safety Data Sheets — Preparation,* American National Standards Institute, Inc., New York, NY.
7. Title 29, Code of Federal Regulations, Part 1910.1200, "Hazard Communication," U.S. Government Printing Office, Washington, DC, July 1, 2001.
8. Title 18, Code of Federal Regulations, Chapter 113, Section 2331, "Crimes and Criminal Procedure," U.S. Government Printing Office, Washington, DC.
9. NRT-1, *Hazardous Materials Emergency Planning Guide*, U.S. National Response Team, Washington, DC 20592.

Additional References

Cote, A., ed., *Fire Protection Handbook®*, 20th ed., National Fire Protection Association, Quincy, MA, 2008.

U.S Air Force Manual, 91-201, *Explosives Safety Standards*, Figure 2.2, released 18 October 2001.

Responder, Fire Engineering, New York, 1989.

Title 28, Code of Federal Regulations, Part 0.85, U.S. Government Printing Office, Washington, DC.

Title 29, Code of Federal Regulations, Part 1910, U.S. Government Printing Office, Washington, DC.

Title 40, Code of Federal Regulations, Part 261, U.S. Government Printing Office, Washington, DC.

Title 49, Code of Federal Regulations, Part 173, U.S. Government Printing Office, Washington, DC.

Title 49, Code of Federal Regulations, Part 178, U.S. Government Printing Office, Washington, DC.

Core Competencies for Operations Level Responders

CHAPTER 5

Core competencies for operations level responders build on those for awareness level personnel. In the 2008 edition of NFPA 472, core competencies for operations level personnel are addressed in Chapter 5, and additional mission-specific competencies are addressed in Chapter 6. When a training entity designs courses for operations level responders, it is up to the authority having jurisdiction (AHJ) to choose the appropriate parts of Chapter 6 to include in that training. For example, many jurisdictions will certainly be interested in personal protective equipment and decontamination as additional operations level competencies, but they might not be as concerned with evidence preservation or sampling, as those duties could be assigned to other personnel.

5.1 General

5.1.1 Introduction.

5.1.1.1* The operations level responder shall be that person who responds to hazardous materials/weapons of mass destruction (WMD) incidents for the purpose of protecting nearby persons, the environment, or property from the effects of the release.

A.5.1.1.1 Operations level responders need only be trained to meet the competencies in Chapter 5. The competencies listed in Chapters 6 (mission-specific competencies) are not required and should be viewed as optional at the discretion of the AHJ based on an assessment of local risks. The purpose of Chapter 6 is to provide a more effective and efficient process so that the AHJ can match the expected tasks and duties of its personnel with the required competencies to perform those tasks. Table A.5.1.1.1 is a sample operations level responder matrix.

Table A.5.1.1.1 is designed to help users of this standard determine which competencies in Chapters 5 and 6 can be utilized to ensure that operations level responders have the appropriate knowledge and skills to perform their expected tasks. These competencies are above the core competencies contained in Chapter 5 and are optional. This matrix is provided only as a sample. The selection of competencies should always be based on the expected mission and tasks, as assigned by the AHJ.

5.1.1.2 The operations level responder shall be trained to meet all competencies at the awareness level (Chapter 4) and the competencies of this chapter.

5.1.1.3* The operations level responder shall receive additional training to meet applicable governmental occupational health and safety regulations.

Some examples of additional training would include programs from the U.S. Department of Transportation (DOT), U.S. Environmental Protection Agency (EPA), Occupational Safety and Health Administration (OSHA) regulatory courses, and any local or state standard operating procedures.

A.5.1.1.3 Operations level responders who are expected to perform additional missions should work under the direction of a hazardous materials technician, a written emergency response plan or standard operating procedures, or an allied professional.

TABLE A.5.1.1.1 *NFPA 472 Operations Level Responder Matrix*

	Competencies						
Responders	***Use PPE***	***Perform Technical or Mass Decontamination****	***Perform Product Control***	***Perform Air Monitoring***	***Perform Victim Rescue and Removal***	***Preserve Evidence and Perform Sampling***	***Respond to Illicit Lab Incident***
Fire fighters expected to perform basic defensive product control measures	X	X	X	—	—	—	—
Emergency responders assigned to a decontamination company or decontamination strike force	X	X	—	—	—	—	—
Emergency responders assigned to a unit tasked with providing rapid rescue and extraction from a contaminated environment	X	X	—	X	X	—	—
Emergency responders assigned to provide staffing or support to a hazardous materials response team	X	X	X	X	X	—	—
Law enforcement personnel involved in investigation of criminal events where hazardous materials are present	X	X	—	X	—	X	X
Law enforcement personnel involved in investigation of incidents involving illicit aboratories	X	X	—	X	—	X	X
Public health personnel involved in the investigation of public health emergencies	X	X	—	—	—	X	—
Environmental health and safety professionals who provide air monitoring support	X	X	—	X	—	—	—

*The scope of the decontamination competencies would be based on whether the mission involves the responder being the "customer" of the decontamination services being provided or is part of those responders who are responsible for the set-up and implementation of the decontamination operation.

5.1.2 Goal.

5.1.2.1 The goal of the competencies at this level shall be to provide operations level responders with the knowledge and skills to perform the core competencies in 5.1.2.2 safely.

Safety, which is the goal of 5.1.2, must be considered in every action taken at an incident scene because risks are always present. By demonstrating the knowledge and skills that follow, a responder should be able to reduce the risk inherent in a hazardous materials incident. A risk-based response process should be put in place by operations level responders to ensure the safety of personnel. This is a systematic process by which responders analyze a problem involving hazardous materials/weapons of mass destruction (WMD), assess the hazards, evaluate the potential consequences, and determine appropriate response actions based upon facts, science, and the circumstances of the incident.

5.1.2.2 When responding to hazardous materials/WMD incidents, operations level responders shall be able to perform the following tasks:

(1) Analyze a hazardous materials/WMD incident to determine the scope of the problem and potential outcomes by completing the following tasks:

Charles Wright notes the following in the *Fire Protection Handbook*® (FPH™):

> "The analysis process begins when a responder receives notification of a problem and continues throughout the incident, typically at the scene threatened by any hazardous materials involved." [1] p.13-121

Outcomes are the direct and indirect results or consequences associated with an emergency. Direct outcomes are considered in terms of people, property, and/or the environment as shown in Exhibit I.5.1.

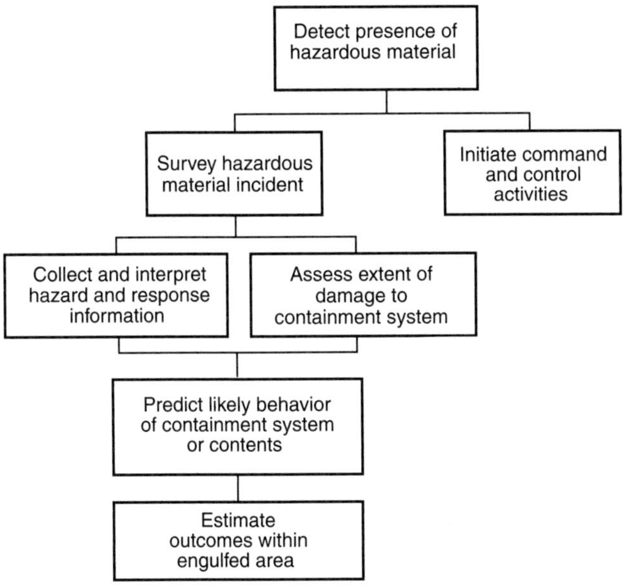

EXHIBIT I.5.1 *This flow diagram displays the tasks associated with analyzing a hazardous material problem.*

(a) Survey a hazardous materials/WMD incident to identify the containers and materials involved, determine whether hazardous materials/WMD have been released, and evaluate the surrounding conditions.

According to Charles Wright, the incident survey should be done from a safe distance so that the responder is not exposed to any released materials [1]. The following steps should be completed during the survey of a hazardous material incident:

1. Identify each containment system by type, identifier, and size. Containment systems (see Commentary Table I.5.1) fit into one of the following three types:
   - Nonbulk
   - Bulk
   - Facility containment systems

   The containment system identifier (i.e., facility or carrier name and number) is used to differentiate one containment system from another and to allow tracking of that container throughout the incident. The quantity within (or capacity of) the containment system can

COMMENTARY TABLE I.5.1 *Types of Nonbulk, Bulk, and Facility Containment Systems*

Transportation		***Facility Containment Systems***
Nonbulk	***Bulk***	
Bags	Bulk bags	Buildings
Bottles	Bulk boxes	Machinery
Boxes	Cargo tanks	Open piles (outdoors and indoors)
Carboys	Covered hopper cars	Piping
Cylinders	Freight containers	Reactors (chemical and nuclear)
Drums	Gondolas	Storage bins, cabinets, or shelves
Jerricans	Intermediate bulk containers (IBCs)	
Multicell	Pneumatic hopper trailers	
Tanks and storage vessels packages	Portable tanks and bins	
Wooden barrels	Protective overpacks for radioactive materials	
	Tank cars	
	Tank containers	
	Ton containers	
	Van trailers	

Source: *Fire Protection Handbook*, 20th edition, Table 13.8.4.

be obtained from markings on the container or entries on the shipping papers or facility documents. The quantity information helps indicate the magnitude of the problem.

2. Identify the name, DOT identification number, or placard applied to each hazardous material containment system. This information provides a means of accessing various sources of hazard and response information.

For facilities, sources for identifying the material include pre-emergency planning documents, markings and color, contact with the facility manager, and review of appropriate material safety data sheets (MSDSs).

In transportation, the identity of the hazardous material can be determined from the DOT identification number, commodity stencil, type of placard or label applied, shipping paper entries, and manufacturer, shipper, or consignee contacts using the 24-hour emergency telephone number on the shipping papers. Pipeline markers provide the name of the commodity or at least the name and telephone number of the pipeline company.

3. Identify leaking containment systems. Clues indicating leakage include material on the outside of the containment system, taste or smell, presence of vapor clouds (see Exhibit I.5.2), or the operation of a safety relief valve. If possible during the survey, the form of the released material (solid, liquid, or gas) and the location of the release should be noted.

EXHIBIT I.5.2 *Leaking containers, smells, vapor clouds (such as the one shown here indicating a leak), fires, or explosions help a responder establish whether a hazardous materials release has occurred at the site.*

4. Identify the surrounding conditions. Surrounding conditions should be noted when surveying hazardous material incidents. These conditions include topography, land use (including utilities and fiber optic cables), accessibility, weather conditions, bodies of water (including recharging ponds), public exposure potential, and the nature and extent of injuries.

If a facility is involved, information about floor drains, ventilation ducts, air returns, and so forth should be gathered as appropriate.

For ease of collection, recording, and interpretation of information obtained during the survey of an incident, a hazardous material incident survey form, such as shown in Exhibit I.5.3, is used.

The information collected in the incident survey should be reviewed to verify its accuracy. For example, if the shipping paper identifies the material as a gas and a solid is being released, some of the information obtained might not be correct.

An understanding of the characteristics of various containment systems helps to verify information about the contents.

(b) Collect hazard and response information from MSDS; CHEMTREC/CANUTEC/SETIQ; local, state, and federal authorities; and shipper/manufacturer contacts.

It is very important for the responder to know what resources are available for providing technical assistance during hazardous materials emergencies and to know how to use those resources.

(c) Predict the likely behavior of a hazardous material/WMD and its container.

The responder is required to be able to assess the potential behavior of a hazardous material/WMD and its container. Some questions the responder should ask include the following. Is it likely to explode? Is it nonflammable? Is it corrosive? Will the container rupture violently?

(d) Estimate the potential harm at a hazardous materials/WMD incident.

As the responder collects the information necessary to predict the behavior of the material and its container, he or she can begin to estimate the potential harm it presents, which is required in 5.1.2.2(1)(d). Generally, initial assessments should be very conservative and should look at the worst possible event that could occur. As more information becomes available, these predictions can be modified.

(2) Plan an initial response to a hazardous materials/WMD incident within the capabilities and competencies of available personnel and personal protective equipment by completing the following tasks:

In 5.1.2.2(2), a plan of action is required to be developed using the information gathered and the estimates of potential harm. This plan establishes the responders' objectives in controlling the incident. Any plan of action must be formulated based on the resources that can be brought to bear on the mitigation process. Operations level responders would perform to the extent their training and equipment allowed and would be required to summon appropriate help.

(a) Describe the response objectives for the hazardous materials/WMD incident.

The response objectives for a hazardous materials/WMD incident focus on controlling events as they occur.

(b) Describe the response options available for each objective.

Response options are those actions the responders can take safely without coming in direct contact with the hazardous materials/WMD involved in the incident. These actions can be

Commonwealth of Virginia Hazardous Materials
Incident Analysis Worksheet - Chemical Hazard Profile

PRODUCT ID

UN NUMBER ________ STCC NUMBER ________ CAS NUMBER ________

CHEMICAL NAME ________ HAZARD CLASS ________

NFPA 704 HEALTH ________ FIRE ________ REACTIVITY ________ SPECIAL ________

HIGH ENERGY EVALUATION

HAZARD	ACTION	NO	YES	PROFILE
EXPLOSIVE		____	____	**EXPLOSIVE**
REACTIVE	IN CONTACT WITH OTHER CHEMICALS	____	____	**REACTIVE**
	REACTIVE WITH OTHER CHEMICALS	____	____	
	WATER REACTIVE	____	____	
	AIR REACTIVE	____	____	
	VIOLENT POLYMERIZATION	____	____	
	CHEMICAL UNSTABLE	____	____	
RADIOACTIVE		____	____	**RADIOACTIVE**

PHYSICAL STATE

AMBIENT TEMP ________ BOILING POINT ________ MELTING POINT ________

EVALUATION	YES	PROFILE
BOILING POINT **BELOW** AMBIENT TEMPERATURE	____	**GAS**
BOILING POINT **BELOW** 300°F BUT ABOVE AMBIENT TEMPERATURE	____	**LIQUID / GAS**
BOILING POINT **ABOVE** AMBIENT TEMPERATURE AND ABOVE 300°F	____	**LIQUID**
MELTING POINT **ABOVE** AMBIENT TEMPERATURE	____	**SOLID**

EVALUATE GAS HAZARDS

					PROFILE
FLASH POINT	____	BELOW 100°F			**FLAMMABLE**
		ABOVE 100°F			**COMBUSTIBLE**
IGN. TEMP	____	LEL ____	UEL	____	
VAPOR DENSITY	____	BELOW 1			**RISE**
		ABOVE 1			**SINK**
CARCINOGEN		NO ____	YES	____	**CARCINOGEN**

			INHALATION HAZARD
LC_{50}		LESS THAN 100 PPM (0.01% v/v)	**HIGH**
PEL	____	101 to 1,000 PPM (0.01% to 0.1% v/v)	**MODERATE**
STEL	____	1,000 to 10,000 PPM (0.1% to 1.0% v/v)	**LOW**
IDLH	____		
pH	____	0 – 3 or 12 – 14	**CORROSIVE**

Revision 02/2003 Page 1

EXHIBIT I.5.3 *This worksheet is a sample Hazardous Materials Incident Survey Form. (Source: Virginia Department of Emergency Management)*

Commonwealth of Virginia Hazardous Materials
Incident Analysis Worksheet - Chemical Hazard Profile

EVALUATE LIQUID HAZARDS

					PROFILE
SOLUBLE IN WATER		NO ______		YES ______	**SOLUBLE**
SPECIFIC GRAVITY	______	BELOW 1			**FLOATS**
		ABOVE 1			**SINK**
CARCINOGEN		NO ______		YES ______	**CARCINOGEN**
					INHALATION HAZARD
LC_{50}		LESS THAN 50 mg/Kg			**EXTREME**
PEL	______	50 mg/Kg to 500 mg/Kg			**HIGH**
STEL	______	500 mg/Kg to 5 g/Kg			**MODERATE**
IDLH	______	5 g/Kg to 15 g/Kg			**LOW**
pH	______	0 – 3 or 12 – 14			**CORROSIVE**

EVALUATE SOLID HAZARDS

					PROFILE
SUBLIME		NO ______		YES ______	EVALUATE GAS
COMBUSTIBLE		NO ______		YES ______	COMBUSTIBLE
CARCINOGEN		NO ______		YES ______	CARCINOGEN
					INHALATION HAZARD
LD_{50}		LESS THAN 50 mg/Kg			**EXTREME**
PEL	______	50 mg/Kg to 500 mg/Kg			**HIGH**
STEL	______	500 mg/Kg to 5 g/Kg			**MODERATE**
IDLH	______	5 g/Kg to 15 g/Kg			**LOW**
pH	______	0 – 3 or 12 – 14			**CORROSIVE**

OTHER INFORMATION

Revision 02/2003 Page 2

EXHIBIT I.5.3 *Continued*

done safely with the protective clothing and equipment available to the responder at this level, providing they have received applicable training. See 6.1.1.1 in Chapter 6.

(c) Determine whether the personal protective equipment provided is appropriate for implementing each option.

Individuals involved in emergency response must understand their limitations, particularly concerning personal protective clothing and equipment. Responders must also understand that clothing and equipment requirements vary depending on the material involved in an incident.

Fire fighters responding to structural fires generally use the same type of protective clothing each time. In hazardous materials incidents, however, what is appropriate for one material could be totally unacceptable for another. Responders using air purifying respirators at one incident might require self-contained breathing apparatus (SCBAs) at another incident.

(d) Describe emergency decontamination procedures.

Emergency decontamination might be necessary if a life-threatening exposure has occurred and the individual needs immediate medical attention, but it is delayed due to contamination. Technical decontamination is not available until responders trained and equipped arrive on the incident scene.

(e) Develop a plan of action, including safety considerations.

A site safety plan or incident action plan (IAP) with safety issues identified must be developed by the incident commander (IC) or unified command for incidents involving the release of hazardous materials.

(3) Implement the planned response for a hazardous materials/WMD incident to favorably change the outcomes consistent with the emergency response plan and/or standard operating procedures by completing the following tasks:

Once responders have analyzed the incident and planned the initial response, they must implement that response in accordance with 5.1.2.2(3). Although some might expect that this process is lengthy, it frequently takes only a few minutes.

For example, a responder might be able to analyze, plan, and implement the response to a small spill of home heating fuel in a nonthreatening location within minutes. However, if the incident involved an 8,000 gal (30,280 L) gasoline tank truck overturned on a congested highway, the process would take a little longer. No matter how severe the incident, the responder should go through the same planning process to ensure that nothing was overlooked.

(a) Establish and enforce scene control procedures, including control zones, emergency decontamination, and communications.

Scene control is critical in keeping both responders and the public safe, so it should be established immediately. Responders can do this by establishing control zones and an exclusion perimeter to keep the public away from the working areas of the emergency responders.

(b) Where criminal or terrorist acts are suspected, establish means of evidence preservation.

(c) Initiate an incident command system (ICS) for hazardous materials/WMD incidents.

When arriving on the scene, operations level responders must begin implementing the incident command system (ICS) in the local emergency response plan. An ICS identifies the roles and responsibilities that help personnel control the incident safely and effectively.

(d) Perform tasks assigned as identified in the incident action plan.

Operations level responders must be familiar with the ICS and incident management system (IMS) and work from the incident action plans developed by the planning section. These plans will address the mission objectives, and the safety issues raised by them.

(e) Demonstrate emergency decontamination.

Operations level responders must be able to properly demonstrate the agency's standard operating procedures for emergency decontamination to both civilians and responders. This process is critical to the health of civilians and responders who may be accidentally contaminated and suffer serious effects.

(4) Evaluate the progress of the actions taken at a hazardous materials/WMD incident to ensure that the response objectives are being met safely, effectively, and efficiently by completing the following tasks:

Part of an effective response is the ongoing evaluation of the actions that have been undertaken and their effectiveness. Operations level responders must not base their actions solely on their initial assessment of conditions at the site because these conditions can, and do, change. For example, the wind could shift, rain could begin to fall, or resources could become unavailable.

(a) Evaluate the status of the actions taken in accomplishing the response objectives.

Some of the questions the responders should ask include the following. Are the actions having the desired results? Is the incident stabilizing or is it intensifying? Responders might find that the actions they chose initially are no longer correct because they are no longer suit the circumstances. The weather could have changed, for example, or the arrival of additional personnel could have been delayed.

(b) Communicate the status of the planned response.

The incident commander must be kept informed of the effectiveness of the actions. An effective response cannot be carried out without frequent status reports, which are required in 5.1.2.2(4)(b). Everyone must be aware of this and provide the necessary information through the appropriate channels.

5.2 Core Competencies — Analyzing the Incident

5.2.1* Surveying Hazardous Materials/WMD Incidents.

Given scenarios involving hazardous materials/WMD incidents, the operations level responder shall survey the incident to identify the containers and materials involved, determine whether hazardous materials/WMD have been released, and evaluate the surrounding conditions and shall meet the requirements of 5.2.1.1 through 5.2.1.6.

A.5.2.1 The survey of the incident should include an inventory of the type of containers involved, identification markings on containers, quantity in or capacity of containers, materials involved, release information, and surrounding conditions. The accuracy of the data should be verified.

The survey of the incident should include an inventory of the type of containers involved, identification markings on containers, quantity in or capacity of containers, materials involved, release information, and surrounding conditions. The accuracy of the data should be verified.

Charles Wright notes the following in FPH:

> After detecting the presence of hazardous materials in an emergency, and while initiating command and control activities, the next task is to survey or inventory the hazardous materials incident. Completion of this task provides an inventory of the containment systems and materials involved, materials released, and surrounding conditions. This incident survey

should be conducted from a safe distance, using aided vision, without exposure to the released materials. [1] p. 13–126

Recognition of the shapes of various containers and knowledge of what each normally holds, which are requirements of 5.2.1, helps the responder verify the presence of hazardous materials and could help in the identification of the particular materials involved in an incident. See Exhibit I.5.4 for an illustration of the parts of a typical tank car.

5.2.1.1* Given three examples each of liquid, gas, and solid hazardous material or WMD, including various hazard classes, operations level personnel shall identify the general shapes of containers in which the hazardous materials/WMD are typically found.

A.5.2.1.1 Examples should include all containers, including nonbulk packaging, bulk packaging, vessels, and facility containers such as piping, open piles, reactors, and storage bins.

The packaging, storage, and transport containers in 5.2.1.1 vary greatly, depending upon the type of material (solid, liquid, or gas), the quantity, and the associated hazards. Liquids can be contained in drums ranging in size from one gallon up to 85 gallon overpack drums, encasing, for example, corrosive liquids (Class 8). Drums can be made from plastic-lined and unlined fiberboard to plastic (poly) steel, stainless steel, and aluminum. There are also carboys for acids and caustics, combination packaging for etiological agents (Class 6), multi-cell packaging, and plastic and glass bottles, holding, for example, organic peroxides (Class 5).

Cylinders can be used for pressurized, liquefied, and dissolved gases, such as aerosol containers with propane (Class 2.1), uninsulated containers with chlorine (Class 2.3), and cryogenic insulated cylinders with cryogenic liquid (Class 2.2).

Many household products, such as poisonous pesticides, insecticides, caustic powders and fertilizers (Class 5), are contained in cloth, burlap, or plastic bags; jugs or jars; or cardboard boxes. Totes, bulk bags, and drums can hold flammable solids, such as calcium carbide and water treatment chemicals (Classes 2, 3, and 4), or combustible (Class 4), toxic (Class 6), and corrosive materials (Class 8).

5.2.1.1.1 Given examples of the following tank cars, the operations level responder shall identify each tank car by type, as follows:

Charles Wright notes the following in the FPH's section on "Rail Transportation Systems": "Tank cars are classed according to their construction, features, and fittings. The tank's spec-

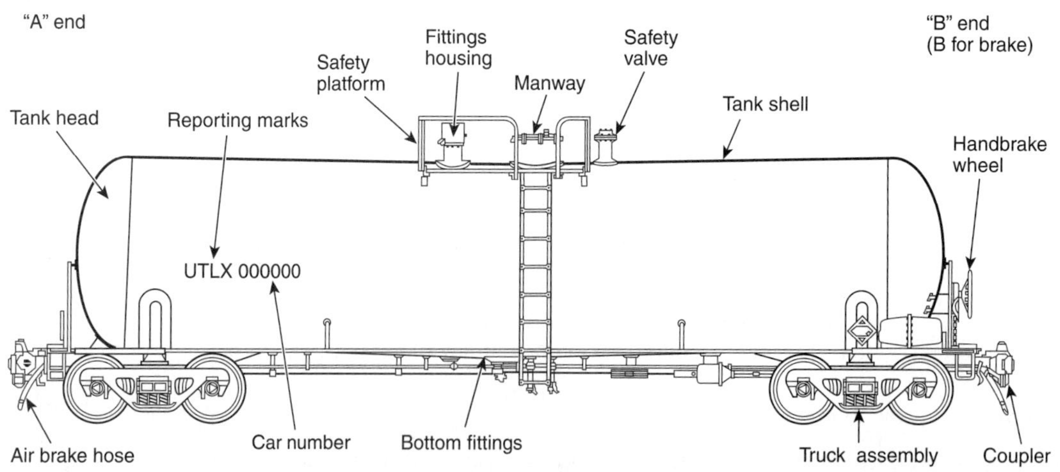

EXHIBIT I.5.4 *The diagram shows a tank car with basic features identified. (Source: Union Tank Car Company)*

ification determines the product it may transport." The responder is required to be able to identify various shapes of containers and have a general idea of the products they contain. See Exhibit I.5.5 for an example of containers that hold a specific product.

EXHIBIT I.5.5 *The containers shown all hold liquefied natural gas, maintained under cryogenic conditions.*

(1) Cryogenic liquid tank cars

Cryogenic liquid tank cars carry low-pressure [25 psi (172 kPa) or lower] liquids refrigerated to –150°F (–101°C) and below. The liquids typically include argon, ethylene, hydrogen, nitrogen, and oxygen. A cryogenic liquid tank car is actually a tank within a tank, and the inner tank is made of stainless steel or nickel. The space between the inner and outer tanks is filled with insulation and is under a vacuum. [1]

(2) Nonpressure tank cars (general service or low pressure cars)

Wright notes the following in the "Rail Transportation Systems" section of FPH:

> Nonpressure tank cars, also known as general-service tank cars or acid-service tank cars, transport a wide variety of hazardous and nonhazardous materials at low pressures.
>
> Nonpressure tank cars transport hazardous materials, such as flammable and combustible liquids , flammable solids, oxidizers, organic peroxides and poison, corrosive materials, and molten solids. They also transport nonhazardous materials, such as tallow, clay slurry, corn syrup, and other food products.
>
> Tank test pressures for nonpressure tank cars range from 430 to 689 kPa (60 to 100 psi). Capacities range from 15 to 181 m^3 (4,000 to 45,000 gal).
>
> Nonpressure tank cars are cylindrical with rounded heads. [1] p. 21–126.

These tanks are shown in Exhibits I.5.6 and I.5.7.

(3) Pressure tank cars

Wright notes the following in the "Rail Transportation Systems" section of the FPH:

> Pressure tank cars typically transport hazardous materials, including flammable, nonflammable, or poisonous gases at higher pressures. However, pressure tank cars can transport other commodities, depending upon the characteristics of the product or the process for loading and unloading the tank.
>
> Other products transported in pressure tank cars are ethylene oxide, pyrophoric liquids, sodium metal, motor fuel anti-knock compounds, bromine, anhydrous hydrofluoric acid, and acrolein.
>
> Tank test pressures for these tank cars range from 100 to 600 psi (689 to 4137 kPa). Pressure tank cars range in capacity from 4,000 to 45,000 gallons (15 to 170 m^3).
>
> Pressure tank cars are cylindrical, non-compartmented steel or aluminum tanks with rounded heads. They are top-loading, with fittings for loading and unloading, pressure relief, and gauging located inside protective housing mounted on a single manway.

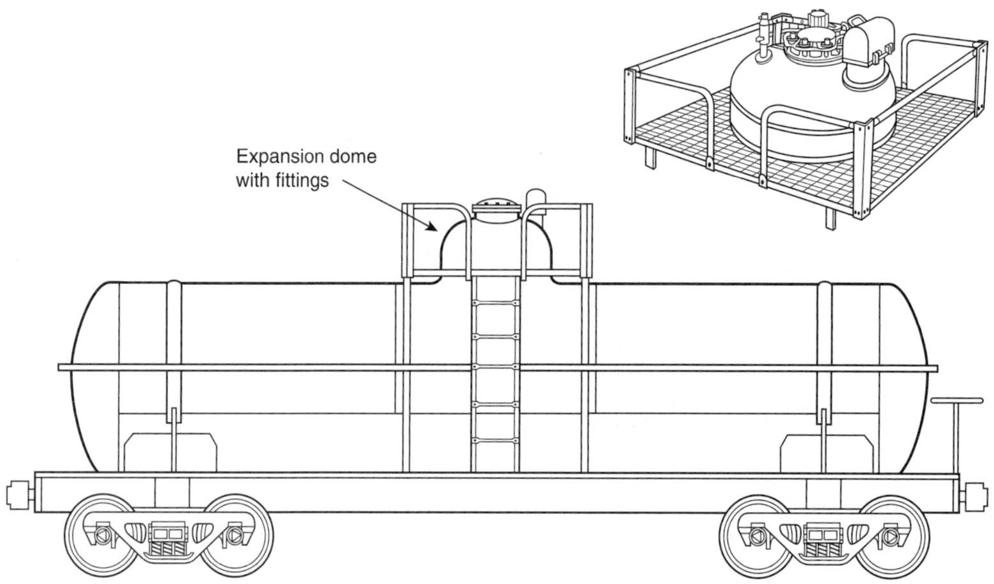

EXHIBIT I.5.6 *The diagram shows a typical nonpressure tank car with an expansion dome. Older models, such as the one shown here, have the dome. (Source: Union Pacific Railroad)*

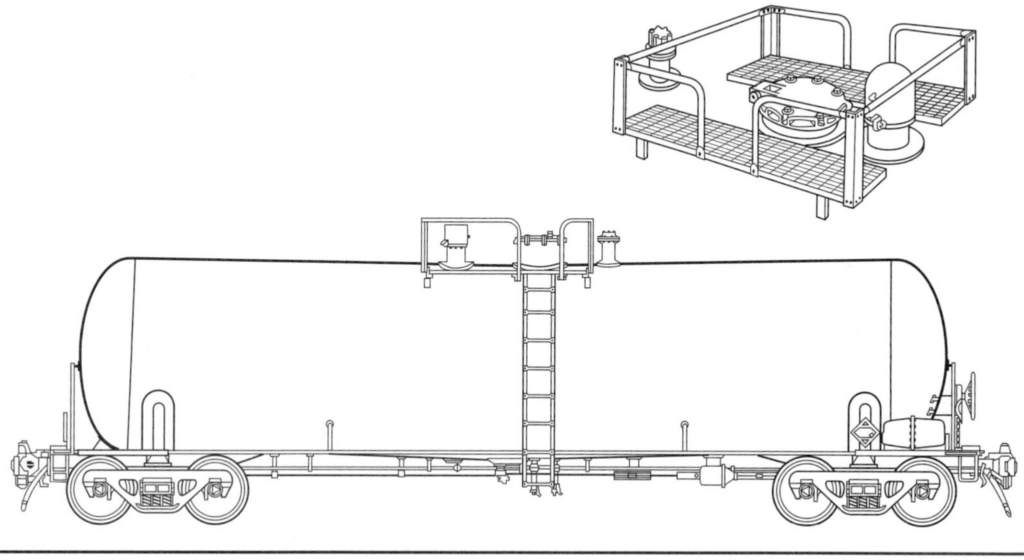

EXHIBIT I.5.7 *The diagram shows a typical nonpressure tank car without an expansion dome, which is generally true of newer cars. (Source: Union Pacific Railroad)*

Pressure tank cars may be insulated and/or thermally protected. The top two-thirds of pressure tank cars without insulation and without jacketed thermal protection will be painted white or another reflective color. [1] p. 21-128 – 21-129

See Exhibit I.5.8.

5.2.1.1.2 Given examples of the following intermodal tanks, the operations level responder shall identify each intermodal tank by type, as follows:

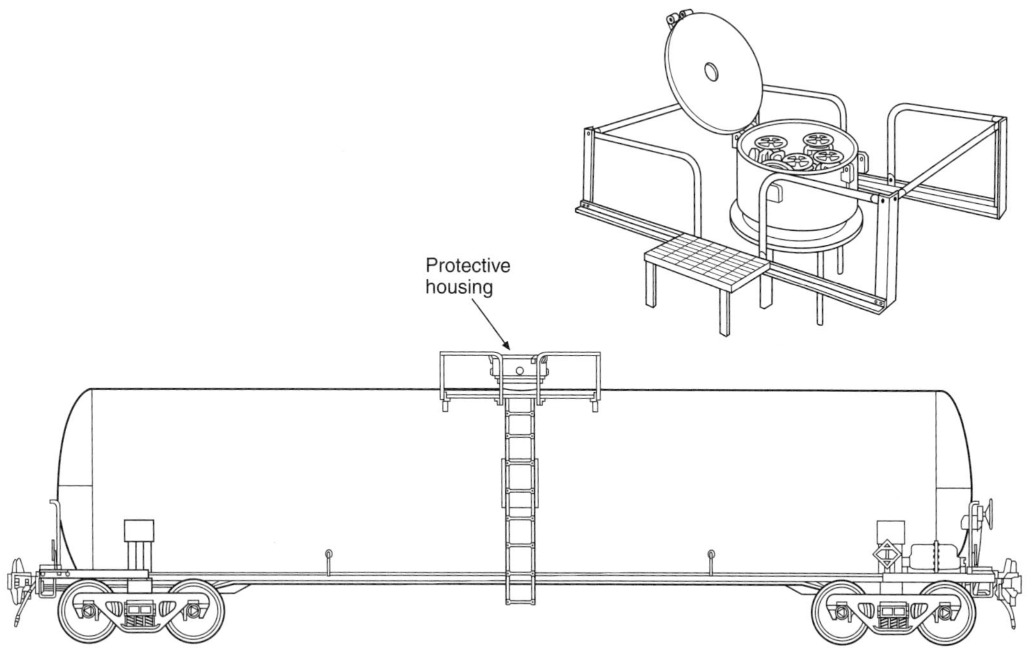

EXHIBIT I.5.8 The diagram shows a typical pressure tank car. (Source: Union Pacific Railroad)

An operations level responder is required to be able to identify intermodal tank containers, which are being used more and more frequently in North America to transport a wide range of commodities, including an increasing number of hazardous materials. Among the factors that account for this increased use are the containers' improved safety, their portability, and the lower transportation costs. Tank containers also offer the benefits of a multimodal transport system. Because the containers consist of a single metal tank mounted inside a sturdy metal supporting structure, they can be used interchangeably on several modes of transport, such as railroad cars, tank trucks, and ships. [1]

(1) Nonpressure intermodal tanks

Nonpressure tank containers, as shown in Exhibit I.5.9, are sometimes referred to as intermodal portable tanks or IM portable tanks. These containers are usually used to transport liquid or solid materials at pressures of up to 100 psi (689 kPa). [1]

In the 2002 edition of NFPA 472, the following two types of tanks were identified in 5.2.1.1.2(1)(a):

1. IM-101 (IMO Type 1 internationally) portable tank, which is built to withstand pressures from 25.4 to 100 psi (175 to 689 kPa)
2. IM-102 (IMO Type 2 internationally) portable tank, which is built to withstand pressures from 14.5 to 25.4 psi (100 to 175 kPa)

These tanks are no longer included in the 2008 edition at the Operations Level and are simply referred to as pressure or nonpressure intermodal tanks. However, this terminology still remains at the hazardous materials technician level in the 2008 edition of NFPA 472 (see 7.2.3.1.1).

(2) Pressure intermodal tanks

Wright notes the following in the FPH section titled, "Rail Transportation Systems":

> Pressure tank containers are designed to accommodate internal pressures of 100 to 500 psi (690 to 3450 kPa) and are generally used to transport gases liquefied under pressure, such as

LP-gas and anhydrous ammonia. Pressure tank containers may also carry liquids such as motor fuel antiknock compounds and aluminum alkyls. [1] pp. 21-132–21-133

See Exhibit I.5.10 for an example of this type of container.

(3) Specialized intermodal tanks, including the following:

(a) Cryogenic intermodal tanks

Cryogenic intermodal tanks, such as the one shown in Exhibit I.5.11, carry refrigerated liquid gases. Internationally they are called IMO Type 7 tank containers.

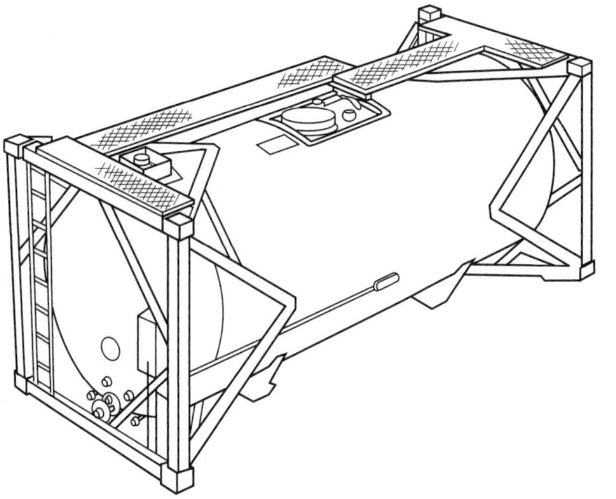

EXHIBIT I.5.9 *The diagram illustrates a nonpressure tank container. (Source: Union Pacific Railroad)*

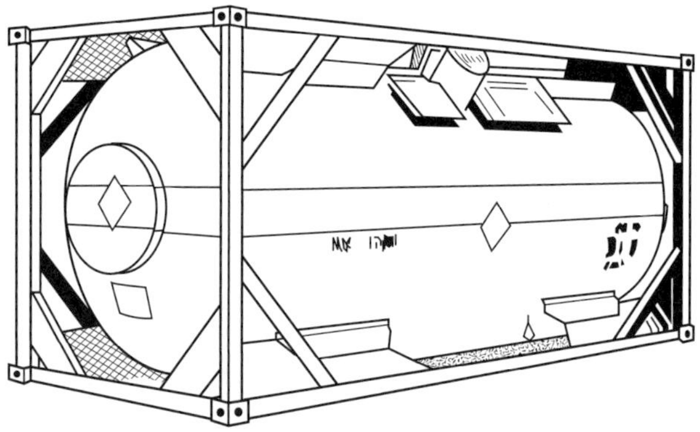

EXHIBIT I.5.10 *The diagram shows a typical pressure tank container. (Source: Union Pacific Railroad)*

EXHIBIT I.5.11 *This is an example of a cryogenic intermodal tank. (Source: Union Pacific Railroad)*

(b) Tube modules

Tube modules, such as the one shown in Exhibit I.5.12, transport gases in high-pressure cylinders permanently mounted within an International Standards Organization (ISO) frame.

5.2.1.1.3 Given examples of the following cargo tanks, the operations level responder shall identify each cargo tank by type, as follows:

Sometimes referred to as tank motor vehicles or tank trucks, cargo tanks are the most common vehicles used to transport combustible, flammable, and corrosive materials as well as flammable and nonflammable compressed gases. Exhibits I.5.13 through I.5.18 illustrate the

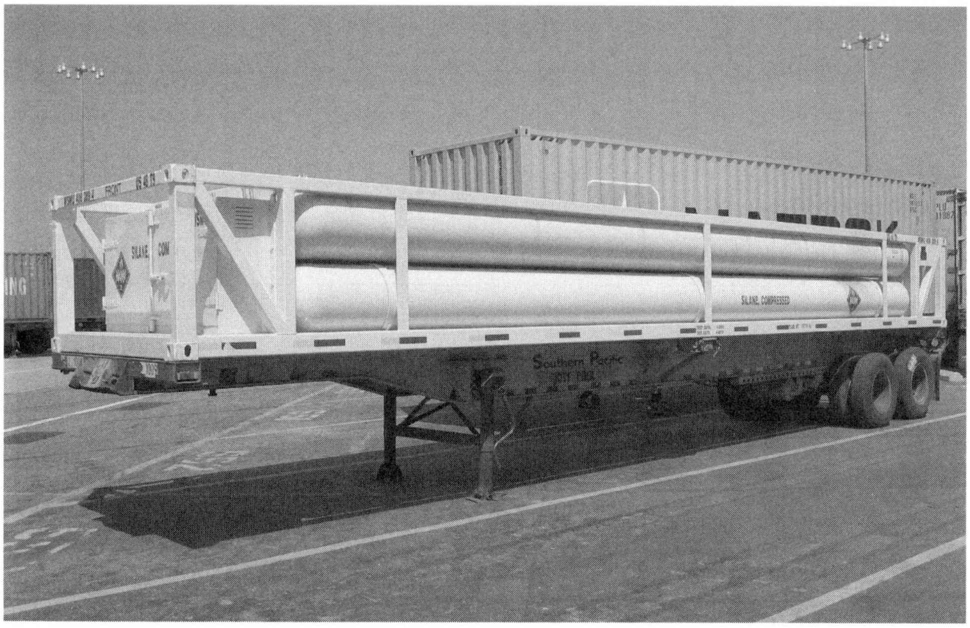

EXHIBIT I.5.12 *This is an example of a tube module. (Source: Union Pacific Railroad)*

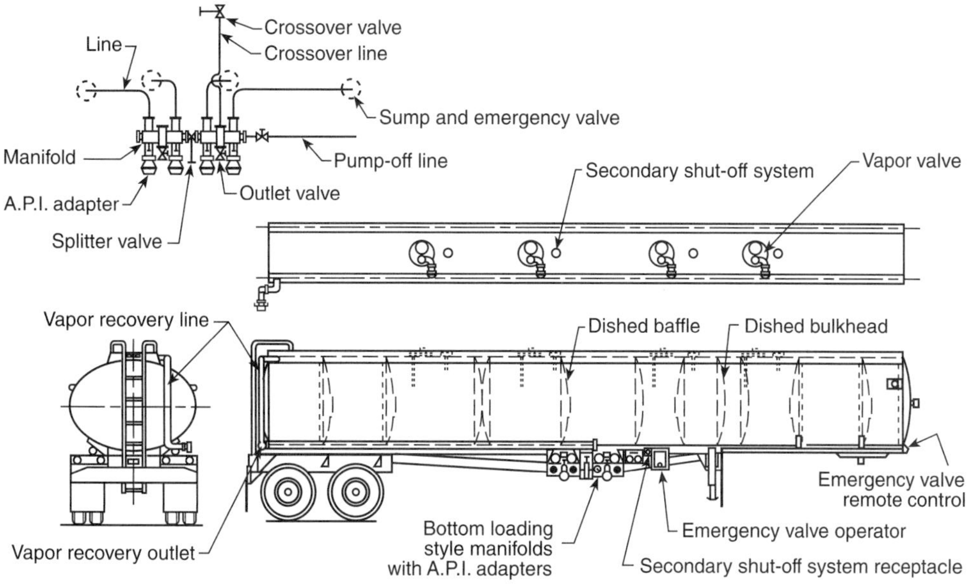

EXHIBIT I.5.13 *The diagram illustrates a typical MC-306 cargo tank and its components.*

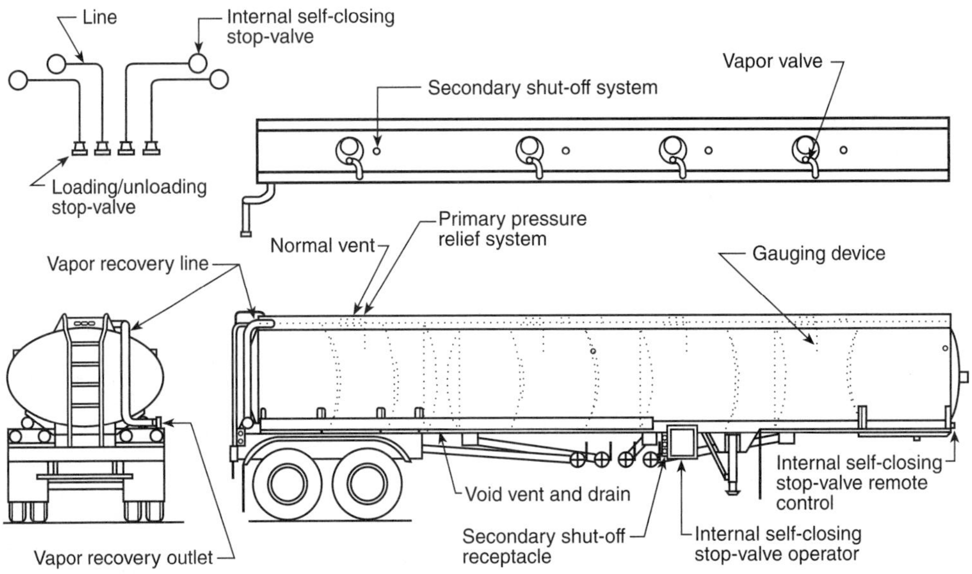

EXHIBIT I.5.14 *The diagram illustrates a typical DOT-406 cargo tank and its components.*

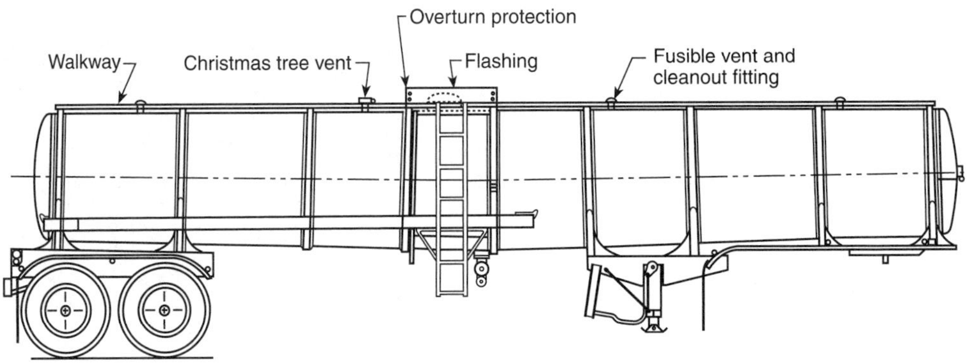

EXHIBIT I.5.15 *The diagram illustrates a typical MC-307 cargo tank and its components.*

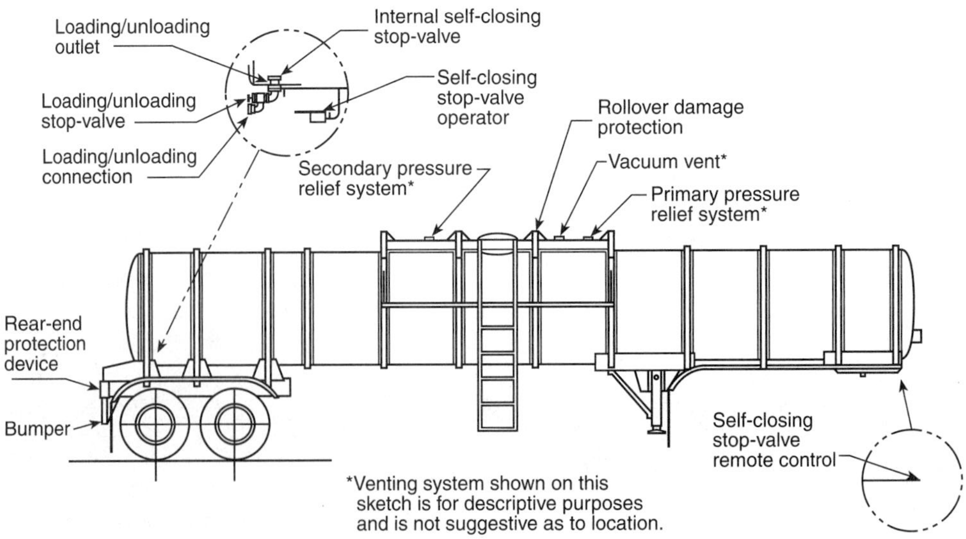

EXHIBIT I.5.16 *The diagram illustrates a typical DOT-407 cargo tank and its components.*

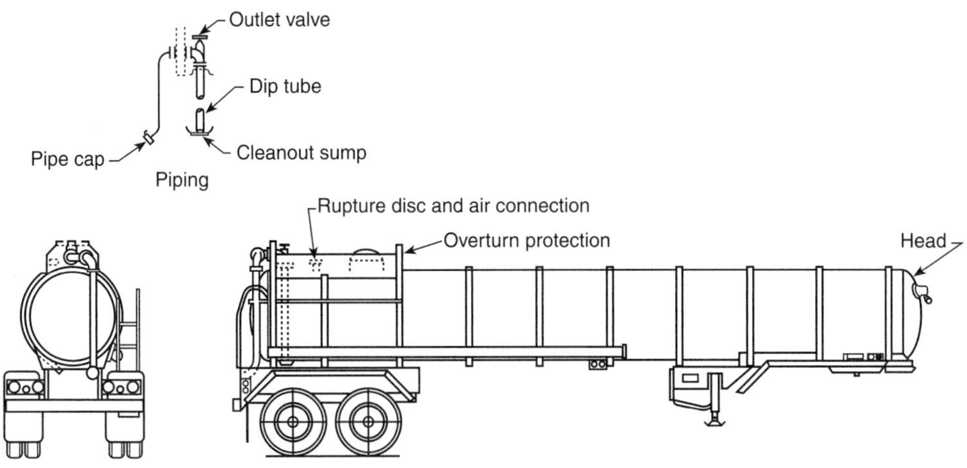

EXHIBIT I.5.17 *The diagram illustrates a typical MC-312 cargo tank and its components.*

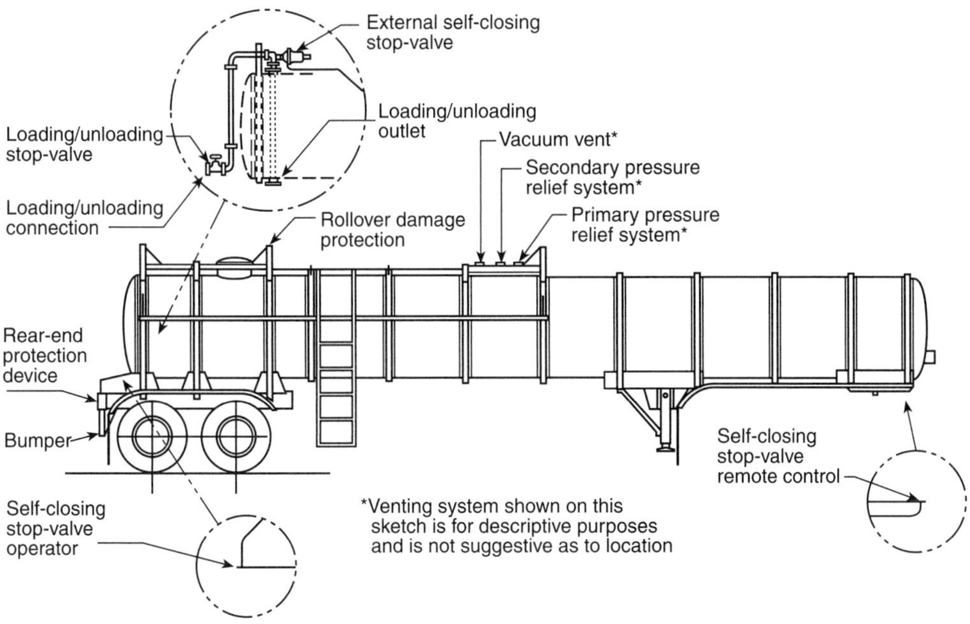

EXHIBIT I.5.18 *The diagram illustrates a typical DOT-412 cargo tank and its components.*

different types of cargo tanks with which the operations level responder must be familiar. As discussed in the commentary to A.5.2.1, the operations level responder should be able to look at a particular type of cargo tank and know something about the nature of the material inside.

(1) Compressed gas tube trailers

On compressed gas tube trailers, multiple cylinders are stacked and manifolded together with controls at the rear. Pressures range up to 5,000 psi (34,470 kPa). These trailers can often be found at construction and industrial sites where a driver may drop off a full trailer and later pick up the empty trailer for refilling.

(2) Corrosive liquid tanks

The corrosive liquid tank cross-section is circular, except for some insulated tanks, with a single compartment. The capacity of a corrosive liquid tank can be up to 7,000 gal (76.5 m^3), and reinforcing ribs are often visible. Overturn and splash protection is at the dome cover and valve locations, which are more often at the rear. The access housing area is often coated with a black, tar-like material to protect the surface from the contents.

(3) Cryogenic liquid tanks

Cyrogenic liquid tanks, such as the one shown in Exhibit I.5.19, are insulated double shell with pressure-relief protection. The product carried is a gas that is cooled to at least –150°F (–101°C) until it becomes a liquid. The space between the shells is placed under a vacuum as part of the cooling process. The tank ends are flat. The piping is usually at the end, contained in a box with double doors. The material is kept in liquid form through refrigeration. Heat from the sun can increase the pressure on the material, and discharging vapors from top rear relief valves is normal.

(4) Dry bulk cargo tanks

Dry bulk cargo tanks are large uninsulated containers with bottom hoppers for unloading. The tanks are used for hauling dry product or sometimes a slurry-like concrete in bulk. Common products carried include fertilizer, grain, and other food products but can include toxic materials.

(5) High pressure tanks

High pressure is considered to be more than 100 psi (689 kPa). The tank with capacities of 2,500 to 11,500 gal (9.46 to 43.5 m^3) has a circular cross-section and rounded ends. The construction is single shell and noninsulated. The upper two-thirds of the tank is painted white or with reflective color to reduce potential heating from the sun. A common product carried in this tank is propane.

(6) Low pressure chemical tanks

Low pressure is considered to be less than 40 psi (275.8 kPa). The most common construction is doubleshell with insulation, featuring one or two compartments with overturn protection. The tank cross-section is circular, except for some insulated tanks. Uninsulated tanks have a single compartment and stiffening rings around the tank. The contents can include a variety of chemicals, and the capacity is between 2,000 and 7,000 gal (7.57 and 26.5 m^3).

(7) Nonpressure liquid tanks

Nonpressurized cargo tanks are the most common on the road and have an elliptical cross-section. Construction is usually of aluminum. The tanks are often used for gasoline and diesel fuel but may contain any liquid.

5.2.1.1.4 Given examples of the following storage tanks, the operations level responder shall identify each tank by type, as follows:

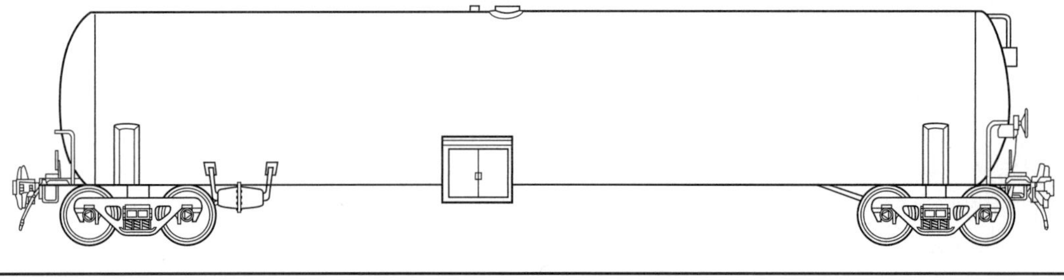

EXHIBIT I.5.19 *The diagram shows a typical cryogenic liquid tank car. (Source: Union Pacific Railroad)*

In the FPH section "Storage of Flammable and Combustible Liquids," Anthony M. Ordile notes the following:

> Tanks can be installed aboveground, underground, or, under certain conditions, inside buildings. [1] p. 7-15

It is important that the responder be able to identify the difference between pressure and nonpressure tanks.

(1) Cryogenic liquid tank

Cryogenic liquid tanks are heavily insulated with a vacuum in the space between the outer and inner shells. The nonpressure tanks are designed to carry refrigerated commodities such as carbon dioxide, nitrogen, argon, hydrogen, and oxygen.

(2) Nonpressure tank

Nonpressure tanks, also called atmospheric tanks, are designed for pressures of 0 to 0.5 psi (4 kPa).

(3) Pressure tank

Pressure tanks are divided into low-pressure storage tanks with pressures of 0.5 to 15 psi (4 to 103 kPa) and pressure vessels, with pressures above 15 psi (103 kPa). Exhibit I.5.20 shows some examples of low-pressure tanks.

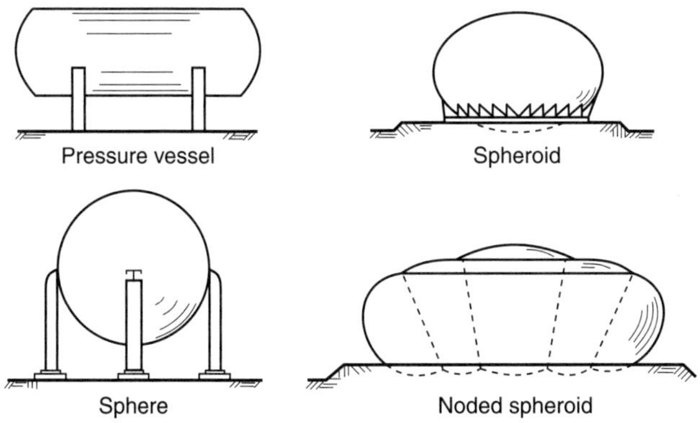

EXHIBIT I.5.20 *The drawing illustrates common types of low-pressure tanks or pressure vessels.*

5.2.1.1.5 Given examples of the following nonbulk packaging, the operations level responder shall identify each package by type, as follows:

(1) Bags

Bags can be many sizes and contain many different types of products, from food to poisons. The construction material can be paper and fiber to plastic and plastic lined. Reinforced sacks can hold very large quantities.

(2) Carboys

A carboy is a glass bottle with a protective cover to keep the bottle from breaking during transportation or should the container be dropped. A common size is 1 gal (3.78 L).

(3) Cylinders

Metal cylinders hold a variety of chemical products under pressure. The cylinders can vary in size from a few pounds to several thousand. A 20 lb (9.1 kg) propane cylinder is common for home barbecues, and larger cylinders up to 250 lb (113.4 kg) are used as a home fuel source. A relief valve or frangible disk provides protection in case of fire or overpressure.

(4) Drums

Drums are often metal cylinders holding 55 gal (208 L) of liquid, but they can be constructed of plastic or fiberboard to hold other products. Drums are often metal cylinders holding liquid, but they can be constructed of plastic or fiberboard to hold other products. A common size in use is the 55-gal (208-L) drum.

(5) Dewar flask (cryogenic liquids)

Dewar flasks are containers within a container. Insulating material and the use of a vacuum space keep the cryogenic material cooled and in a liquid state.

5.2.1.1.6* Given examples of the following radioactive material packages, the operations level responder shall identify the characteristics of each container or package by type, as follows:

(1) Excepted

Excepted packages are for materials with extremely low levels of radioactivity. Due to the very limited hazard of the contents, packaging requirements include ease of handling as well as reasonable strength for transportation. Packaging can range from a fiberboard box to a more-sturdy wooden or steel crate. Packages are not identified as such by package markings or on shipping papers. Excepted packages, such as the one shown in Exhibit I.5.21, are used for transporting limited quantities of radioactive material that would pose very low hazard if released in an accident.

*<u>**EXHIBIT I.5.21**</u> This package is a shipment of low specific activity material en route to a disposal facility. (Source: Department of Energy Transportation Emergency Preparedness Program)*

(2) Industrial

Industrial packages, such as the ones shown in Exhibit I.5.22, are intended for materials with a low concentration of radioactivity that poses a limited hazard to the public and the environment. The radioactive material can be liquid or solidified in such materials as concrete or glass. Industrial packages are not identified as such by package markings or on shipping papers. The following three categories are based on strength:

1. IP-l packages must meet the same design requirements as excepted packaging.
2. IP-2 packages must pass the same tests as Type A for free-drop and stacking.
3. IP-3 packages must pass IP-2 tests and the water spray and penetration tests for Type A shipment of solid contents.

EXHIBIT I.5.22 *These industrial packages contain low activity material and contaminated objects that are categorized as radioactive waste. (Source: Department of Energy Transportation Emergency Preparedness Program)*

(3) Type A

Type A packages, such as the one shown in Exhibit I.5.23, are used to transport radioactive material with higher concentrations of radioactivity than those allowed in excepted and industrial packages. They often have an inner containment vessel made of glass, plastic, or metal surrounded by packaging material of polyethylene, rubber, vermiculite, or wood. The packaging might be an absorbent in a fiberboard, wood, or metal outer container. This packaging must be able to withstand heavy rain equivalent to 2 in. (5. 1 cm) per hour, free-dropping from 4 ft (1.22 m), stacking (compression equal to the weight of the package for at least 24 hr), vibration [1 hr, strong enough to raise the package 0.063 in. (1.6 cm)], and penetration by a dropped weight [1.25 in. (3.18 cm) in diameter and 13.2 lb (5.99 kg) dropped from 40 in. (1.02 m)].

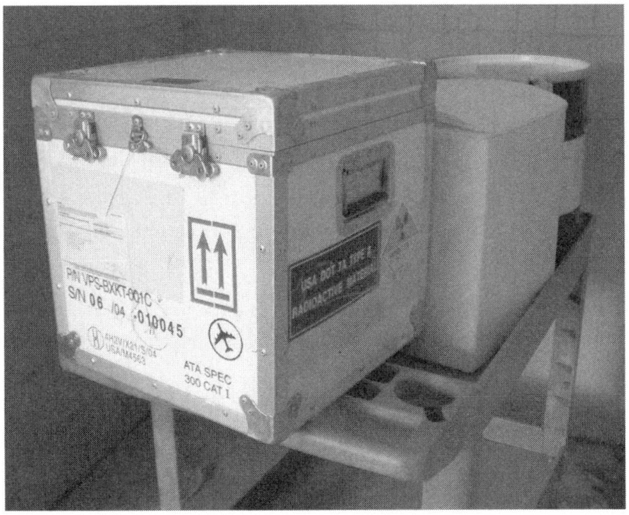

EXHIBIT I.5.23 *This Type A package is designed to be a reusable container. (Source: Department of Energy Transportation Emergency Preparedness Program)*

(4) Type B

Type B protects materials with higher radioactivity levels, including spent nuclear fuel, so it is substantially constructed to retain the contents under normal transport conditions and also under severe accident conditions (see Exhibit I.5.24). Size ranges from small handheld radiography cameras to small drums [55 gal (208 L)] to heavily shielded steel casks that can weigh more than 100 tons (101.6 kilotons). This packaging must be strong enough to withstand tests for dropping 30 ft (9.1 m) so that the package's weakest point is hit; puncture, dropped 40 in. (1.02 m) onto a 6 in. (15.2 cm) diameter steel rod 8 in. (20.3 cm) high, again hitting the package's weakest point; heat, 1475°F (802°C) for 30 min; crush, for some lightweight packages, a drop of 1,100 lb (499 kg) mass 30 ft (9.1 m) onto the package, and immersion under 50 ft (15.2 m) of water. Packages are identified as *Type B* by markings on the package and shipping papers.

EXHIBIT I.5.24 *This RH-72B Type B shipping cask is used to provide double containment for shipment of transuranic waste material. (Source: Department of Energy Transportation Emergency Preparedness Program)*

(5) Type C

Type C packages are rarely used. They contain the most hazardous amounts of radiation. Life-threatening conditions can exist only if the contents are released or package shielding fails.

A.5.2.1.1.6 See A.3.3.43.3.

5.2.1.2 Given examples of containers, the operations level responder shall identify the markings that differentiate one container from another.

Containers at fixed facilities can be marked with the NFPA 704, *Standard System for the Identification of the Hazards of Materials for Emergency Response*, system, and transportation vehicles can be marked with DOT placards or identification numbers [2]. Responders are required to know how to differentiate both markings. Particular tanks or storage areas at fixed facilities can also be identified by labeling and pre-incident planning documents, as can the products they contain.

5.2.1.2.1 Given examples of the following marked transport vehicles and their corresponding shipping papers, the operations level responder shall identify the following vehicle or tank identification marking:

The identification marking on each transport vehicle is included on the shipping papers. This system allows the responders to ensure that the shipping papers and vehicles match. The identification number also provides a way to contact the shipper for information about a specific vehicle.

(1) Highway transport vehicles, including cargo tanks

In highway transportation, the shipping papers are called a bill of lading or shipping papers. Visible markings include company names and logos, vehicle identification numbers, the manufacturer's specification plate, and tank color for specific tanks.

The identification marking on each transport vehicle is included on the shipping papers, which allows the responders to contact the shipper for information about a specific vehicle. Identification numbers, which are assigned to each proper hazardous material shipping name, are required on or near bulk transport container placards and on shipping papers. The numbers begin with the prefix UN for United Nations or NA for North America. NA is used only between the United States and Canada for those not covered by the UN system.

(2) Intermodal equipment, including tank containers

Reporting marks and the tank number on intermodal portable tanks are registered with the International Container Bureau in France. These markings detail the ownership of the tank by the initials and the specific tank by the tank number. The standards to which a portable tank was built are shown by the specification markings. Other markings include DOT exemption markings, Association of American Railroads (AAR) 600 markings for interchange purposes, the permanently attached data plate, and the size, type, and country codes.

(3) Rail transport vehicles, including tank cars

For rail, shipping papers for a train are called the train consist. Individual waybills for a specific rail car can be generated if requested. A seven-digit number, starting with either a 48 or a 49, indicates that the material is hazardous, as per the Standard Transportation Commodity Code (STCC), or stick number. A 48 indicates hazardous waste, while a 49 refers to an uncontaminated product. Railroad hopper cars can display the four-digit identification numbers. Other markings include a commodity stencil for specific tank cars, reporting marks and numbers for the contents, capacity stencil for volume, and specification markings. Rail industry standards previously recommended a vertical red stripe painted the length of an entirely white railroad tank car and a vertical red stripe 4 ft from each end to identify hydrogen cyanide rail tank cars. This is being phased out over time as each tank car receives maintenance.

5.2.1.2.2 Given examples of facility containers, the operations level responder shall identify the markings indicating container size, product contained, and/or site identification numbers.

Containers at fixed facilities are often stenciled with a product name or some type of identification number that refers to a site plan or an emergency operations plan that identifies the product and the quantity stored. Containers at fixed facilities can be marked with the NFPA 704 marking system.

5.2.1.3 Given examples of hazardous materials incidents, the operations level responder shall identify the name(s) of the hazardous material(s) in 5.2.1.3.1 through 5.2.1.3.3.

In order to develop the appropriate action options, the operations level responder is required to gather whatever information is necessary to identify the hazardous materials at an incident. When the responder can ascertain the material involved, he or she can identify the hazards of the material and the routes of exposure. When they know the routes of exposure, they can determine if the personal protective equipment (PPE) they have will protect them from the hazards of the material.

5.2.1.3.1 The operations level responder shall identify the following information on a pipeline marker:

(1) Emergency telephone number
(2) Owner
(3) Product

Pipelines that carry hazardous materials must be identified, and the information is routinely provided on the pipeline marker, such as the one shown in Exhibit I.5.25. The term *product* refers to product class (petroleum), not a specific product identification (gasoline). The responder should be aware that markers are not always exact indicators of pipeline location and

EXHIBIT I.5.25 *This pipeline marker displays typical markings.*

that pipelines do not always follow a straight line between markers. The responder should look for a pipeline marker where it intersects with a street or railroad.

5.2.1.3.2 Given a pesticide label, the operations level responder shall identify each of the following pieces of information, then match the piece of information to its significance in surveying hazardous materials incidents:

According to William J. Keffer and Matthew Woody's section on "Pesticides" in the *Fire Protection Handbook*, 19th edition, pesticides are classified according to their primary or specific control purposes or to reflect the manner in which they are used. Among the pesticides classified by control purposes are insecticides, fungicides, herbicides, nematocides, and rodenticides. Among those pesticides classified by the manner in which they are used are fumigants. [3]

(1) Active ingredient

Each active ingredient in the pesticide is identified and the percentage is indicated. Inert ingredients are also shown but only by percentage.

(2) Hazard statement

A hazard statement typically indicates that the product poses an environmental hazard and advises against contaminating water supplies.

(3) Name of pesticide

The label contains the manufacturer's name for the pesticide, which the responder is required to identify.

(4) Pest control product (PCP) number (in Canada)

In Canada, labels carry a pest control product number to acquire additional information regarding a specific product. In the United States, labels carry an EPA registration number.

(5) Precautionary statement

Labels also carry a precautionary statement indicating the care that must be taken when using the product. Such statements include "Keep Out of Reach of Children," "Restricted Use Pesticide," or "Hazard to Humans and Domestic Animals."

(6) Signal word

Pesticide labels must have a signal word that indicates the relative hazard of the product. Commentary Table I.5.2 lists the current warnings the EPA requires, based on the hazard of the active ingredient.

COMMENTARY TABLE I.5.2 EPA Toxicity Categories

Category	Description
Category 1: Poison/Danger	All pesticide products meeting the following criteria: • Oral LD_{50} up to and including 50 mg/kg • Inhalation LD_{50} up to and including 0.2 mg/liter • Dermal LD_{50} up to and including 200 mg/kg • Eye effects—corneal opacity not reversible within 7 days, and skin effects corrosive. Must bear on the front panel the signal word "Danger." In addition, if the product was assigned to Category 1 on the basis of its oral, inhalation, or dermal toxicity (as distinct from skin and eye local effects), the word "poison" must appear in red on a background of distinctly contrasting color, and the skull and crossbones must appear in immediate proximity to the word "Poison."
Category 2: Warning	All pesticides meeting the following criteria: • Oral LD_{50} from 50 through 500 mg/kg • Inhalation LD_{50} from 0.2 through 2.0 mg/liter • Dermal LD_{50} from 200 through 2000 mg/kg • Eye effects—corneal opacity reversible within 7 days (irritation persisting for 7 days) • Skin effects—severe irritation at 72 hr Must bear on the front panel the signal word "Warning."
Categories 3 and 4: Caution	All pesticide products meeting the following criteria: • Oral LD_{50} greater than 500 mg/kg • Inhalation LD_{50} greater than 2.0 mg/liter • Dermal LD_{50} greater than 2000 mg/kg • Eye effects—no corneal opacity • Skin effects—moderate irritation at 72 hr Must bear on the front panel the signal word "Caution."

Source: Adapted from *Fire Protection Handbook®*, 19th edition, Table 8.11.2.

5.2.1.3.3 Given a label for a radioactive material, the operations level responder shall identify the type or category of label, contents, activity, transport index, and criticality safety index as applicable.

Each label covered in 5.2.1.3.3 has from one to three red vertical bars used to identify the label category. Each label provides a space where the shipper notes the contents of the package and the activity level of the radioactive material inside the package. Additionally, the Radioactive Yellow-II and Radioactive Yellow-III labels provide a space for the transport index. The following radioactive labels are applied to a package based on external radiation levels:

- Radioactive White-I label indicates low external radiation levels.
- Radioactive Yellow-II label indicates medium levels of radiation on the external surface of the package.
- Radioactive Yellow-III label indicates the highest levels of radiation on the external surface of the package. The criticality safety index for each package will be noted on the label. The criticality safety index is displayed on the label to assist the shipper in controlling how many fissile packages can be grouped together on a conveyance.

5.2.1.4* The operations level responder shall identify and list the surrounding conditions that should be noted when a hazardous materials/WMD incident is surveyed.

A.5.2.1.4 The list of surrounding conditions should include topography; land use; accessibility; weather conditions; bodies of water; public exposure potential; overhead and underground wires and pipelines; storm and sewer drains; possible ignition sources; adjacent land use such as rail lines, highways, and airports; and nature and extent of injuries. Building information, such as floor drains, ventilation ducts, and air returns, also should be included where appropriate.

Surrounding conditions are important to the responder in that they influence the options available. The responder should think very broadly when considering the surrounding conditions.

5.2.1.5 The operations level responder shall give examples of ways to verify information obtained from the survey of a hazardous materials/WMD incident.

Responders should continuously collect information about an incident so that they can validate the information collected earlier. This information can be verified by, among other things, contacting CHEMTREC/CANUTEC/SETIQ or any of the many online websites that store chemical data sheets to verify the hazard information found in emergency response guides, contacting the shipper to verify the products listed on shipping papers, and using additional references to confirm the emergency handling procedures.

5.2.1.6* The operations level responder shall identify at least three additional hazards that could be associated with an incident involving terrorist or criminal activities.

A.5.2.1.6 The following are examples of such hazards:

(1) Secondary events intended to incapacitate or delay emergency responders
(2) Armed resistance
(3) Use of weapons
(4) Booby traps
(5) Secondary contamination from handling patients

5.2.2 Collecting Hazard and Response Information.

Given scenarios involving known hazardous materials/WMD, the operations level responder shall collect hazard and response information using MSDS, CHEMTREC/CANUTEC/SETIQ, governmental authorities, and shippers and manufacturers and shall meet the following requirements:

In the *Fire Protection Handbook*, Charles Wright notes the following:

> Once a hazardous material is identified, information about the material's hazards, behavior characteristics, and suggested response options is collected. This information, which may be

> collected simultaneously with determining the extent of containment system damage, is used to predict the behavior of that material. The information to be collected is divided into six basic groups:
>
> **1.** Material identification information
> **2.** Physical properties
> **3.** Chemical properties
> **4.** Physical hazards
> **5.** Health hazards
> **6.** Response information
>
> The task of obtaining, recording, and interpreting hazardous material information can be lengthy and rigorous. Various forms are being used to record hazard and response information. [1] p. 13-127

The information the operations level responder is required to collect by 5.2.2 allows the responder to determine the defensive options that can be performed safely, given the personnel and equipment available. At the technician level, this same process continues and the responder uses the information collected to determine whether conducting offensive operations is feasible.

(1) Match the definitions associated with the UN/DOT hazard classes and divisions of hazardous materials/WMD, including refrigerated liquefied gases and cryogenic liquids, with the class or division.

The responder is required to match the hazard class or division of a hazardous material with the appropriate definition of that material. For example, the responder should be able to match the definition of a flammable gas with a Class 2, Division 1 material.

(2) Identify two ways to obtain an MSDS in an emergency.

MSDSs are available at fixed facilities and can be found in transporting vehicles as well. A responder can obtain MSDSs from CHEMTREC or from the shipper, who can fax them to dispatch offices or to portable fax machines available to field personnel. Many websites are also available to access data sheets including many of the shippers and manufacturers sites.

(3) Using an MSDS for a specified material, identify the following hazard and response information:

In the United States, MSDSs are required by OSHA. Although OSHA provides a standard form, manufacturers can use similar forms of their own design that are approved by OSHA. Certain basic information specified by OSHA must appear on all forms. Responders should be familiar with this information and know how to locate it on an MSDS, which is a valuable source of product information. The responder should understand that, although the location of certain information on the MSDS might vary, the basic information outlined in 5.2.2(3) can be found somewhere on the sheet. For example, information relating to storage found on one form in a section called *Precautions for Safe Handling and Use* might be found on another form in a section called *Special Protection Information.*

(a) Physical and chemical characteristics

The MSDS information in 5.2.2(3)(a) provides the responder with information about physical characteristics of the hazardous material such as its vapor density, boiling point, specific gravity, water solubility, pH, and physical appearance. For example, a substance might be described as white to pale yellow sticks, granules, or powder, no odor.

(b) Physical hazards of the material

The MSDS information includes information about a material's fire and explosion hazards, including its flash point, autoignition temperature, and flammability limits as well as information about the extinguishing agents that might be used on the material. This section might also provide information about hazards associated with fire control operations.

For example, an entry might read, "A water stream directed at molten material can scatter the material, increasing the flammability of any combustible material it contacts." This section might also recommend appropriate personal protective clothing and respiratory protection.

A separate reactivity section generally provides information about the material's stability and indicates what the material will react with. For instance, an entry might read, "is a strong oxidizing agent that will increase the flammability of all combustible materials it contacts."

(c) Health hazards of the material
(d) Signs and symptoms of exposure
(e) Routes of entry
(f) Permissible exposure limits

Information on permissible exposure limits is generally contained in the health hazard section, which provides the responder with important data on the health hazards a material presents including the threshold limit value (TLV), the routes of exposure, and the material's effects. Also provided in the section in 5.2.2(3)(f) is information about emergency first-aid measures.

(g) Responsible party contact

Information about the material's manufacturer and possibly the names of its distributors is also provided as are the telephone numbers for emergency contacts.

(h) Precautions for safe handling (including hygiene practices, protective measures, and procedures for cleanup of spills and leaks)

The MSDS also tells the responder what steps should be taken in the event of a spill or leak and how to dispose of such spilled or leaked material. For example, this MSDS section might instruct the responder to remove ignition sources or to suppress the materials vapors with foam.

(i) Applicable control measures, including personal protective equipment

This MSDS information helps the responder choose the appropriate respiratory protection, eye protection, protective gloves, and so forth for working with the hazardous material. The information might also indicate how the material should be stored and how to recognize improperly stored materials.

(j) Emergency and first-aid procedures

Information about emergency and first-aid procedures is often found with other health-related data. This MSDS section details the actions that should be taken immediately if an individual is exposed to a hazardous material and recommend when to seek additional medical attention.

(4) Identify the following:

(a) Type of assistance provided by CHEMTREC/CANUTEC/SETIQ and governmental authorities

CHEMTREC stands for Chemical Transportation Emergency Center and is a public service of the Chemical Manufacturers Association. CHEMTREC provides the on-scene commander with immediate advice by telephone and contacts the involved shipper for detailed assistance and response follow-up. The organization can also notify the National Response Center (NRC) of significant incidents and bridge a caller to the NRC to report a spill. CHEMTREC operates 24 hours a day and can be contacted throughout the United States and Canada.

CHEMTREC can usually provide hazard information warnings and guidance when given a material's four-digit identification number, the name of the product, and the nature of the problem. If the product is unknown or more detailed information and assistance is needed, the caller should attempt to provide as much of the following information as possible:

- Caller's name and a call-back number
- Guide number being used
- Name of the shipper or manufacturer
- Rail car or truck number
- Carrier's name
- Consignee
- Local conditions

At an incident, the caller should try to keep a phone line open to CHEMTREC so that they can provide guidance and assistance. CHEMTREC can also provide a teleconferencing bridge that allows them to connect technical experts to the caller's line as necessary.

CANUTEC is the Canadian Transport Emergency Center, which is operated by the Transport Dangerous Goods Directorate of Transport Canada. The organization provides technical assistance to emergency responders much the same as CHEMTREC. Personnel provide technical information regarding the physical, chemical, toxicological, and other properties of the products involved in an incident; recommend remedial actions for fires, spills, or leaks; provide advice on protective clothing and emergency first aid; and contact the shipper, manufacturer, or others who are deemed necessary.

SETIQ is the Emergency Transportation System for the Chemical Industry in Mexico and provides the same services as CHEMTREC and CANUTEC. Local and state authorities might have a specific role in receiving incident information or providing assistance. This role should be covered in the local emergency response plan, which should be consulted as needed for the agencies and phone numbers involved. Federal authorities might also provide information and other assistance. The responder might be required to notify federal authorities according to the emergency at hand.

The responder should collect and provide the following information, as much as can safely be obtained, to the chain-of-command and specialists listed in the *Emergency Response Guidebook* (ERG) who might be contacted for technical guidance [4]:

- Responder's name, call-back telephone number, and fax number
- Location and nature of the problem
- Name and ID number of the material involved
- Shipper, consignee, and point of origin
- Carrier name, rail car number, or truck number
- Container type and size
- Quantity of material transported and released
- Local conditions of weather, terrain, and proximity to schools, hospitals, and waterways
- Injuries and exposures
- Local emergency services that have been notified

The ERG lists emergency response telephone numbers for CHEMTREC, CANUTEC, military shipments, and the NRC. The NRC receives reports required when dangerous goods and hazardous substances are spilled. After receiving notification of an incident, the NRC will immediately notify the appropriate federal on-scene coordinator and concerned federal agencies. Federal law requires that anyone who releases a reportable quantity of a hazardous substance into the environment must immediately notify the NRC. Calling other agencies does not constitute compliance with the requirement to call the NRC.

(b) Procedure for contacting CHEMTREC/CANUTEC/SETIQ and governmental authorities
(c) Information to be furnished to CHEMTREC/CANUTEC/SETIQ and governmental authorities

(5) Identify two methods of contacting the manufacturer or shipper to obtain hazard and response information.

The two methods that the responder can use to contact the manufacturer or shipper are through CHEMTREC/CANUTEC/SETIQ, as described in the commentary to 5.2.2(4), or by using information provided on the shipping papers or on the MSDS.

(6) Identify the type of assistance provided by governmental authorities with respect to criminal or terrorist activities involving the release or potential release of hazardous materials/WMD.

Criminal or terrorist activities can involve the use of a variety of hazardous materials, but all such incidents involve law enforcement response. Notifying the local law enforcement agency initiates the needed state and federal agency notification and response. This assistance support includes scene security, crime scene preservation, evidence collection, and other law enforcement missions.

(7) Identify the procedure for contacting local, state, and federal authorities as specified in the emergency response plan and/or standard operating procedures.

The key to accomplishing this requirement is becoming familiar with the local emergency response plan (ERP) or organization's standard operating procedures (SOP) and making detailed preparation before the emergency. Planning documents have given consideration to which authorities need to be contacted and under what circumstances. From that list the relevant phone numbers are available or can be compiled.

The methods for keeping the list of contacts and phone numbers updated needs to be thought through. When contacted, the governmental authorities need as much accurate information as can be provided. At that time differentiating between known facts and other reported information that cannot be verified is very important.

(8)* Describe the properties and characteristics of the following:

(a) Alpha radiation

Radioactive materials transmit energy through space in the form of particles and rays. The energy is the result of spontaneous disintegration of atomic nuclei by the emission of subatomic particles. Alpha particles, listed in 5.2.2(8)(a), are positively charged nuclear particles consisting of two protons bound to two neutrons that are ejected from the core of a radioactive element. Alpha particles are a slow-moving, relatively heavy, but relatively weak form of energy with a very short range [3 in. (7.6 cm)] and little penetration. Because of the alpha particle's short range and limited penetrating ability, external shielding is not required. The particles can be stopped by clothing or even sheets of paper. Alpha particles cannot penetrate the skin, but they can be harmful if inhaled or ingested into the body where they continue to transmit energy at closer range where they can damage body tissue. Inside the body, alpha particles can be the most serious internal radiation hazard. Alpha particles are 7000 times larger than beta particles.

(b) Beta radiation

Beta particles are fast-moving, positively or negatively (more common) charged electrons emitted as energy from the nucleus during radioactive decay. Beta particles are formed when a neutron breaks up into its component parts, which is one proton and one neutron held together by a binding energy. When this energy is released, the electron is ejected as a beta particle. High-speed beta particles can travel farther [a mean range of about 7 ft (2.1 m)] and are more penetrating than alpha particles, but are less damaging over the same distances. Some beta particles can be stopped by layers of clothing (fire fighter personal protective clothing), plastic (SCBA face shield), aluminum, thick cardboard, a building wall, or other shielding. Some beta particles are capable of penetrating the skin from 0.25 in. (0.64 cm) or closer and causing radiation damage. Eye and skin damage is possible if the source is strong. As with alpha particles, beta particles are generally more hazardous when the radioactive material emitting them is inhaled or ingested but less so than alpha particles. Internally, beta radiation is less hazardous than alpha radiation because beta particles travel farther. As a result, the energy deposited by beta radiation is spread out over a larger area, which causes less harm to individual cells or organs.

(c) Gamma radiation

Gamma radiation, like x-rays, is electromagnetic radiation consisting not of particles but waves of energy with no mass and no electrical charge. Gamma rays are similar to x-rays and are weightless packets of energy called photons. They may accompany the emission of alpha or beta particles from a decaying nucleus, but they travel at nearly the speed of light. Because gamma rays have no mass and no electrical charge, they travel great distances and are more penetrating and easily pass through the human body from several hundred feet away, or they can be absorbed by tissue. This movement produces a radiation risk for the entire body and can be lethal. Shielding to stop the more energetic gamma rays requires several feet of concrete or a few inches of lead.

(d) Neutron radiation

Neutron radiation consists of neutron particles that are ejected from an atom's nucleus. Neutrons are a basic part of an atom. An atom consists of a small positively charged core nucleus surrounded by electrons. This nucleus contains most of the mass of the atom and is composed of neutrons bound tightly with protons by a nuclear force much greater than the electrical forces binding electrons to the nucleus. Many nuclei are unstable and subject to slow disintegration and release of radiation energy, or a nucleus can be made to rapidly disintegrate. If the binding nuclear force of neutrons and protons is intentionally overcome, the atom breaks up (fission) and releases energy.

Fission requires an atomic particle with sufficient energy and mass to penetrate the nucleus. Neutrons are capable of this penetration because they are electrically neutral and are not repelled by the positive charge of a nucleus. This fission can produce an explosion or, if controlled, produce energy for a power supply. Neutron radiation is best shielded with high hydrogen content material (i.e., water, plastic). In transportation situations, neutron radiation is not commonly encountered. Neutron radiation is usually associated with operating nuclear power plants.

A.5.2.2(8) Radioactive materials transmit energy through space in the form of particles and rays. The energy is the result of spontaneous disintegration of atomic nuclei by the emission of subatomic particles. Alpha particles are positively charged nuclear particles consisting of two protons bound to two neutrons that are ejected from the nucleus of a radioactive atom. Alpha particles travel at about 1/20th the speed of light but have a very short range [7.6 cm (3 in.)] and little penetration power. Because of the alpha particle's short range and limited penetrating ability, external shielding is not required. The particles can be stopped by clothing or even sheets of paper. Alpha particles cannot penetrate the skin, but they can be harmful if the radioactive material emitting the alpha particles is inhaled or ingested into the body, where they continue to emit alpha particles; at closer range, they can damage body tissue. Inside the body, alpha particles can be the most serious internal radiation hazard. Alpha particles are 7000 times larger than beta particles.

5.2.3* Predicting the Likely Behavior of a Material and Its Container.

Given scenarios involving hazardous materials/WMD incidents, each with a single hazardous material/WMD, the operations level responder shall predict the likely behavior of the material or agent and its container and shall meet the following requirements:

A.5.2.3 Predicting the likely behavior of a hazardous material and its container requires the ability to identify the types of stress involved and the ability to predict the type of breach, release, dispersion pattern, length of contact, and the health and physical hazards associated with the material and its container. References can be made to the National Fire Academy program, *Hazardous Materials Incident Analysis,* or the *Fire Protection Handbook* chapter titled "Managing the Response to Hazardous Material Incidents."

In *A Textbook for Use in the Study of Hazardous Materials Emergencies*, Ludwig Benner, Jr. describes the process of predicting hazardous materials behavior as the following:

> . . . visualization of an event's sequences in a 'mental movie' framework. . . . The responder needs to think in terms of events and then relate them to the prediction of the emergency events." The responder ". . . needs to focus on what the hazardous material is going to do . . . in order to influence the sequence of events. [5]

(1) Interpret the hazard and response information obtained from the current edition of the DOT *Emergency Response Guidebook,* MSDS, CHEMTREC/CANUTEC/SETIQ, governmental authorities, and shipper and manufacturer contacts, as follows:

Not only must the responder know where to find response information, but he or she must also be able to interpret that information in order to decide what actions are appropriate. The responder must also recognize that different emergency response guides might present conflicting information or emphasize one area more than another. The information that is most appropriate for a given situation must be gathered, interpreted, and chosen.

(a) Match the following chemical and physical properties with their significance and impact on the behavior of the container and its contents:

i. Boiling point

Boiling point is the temperature of a substance when the vapor pressure exceeds atmospheric pressure and the liquid turns into a gas at the surface.

ii. Chemical reactivity

Chemical reactivity is the ability of a material to undergo a chemical change. The catalyst for the chemical reaction could be exposure to light, heat, shock or contact with other chemicals. Undesirable effects such as pressure buildup and/or increasing temperature can result in catastrophic failure of the container or the formation of other hazardous materials.

iii. Corrosivity (pH)

Corrosivity is a measure of a substance's tendency to deteriorate in the presence of another substance or in a particular environment. The U.S. Code of Federal Regulations defines a corrosive material as a liquid or solid that causes visible destruction or irreversible alterations in human skin tissue at the site of contact or that causes steel to corrode at a severely accelerated rate [16]. The degree of corrosiveness is measured by pH, which ranges from 1 to 14. A pH of 7 is neutral, while a pH below 7 is acidic and a pH above 7 represents a base.

iv. Flammable (explosive) range [lower explosive limit (LEL) and upper explosive limit (UEL)]

A material's flammable, or explosive, range is the difference between its upper and lower explosive limits. The lower explosive limit (LEL) is the minimum concentration of vapor to air below which a flame does not propagate in the presence of an ignition source. The upper explosive limit (UEL) is the maximum vapor-to-air concentration above which a flame does not propagate. If a vapor-to-air mixture is below the LEL, the mixture is described as being "too lean" to burn; if it is above the UEL, it is "too rich" to burn. When the vapor-to-air ratio is somewhere between the LEL and the UEL, fires and explosions can occur, and the mixture is said to be in the flammable range. The flammable range for gasoline is 1.4 percent to 7.6 percent, and the flammable range for carbon monoxide is 12.5 percent to 74 percent. Awareness of the range as well as the upper and lower flammable limits is important for the responder. If the responder suspects or knows that flammable vapors are present, he or she must determine the concentration of vapor in air. Combustible gas detection instruments are used for this purpose.

v. Flash point

The flash point of a liquid is the minimum temperature at which the liquid gives off vapor in sufficient concentration to form an ignitable mixture with air. A liquid's flash point is the primary property or characteristic used to determine its relative degree of flammability. Since the vapors of flammable liquids are what burn, vapor generation is a primary factor in determining the liquid's fire hazard.

vi. Ignition (autoignition) temperature

The terms *ignition temperature* and *autoignition temperature* are interchangeable. The ignition temperature of a substance, whether solid, liquid, or gaseous, is the minimum temperature required to cause self-sustained combustion in the absence of any source of ignition. The responder should look upon assigned ignition temperatures as approximations.

Ignition temperatures can be quite high, especially in relation to a liquid's flash point. For example, the flash point of gasoline is –45°F (–43°C), while its ignition temperature is well over 500°F (260°C).

vii. Particle size

Particle size refers to solids, and it is expressed in microns or percent passing through a meshed screen.

viii. Persistence

The term *persistence* refers to a material's ability to stay within the area of release for long periods of time. This is generally considered to be more than 24 hours and is intended to prevent personnel from re-entering the area due to concentrations that remain high.

ix. Physical state (solid, liquid, gas)

Hazardous materials are either solids, liquids, or gases, and the responder should understand the difference physical form makes on the hazards a material presents. For example, gases present significantly different hazards than solids.

x. Radiation (ionizing and non-ionizing)

Radiation can be grouped into two categories: ionizing radiation (i.e., alpha, beta, gamma, neutron) and non-ionizing radiation (i.e., microwaves, radio waves, visible light).

Ionizing radiation consists of high-energy rays (gamma rays, x-rays) or particles (alpha particles, beta particles). The term *ionizing* is used because of the mode of action. The result is that the high energy impacts an atom and has the ability to cause a physical change in atoms by making them electrically charged (ionized). This ability makes ionizing radiation hazardous. The damage can be immediate and physical and also lead to genetic mutations. Ionizing radiation's ability to cause a physical change in atoms does not have an effect on the container or its contents. However, as the hazard level of the radioactive material increases so does the strength of the package.

Non-ionizing radiation consists of ultraviolet and visible light, sound waves, microwaves, and magnetic fields. Ultraviolet radiation exposure causes familiar sunburn. Nonionizing radiation from the sun can cause genetic damage to skin cells and result in skin cancer. This radiation does not have the energy required to impact an atom enough to eject an orbital electron. Non-ionizing radiation is generally beyond the control of hazardous materials responders.

xi. Specific gravity

Specific gravity is the ratio of the weight of a volume of liquid or solid to the weight of an equal volume of water, with the gravity of water being 1. A substance with a specific gravity of less than 1 will float on water, and a substance with a specific gravity greater than 1 will sink.

xii. Toxic products of combustion

All products of combustion should be considered toxic, but those produced by hazardous materials might be more toxic than those produced by nonhazardous materials. The danger that the products of combustion of hazardous materials are contaminated to a greater degree with higher levels of toxins than one would find in a regular structural fire increases the need for respiratory protection and for evacuation downwind of the fire.

However, there are instances when burning hazardous materials, such as pesticides, can destroy the hazardous materials during the combustion process. Fire fighters must be aware that this is as true for outside fires as it is for structural fires.

xiii. Vapor density

Vapor density measures the weight of a given vapor as compared with an equal volume of air, with air having a value of 1.0. A vapor density greater than 1.0 indicates it is heavier than air, and a value less than 1.0 indicates it is lighter. Vapor density can be important to the responder, because it determines the behavior of free vapor at the scene of a liquid spill or gas release.

xiv. Vapor pressure

Vapor pressure is the pressure at any given temperature at which the vapor and liquid phases of the substance are in equilibrium in a closed container. A change in temperature or atmospheric pressure can increase the pressure inside the container, increasing the stress on the container. If the container breaches, the material will spread quickly. For example, the vapor pressure of diesel fuel increases with elevation due to reduced barometric pressure (lower atmospheric pressure).

xv. Water solubility

Water solubility, or the degree to which a substance is soluble in water, can be useful in determining the effectiveness of water as an extinguishing agent in dilution and in a decontamination process.

(b) Identify the differences between the following terms:

i. *Contamination* and *secondary contamination*

Contamination is a direct transfer of a hazardous material, as described in the commentary to 5.2.3(l)(b)ii. Secondary contamination is an indirect transfer when, for example, contaminated personnel or equipment carries a contaminant away from a hot zone and transfers it to another person. Responders working in the hot zone may become contaminated during control operations. If the responders carry that contamination outside the hot zone on their equipment clothing, skin, or hair in sufficient quantities and are not adequately decontaminated, they could contaminate others.

ii. *Exposure* and *contamination*

Exposure occurs when a material comes directly in contact with or is taken into a person's body through a route of exposure. It might or might not result in contamination. An example of this is carbon monoxide that is breathed in causes an exposure but does not lead to contamination of the person.

A person can be exposed to radiation and not become contaminated. On the other hand, radioactive contamination emits radiation. If a person is contaminated with radioactive material, the person continues to be exposed to radiation until the contamination is removed. Radioactive materials exist in nature or can be man-made. Natural sources of low-level radioactive material can include the soil or the air, and through these sources radioactive material can enter food. Man-made radiation might be intentionally directed at us from television tubes or in the form of medical procedures such as x-rays, which constitutes a normal exposure. Any of these exposures can go from external to internal if the material emitting the radiation is inhaled or ingested. The term *exposure* means being exposed to ionizing radiation or to radioactive material. The risk from short duration, low levels of radiation exposure is small. Management of any risk from radiation exposure includes attention to the following three radiation protection principles:

1. Minimize your time in a field of radiation.
2. Maintain as much distance between you and the source of radiation.
3. Use available shielding whenever possible. This could involve standing behind a vehicle, building, or any other object you can place between you and the source of radiation.

Personnel and equipment can become contaminated if they come in contact with radioactive material that has been released from its containment or package. When individuals (accident victims or response personnel), PPE, or equipment become contaminated, the contamination can easily be spread by secondary contamination to other persons, equipment, or surfaces. Care should be taken to avoid secondary contamination.

Radioactive contamination can be determined by direct or indirect measurement. Direct measurement is possible with portable detection instruments for fixed and removable contamination when background radiation levels are negligible and the detector has sufficient sensitivity. Indirect measurement only detects removable contamination by wipe tests.

iii. *Exposure* and *hazard*

A hazard is something capable of posing an unreasonable risk to health and safety, while an exposure is the process by which people, animals, and equipment come in contact with a hazardous material. An exposure is affected by duration and concentration of the hazardous material. A person might be exposed to large quantities of a hazardous material in concentrations that do not present a hazard or to small amounts of a hazardous material that present a very high hazard.

iv. *Infectious* and *contagious*

Contagious means a substance is capable of being transmitted from one individual (animal or human) to another through contact, typically by bodily fluids or secretions. Infectious means disease is caused by exposure to harmful microorganisms. Microorganisms multiply and typically attack other organs or cells in the body. Not all infectious diseases are contagious.

v. *Acute effects* and *chronic effects*

Acute effects present symptoms immediately while chronic effects manifest at a later time, which can even be years later.

vi. *Acute exposures* and *chronic exposures*

Acute exposures are generally considered to be large concentrations over a short period of time. An acute exposure can have effects that are both immediate and/or long term. Chronic exposures are over long periods of time, over several days or longer with repeated periods of contact at relatively low levels of concentrations.

(2)*Identify three types of stress that can cause a container system to release its contents.

A.5.2.3(2) The three types of stress that could cause a container to release its contents are thermal stress, mechanical stress, and chemical stress.

(3)*Identify five ways in which containers can breach.

A.5.2.3(3) The five ways in which containers can breach are disintegration, runaway cracking, closures opening up, punctures, and splits or tears. The performance objectives contained in 5.2.3(3) through 5.2.3(5) should be taught in a manner and language understandable to the audience. The intent is to convey the simple concepts that containers of hazardous materials/WMD under stress can open up and allow the contents to escape. This refers to both pressurized and nonpressurized containers. This content release will vary in type and speed. A pattern will be formed by the escaping product that will possibly expose people, the environment, or property, creating physical and/or health hazards. This overall concept is often referred to as a *general behavior model* and is used to estimate the behavior of the container and its contents under emergency conditions.

When a hazardous materials container loses its integrity, the incident often escalates. The timing of such a release cannot always be predicted, and the release will vary with the duration, intensity, and type of stress to which the container is subjected.

The responder must understand that a container or its contents might be stressed and that a difference exists between the two phenomena. A container can degrade under stress. The material inside a container can either degrade the container or breach a container that has not been degraded [5].

According to Charles Wright in FPH, factors that affect the intensity of the breach include the following:

> Type and duration of the stress being applied, behavior of the containment system under the stresses applied, behavior of the contents, location of the applied stresses, force of opening of the containment system, size of breach, and speed of breach event. [1] p. 13-132

(4)* Identify four ways in which containers can release their contents.

A.5.2.3(4) The four ways in which containment systems can release their contents are detonation, violent rupture, rapid relief, and spill or leak.

The types of release include detonation, disintegration of the container, and/or detonation of the contents; violent massive failure behavior, the runaway cracking of the container, and rapid-acceleration polymerization or oxidizing hazardous materials reactions that burst the container abruptly; rapid relief behavior, including pressure ruptures or safety valve operation; and spill or leak behavior, including gradual flow through openings, tears or splits, and punctures [5].

(5)* Identify at least four dispersion patterns that can be created upon release of a hazardous material.

A.5.2.3(5) Seven dispersion patterns can be created upon release of agents: hemisphere, cloud, plume, cone, stream, pool, and irregular.

Knowing how hazardous materials behave when they are released is important in determining the area potentially endangered. Dispersion patterns are influenced by the way the material is released, physical behavior of the material, and weather conditions.

(6)* Identify the time frames for estimating the duration that hazardous materials/WMD will present an exposure risk.

A.5.2.3(6) The three general time frames for predicting the length of time that an exposure can be in contact with hazardous materials/WMD in an endangered area are short-term (minutes and hours), medium-term (days, weeks, and months), and long-term (years and generations).

The factors that influence the length of time an exposure might last in an endangered area include the quantity of the material released, the method of dispersion, and the speed at which it is released. For example, did the container leak or did it detonate? The presence of secondary reactions also influences the length of an exposure.

(7)* Identify the health and physical hazards that could cause harm.

A.5.2.3(7) The health and physical hazards that could cause harm in a hazardous materials/WMD incident are thermal, mechanical, poisonous, corrosive, asphyxiating, radiological, and etiologic.

Harm is the injury or damage caused by being exposed to the hazards of the released contents. The hazards associated with the released contents are directly proportional to the concentrations of released materials and to the duration of contact.

(8)*Identify the health hazards associated with the following terms:

(a) Alpha, beta, gamma, and neutron radiation

The fact sheet "Potential Health Hazards of Radiation" from the U.S. Department of Energy explains the following:

> Radiation is energy emitted by unstable (radioactive) atoms. Unstable atoms contain extra energy that is released as invisible particles or waves as the atoms change, or decay, into more stable forms. Particles and waves are referred to as radiation and their emission is called radioactivity. People are exposed to radiation from natural and man-made sources. [6]

Exhibit I.5.26 illustrates sources of radiation.

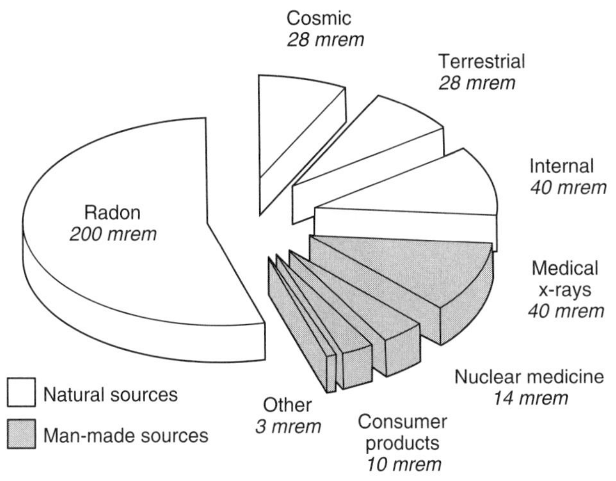

EXHIBIT I.5.26 *This diagram illustrates that there are various sources of radiation and those sources can be natural or man-made. (Source: Potential Health Hazards of Radiation, U.S. Department of Energy)*

Although alpha particles travel only a few inches (centimeters) through air and are not strong enough to penetrate the outer layer of skin, if radioactive material emitting alpha particles is inhaled or ingested, the alpha particles pose a much greater hazard in close proximity to internal tissue.

Low levels of exposure to beta particles pose only a slight hazard outside the body, but they are capable of penetrating the skin and can travel longer distances [mean range of 7 ft (2.1 m)] than alpha particles. Exposure to high levels of beta particles can cause damage to the skin and eyes. If material emitting beta particles is inhaled or ingested, the beta particles can cause more internal damage than from external exposure.

Both gamma and neutron radiation pose a hazard to the entire body because they can penetrate human tissue. Gamma rays contain more energy than alpha or beta particles and can pass through the human body or be absorbed by tissue. Shielding requires several feet (meters) of concrete or a few inches (centimeters) of lead.

Depending on the level of exposure, radiation can be a health risk. The damage can be acute or late. A high dose for even a short time is acute and can cause tissue damage or even death. Late damage effects are not seen until years after exposure. Late effects can range from a previous high acute dose or a low but chronic dose. Late effects can include genetic consequences.

Acute effects are mainly at the cell function level of skin, blood, gastrointestinal tissue, reproductive tissue, and brain tissue. At a high dose the cell membrane might be ruptured, killing the cell, interrupting the cell's energy supply, or changing its function in a harmful way. The most radiosensitive cells are those that have a high division rate, have a high metabolic rate, are of a nonspecialized type, and are well nourished.

Symptoms of radiation sickness include gastrointestinal disorders, bacterial infections, hemorrhaging, anemia, loss of body fluids, and electrolyte imbalance.

Late effects include cancer, genetic effects, and life shortening. If a cell survives an acute dose that causes damaged DNA, the cell can become cancerous and multiply uncontrollably. A cell exposed to high radiation might survive with altered DNA, which might be passed to offspring. Problems in this case can be caused from missing essential parts of DNA, the addition of harmful extra DNA, or rearrangement of the strands of DNA.

Life-shortening effects of radiation, in addition to cancer, can include an acceleration of the aging process.

Neutron radiation is most commonly seen in conjunction with nuclear power generation and medical treatment. A relatively new medical procedure is the use of higher-energy neutron radiotherapy. This use is spreading because of the effectiveness of neutrons. With neutrons, the required tumor dose is about one-third the dose required with other radiation therapy. This medical use is intended for highly controlled circumstances, but accidents can occur. One commonly used and transported neutron source is a density gauge. These gauges often have an americium/beryllium source (AmBe) that emits neutron radiation. The gauges are commonly used in the construction industry to check the moisture content of soil. The health hazard of neutron radiation is similar to that of other forms of radiation.

(b) Asphyxiant

An asphyxiant is a substance that can cause unconsciousness or death by suffocation. Asphyxiation is an extreme hazard when working in enclosed spaces. Asphyxiants are usually classified as either simple or chemical. A simple asphyxiant will displace available oxygen in air while a chemical asphxiant will prevent oxygen in the breathed air from being absorbed into the body.

(c)* Carcinogen

A carcinogen is a chemical or material that is known to, or is suspected to cause cancer and that falls within any of the following categories:

1. It has been evaluated by the International Agency for Research on Cancer (IARC) and found to be a carcinogen or potential carcinogen.
2. It is listed as a carcinogen or potential carcinogen in the latest edition of the "Annual Report on Carcinogens" published by the National Toxicology Program (NTP).
3. It is federally regulated by OSHA as a carcinogen (can be regulated additionally by states).

A.5.2.3(8)(c) Chronic health hazards include carcinogen, mutagen, and teratogen.

(d) Convulsant

A convulsant is a drug or chemical that causes convulsions. Convulsions are a violent shaking of the body or limbs caused by uncontrollable muscle contractions, which can be a symptom of brain disorders and other conditions such as exposure to a hazardous material. A person having convulsions can receive additional injury from falling and striking an object or the ground. The convulsion can also cause the exposure to the substance to be lengthened if the person cannot escape from a hazardous area.

(e) Corrosive

A corrosive is a chemical that causes visible destruction or irreversible alterations to living tissue by chemical action at the site of contact.

(f) Highly toxic

A highly toxic chemical is one that falls within any of the following categories:

1. A chemical that has a median lethal dose (LD50) of 50 mg or less per kg of body weight when administered orally to albino rats weighing between 200 g and 300 g each
2. A chemical that has an LD50 of 200 mg or less per kg of body weight when administered by continuous contact for 24 hours (or less if death occurs within 24 hours) with the bare skin of albino rabbits weighing between 2 kg and 3 kg each
3. A chemical that has a median lethal concentration also LD50) in air of 200 parts per million by volume or less of gas or vapor, or 2 mg/L or less of mist, fume, or dust, when administered by continuous inhalation for 1 hour (or less if death occurs within one hour) to albino rats weighing between 200 g and 300 g each

(g) Irritant

An irritant is a chemical that is not corrosive but one that causes a reversible inflammatory effect on living tissue by chemical action at the site of contact.

(h) Sensitizer, allergen

An allergen in 5.2.3(8)(h) is an allergy-provoking substance. Allergens are substances that, in some people, the immune system recognizes and attacks as "foreign" or "dangerous" but cause no response for most people. Allergic reactions vary. Reactions can be mild or serious and can be confined to a small area of the body or affect the entire body. Most occur within seconds or minutes after exposure to the allergen, but some can occur after days or weeks. Anaphylaxis is a sudden and severe allergic reaction that occurs within minutes of exposure.

Common allergens include certain contactants (such as chemicals, plants), drugs (such as antibiotics, serums), foods (such as milk, chocolate, strawberries, wheat), infectious agents (such as bacteria, viruses, animal parasites), inhalants (such as dust, pollen, perfumes, smoke), and physical agents (such as heat, light, friction, radiation).

A sensitizer is defined by OSHA as the following:

> A chemical that causes a substantial proportion of exposed people or animals to develop an allergic reaction in normal tissue after repeated exposure to the chemical. [16]

(i) Target organ effects

Target organ effects are signs and symptoms from chemical exposure that affect specific organs. The following examples are not intended to be all-inclusive but do illustrate the range and diversity of effects and hazards that can be encountered:

1. Hepatotoxins. Chemicals that produce liver damage (signs and symptoms: jaundice, liver enlargement; chemicals: carbon tetrachloride, nitrosamines)
2. Nephrotoxins. Chemicals that produce kidney damage (signs and symptoms: edema protein urea; chemicals: halogenated hydrocarbons, uranium)
3. Neurotoxins. Chemicals that produce their primary toxic effects on the nervous system. Both the central nervous system (CNS) and the peripheral nervous system (PNS) can be affected. The following are the symptoms associated with CNS and PNS exposure to neurotoxins:
   - Central nervous system hazards are chemicals that cause depression or stimulation of consciousness or otherwise injure the brain (signs and symptoms: drooping of upper eyelids, respiratory difficulty, seizures, unconsciousness)
   - Peripheral nervous system hazards are chemicals that damage the nerves that transmit messages to and from the brain and the rest of the body (signs and symptoms: numbness, tingling, decreased sensation, change in reflexes, decreased motor strength; examples: arsenic, lead, toluene, styrene)
4. Agents that decrease the function of hemoglobin in the blood: deprive the hematopoietic body tissues of oxygen (signs and symptoms: cyanosis, loss of consciousness; chemicals: carbon monoxide, benzene)

5. Agents that irritate the lung or damage the pulmonary tissue (signs and symptoms: cough, tightness in chest, shortness of breath; chemicals: silica, asbestos, HCl)
6. Reproductive Toxins. Chemicals that affect the reproductive capabilities including chromosomal damage (mutations) and affect fetuses (teratogenesis) (signs and symptoms: birth defects, sterility; chemicals: lead, DBCP)
7. Cutaneous Hazards. Chemicals that affect the dermal layer of the body (signs and symptoms: defatting of the skin, rashes, irritation; chemicals: ketones, chlorinated compounds)
8. Eye Hazards. Chemicals that affect the eye or visual capacity (signs and symptoms: conjunctivitis, corneal damage; chemicals: organic solvents, acids)

(j) Toxic

A toxic chemical is one that falls within any of the following categories:

1. Has a median lethal dose (LD50) or more than 50 mg per kg but not more than 500 mg per kg of body weight when administered orally to albino rats weighing between 200 g and 300 g each
2. Has an LD50 of more than 200 mg per kg but not more than 1000 mg per kg of body weight when administered by continuous contact for 24 hours (or less if death occurs within 24 hours) with the bare skin of albino rabbits weighing between 2 kg and 3 kg each.
3. Has aLD50 concentration in air of more than 200 parts per million but not more than 3000 parts per million by volume of gas or vapor, or more than 2 mg/L but not more than 200 mg/L of mist, fume, or dust, when administered by continuous inhalation for one hour (or less if death occurs within 1 hour) to albino rats weighing between 200 g and 300 g each.

A.5.2.3(8) Terms used to explain health hazards are defined as follows:

(1) *Carcinogen.* A chemical that falls within any of the following categories:
   (a) A chemical that has been evaluated by the International Agency for Research on Cancer (IARC) and found to be a carcinogen or potential carcinogen
   (b) A chemical that is listed as a carcinogen or potential carcinogen in the latest edition of the National Toxicology Program (NTP) "Annual Report on Carcinogens."
   (c) A chemical that is regulated by the Occupational Safety and Health Administration (OSHA) as a carcinogen (can be regulated additionally by states)
(2) *Corrosive.* A chemical that causes visible destruction of or irreversible alterations in living tissue by chemical action at the site of contact.
(3) *Highly toxic.* A chemical that falls within any of the following categories:
   (a) A chemical that has a median lethal dose (LD_{50}) of 50 mg or less per kilogram of body weight when administered orally to albino rats weighing between 200 g and 300 g each
   (b) A chemical that has a median lethal dose (LD_{50}) of 200 mg or less per kilogram of body weight when administered by continuous contact for 24 hours (or less if death occurs within 24 hours) with the bare skin of albino rabbits weighing between 2 kg and 3 kg each
   (c) A chemical that has a median lethal concentration (LD_{50}) in air of 200 parts per million by volume or less of gas or vapor, or 2 mg per liter or less of mist, fume, or dust, when administered by continuous inhalation for 1 hour (or less if death occurs within 1 hour) to albino rats weighing between 200 g and 300 g each
(4) *Irritant.* A chemical that is not corrosive but that causes a reversible inflammatory effect on living tissue by chemical action at the site of contact.
(5) *Sensitizer.* A chemical that causes a substantial proportion of exposed people or animals to develop an allergic reaction in normal tissue after repeated exposure to the chemical.
(6) *Toxic.* A chemical that falls within any of the following categories:

(a) A chemical that has a median lethal dose (LD_{50}) of more than 50 mg per kilogram but not more than 500 mg per kilogram of body weight when administered orally to albino rats weighing between 200 g and 300 g each
(b) A chemical that has a median lethal dose (LD_{50}) of more than 200 mg per kilogram but not more than 1000 mg per kilogram of body weight when administered by continuous contact for 24 hours (or less if death occurs within 24 hours) with the bare skin of albino rabbits weighing between 2 kg and 3 kg each
(c) A chemical that has a median lethal concentration (LD_{50}) in air of more than 200 parts per million but not more than 3000 parts per million by volume of gas or vapor or more than 2 mg per liter but not more than 200 mg per liter of mist, fume, or dust, when administered by continuous inhalation for 1 hour (or less if death occurs within 1 hour) to albino rats weighing between 200 g and 300 g each

(7) *Target organ effects.* A target organ categorization of effects that can occur, including examples of signs and symptoms and chemicals that have been found to cause such effects. The following examples illustrate the range and diversity of effects and hazards that can be encountered and are not intended to be all-inclusive:

(a) *Hepatotoxins.* Chemicals that produce liver damage (signs and symptoms: jaundice, liver enlargement; examples: carbon tetrachloride, nitrosamines)
(b) *Nephrotoxins.* Chemicals that produce kidney damage (signs and symptoms: edema, protein urea; examples: halogenated hydrocarbons, uranium)
(c) *Neurotoxin.* Chemicals that produce their primary toxic effects on the nervous system:

i. *Central nervous system hazards.* Chemicals that cause depression or stimulation of consciousness or otherwise injure the brain (signs and symptoms: drooping of upper eyelids, respiratory difficulty, seizures, unconsciousness)
ii. *Peripheral nervous system hazards.* Chemicals that damage the nerves that transmit messages to and from the brain and the rest of the body (signs and symptoms: numbness, tingling, decreased sensation, change in reflexes, decreased motor strength; examples: arsenic, lead, toluene, styrene)

(d) Agents that decrease hemoglobin in the blood of function and deprive the hematopoietic body tissues of oxygen (signs and symptoms: cyanosis, loss of consciousness; examples: carbon monoxide, benzene)
(e) Agents that irritate the lung or damage the pulmonary tissue [signs and symptoms: cough, tightness in chest, shortness of breath; examples: silica, asbestos, hydrochloric acid (HCl)]
(f) *Reproductive Toxins.* Chemicals that affect the reproductive capabilities, including chromosomal damage (mutations) and effects on fetuses (teratogenesis) (signs and symptoms: birth defects, sterility; examples: lead, DBCP)
(g) *Cutaneous hazards.* Chemicals that affect the dermal layer of the body (signs and symptoms: defatting of the skin, rashes, irritation; examples: ketones, chlorinated compounds)
(h) *Eye hazards.* Chemicals that affect the eye or visual capacity (signs and symptoms: conjunctivitis, corneal damage; examples: organic solvents, acids)

(9)* Given the following, identify the corresponding UN/DOT hazard class and division:

(a) Blood agents
(b) Biological agents and biological toxins
(c) Choking agents
(d) Irritants (riot control agents)
(e) Nerve agents
(f) Radiological materials
(g) Vesicants (blister agents)

A.5.2.3(9) Some examples of hazard class are given in Table A.5.2.3(9).

TABLE A.5.2.3(9) Examples of Hazard Class

Common Name	*Military Abbreviation*	*UN/DOT Hazard Class*
Nerve agents		
Tabun	GA	6.1
Sarin	GB	6.1
Soman	GD	6.1
V agent	VX	6.1
Vesicants (blister agents)		
Mustard	H	6.1
Distilled mustard	HD	6.1
Nitrogen mustard	HN	6.1
Lewisite	L	6.1
Blood agents		
Hydrogen cyanide	AC	6.1
Cyanogen chloride	CK	2.3
Choking agents		
Chlorine	CL	2.3
Phosgene	CG	2.3
Irritants		
Tear gas	CS	6.1
Dibenzoxazepine	CR	6.1
Chloroacetophone	CN	6.1
Pepper spray, Mace	OC	2.2 (subsequent risk 6.1)
Mace, phenylchloro-methylketone, chloropicrin	PS	6.1
Biological agents and toxins	—	
Anthrax	—	6.2
Mycotoxin	—	6.1 or 6.2
Plague	—	6.2
Viral hemorrhagic fevers	—	6.2
Smallpox	—	6.2
Ricin	—	6.2

5.2.4* Estimating Potential Harm.

Given scenarios involving hazardous materials/WMD incidents, the operations level responder shall estimate the potential harm within the endangered area at each incident and shall meet the following requirements:

A.5.2.4 The process for estimating the potential outcomes within an endangered area at a hazardous materials/WMD incident includes determining the dimensions of the endangered area; estimating the number of exposures within the endangered area; measuring or predicting concentrations of materials within the endangered area; estimating the physical, health, and safety hazards within the endangered area; identifying the areas of potential harm within the endangered area; and estimating the potential outcomes within the endangered area.

The operational level responder must be able to estimate the area of potential harm. Responders at this level are not expected to engage in intricate calculations or even to use sophisticated computer modeling, but they should be able to make general estimates based on the hazard information collected so that they can begin evacuations and safely implement defensive operations. As an incident becomes more complex, technician level responders would probably become involved, and they would be able to make more definitive predictions about the endangered area.

(1)* Identify a resource for determining the size of an endangered area of a hazardous materials/WMD incident.

A.5.2.4(1) Resources for determining the size of an endangered area of a hazardous materials/WMD incident are the current edition of the DOT *Emergency Response Guidebook* and plume dispersion modeling results from facility pre-incident plans.

The ERG recommends isolation and evacuation distances for some hazardous materials. CHEMTREC, CANUTEC, or SETIQ can provide additional information. Plume modeling data is now available for many areas and real time monitoring and detection information are also valuable resources.

(2) Given the dimensions of the endangered area and the surrounding conditions at a hazardous materials/WMD incident, estimate the number and type of exposures within that endangered area.

Exposures include people, the environment, and property. Once the responder has estimated the size and location of the endangered area, he or she can determine what is inside the perimeter. Factors influencing the decision include the time of day, the type of occupancies involved, and the type of area involved. Is it rush hour, for example? Is the occupancy a petrochemical plant? Is the area congested or is it rural?

(3) Identify resources available for determining the concentrations of a released hazardous material/WMD within an endangered area.

Various types of monitoring equipment and dispersion modeling programs are available for determining the concentrations of a released hazardous material. In most cases, however, operations level responders would not have this type of equipment. Nonetheless, they are required to understand what can be used, how to acquire it, and the technical assistance needed to operate it. For example, responders should be aware of any company in their area that handles hazardous materials and know whether the company has personnel and equipment available to monitor hazardous conditions.

Responders should also know whether they can ask a regional hazardous materials response team to assist them. State or county environmental agencies or health departments can often provide monitoring equipment and personnel. However, responders should plan ahead and make these arrangements before an incident occurs, not at 3:00 a.m. on a rainy night when an overturned cargo tanker begins leaking hazardous materials all over the highway.

(4)* Given the concentrations of the released material, identify the factors for determining the extent of physical, health, and safety hazards within the endangered area of a hazardous materials/WMD incident.

A.5.2.4(4) The factors for determining the extent of physical, health, and safety hazards within an endangered area at a hazardous materials/WMD incident are surrounding conditions, an indication of the behavior of the hazardous materials/WMD and its container, and the degree of hazard.

Responders must determine the extent of physical harm that can be expected in an endangered area and compare the gains they could receive from intervening, or not intervening, before they can determine the actions they should or should not take. The factors that influence this decision are the quantity and concentration of the hazardous materials released, the number of exposures in the endangered area, and the manner in which the exposures are subjected to the hazardous material. Is the material a liquid or a gas, for example? How far away from the source are the exposures? How fast is the material being released?

(5) Describe the impact that time, distance, and shielding have on exposure to radioactive materials specific to the expected dose rate.

The three basic principles in 5.2.4(5) that affect an exposure to external radiation are time, distance, and shielding. Increased protection occurs when you decrease the time of the exposure, increase the distance from the radiation source, and increase shielding between you and the source.

Spend as little *time* as possible near the radiation source. Reducing the time of exposure reduces the relative amount of radiation energy absorbed under any conditions. Except for someone trapped, this variable can be controlled in an emergency response incident.

Maintain *distance* from the radiation source. Increasing the distance from the radiation source is an effective way of minimizing exposure. Radiation intensity varies inversely with the square of the distance, which means that when a 4000 milliroentgen per hour (mR/hour) exposure rate is present at 2 ft (0.61 m) from a source, the exposure rate at 4 ft (1.22 m) would be 1,000 mR/hour. Doubling the distance reduced the exposure four times. A practical application of this would be the use of forceps, tongs, and other tools to extend an individual's reach to pick up or move a radiation source if necessary. Decreasing time and increasing distance are often enough to assure safe operations with radiation.

Use available *shielding*. For large radiation sources, controlling time and distance might not be enough. In other situations, time and distance might be safe but not efficient, such as when closer work may be required for rescue. Then shielding can be more of a factor.

Shielding is a barrier between the source radiation and other objects that might be affected. The shield absorbs the radiation energy. Different materials are used to shield different types of radiation. A dense material is more effective at shielding than a less-dense material. However, matter is mostly empty space, and some rays penetrate any barrier without hitting anything and being absorbed or slowed. Shielding is rated as half-value layers; that is, the amount of material required to stop half the radiation. An additional layer of the same-rated shielding would stop another half of the remaining penetrating radiation. The layers are intended to reduce the penetrating radiation to acceptable levels, in addition to time and distance factors. If the half-value layer value is not known, remember that even similar materials can have different values. For example, many different types of concrete with substantial differences in densities are available. Many objects, including vehicles, a mound of dirt, or a piece of heavy equipment, can be used to diminish the exposure level if they are located between the responder and the radiation source.

Shielding is not always practical during emergency field operations, and administering emergency care should never be delayed in the interest of seeking shielding material. In emergency care situations, the responder can still reduce radiation exposure through the factors of time and distance.

5.3 Core Competencies — Planning the Response

5.3.1 Describing Response Objectives.

Given at least two scenarios involving hazardous materials/WMD incidents, the operations level responder shall describe the response objectives for each example and shall meet the following requirements:

Up to this point, NFPA 472 has addressed the competencies that allow the responder to gather the information necessary to make appropriate decisions when determining response objectives. Section 5.1 outlines the competencies necessary for using this information to develop defensive options that can be safely implemented to influence the outcomes of the incident. Responders need to remember that actions are limited by the resources available. Responders should not take any action until they have a clear idea of what they are trying to accomplish. In FPH Wright notes the following:

> The planning process begins as part of the pre-emergency response planning efforts prior to the incident and continues on the scene of the incident, again from a safe location. Federal, state, and local agencies, industry, and carrier personnel may be called upon to help. The process of response planning is based on the following tasks:
>
> - Determine the response objectives.
> - Determine the available response options that could favorably change the outcomes.
> - Identify the personal protective equipment for the response options.
> - Identify an appropriate decontamination process for each response option.
> - Select the response options within the response community's capabilities that will most favorably change the outcomes.
> - Develop a plan of action including safety considerations. [1] p. 13-137

(1) Given an analysis of a hazardous materials/WMD incident and the exposures, determine the number of exposures that could be saved with the resources provided by the AHJ.

The key component of determining the exposures that could be saved lies within the *hazard risk assessment* of the situation. This assessment should include the hazardous materials involved, the chemical and physical properties of those materials, the population at risk for exposure, and the resources available to either mitigate the release of the material or isolate the population.

As an example, a breach of a DOT105 chlorine rail tank car upwind of a school presents a greater exposure risk than a Ricin biological toxin release in an isolated office. Each situation must be assessed for its potential risk to the population, and resources should be ordered to meet the threat. Incidents involving high-risk materials, such as toxic by inhalation chemicals, can quickly involve local, state, and federal assets.

(2) Given an analysis of a hazardous materials/WMD incident, describe the steps for determining response objectives.

Wright also noted the following:

> Response objectives, based on the stage of the incident, are the strategic goals for stopping the event now occurring or keeping future events from occurring . . . Decisions should focus on changing the actions of the stressors, the containment system, and the hazardous material. [1] p. 13-137

As part of the planning process during the development of the incident action plan, the development of objectives is one of the first actions that should be taken by the IC/unified command. Every facet of the incident must key off of the objectives that are developed. Specifically, the initial objectives should address the incident priorities of life safety, incident stabilization, and preservation of property. Incident objectives should be specific, measurable, and realistic based upon the resources available.

(3) Describe how to assess the risk to a responder for each hazard class in rescuing injured persons at a hazardous materials/WMD incident.

The risk of harm to a responder from hazardous materials was defined by Ludwig Benner as part of his "General Behavior Model of Hazardous Materials." He coined the acronym *TRACEM*, for Thermal, Radioactive, Asphyxia, Chemical, Etiological, and Mechanical. The specific risks listed in the 9 DOT hazard classes that follow are all represented in Benner's Model [5]:

- **Class 1 Risk.** Explosives that may involve thermal injury, due to the heat generated by the detonation, mechanical injury from the shock, blast overpressure, fragmentation, shrapnel or structural damage, and secondary and tertiary contact with objects. Chemical injuries may result from associated contamination. During a rescue, contact with the blood or other bodily fluids from the victim can result in etiological harm. Because burning depletes oxygen, asphyxiation should always be a consideration.

- **Class 2 Risk.** Gases that are stored in containers that might be under pressure can rupture violently and, depending upon their contents, could create a thermal, asphyxiant, chemical or mechanical hazard. Cryogenic materials (Class 2.2), such as liquid oxygen, can cause thermal harm because of its extremely cold temperature. Asphyxiation is always a concern with chemical vapors in a confined space, because chemical reactions can deplete oxygen, affect the body's ability to use oxygen, or create gases that displace oxygen, such as carbon dioxide.
- **Class 3 Risk.** Flammable liquids can freely burn, disseminate heat and shrapnel caused by a forceful explosion, resulting in thermal hazards from heat and fire as well as associated chemical and mechanical injuries.
- **Class 4 Risk.** Flammable solids, spontaneously combustible materials, and materials that are dangerous when wet could cause thermal harm from heat and flammability. Many of these materials burn at extremely high temperatures. Mechanical harm could occur as some materials react spontaneously and create slip, trip, and fall hazards, while other water-reactive, toxic, and/or corrosive materials can cause chemical harm.
- **Class 5 Risk.** Oxidizers and organic peroxides can create thermal, chemical, and mechanical harm because they supply oxygen to support combustion and are sensitive to heat shock, friction, and contamination.
- **Class 6 Risk.** Toxic/poisonous materials and infectious substances that cause chemical harm due to toxicity by inhalation, ingestion, and skin and eye contact, and in rare circumstances injection. Etiological harm can come from biological agents like bacteria and viruses or toxins derived from living organisms. Because these products may be flammable, thermal injuries are a potential hazard.
- **Class 7 Risk.** Radioactive material can cause radiological harm from alpha, beta, gamma, and neutron radiation. The hazard to humans is relative to the radiation dose rate being received over time. Limiting the time of exposure, the distance to the material, and placing shielding between the material and the responder reduces the harm. Radiological substances can also be associated with chemical hazards, for example uranium hexafluoride, which is both radioactive and corrosive.
- **Class 8 Risk.** Corrosive materials that have chemical and thermal hazards associated with damage to contacted tissues and the recognition that chemical reactions create heat especially if the material is fuming and/or water reactive. Acids tend to cause direct immediate damage to tissue, and alkalis can emulsify human tissue. Corrosive chemicals, like strong acids, can weaken structural elements, causing the potential for mechanical harm.
- **Class 9 Risk.** Miscellaneous materials present unspecified danger from harm due to the wide variety of materials that fall within the hazard class. Hazardous wastes, hot materials, and many other unique materials fall in this class. Responders should anticipate that these materials could cause harm under any of the TRACEM components.

(4)*Assess the potential for secondary attacks and devices at criminal or terrorist events.

Consideration should be given to the possibility that criminal suspects might still be present on scene during hazardous materials/WMD incidents. Law enforcement must address the potential threat from adversaries during the tactical phase of the incident, prior to any life saving or mitigation activities.

Additionally, improvised explosive devices (IEDs) might be present and set to initiate after a set amount of time, upon trip of an anti-personnel switch (booby trap) or via a remote triggering device such as a radio or cell phone. These IEDs might be placed to target emergency responders or onlookers, which would place increased demands on the emergency response. Hazardous devices technicians (bomb technicians) must clear all working points at a bombing incident for IEDs prior to responders entering the area.

A.5.3.1(4) Consideration should be given to the possibility that criminal suspects may still be on scene during hazardous materials/WMD incidents. The potential hazards presented by human threats or secondary explosive devices demonstrate the need for multiple response disciplines to prioritize, plan, and conduct response operations concurrently.

5.3.2 Identifying Action Options.

Given examples of hazardous materials/WMD incidents (facility and transportation), the operations level responder shall identify the options for each response objective and shall meet the following requirements:

After responders have determined what or who to protect, they must determine how to protect them. Operational level responders must remember that their actions are expected to be defensive in nature.

(1) Identify the options to accomplish a given response objective.

The available defensive options of 5.3.2(1) fall into the following two categories:

1. Evacuation
2. Recognition, identification, notification and isolation

(2) Describe the prioritization of emergency medical care and removal of victims from the hazard area relative to exposure and contamination concerns.

Prioritization of emergency medical care and victim removal will depend on local medical procedures and standard operating procedures.

5.3.3 Determining Suitability of Personal Protective Equipment.

Given examples of hazardous materials/WMD incidents, including the name of the hazardous material/WMD involved and the anticipated type of exposure, the operations level responder shall determine whether available personal protective equipment is applicable to performing assigned tasks and shall meet the following requirements:

Responders at this level are not expected to use specialized chemical-protective clothing. Rather, they are expected to use the type of protective clothing they normally wear in their working environment. For example, fire fighters would wear structural fire-fighting protective clothing, while the employee of an industrial facility might use liquid splash-protective clothing.

Responders are required in 5.3.3 to understand the differences among the various types of protective clothing and the levels of protection they afford. The level of protective clothing available to operational level responders is a significant factor when considering what type of defensive operations they can safely undertake and when they should request additional expertise and specialized equipment.

(1)*Identify the respiratory protection required for a given response option and the following:

A.5.3.3(1) The minimum requirement for respiratory protection at hazardous materials/ WMD incidents (emergency operations until concentrations have been determined) is positive pressure self-contained breathing apparatus (SCBA).

The respiratory hazards presented by the hazardous materials to which the first responder at the operational level might be exposed can vary widely. A risk-based method of selecting respiratory protection is therefore needed.

For most materials, positive pressure SCBA is appropriate and readily available. However, lower-risk incidents such as a powder spilled from an envelope might warrant downgrading respiratory protection to air-purifying respirators, in accordance with protocols set out by the AHJ.

Similarly, long-duration reduced-risk activities such as mass decontamination might warrant downgrading respiratory protection to powered air-purifying respirators or supplied-air respirators. Choices in respiratory protection are many and must be matched to the risk faced by the responder.

In all cases, the respiratory protective device should be approved under the applicable respiratory protection program legislation such as 29 CFR 1910.134 or local equivalent. Where exposure to chemical, biological, or radiological warfare agents is possible, the respiratory protective device should have CBRN certification under NIOSH or under a local equivalent agency in jurisdictions where NIOSH does not apply.

Referenced in A.5.3.3 (1), OSHA 1910.120, Subpart L, 4iiiD requires the following:

> Employees engaged in emergency response and exposed to hazardous substances shall wear positive pressure self-contained breathing apparatus while engaged in emergency response until such time that the individual in charge of the incident command system (ICS), the incident commander, determines through the use of air monitoring that a decreased level of respiratory protection will not result in hazardous exposure to employees. [8]

This requirement does not mean that the responder at the operational level cannot use other types of respiratory protection. Once the material is identified and an appropriate type of protection is determined, the protection can be worn for the emergency. The incident commander should make this decision only after consulting with the hazardous materials safety officer.

(a) Describe the advantages, limitations, uses, and operational components of the following types of respiratory protection at hazardous materials/WMD incidents:

The six types of respiratory protection are the air purifying respirator (APR), the supplied-air respirator (SAR), the self-contained breathing apparatus (SCBA), powered air-purifying respirator (PAPR), particulate respirator, and closed circuit breathing respirator. See Commentary Table I.5.3 for the advantages and disadvantages of emergency response respirators.

The use of SCBA has specific advantages and user limitations (see Commentary Table I.5.4). Positive pressure SCBA can place a strain on the wearer's cardiovascular system because of the additional weight and airflow resistance to breathing created by the SCBA and the physical fitness of the user. In general, the better the aerobic capacity of the wearer, the longer a person can breathe/operate before egressing or escaping from a workplace or running out of air. This physiological capacity difference can be substantial between different users/wearers. In addition, some individuals are claustrophobic or have facial dimensions that prohibit safe use or wear of a full-face, tight-fitting, man-packed SCBA and simply cannot wear one. OSHA regulations require that an individual be medically certified to wear respiratory protective devices safely.

NFPA 1500, *Standard on Fire Department Occupational Safety and Health Program*, requires that all responders using SCBA be medically certified annually by a physician [9]. The certifying physician should consult ANSI Z88.6, *Standard for Respiratory Protection — Respirator Use — Physical Qualifications for Personnel*, to determine which medical review is appropriate [10]. NFPA 1500 also requires that SCBA users be trained, tested, and certified regularly in the safe and proper use of the equipment. NFPA 1404, *Standard for Fire Service Respiratory Protection Training*, states the following:

> SCBA used in training evolutions simulating exposure to weapons of mass destruction (WMD) shall meet the appropriate sections of 42 CFR Part 84 and shall be marked with a chemical, biological, radiological and nuclear (CBRN) rating. [11, 12]

Additionally, NFPA 1404 requires the training program of the AHJ to "evaluate the ability of members to determine if a given SCBA has been tested and certified by NIOSH for use by emergency responders in CBRN environments.

i. Positive pressure self-contained breathing apparatus (SCBA)

COMMENTARY TABLE I.5.3 *Advantages and Disadvantages of Emergency Response Respirators*

Type	*Advantages*	*Limitations*
Open-Circuit, Pressure-Demand, Positive Pressure SCBA	**Movement:** Allows free walking movement of the protected user over a specific work area. **Protection:** Provides highest level of respiratory protection against airborne contaminants and oxygen deficiency when properly fitted and maintained. **CBRN Protections:** CBRN certification offers novel verification of new designs available to emergency responders, which would not otherwise be available. **Live Agent Testing (LAT) & Laboratory Respirator Protection Level (LRPL):** CBRN SCBA are laboratory evaluated and determined certified against CASARM grade CWA (GB and HD) and quantitatively evaluated for manufacturer specified face-to-facepiece fit factors and sizing using corn-oil particulate on human test subjects (Laboratory Respirator Protection Level (LRPL) testing). **Breathing Air:** SCBA is a man-packed portable source of breathable CGA specification G-7.1 Grade D air. **PASS:** Integrated and stand alone PASS devices alert fellow users of downed responder(s). Certain passive tactile or silent alarms allow users to maintain noise discipline when mission dictates. **End of Service Time Indicators (EOSTI):** Heads-Up-Display and redundant EOSTI indicate breathing air cylinder consumption rates over time via visual or audio codes to user. **Full Facepiece:** Tight fitting, full facepiece protects the entire frontal facial dimensions of a user, has an integrated impact resistant visor and is available in multiple sizes. **Rules of Air Management (ROAM):** SCBA can be donned and user seal checked while user is standing by to respond. Breathing air can be conserved through use of available SCBA technologies and application of local rules of air management. **Fit Evaluation/Testing:** Annual facepiece seal-to-face fit evaluation/testing using recognized quantitative methods increases user confidence in respirator protective qualities and validates that the issued SCBA provides the highest level of respiratory protection afforded from a CBRN SCBA or a NON-CBRN SCBA. **Ease of Use:** Hologram sights are recommended for tactical operators that wear SCBA and have to sight a weapon while conducting criminal apprehensions under CBRN or hazardous materials incident response. Ballistic protection	**Complicated:** One of the most complicated respirators used by trained emergency responders. Extensive special technical training and maintenance required. **Ergonomics:** User has to adapt to the weight and backframe designs of SCBA to ensure adequate proficiency of use. Some SCBA may be considered bulky, awkward, or heavy to untrained users. **Air Requirement:** A finite air supply requires refilling or replacement. **Duration:** Available air cylinder and re-supply defines work duration cycles. **Profile:** SCBA hardware may impair movement in confined spaces, structural collapse, or low/non-visibility conditions. **Non-CBRN:** Use of NON-CBRN compliant SCBA may severely impact respiratory protection of a user in CBRN incidents. **Cylinder Valve:** Cylinder and neck valve assemblies require special emphasis compliance to specified manufacturer maintenance instructions regarding torque, use, and cleanliness. 4500, 3000 and 2216 psig pressure ranges exist and require like pressure rated hardware for compatibility. **Dermal Protection:** SCBA must be integrated and worn with appropriate dermal protection ensembles to attain recognized level of total user protection per hazard type. **CBRN Respirator In-Use Life (CRUL):** CBRN SCBA have a maximum use life of six hours when contaminated with chemical warfare agents. NON-CBRN SCBA have been shown to catastrophically fail or allow levels of CWA contamination in the SCBA breathing zone within minutes of contamination. Decontamination is difficult. **Head Harness:** SCBA head harness assembly that holds the facepiece to the user's head can become entangled, torn or fail during use and cause facepiece seal breakage, slippage, or dumping of air. **User Seal Checks:** User must be trained and remember to conduct facepiece user seal check and regulator check before "going on air". **Survival Skills:** Use of SCBA requires ingrained training repetetiveness for effective use. Communications is hindered. MAYDAY messages require realistic training and the creation of unique survival skills. **Donning Time:** Routine and non-routine donning time standards for SCBA are required to be defined, validated, and trained. **Variation of Models:** Multiple types and styles of SCBA are fielded and must be maintained to applicable user instructions and edition-specific performance standards.

(continues)

COMMENTARY TABLE I.5.3 Continued

Type	*Advantages*	*Limitations*
	of cylinder and high pressure hose lines is best accomplished by use of stealth and concealment measures. Integration of ballistic protection plates adds weight.	**Sight Picture:** Facepiece design may challenge a user's ability to attain a consistent sight picture using chin-to-stock.
Positive pressure SAR	**Duration:** Longer work periods than with SCBA **Ergonomics:** Less bulky and lighter than most SCBA **Protections:** Protects against airborne industrial contaminants **Like Parts:** Similar in operation to the demand and pressure-demand SCBA regulators. **Comfort:** Select models of SAR provide vortex tubes that cool or heat the respirable air for user comfort or physiological support. **Types of Masks:** Full face and half masks air line respirators are available. **Construction Specific:** Abrasive-blasting air line respirators are available for structural collapse work. **Ensembles:** Full suit air line respirators may be available. **Air Supply:** Requires same quality of breathing air as SCBA. **Escape Trait:** SAR with escape bottle can be disengaged from air line source and user can egress while on escape bottle air.	**Approval:** Not approved for structural fire or CBRN incident work. Use in a CBRN warm zone requires strict delineation of use and incident commander risk assessment. **Air Supply:** SAR require an air supply hose that will restrict or impair mobility **Maximum Use Distance:** MSHA/NIOSH hose length is a maximum of 300 ft (91 m) to ensure proper air flow. **Air Flow:** As length of hose is increased, supply of minimum approved airflow may require adjustment to maintain proper air flow at the facepiece **Decontamination and Damage:** Air line is vulnerable to damage, chemical contamination, and degradation; decontamination of hoses may be difficult. **Air Pressure Range:** Air pressure is limited to 125 psig and therefore there is only a single stage reduction. **Restriction on Movement:** Worker must retrace steps to leave work area **Air Line Protection:** Requires protection of the air line and supervision/monitoring of the air supply status and line condition. **IDLH:** Select models are not approved for use in atmospheres immediately dangerous to life or health (IDLH) or in oxygen-deficient atmospheres unless equipped with an emergency egress only SCBA that can provide immediate emergency respiratory protection in case of air line failure. **Demand Mode:** Air line respirators that are demand mode have negative pressure in the facepiece.
Air-purifying respirators (APR), including powered air purifying respirators (PAPR), air-purifying escape respirators (APER), and elastomeric half-masks	**CBRN:** Select models of APR, PAPR, and APER are compliant to NIOSH-certified CBRN protection standards. **TRA:** CBRN Cap 1 rating signifies NIOSH approval of the canister at 15 minute laboratory test times at known concentration gradients of 11 Test Representative Agents covering 139 TIC/TIM. **LAT:** CBRN rating equates to 8 hours of respirator protection against a known concentration of aerosolized GB (Sarin) and HD (Blister) and two hours of protection against liquid HD droplets. **Ease of Use:** Overall, easier to use because the threat is defined and toxicity is known. **Canister Life:** Each canister is vacuum packaged to maintain integrity of performance in storage and reaction with contaminants or air starts as soon as canister is exposed to atmosphere.	**APF:** Lowest assigned protection factors & levels. **INCIDENT USE:** Per response, a new canister must be taken out of vacuum package and correctly applied to the facepiece or powered air source to gain maximum protection afforded by the APR or PAPR. Canisters left on facepieces while in storage will lose approved efficiency of carbon bed but retain particulate protection efficiency provided they are not destructively probed or damaged. **IDLH:** Filtration media subject to saturation when respirator is inappropriately or distinctly used in unknown, greater than IDLH, or IDLH respiratory hazardous conditions. **Ensembles:** Requires dermal protection appropriate to the hazard. **Hybrids:** SCBA-PAPR/APR combination-hybrid respirators are not NIOSH-approved as a system configuration. NIOSH standards are under development.

COMMENTARY TABLE I.5.3 Continued

Type	*Advantages*	*Limitations*
	Less Parts: Less moving/pressurized parts. **Movement:** Unlimited movement of user. **Hydration:** Hydration device may be integral. **Protection:** Combination P100 and carbon filtration canisters and cartridges are primary mechanisms for filtering characterized contaminated air. **Air:** Breathable air concentrations of oxygen at 19.5% are required for use. **Particulate Media:** P100 particulate media retains longer service life than adsorbent or absorbent carbon media, when used to protect against respiratory hazards consisting of only particulate in less than IDLH conditions (riot control agent/CS/CN/OC are examples). **Weight:** Exceedingly lighter in weight than SCBA. **Integration:** PAPR can be integrated with SCBA to create a specified design unique to the respirator manufacturer. **Eye and Face:** Eye and face protection attained is contingent on the type of respirator used and the compliance of the respirator to current ANSI and equivalent performance standards. **Fit Testing:** Loose-fitting PAPR and APER do not require annual fit testing.	**Detection:** Sampling and monitoring/characterization of workplace required before authorization to use APR or PAPR is recommended. **Logistics:** Requires a single or dual filtration media and replacement supplies that attach or are integral to the full facepiece, half mask, or hooded respirator. **Break Thru:** Requires filtration media change-out schedule for mandatory removal and replacement of approved filtration media prior to contaminant break-thru. **Air:** Cartridge filtration media cannot be used in IDLH or oxygen-deficient atmosphere (less than 19.5 percent oxygen at sea level.) **Escape Use:** Canister filtration media cannot be used to enter IDLH but when used in less than IDLH conditions and a secondary device creates IDLH conditions, canister media can be used to escape from IDLH conditions in a single use. **Duration:** Limited duration of protection. May be hard to gauge safe operating time in field conditions. **Limits:** Only protects against specific chemicals and up to specific concentrations **CBRN Use:** In CBRN/WMD incidents, use requires continuous sampling and monitoring of contaminants and oxygen levels. **Restrictions:** Can only be used when/for: 1. Workplace respiratory hazards are characterized and appropriate respiratory protection is available. 2. Specific gases, vapors, dusts, fumes or mists provided that the service life is known via a change-out schedule that is validated and enforced by use of a calculated formula or an integrated end-of-service-life indicator (ESLI).

COMMENTARY TABLE I.5.4 Advantages and Disadvantages of SCBA

Type	*Description*	*Advantages*	*Disadvantages*	*Comments*
Entry-and-Escape SCBA or Open-Circuit SCBA, Pressure-Demand, Positive Pressure Respirator	**Air:** Supplies breathable air to wearer from cylinder. **Exhalation and Indicators:** Wearer exhales air directly to atmosphere while monitoring all visual and manual indicators of the respirator.	**Alarms:** Primary and redundant warning alarms and gauges signal when 20 to 25 percent of air supply remains. **Use:** Allows full walking range of movement for user, within restrictions of workplace and duration of cylinder. **Air:** Grade D air supply. **CBRN:** CBRN protection compliance is for the entire SCBA respirator	**Industrial:** Industrial models are not approved for structural fire or CBRN incident response. **Duration:** Shorter operating time (30, 45, 60 minutes), heavier weight [up to 35 lb (15.9 kg)] than closed-circuit SCBA. **Fit:** Fit testing required. **Maintenance:** Extensive technician trained logistics support program required.	Operating time may vary depending on size of air tank, physical fitness, and work rate of individual.

(continues)

COMMENTARY TABLE 5.4 Continued

Type	*Description*	*Advantages*	*Disadvantages*	*Comments*
		system, independent of the protection afforded by any protective ensemble rated for CBRN incident response or proximity fire fighting. **Respirator of Choice:** A CBRN SCBA protected by an encapsulating ensemble will assume the protective qualities of the ensemble until the suit is compromised. The CBRN SCBA is then the respirator of choice when the respiratory route is the most likely route of entry for a CBRN agent due to the penetration or failure of protective ensembles and the possible variations in protective equipment ensemble interfaces, types, models, and break-thru times.	**Cylinder:** Compliance review every 5 years. **Service:** Annual service flow checks. **Checks:** Monthly and weekly maintenance checks. **Shared Hardware:** Facepieces are issued individually but certain responders may not have an individual regulator and thus may share regulators and SCBA hardware. **Sanitization:** Regulator sanitization process per UI. **Adapters:** Select facepieces require adaptor configuration to convert SCBA facepiece to a negative-pressure facepiece. **Exhalation Valves:** Select SCBA have unique exhalation valves that prevent the facepiece from being converted to negative pressure use. **Cylinder:** Compressed air vessel regulated by separate agency that regulates respirator breathing zone performance. **Use in Fire:** Proximity fire fighting requires SCBA to be protected by fire hardened ensemble. APR integrated in SCBA is not fire resistant.	
Closed-Circuit SCBA (Rebreather), Pressure-Demand, Positive Pressure Respirator	**Air:** Supplies regenerated dry breathing air from man-packed device. **Recycle:** Recycles exhaled gases (CO_2, O_2, nitrogen) by removing CO_2 with alkaline scrubber and replenishing consumed oxygen with oxygen from liquid or gaseous source.	**Air-Supplied Respirator:** With future CBRN protections compliance, it is expected to be one of the respirator ASR types that provides a higher respiratory protection level and type to the user.	**Dry Air:** Hot dry air is generated for the user over the entire duration of use. **Approvals:** Not approved for structural fire or CBRN incident response. **Mouthpiece:** User may have to use teeth to bite down on mouthpiece to keep mouthpiece sealed to lips.	**Types:** Positive pressure units offer more protection than negative-pressure closed-circuit SCBA, which are not recommended on hazardous waste sites. **Design:** While these devices may be certified as closed-circuit SCBAs, manufacturers do not

COMMENTARY TABLE 5.4 Continued

Type	*Description*	*Advantages*	*Disadvantages*	*Comments*
	Facepiece: Self-contained back packed respirator capable of interfacing with compatible or different facepieces of like origin-at the risk of voiding NIOSH approval. **Duration:** 4 hours is normal service life for one closed circuit SCBA **Use:** If user can withstand quality of breathing air generated, closed circuit SCBA are ideal for underground hazardous materials incidents where fire hazard is minimal to none and longest duration of use is required.	**Duration:** Longer operating time (up to 4 hours), lighter weight [21 to 30 lb (9.5 to 13.6 kg)] than open-circuit apparatus **Alarms:** Warning alarm signals when 20 to 25 percent of oxygen supply remains. **Air:** Oxygen supply is depleted before CO_2 sorbent scrubber supply, protecting wearer from CO_2 breakthrough.	**Cold Temp:** At very cold temperatures, scrubber efficiency may be reduced, CO_2 breakthrough may occur. **Heat Stress:** Units retain heat exchanged in exhalation, generate heat in CO_2 scrubbing operations, adding to danger of heat stress Auxiliary cooling devices may be required. **Permeation:** When worn outside encapsulating suit, breathing bag may be permeated by industrial chemicals, contaminating breathing apparatus and air. **Breathing Bag:** Decontamination of breathing bag may be difficult **Size:** Ergonomic features may seem more bulky compared to open-circuit SCBA. **Vision:** Dual hose assembly interface with facepiece inhibits downward vision. **Hydration and Communication:** Hydration devices and communication devices required.	design closed-circuit SCBA as positive pressure devices due to limitations in design and performance standards. **Mil-Spec:** Variations of the closed-circuit SCBA used in special military specified operations exist and all variations are based on the same traditional technology.
Escape-Only SCBA/Self-Contained Escape Respirator (SCER)	**Air:** Supplies clean air from an air cylinder or from oxygen-generating chemical reaction vessel. **Approval:** Approved for escape only uses.	**Ergonomics:** Lightweight [10 lb (4.5 kg) or less], low bulk, easy to carry-one time use. **Styles:** Available in pressure-demand, continuous flow modes with industrial protections. **CBRN:** CBRN protections available for NIOSH-approval holders that can meet existing technical performance standards.	**Use:** Cannot be used for entry **Mis-Use:** Untrained or uninformed user may attempt to rely on a SCER for entry, or sustainment of the user or users in situations where escape is viewed as one of two or three user options.	**Duration:** Provides 5 to 15 minutes of respiratory protection, depending on model, wearer breathing rate, and other critical respiratory protection factors.

Positive pressure SCBA provides the highest level of respiratory protection available. The equipment can be used in oxygen-deficient atmospheres and does not restrict the distance the wearer can travel or the path the wearer can take. However, the positive pressure SCBA's limited air supply, which is 2 hours at the most, limits the length of time it can be used, thus controlling the distance the wearer can travel. This limitation is even greater for individuals not in good physical condition. In addition, the weight of the SCBA might cause the wearer to exert additional physical effort, which may also shorten the period of effectiveness. See Commentary Table I.5.4 for advantages and disadvantages of SCBA.

ii. Positive pressure air-line respirator with required escape unit

Supplied air respirators are supplied with air by an external source, usually a compressor or compressed air cylinders located away from the actual work site. SARs can be used in oxygen-deficient atmospheres, they can operate longer than SCBA, and they are lighter to wear than SCBA. However, SAR hoses limit the distance the wearer can travel and can become tangled or twisted. In addition, the wearer must enter and leave the site at the same point.

iii. Closed-circuit SCBA

iv. Powered air-purifying respirator (PAPR)

PAPRs operate similarly to APRs in that they use filtration canisters to filter the air of contaminants. It cannot be used in an oxygen-deficient atmosphere and must only be worn in atmospheres in which the hazard has been identified and the concentrations are within allowable limits. It has a battery-powered fan that pulls the air through the filtration canisters minimizing the stress of the user.

v. Air-purifying respirator (APR)

APRs are lightweight, and they provide the user with more movement and travel distance than many of the other types of respiratory protection. However, APRs cannot be used in oxygen-deficient atmospheres and must be worn only in atmospheres in which the hazard has been identified and the concentrations are within allowable limits.

vi. Particulate respirator

(b) Identify the required physical capabilities and limitations of personnel working in respiratory protection.

The use of SCBA has certain limitations, as listed in Commentary Table I.5.4. Positive pressure SCBA places a strain on the wearer's cardiovascular system because of the additional weight. In general, the better the aerobic capacity, the longer a person can operate before running out of air. This difference can be substantial. In addition, some individuals are claustrophobic and simply cannot wear SCBA. OSHA regulations require that an individual be medically certified to wear respiratory protection safely.

NFPA 1500 requires that all responders using SCBA be medically certified annually by a physician. The certifying physician should consult ANSI Z88.6 to determine which medical review is appropriate. NFPA 1500 also requires that SCBA users be trained, tested, and certified regularly in the safe and proper use of the equipment.

(2) Identify the personal protective clothing required for a given option and the following:

Given the name of the material involved in an incident and the type of exposure, the responder is required to be able to determine what type of personal protective equipment is required when implementing options.

(a) Identify skin contact hazards encountered at hazardous materials/WMD incidents.

A number of skin contact hazards, including burns, rashes, and absorption of toxic substances can occur, so protecting exposed skin is critical.

(b) Identify the purpose, advantages, and limitations of the following types of protective clothing at hazardous materials/WMD incidents:

i. Chemical-protective clothing: liquid splash–protective clothing and vapor-protective clothing
ii. High temperature–protective clothing: proximity suit and entry suits
iii. Structural fire-fighting protective clothing

A responder needs to remember that no single type of protective clothing protects the wearer against all possible hazards, which is true even of chemical-protective clothing.

5.3.4* Identifying Decontamination Issues.

Given scenarios involving hazardous materials/WMD incidents, operations level responders shall identify when emergency decontamination is needed and shall meet the following requirements:

(1) Identify ways that people, personal protective equipment, apparatus, tools, and equipment become contaminated.

Contamination occurs when responders come in contact with hazardous substances at an incident. If the responders are adequately protected, only the garments they are wearing or the equipment they are using in the hot zone will be contaminated. All contaminants must be removed from the responder's personal protective clothing before the responder removes the clothing.

(2) Describe how the potential for secondary contamination determines the need for decontamination.

Operations level personnel who inadvertently or in the performance of an emergency rescue enter the hot zone can carry contaminants out of the hot zone on their protective clothing or equipment. Emergency decontamination procedures must be implemented immediately to minimize the threat of secondary contamination.

(3) Explain the importance and limitations of decontamination procedures at hazardous materials incidents.

The purpose of following decontamination procedures is to limit exposure, prevent the spread of the hazardous materials/WMD, and protect the environment. The limitations of these procedures include knowing the material, weather conditions, equipment available, topography, etc.

(4) Identify the purpose of emergency decontamination procedures at hazardous materials incidents.

Emergency decontamination is the physical process of immediately reducing contamination of individuals in potentially life-threatening situations with or without the formal establishment of a decontamination corridor. Exhibit I.5.27 shows the process of emergency decontamination.

(5) Identify the factors that should be considered in emergency decontamination.

Factors that should be considered in emergency decontamination include the following:

1. Identification of the material
2. Physical state of the material
3. Health condition of the patient contaminated
4. Potential health effects on the responder
5. Available PPE and equipment to the responder

EXHIBIT I.5.27 *A hazmat team conducts training in performing an emergency decontamination. (Courtesy of Hildebrand and Noll Technical Resources, LLC)*

(6) Identify the advantages and limitations of emergency decontamination procedures.

Emergency decontamination can be implemented without establishing a formal decontamination corridor. Only gross decontamination is provided, however, so the victim can still be exposed to contaminants and may pose a threat of secondary contamination.

A.5.3.4 Refer to *Hazardous Materials/Weapons of Mass Destruction Response Handbook*, 2008 ed.

Decontamination or contamination reduction procedures are critical to health and safety at a hazardous materials incident. It protects responders from hazardous substances that can contaminate and eventually permeate their personal protective equipment and the other equipment they use at the incident. Decontamination also minimizes the transmission of harmful substances from one control zone to another and helps protect the environment. Decontamination procedures must be specific for the type of hazard encountered.

5.4 Core Competencies — Implementing the Planned Response

5.4.1 Establishing and Enforcing Scene Control Procedures.

Given two scenarios involving hazardous materials/WMD incidents, the operations level responder shall identify how to establish and enforce scene control, including control zones and emergency decontamination, and communications between responders and to the public and shall meet the following requirements:

(1) Identify the procedures for establishing scene control through control zones.

Scene control must be put into place very quickly at each hazardous materials/WMD incident to maintain control of the scene. The size of the control zones is based on the degree of hazard present and /or any potential hazards that might occur.

(2) Identify the criteria for determining the locations of the control zones at hazardous materials/WMD incidents.

Initially, the responder probably determines where to locate the control zones using recommendations from emergency response guides, such as the initial isolation and evacuation distances in the ERG, or the advice of such organizations as CHEMTREC or CANUTEC.

In addition, the responder relies on his or her own observations of the incident and on an assessment of related information. As the incident progresses, the responder can adjust the zones based on sampling and monitoring results, on evaluations of the extent of contamination and the path it might take in case of a leak, and on the space needed to support control operations.

(3) Identify the basic techniques for the following protective actions at hazardous materials/WMD incidents:

(a) Evacuation

Evacuation is defined by the ERG as the process of moving people at risk from the area threatened to safety. Opportunity and enough time must be available to warn the people in the affected area, and enough time is also needed for them to get ready and leave that area.

Evacuees should be sent upwind of the threatened area by a specific route to a place far enough away from the incident so that they will not have to be moved again if the conditions change. Contaminated evacuees must be kept in a safe refuge area until they can be decontaminated and receive medical treatment if necessary.

(b) Sheltering-in-place

Sheltering-in-place means that people should seek shelter inside a building and remain inside until the danger passes. Sheltering-in-place is used when evacuating the public would cause greater risk then staying in place, or when an evacuation cannot be performed.

(4)* Demonstrate the ability to perform emergency decontamination.

Even though emergency decontamination is the immediate reduction of contamination on individuals in potentially life-threatening situations, these procedures must be specific for the type of hazard encountered. Decontamination procedures should be undertaken expeditiously without compromising the health and safety of those involved.

The victim(s) must be evacuated from the source of contamination. With copious amounts of water, the responder should immediately begin flushing all of the exposed body parts that might have been contaminated. As the victim's clothing is removed, the responder should ensure that the victim does not become secondarily contaminated by cross-contact with their contaminated clothing. Following this flushing, the victim should be transferred to a clean area for first aid and medical treatment. Prior to transport, the receiving hospital personnel should be informed of the contaminant involved.

A.5.4.1(4) Refer to *Hazardous Materials/Weapons of Mass Destruction Response Handbook*, 2008 ed.

(5)* Identify the items to be considered in a safety briefing prior to allowing personnel to work at the following:

The following items should be presented during a safety briefing:

1. Preliminary evaluation
2. Hazard identification
3. Description of the site
4. Task(s) to be performed
5. Length of time for task(s)
6. Required personal protective clothing
7. Monitoring requirements
8. Notification of identified risks

The factors in this list are some of the most important factors to consider, but others may be included depending on the type of incident.

(a) Hazardous material incidents
(b)* Hazardous materials/WMD incidents involving criminal activities

A.5.4.1(5)(b) The following are examples of such hazards:

(1) Secondary events intended to incapacitate or delay emergency responders
(2) Armed resistance
(3) Use of weapons
(4) Booby traps
(5) Secondary contamination from handling patients

A.5.4.1(5) Refer to NIOSH/OSHA/USCG/EPA, *Occupational Safety and Health Guidance Manual for Hazardous Waste Site Activities.*

(6) Identify the procedures for ensuring coordinated communication between responders and to the public.

5.4.2* Preserving Evidence.

Given two scenarios involving hazardous materials/WMD incidents, the operations level responder shall describe the process to preserve evidence as listed in the emergency response plan and/or standard operating procedures.

A.5.4.2 Preservation of evidence is essential to the integrity and credibility of an incident investigation. Preservation techniques must be acceptable to the law enforcement agency having jurisdiction; therefore, it is important to get their agreement ahead of time for the techniques that are set out in the local emergency response plan or the organization's standard operating procedures.

General procedures to follow for these types of incidents include the following:

(1) Secure and isolate any incident area where evidence is located. This can include discarded personal protection equipment, specialized packaging (shipping or workplace labels and placards), biohazard containers, glass or metal fragments, containers (e.g., plastic, pipes, cylinders, bottles, fuel containers), and other materials that appear relevant to the occurrence, such as roadway flares, electrical components, fluids, and chemicals.
(2) Leave fatalities and body parts in place and secure the area in which they are located.
(3) Isolate any apparent source location of the event (e.g., blast area, spill release point).
(4) Leave in place any explosive components or housing materials.
(5) Place light-colored tarpaulins on the ground of access and exit corridors, decontamination zones, treatment areas, and rehabilitation sectors to allow possible evidence that might drop during decontamination and doffing of clothes to be spotted and collected.
(6) Secure and isolate all food vending locations in the immediate area. Contaminated food products will qualify as primary or secondary evidence in the event of a chemical or biological incident.

The collection (as opposed to preservation) of evidence is usually conducted by law enforcement personnel, unless other protocols are in place. If law enforcement personnel are not equipped or trained to enter the hot zone, hazardous materials technicians should be trained to collect samples in such a manner as to maintain the integrity of the samples for evidentiary purposes and to document the chain of evidence.

5.4.3* Initiating the Incident Command System.

Given scenarios involving hazardous materials/WMD incidents, the operations level responder shall initiate the incident command system specified in the emergency response plan and/or standard operating procedures and shall meet the following requirements:

One of the key elements of conducting a safe and effective control operation at a hazardous materials incident is the implementation of an effective IMS. For an IMS to be effective, the system must be part of a pre-incident plan and be adopted as a standard operating procedure for emergency responders in a given area.

NFPA 1561, *Standard on Emergency Services Incident Management System*, provides guidance on what an effective IMS should provide in order to operate effectively [13]. The National Incident Management System (NIMS) lists the following components of its ICS:

- Common terminology
- Modular organizations
- Integrated communications
- Unified command structure
- Consolidated action plans
- Manageable span of control

(1) Identify the role of the operations level responder during hazardous materials/WMD incidents as specified in the emergency response plan and/or standard operating procedures.

The local emergency response plan should outline the role of emergency responders at the operational level. Primarily, that role includes responding to an emergency, assessing the nature of the incident, implementing initial actions, notifying other involved parties, and asking for additional assistance when needed.

(2) Identify the levels of hazardous materials/WMD incidents as defined in the emergency response plan.

The book, *Hazardous Materials — Managing the Incident,* describes levels of hazardous materials incidents as the following:

- **Level I: Potential Emergency Condition.** An incident or threat of a release, which can be controlled by the first responder. It does not require evacuation, beyond the involved structure or immediate outside area. The incident is confined to a small area and poses no immediate threat to life and property.
- **Level II: Limited Emergency Conditions.** An incident involving a greater hazard or larger area than Level I which poses a potential threat to life and property. It may require a limited protective action of the surrounding area.
- **Level III: Full Emergency Condition.** An incident involving a severe hazard or a large area which poses a significant threat to life and property and which may require a large-scale protective action [14].

(3) Identify the purpose, need, benefits, and elements of the incident command system for hazardous materials/WMD incidents.

The ICS is an organized structure of roles, responsibilities, and procedures for the command and control of emergency operations. Because the ICS is modular and can expand and contract based on the size and nature of the incident, it enables multiple disciplines and multiple jurisdictions to work together safely and effectively.

Three management concepts of ICS include unity of command, span of control, and functional positions. Unity of command stipulates that only one incident commander or a unified command is ultimately responsible for the entire incident. The command structure encompasses clearly defined lines of authority in which everyone is responsible to, and directed by, one position. Span of control establishes that only three to seven individuals report to one position so that no one position becomes overloaded, with the optimum span of control at five. Under the functional positions concept, all resources at the scene (command, planning, operations, logistics, and administration and finance) are assigned to one functional position in the ICS and should remain in that position until reassigned or released from the incident.

(4) Identify the duties and responsibilities of the following functions within the incident management system:

(a) Incident safety officer

The incident safety officer should be designated specifically at all hazardous material incidents per 29 CFR 1910. 120 [15]. The incident safety officer has the following responsibilities:

1. Obtains a briefing from the incident commander (IC)
2. Participates in the preparation of and monitors the implementation of the incident safety considerations (including medical monitoring of entry team personnel before and after entry)
3. Advises the incident commander/sector officer of deviations from the incident safety considerations and of any dangerous situations
4. Alters, suspends, or terminates any activity that is judged to be unsafe

The position of incident safety officer is critical. The IC can be responsible for implementing tasks at an incident, but the safety officer has the authority to see that they are accomplished safely in the hot and warm zones. At most incidents, only one safety officer is present. At more complex incidents, however, additional assistants may be available.

(b) Hazardous materials branch or group

(5) Identify the considerations for determining the location of the incident command post for a hazardous materials/WMD incident.

The initial command post could be the first-arriving unit at the incident. As more equipment and personnel arrive, however, the IC can use a designated command post, which might be a specially designed vehicle. The IC can also use radios or telephones to disseminate information.

Each incident should have only one command post. That post should be clearly marked and access to the post should be controlled. The responder should consider locating the command post in an area where it will not have to be moved.

(6) Identify the procedures for requesting additional resources at a hazardous materials/WMD incident.

Responders at every level are required to know what types of resources are available and understand how to request them.

(7) Describe the role and response objectives of other agencies that respond to hazardous materials/WMD incidents.

A.5.4.3 Jurisdictions that have not developed an emergency response plan can refer to the National Response Team document NRT-1, *Hazardous Materials Emergency Planning Guide*.

The National Response Team, composed of 16 federal agencies having major responsibilities in environmental, transportation, emergency management, worker safety, and public health areas, is the national body responsible for coordinating federal planning, preparedness, and response actions related to oil discharges and hazardous substance releases.

Under the Superfund Amendments and Reauthorization Act of 1986, the NRT is responsible for publishing guidance documents for the preparation and implementation of hazardous substance emergency plans.

5.4.4 Using Personal Protective Equipment.

The operations level responder shall describe considerations for the use of personal protective equipment provided by the AHJ, and shall meet the following requirements:

Not only should responders know all about the protective clothing and equipment they are given, but they must also be able to use it.

(1) Identify the importance of the buddy system.

OSHA has dictated that employees working in immediately dangerous to life and health (IDLH) environments work in pairs or teams. No one involved in operations at an incident works alone. Everyone either is part of a team or has a partner, which is shown in Exhibit I.5.28.

EXHIBIT I.5.28 *Nurses from a local hospital demonstrate the buddy system. (Courtesy of Ken deBolt, Geneva NY Fire Department)*

(2) Identify the importance of the backup personnel.

During emergency operations, responders often work in conditions that deteriorate rapidly and unexpectedly, even with the best of planning. Thus, it is vitally important that backup personnel are available, that they are equipped with the same level of personal protective equipment as the personnel they are backing up, and that they can be deployed immediately in the event of an emergency.

(3) Identify the safety precautions to be observed when approaching and working at hazardous materials/WMD incidents.

An incident should be approached from upwind and uphill whenever possible, and the approach should be calculated and deliberate. Binoculars can help responders identify the material involved, and monitoring equipment can help them assess the hazard and determine what protective equipment they should use from a safe distance.

Responders working at an incident should be aware of what is happening around them. The following are some examples of what the responders should ask themselves while working at a hazardous materials incident:

- Is there an increase in the rate of venting?
- Does the fire seem bigger than it was upon arrival?
- Does the problem seem to be lessening?
- Is there a change in atmospheric conditions? Is it getting windier?
- Has the wind changed direction?

(4) Identify the signs and symptoms of heat and cold stress and procedures for their control.

The two most serious heat problems for responders are heat exhaustion and heat stroke. Heat exhaustion is characterized by fatigue, headache, nausea, dizziness, and profuse sweating and usually occurs when a person is dehydrated from not having adequate liquid intake. Heat stroke is characterized by hot, dry skin because of the inability to sweat. This condition can occur quickly, and the victim will exhibit confusion and impaired judgment.

Cold stress can occur if responders are exposed to prolonged cold temperatures. Hypothermia is one of the more severe manifestations of cold stress. Victims exhibit shivering, apathy, listlessness, drowsiness, slow pulse, a low respiratory rate, and possible freezing of the extremities. If the victim is exposed to water, either by sweating or if clothing becomes wet, the possibility of cold stress increases.

(5) Identify the capabilities and limitations of personnel working in the personal protective equipment provided by the AHJ.

The physical and medical requirements for personnel wearing respiratory protection were discussed in the commentary to 5.3.3(1)(b). In addition to those requirements, responders must have the physical stamina required in 5.4.4(5) to wear the protective clothing and equipment provided and to work under strenuous conditions. These criteria are why conducting realistic training exercises that require responders to demonstrate their true abilities are so important.

A job-oriented physical fitness test should be given to new responders and administered annually to existing responders. This test is the only way to determine fitness for duty before placing personnel into physically strenuous situations.

(6) Identify the procedures for cleaning, disinfecting, and inspecting personal protective equipment provided by the AHJ.
(7) Describe the maintenance, testing, inspection, and storage procedures for personal protective equipment provided by the AHJ according to the manufacturer's specifications and recommendations.

5.5 Core Competencies — Evaluating Progress

5.5.1 Evaluating the Status of Planned Response.

Given two scenarios involving hazardous materials/WMD incidents, including the incident action plan, the operations level responder shall evaluate the status of the actions taken in accomplishing the response objectives and shall meet the following requirements:

All responders should understand why their efforts must be evaluated. If they are not making progress, the plan must be re-evaluated to determine why progress is not being made.

(1) Identify the considerations for evaluating whether actions taken were effective in accomplishing the objectives.

To decide whether the actions being taken at an incident are effective and the objectives are being achieved, the responder must determine whether the incident is stabilizing or increasing in intensity. Factors to be considered include reduction of potential impact to persons or the environment and status of resources available to manage the incident. This evaluation should take place upon initiation of the incident action plan, and the IC/unified command and general staff chiefs should constantly monitor the status of the incident. The actions taken should be leading to a desirable outcome, with minimal loss of life and property. Changes in the status of the incident should influence the development of the incident action plan for the next operational period.

(2) Describe the circumstances under which it would be prudent to withdraw from a hazardous materials/WMD incident.

Remaining in the immediate vicinity of an incident when nothing can be done to mitigate it and the situation is about to deteriorate is pointless. If flames are impinging on an LP-gas vessel, for example, and providing the necessary volume of water to cool it is impossible, it would be prudent to withdraw to a safe distance. ICs should always evaluate the benefit of operations in contrast to the risk taken to implement those operations.

5.5.2 Communicating the Status of the Planned Response.

Given two scenarios involving hazardous materials/WMD incidents, including the incident action plan, the operations level responder shall communicate the status of the planned response through the normal chain of command and shall meet the following requirements:

(1) Identify the methods for communicating the status of the planned response through the normal chain of command.

The proper method for communicating the status of the planned response lies within the guidelines of the incident command system (ICS) as dictated by an incident-specific incident action plan. The ICS identifies two types of communication at incidents, formal and informal. Formal communication should be used for all policy related communication, using the ICS principles of unity of command (one supervisor), chain of command (reporting only through a supervisor), while maintaining span of control (between 3 and 7 persons per supervisor). All critical information should ideally be communicated face-to-face if possible. All situations involving transfer of command must be performed face-to-face.

The format for communications within the ICS must be established by the IC/unified command with input from the general staff chiefs. The incident action plan must clearly reflect this format.

(2) Identify the methods for immediate notification of the incident commander and other response personnel about critical emergency conditions at the incident.

A procedure should be established to allow responders to notify the incident commander immediately when conditions become critical and personnel are threatened. This notification can take the form of a pre-established emergency radio message or tone that signifies danger, for example, or it might be repeated blasts on an air horn. The message should not be delayed while responders try to locate a specific person in the chain of command.

5.6* Competencies — Terminating the Incident

(Reserved)

A.5.6 No competencies are currently required at this level.

REFERENCES CITED IN COMMENTARY

1. Cote, A., ed. *Fire Protection Handbook®*, 20th ed., National Fire Protection Association, Quincy, MA, 2008.
2. NFPA 704, *Standard System for the Identification of the Hazards of Materials for Emergency Response*, National Fire Protection Association, Quincy, MA, 2007.
3. Cote, A., ed. *Fire Protection Handbook®*, 19th ed., National Fire Protection Association, Quincy, MA, 2003.
4. *Emergency Response Guidebook*, U.S. Department of Transportation, Washington, DC, 2004.
5. Benner, Jr, L., *A Textbook for Use in the Study of Hazardous Materials Emergencies*, 2nd ed., Lufred Industries, Inc., Oakton, VA, 1978.
6. Fact Sheet, "Potential Health Hazards of Radiation" U. S. Department of Energy, Grand Junction Projects Office, August 1995.
7. Benner, L., "The Story of GEBMO: How the General Hazardous Materials Behavior Model (GEBMO) for Hazardous Materials Emergency Responders Evolved," 2001. Available online at http://www.iprr.org/HazMatdocs/GEBMO/GEBMO.html.

8. OSHA 1910. 120, Subpart L, 4iiiD, U.S. Government Printing Office, Washington, DC.
9. NFPA 1500, *Standard on Fire Department Occupational Safety and Health Program*, National Fire Protection Association, Quincy, MA, 2007.
10. ANSI Z88.6, *Standard for Respiratory Protection — Respirator Use — Physical Qualifications for Personnel*, American National Standards Institute, Inc., New York, NY, 1984.
11. NFPA 1404, *Standard for Fire Service Respiratory Protection Training*, National Fire Protection Association, Quincy, MA, 2006.
12. Title 42, Code of Federal Regulations, Part 84, U.S. Government Printing Office, Washington, DC.
13. NFPA 1561, *Standard on Emergency Services Incident Management System,* National Fire Protection Association, Quincy, MA, 2008.
14. Noll, G., Hildebrand, M., and Yvorra, J. *Hazardous Materials — Managing the Incident*, 2nd ed., Fire Protection Publications, Oklahoma State University, 1995.
15. Title 29, Code of Federal Regulations, 1910.120, U.S. Government Printing Office, Washington, DC.
16. Title 29, Code of Federal Regulations, 1910.1200, Appendix A, U.S. Government Printing Office, Washington, DC.

Competencies for Operations Level Responders Assigned Mission-Specific Responsibilities

CHAPTER 6

The competencies for operations level responders who are assigned mission-specific responsibilities are discussed in this chapter. It is important to note that the authority having jurisdiction (AHJ) might only require some of these competencies for training to the operations level, as some of the competencies might not be applicable to their local jurisdiction. The 2008 edition of NFPA 472 separates out what is now referred to as *core competencies for operations level responders* in Chapter 5. Chapter 6 now describes of the various competencies from which the AHJ can select for training operations level responders.

6.1 General

6.1.1 Introduction.

6.1.1.1* This chapter shall address competencies for the following operations level responders assigned mission-specific responsibilities at hazardous materials/WMD incidents by the authority having jurisdiction beyond the core competencies at the operations level (Chapter 5):

(1) Operations level responders assigned to use personal protective equipment
(2) Operations level responders assigned to perform mass decontamination
(3) Operations level responders assigned to perform technical decontamination
(4) Operations level responders assigned to perform evidence preservation and sampling
(5) Operations level responders assigned to perform product control
(6) Operations level responders assigned to perform air monitoring and sampling
(7) Operations level responders assigned to perform victim rescue/recovery
(8) Operations level responders assigned to respond to illicit laboratory incidents

A.6.1.1.1 Operations level responders need only be trained to meet the competencies in Chapter 5. All of the competencies listed in Chapters 6 (mission-specific competencies) are not required and should be viewed as optional at the discretion of the AHJ based on an assessment of local risks. The purpose of Chapter 6 is to provide a more effective and efficient process so that the AHJ can match the expected tasks and duties of its personnel with the required competencies to perform those tasks.

6.1.1.2 The operations level responder who is assigned mission-specific responsibilities at hazardous materials/WMD incidents shall be trained to meet all competencies at the awareness level (Chapter 4), all core competencies at the operations level (Chapter 5), and all competencies for the assigned responsibilities in the applicable section(s) in this chapter.

6.1.1.3* The operations level responder who is assigned mission-specific responsibilities at hazardous materials/WMD incidents shall receive additional training to meet applicable governmental occupational health and safety regulations.

A.6.1.1.3 Additional training opportunities can be available through local and state law enforcement, public health agencies, the Federal Bureau of Investigation (FBI), the Drug Enforcement Administration (DEA), and the Environmental Protection Agency (EPA).

6.1.1.4 The operations level responder who is assigned mission-specific responsibilities at hazardous materials/WMD incidents shall operate under the guidance of a hazardous materials technician, an allied professional, an emergency response plan, or standard operating procedures.

6.1.1.5 The development of assigned mission-specific knowledge and skills shall be based on the tools, equipment, and procedures provided by the AHJ for the mission-specific responsibilities assigned.

6.1.2 Goal.

The goal of the competencies in this chapter shall be to provide the operations level responder assigned mission-specific responsibilities at hazardous materials/WMD incidents by the AHJ with the knowledge and skills to perform the assigned mission-specific responsibilities safely and effectively.

6.1.3 Mandating of Competencies.

This standard shall not mandate that the response organizations perform mission-specific responsibilities.

6.1.3.1 Operations level responders assigned mission-specific responsibilities at hazardous materials/WMD incidents, operating within the scope of their training in this chapter, shall be able to perform their assigned mission-specific responsibilities.

6.1.3.2 If a response organization desires to train some or all of its operations level responders to perform mission-specific responsibilities at hazardous materials/WMD incidents, the minimum required competencies shall be as set out in this chapter.

6.2 Mission-Specific Competencies: Personal Protective Equipment

6.2.1 General.

6.2.1.1 Introduction.

6.2.1.1.1 The operations level responder assigned to use personal protective equipment shall be that person, competent at the operations level, who is assigned to use of personal protective equipment at hazardous materials/WMD incidents.

6.2.1.1.2 The operations level responder assigned to use personal protective equipment at hazardous materials/WMD incidents shall be trained to meet all competencies at the awareness level (Chapter 4), all core competencies at the operations level (Chapter 5), and all competencies in this section.

6.2.1.1.3 The operations level responder assigned to use personal protective equipment at hazardous materials/WMD incidents shall operate under the guidance of a hazardous materials technician, an allied professional, or standard operating procedures.

Although some of the mission-specific competencies in this section are taken from Chapter 7, "Competencies for Hazardous Materials Technicians," the technical committee wants to clearly state that operations level responders with a mission-specific competency are not replacements for a hazardous materials technician (HMT). Operations level responders with a mission-specific competency can perform some technician level skills, but do not have the broader skills and competencies required of an HMT, particularly regarding risk assessment and the selection of control options. The following two options are examples of how guidance can be provided to ensure that operations level responders do not go beyond their level of training and equipment:

- Direct Guidance: Operations level responders are working under the control of an HMT or allied professional who has the ability to (1) continuously assess and/or observe their actions and (2) provide immediate feedback. Guidance by an HMT or an allied professional can be provided through direct visual observation or through assessment reports communicated by the operations level responder to them.
- Written Guidance: Written standard operating procedures or similar guidance clearly states the "rules of engagement" for operations level responders with the mission-specific competency. Emphasis should be placed upon (1) tasks expected of operations level responders, (2) tasks beyond the capability of operations level responders, (3) required personal protective equipment (PPE) and equipment to perform these expected tasks, and (4) procedures for ensuring coordination within the local ICS.

6.2.1.1.4* The operations level responder assigned to use personal protective equipment shall receive the additional training necessary to meet specific needs of the jurisdiction.

If the AHJ has supplied specialized hazardous materials/WMD PPE, the personnel must be trained in its appropriate use. For example, law enforcement personnel should be required to conduct firearms training while using respiratory equipment, fire fighters should conduct victim decontamination exercises while wearing hazardous materials/WMD PPE, and emergency medical services personnel should use hazardous materials/WMD PPE while doing victim triage.

A.6.2.1.1.4 See A.6.1.1.3.

6.2.1.2 Goal. The goal of the competencies in this section shall be to provide the operations level responder assigned to use personal protective equipment with the knowledge and skills to perform the following tasks safely and effectively:

(1) Plan a response within the capabilities of personal protective equipment provided by the AHJ in order to perform mission specific tasks assigned.
(2) Implement the planned response consistent with the standard operating procedures and site safety and control plan by donning, working in, and doffing personal protective equipment provided by the AHJ.
(3) Terminate the incident by completing the reports and documentation pertaining to personal protective equipment.

6.2.2 Competencies — Analyzing the Incident

(Reserved)

6.2.3 Competencies — Planning the Response.

6.2.3.1 Selecting Personal Protective Equipment. Given scenarios involving hazardous materials/WMD incidents with known and unknown hazardous materials/WMD, the operations level responder assigned to use personal protective equipment shall select the personal protective equipment required to support mission-specific tasks at hazardous materials/WMD incidents based on local procedures and shall meet the following requirements:

(1)*Describe the types of protective clothing and equipment that are available for response based on NFPA standards and how these items relate to EPA levels of protection.

A.6.2.3.1(1) A written personal protective equipment program should be established in accordance with 29 CFR 1910.120. Elements of the program should include personal protective equipment (PPE) selection and use; storage, maintenance, and inspection; and training consideration.

Proper selection of PPE for individual responders during a specific emergency must be based on a careful assessment of two factors:

(1) The hazards anticipated to be present at the scene
(2) The probable impact of those hazards, based on the mission role of the individual

The emergency responder must be provided with appropriate respiratory and dermal protection from suspect or known hazards. The amount of protection required is material and hazard specific. The protective ensembles must be sufficiently strong and durable to maintain protection during operations. According to 29 CFR 1910.120(q)(3)iii, the individual in charge of the ICS ensures that the personal protective ensemble worn is appropriate for the hazards to be encountered.

Currently, no single personal protective ensemble can protect the wearer from exposure to all hazards. It is important that the appropriate combination of respirator, ensemble, and other equipment be selected based on a hazard assessment at the scene.

The OSHA/EPA categories of personal protective equipment are defined in 29 CFR 1910.120, "Hazardous Waste Operations and Emergency Response" (HAZWOPER), Appendix B, as follows:

(1) Level A — To be selected when the greatest level of skin, respiratory, and eye protections is required
(2) Level B — To be selected when the highest level of respiratory protection is necessary but a lesser level of skin protection is needed
(3) Level C — To be selected when the concentration(s) and type(s) of airborne substances are known and the criteria for using air-purifying respirators (APRs) are met

Except for the inflation and inward leakage tests on Level A garments, HAZWOPER does not specify minimum performance criteria of protective clothing and respirators required for specific threats, such as chemical permeation resistance and physical property characteristics. The use of these general levels of protection does not ensure that the wearer is adequately protected from CBRN-specific hazards.

Relying solely on OSHA/EPA nomenclatures in selection of personal protective equipment could result in exposure above acceptable limits or an unnecessary reduction in operational effectiveness through lack of mobility, decreased dexterity, or reduced operational mission duration.

The clothing and ensemble standards developed by the NFPA Technical Committee on Hazardous Materials Protective Clothing and Equipment establish minimum performance requirements for physical and barrier performance during hazardous materials emergencies, including those involving chemical, biological, and radioactive terrorism materials. These standards are integrated with the NIOSH and NFPA standards on respiratory equipment.

Table A.6.2.3.1(1) is provided to assist emergency response organizations in transitioning from the OSHA/EPA Levels A, B, and C to protection-based standards terminology. Because the OSHA/EPA levels are expressed in more general terms than the standards and do not include testing to determine protection capability, it is not possible to "map" those levels to specific standards. However, it is possible to look at specific configurations and infer their OSHA/EPA levels based on the definitions of those levels. Examples of ensembles and conservative interpretations of their corresponding levels are provided in Table A.6.2.3.1(1).

All purchasers of personal protective equipment are cautioned to examine their hazard and mission requirements closely and to select appropriate performance standards. All personal protective equipment must be used in accordance with 29 CFR 1910.120 (or equivalent EPA or state regulations). Also applicable in states with OSHA-approved health and safety programs and for Federal employers is 29 CFR 1910.134, "Respiratory Protection" (or an equivalent EPA or state regulation). Both 29 CFR 1910.120 and 29 CFR 1910.134 include requirements for formal plans, medical evaluation, and training to ensure the safety and health of emergency responders. Additional information, a list of allowable equipment, and information on related standards, certifications, and products are available on the Department of Homeland Security (DHS)–sponsored Responder Knowledge Base (http://www.rkb.mipt.org).

TABLE A.6.2.3.1(1) Protective Clothing Standards That Correspond to OSHA/EPA Levels

Ensemble Description Using Performance-Based Standard(s)[a]	*OSHA/EPA Level*
NFPA 1991 worn with NIOSH CBRN SCBA	A
NFPA 1994 Class 2 worn with NIOSH CBRN SCBA[b]	B
NFPA 1971 with CBRN option worn with NIOSH CBRN SCBA[c]	B
NFPA 1994 Class 3 worn with NIOSH CBRN APR[b]	C
NFPA 1994 Class 4 worn with NIOSH CBRN APR	C

[a] The 2007 edition of NFPA 1994 (effective on August 17, 2006) eliminated the Class 1 requirements, relying instead on NFPA 1991 as the standard for vapor protective ensembles. The 2007 edition of NFPA 1994 also included a new Class 4 requirement for biological and radiological particulate protective ensembles.

[b] Vapor protection for NFPA 1994, Class 2 and Class 3, is based on challenge concentrations established for certification of CBRN open-circuit SCBA and APR respiratory equipment. Class 2 and Class 3 do not require the use of totally encapsulating garments.

[c] The 2007 edition of NFPA 1971 (effective August 17, 2006) included options for protection from CBRN hazards. Only complete ensembles certified against these additional optional requirements provide this protection. The protection levels set in the NFPA 1971 CBRN option are based on the Class 2 requirements contained in NFPA 1994.

Commentary Table I.6.1 compares the NFPA hazardous materials protective clothing standards, the Occupational Safety and Health Administration (OSHA)/Environmental Protection Agency (EPA) Levels A, B, and C, and the new National Institute for Occupational Safety and Health (NIOSH)-certified respirator with chemical, biological, radiological, and nuclear (CBRN) protection standards.

(2) Describe personal protective equipment options for the following hazards:
   (a) Thermal
   (b) Radiological
   (c) Asphyxiating
   (d) Chemical
   (e) Etiological/biological
   (f) Mechanical

(3) Select personal protective equipment for mission-specific tasks at hazardous materials/WMD incidents based on local procedures.
   (a) Describe the following terms and explain their impact and significance on the selection of chemical-protective clothing:
      i. Degradation
      ii. Penetration
      iii. Permeation
   (b) Identify at least three indications of material degradation of chemical-protective clothing.
   (c) Identify the different designs of vapor-protective and splash-protective clothing and describe the advantages and disadvantages of each type.
   (d)*Identify the relative advantages and disadvantages of the following heat exchange units used for the cooling of personnel operating in personal protective equipment:

A.6.2.3.1(3)(d) Phase change technology creates a constant temperature vest and is a completely unique body management device. The unique cooling formulation encapsulated in an anatomically designed device makes a change in minutes from a clear liquid to a semisolid, white waxy form and maintains a temperature of 59°F (15°C). Unlike the extremely cold temperatures of ice and gel, the higher temperature formulation in these devices works in

COMMENTARY TABLE I.6.1 *Comparison of NFPA Standards and OSHA/EPA Levels for Respiratory Protection*

				Expected Dermal Protection from Suit(s)			
NFPA Standard	**Minimum OSHA/EPA Level**	**Respirator**	**NFPA Chemical Barrier Protection Method**	**Chemical Vapor[5]**	**Chemical Liquid**	**Particulate**	**Liquid-borne Viral Penetration**
1991 (2005)	A	SCBA (open-circuit) — NFPA 1981 and NIOSH CBRN Compliant	Permeation against 25 industrial chemicals, HD and GB[1]	X	X	X	X
1992 (2005)	B	NFPA 1981 Open circuit SCBA	Penetration against 5 liquids[2]		X		
1994 (2007), Class 1	(NOTE: NFPA 1994, Class 1 ensemble was removed in 2007)						
1994, Class 2	B	NIOSH CBRN and NFPA 1981 Compliant CBRN Open Circuit SCBA	Permeation[3]	X	X	X	X
1994, Class 3	C	NIOSH CBRN Compliant APR or PAPR	Permeation[4] No Chemical Testing	X NA	X NA	X X	X X
	C	NIOSH CBRN Compliant APR or PAPR					

[1] NFPA 1991 permeation testing conducted with excess 100% HD, GB, solvents and gases, 98% sulfuric acid, or saturated sodium hydroxide

[2] 100% acetone, ethyl acetate, tetrahydrofuran, 50% sodium hydroxide and 98% sulfuric acid at 2 psi applied pressure.

[3] For Class 2, 350 ppm ammonia and chlorine gas and as droplets, 10 g/m^2 acrolein, acrylonitrile, soman, sulfur mustard and dimethyl sulfate.

[4] For Class 2, 350 ppm ammonia and chlorine gas and as droplets, 10 g/m^3 acrolein, acrylonitrile, soman, sulfur mustard, and dimethyl sulfate.

[5] NFPA 1991 ensemble vapor protection based on < 0.2% sulfur hexafluoride inward leakage. Vapor protection for NFPA 1994 based on MIST testing at challenge concentrations established for NIOSH certification of CBRN open-circuit SCBA and APR/PAPR respiratory equipment, which are 320 ppm GB and 40 ppm for Class 2 and 34 ppm GB and 7 ppm HD for Class 3.

Note: See references at the end of the chapter for more complete information on the NFPA standards cited.

harmony with the body. When an energized cool vest is worn, the cool phase change material absorbs the excessive heat the body creates when wearing protective clothing or encapsulating suits.

i. Air cooled
ii. Ice cooled
iii. Water cooled
iv. Phase change cooling technology

(e) Identify the physiological and psychological stresses that can affect users of personal protective equipment.
(f) Describe local procedures for going through the technical decontamination process.

6.2.4 Competencies — Implementing the Planned Response.

6.2.4.1 Using Protective Clothing and Respiratory Protection. Given the personal protective equipment provided by the AHJ, the operations level responder assigned to use personal protective equipment shall demonstrate the ability to don, work in, and doff the equipment provided to support mission-specific tasks and shall meet the following requirements:

(1) Describe at least three safety procedures for personnel wearing protective clothing.
(2) Describe at least three emergency procedures for personnel wearing protective clothing.
(3) Demonstrate the ability to don, work in, and doff personal protective equipment provided by the AHJ.
(4) Demonstrate local procedures for responders undergoing the technical decontamination process.
(5) Describe the maintenance, testing, inspection, storage, and documentation procedures for personal protective equipment provided by the AHJ according to the manufacturer's specifications and recommendations.

6.2.5 Competencies — Terminating the Incident.

6.2.5.1 Reporting and Documenting the Incident. Given a scenario involving a hazardous materials/WMD incident, the operations level responder assigned to use personal protective equipment shall identify and complete the reporting and documentation requirements consistent with the emergency response plan or standard operating procedures regarding personal protective equipment.

6.3 Mission-Specific Competencies: Mass Decontamination

6.3.1 General.

When a release of hazardous materials has potentially contaminated a large number of people, a different approach needs to be taken to decontamination. What constitutes a large number will depend on the response capabilities of the locality where the incident has occurred. If the use of technical decontamination procedures is not feasible due to the volume of victims, then mass decontamination will need to be implemented. Ideally, mass decontamination will use methods that can be readily applied, with existing resources and techniques consistent with current responder training, and will take into account resource limits and other constraints such as human nature.

There are a number of ways to perform mass decontamination, but the procedure is dependent on the number and types of fire apparatus available locally. The first action to be instigated is disrobing of the victims down to their undergarments, as illustrated in Exhibit I.6.1. According to studies conducted in 2001 by the U.S. Army Soldier and Biological Command (SBCCOM, now known as RDECOM) when harmless nerve agent simulants were used to track contaminants, 80 percent of the contaminants were removed by simply undressing. In most cases, speed is a critical component in limiting victim exposure to contaminant hazards and disrobing achieves speedy contaminant reduction. There could be modesty issues for reasons such as gender, religious beliefs, and other societal norms; therefore provision should be made to preserve victim modesty and not offend their dignity. Those victims refusing to disrobe should be flushed while clothed: failure to remove outer garments should never be a cause for excluding victims from the mass decontamination process.

Victims should be flushed with large volumes of low-pressure water applied from fire apparatus through fog lines directed horizontally and vertically down. Where aerial devices are available, a water-tower based "shower" is easily implemented, but fog nozzles can also be attached to one or more ground ladders secured on top of two engines, then aimed downward. Nozzles should be placed into side outlets, trashlines, and deck guns on apparatus to provide a horizontal spray without the need for pump operators or fire fighters to be in the decontamination path. Pressures should be regulated so as to produce approximately 50 psi (350 kPa).

EXHIBIT I.6.1 *Having victims remove their outer garments will quickly remove up to 80 percent of contaminants. (Source: Hildebrand and Noll Technical Resources, LLC)*

The objective is to gain volume but not to risk hurting or knocking over the people being decontaminated. (See Exhibit I.6.2.)

Where multiple decontamination paths can be established, there will be a benefit in various ways: (a) It could resolve modesty issues if gender-based paths were established; (b) it could increase the number of victims decontaminated in a given time; (c) or it could be used to provide one path for ambulatory victims, one for nonambulatory people requiring assis-

EXHIBIT I.6.2 *Two engines plus an elevated stream will create a large volume of low-pressure water. (Source: Rem Gaade, Gaade & Associates)*

tance from fire fighters or emergency medical technicians (EMT), and one for critically ill patients who will need immediate emergency medical services (EMS) attention even while being decontaminated. If multiple decontamination paths are established, it is important to ensure that children are not separated from those responsible for them.

It is generally considered acceptable for there to be a separate decontamination corridor for emergency responders, who will undergo technical (rather than mass) decontamination.

It is important to ensure that following their flush-down, victims are marshaled to a controlled area where they can be medically assessed, be given dry garments, and be provided psychological support. Law enforcement might wish to conduct interviews, and where available, social services should be in attendance to provide support to those with logistical problems in transportation, housing, and reuniting of families. It is essential that all potential participants in the mass decontamination process, including any required follow-up after the wash-down, establish their roles and responsibilities during local pre-planning for these events.

When responders suspect that a hazardous materials release has been caused through criminal or terrorist activity, triaging needs to be done before mass decontamination is initiated. First, victims should be observed to see if any are symptomatic; those who are should have treatment initiated and continued through decontamination. Second, if everyone appears asymptomatic, a determination needs to be made as to whether actual chemical agent exposure has occurred: where exposure appears unlikely, decontamination should be deferred pending on-scene investigation. (Should signs or symptoms appear, appropriate treatment of patients and decontamination of all should be started.) The decision to decontaminate should be based on the probability of agent exposure, the environmental conditions at the time, and the age and health status of the victims. If asymptomatic victims are clamoring for decontamination, this is likely more for psychological reasons, and incident commanders need to weigh the risks inherent in the decontamination process against the benefits gained.

Several factors need to be taken into consideration during decontamination in cold weather: location for wash-down, temperature and windchill, condition of the person being decontaminated, and the availability of shelter. In cold weather, decontamination of individuals (technical decontamination) and of groups of people (mass decontamination) can be done either indoors or outdoors; the temperature will be the deciding factor.

According to a SBCCOM report on cold weather decontamination procedures, in ambient temperatures over 65°F (18°C), decontamination and follow-up can be done outdoors; when temperatures are above 35°F (2°C), the outdoor decontamination should be done in a covered, sheltered decontamination corridor with follow-up indoors; and where temperatures are below 35°F (2°C), decontamination can be done either indoors or outdoors as long as a heated, sheltered decontamination corridor is used. If during outdoor decontamination the runoff water forms ice and creates slip-and-fall hazards, both the decontamination and follow-up should be done indoors. Note that by definition, emergency decontamination is a critical activity that should be done immediately where and when necessary, without regard for ambient temperature considerations.

Showers can be in indoor locations suitable for technical decontamination: both the regular domestic type found in schools, fitness centers, executive bathrooms, and industrial change rooms, and the industrial safety showers located in many facilities where hazardous materials incidents can occur. For mass decontamination a variety of other options are possible for indoor decontamination: buildings such as community centers with shower or swimming pool facilities, sprinkler systems in hallways or warehouses (when water damage to building fabric and contents can be minimized), and indoor open areas such as hockey arenas, where people can be hosed off using fog lines yet the runoff can be controlled and the area heated.

Consideration needs to be given to the proximity of available facilities. If victims have to be transported to a location away from the incident area to achieve warmer conditions for decontamination, they should disrobe before transport and any remaining visible liquids be blotted off them to minimize chemical exposure during the time prior to decontamination.

The interior of any vehicles being used for such transport should be able to be hosed down afterward: Transit buses and trucks would be better choices than automobiles and vans.

When mass decontamination has taken place, the victims need to be given shelter from the cold and issued with temporary covering until suitable clothing can be issued. Large towels, sheets, and blankets can be used, and arrangements should be established with local hotels, hospitals, college residences and dormitories, and industrial laundries prior to any event for the loan of such items in an emergency. Post-decontamination observation of victims is necessary to ensure they do not suffer ill effects from cold shock or hypothermia and are kept away from areas where windchill could affect them. Victims who are shivering are displaying a normal physiological response and only need to be attended to if the shivering should stop and they start to display signs of hypothermia.

Whereas emergency responders who are exposed to chemical contamination can undergo technical decontamination in cold weather with minimal effect on their body systems thanks to temperature protection provided by their PPE, citizens being put through mass decontamination in low-temperature (or high windchill) conditions can suffer physiological effects ranging from discomfort to hypothermic reactions. Nevertheless, as in emergency decontamination situations, where people have been in contact with life-threatening levels of contamination, they should disrobe and be flushed with copious amounts of low-pressure, high-volume water regardless of the ambient temperature. They should then be sheltered and warmed as expeditiously as possible.

6.3.1.1 Introduction.

6.3.1.1.1 The operations level responder assigned to perform mass decontamination at hazardous materials/WMD incidents shall be that person, competent at the operations level, who is assigned to implement mass decontamination operations at hazardous materials/WMD incidents.

6.3.1.1.2 The operations level responder assigned to perform mass decontamination at hazardous materials/WMD incidents shall be trained to meet all competencies at the awareness level (Chapter 4), all core competencies at the operations level (Chapter 5), all mission-specific competencies for personal protective equipment (Section 6.2), and all competencies in this section.

6.3.1.1.3 The operations level responder assigned to perform mass decontamination at hazardous materials/WMD incidents shall operate under the guidance of a hazardous materials technician, an allied professional, or standard operating procedures.

Although some of the mission-specific competencies in this section are taken from Chapter 7, "Competencies for Hazardous Materials Technicians," the technical committee wants to clearly state that operations level responders with a mission-specific competency are not replacements for a hazardous materials technician (HMT). Operations level responders with a mission specific competency can perform some technician-level skills but do not have the broader skills and competencies required of a HMT, particularly regarding risk assessment and the selection of control options. The following two options are examples of how guidance can be provided to ensure that operations level responders do not go beyond their level of training and equipment:

- Direct Guidance: Operations level responders are working under the control of an HMT or allied professional who has the ability to (1) continuously assess and/or observe their actions, and (2) provide immediate feedback. Guidance by an HMT or an allied professional can be provided through direct visual observation or through assessment reports communicated by the operations level responder to them.
- Written Guidance: Written standard operating procedures or similar guidance clearly state the "rules of engagement" for operations level responders with the mission-specific competency. Emphasis should be placed on the following:

1. Tasks expected of operations level responders,
2. Tasks beyond the capability of operations level responders,
3. Required PPE and equipment to perform these expected tasks, and
4. Procedures for ensuring coordination within the local ICS.

6.3.1.1.4* The operations level responder assigned to perform mass decontamination at hazardous materials/WMD incidents shall receive the additional training necessary to meet specific needs of the jurisdiction.

A.6.3.1.1.4 Additional training opportunities can be available through local and state law enforcement, public health agencies, the Federal Bureau of Investigation (FBI), the Drug Enforcement Administration (DEA), and the Environmental Protection Agency (EPA).

6.3.1.2 Goal.

6.3.1.2.1 The goal of the competencies in this section shall be to provide the operations level responder assigned to perform mass decontamination at hazardous materials/WMD incidents with the knowledge and skills to perform the tasks in 6.3.1.2.2 safely and effectively.

6.3.1.2.2 When responding to hazardous materials/WMD incidents, the operations level responder assigned to perform mass decontamination shall be able to perform the following tasks:

(1) Plan a response within the capabilities of available personnel, personal protective equipment, and control equipment by selecting a mass decontamination process to minimize the hazard.
(2) Implement the planned response to favorably change the outcomes consistent with standard operating procedures and the site safety and control plan by completing the following tasks:
  (a) Perform the decontamination duties as assigned.
  (b) Perform the mass decontamination functions identified in the incident action plan.
(3) Evaluate the progress of the planned response by evaluating the effectiveness of the mass decontamination process.
(4) Terminate the incident by providing reports and documentation of decontamination operations.

6.3.2 Competencies — Analyzing the Incident. (Reserved)

6.3.3 Competencies — Planning the Response.

6.3.3.1 Selecting Personal Protective Equipment. Given an emergency response plan or standard operating procedures, the operations level responder assigned to mass decontamination shall select the personal protective equipment required to support mass decontamination at hazardous materials/WMD incidents based on local procedures *(see Section 6.2)*.

6.3.3.2 Selecting Decontamination Procedures. Given scenarios involving hazardous materials/WMD incidents, the operations level responder assigned to mass decontamination operations shall select a mass decontamination procedure that will minimize the hazard and spread of contamination, determine the equipment required to implement that procedure, and meet the following requirements:

(1) Identify the advantages and limitations of mass decontamination operations.
(2) Describe the advantages and limitations of each of the following mass decontamination methods:
  (a) Dilution
  (b) Isolation
  (c) Washing

(3) Identify sources of information for determining the correct mass decontamination procedure and identify how to access those resources in a hazardous materials/WMD incident.
(4) Given resources provided by the AHJ, identify the supplies and equipment required to set up and implement mass decontamination operations.
(5) Identify procedures, equipment, and safety precautions for communicating with crowds and crowd management techniques that can be used at incidents where a large number of people might be contaminated.

The following is according to the Fairfield (CT) Police Department, Special Services Division:

> Emergency responders should understand that when responding to incidents which may involve the release of a hazardous material, they may encounter large crowds requiring direction, evaluation, and possibly mass decontamination. Factors affecting crowd size can include location, time of day, day of week, weather, and possible events occurring at the time. During response, personnel should assess these and other relevant factors in order to properly estimate possible crowd size or number of victims and resulting response needs. Generally, law enforcement personnel provide for the protection of the scene, limiting of access, and crowd management. All first responders should be aware of initial steps to control a crowd:
>
> Containment + Communication = Control
>
> Containment is determined by factors including the size and condition of crowds, potential presence of hazardous materials, and number of emergency response personnel present. Responders should realize a crisis situation contained is far better than a crisis situation uncontained. In addition to identifying the hot zone, responders should identify and establish a containment area (perimeter) with controlled access. It helps to locate and utilize existing natural barriers (bodies of water, trees, etc.), other landscape barriers including buildings, parked vehicles, and fencing already present to assist in defining the containment area. Responders may also utilize wooden barricades, yellow barrier tape, or other visible barriers to clearly identify the containment area. Within the containment area locations should be identified where victims may be directed. As more resources arrive, the outer perimeter and hot zones should be clearly communicated to prevent responder contamination. These responders, once they are properly protected, should be utilized to assist in establishing and fortifying containment areas.
>
> Communication: Victims and others within the containment area will seek and clearly need re-assurance, information and direction. It is important to gain their trust and confidence by maintaining a professional controlling demeanor. Responders should not approach those individuals with possible exposure, until they themselves are properly protected. Generally, the quicker containment is established, and the quicker responders communicate with the victims, the quicker the situation is likely to be controlled. Utilizing bullhorns, vehicle public address equipment, signs or other written displays, or interior building intercom / PA equipment if the situation is located indoors, responders should assume control by identifying themselves or their agency and informing the crowd of their intention to provide help. Regardless of the device or means utilized to communicate, responders should maintain a command presence by providing specific, authoritative instruction.
>
> Communications should be in short three- to five-word statements repeated over and over again as needed. The message should include instructions to remain calm, confirmation that help has been dispatched, and a statement stressing the need for them to follow these instructions. As the scene becomes more controlled and more resources arrive, more detailed instructions should be included.
>
> First responders should realize that depending on the time that it takes them to get on the scene, a number of the crowd intent on leaving will have done so. Of course this depends on the size of the crowd or the venue where the incident occurs (a concert hall or stadium cannot be emptied in a few minutes). Those left behind may be more likely to listen to and follow instructions, reinforcing the need for the communication process to express competence and authority.

6.3.4 Competencies — Implementing the Planned Response.

6.3.4.1 Performing Incident Management Duties. Given a scenario involving a hazardous materials/WMD incident and the emergency response plan or standard operating procedures, the operations level responder assigned to mass decontamination operations shall demonstrate the mass decontamination duties assigned in the incident action plan by describing the local procedures for the implementation of the mass decontamination function within the incident command system.

6.3.4.2 Performing Decontamination Operations Identified in Incident Action Plan. The operations level responder assigned to mass decontamination operations shall demonstrate the ability to set up and implement mass decontamination operations for ambulatory and nonambulatory victims.

6.3.5 Competencies — Evaluating Progress.

6.3.5.1 Evaluating the Effectiveness of the Mass Decontamination Process. Given examples of contaminated items that have undergone the required decontamination, the operations level responder assigned to mass decontamination operations shall identify procedures for determining whether the items have been fully decontaminated according to the standard operating procedures of the AHJ or the incident action plan.

6.3.6 Competencies — Terminating the Incident.

6.3.6.1 Reporting and Documenting the Incident. Given a scenario involving a hazardous materials/WMD incident, the operations level responder assigned to mass decontamination operations shall complete the reporting and documentation requirements consistent with the emergency response plan or standard operating procedures and shall meet the following requirements:

(1) Identify the reports and supporting documentation required by the emergency response plan or standard operating procedures.
(2) Describe the importance of personnel exposure records.
(3) Identify the steps in keeping an activity log and exposure records.
(4) Identify the requirements for filing documents and maintaining records.

6.4 Mission-Specific Competencies: Technical Decontamination

6.4.1 General.

6.4.1.1 Introduction.

6.4.1.1.1 The operations level responder assigned to perform technical decontamination at hazardous materials/WMD incidents shall be that person, competent at the operations level, who is assigned to implement technical decontamination operations at hazardous materials/WMD incidents.

6.4.1.1.2 The operations level responder assigned to perform technical decontamination at hazardous materials/WMD incidents shall be trained to meet all competencies at the awareness level (Chapter 4), all core competencies at the operations level (Chapter 5), all mission-specific competencies for personal protective equipment (Section 6.2), and all competencies in this section.

6.4.1.1.3 The operations level responder assigned to perform technical decontamination at hazardous materials/WMD incidents shall operate under the guidance of a hazardous materials technician, an allied professional, or standard operating procedures.

Although some of the mission-specific competencies in this section are taken from Chapter 7, "Competencies for Hazardous Materials Technicians," the technical committee wants to clearly state that operations level responders with a mission-specific competency are not replacements for an HMT. Operations level responders with a mission specific competency may perform some technician-level skills but do not have the broader skills and competencies required of an HMT, particularly regarding risk assessment and the selection of control options. The following two options are examples of how guidance can be provided to ensure that operations level responders do not go beyond their level of training and equipment:

- Direct Guidance: Operations level responders are working under the control of an HMT or allied professional who has the ability to (1) continuously assess and/or observe their actions and (2) provide immediate feedback. Guidance by an HMT or an allied professional can be provided through direct visual observation or through assessment reports communicated by the operations level responder to them.
- Written Guidance: Written standard operating procedures or similar guidance clearly state the "rules of engagement" for operations level responders with the mission-specific competency. Emphasis should be placed on the following:

  1. Tasks expected of operations level responders,
  2. Tasks beyond the capability of operations level responders,
  3. Required PPE and equipment to perform these expected tasks, and
  4. Procedures for ensuring coordination within the local ICS.

6.4.1.1.4* The operations level responder assigned to perform technical decontamination at hazardous materials/WMD incidents shall receive the additional training necessary to meet specific needs of the jurisdiction.

A.6.4.1.1.4 See A.6.3.1.1.4.

6.4.1.2 Goal.

6.4.1.2.1 The goal of the competencies in this section shall be to provide the operations level responder assigned to perform technical decontamination at hazardous materials/WMD incidents with the knowledge and skills to perform the tasks in 6.4.1.2.2 safely and effectively.

6.4.1.2.2 When responding to hazardous materials/WMD incidents, the operations level responder assigned to perform technical decontamination shall be able to perform the following tasks:

(1) Plan a response within the capabilities of available personnel, personal protective equipment, and control equipment by selecting a technical decontamination process to minimize the hazard.

At every incident involving hazardous materials, there is a possibility that personnel, their equipment, and members of the general public will become contaminated. The contaminant poses a threat, not only to the persons contaminated but to other personnel who might subsequently come into contact with the contaminated personnel and equipment. The entire process of decontamination should be directed toward confinement of the contaminant within the hot zone and the decontamination corridor to maintain the safety and health of response personnel, the general public, and the environment. Sound judgment should be exercised, and the potential effects of the decontamination process upon personnel should be considered when developing the decontamination plan.

Although decontamination is typically performed following site entry, the determination of applicable decontamination methods and procedures needs to be considered before the incident as part of the overall pre-incident planning and hazard and risk evaluation process. No entry into the hot zone should be permitted until decontamination methods are determined and established based on the hazards present, except in those situations where a rescue might be possible and emergency decontamination is available.

Personnel and their equipment can be decontaminated by removing or neutralizing the contaminants that have accumulated on them or by disposing of their protective clothing. Decontamination requires an organized and well-ordered procedure, hence the need for a plan for successful execution. The plan must take into account measures that minimize contamination as a line of first defense.

The decontamination plan should address such factors as the following:

1. Site layout
2. Decontamination methods to be used and equipment needed
3. Number of personnel needed
4. Level of protective clothing and equipment that have to be processed
5. Disposal methods
6. Runoff control
7. Emergency medical requirements
8. Methods for collecting and disposing of contaminated clothing and equipment

(2) Implement the planned response to favorably change the outcomes consistent with standard operating procedures and the site safety and control plan by completing the following tasks:

(a) Perform the technical decontamination duties as assigned.

(b) Perform the technical decontamination functions identified in the incident action plan.

(3) Evaluate the progress of the planned response by evaluating the effectiveness of the technical decontamination process.

The process of evaluating the effectiveness of decontamination in the field is mainly limited to visually determining whether the contaminants have been removed from PPE and by using monitoring instruments available that might be available to emergency responders.

(4) Terminate the incident by completing the providing reports and documentation of decontamination operations.

6.4.2 Competencies — Analyzing the Incident. (Reserved)

6.4.3 Competencies — Planning the Response.

6.4.3.1 Selecting Personal Protective Equipment. Given an emergency response plan or standard operating procedures, the operations level responder assigned to technical decontamination operations shall select the personal protective equipment required to support technical decontamination at hazardous materials/WMD incidents based on local procedures *(see Section 6.2).*

Typically, the PPE worn by personnel performing technical decontamination duties is one level down from the persons they are decontaminating, but as a minimum it will include liquid-splash-protective clothing (or bunker gear where fire fighters are involved) with face and respiratory protection such as offered by a self-contained breathing apparatus (SCBA) or supplied air respirator (SAR). See Exhibit I.6.3.

6.4.3.2 Selecting Decontamination Procedures. Given scenarios involving hazardous materials/WMD incidents, the operations level responder assigned to technical decontamination operations shall select a technical decontamination procedure that will minimize the hazard and spread of contamination and determine the equipment required to implement that procedure and shall meet the following requirements:

EXHIBIT I.6.3 *This photograph shows typical PPE for personnel performing technical decontamination. (Source: Hildebrand and Noll Technical Resources, LLC)*

Decontamination consists of reducing and preventing the spread of contamination from persons and equipment used at a hazardous materials incident by physical and/or chemical processes. Emergency response personnel should implement a thorough, technically sound decontamination procedure until it is determined by technically knowledgeable staff at the scene to be no longer necessary.

The two basic ways to decontaminate something are physical and chemical. Physical methods manually separate the chemical from the material being decontaminated by scrubbing or washing the material, or both. Physical decontamination is often easier than chemical decontamination, but it might not completely remove all the contaminants. Chemical methods involve adding another chemical that changes the physical or chemical properties of one chemical into another or into a form that facilitates its removal. Unfortunately, the chemical process involved could introduce other hazards. Care must be taken to collect all the contamination that has been removed by either method and to dispose of it properly.

Emergency response personnel should have an established procedure to minimize contamination or contact, to limit migration of contaminants, and to dispose of contaminated materials. The primary objective of decontamination is to avoid becoming contaminated or contaminating other personnel or equipment outside the hot zone. If contamination is suspected, decontamination of personnel, equipment, and apparatus should be performed.

Procedures for all types of decontamination need to be developed and implemented to reduce the possibility of contamination to personnel and equipment. Initial procedures should be upgraded or downgraded as additional information is obtained concerning the type of hazardous materials involved, the degree of hazard, and the probability of exposure of response personnel. Assuming protective equipment is contaminated, decontamination methods should be used appropriate to the hazards presented by the chemicals encountered.

The decision to implement all or part of a decontamination procedure should be based upon a field analysis of the hazards and risks involved. This analysis generally consists of referring to technical reference sources to determine the general hazards, such as flammability and toxicity, and then evaluating the relative risks. Decontamination procedures should be implemented upon arrival at the scene, should provide an adequate number of decontamination stations and personnel, and should continue until the incident commander determines that decontamination procedures are no longer required.

There are occasions when an apparently normal alarm response turns into a hazardous materials incident. Frequently, most of the initial assignment crews will have already gone into the incident area and exposed themselves to the contamination threat.

It is essential that all members so involved remove themselves from the area at once, call for decontamination capability, and stay together in one location. They must not wander around, climb into and out of vehicles, and mix with other personnel since there is a potential for them to be contaminated.

Responders so exposed should be given gross decontamination as a precautionary measure. Knowledgeable hazardous materials personnel, such as the hazardous materials officer, in conjunction with the incident commander, should determine whether technical decontamination is necessary.

The primary objective of decontamination is to avoid contaminating anyone or anything beyond the hot zone. When in doubt about contamination, decontaminate all involved personnel, equipment, and apparatus.

(1) Identify the advantages and limitations of technical decontamination operations.
(2) Describe the advantages and limitations of each of the following technical decontamination methods:
   (a) Absorption

Absorption is the process by which materials hold liquids. Many types of commercial absorbents are available. Sand or soil can also be used for this purpose, although they are more suited for decontaminating equipment or the area surrounding a spill than they are for decontaminating personnel. Absorbents are often readily available, but they must be disposed of properly because the absorbent substance retains the properties of the material absorbed.

   (b) Adsorption

Adsorption is a chemical method of decontamination involving the interaction of a hazardous liquid and a solid sorbent surface. Examples of adsorbents are activated charcoal, silica or aluminum gel, fuller's earth, and other clays. Adsorption produces heat and can cause spontaneous combustion. Adsorbents must be disposed of properly.

   (c) Chemical degradation

Chemical degradation is the natural breakdown of the contaminants as they age. An example of chemical degradation is the evaporation of a flammable liquid spill. The decontamination of an oil spill on a beach because of manual (pressure washer) or natural (wave action) action is an example of physical degradation. Either of the two methods has limitations depending on factors such as the location of the spill and the toxicity of the material. In some cases, however, these methods are the most practical.

   (d) Dilution

Dilution, which simply reduces the concentration of a contaminant, is best used on materials that are soluble or miscible in water, such as chlorine and ammonia. An advantage of dilution is that solutes, especially water, are generally available in large quantities. A disadvantage is that the runoff must be collected and disposed of.

   (e) Disinfection

Disinfection is a process to kill most (but not all) pathogenic microorganisms. The two types of disinfectants are chemical disinfectants and antiseptic disinfectants. The limitations and capabilities of each disinfectant should be known.

Proper disinfection results in a reduction in the number of viable organisms to some acceptable level. It might not totally destroy 100 percent of the microorganisms.

   (f) Evaporation

In some cases, responders can allow a hazardous material to evaporate, particularly if the vapors do not present a hazard. A small spill of gasoline, for example, might be allowed to evaporate as long as it does not present a vapor problem. Evaporation is an easy operation and requires minimal personnel. This method is not as effective on porous surfaces as it is on nonporous surfaces, however, and it could take quite a while, depending on the quantity of the chemical involved.

(g) Isolation and disposal

Disposal is the direct removal of a contaminant from a carrier. An example of this method is the removal of a contaminated object from a piece of equipment. This type of decontamination might not entirely remove all contamination.

(h) Neutralization

Neutralizers alter a contaminant chemically so that the resulting chemical is harmless. For example, the addition of soda ash to an acidic solution can increase the pH, making it a chemically harmless substance. Many neutralizing chemicals present their own hazards, however, and should only be used by hazardous materials technicians who are fully aware of the consequences. One advantage of neutralizers is that by rendering the remaining material harmless, they reduce the problem of disposal.

(i) Solidification

Commercial products are available that cause certain liquids to solidify. One advantage of solidification is that it allows responders to confine a small spill relatively quickly. As with other decontaminants, however, the resulting solid must be disposed of properly when the incident is over.

(j) Sterilization

Sterilization destroys all microorganisms through the use of steam, concentrated chemical agents, or ultraviolet light radiation. Although some liquid disinfectants can be concentrated enough to sterilize, the products then have side effects. The field use of sterilization is therefore limited.

(k) Vacuuming

Vacuuming allows for the collection of materials, either liquid or solid, into containers. The equipment being used must be appropriate for the material being vacuumed. If the material is corrosive or flammable, for example, specialized equipment is needed.

(l) Washing

A very effective decontamination process for many materials involves washing the contaminated person, building, or equipment. Materials that are not soluble in water, such as oil-based contaminants, can be washed with detergent solutions. Washing equipment, protective clothing, and personnel is one of the easiest methods of decontamination. However, collecting and properly disposing of any runoff is necessary. See Exhibit I.6.4.

Summary of Decontamination Approaches

Physical Methods. Physical methods generally involve the physical removal of the contaminant from the contaminated person or object and containment of the contaminant for disposal. While these methods can reduce the contaminant's concentration, generally the containment remains chemically unchanged. Examples of physical decontamination methods include the following:

1. Absorption
2. Brushing and scraping
3. Isolation and disposal
4. Vacuuming
5. Washing

EXHIBIT I.6.4 *The method of decontamination must take into account the state of the contaminant and the type of PPE worn. (Source: Catherine Blair, Gaade & Associates)*

Placing contaminated outer clothing on a plastic drop cloth within the decontamination area of the warm zone is a common and useful procedure when using the physical methods of decontamination. Contaminated clothing and equipment should be packed in hazardous waste containers. Lined containers should be on hand in which to pack the contaminated clothing and equipment.

If time is not an overriding factor, and if technicians are not on site or immediately available, the detergent–water solution method of preliminary decontamination is the safest and most appropriate approach. Metal or plastic drums are effective for storing washing or rinsing solutions used in decontamination work.

For radioactive materials decontamination, use of traditional hazardous materials decontamination procedures might not be necessary if radioactive material is the only hazard present. While use of traditional hazardous material decontamination processes are effective for radioactive material, their use can generate large quantities of wastewater. Consideration should be given to methods that minimize the amount of waste generated. Simpler methods are available for decontamination that are less time consuming, require fewer resources, and generate less waste.

Removing all clothing can dramatically reduce the radioactive contaminants on a person's body. After performing a gross decontamination, clothing should be left inside the hot zone. This clothing should be contained and controlled until surveyed. Minimizing the accumulation of contaminated or radioactive material, such as removed clothing, packages, and so forth, in the area helps keep area radiation dose rates low.

Radiological survey instruments can be used to locate radioactive contamination on personnel and to determine the effectiveness of decontamination efforts. Personnel decontamination can be accomplished by using conventional cleansing techniques on localized contaminated body surfaces (i.e., gentle washing and flushing that does not abrade the skin surface). When washing and flushing skin surfaces, mild soap and lukewarm water are preferred. Lukewarm water is preferred because cold water can cause skin pores to close, fixing the contamination into the skin. Hot water can cause skin pores to open, allowing the contamination to go deeper into the skin. Any water or material used in this process needs to be contained and considered as radioactive waste. Techniques beyond gross decontamination should only be performed by properly trained personnel.

Chemical Methods. Chemical methods are used on equipment, not people, and generally involve decontamination by changing the contaminant through some type of chemical reaction in an effort to render the contaminant less harmful. In the case of etiologic contaminants, chemical methods are actually biologically "killing" the organism. Examples of chemical methods include the following:

1. Adsorption
2. Chemical degradation
3. Disinfection or sterilization
4. Neutralization
5. Solidification

The precaution on limiting the use of chemical methods to decontaminate equipment deserves emphasis. Chemical degradation, for example, is used to alter the chemical structure of the hazardous material. Commonly used agents, including sodium hypochlorite (household bleach), sodium hydroxide as a saturated solution (household drain cleaner), and calcium oxide slurry (lime), are harmful to skin and never should be applied directly to skin.

Some decontamination procedures can actually present additional hazards. For example, the decontamination solution could react with the chemical to which the clothing was exposed, it could permeate or degrade some protective clothing, or it could emit harmful vapors. Compatibility should be determined before use: Technical advice might be needed to ensure that any solution used on equipment is not reactive with the contaminant.

Prevention Methods. If contact with a contaminant can be controlled, the risk of exposure is reduced and the need for decontamination can be minimized. The following points should be considered to prevent contamination:

1. Stress work practices that minimize contact with hazardous substances.
2. Wear limited-use or disposable protective clothing and equipment, where appropriate.

(3) Identify sources of information for determining the correct technical decontamination procedure and identify how to access those resources in a hazardous materials/WMD incident.

Technical reference sources such as books and databases are the best references for hazard information on contaminants, but they often are lacking in details concerning the best methods for safely and effectively removing the contaminant from people and equipment. The best sources for this information come from the manufacturer of the product. Manufacturers have years of experience handling product emergencies and usually have current information on the best techniques as well as emergency medical treatment. CHEMTREC/CANUTEC/SETIQ can facilitate access to manufacturers. Local and regional poison control centers can frequently be a source of technical information to aid in determining the correct decontamination procedure.

An HMT or allied professional can also provide information regarding decontamination methods.

The ultimate responsibility for implementing a decontamination plan falls to the incident commander. The hazardous materials officer often oversees the implementation of the decontamination procedures.

(4) Given resources provided by the AHJ, identify the supplies and equipment required to set up and implement technical decontamination operations.

A standard complement for decontamination may include the following:

- Containment to prevent runoff
- A water source
- Tarps and plastic sheeting suitable to prevent surface contamination
- Brushes and sponges
- Bucket
- Containment and disposal vessels for contaminants

(5) Identify the procedures, equipment, and safety precautions for processing evidence during technical decontamination operations at hazardous materials/WMD incidents.

A hazard risk assessment is performed to determine the method for decontamination of items of evidence brought from the exclusion zone. A separate decontamination line needs to be established for evidence decontamination at the entrance to the decontamination corridor. Evidence decontamination is designed to remove contamination from the exterior evidence packaging container only. At no time is the exterior evidence container to be breeched for the purposes of decontaminating the interior evidence packaging; care must be taken to preserve the integrity of forensic evidence, such as fingerprints, during the decontamination process. Once the exterior evidence packaging is decontaminated, the evidence is moved following law enforcement AHJ procedures for chain of custody to an evidence custodian, for documentation into the evidence chain.

(6) Identify procedures, equipment, and safety precautions for handling tools, equipment, weapons, criminal suspects, and law enforcement/search canines brought to the decontamination corridor at hazardous materials/WMD incidents.

Procedures for the decontamination of law enforcement tools, equipment, and weapons must be designed with cooperation between local or state and federal public safety and law enforcement agencies. Law enforcement personnel, whether patrol officers or tactical team members, could be wearing several layers of law enforcement–specific equipment outside their chemical-protective clothing. Procedures for the decontamination of law enforcement personnel should focus on systematic removal of these layers of external equipment, isolation and security of that equipment pending a hazard risk assessment of the potential contamination to the equipment, followed by standard technical decontamination procedures for the personnel. The law enforcement AHJ must have a procedure in place for the clearing of ammunition from weapons and securing of distraction devices, to ensure the safety of the personnel in the decontamination corridor.

Procedures for the decontamination of criminal subjects must meet the added challenge of positive custodial control over the subjects by law enforcement personnel. Non–law enforcement responders should not be placed in a position where they are the sole custodian of criminal suspects. Law enforcement authorities must follow their procedures regarding the control of criminal suspects in the decontamination corridor. It might be necessary for law enforcement personnel in chemical protective clothing to accompany the suspect. Decontamination of suspects should follow procedures developed by the AHJ for emergency decontamination of civilians.

The use of law enforcement or search canines at incidents involving the known use of hazardous materials/WMD is cautioned. Development of procedures for decontamination of canines must be drafted through cooperation between the AHJ and their veterinary medicine specialist.

6.4.4 Competencies — Implementing the Planned Response.

6.4.4.1 Performing Incident Management Duties. Given a scenario involving a hazardous materials/WMD incident and the emergency response plan or standard operating procedures, the operations level responder assigned to technical decontamination operations shall demonstrate the technical decontamination duties assigned in the incident action plan and shall meet the following requirements:

(1) Identify the role of the operations level responder assigned to technical decontamination operations during hazardous materials/WMD incidents.

During doffing of PPE, the clothing should be removed in such a manner that the outside surfaces do not touch or make contact with the wearer. A log of PPE used during the incident should be maintained. Personnel wearing disposable PPE should proceed through the decontamination process setup in the decontamination area, and the disposable protective equipment should be containerized and identified for disposal in accordance with established procedures.

The log should record the type of equipment used, the duration of use of the equipment and exposure to the chemicals, the decontamination procedures used, and the types of chemicals to which the equipment was exposed. The log should also record the name of the person using the equipment.

Disposable protective equipment should be placed in plastic bags or plastic trash cans pending final disposal.

The physical and chemical compatibility of decontamination solutions needs to be determined before they are used. Any decontamination method that permeates, degrades, damages, or otherwise impairs the safe function of PPE should not be used unless there are plans to isolate and dispose of the PPE.

Water or other solutions for washing or rinsing may have to be confined, collected, containerized, and analyzed for treatment and disposal. Consult with environmental and public health agencies or other reference sources and guidelines to determine the need for confinement and the disposal methods for collected decontamination fluids and PPE.

Decontamination methods vary in their effectiveness for removing different substances. The effectiveness of any decontamination method should be assessed throughout the decontamination operation. If decontamination does not appear to be effective, a different method should be selected and implemented. Before initiating decontamination, the following questions should be considered:

1. Can decontamination be conducted safely?
2. Are existing resources adequate and immediately available to perform decontamination of personnel and equipment? If not, where can they be obtained, and how long will it take to get them?

If the decontamination method being used is not effectively removing contaminants, different and additional measures must be taken to prevent further contamination. If necessary, technical specialists should be consulted.

The decontamination plan might have to be changed whenever the type of protective clothing and equipment changes, when site conditions change, or when new information is received. If necessary, the manufacturers of the protective clothing and equipment should be consulted.

Criteria that can be used for evaluating decontamination effectiveness during field operations include the following:

1. Contamination levels are reduced as personnel move through the decontamination area.
2. Contamination is confined to the hot zone and decontamination area.
3. Contamination is reduced to a level that is as low as reasonably achievable.

Large items of equipment, such as vehicles and trucks, should be subjected to decontamination by washes, high-pressure washes, steam, or special solutions. Water or other solutions used for washing or rinsing might have to be confined, collected, containerized, analyzed, and treated prior to disposal. Consult with environmental and public health agencies to determine the appropriate disposition.

Vehicles can become contaminated in several ways. These situations include placing them downwind where smoke or vapors can contaminate them, driving through a spill, placing the vehicle too close to the isolation area where it can be splashed or sprayed by the hazardous material, and placing contaminated equipment back in the vehicle. Placing contaminated persons in police and EMS vehicles has been a source of contamination that has led to extensive downtime while the vehicle was decontaminated. To prevent contamination of an ambulance, a victim must undergo gross decontamination prior to hand-off to EMS, as shown in Exhibit I.6.5.

If a large number of vehicles need to be decontaminated, the following recommendations should be considered:

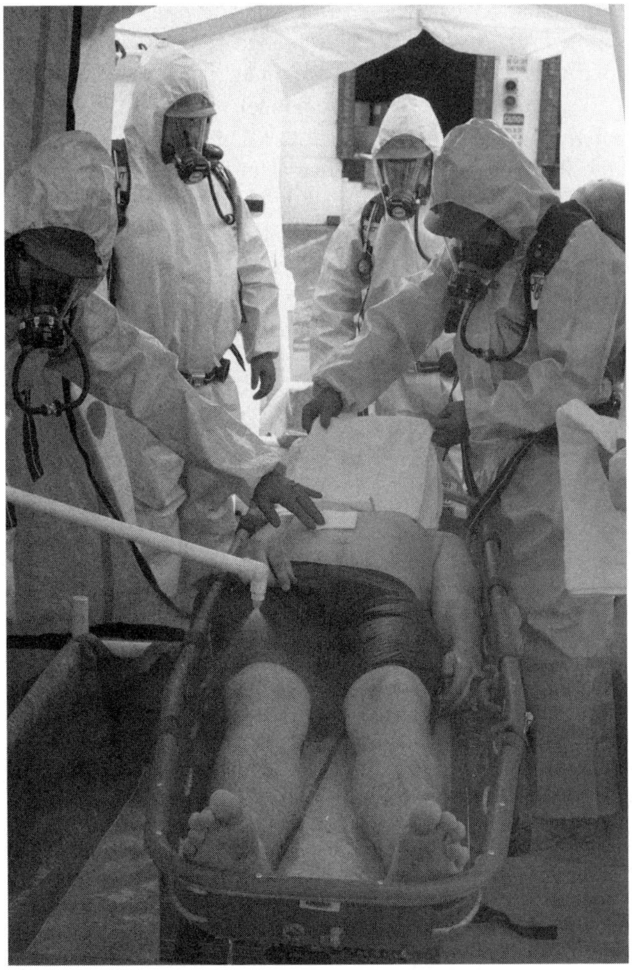

EXHIBIT I.6.5 *This victim is being decontaminated to prevent contamination of an ambulance. (Source: Hildebrand and Noll Technical Resources, LLC)*

1. Establish a decontamination pad as a primary wash station. The pad might be a coarse gravel pad, a concrete slab, or a pool liner. It might be necessary to collect these decontamination fluids, and the decontamination pad can be bermed or diked with a sump or some form of water recovery system.
2. Completely wash and rinse vehicles several times with detergent. Pay particular attention to wheel wells, radiators, engines, and chassis. Depending on the nature of the contaminant, it might be necessary to collect all runoff water from the initial gross rinse, particularly if there is contaminated mud and dirt on the underside of the chassis.
3. Inspect vehicles thoroughly for possible mechanical or electrical damage. Areas of concern include air intakes, filters, cooling systems, and air-operated systems.
4. Empty completely and thoroughly wash any outside compartments that were opened. The equipment should be washed and rinsed prior to being replaced.
5. Equipment sprayed with acids should be flushed or washed as soon as possible with a neutralizing agent such as baking soda and then flushed again with rinse water.
6. Decontaminate vehicles on-site if they have been exposed to minimal contaminants such as smoke and vapors. They can then be driven to an off-site car wash for a second, more thorough washing. Car washes can be suitable if the drainage area is fully contained and all runoff drains into a holding tank.
7. Verify adequacy of decontamination, which can consist of samples collected from the cab and exterior surfaces that are analyzed in an off-site laboratory.

Personnel assigned to the decontamination team should wear an appropriate level of PPE and could require decontamination themselves. PPE can be upgraded or downgraded as additional information is obtained concerning the type of hazardous materials involved, the degree of hazard, and the probability of exposure of response personnel.

The members of the decontamination team closest to the hot zone might require a higher level of protective clothing than those closest to the cold zone. The level of protection required varies with the decontamination equipment in use. Protective clothing should be selected by a hazardous materials technician under the direction of the hazardous materials officer. The selection should be approved by the incident commander.

If personnel display any symptoms of heat exhaustion or possible exposure, emergency measures need to be implemented to doff PPE, while protecting the individual from contaminants and preventing the spread of any contaminants. These individuals should be transferred to the care of emergency medical services personnel who have completed training in accordance with applicable standards such as NFPA 473, *Standard for Competencies for EMS Personnel Responding to Hazardous Materials/Weapons of Mass Destruction Incidents* [1].

The responder in PPE must be aware of the signs and symptoms of heat exhaustion or other adverse effects and be able to request assistance at the first indication of a problem. For the wearer to doff the PPE quickly and be decontaminated properly, sufficient time must be allowed for the process.

A debriefing should be held for those involved in decontamination as soon as practical. Exposed persons should be provided with as much information as possible about the delayed health effects of the hazardous materials involved in the incident. If necessary, follow-up examinations should be scheduled with medical personnel.

Exposure records should be maintained for future reference by the individual's personal physician and employer.

(2) Describe the procedures for implementing technical decontamination operations within the incident command system.

6.4.4.2 Performing Decontamination Operations Identified in Incident Action Plan. The responder assigned to technical decontamination operations shall demonstrate the ability to set up and implement the following types of decontamination operations:

(1) Technical decontamination operations in support of entry operations
(2) Technical decontamination operations for ambulatory and nonambulatory victims

Responders might encounter ambulatory and nonambulatory victims who need to be decontaminated. The primary methodology will be gross decontamination or emergency decontamination sufficient to safely forward the patients to emergency medical services for treatment.

6.4.5 Competencies — Evaluating Progress.

6.4.5.1 Evaluating the Effectiveness of the Technical Decontamination Process. Given examples of contaminated items that have undergone the required decontamination, the operations level responder assigned to technical decontamination operations shall identify procedures for determining whether the items have been fully decontaminated according to the standard operating procedures of the AHJ or the incident action plan.

Methods that can be useful in assessing the effectiveness of decontamination include the following:

1. Visual observation (stains, discolorations, corrosive effects, etc.)
2. Monitoring devices [Devices such as photoionization detectors (PIDs), detector tubes, radiation monitors, and pH paper strips/meters can show that contamination levels are at least below the device's detection limit.]
3. Wipe sampling. Such sampling provides after-the-fact information on the effectiveness of decontamination. Once a wipe swab is taken, it is analyzed by chemical means, usually in a laboratory. Protective clothing, equipment, and skin can be tested using wipe samples.

6.4.6 Competencies — Terminating the Incident.

6.4.6.1 Reporting and Documenting the Incident. Given a scenario involving a hazardous materials/WMD incident, the operations level responder assigned to technical decontamination operations shall complete the reporting and documentation requirements consistent with the emergency response plan or standard operating procedures and shall meet the following requirements:

(1) Identify the reports and supporting technical documentation required by the emergency response plan or standard operating procedures.
(2) Describe the importance of personnel exposure records.
(3) Identify the steps in keeping an activity log and exposure records.
(4) Identify the requirements for filing documents and maintaining records.

6.5 Mission-Specific Competencies: Evidence Preservation and Sampling

6.5.1 General.

6.5.1.1 Introduction.

6.5.1.1.1 The operations level responder assigned to perform evidence preservation and sampling shall be that person, competent at the operations level, who is assigned to preserve forensic evidence, take samples, and/or seize evidence at hazardous materials/WMD incidents involving potential violations of criminal statutes or governmental regulations.

This mission-specific competency is designed to prepare the operations level responder to support law enforcement operations in providing evidence collection and evidence preservation tasks at a crime scene involving hazardous materials. The competency is designed to raise the awareness of the responder during crime scene operations and to prepare them to assist crime scene investigators during the performance of their jobs. See Exhibit I.6.6.

Additional guidance in evidence preservation and sampling can be obtained from the local office of the Federal Bureau of Investigation's (FBI) Weapons of Mass Destruction Coordinator, the local office of the Drug Enforcement Agency (DEA), state or local law enforcement, or governmental sponsored training programs.

6.5.1.1.2 The operations level responder assigned to perform evidence preservation and sampling at hazardous materials/WMD incidents shall be trained to meet all competencies at the awareness level (Chapter 4), all core competencies at the operations level (Chapter 5), all mission-specific competencies for personal protective equipment (Section 6.2), and all competencies in this section.

6.5.1.1.3 The operations level responder assigned to perform evidence preservation and sampling at hazardous materials/WMD incidents shall operate under the guidance of a hazardous materials technician, an allied professional, or standard operating procedures.

Although some of the mission-specific competencies in this section are taken from Chapter 7, "Competencies for Hazardous Materials Technicians," the technical committee wants to clearly state that operations level responders with a mission-specific competency are not replacements for an HMT. Operations level responders with a mission-specific competency can perform some technician-level skills but do not have the broader skills and competencies required of an HMT, particularly regarding risk assessment and the selection of control options. The following two options are examples of how guidance can be provided to ensure that operations level responders do not go beyond their level of training and equipment:

- Direct Guidance: Operations level responders are working under the control of an HMT or allied professional who has the ability to (1) continuously assess and/or observe their

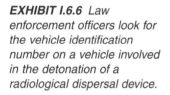

EXHIBIT I.6.6 *Law enforcement officers look for the vehicle identification number on a vehicle involved in the detonation of a radiological dispersal device.*

actions and (2) provide immediate feedback. Guidance by an HMT or an allied professional can be provided through direct visual observation or through assessment reports communicated by the operations level responder to them.

- Written Guidance: Written standard operating procedures or similar guidance clearly state the "rules of engagement" for operations level responders with the mission-specific competency. Emphasis should be placed on the following:
  1. Tasks expected of operations level responders
  2. Tasks beyond the capability of operations level responders
  3. Required PPE and equipment to perform these expected tasks
  4. Procedures for ensuring coordination within the local ICS

6.5.1.1.4* The operations level responder assigned to perform evidence preservation and sampling at hazardous materials/WMD incidents shall receive the additional training necessary to meet specific needs of the jurisdiction.

A.6.5.1.1.4 See A.6.3.1.1.4.

6.5.1.2 Goal.

6.5.1.2.1 The goal of the competencies in this section shall be to provide the operations level responder assigned to evidence preservation and sampling at hazardous materials/WMD incidents with the knowledge and skills to perform the tasks in 6.5.1.2.2 safely and effectively.

6.5.1.2.2 When responding to hazardous materials/WMD incidents involving potential violations of criminal statutes or governmental regulations, the operations level responder assigned to perform evidence preservation and sampling shall be able to perform the following tasks:

(1) Analyze a hazardous materials/WMD incident to determine the complexity of the problem and potential outcomes by completing the following tasks:
   (a) Determine if the incident is potentially criminal in nature and identify the law enforcement agency having investigative jurisdiction.
   (b) Identify unique aspects of criminal hazardous materials/WMD incidents.

(2) Plan a response for an incident where there is potential criminal intent involving hazardous materials/WMD within the capabilities and competencies of available personnel, personal protective equipment, and control equipment by completing the following tasks:
   (a) Determine the response options to conduct sampling and evidence preservation operations.
   (b) Describe how the options are within the legal authorities, capabilities, and competencies of available personnel, personal protective equipment, and control equipment.

(3) Implement the planned response to a hazardous materials/WMD incident involving potential violations of criminal statutes or governmental regulations by completing the following tasks under the guidance of law enforcement:
   (a) Preserve forensic evidence.
   (b) Take samples.
   (c) Seize evidence.

6.5.2 Competencies — Analyzing the Incident.

6.5.2.1 Determining If the Incident Is Potentially Criminal in Nature and Identifying the Law Enforcement Agency That Has Investigative Jurisdiction. Given examples of hazardous materials/WMD incidents involving potential criminal intent, the operations level responder assigned to evidence preservation and sampling shall describe the potential criminal violation and identify the law enforcement agency having investigative jurisdiction and shall meet the following requirements:

Law enforcement jurisdiction, investigative guidelines, and investigative priorities are complex and dynamic. Specific jurisdictional situations should be identified with the responder's local or state law enforcement authorities and federal investigative agencies such as the FBI, DEA, the United States Postal Inspection Service, and the Environment Protection Agency (EPA).

(1) Given examples of the following hazardous materials/WMD incidents, the operations level responder shall describe products that might be encountered in the incident associated with each situation:

Response agencies should maintain situational awareness by receiving a threat briefing from law enforcement officials concerning the anticipated threats from hazardous materials/WMD incidents and being aware of potential illicit uses for hazardous materials in their community.

(a) Hazardous materials/WMD suspicious letter
(b) Hazardous materials/WMD suspicious package

There have been actual cases of hazardous materials/WMD sent in letters and packages that contained explosive materials, explosive devices, chemicals, toxins, biological materials, and radioactive materials.

(c) Hazardous materials/WMD illicit laboratory

Illicit laboratories can include the illegal manufacture of drugs, toxins, nontraditional chemical weapons, and biologic agents. Specific guidance on the products that might be encountered must be obtained from local or state law enforcement or federal agencies, based upon current threats and trends.

(d) Release/attack with a WMD agent

Identifying the products and methods that can be used during a WMD agent attack is complex. Special attention should be given to materials that are commonly found or acquired either in transportation or at fixed facilities, such as industrial chemicals and toxins.

(e) Environmental crimes

Illegal disposal of either hazardous materials or hazardous waste is incident specific. The potential threat from chemical, biological, or radiological materials or waste materials must be assessed in cooperation with local or state and federal agencies.

(2) Given examples of the following hazardous materials/WMD incidents, the operations level responder shall identify the agency(s) with investigative authority and the incident response considerations associated with each situation:

Law enforcement jurisdiction, investigative guidelines, and investigative priorities are complex and dynamic. The specific jurisdictional situations should be identified with the responder's local or state law enforcement authorities and federal investigative agencies.

(a) Hazardous materials/WMD suspicious letter
(b) Hazardous materials/WMD suspicious package
(c) Hazardous materials/WMD illicit laboratory
(d) Release/attack with a WMD agent
(e) Environmental crimes

6.5.3 Competencies — Planning the Response.

6.5.3.1 Identifying Unique Aspects of Criminal Hazardous Materials/WMD Incidents. The operations level responder assigned to evidence preservation and sampling shall be capable of identifying the unique aspects associated with illicit laboratories, hazardous materials/WMD incidents, and environmental crimes and shall meet the following requirements:

When responding to hazardous materials/WMD incidents, the operations level responder should be observant for signs of criminal activity involving chemical, biological, radiological, or explosive materials or devices. The operations level responder should be familiar with the jurisdictional procedures for operation within a crime scene, including investigative law enforcement leadership, search warrant requirements, rules of evidence, crime scene documentation, policies regarding photography, evidence custodial requirements, chain of custody, and specific requirements set forth by the prosecuting attorney.

(1) Given an incident involving illicit laboratories, a hazardous materials/WMD incident, or an environmental crime, the operations level responder shall perform the following tasks:

(a) Describe the procedure to secure, characterize, and preserve the scene.

The operations level responder should follow local or state and federal jurisdictional procedures for crime scene security, at the direction of law enforcement. These activities can include full accountability and identification of all personnel in the crime scene, documentation of any items disturbed within the crime scene, and protection of evidence from potential damage or destruction.

(b) Describe the procedure to document personnel and scene activities associated with the incident.

Local procedures vary for crime scene documentation. For example, some jurisdictions do not allow the use of video documentation. The operations level responder should become familiar with the local or state and federal procedures for documentation at a crime scene.

(c) Describe the procedure to determine whether the operations level responders are within their legal authority to perform evidence preservation and sampling tasks.

It is very important for the operations level responder to coordinate all evidence preservation and sampling tasks with the law enforcement AHJ. All evidentiary tasks are required to follow rules of evidence, Fourth Amendment guarantees, and judicial precedents. Current information involving judicial case law is available from the law enforcement AHJ, and it is absolutely critical that the operations level responder be aware of this information to avoid problems with inadmissible evidence.

(d) Describe the procedure to notify the agency with investigative authority.

The elements of the criminal offense will drive the creation of an evidence sampling plan, which will identify the items to be collected or sampled during crime scene operations. The law enforcement AHJ will be in contact with the prosecuting attorney's office to determine the elements that are required to prove the criminal offense.

(e) Describe the procedure to notify the explosive ordnance disposal (EOD) personnel.
(f) Identify potential sample/evidence.
(g) Identify the applicable sampling equipment.

The operations level responder should be familiar with any existing local or state and federal requirements for the selection of appropriate sampling equipment.

(h) Describe the procedures to protect samples and evidence from secondary contamination.

The operations level responder needs to ensure that all sampling containers and equipment are free of contamination. This could involve the use of sterile containers for the collection of biological materials, certified clean containers for chemicals and radiological materials, and control blanks. Control blanks are a collection of each type and lot of sampling equipment used during crime scene sampling operations, isolated in their clean or sterile packaging, and entered into evidence.

(i) Describe documentation procedures.

Specific documentation procedures vary among law enforcement agencies. Specific guidance must be obtained to ensure that the documentation requirements of the law enforcement AHJ have been met.

(j) Describe evidentiary sampling techniques.

Prior to sampling operations, a sampling plan will be developed by the criminal investigator in cooperation with the crime scene investigation team. This plan will identify specific sampling techniques required to obtain forensic evidence, ensure appropriate documentation, and maintain sample integrity. A crime scene investigation team could include a three-person sampling team: a sampler, an assistant, and a documenter.

(k) Describe field screening protocols for collected samples and evidence.

The operations level responder must field screen samples taken into evidence to rule out specific hazards prior to the introduction of those samples into a laboratory facility. The field screening shall include, at a minimum, nondestructive testing to identify the presence of explosive ordnance disposal (EOD) devices, radiological materials, flammable materials, toxic materials, strong oxidizers, and the corrosive properties of liquid samples.

(l) Describe evidence labeling and packaging procedures.

The law enforcement AHJ, in cooperation with the receiving laboratory, is responsible for establishing the procedures for labeling hazardous materials/WMD evidence. The labeling system must be clear to all personnel sampling, collecting, and packaging evidence. Labeling of evidence packaging materials can be best performed prior to entry to the exclusion zone. Selection of packaging materials and procedures will be guided by the evidence sampling plan, in accordance with the law enforcement AHJ and in cooperation with the receiving laboratory.

(m) Describe evidence decontamination procedures.

A hazard risk assessment should be performed to determine the method for decontamination of items of evidence brought from the exclusion zone. A separate decontamination line should be established for evidence decontamination at the entrance to the decontamination corridor. Evidence decontamination is designed to remove contamination from the exterior evidence packaging container only. At no time shall the exterior evidence container be breeched for the purposes of decontaminating the interior evidence packaging. Care must be taken to preserve the integrity of forensic evidence, such as fingerprints, during the decontamination process. Once the exterior evidence packaging is decontaminated, the evidence is moved following law enforcement AHJ procedures for chain of custody to an evidence custodian for documentation into the evidence chain.

(n) Describe evidence packaging procedures for evidence transportation.

Procedures for the safe external packaging and transportation of evidence are the responsibility of the law enforcement AHJ, in cooperation with the receiving laboratory and the operator of the transport vehicle. Packaging for transportation must ensure the safe transit of the evidence, prevent release of the hazardous material during transport, and follow any applicable regulations or AHJ policy. Whenever possible, the external evidence container should fit within a standard bio-safety protection cabinet or chemical hood.

(o) Describe chain-of-custody procedures.

Chain-of-custody is the practice of maintaining positive visual or physical control over evidence from collection at a site until presentation in court. Each person encountering positive control over the evidence must be entered into the chain-of-custody documentation and as such is a candidate for subpoena to court. The chain-of-custody procedures are the responsibility of the law enforcement AHJ.

(2) Given an example of an illicit laboratory, the operations level responder assigned to evidence preservation and sampling shall be able to perform the following tasks:

Illicit laboratories take several forms and involve numerous types of hazards. It is important for the operations level responder to understand that the potential for chemical, biological, radiological, and explosive materials is present at all illicit laboratories. Consultation with local or state and federal law enforcement about current intelligence, recent incidents, and local trends is critical for the operations level responder.

(a) Describe the hazards, safety procedures, decontamination, and tactical guidelines for this type of incident.

Illicit laboratories can be designed for the production of many different end products, including illegal drugs such as methamphetamine, chemical modification such as distillation of pesticides, and biological toxins or pathogens. Operators of these illicit laboratories often have a vested interest in the protection of their product and evading law enforcement. There is a risk that the operator will be present within the laboratory, with access to weapons. The human element must be addressed by law enforcement tactical teams specifically trained to operate within a hazardous environment. The level of protection for the tactical team should be based upon an assessment of the intelligence and information available on the intent of the laboratory, and it should also include any protective clothing used by the operator, activity of animals in the laboratory, interviews with neighbors, and so forth. Additionally, there is a risk of anti-personnel devices (booby traps) around and within the laboratory. The operations level responder must leave clearance of potential anti-personnel devices to EOD personnel trained for these procedures. Once the human and EOD hazards have been cleared, the operations level responder should conduct a hazard risk assessment to determine the potential harm from hazardous materials/WMD within the illicit laboratory. Decontamination procedures should be based upon the results of the hazard risk assessment.

(b) Describe the factors to be evaluated in selecting the personal protective equipment, sampling equipment, detection devices, and sample and evidence packaging and transport containers.

Selection of PPE and detection devices are based upon assessment of the intelligence, outward warning signs, and detection clues. The detection and monitoring equipment selected must be capable of assisting in the performance of the hazard risk assessment and, at a minimum, should include a combustible gas indicator, oxygen level meter, photoionization meter, pH paper, and radiological monitoring equipment. The selection of sampling devices, packaging, and transport containers is predicated upon the evidence sampling plan, as developed by the law enforcement AHJ in coordination with the receiving laboratory.

(c) Describe the sampling options associated with liquid and solid sample and evidence collection.

The decision on sampling techniques and collection tools will be guided by the evidence sampling plan. Potential collection tools for liquid samples include syringes, pipettes, composite liquid waste sampler (coliwasa) tubes, drum thieves, certified clean jars, and sterile vials. Potential collection tools for solid samples include swabs, scoops, spatulas, certified clean jars, and sterile vials.

(d) Describe the field screening protocols for collected samples and evidence.

The operations level responder must field screen samples taken into evidence to rule out specific hazards prior to the introduction of those samples into a laboratory facility. The field screening shall include, at a minimum, nondestructive testing to identify the presence of EOD devices, radiological materials, flammable materials, toxic materials, strong oxidizers, and the corrosive properties of liquid samples. See Exhibit I.6.7.

(3) Given an example of an environmental crime, the operations level responder assigned to evidence preservation and sampling shall be able to perform the following tasks:

(a) Describe the hazards, safety procedures, decontamination, and tactical guidelines for this type of incident.

Environmental crime sites could involve the illegal use and disposal of hazardous materials and hazardous waste. Operators and/or owners of environmental crime sites might be evading law enforcement. There is a risk that the operator and/or owner will be present within the site, with access to weapons. The human element must be addressed by law enforcement personnel. The level of protection for the site entry should be based upon an assessment of the intelligence and information on the intent of the environmental crime site. This information should include any protective clothing used by the operator, activity of animals in the area, and interviews with neighbors and employees. Once the human and any EOD hazards have been cleared, the operations level responder should conduct a hazard risk assessment to determine the potential harm from hazardous materials/WMD within the environmental crime site. Decontamination procedures should be based upon the results of the hazard risk assessment.

(b) Describe the factors to be evaluated in selecting the personal protective equipment, sampling equipment, detection devices, and sample and evidence packaging and transport containers.

Selection of PPE and detection devices is based upon assessment of the intelligence, outward warning signs, and detection clues. The detection and monitoring equipment selected must be capable of assisting in the performance of the hazard risk assessment, and at a minimum, it should include a combustible gas indicator, oxygen level meter, photoionization meter, pH

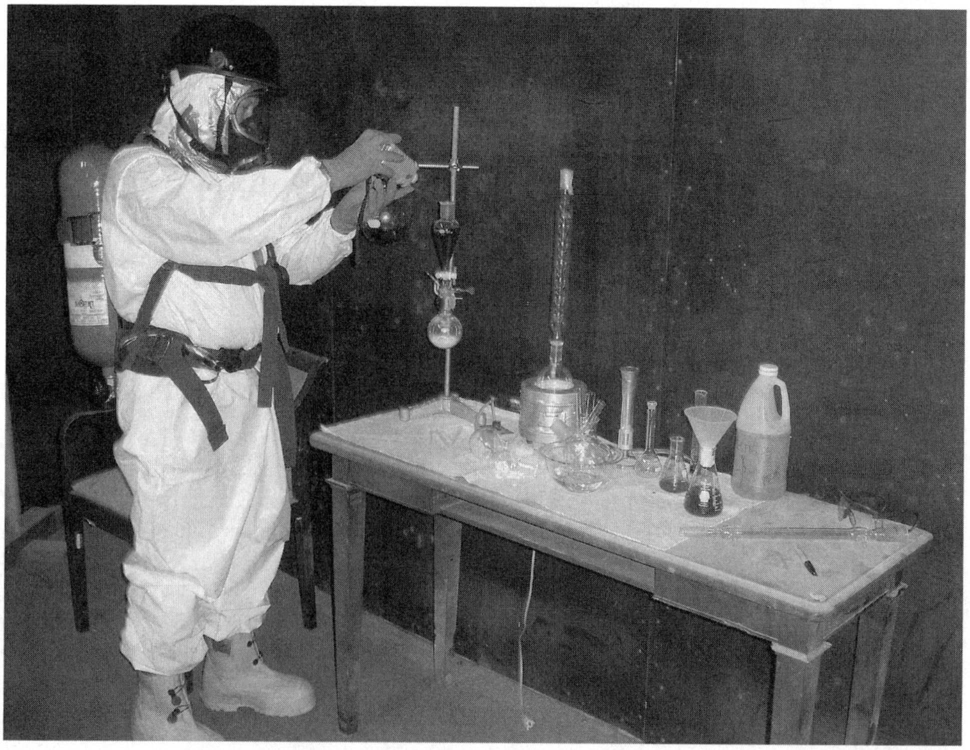

EXHIBIT I.6.7 *A law enforcement officer in Level B chemical protective clothing performs field screening in a suspected illicit chemical laboratory.*

paper, and radiological monitoring equipment. The selection of sampling devices, packaging, and transport containers is predicated upon the evidence sampling plan, as developed by the law enforcement AHJ in coordination with the receiving laboratory.

(c) Describe the sampling options associated with the collection of liquid and solid samples and evidence.

The decision on sampling techniques and collection tools will be guided by the evidence sampling plan. Potential collection tools for liquid samples include syringes, pipettes, coliwasa tubes, drum thieves, certified clean jars, and sterile vials. Potential collection tools for solid samples include swabs, scoops, spatulas, certified clean jars, and sterile vials.

(d) Describe the field screening protocols for collected samples and evidence.

The operations level responder must field screen samples taken into evidence to rule out specific hazards prior to the introduction of those samples into a laboratory facility. At a minimum, the field screening must include nondestructive testing to identify the presence of EOD devices, radiological materials, flammable materials, toxic materials, strong oxidizers, and the corrosive properties of liquid samples.

(4) Given an example of a hazardous materials/WMD suspicious letter, the operations level responder assigned to evidence preservation and sampling shall be able to perform the following tasks:

(a) Describe the hazards, safety procedures, decontamination, and tactical guidelines for this type of incident.

Hazmat/WMD suspicious letters have the potential to contain explosive devices, explosive materials, chemicals, biological materials, or radioactive materials. The operations level responder should participate in a threat assessment with local or state and federal law enforcement, hazardous materials response authorities, and public health officials to evaluate the articulated or implied threat from the suspicious letter. The threat assessment should include evaluation of the behavioral resolve of the originator of the letter, the technical feasibility that hazardous materials are present, and the operational practicality of successful delivery of hazardous materials in the letter.

Once the letter has been cleared for EOD hazards, the operations level responder should conduct a hazard risk assessment. The assessment should include field screening to determine the potential harm from hazardous materials/WMD within the suspicious letter. In the presence of an articulated or implied threat, the operations level responder must ensure that the bulk of any potential hazardous materials found within the letter, as well as the letter and envelope containing the letter, are packaged for law enforcement–supervised transport to a receiving laboratory. Decontamination procedures should be based upon the results of the hazard risk assessment.

(b) Describe the factors to be evaluated in selecting the personal protective equipment, sampling equipment, detection devices, and sample and evidence packaging and transport containers.

Selection of PPE and detection devices is based on assessment of the intelligence, outward warning signs, and detection clues. The detection and monitoring equipment selected must be capable of assisting in the performance of the hazard risk assessment, and at a minimum should include a combustible gas indicator, oxygen level meter, photoionization meter, pH paper, and radiological monitoring equipment. The selection of sampling devices, packaging, and transport containers is predicated upon the evidence sampling plan, as developed by the law enforcement AHJ in coordination with the receiving laboratory.

(c) Describe the sampling options associated with the collection of liquid and solid samples and evidence.

The decision on sampling techniques and collection tools will be guided by the evidence sampling plan. Potential collection tools for liquid samples include syringes, pipettes, certified clean jars, and sterile vials. Potential collection tools for solid samples include swabs, scoops, spatulas, certified clean jars, and sterile vials. The operations level responder should make every effort to document any text or graphics present on the letter or envelope containing the letter. This could include either law enforcement–approved photographs, notes, or packaging of the letter and envelope in clear packaging bags.

(d) Describe the field screening protocols for collected samples and evidence.

The operations level responder must field screen samples taken into evidence to rule out specific hazards prior to the introduction of those samples into a laboratory facility. The field screening shall include, at a minimum, non-destructive testing to identify the presence of EOD devices, radiological materials, flammable materials, toxic materials, strong oxidizers, and the corrosive properties of liquid samples.

(5) Given an example of a hazardous materials/WMD suspicious package, the operations level responder assigned to evidence preservation and sampling shall be able to perform the following tasks:

(a) Describe the hazards, safety procedures, decontamination, and tactical guidelines for this type of incident.

Hazmat/WMD suspicious packages have the potential to contain explosive devices, explosive materials, chemicals, biological materials, or radioactive materials. The operations level responder should participate in a threat assessment with local or state and federal law enforcement, hazardous materials response authorities, and public health officials to evaluate the articulated or implied threat from the suspicious package. The threat assessment should include evaluation of the behavioral resolve of the originator of the package, the technical feasibility that hazardous materials are present, and the operational practicality of successful delivery of hazardous materials in the package. Once the package has been cleared for EOD hazards, the operations level responder should conduct a hazard risk assessment, to include field screening, to determine the potential harm from hazardous materials/WMD within the suspicious package. In the presence of an articulated or implied threat, the operations level responder must ensure that the bulk of any potential hazardous materials found within the package, as well as the outer package, are packaged for law enforcement supervised transport to a receiving laboratory. Decontamination procedures should be based upon the results of the hazard risk assessment.

(b) Describe the factors to be evaluated in selecting the personal protective equipment, sampling equipment, detection devices, and sample and evidence packaging and transport containers.

Selection of PPE and detection devices is based on assessment of the intelligence, outward warning signs, and detection clues. The detection and monitoring equipment selected must be capable of assisting in the performance of the hazard risk assessment, and at a minimum should include a combustible gas indicator, oxygen level meter, photoionization meter, pH paper, and radiological monitoring equipment. The selection of sampling devices, packaging, and transport containers is predicated upon the evidence sampling plan, as developed by the law enforcement AHJ in coordination with the receiving laboratory.

(c) Describe the sampling options associated with liquid and solid sample/evidence collection.

The decision on sampling techniques and collection tools will be guided by the evidence sampling plan. Potential collection tools for liquid samples include syringes, pipettes, certified clean jars, and sterile vials. Potential collection tools for solid samples include swabs, scoops, spatulas, certified clean jars, and sterile vials. The operations level responder should make every effort to document any text or graphics present on any materials within the package, and on the package itself, including law enforcement–approved photographs, notes, or packaging of any written materials found within the package in clear packaging bags.

(d) Describe the field screening protocols for collected samples and evidence.

The operations level responder must field screen samples taken into evidence to rule out specific hazards prior to the introduction of those samples into a laboratory facility. The field screening must include, at a minimum, nondestructive testing to identify the presence of EOD devices, radiological materials, flammable materials, toxic materials, strong oxidizers, and the corrosive properties of liquid samples.

(6) Given an example of a release/attack involving a hazardous material/WMD agent, the operations level responder assigned to evidence preservation and sampling shall be able to perform the following tasks:

(a) Describe the hazards, safety procedures, decontamination and tactical guidelines for this type of incident.

Potential attack scenarios involving hazardous materials/WMD include explosive devices, biological toxins, release or burning of toxic industrial chemicals, biological pathogens, radioactive sources, chemical warfare agents, or nuclear devices. In the event that the release/attack has already occurred, operations level responders should perform their duties as prescribed in the local emergency response plan. The operations level responder should participate in a threat assessment with local or state and federal law enforcement, hazardous materials response authorities, and public health officials to evaluate the articulated or implied threat from the release/attack. The threat assessment should include evaluation of the behavioral resolve of the originator of the release/attack, the technical feasibility that hazardous materials are present, and the operational practicality of successful delivery of hazardous materials during the release/attack. Once the release/attack site has been cleared for EOD hazards, the operations level responder should conduct a hazard risk assessment, including field screening to determine the potential harm from any hazardous materials/WMD involved in the release/attack. Decontamination procedures should be based upon the results of the hazard risk assessment.

(b) Describe the factors to be evaluated in selecting the personal protective equipment, sampling equipment, detection devices, and sample and evidence packaging and transport containers.

Selection of PPE and detection devices is based on assessment of the intelligence, outward warning signs, and detection clues. The detection and monitoring equipment selected must be capable of assisting in the performance of the hazard risk assessment and, at a minimum, should include a combustible gas indicator, oxygen level meter, photoionization meter, pH paper, and radiological monitoring equipment. The selection of sampling devices, packaging, and transport containers is predicated upon the evidence sampling plan, as developed by the law enforcement AHJ in coordination with the receiving laboratory.

(c) Describe the sampling options associated with the collection of liquid and solid samples and evidence.

The decision on sampling techniques and collection tools will be guided by the evidence sampling plan. Potential collection tools for liquid samples include syringes, pipettes, certified clean jars, and sterile vials. Potential collection tools for solid samples include swabs, scoops, spatulas, certified clean jars, and sterile vials.

(d) Describe the field screening protocols for collected samples and evidence.

The operations level responder must field screen samples taken into evidence to rule out specific hazards prior to the introduction of those samples into a laboratory facility. The field screening shall include, at a minimum, nondestructive testing to identify the presence of EOD devices, radiological materials, flammable materials, toxic materials, strong oxidizers, and the corrosive properties of liquid samples.

(7) Given examples of different types of potential criminal hazardous materials/WMD incidents, the operations level responder shall identify and describe the application, use, and limitations of the various types field screening tools that can be utilized for screening the following:

Refer to the mission-specific competencies in Section 6.7, Mission-Specific Competencies: Air Monitoring and Sampling.

(a) Corrosivity
(b) Flammability
(c) Oxidation
(d) Radioactivity
(e) Volatile organic compounds (VOC)

(8) Describe the potential adverse impact of using destructive field screening techniques.
(9) Describe the procedures for maintaining the evidentiary integrity of any item removed from the crime scene.

6.5.3.2 Selecting Personal Protective Equipment. The operations level responder assigned to evidence preservation and sampling shall select the personal protective equipment required to support evidence preservation and sampling at hazardous materials/WMD incidents based on local procedures *(see Section 6.2)*.

Refer to the mission-specific competencies for PPE in Section 6.2.

6.5.4 Competencies — Implementing the Planned Response.

6.5.4.1 Implementing the Planned Response. Given the incident action plan for a criminal incident involving hazardous materials/WMD, the operations level responder assigned to evidence preservation and sampling shall implement or oversee the implementation of the selected response actions safely and effectively and shall meet the following requirements:

(1) Secure, characterize, and preserve the scene.
(2) Document personnel and scene activities associated with the incident.
(3) Describe whether the responders are within their legal authority to perform evidence preservation and sampling tasks.
(4) Notify the agency with investigative authority.
(5) Notify the EOD personnel.
(6) Identify potential samples and evidence to be collected.
(7) Demonstrate the procedures to protect samples and evidence from secondary contamination.
(8) Demonstrate the correct techniques to collect samples utilizing the equipment provided.
(9) Demonstrate the documentation procedures.
(10) Demonstrate the sampling protocols.
(11) Demonstrate field screening protocols for samples and evidence collected.
(12) Demonstrate evidence labeling and packaging procedures.
(13) Demonstrate evidence decontamination procedures.
(14) Demonstrate evidence packaging procedures for evidence transportation.

6.5.4.2 The operations level responder assigned to evidence preservation and sampling shall describe local procedures for the technical decontamination process.

Refer to the mission-specific competencies for decontamination in Section 6.3.

6.5.5 Competencies — Implementing the Planned Response. (Reserved)

6.5.6 Competencies — Terminating the Incident. (Reserved)

6.6 Mission-Specific Competencies: Product Control

6.6.1 General.

6.6.1.1 Introduction.

Product control has been moved to the mission-specific competencies in the 2008 edition of the standard. Most law enforcement responders do not need to be trained to implement product control as a core competency, but it is expected that most AHJs will require fire fighters to be trained to mission-specific competencies product control.

6.6.1.1.1 The operations level responder assigned to perform product control shall be that person, competent at the operations level, who is assigned to implement product control measures at hazardous materials/WMD incidents.

6.6.1.1.2 The operations level responder assigned to perform product control at hazardous materials/WMD incidents shall be trained to meet all competencies at the awareness level (Chapter 4), all core competencies at the operations level (Chapter 5), all mission-specific competencies for personal protective equipment (Section 6.2), and all competencies in this section.

6.6.1.1.3 The operations level responder assigned to perform product control at hazardous materials/WMD incidents shall operate under the guidance of a hazardous materials technician, an allied professional, or standard operating procedures.

Although some of the mission-specific competencies in this section are taken from Chapter 7, which covers the hazardous materials technician (HMT), the technical committee wants to clearly state that first responder operational (FRO) level personnel with a mission-specific competency is not a replacement for an HMT. FRO personnel with a mission-specific competency can perform some technician-level skills but do not have the broader skills and competencies required of an HMT, particularly regarding risk assessment and the selection of control options. The following two options are examples of how guidance can be provided to ensure that FRO personnel do not go beyond their level of training and equipment:

1. Direct Guidance: FRO personnel are working under the control of an HMT or allied professional who has the ability (1) to continuously assess and/or observe their actions and (2) provide immediate feedback. Guidance by an HMT or an allied professional can be provided through direct visual observation or through assessment reports communicated by the FRO to them.
2. Written Guidance: Written standard operating procedures or similar guidance clearly states the "rules of engagement" for FRO personnel with the mission-specific competency. Emphasis should be placed upon (1) tasks expected of FRO personnel, (2) tasks beyond the capability of FRO personnel, (3) required PPE and equipment to perform these expected tasks, and (4) procedures for ensuring coordination within the local ICS.

6.6.1.1.4* The operations level responder assigned to perform product control at hazardous materials/WMD incidents shall receive the additional training necessary to meet specific needs of the jurisdiction.

A.6.6.1.1.4 See A.6.3.1.1.4.

6.6.1.2 Goal.

6.6.1.2.1 The goal of the competencies in this section shall be to provide the operations level responder assigned to product control at hazardous materials/WMD incidents with the knowledge and skills to perform the tasks in 6.6.1.2.2 safely and effectively.

6.6.1.2.2 When responding to hazardous materials/WMD incidents, the operations level responder assigned to perform product control shall be able to perform the following tasks:

(1) Plan an initial response within the capabilities and competencies of available personnel, personal protective equipment, and control equipment and in accordance with the emergency response plan or standard operating procedures by completing the following tasks:
   (a) Describe the control options available to the operations level responder.
   (b) Describe the control options available for flammable liquid and flammable gas incidents.
(2) Implement the planned response to a hazardous materials/WMD incident.

6.6.2 Competencies — Analyzing the Incident. (Reserved)

6.6.3 Competencies — Planning the Response.

6.6.3.1 Identifying Control Options. Given examples of hazardous materials/WMD incidents, the operations level responder assigned to perform product control shall identify the options for each response objective and shall meet the following requirements as prescribed by the AHJ:

(1) Identify the options to accomplish a given response objective.
(2) Identify the purpose for and the procedures, equipment, and safety precautions associated with each of the following control techniques:
   (a) Absorption
   (b) Adsorption
   (c) Damming
   (d) Diking
   (e) Dilution
   (f) Diversion
   (g) Remote valve shutoff
   (h) Retention
   (i) Vapor dispersion
   (j) Vapor suppression

Operations level responders with this mission-specific competency need to understand the advantages and limitations of each of the techniques described earlier. Their application is dependent on the chemical and physical properties of the product as well as the environmental factors. Each of the methods has a specific methodology to be used when making the application. A good example of this would be the appropriate type, use, and application of Class B foam. There is a concern today about the use of the wrong foam on ethanol blended fuels versus on standard hydrocarbon fires.

6.6.3.2 Selecting Personal Protective Equipment. The operations level responder assigned to perform product control shall select the personal protective equipment required to support product control at hazardous materials/WMD incidents based on local procedures *(see Section 6.2)*.

Personnel who are trained to this competency should understand that the type of PPE worn at an incident requiring product control is dependent on the proximity to the release. For example, if the responder is constructing underflow dams, overflow dams, or retention basins and is far enough away from the release, he or she should only wear the appropriate safety equipment for the hazards. Whereas if a responder is dealing with a point release and using adsorbents or absorbent, the proximity to the spill would require the need for PPE, depending on the product and the equipment provided by the AHJ.

6.6.4 Competencies — Implementing the Planned Response.

6.6.4.1 Performing Control Options. Given an incident action plan for a hazardous materials/WMD incident, within the capabilities and equipment provided by the AHJ, the operations level responder assigned to perform product control shall demonstrate control functions set out in the plan and shall meet the following requirements as prescribed by the AHJ:

(1) Using the type of special purpose or hazard suppressing foams or agents and foam equipment furnished by the AHJ, demonstrate the application of the foam(s) or agent(s) on a spill or fire involving hazardous materials/WMD.
(2) Identify the characteristics and applicability of the following Class B foams if supplied by the AHJ:
    (a) Aqueous film-forming foam (AFFF)
    (b) Alcohol-resistant concentrates
    (c) Fluoroprotein
    (d) High-expansion foam
(3) Given the required tools and equipment, demonstrate how to perform the following control activities:
    (a) Absorption
    (b) Adsorption
    (c) Damming
    (d) Diking
    (e) Dilution
    (f) Diversion
    (g) Retention
    (h) Remote valve shutoff
    (i) Vapor dispersion
    (j) Vapor suppression
(4) Identify the location and describe the use of emergency remote shutoff devices on MC/DOT-306/406, MC/DOT-307/407, and MC-331 cargo tanks containing flammable liquids or gases.
(5) Describe the use of emergency remote shutoff devices at fixed facilities.

6.6.4.2 The operations level responder assigned to perform product control shall describe local procedures for going through the technical decontamination process.

6.6.5 Competencies — Evaluating Progress. (Reserved)

6.6.6 Competencies — Terminating the Incident. (Reserved)

6.7 Mission-Specific Competencies: Air Monitoring and Sampling

6.7.1 General.

6.7.1.1 Introduction.

6.7.1.1.1 The operations level responder assigned to perform air monitoring and sampling shall be that person, competent at the operations level, who is assigned to implement air-monitoring and sampling operations at hazardous materials/WMD incidents.

6.7.1.1.2 The operations level responder assigned to perform air monitoring and sampling at hazardous materials/WMD incidents shall be trained to meet all competencies at the awareness level (Chapter 4), all core competencies at the operations level (Chapter 5), all mission-specific competencies for personal protective equipment (Section 6.2), and all competencies in this section.

6.7.1.1.3 The operations level responder assigned to perform air monitoring and sampling at hazardous materials/WMD incidents shall operate under the guidance of a hazardous materials technician, an allied professional, or standard operating procedures.

Although some of the mission-specific competencies in this section are taken from Chapter 7, which covers the hazardous materials technician, the technical committee wants to clearly state that FRO personnel with a mission-specific competency is not a replacement for an HMT. FRO personnel with a mission-specific competency can perform some technician-level skills but do not have the broader skills and competencies required of an HMT, particularly regarding risk assessment and the selection of control options. The following two options are examples of how guidance can be provided to ensure that FRO personnel do not go beyond their level of training and equipment:

1. Direct Guidance: FRO personnel are working under the control of an HMT or allied professional who has the ability (1) to continuously assess and/or observe their actions and (2) provide immediate feedback. Guidance by an HMT or an allied professional can be provided through direct visual observation or through assessment reports communicated by the FRO to them.
2. Written Guidance: Written standard operating procedures or similar guidance clearly states the "rules of engagement" for FRO personnel with the mission-specific competency. Emphasis should be placed upon (1) tasks expected of FRO personnel, (2) tasks beyond the capability of FRO personnel, (3) required PPE and equipment to perform these expected tasks, and (4) procedures for ensuring coordination within the local ICS.

6.7.1.1.4* The operations level responder assigned to perform air monitoring and sampling at hazardous materials/WMD incidents shall receive the additional training necessary to meet specific needs of the jurisdiction.

Air monitoring equipment is equipment used to detect or measure amounts of hazardous materials/WMD agents. The equipment that could be expected to be used by operations trained responders include the following:

1. Carbon monoxide meter
2. Colorimetric tubes
3. Combustible gas indicator
4. Oxygen meter
5. Passive dosimeters
6. pH indicators and/or pH meters
7. Photoionization and/or flame ionization detectors
8. Radiation detection instruments
9. Reagents
10. Test strips
11. WMD detectors (chemical and/or biological)
12. Other equipment provided by the AHJ

Sampling equipment that can be used under this competency is intended to be used for environmental sampling and not for evidence collection. If it is intended for evidence sampling and collection, the FRO must be trained to mission-specific competency of Section 6.5, which covers evidence preservation and sampling. The sampling equipment that can be expected to be used by operations trained responders, based on AHJ requirements, includes but is not limited to the following:

1. Any tool designated to remove liquid or solid product from a container for the purpose of environmental sampling and testing
2. Any container suitable for the collection of a liquid or solid sample based on the type and quantity

A.6.7.1.1.4 See A.6.3.1.1.4.

6.7.1.2 Goal.

Mission-specific competencies for air monitoring and sampling were added to this standard with the thought that operations trained responders are now expected to operate air monitoring equipment, based on AHJ requirements, under the guidance of HMTs, allied professionals, or standard operating procedures. Prior to this update, operations trained responders had no standardized competencies when the AHJ required air monitoring and/or sampling at hazardous materials incidents.

6.7.1.2.1 The goal of the competencies in this section shall be to provide the operations level responder assigned to air monitoring and sampling at hazardous materials/WMD incidents with the knowledge and skills to perform the tasks in 6.7.1.2.2 safely and effectively.

6.7.1.2.2 When responding to hazardous materials/WMD incidents, the operations level responder assigned to perform air monitoring and sampling shall be able to perform the following tasks:

(1) Plan the air monitoring and sampling activities within the capabilities and competencies of available personnel, personal protective equipment, and control equipment and in accordance with the emergency response plan or standard operating procedures describe the air monitoring and sampling options available to the operations level responder.
(2) Implement the air monitoring and sampling activities as specified in the incident action plan.

Commentary Table I.6.2 can assist the responder in selecting the appropriate air monitoring equipment and sampling protocols.

6.7.2 Competencies — Analyzing the Incident. (Reserved)

6.7.3 Competencies — Planning the Response.

6.7.3.1 Given the air monitoring and sampling equipment provided by the AHJ, the operations level responder assigned to perform air monitoring and sampling shall select the detection or monitoring equipment suitable for detecting or monitoring solid, liquid, or gaseous hazardous materials/WMD.

6.7.3.2 Given detection and monitoring device(s) provided by the AHJ, the operations level responder assigned to perform air monitoring and sampling shall describe the operation, capabilities and limitations, local monitoring procedures, field testing, and maintenance procedures associated with each device.

6.7.3.3 Selecting Personal Protective Equipment. The operations level responder assigned to perform air monitoring and sampling shall identify the local procedures for selecting personal protective equipment to support air monitoring and sampling at hazardous materials/WMD incidents.

6.7.3.4 Selecting Personal Protective Equipment. The operations level responder assigned to perform air monitoring and sampling shall select the personal protective equipment required to support air monitoring and sampling at hazardous materials/WMD incidents based on local procedures *(see Section 6.2)*.

It is expected that operations level responders will, as a prerequisite, be trained to meet the requirements of Section 6.2, Mission-Specific Competencies: Personal Protective Equipment.

COMMENTARY TABLE I.6.2 *Air Monitoring Equipment and Sampling Protocols Selection Table*

	Hazard Class												
Characteristics	***1***	***2***			***3***	***4***			***5***		***6***	***7***	***8***
Division	All	2.1	2.2	2.3	—	4.1	4.2	4.3	5.1	5.2	—	—	—
Can it be detected	No	Yes	Yes Gas Specific	Yes Gas Specific	Yes Gas Specific						Yes	Yes	Yes
Can it be monitored	No	Yes	Yes	Yes	Yes						Yes	Yes	Yes
Units, Note*: Also found as micro, mille, or sieverts	—	% LEL	O_2 - % Others – PPM	PPM	% LEL						PPM	Atmospheric* •R/hr mR/hr R/hr Individual dose* •R mR R	pH 0 - 14
Equipment used to detect/ monitor	—	CGI PID	Electro – Chemical cell Colorimetric Tube Specific Sensor	Electro – Chemical cell Colorimetric Tube Specific Sensor	Electro – Chemical cell Colorimetric Tube Specific Sensor						Specific test kits Pesticide PCB Chlorine	G/M tube Alpha scintillators Gamma scintillators	pH paper pH meter

6.7.4 Competencies — Implementing the Planned Response.

6.7.4.1 Given a scenario involving hazardous materials/WMD and detection and monitoring devices provided by the AHJ, the operations level responder assigned to perform air monitoring and sampling shall demonstrate the field test and operation of each device and interpret the readings based on local procedures.

The personnel assigned to this task—when given three hazardous materials/WMD, one of which is a solid, one a liquid, and one a gas, and using the following monitoring equipment, test strips, and reagents—need to be able to select the equipment and demonstrate the correct techniques to identify the hazards (corrosivity, flammability, oxygen content, oxygen deficiency, radioactivity, toxicity, and pathogenicity) using the following equipment:

1. Carbon monoxide meter
2. Colorimetric tubes
3. Combustible gas indicator
4. Oxygen meter
5. Passive dosimeters
6. pH indicators and/or pH meters
7. Photoionization
8. Flame ionization detectors
9. Radiation detection instruments
10. Reagents
11. Test strips

12. WMD detectors (chemical and/or biological)
13. Other equipment provided by the AHJ

6.7.4.2 The operations level responder assigned to perform air monitoring and sampling shall describe local procedures for decontamination of themselves and their detection and monitoring devices upon completion of the air monitoring mission.

Operations level trained responders will also be expected to meet the requirements of Section 6.4, Mission-Specific Competencies: Technical Decontamination, which will train them on the proper decontamination of themselves and equipment.

6.7.5 Competencies — Evaluating Progress. (Reserved)

6.7.6 Competencies — Terminating the Incident. (Reserved)

6.8 Mission-Specific Competencies: Victim Rescue and Recovery

6.8.1 General.

6.8.1.1 Introduction.

6.8.1.1.1 The operations level responder assigned to perform victim rescue and recovery shall be that person, competent at the operations level, who is assigned to rescue and recover exposed and contaminated victims at hazardous materials/WMD incidents.

Operations level responders conducting victim rescue missions might be fire fighters, HMTs, EMS personnel, law enforcement officers, other trained personnel (e.g., industrial or transportation carrier employees), or a combination of those personnel assembled into a team under a predetermined response plan by the AHJ. Victim rescue missions can have a different meaning for each discipline based upon the given hazardous materials/WMD incident, but should all be conducted on a risk-based model.

6.8.1.1.2 The operations level responder assigned to perform victim rescue and recovery at hazardous materials/WMD incidents shall be trained to meet all competencies at the awareness level (Chapter 4), all core competencies at the operations level (Chapter 5), all mission-specific competencies for personal protective equipment (Section 6.2), and all competencies in this section.

6.8.1.1.3 The operations level responder assigned to perform victim rescue and recovery at hazardous materials/WMD incidents shall operate under the guidance of a hazardous materials technician, an allied professional, or standard operating procedures.

Although some of the mission-specific competencies in this section are taken from Chapter 7, which covers the hazardous materials technician, the technical committee wants to clearly state that FRO personnel with a mission-specific competency is not a replacement for an HMT. FRO personnel with a mission-specific competency can perform some technician-level skills but do not have the broader skills and competencies required of an HMT, particularly regarding risk assessment and the selection of control options. The following two options are examples of how guidance can be provided to ensure that FRO personnel do not go beyond their level of training and equipment:

1. Direct Guidance: FRO personnel are working under the control of an HMT or allied professional who has the ability (1) to continuously assess and/or observe their actions, and (2) provide immediate feedback. Guidance by a HMT or an allied professional can be provided through direct visual observation or through assessment reports communicated by the FRO to them.
2. Written Guidance: Written standard operating procedures or similar guidance clearly states the "rules of engagement" for FRO personnel with the mission-specific competency.

Emphasis should be placed upon (1) tasks expected of FRO personnel, (2) tasks beyond the capability of FRO personnel, (3) required PPE and equipment to perform these expected tasks, and (4) procedures for ensuring coordination within the local ICS.

6.8.1.1.4* The operations level responder assigned to perform victim rescue and recovery at hazardous materials/WMD incidents shall receive the additional training necessary to meet specific needs of the jurisdiction.

A.6.8.1.1.4 See A.6.3.1.1.4.

6.8.1.2 Goal.

6.8.1.2.1 The goal of the competencies in this section shall be to provide the operations level responder assigned victim rescue and recovery at hazardous materials/WMD incidents with the knowledge and skills to perform the tasks in 6.8.1.2.2 safely and effectively.

6.8.1.2.2 When responding to hazardous materials/WMD incidents, the operations level responder assigned to perform victim rescue and recovery shall be able to perform the following tasks:

(1) Plan a response for victim rescue and recovery operations involving the release of hazardous materials/WMD agent within the capabilities of available personnel and personal protective equipment.
(2) Implement the planned response to accomplish victim rescue and recovery operations within the capabilities of available personnel and personal protective equipment.

The personnel involved in victim rescue/recovery should be proficient in the use of the appropriate type of protective clothing offered and should meet the requirements of Section 6.2.

6.8.2 Competencies — Analyzing the Incident. (Reserved)

6.8.3 Competencies — Planning the Response.

6.8.3.1 Given scenarios involving hazardous materials/WMD incidents, the operations level responder assigned to victim rescue and recovery shall determine the feasibility of conducting victim rescue and recovery operations at an incident involving a hazardous material/WMD and shall be able to perform the following tasks:

(1) Determine the feasibility of conducting rescue and recovery operations.
(2) Describe the safety procedures, tactical guidelines, and incident response considerations to effect a rescue associated with each of the following situations:
    (a) Line-of-sight with ambulatory victims
    (b) Line-of-sight with nonambulatory victims
    (c) Non-line-of-sight with ambulatory victims
    (d) Non-line-of-sight with nonambulatory victims
    (e) Victim rescue operations versus victim recovery operations

Personnel planning a victim rescue operation should consider all hazards that are or could potentially be present during the incident. Efforts should be made to identify whether hostile human threats, improvised explosive devices, or other devices utilized in an intentional release incident are still present, simultaneously with evaluation of the agent type and potential harm. Additional consideration should also include the following precautions:

1. Emergency responders will enter potentially contaminated areas only to perform rescue of known live victims or to perform an immediate reconnaissance to determine if live victims exist.
2. Emergency responders will immediately exit any area where they encounter evidence of chemical contamination and cannot identify any living victims.
3. Emergency responders will avoid contact with any unidentified materials.

4. Emergency responders and rescued victims will undergo an emergency decontamination immediately upon exit from the potentially hazardous area.
5. Immediate medical assistance such as that provided by EMS providers is immediately available.
6. Emergency responders, when finding conditions in excess of immediately dangerous to life or health (IDLH) should attempt to change the environment (ventilation, vapor dispersion/suppression, etc.) to enable others to respond to assist.
7. While reducing the hazards to create a safer environment in which to operate is always a good work practice, it is essential when performing victim recovery.

(3) Determine if the options are within the capabilities of available personnel and personal protective equipment.
(4) Describe the procedures for implementing victim rescue and recovery operations within the incident command system.

6.8.3.2 Selecting Personal Protective Equipment. The operations level responder assigned to perform victim rescue and recovery shall select the personal protective equipment required to support victim rescue and recovery at hazardous materials/WMD incidents based on local procedures *(see Section 6.2).*

Personnel who are to perform duties as described in this competency must have completed the mission-specific competencies for PPE in Section 6.2.

6.8.4 Competencies — Implementing the Planned Response.

6.8.4.1 Given a scenario involving a hazardous material/WMD, the operations level responder assigned to victim rescue and recovery shall perform the following tasks:

(1) Identify the different team positions and describe their main functions.
(2) Select and use specialized rescue equipment and procedures provided by the AHJ to support victim rescue and recovery operations.
(3) Demonstrate safe and effective methods for victim rescue and recovery.
(4) Demonstrate the ability to triage victims.
(5) Describe local procedures for performing decontamination upon completion of the victim rescue and removal mission.

6.8.5 Competencies — Evaluating Progress. (Reserved)

6.8.6 Competencies — Terminating the Incident. (Reserved)

6.9 Mission-Specific Competencies: Response to Illicit Laboratory Incidents

6.9.1 General.

6.9.1.1 Introduction.

6.9.1.1.1 The operations level responder assigned to respond to illicit laboratory incidents shall be that person, competent at the operations level, who, at hazardous materials/WMD incidents involving potential violations of criminal statutes specific to the illegal manufacture of methamphetamines, other drugs, or WMD, is assigned to secure the scene, identify the laboratory or process, and preserve evidence at hazardous materials/WMD incidents involving potential violations of criminal statutes specific to the illegal manufacture of methamphetamines, other drugs, or WMD.

This mission-specific competency is designed to prepare the operations level responder to support law enforcement operations in the identification of illicit laboratories and evidence preservation tasks at a crime scene. The competency is designed to raise the awareness of responders and to prepare them to assist crime scene investigators during the performance of their jobs.

Additional guidance in response to illicit laboratories can be obtained from the local office of the FBI's Weapons of Mass Destruction Coordinator, the local office of the DEA, state or local law enforcement, or governmental sponsored training programs.

6.9.1.1.2 The operations level responder who responds to illicit laboratory incidents shall be trained to meet all competencies at the awareness level (Chapter 4), all core competencies at the operations level (Chapter 5), all mission-specific competencies for personal protective equipment (Section 6.2), and all competencies in this section.

6.9.1.1.3 The operations level responder who responds to illicit laboratory incidents shall operate under the guidance of a hazardous materials technician, an allied professional, or standard operating procedures.

Although some of the mission-specific competencies in this section are taken from Chapter 7, Competencies for Hazardous Materials Technicians, the technical committee wants to clearly state that operations level responders with a mission-specific competency are not replacements for an HMT. Operations level responders with a mission-specific competency can perform some technician-level skills but do not have the broader skills and competencies required of an HMT, particularly regarding risk assessment and the selection of control options. The following two options are examples of how guidance can be provided to ensure that operations level responders do not go beyond their level of training and equipment:

- Direct Guidance: Operations level responders are working under the control of an HMT or allied professional who has the ability to (1) continuously assess and/or observe their actions and (2) provide immediate feedback. Guidance by an HMT or an allied professional can be provided through direct visual observation or through assessment reports communicated by the operations level responder to them.
- Written Guidance: Written standard operating procedures or similar guidance clearly states the "rules of engagement" for operations level responders with the mission-specific competency. Emphasis should be placed on the following:
  1. Tasks expected of operations level responders
  2. Tasks beyond the capability of operations level responders
  3. Required PPE and equipment to perform these expected tasks
  4. Procedures for ensuring coordination within the local ICS

6.9.1.1.4* The operations level responder who responds to illicit laboratory incidents shall receive the additional training necessary to meet specific needs of the jurisdiction.

A.6.9.1.1.4 See A.6.3.1.1.4.

6.9.1.2 Goal.

6.9.1.2.1 The goal of the competencies in this section shall be to provide the operations level responder assigned to respond to illicit laboratory incidents with the knowledge and skills to perform the tasks in 6.9.1.2.2 safely and effectively.

6.9.1.2.2 When responding to hazardous materials/WMD incidents, the operations level responder assigned to respond to illicit laboratory incidents shall be able to perform the following tasks:

(1) Analyze a hazardous materials/WMD incident to determine the complexity of the problem and potential outcomes and whether the incident is potentially a criminal illicit laboratory operation.
(2) Plan a response for a hazardous materials/WMD incident involving potential illicit laboratory operations in compliance with evidence preservation operations within the capabilities and competencies of available personnel, personal protective equipment, and control equipment after notifying the responsible law enforcement agencies of the problem.
(3) Implement the planned response to a hazardous materials/WMD incident involving potential illicit laboratory operations utilizing applicable evidence preservation guidelines.

6.9.2 Competencies — Analyzing the Incident.

6.9.2.1 Determining If a Hazardous Materials/WMD Incident Is an Illicit Laboratory Operation. Given examples of hazardous materials/WMD incidents involving illicit laboratory operations, the operations level responder assigned to respond to illicit laboratory incidents shall identify the potential drugs/WMD being manufactured and shall meet the following related requirements:

Illicit laboratories can be designed for the production of many different end products. Examples include illegal drugs such as methamphetamine, chemical modification such as distillation of pesticides, and biological toxins or pathogens. Identification of the specific materials being produced is based upon a scientific assessment of the tools and materials utilized in the process combined with law enforcement intelligence. Various specialized teams exist to assist in this process and support these types of operations. Examples of these teams include DEA Clandestine Lab Teams, local or state law enforcement lab teams for illegal manufacture of drugs, and the FBI Hazardous Materials Response Unit for manufacture of WMD materials.

Operators of these illicit laboratories often have a vested interest in the protection of their product and eluding law enforcement. There is a risk that the operator will be present within the laboratory, with access to weapons. The human element must be addressed by law enforcement tactical teams specifically trained to operate within a hazardous environment. The level of protection for the tactical team should be based upon an assessment of the intelligence, information on the materials being produced, and information including any protective clothing used by the operator, activity of animals in the laboratory, and interviews with neighbors. Additionally, there is a risk of anti-personnel devices (booby traps) around and within the laboratory. The operations level responder must leave clearance of potential anti-personnel devices to EOD personnel trained for these procedures. Once the human and EOD hazards have been cleared, the operations level responder should rely on the hazard risk assessment provided by the HMT or allied professional to determine the potential harm from hazardous materials/WMD within the illicit laboratory. Evidence recovery and forensic operations must be conducted by a specialized team trained for these specific operations. All site activities are guided by the law enforcement agency having jurisdiction, within the parameters established by the search warrant or protocol in effect authorizing the seizure of the illicit laboratory. Decontamination procedures should be based upon the results of the hazard risk assessment.

(1) Given examples of illicit drug manufacturing methods, describe the operational considerations, hazards, and products involved in the illicit process.

The processes involved in illicit drug production change frequently, and as such response agencies must maintain frequent interaction with law enforcement drug response teams. Response agencies should maintain situational awareness by receiving a threat briefing from law enforcement concerning the anticipated threats from hazardous materials/WMD incidents. These agencies should be aware of potential hazards involving illicit drug production, including but not limited to flammable gases, flammable solvents, and toxic by inhalation and/or by dermal contact materials.

(2) Given examples of illicit chemical WMD methods, describe the operational considerations, hazards, and products involved in the illicit process.

Illicit chemical WMD methods include such procedures as distillation of organophosphate pesticides or the use of acids and cyanides. Identifying the products and methods that can be utilized in a WMD chemical process is complex. Special attention should be given to materials that are commonly found or acquired either in transportation or at fixed facilities, such as industrial chemicals and toxins.

(3) Given examples of illicit biological WMD methods, describe the operational considerations, hazards, and products involved in the illicit process.

Illicit biological WMD methods include such procedures as the processing of castor beans to refine the protein toxin ricin and the culture of bacterial or viral agents. Identifying the products and methods that might be utilized in a WMD biological process is complex. Special attention should be given to scenes involving the discovery of potential biological laboratory equipment, such as incubators, fermentors, and agar growth plates.

(4) Given examples of illicit laboratory operations, describe the potential booby traps that have been encountered by response personnel.

Specific information on potential booby traps should be obtained from EOD or bomb squad teams. The information should be utilized for awareness only. Searching for and dismantling booby traps should only be done by EOD teams specifically trained in this task.

(5) Given examples of illicit laboratory operations, describe the agencies that have investigative authority and operational responsibility to support the response.

Law enforcement jurisdiction, investigative guidelines, and investigative priorities are complex and dynamic. Specific jurisdictional situations should be identified with the responder's local or state law enforcement authorities and federal investigative agencies such as the FBI, DEA, the United States Postal Inspection Service, and the EPA.

6.9.3 Competencies — Planning the Response.

6.9.3.1 Determining the Response Options. Given an analysis of hazardous materials/WMD incidents involving illicit laboratories, the operations level responder assigned to respond to illicit laboratory incidents shall identify possible response options.

6.9.3.2 Identifying Unique Aspects of Criminal Hazardous Materials/WMD Incidents.

6.9.3.2.1 The operations level responder assigned to respond to illicit laboratory incidents shall identify the unique operational aspects associated with illicit drug manufacturing and illicit WMD manufacturing.

6.9.3.2.2 Given an incident involving illicit drug manufacturing or illicit WMD manufacturing, the operations level responder assigned to illicit laboratory incidents shall describe the following tasks:

(1) Law enforcement securing and preserving the scene

These tasks include neutralization of any tactical threat, rendering safe any EOD or booby traps, full accountability and identification of all personnel in the crime scene, documentation of any items disturbed within the crime scene, and protection of evidence from potential damage or destruction.

(2) Joint hazardous materials and EOD personnel site reconnaissance and hazard identification

Hazmat and EOD teams will need to work together to resolve situations found within illicit drug or WMD laboratories—for example, explosive/chemical, explosive/radiological, or explosive/biological devices and materials; clearance of booby traps; and characterization of

EOD and hazmat hazards. Responders should develop liaison between EOD and hazardous materials personnel in preparation for such events.

(3) Determining atmospheric hazards through air monitoring and detection

Assessment of atmospheric hazards is a primary task performed as part of the site hazard risk assessment. The detection and monitoring equipment selected to determine atmospheric hazards must be capable of assisting in the performance of the hazard risk assessment and, at a minimum, should include a combustible gas indicator, oxygen level meter, photoionization meter, pH paper, and radiological monitoring equipment.

(4) Mitigation of immediate hazards while preserving evidence

Standard priorities of scene operations apply during responses to illicit drug or WMD laboratories, specifically, life safety, incident stabilization, and preservation of property. In the course of reconnaissance of the laboratory, responders should assess immediate hazards to life and the safety of responders and investigators. Responders should liaison with the appropriate law enforcement agency prior to reconnaissance operations in order to be briefed on intelligence involving the laboratory or site.

(5) Coordinated crime scene operation with the law enforcement agency having investigative authority

When responding to potential illicit drug or WMD laboratories, the operations level responder should be observant for signs of criminal activity involving chemical, biological, radiological, or explosive materials or devices. The operations level responder should be familiar with the jurisdictional procedures for operation within a crime scene, to include investigative law enforcement leadership, search warrant requirements, rules of evidence, crime scene documentation, policies regarding photography, evidence custodial requirements, chain of custody, and specific requirements set forth by the prosecuting attorney.

The operations level responder should be an advocate for the law enforcement operation, and support the operation within their scope of training.

(6) Documenting personnel and scene activities associated with incident

Local procedures vary for crime scene documentation. For example, some jurisdictions do not allow the use of video documentation. The operations level responder should become familiar with the local or state and federal procedures for documentation at a crime scene.

6.9.3.3 Identifying the Law Enforcement Agency That Has Investigative Jurisdiction. The operations level responder assigned to respond to illicit laboratory incidents shall identify the law enforcement agency having investigative jurisdiction and shall meet the following requirements:

Law enforcement jurisdiction, investigative guidelines, and investigative priorities are complex and dynamic. Specific jurisdictional situations should be identified with the responder's local or state law enforcement authorities and federal investigative agencies such as the FBI, DEA, the United States Postal Inspection Service, and the EPA.

(1) Given scenarios involving illicit drug manufacturing or illicit WMD manufacturing, identify the law enforcement agency(s) with investigative authority for the following situations:
   - (a) Illicit drug manufacturing
   - (b) Illicit WMD manufacturing
   - (c) Environmental crimes resulting from illicit laboratory operations

6.9.3.4 Identifying Unique Tasks and Operations at Sites Involving Illicit Laboratories.

6.9.3.4.1 The operations level responder assigned to respond to illicit laboratory incidents shall identify and describe the unique tasks and operations encountered at illicit laboratory scenes.

Unique tasks include clearance of hostile suspects, clearance of EOD devices or materials, clearance of anti-personnel devices (booby traps), performance of tactical decontamination, acquisition of warrants and affidavits, isolation of chemical reactions, collection of evidence, and site remediation.

6.9.3.4.2 Given scenarios involving illicit drug manufacturing or illicit WMD manufacturing, describe the following:

(1) Hazards, safety procedures, and tactical guidelines for this type of emergency

Illicit laboratories might be designed for the production of many different end products, for example, illegal drugs such as methamphetamine, chemical modification such as distillation of pesticides, and biological toxins or pathogens. Operators of these illicit laboratories often have a vested interest in the protection of their product and evading law enforcement. There is a risk that the operator will be present within the laboratory and have access to weapons. The human element must be addressed by law enforcement tactical teams specifically trained to operate within a hazardous environment. Additionally, there is a risk of anti-personnel devices (booby traps) around and within the laboratory. The operations level responder must leave clearance of potential anti-personnel devices to EOD personnel trained for these procedures. Once the human and EOD hazards have been cleared, the operations level responder should conduct a hazard risk assessment to determine the potential harm from hazardous materials/WMD within the illicit laboratory.

(2) Factors to be evaluated in selection of the appropriate personal protective equipment for each type of tactical operation

Selection of PPE is based upon assessment of available intelligence, outward warning signs, and detection clues and includes any protective clothing used by the operator, activity of animals in the laboratory, and interviews with neighbors. Law enforcement activities may require PPE designed for tactical law enforcement operations; however, the PPE must be evaluated for appropriate protection from the anticipated hazards as identified during the hazard risk assessment. EOD operations will require the appropriate level of EOD protective garment, augmented by chemical protective clothing appropriate for the anticipated hazard as identified during the hazard risk assessment.

(3) Factors to be considered in selection of appropriate decontamination procedures

Decontamination procedures should be based upon the results of the hazard risk assessment. Dynamic tactical entries may require the use of tactical decontamination procedures. Tactical decontamination is predicated upon rapid deployment and a focus on reception of four potential situations: uninjured tactical operators and their equipment, injured tactical operators, uninjured suspects, and injured suspects. Operations level responders must coordinate decontamination procedures with law enforcement tactical teams to resolve potential issues, such as live ammunition, pyrotechnic devices, and custody of suspects.

(4) Factors to be evaluated in the selection of detection devices

The detection and monitoring equipment selected must be capable of assisting in the performance of the hazard risk assessment, and at a minimum should include a combustible gas indicator, oxygen level meter, photoionization meter, pH paper, and radiological monitoring equipment.

(5) Factors to be considered in the development of a remediation plan

The operations level responder must become familiar with local or state and federal agency policies concerning the remediation of illicit drug/WMD scenes. Potential sources of assistance include local or state health and/or environmental departments, emergency management agencies, and federal agencies such as the DEA or the EPA.

6.9.3.5 Selecting Personal Protective Equipment. The operations level responder assigned to respond to illicit laboratory incidents shall select the personal protective equipment required to respond to illicit laboratory incidents based on local procedures.

Refer to mission-specific competencies for PPE in Section 6.2.

6.9.4 Competencies — Implementing the Planned Response.

6.9.4.1 Implementing the Planned Response. Given scenarios involving an illicit drug/WMD laboratory operation involving hazardous materials/WMD, the operations level responder assigned to respond to illicit laboratory incidents shall implement or oversee the implementation of the selected response options safely and effectively.

6.9.4.1.1 Given a simulated illicit drug/WMD laboratory incident, the operations level responder assigned to respond to illicit laboratory incidents shall be able to perform the following tasks:

(1) Describe safe and effective methods for law enforcement to secure the scene.
(2) Demonstrate decontamination procedures for tactical law enforcement personnel (SWAT or K-9) securing an illicit laboratory.
(3) Demonstrate methods to identify and avoid potential unique safety hazards found at illicit laboratories such as booby traps and releases of hazardous materials.
(4) Demonstrate methods to conduct joint hazardous materials/EOD operations to identify safety hazards and implement control procedures.

6.9.4.1.2 Given a simulated illicit drug/WMD laboratory entry operation, the operations level responder assigned to respond to illicit laboratory incidents shall demonstrate methods of identifying the following during reconnaissance operations:

(1) The potential manufacture of illicit drugs
(2) The potential manufacture of illicit WMD materials
(3) Potential environmental crimes associated with the manufacture of illicit drugs/WMD materials

6.9.4.1.3 Given a simulated illicit drug/WMD laboratory incident, the operations level responder assigned to respond to illicit laboratory incidents shall describe joint agency crime scene operations, including support to forensic crime scene processing teams.

6.9.4.1.4 Given a simulated illicit drug/WMD laboratory incident, the operations level responder assigned to respond to illicit laboratory incidents shall describe the policy and procedures for post–crime scene processing and site remediation operations.

6.9.4.1.5 The operations level responder assigned to respond to illicit laboratory incidents shall be able to describe local procedures for performing decontamination upon completion of the illicit laboratory mission.

6.9.5 Competencies — Evaluating Progress. (Reserved)

6.9.6 Competencies — Terminating the Incident. (Reserved)

REFERENCES CITED IN COMMENTARY

1. NFPA 473, *Standard for Competencies for EMS Personnel Responding to Hazardous Materials/Weapons of Mass Destruction Incidents*, National Fire Protection Association, Quincy MA, 2008.

Additional References

NFPA 1981, *Standard on Open-Circuit Self-Contained Breathing Apparatus (SCBA) for Emergency Services*, 2007.

NFPA 1991, *Standard on Vapor-Protective Ensembles for Hazardous Materials Emergencies,* National Fire Protection Association, Quincy MA, 2005.

NFPA 1992, *Standard on Liquid Splash-Protective Ensembles and Clothing for Hazardous Materials Emergencies*, National Fire Protection Association, Quincy MA, 2005.

NFPA 1994, *Standard on Protective Ensembles for First Responders to CBRN Terrorist Incidents*, National Fire Protection Association, Quincy MA, 2007.

Chemical Weapons Improved Response Program Report, U.S. Army Soldier and Biological Command (SBCCOM, now known as RDECOM), Aberdeen Proving Ground, MD.

Competencies for Hazardous Materials Technicians

CHAPTER 7

The technician level competencies are achieved through more advanced training and build upon the skills required at the awareness and operations levels. Training programs at this level are required to ensure that trainees have first mastered the competencies in Chapter 4 and Chapter 5 of NFPA 472. The hazardous materials technician is referred to as an HMT throughout the commentary of this chapter.

7.1 General

7.1.1 Introduction.

7.1.1.1 The hazardous materials technician shall be that person who responds to hazardous materials/WMD incidents using a risk-based response process by which he or she analyzes a problem involving hazardous materials/WMD, selects applicable decontamination procedures, and controls a release using specialized protective clothing and control equipment *[see 7.1.2.2(1)]*.

7.1.1.2 The hazardous materials technician shall be trained to meet all competencies at the awareness level (Chapter 4), all core competencies at the operations level (Chapter 5), and all competencies of this chapter.

7.1.1.3* The hazardous materials technician shall receive additional training to meet applicable governmental occupational health and safety regulations.

A.7.1.1.3 Additional training sources might include, but are not limited to, local and state public health agencies and the Centers for Disease Control and Prevention (CDC). Additional training options include, but are not limited to, programs offered at the Center for Domestic Preparedness in Anniston, Alabama, and at the U.S. Army Dugway Proving Grounds in Utah.

7.1.1.4 The hazardous materials technician shall be permitted to have additional competencies that are specific to the response mission, expected tasks, and equipment and training as determined by the AHJ.

The HMT, in addition to the basic competencies, could have specific mission requirements based on the employer's emergency response plan and standard operating procedures. The intent of this statement is that the HMT must be properly trained and equipped for any mission to be undertaken as determined by the authority having jurisdiction (AHJ).

7.1.2 Goal.

7.1.2.1 The goal of the competencies at this level shall be to provide the hazardous materials technician with the knowledge and skills to perform the tasks in 7.1.2.2 safely.

7.1.2.2 In addition to being competent at both the awareness and the operations levels, the hazardous materials technician shall be able to perform the following tasks:

(1) Analyze a hazardous materials/WMD incident to determine the complexity of the problem and potential outcomes by completing the following tasks:

The HMT is required in 7.1.2.2(1) to have an in-depth knowledge of containers and to be able to operate monitoring equipment in order to identify the presence and concentrations of hazardous materials. The HMT should be able to access and interpret a wide variety of resources for hazard and response information, as well as find ways to identify hazards based on minimal information at the scene. The HMT is expected to be able to predict more definitively than the responder at the operations level how the materials behave when released. Finally, the HMT should have the ability to utilize or access computer-based plume dispersion modeling using real-time weather and geospatial information to more definitively identify endangered areas for use in the risk-based decision process.

(a) Survey the hazardous materials/WMD incident to identify special containers involved, to identify or classify unknown materials, and to verify the presence and concentrations of hazardous materials through the use of monitoring equipment.
(b) Collect and interpret hazard and response information from printed and technical resources, computer databases, and monitoring equipment.
(c) Describe the type and extent of damage to containers.
(d) Predict the likely behavior of released materials and their containers when multiple materials are involved.
(e) Estimate the size of an endangered area using computer modeling, monitoring equipment, or specialists in this field.

(2) Plan a response within the capabilities of available personnel, personal protective equipment, and control equipment by completing the following tasks:

(a) Describe the response objectives for hazardous materials/WMD incidents.

The difference between the competency in 7.1.2.2(2)(a) and the corresponding competency at the operations level is that the HMT is expected to be able to utilize a risk-based decision process to identify and recommend response objectives. The higher level of knowledge and skills allows the HMT to evaluate and determine the proper course of action to recommend to the incident commander.

(b) Describe the potential response options available by response objective.

At this level, the HMT is required to identify the options available in order to achieve the response objectives that have been established. This task is done by considering the choices—given resources, time, and risk analysis. Some of these options could include separating the hazard from potential exposures, establishing barriers or dikes, diluting the material, or transferring the product to an undamaged container.

(c) Select the personal protective equipment required for a given action option.

The HMT is required in 7.1.2.2(2)(c) to know more about personal protective equipment (PPE) because technicians are expected to be involved in product control operations. HMTs might have to work in specialized chemical-protective clothing such as a vapor-protective suit, which is shown in Exhibit I.7.1.

(d) Select a technical decontamination process to minimize the hazard.

Selecting the appropriate decontamination procedures is imperative for technician level responders because they could come in contact with hazardous materials/WMD while conducting such product control operations as plugging or patching.

(e) Develop an incident action plan for a hazardous materials/WMD incident, including a site safety and control plan, consistent with the emergency response plan or stan-

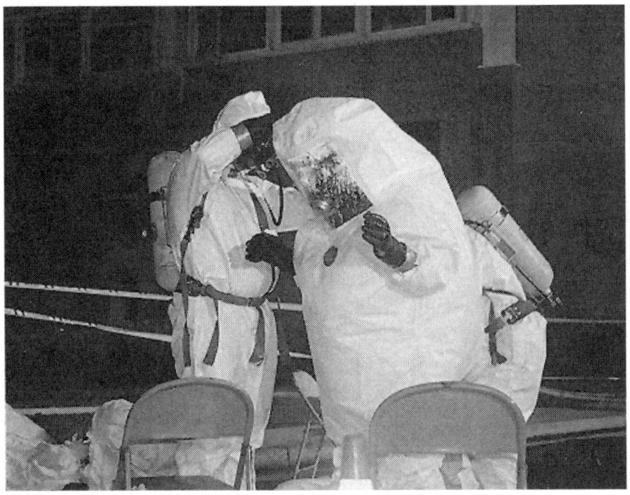

EXHIBIT I.7.1 *These are typical vapor-protective suits. (Courtesy of Michael Ratcliffe)*

dard operating procedures and within the capability of the available personnel, personal protective equipment, and control equipment

After the response objectives are identified in 7.1.2.2(2)(a), the incident action plan developed in 7.1.2.2(2)(e) defines the strategy and tactics and identifies the personnel and equipment necessary to accomplish them.

(3)*Implement the planned response to favorably change the outcomes consistent with the standard operating procedures and site safety and control plan by completing the following tasks:

A.7.1.2.2(3) The following site safety and control plan considerations are from the NIMS Site Safety and Control Plan (form ICS 208 HM):

(1) Site description
(2) Entry objectives
(3) On-site organization
(4) On-site control
(5) Hazard evaluation
(6) Personal protective equipment
(7) On-site work plans
(8) Communication procedures
(9) Decontamination procedures
(10) Site safety and health plan

(a) Perform the duties of an assigned hazardous materials branch or group position within the local incident management system (IMS).

At this level, the HMT is required in 7.1.2.2(3)(a) to be able to perform assigned roles in the incident management system. The incident command system (ICS) is a part of the overall incident management system (IMS), and the technician must know and understand the assigned roles and the chain of command to ensure accountability and safety. Additionally, at large events the IMS could have a multi-agency coordination entity that would be accessed by the on-scene ICS and support the needs of the incident. In most areas this entity will be the local emergency operations center (EOC).

(b) Don, work in, and doff personal protective clothing, including, but not limited to, both liquid splash– and vapor–protective clothing with correct respiratory protection.
(c) Perform the control functions identified in the incident action plan.

Responders at the technician level are required in 7.1.2.2(3)(c) to be able to demonstrate their ability to select and use the appropriate equipment and to implement the tactical options identified in the incident action plan.

(d) Perform the decontamination functions identified in the incident action plan

(4) Evaluate the progress of the planned response by completing the following tasks:
(a) Evaluate the effectiveness of the control functions.

This evaluation should answer the question, "Are the strategy and tactics that have been implemented accomplishing the incident objectives identified in 7.1.2.2(2)(a)?"

(b) Evaluate the effectiveness of the decontamination process.

(5) Terminate the incident by completing the following tasks:
(a) Assist in the incident debriefing.
(b) Assist in the incident critique.
(c) Provide reports and documentation of the incident.

Occupational Safety and Health Administration (OSHA) regulation 29 CFR 1910.120(q) mandates incident debriefing, critique, reporting, and documentation [1]. While the technician might not have ultimate responsibility for ensuring the completion of these tasks, the technician could be required to provide information regarding actions taken and decisions made.

7.2 Competencies — Analyzing the Incident

7.2.1 Surveying Hazardous Materials/WMD Incidents.

Given examples of hazardous materials/WMD incidents, the hazardous materials technician shall identify containers involved and, given the necessary equipment, identify or classify unknown materials involved, verify the identity of the hazardous materials/WMD involved, determine the concentration of hazardous materials, and shall meet the requirements of 7.2.1.1 through 7.2.1.5.

At this level, the HMT is required in 7.2.1 to be able to identify the characteristics of special containers that might indicate the presence of hazardous materials. The HMT is also expected to be able to conduct an analysis that identifies and classifies unknown materials based on container and type of material that could be within. Special containers include high-pressure containers (identified by their rounded ends), cryogenic cargo tanks or cylinders, and casks for radioactive materials.

7.2.1.1 Given examples of various containers for hazardous materials/WMD, the hazardous materials technician shall identify each container by name and specification and identify the typical contents by name and hazard class.

7.2.1.1.1 Given examples of the following railroad cars, the hazardous materials technician shall identify the container by name and specification and identify the typical contents by name and hazard class:

(1) Cryogenic liquid tank cars

Cryogenic tank cars in 7.2.1.1.1(1) carry low-pressure liquids, usually at 25 psi (172 kPa) or lower, that are refrigerated to –155°F (–104°C) or below. The cryogenic tank car is a tank within a tank. The space between the inner and outer tanks is filled with insulation and normally maintained under a vacuum.

Cryogenic cars are distinguished by the absence of top fittings; the fittings are enclosed in cabinets either at ground level on both sides or at the end of the car. Among the materials that can be shipped in cryogenic tank cars are argon, ethylene, hydrogen, nitrogen, and oxygen. Another type of cryogenic tank car is the box tank. This type of tank is built inside a 40-ft (12-m) box car, and its fittings are located inside the doors on both sides.

(2) Nonpressure tank cars

Nonpressure tank cars, also known as general-service tank cars or acid-served tank cars, transport a wide variety of hazardous and nonhazardous materials at low pressures. Nonpressure tank cars transport hazardous materials, such as flammable and combustible liquids, flammable solids, oxidizers, organic peroxides and poison, corrosive materials, and molten solids. They also transport nonhazardous materials, such as tallow, clay, slurry, corn syrup, and other food products. Tank test pressures for nonpressure tank cars range from 60 psi to 90 psi (413 kPa to 620 kPa). Capacities range from 4000 gal to 45,000 gal (15 m^3 to 170 m^3). The tanks themselves are cylindrical with rounded heads. Newer nonpressure tank cars have at least one manway to allow access to the car's interior. Some older nonpressure tank cars have at least one expansion dome with a manway. Fittings for loading and unloading, pressure and/or vacuum relief, gauging, and other purposes are visible at the top and/or bottom of the car.

(3) Pneumatically unloaded hopper cars

Charles Wright notes the following in the *Fire Protection Handbook*®:

> Pneumatically unloaded covered hopper cars are covered hopper cars, which are unloaded through pressure differential, or pneumatics, applying air pressure. Even though the pressure is used only during unloading, tank test pressures for the car range from 20 psi to 80 psi (138 kPa to 551 kPa). Dry caustic soda is one commodity transported in this type of car. [2] p. 21-129

(4) Pressure tank cars

Wright notes the following in the *Fire Protection Handbook*:

> The high pressure tube car is a 40-ft (12-m) box-type open-frame car. Inside the frame is a visible cluster of 30 stainless steel, uninsulated cylinders, arranged horizontally and permanently attached to the car. Tank test pressures of these tube cars range up to 5,000 psi (345,000 kPa). These tube cars transport helium, hydrogen, or oxygen in the gaseous form. . . . A steel case encloses each end of the tube car. Loading and unloading fittings and safety devices are located in a walk-in cabinet at one end of the car. Safety relief devices, either safety relief valves or safety vents, are found on each cylinder. For cars transporting flammable gases, safety relief devices are equipped with ignition devices to burn off any released material. [3] pp.14-65–14-66

7.2.1.1.2 Given examples of the following intermodal tanks, the hazardous materials technician shall identify the container by name and specification and identify the typical contents by name and hazard class:

(1) Nonpressure intermodal tanks
    (a) IM-101 portable tanks (IMO Type 1 internationally)

The tanks in 7.2.1.1.2(1)(a) are built to withstand maximum allowable working pressures (MAWP) of 25.4 psi to 100 psi (175 kPa to 6890 kPa). The tanks coould be used to transport both nonhazardous and hazardous materials, including toxic, corrosive, and flammable materials. Internationally, an IM-101 portable tank is called an IMO Type 1 tank container.

(b) IM-102 portable tanks (IMO Type 2 internationally)

Wright also notes the following:

> These tanks are designed to handle lower maximum allowable working pressures (MAWP)—that is, from 14.5 psi to 25.4 psi (100 kPa to 175 kPa). They transport materials such as liquor, alcohols, some corrosives, pesticides, insecticides, resins, industrial solvents, and flammables with flash points between 32°F and 140°F (0°C and 60°C). More often, they transport various nonregulated materials, such as food commodities. Internationally, an IM-102 tank is called an IMO Type 2 tank container. [2] p. 21-132

(2) Pressure intermodal tank (DOT Specification 51; IMO Type 5 internationally)

Also known as DOT Specification 51 portable tanks, these containers are less common in transport. Designed to handle internal pressures ranging from 100 psi to 500 psi (690 kPa to 3450 kPa), they generally transport gases liquefied under pressure, such as LP-gas and anhydrous ammonia. They can also carry liquids, such as motor fuel anti-knock compounds or aluminum alkyls. Internationally, they are called IMO Type 5 tank containers.

(3) Specialized intermodal tanks

(a) Cryogenic intermodal tanks (DOT Specification 51; IMO Type 7 internationally)

These containers carry refrigerated liquid gases, such as liquefied argon, oxygen, helium, ethylene, and nitrogen, and they are built to DOT specifications for portable tanks intended for the transportation of refrigerated gases. The tanks consist of a tank within a tank with insulation between the inner and outer tanks. The space between the tanks is normally maintained under a vacuum. Internationally, they are called IMO Type 7 tank containers.

(b) Tube modules

These containers transport nonliquefied gases such as helium, nitrogen, and oxygen in high-pressure 3T cylinders mounted within a full- or half-height ISO frame. This rigid bulk packaging consists of several horizontal seamless steel cylinders, from 9 in. to 48 in. (229 mm to 1219 mm) in diameter, that are permanently mounted inside an open frame with a boxlike compartment at one end enclosing the loading and unloading fittings and safety devices. Service pressures range from 3000 psi to 5000 psi (20,670 kPa to 34,450 kPa)

7.2.1.1.3 Given examples of the following cargo tanks, the hazardous materials technician shall identify the container by name and specification and identify the typical contents by name and hazard class:

(1) Compressed gas tube trailers
(2) Corrosive liquid tanks
(3) Cryogenic liquid tanks
(4) Dry bulk cargo tanks
(5) High-pressure tanks
(6) Low-pressure chemical tanks
(7) Nonpressure liquid tanks

The shape of the vehicle can offer clues to its cargo in an emergency. Often the shape is the only key to identifying the cargo of a trailer truck at an accident scene. Exhibit I.7.2 shows the various shapes of road trailers for quick reference.

7.2.1.1.4 Given examples of the following facility storage tanks, the hazardous materials technician shall identify the container by name and identify the typical contents by name and hazard class:

(1) Cryogenic liquid tank
(2) Nonpressure tank
(3) Pressure tank

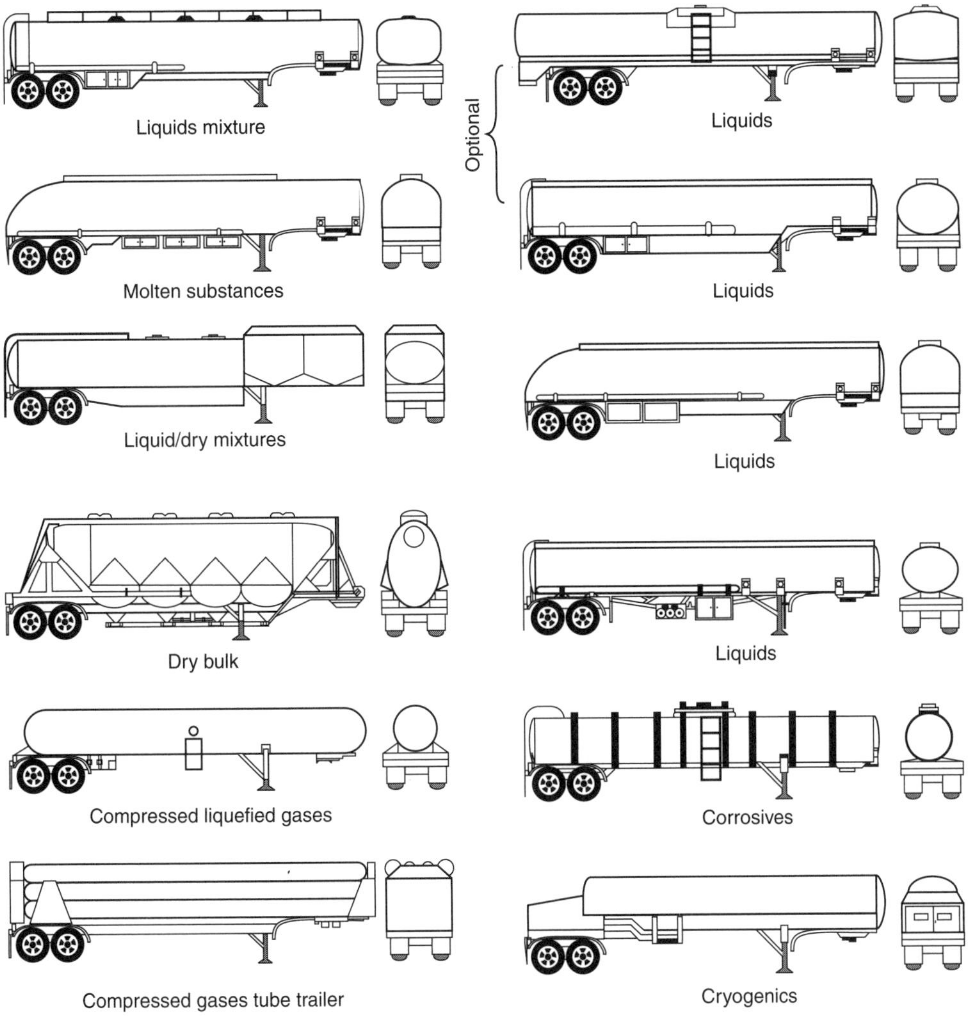

EXHIBIT I.7.2 *The shapes of the most common vehicles are illustrated.*

7.2.1.1.5 Given examples of the following nonbulk packaging, the hazardous materials technician shall identify the package by name and identify the typical contents by name and hazard class:

The specifications for packaging are defined in Title 49 CFR Part 178 [4]. In order to understand the risk posed by packaging, the HMT needs to be able to anticipate the hazards of materials shipped in the different non-bulk packaging.

(1) Bags
(2) Carboys
(3) Cylinders
(4) Drums

7.2.1.1.6 Given examples of the following radioactive materials packages, the hazardous materials technician shall identify the container/package by name and identify the typical contents by name:

(1) Excepted

Excepted packaging in 7.2.1.1.6(1) is used to transport material with extremely low levels of radioactivity. Excepted packages range from a product's fiberboard box to a sturdy wooden or steel crate. Typical shipments in excepted packages include limited quantities of materials, instruments, and articles such as smoke detectors.

(2) Industrial

Industrial packaging in 7.2.1.1.6(2) is used to transport material that presents a limited hazard to the public and environment. Examples of material transported in industrial packages include contaminated equipment and radioactive waste solidified in materials such as concrete. Industrial packages are grouped into three categories (IP-1, IP-2, IP-3), based on the strength of the package.

(3) Type A

Type A packages in 7.2.1.1.6(3) have an inner containment vessel made of glass, plastic, or metal, and the packing material is made of polyethylene, rubber, or vermiculite. Examples of materials shipped in Type A packages include radiopharmaceuticals and low-level radioactive waste. Type A packages containing non-life-endangering amounts of radioactive material are identified with the words *Type A* on the package and also on the shipping papers.

(4) Type B

Type B packaging in 7.2.1.1.6(4) is used to transport material with radioactivity levels higher than those allowed in Type A packages, such as spent fuel and high-level radioactive waste. Limits on activity contained in Type B packages are provided in 49 CFR Part 173.431 [5]. Type B packages range from small drums [55 gal (208 L)] to heavily shielded steel casks that sometimes weigh more than 100 tons. Type B packages are identified with the words *Type B* on the package and also on the shipping papers.

(5) Type C

Type C packages in 7.2.1.1.6(5) are used for high-activity materials (including plutonium) transported by aircraft. They are designed to withstand severe accident conditions associated with air transport without loss of containment or significant increase in external radiation levels. The Type C package performance requirements are significantly more stringent than those for Type B packages. Type C packages are not authorized for domestic use but are authorized for international shipments of radioactive material. The fact that they are not authorized for domestic use is why one does not hear much about them in the United States. The *Emergency Response Guidebook* (ERG) includes information on Type C because it is more of an international document [6].

Regulations require that both Type B and Type C packages be marked with a trefoil symbol to ensure that the package can be positively identified as carrying radioactive material after a severe accident. The trefoil symbol must be resistant to the effects of both fire and water.

The performance requirements for Type C packages include those applicable to Type B packages, with enhancements on some tests that are significantly more stringent than those for Type B packages. For example, a 200 mi (322 km) per hour impact onto an unyielding target is required instead of the 30-ft (9-m) drop test required of a Type B package; a 60-minute fire test instead of the 30-minute for Type B packages; and a 650-ft (198-m) immersion test instead of the 50-ft (15-m) test required for Type B package. These stringent tests are expected to result in package designs that will survive more severe aircraft accidents than Type B package designs.

7.2.1.2 Given examples of three facility and three transportation containers, the hazardous materials technician shall identify the approximate capacity of each container.

Facility containers are often marked with their capacities, and transportation containers are required to be marked with their capacities.

7.2.1.2.1 Using the markings on the container, the hazardous materials technician shall identify the capacity (by weight or volume) of the following examples of transportation vehicles:

(1) Cargo tanks

A data plate and a specification plate must be affixed to cargo tanks. These plates provide the volume and/or weight of the container, which the HMT is required in 7.2.1.2.1(1) to know.

(2) Tank cars

Tank cars in 7.2.1.2.1(2) must be marked by the U.S. Department of Transportation (DOT). DOT requires that tank cars have standardized markings. Exhibit I.7.3 shows an example of tank car markings.

(3) Tank containers

DOT also requires that tank containers have standardized markings as shown in Exhibit I.7.4.

7.2.1.2.2 Using the markings on the container and other available resources, the hazardous materials technician shall identify the capacity (by weight or volume) of each of the following facility containers:

(1) Cryogenic liquid tank
(2) Nonpressure tank (general service or low-pressure tank)
(3) Pressure tank

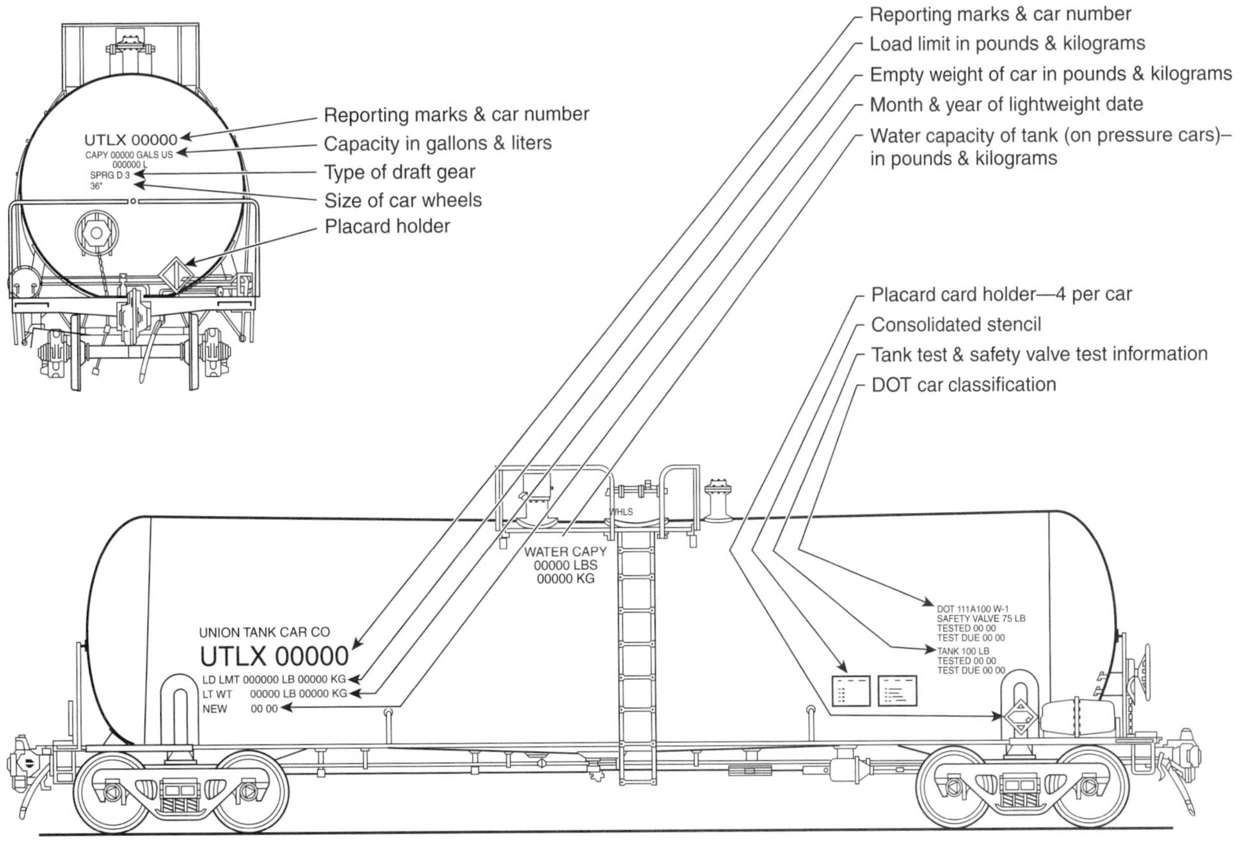

EXHIBIT I.7.3 *This illustration identifies sample DOT markings and their locations on a tank car. (Courtesy of Union Tank Car Company)*

EXHIBIT I.7.4 *Tank container standardized markings are illustrated.*

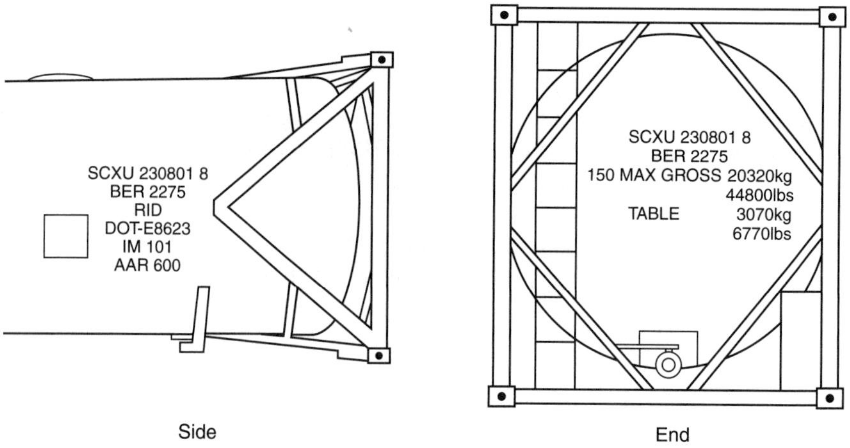

7.2.1.3* Given at least three unknown hazardous materials/WMD, one of which is a solid, one a liquid, and one a gas, the hazardous materials technician shall identify or classify by hazard each unknown material.

The technician level responder is required in 7.2.1.3 to be able to identify unknown materials in the event that a container's shipping papers, placards, material safety data sheet (MSDS), or other identifying items have been destroyed or are unavailable.

A.7.2.1.3 Suggested materials to identify can include the most commonly released materials that are identified annually on several lists, such as those from the federal EPA or the California Environmental Protection Agency (Cal/EPA).

7.2.1.3.1 The hazardous materials technician shall identify the steps in an analysis process for identifying unknown solid and liquid materials.

Unknown materials must be identified cautiously. Monitoring equipment can help identify the hazard class and, in some cases, even specific chemicals. Kits are also available to provide assistance in identification.

If the technician must take samples in order to conduct an analysis or approach the site to monitor it, the HMT must use the proper level of protective clothing. Responders must approach the site from upwind, wearing the appropriate level of protective clothing and working in pairs at least, with adequate backup personnel.

7.2.1.3.2 The hazardous materials technician shall identify the steps in an analysis process for identifying an unknown atmosphere.

Kenneth York and Gerald Grey, in *Hazardous Materials/Waste Handling for the Emergency Responder*, recommend the following:

> Unless you are certain, beyond any doubt, as to what your hazard is, you should approach the incident employing the information listed below. Assuming that you are not certain what all the hazards are, you should measure, in the order shown below, for:
>
> - Radioactivity
> - Combustibility
> - Oxygen availability (deficiency)
> - pH (if liquid)
> - Hydrogen sulfide (if in areas of, or adjacent to petroleum refining activities)
> - Carbon monoxide (in a fire or post-fire incident)
> - Organic vapor [7] p. 281

7.2.1.3.3 The hazardous materials technician shall identify the type(s) of monitoring technology used to determine the following hazards:

Hazardous materials response teams should develop protocols for monitoring the air during operations at hazardous materials incidents. A wide variety of monitoring equipment is available, and response teams must determine what is appropriate for the type of hazards they are likely to encounter. The response teams must also become proficient in calibrating, operating, and maintaining this equipment. Each instrument has its own operating characteristics and limitations, and responders must be familiar with them.

(1) Corrosivity

Corrosivity is measured by determining the pH of a material. This task can be done using pH paper or pH meters. On pH paper, the color changes, indicating the pH level. By comparing the color of the pH paper with the color chart provided, the HMT can determine the pH of the material in question. Care must be taken in the use of color-indicating detectors when used in an atmosphere or liquid with bleaching characteristics. Commercially available pH meters have a probe that is inserted into the material; the pH is indicated on the meter's display. These pH meters provide a more accurate reading than pH paper. See Exhibit I.7.5 for an example.

(2) Flammability

Combustible gas indicators (CGI) can be used to determine the presence of flammable vapors of hydrocarbon products. Certain instruments are specifically designed to monitor methane vapors only; they measure the flammable vapors as a percentage of the lower explosive limit. Flash point instruments are also available for field use. The instruments allow the responder to determine the flammability of an unknown material fairly accurately and the class of flammable or combustible with which they are dealing. Many CGIs are combination instruments, that is, they can measure oxygen content and several toxic substances as well as combustible gas vapors (see Exhibit I.7.6).

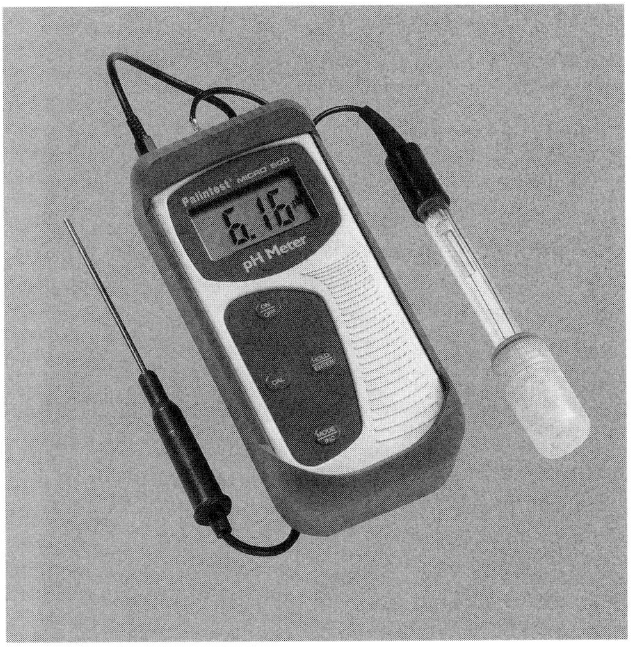

EXHIBIT I.7.5 *A pH meter such as this one is used to determine whether a material is acidic, base, or neutral. (Source: Palintest Company)*

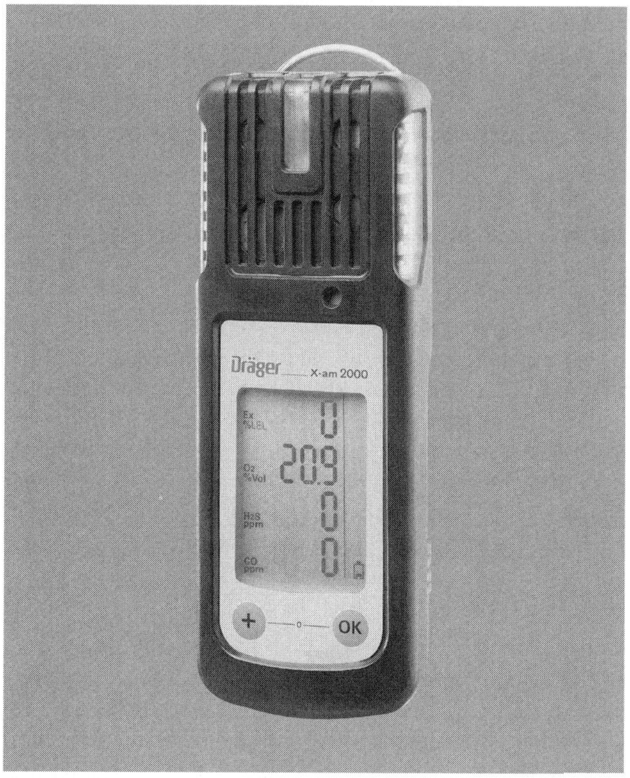

EXHIBIT I.7.6 *Combination instruments typically measure oxygen content, flammable vapor concentration, and specific toxic concentrations. (Source: Draeger Safety, Inc.)*

(3) Oxidation potential

Anytime oxygen is present, substances can become oxidized. The ability of a substance to oxidize is a measure of its propensity to yield oxygen. Oxygen is easily released, especially when heated, and it accelerates the burning of combustible materials. The more readily a material gives up its oxygen molecule, the greater the hazard it presents, which is the case with oxidizing agents. Oxidizer paper can be used to test for the presence of oxidizers, and an oxygen meter can be used to test for oxygen-enriched atmospheres.

(4) Oxygen deficiency

Equipment used to monitor oxygen concentrations generally measures over a range of 0 to 25 percent oxygen in air (see Exhibit I.7.7). Some models of oxygen meters contain an alarm that sounds if the oxygen level drops below 19.5 percent, which is the minimum permissible percentage of oxygen as established by OSHA. For work in areas with oxygen measurements below this level, the responder must have a supplied-air respirator.

(5) Pathogenicity

Pathogenicity is the virulence of a pathogen. Virulence refers to how ill an individual may become. The presence of pathogens such as viruses, bacteria, and fungi can only be determined in the laboratory. There are arguably several methods to determine the presence and identity on the scene. However, the "gold standard" continues to be a laboratory capable of identifying pathogens. In the United States, this should be a Centers for Disease Control and Prevention (CDC)-certified Laboratory Response Network facility. On-scene tentative methods include, with varying success, biological immunoassay indicators, DNA fluoroscopy, and polymerase chain reaction (PCR) hand-held detectors. Proper sampling, collection techniques, and evidence preservation and control must be observed.

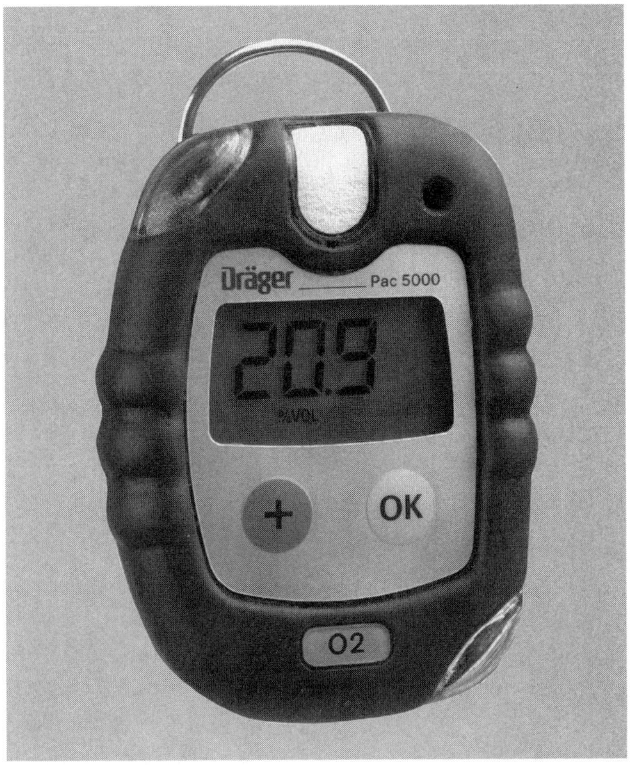

EXHIBIT I.7.7 *An oxygen meter measures the amount of oxygen in the atmosphere. (Source: Draeger Safety, Inc.)*

(6) Radioactivity

Radiation detectors in 7.2.1.3.3(6) are available to monitor alpha particles, beta particles, gamma rays, and neutron particles. Generally, such detectors measure two or more types of radiation. For example, the Geiger counter, probably the most common type of radiation detector, can detect both gamma and beta radiation. However, the detector cannot effectively measure the amounts of beta radiation. Ion chambers can also measure radiation, but they are not effective for low-level radiation. Exhibit I.7.8 shows a typical radiation detector.

(7) Toxicity

Several types of instruments can be used to measure toxic exposures. The instruments include photoionization detectors (see Exhibit I.7.9), flame ionization detectors, infrared spectrophotometers, and detector tubes (see Exhibit I.7.10). Some of these instruments are designed to measure specific chemicals, such as hydrogen sulfide, and some may measure more than one chemical (see Exhibit I.7.11). Detector tubes, which allow responders to evaluate potential hazards quickly, operate by drawing an air sample through a small glass tube. This process causes the material inside the tube to change color, indicating the concentration of the material in the air.

7.2.1.3.4* The hazardous materials technician shall identify the capabilities and limiting factors associated with the selection and use of the following monitoring equipment, test strips, and reagents:

The HMT must understand that with advances in technology and developments in usability and functionality, monitoring and sampling devices and techniques will evolve and change. It is incumbent upon the AHJ and the technician to remain current and knowledgeable of these evolutions and changes. The items associated with 7.2.1.3.4 represent the most current

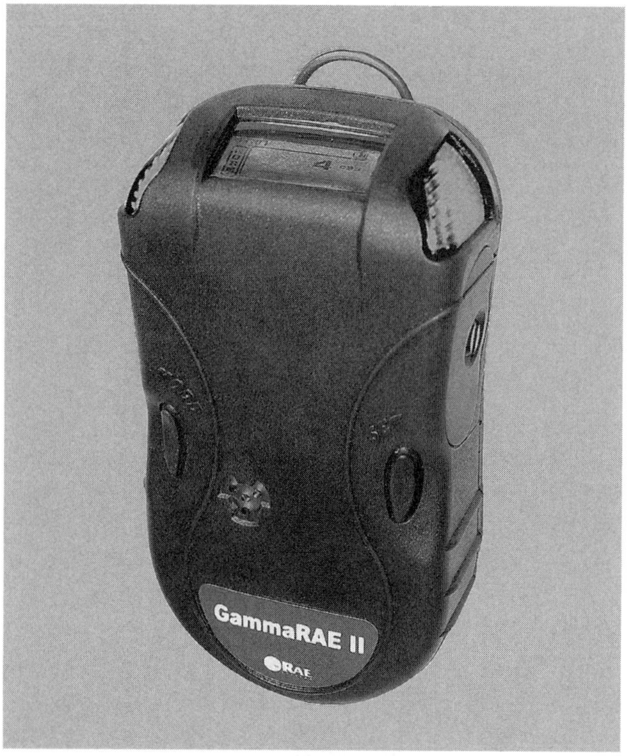

EXHIBIT I.7.8 *Radiation detectors measure radiation that may be present. (Source: RAE Systems)*

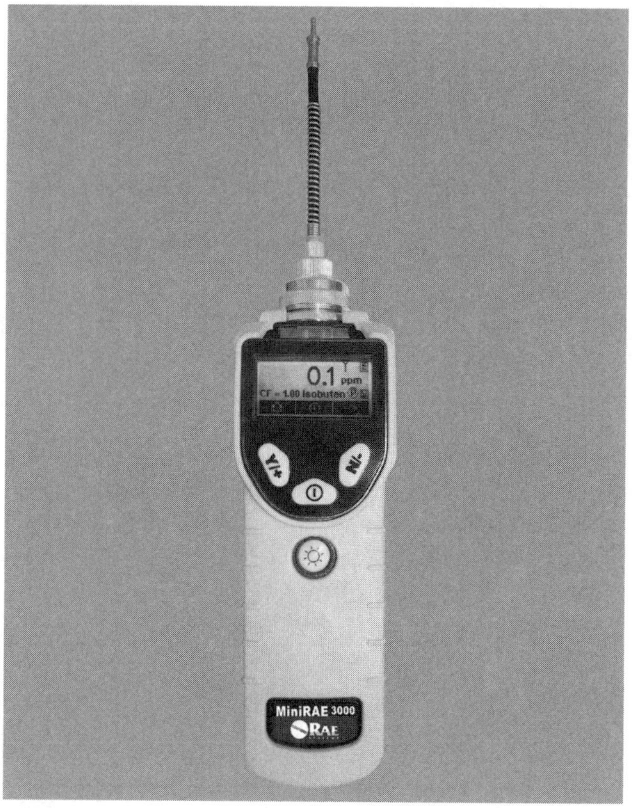

EXHIBIT I.7.9 *A photoionization detector is used to measure the amount of toxic vapors that may be present. (Source: RAE Systems)*

available at the time this handbook was printed. The technician needs only to be proficient in the use of the equipment supplied by the AHJ. However, the technician must be knowledgeable about all of the technologies in order to know what may be available for a specific purpose.

(1) Biological immunoassay indicators
(2) Chemical agent monitors (CAMs)
(3) Colorimetric indicators [colorimetric detector tubes, indicating papers (pH paper and meters), reagents, test strips]
(4) Combustible gas indicator
(5) DNA fluoroscopy
(6) Electrochemical cells (carbon monoxide meter, oxygen meter)
(7) Flame ionization detector
(8) Gas chromatograph/mass spectrometer (GC/MS)
(9) Infrared spectroscopy
(10) Ion mobility spectroscopy
(11) Mass channel analyzer
(12) Metal oxide sensor
(13) Photoionization detectors
(14) Polymerase chain reaction (PCR)
(15) Radiation detection and measurement instruments
(16) Raman spectroscopy

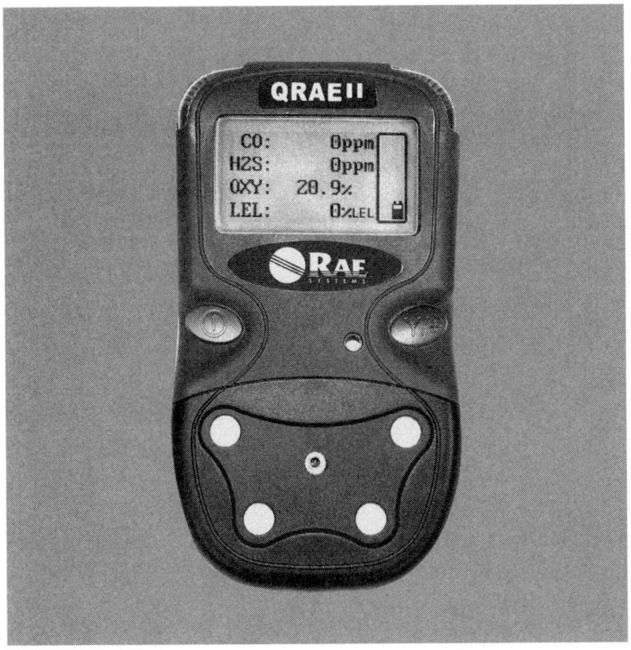

EXHIBIT I.7.10 *Carbon monoxide is one of the substances that can be tested for by using this combination instrument. (Source: RAE Systems)*

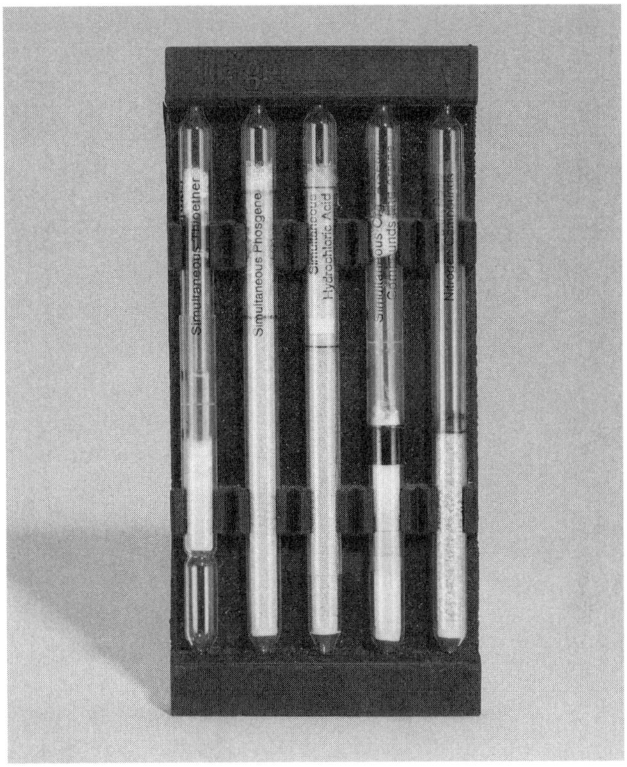

EXHIBIT I.7.11 *Colorimetric detector tubes are glass sampling tubes used to test for specific toxic materials. (Source: Draeger Safety, Inc.)*

(17) Surface acoustical wave (SAW)
(18) Wet chemistry

A.7.2.1.3.4 These factors include, but are not limited to, operation, calibration, response time, detection range, relative response, sensitivity, selectivity, inherent safety, environmental conditions, and nature of hazard. Also refer to NIOSH/OSHA/USCG/EPA, *Occupational Safety and Health Guidance Manual for Hazardous Waste Site Activities.*

7.2.1.3.5* Given three hazardous materials/WMD, one of which is a solid, one a liquid, and one a gas, and using the following monitoring equipment, test strips, and reagents, the hazardous materials technician shall select from the following equipment and demonstrate the correct techniques to identify the hazards (corrosivity, flammability, oxidation potential, oxygen deficiency, radioactivity, toxicity, and pathogenicity):

(1) Carbon monoxide meter
(2) Colorimetric tubes
(3) Combustible gas indicator
(4) Oxygen meter
(5) Passive dosimeters
(6) pH indicators and pH meters
(7) Photoionization and flame ionization detectors
(8) Radiation detection instruments
(9) Reagents
(10) Test strips

(11) WMD detectors (chemical and biological)
(12) Other equipment provided by the AHJ

A.7.2.1.3.5 For example, the techniques for use of the monitoring equipment should include monitoring for lighter-than-air gases in a confined area, heavier-than-air gases and vapors in a confined area, and heavier-than-air gases and vapors in an unconfined area.

The purpose of the competency in 7.2.1.3.5 is to have responders at this level demonstrate their ability to select the proper instrument and demonstrate its use.

7.2.1.3.6 Given monitoring equipment, test strips, and reagents provided by the AHJ, the hazardous materials technician shall demonstrate the field maintenance and testing procedures for those items.

Responders who operate monitoring equipment required in 7.2.1.3.6 must know how to field-check it in order to get accurate readings and must be able to perform minor field maintenance, such as replacing sensors in an oxygen meter or a combustible gas indicator.

7.2.1.4* Given a label for a radioactive material, the hazardous materials technician shall identify the type or category of label, contents, activity, transport index, and criticality safety index as applicable, then describe the radiation dose rates associated with each label.

The radioactive label in 7.2.1.4 provides responders with information about the level of radiation hazard a material presents. Radioactive White I, with a single vertical bar, is the lowest level hazard. Radioactive Yellow II, with two vertical bars, is the next higher level, and Radioactive Yellow III, with three vertical bars, is the highest level of hazard. These levels are based on the radiation activity of the material. The radiation level must be written on the label. On vehicles, these labels might not be easily accessible, so recognition of the special shape of the transport vehicle is vital.

A.7.2.1.4 Examples of radioactive material labels include:

(1) *Radioactive White I.* The Radioactive White I label is attached to packages with extremely low levels of external radiation. The maximum contact radiation level associated with this level is 0.5 mrem/hour.
(2) *Radioactive Yellow II.* The Radioactive Yellow II label is attached to packages with external contact radiation levels ranging from greater than 0.5 mrem/hour to no more than 50 mrem/hour. The Radioactive II level also has a box for the transport index. The maximum allowable transport index for this label is 1.
(3) *Radioactive Yellow III.* The Radioactive Yellow III label is attached to packages with external contact radiation levels ranging from greater than 50 mrem/hour to a maximum of 200 mrem/hour.
(4) *Empty.* Applied to packages that have been emptied of their contents as far as practical but that might still contain regulated amounts of internal contamination and radiation levels of less than 0.5 mrem/hour detectable outside the package.
(5) *Fissile.* Applied to packages that contain fissile materials. The criticality safety index for each package will be noted on the label. The criticality safety index is displayed on the label to assist the shipper in controlling how many fissile packages can be grouped on a conveyance. Where applicable, the fissile label will appear adjacent to the Radioactive White I, Radioactive Yellow II, or Radioactive Yellow III label.

7.2.1.5 The hazardous materials technician shall demonstrate methods for collecting samples of the following:

(1) Gas
(2) Liquid
(3) Solid

7.2.2 Collecting and Interpreting Hazard and Response Information.

Given access to printed and technical resources, computer databases, and monitoring equipment, the hazardous materials technician shall collect and interpret hazard and response information not available from the current edition of the DOT *Emergency Response Guidebook* or an MSDS and shall meet the requirements of 7.2.2.1 through 7.2.2.6.

At the operations level, the responder must be able to collect hazard and response information from sources that, in general, are readily available, such as the ERG or MSDSs. At this level, however, the technician responder is required in 7.2.2 to be able to collect and interpret information from a number of sources. Technician level responders must understand the importance of using multiple resources for gathering hazard and response information. HMTs should collect information from a variety of resources, compare that information, and then make their decisions based on this comparison. Responders generally give more weight to the more conservative information and base their actions on that information. HMTs need to remember that different information is available, and some of it could conflict.

7.2.2.1* The hazardous materials technician shall identify and interpret the types of hazard and response information available from each of the following resources and explain the advantages and disadvantages of each resource:

A.7.2.2.1 For example, the significance of high concentrations of three airborne hazardous materials/WMD readings at scenarios relative to the hazards and harmful effects of the hazardous materials/WMD on the responders and the general public should be known.

A number of databases are available. The Computer-Aided Management of Emergency Operations (CAMEO) is probably the most widely used and is available for both Windows and Macintosh platforms. Other public-domain databases are also available to the responder, including the Oil and Hazardous Materials Technical Assistance Database (OHM/TADS), the Registry of Toxic Effects of Chemical Substances (RTECS), and the Chemical Hazard Response Information System (CHRIS). Each database presents information a little differently and concentrates on different areas. Technician level responders should understand the differences among them and use the databases that are the most appropriate for handling a specific incident.

(1) Hazardous materials databases
(2) Monitoring equipment

Monitoring equipment in 7.2.2.1(2) provides a source of information regarding the hazards that are present. The responder should not rely on a single means of monitoring at any incident because the equipment could be affected by unknown materials and give a false reading. Technician level responders should have a sound understanding of the instruments they are using and recognize the value of not relying on a single source for determining the level of hazard present.

(3) Reference manuals

As with hazardous materials databases, a number of reference manuals are available. Each manual tends to emphasize information a little differently. For example, some manuals present quite a bit of information about medical hazards but very little about handling emergency incidents. The responder must understand this and recognize the value of gathering information from more than one source, comparing that information for a broader picture of the potential hazards.

(4) Technical information centers (i.e., CHEMTREC/CANUTEC/SETIQ and local, state, and federal authorities)

Technical information centers, such as CHEMTREC, can provide the responder with valuable information during hazardous materials incidents, and the HMT's knowledge of the assistance that is available from these sources is important. Using its database, the Chemical Transportation Emergency Response Center (CHEMTREC) can provide initial response information on more than one million product-specific MSDSs. CHEMTREC can put the responder at the scene in contact with the shipper and can help the responder identify the materials involved in an incident using waybill numbers and other sources. In addition to contacting shippers, CHEMTREC can help the responder contact manufacturers and other technical specialists. If needed, CHEMTREC also can activate its emergency response mutual aid network, composed of more than 250 emergency response teams from chemical companies and private contractors that can respond to the scene of an incident and help those on site.

(5) Technical information specialists

Technical information specialists in 7.2.2.1(5) can be a valuable asset to the responder. The HMT may want to keep a directory of individuals who are able to provide technical assistance. However, the responder must remember that no individual is likely to have all the answers. For example, there are chemists who specialize in formulating perfumes and chemists who specialize in formulating explosives. Depending on the type of incident the responder is involved in, the assistance of one would be more appropriate than the assistance of the other.

Developing a network of people with technical knowledge is one of the most important things a responder can do. Because incidents differ, responders cannot rely on finding books or databases that address all the possibilities.

7.2.2.2 The hazardous materials technician shall describe the following terms and explain their significance in the analysis process:

(1) Acid, caustic

Acids, or caustics, can cause the pressure within a container to rise, particularly if they become contaminated. The increased pressure can exceed the design load, increasing the risk of container failure.

(2) Air reactivity

Materials that are potentially air-reactive can ignite if they are exposed to air. The potential for container failure due to overpressurization exists.

(3) Autorefrigeration

Autorefrigeration is a phenomenon that occurs during the rapid release (boiling) of a liquefied gas that causes it to temporarily remain in a liquid state through rapid cooling. This situation could lead the technician to falsely assume product elimination until the product resumes boiling and subsequent release.

(4) Biological agents and biological toxins

Biological agents include bacteria, viruses, fungi, other microorganisms, and their associated toxins. They have the ability to adversely affect human health in a variety of ways, ranging from relatively mild, allergic reactions to serious medical conditions, even death. These organisms are widespread in the natural environment; they are found in water, soil, plants, and animals. Because many microbes reproduce rapidly and require minimal resources for survival, they are a potential danger in a wide variety of occupational settings.

(5) Blood agents

A blood agent or cyanogen agent is a chemical compound containing the cyanide group that prevents the body from utilizing oxygen. The term *blood agent* is a misnomer, however, because these agents do not actually affect the blood in any way. Rather, they exert their toxic effect at the cellular level, by interrupting the electron transport chain in the inner membranes of mitochondria. Some other blood agents are carbon monoxide and arsine, although they have different toxic syndromes that affect the blood.

(6) Boiling point

The boiling point is the temperature at which the transition from liquid to gas occurs. At this temperature, the vapor pressure of a liquid equals the surrounding atmospheric pressure so that the liquid rapidly becomes a vapor. Flammable materials with low boiling points generally present greater problems than those with high boiling points. For example, the boiling point of acetone is 133°F (56°C), and the boiling points for jet fuels range from 400°F to 550°F (204°C to 288°C).

(7) Catalyst

Catalysts are used to control the rate of a chemical reaction by either speeding it up or slowing it down. If used improperly, catalysts can speed up a reaction and cause failure of a container that cannot withstand either the pressure or the heat buildup.

(8) Chemical change
(9) Chemical interactions

The chemical interaction of materials in a container can result in a buildup of heat that, in turn, causes an increase in pressure. The combined materials could be more corrosive than the material the container was originally designed to withstand, and the container might fail.

(10) Compound, mixture

Compounds have a tendency to break down into their component parts, sometimes in an explosive manner. If the compound nitroglycerine is contaminated, for example, it can decompose explosively when heated or shocked.

(11) Concentration

When dealing with corrosives, the amount of acid or base is compared with the amount of water present. Concentrated acids are not the same as strong acids. A high concentration of a weak acid and a low concentration of a strong acid are both possible.

(12) Critical temperature and pressure

Critical temperature and pressure relate to the process of liquefying gases. The critical temperature is the minimum temperature at which a gas can be liquefied no matter how much pressure is applied. The critical pressure is the pressure that must be applied to bring a gas to its liquid state.

A gas cannot be liquefied above its critical temperature. The lower the critical temperature, the less pressure is required to bring a gas to its liquid state. If a liquefied gas container exceeds its critical temperature, the liquid converts instantaneously to gas, which can cause the container to fail violently.

(13) Dissociation and corrosivity

Corrosivity is measured by pH, which indicates the concentration of hydrogen ions in the material being tested. Title 40 CFR Part 261.22 defines a corrosive material as one with a pH of 2 or less or of 12.5 or more [8]. Those materials with a pH of 2 or less are acidic, and those materials with a pH at or over 12.5 are bases.

(14) Dose

Radiation dose is the amount of radiation energy deposited in the body. Radiation dose rate is a measure of the rate at which radiation energy is deposited in the body. Radiation dose rate is often measured in terms of exposure per unit of time. These measurements are similar to the speedometer and odometer in a car. The speedometer measures the rate of speed, like dose rate; the odometer measures the total distance traveled, like total dose received.

Radiation dose is usually measured in terms of millirem (mrem), and radiation dose rate is usually measured in terms of millirem per hour. In the United States, the annual average radiation dose per person from all sources is about 360 mrem. However, receiving far more than that in a given year (largely due to medical procedures) is not uncommon. As an example, workers at nuclear facilities are allowed up to 5000 mrem of radiation exposure each year.

(15) Dose response

Dose–response relationship is the biological reaction caused by the dose in the body. This can relate to chemical, biological, or radiological doses. The degree of harm is directly related to the dose (time and amount) and its impact on bodily functions.

(16) Expansion ratio

The expansion ratio is the amount of gas produced by a given volume of liquid at a given temperature. For instance, liquid propane has an expansion ratio of liquid to gas of 270 to 1, while liquefied natural gas has an expansion ratio of 635 to 1. Obviously, the greater the expansion ratio, the more gas is produced, and the larger the endangered area becomes.

(17) Fire point

Several important factors need to be remembered about the flammable range of a material. One characteristic is the lower limit, and another is the width of the range. For example, gasoline has a lower explosive limit (LEL) of 1.4 percent and an upper explosive limit (UEL) of 7.6 percent. Carbon monoxide, however, has an LEL of 12.5 percent and a UEL of 74 percent.

(18) Flammable (explosive) range (LEL and UEL)

When flammable and combustible liquids continue to be heated above their flash points, they reach a temperature at which their output of flammable vapors is in balance with air, so that their vapors continue to burn even after the source of ignition is removed. That temperature is known as the fire point, and it is always a few degrees above a liquid's flash point.

(19) Flash point

Flash point is important to take into account when determining the level of hazard of flammable liquids. Aviation-grade gasoline (100 to 130 octane) has a flash point of –50°F (–45.5°C), and kerosene has a flash point of 100°F (37.8°C). A kerosene spill that occurs when the temperature is 25°F (–3.8°C) presents less danger than a spill that occurs when the temperature is 110°F (43°C). Gasoline would present a serious hazard even at 25°F (–3.8°C).

(20) Half-life

The term *half-life* refers to the time it takes for one-half of the radioactive atoms in a sample to decay to another form. Some radioactive materials decay rapidly, with half-lives of seconds, hours, or days. Some radioactive materials have half-lives of billions of years. For example, a radioactive form of technetium commonly used in medical diagnostic studies has a half-life of only 6 hours. Thorium, the radioactive material found in many foreign-made lantern mantles, however, has a half-life of 14 billion years.

(21) Halogenated hydrocarbon

A hydrocarbon with halon atoms attached is a halogenated hydrocarbon. Halogenated hydrocarbons are used to produce liquids such as flammable liquids, combustible liquids, and liquids used as extinguishing agents. They are often more toxic than naturally occurring organic chemicals, and they all decompose into smaller, more harmful elements when exposed to high temperatures for long periods of time.

(22) Ignition (autoignition) temperature

The ignition temperature is the minimum temperature a material must achieve before it ignites; the ignition source must also be this temperature. Carbon disulfide has an ignition temperature of 194°F (90°C), while ammonia has an ignition temperature of 1204°F (651°C). Products with lower ignition temperatures are in greater danger of igniting than those with higher ignition temperatures.

(23) Inhibitor

Inhibitors are added to products to control their chemical reaction with other products. For example, an inhibitor is added to a monomer, such as ethylene, to keep it from polymerizing when the material is being shipped. If the inhibitor is not added or escapes during an incident, the material will begin to polymerize, which creates a very dangerous situation. The final result may be a violent rupture of the container.

(24) Instability

Materials that decompose spontaneously, polymerize, or otherwise self-react are generally considered unstable. They do not need to mix with other chemicals to react. Organic peroxides, for example, exhibit this characteristic. At low temperatures, these materials can be fairly stable, but they begin to decompose rapidly when exposed to higher temperatures, and once this reaction has begun, it cannot be stopped. The term *instability* is often used interchangeably with the term *reactivity*.

(25) Ionic and covalent compounds

Ionic bonding occurs when ions are formed by the transfer of one or more electrons such as sodium chloride. Covalent bonding occurs when compounds share electrons by a pair of atoms, such as is the case with most hydrocarbons.

(26) Irritants (riot control agents)

The sole purpose of irritants, also known as tear gas, riot control agents, and lachrymators, is to produce immediate discomfort and eye closure to render the victim incapable of fighting or resisting. Riot control agents are solids with low vapor pressures that are dispersed as fine particles or in solution. CS gas and chloracetophenone (CN) are SN_2 (bimolecular nucleophilic substitution) alkylating agents and react at nucleophilic sites.

(27) Maximum safe storage temperature (MSST)

The maximum safe storage temperature is the temperature at which organic peroxides should be stored to ensure that they do not reach the self-accelerating decomposition temperature. The maximum safe storage temperature for organic peroxides should not be exceeded.

(28) Melting point and freezing point

The melting point is the temperature at which a solid becomes a liquid. Materials with low melting points present problems because they easily become liquid and spread more readily. The reverse is also true, however. If the temperature of a liquid can be lowered, the responder might be able to convert it to a solid.

(29) Miscibility

The term *miscible* refers to the tendency or ability of two or more liquids to form a uniform blend, or to dissolve in each other. Liquids may be totally miscible, partially miscible, or not at all miscible.

(30) Nerve agents

Nerve agents (also known as nerve gases, even though these chemicals are liquid at room temperature) are a class of phosphorus-containing organic chemicals (organophosphates) that disrupt the mechanism by which nerves transfer messages to organs. The disruption is caused by blocking acetylcholinesterase, an enzyme that normally relaxes the activity of acetylcholine, a neurotransmitter.

(31) Organic and inorganic

Organic materials are derived from materials that are living or were once living, such as plants or decayed products, and they contain chains of two or more carbon atoms. An example of an organic material is methane (CH_4). Inorganic materials lack carbon chains, but they can contain a carbon atom. An example of an inorganic material that contains a carbon atom is carbon dioxide (CO_2). Knowledge of whether a material is organic or inorganic can be helpful in choosing the proper instrumentation. Some organic materials are reactive with oxidizers. In addition, inorganic acids are generally stronger than organic acids at the same concentration. Organic acids are generally flammable and; as a rule, they are also the toxic and explosive acids.

(32) Oxidation potential

Anytime oxygen is present, substances can become oxidized. The rusting of steel is slow oxidation, while the burning of wood is rapid oxidation. The ability of a substance to oxidize is a measure of its propensity to yield oxygen. Oxygen is easily released, especially when heated, and it accelerates the burning of combustible materials. The more readily a material gives up its oxygen molecule, the greater the hazard it presents, which is the case with oxidizing agents.

(33) Persistence

Persistence refers to a chemical's ability to remain in the environment. The more persistent, the greater the propensity for it to remain harmful over a period of time. The term is used by the military to describe chemical agent volatility.

(34) pH

The pH of a substance is a numerical measure of its relative acidity or alkalinity. An accurate determinant of a solution's hydrogen ion concentration is pH. The level of pH determines the type of container in which a material can be stored or transported. With a low pH of 2 or less, or a high pH of 12.5 or more, special containers are needed. Containers that do not meet the requirements will fail and then the product will be released.

(35) Physical change
(36) Physical state (solid, liquid, gas)

The physical state of a hazardous material plays a significant role in determining not only the measures that are used to control a spill but often the hazards that it presents. A solid is much easier to contain than a liquid, and liquids that have to be diked or dammed while in a gaseous state generally cannot be contained.

(37) Polymerization

Polymerization is a chemical reaction in which small molecules combine to form larger molecules. A hazardous polymerization takes place at a rate that releases large amounts of energy, which can cause a fire or explosion or burst a container. Materials that polymerize usually contain inhibitors that delay the reaction.

(38) Radioactivity

The process of an unstable (radioactive) atom-emitting radiation is called *radioactivity.* Radioactive atoms can be generated through nuclear processes, but they also exist naturally in uranium ore, thorium rock, and some forms of potassium. When a radioactive atom goes through the process of radioactivity, also called radioactive decay, the atom changes to another type of atom. In fact, a radioactive atom can change from one element to another element during the decay process. For example, the element uranium eventually changes to lead through radioactive decay. The length of time for this stabilizing process can take from a fraction of a second to billions of years, depending on the particular type of atom.

The level and type of radioactivity determines the type of packaging a radioactive material needs. The greater the radioactivity, the greater are the packaging requirements.

(39) Reactivity

Chemical reactivity describes a substance's propensity to release energy or undergo change. Some materials are self-reactive or can polymerize. Others undergo violent reactions if they come in contact with other materials. Substances that are air-reactive ignite or release energy when exposed to air. Organic peroxides are examples of highly reactive materials. Other examples are corrosives, radioactive materials, oxidizing materials, pyrophoric substances, explosives, and water-reactive materials.

Water reactivity describes the sensitivity of a material to water without the addition of heat or confinement. The more sensitive materials release heat or flammable or toxic gases. Some materials even react explosively when they are exposed to water. Examples of water-reactive substances are sulfuric acid and sodium and aluminum chloride.

(40) Riot control agents

See Irritants 7.2.2.2(26).

(41) Saturated, unsaturated (straight and branched), and aromatic hydrocarbons

Saturated hydrocarbons *(alkanes)* are those in which the carbon atoms are linked by only single covalent bonds. In saturated hydrocarbons, all the carbon atoms are saturated with hydrogen, such as methane (CH_4) and ethane (C_2H_6). Unsaturated hydrocarbons (alkenes and alkynes) have at least one multiple bond between two carbon atoms somewhere in the molecule, such as ethylene (C_2H_4) and acetylene (C_2H_2).

Unsaturated hydrocarbons are generally more active chemically than saturated hydrocarbons. As a result, these hydrocarbons are considered more hazardous.

Aromatic hydrocarbons contain the benzene "ring," which is formed by six carbon atoms and contains double bonds, as shown in Exhibit I.7.12. Examples of aromatic hydrocarbons are benzene (C_6H_6) and toluene (C_7H_8).

(42) Self-accelerating decomposition temperature (SADT)

Once an organic peroxide exceeds the maximum safe storage temperature, it can reach the self-accelerating decomposition temperature (SADT). Once the organic peroxide reaches the SADT, there is no intervention possible to stop the decomposition.

(43) Solubility

The ability of a substance to form a solution with water can be important when determining control methods. For example, gasoline is insoluble, while anhydrous ammonia is soluble.

(44) Solution and slurry

A solution is a mixture in which all of the ingredients are completely dissolved. It is a homogeneous mixture of the molecules, atoms, or ions of two or more different substances. A slurry is a pourable mixture of a solid and a liquid.

(45) Specific gravity

Specific gravity is the weight of a solid or liquid compared to an equal volume of water. If a material has a specific gravity greater than 1.0 and it does not dissolve in water, it will sink. If its specific gravity is less than 1.0, it will float on water. Specific gravity becomes important when conducting some types of damming or booming operations and when dealing with

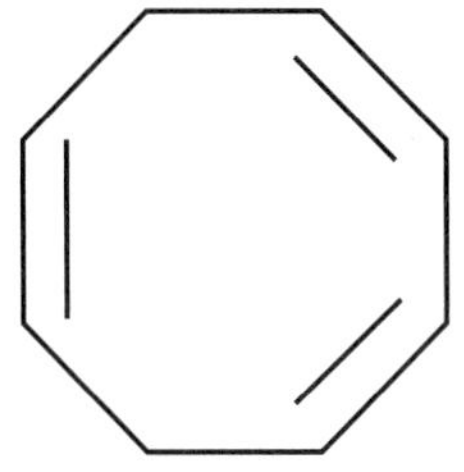

EXHIBIT I.7.12 *This sketch presents the chemical structure of benzene.*

flammable liquids, which generally have specific gravities less than 1.0 and spread around when water is applied.

(46) Strength

Strength is a term used to describe the concentration of a solution. In corrosives, strength refers to the degree of ionization of the acid or base in water. Hydrochloric acid is a strong acid, for example, and acetic acid is a weak acid.

(47) Sublimation

In sublimation, a substance passes directly from the solid state to the vapor state without passing through the liquid state. Solids such as dry ice and naphthalene (used in mothballs) are two examples. An increase in temperature increases the rate of sublimation. During an incident, a responder should assess the toxicity and flammability of the vapors of any spilled material that sublimes. The opposite of sublimation is deposition.

(48) Temperature of product

The temperature of a product influences the measures taken to control an incident that involves that product. A product's temperature may also present hazards. An incident involving molten sulfur, for example, raises a different set of concerns than one involving a cryogenic material such as liquefied natural gas.

(49) Toxic products of combustion

All products of combustion, from cigarette smoke to the smoke from a fire involving pesticides, have some toxic effects. Some materials generate more highly toxic products of combustion than others, and appropriate levels of protective clothing and equipment must be used to counter them.

(50) Vapor density

Vapor density is the relative density of a vapor compared to air. The vapor density of air is 1.0. If a material has a vapor density higher than 1.0, it is heavier than air and settles. Toluene, for example, has a vapor density of 3.14, and it settles and pools in low-lying areas. If a vapor's density is less than 1.0, it is lighter than air and rises and tends to dissipate.

(51) Vapor pressure

Vapor pressure is the pressure exerted on the inside of a closed container by the vapor in the space above the liquid in the container. Products with high vapor pressures have a greater potential to breach their containers when heated, because the pressure increases as the temperature rises. Products with high vapor pressures are more volatile. Vapor pressure is measured in millimeters of mercury. The vapor pressure of water is 21 mm mercury, and the vapor pressure of chlorine is 4800 mm mercury.

(52) Vesicants (blister agents)

Vesicants, also known as blister and mustard agents, are used in warfare to produce casualties, degrade fighting efficiency, and force opposing troops to wear full protective equipment. Mustard agents include nitrogen mustards (HN-1, HN-2, HN-3), sulfur mustards (H, HD, HT), and mustard-lewisite (HL). Mustard agents are oily liquids ranging from colorless (in pure state) to pale yellow to dark brown, depending on the type and purity. They have a faint odor of mustard, onion, garlic, or horseradish, but because of olfactory fatigue, odor cannot be relied on for detection. Volatility varies with the particular compound. Mustard agents are only slightly soluble in water and may persist for long periods. HN-1 is more volatile and less persistent than HD, but it is only one-fifth as potent a vesicant to the skin. HN-3 is less volatile and more persistent than HD and has equal vesicant effects.

(53) Viscosity

Viscosity, a measure of the thickness of a liquid, determines how easily it flows. Liquids with high viscosity, such as heavy oils, must be heated to increase their fluidity. Liquids that are more viscous tend to flow more slowly, while those that are less viscous spread more easily. During an incident, liquids that are less viscous are likely to flow away from a leaking container, expanding the endangered area.

(54) Volatility

Volatility describes the ease with which a liquid or solid can pass into the vapor state. The higher a material's volatility, the greater its rate of evaporation. Vapor pressure is a measure of a liquid's propensity to evaporate. Thus, the higher a liquid's vapor pressure, the more volatile it is. During an incident, a volatile material disperses in air and expands the endangered area.

7.2.2.3 The hazardous materials technician shall describe the heat transfer processes that occur as a result of a cryogenic liquid spill.

Because cryogenic liquids are kept at temperatures below –150°F (–101°C), cryogenic liquid spills vaporize rapidly when exposed to the higher ambient temperatures of the atmosphere outside the tank. Expansion ratios for common cryogenics range from 560 to 1445 to 1.

7.2.2.4* Given five hazardous materials/WMD scenarios and the associated reference materials, the hazardous materials technician shall identify the signs and symptoms of exposure to each material and the target organ effects of exposure to that material.

The responder at this level is required to be able to use various references to determine what effect various chemicals have on target organs. For example, *Emergency Care for Hazardous Materials Exposures* is a good source for signs and symptoms and notes the chemicals that may have caused the effects [9].

A.7.2.2.4 The selection of scenarios to test the knowledge and ability to identify exposure symptoms should include the following:

(1) Select materials common to the jurisdiction. This selection can be based on historical local records or any of the materials listed in Table A.5.2.3(9) that are commonly spilled throughout the country (i.e., chlorine, anhydrous ammonia, mineral acids, bases, and aliphatic and aromatic solvents).
(2) Select concentrations and formulation of the materials common to the jurisdiction. It is especially important with pesticides to select realistic scenarios because the state of matter, behavior, and exposure routes can vary considerably from technical-grade materials to common-use formulations.
(3) Select weather conditions and release conditions appropriate to the jurisdiction because the behavior and the exposure hazards can vary considerably from summer conditions in the deep south to winter conditions in the north.

The setting for scenarios and simulations should be as realistic as possible and should be locally oriented to increase the realism. Fictionalizing of an event or incident may be necessary to eliminate legal or political implications, but should replicate the local environment as closely as possible.

7.2.2.5 The hazardous materials technician shall identify two methods for determining the pressure in bulk packaging or facility containers.

7.2.2.6 The hazardous materials technician shall identify one method for determining the amount of lading remaining in damaged bulk packaging or facility containers.

7.2.3* Describing the Condition of the Container Involved in the Incident.

Given examples of container damage, the hazardous materials technician shall describe the damage and shall meet the related requirements of 7.2.3.1 through 7.2.3.5.

A.7.2.3 The condition of the container should be described using one of the following terms:

(1) Undamaged, no product release
(2) Damaged, no product release
(3) Damaged, product release
(4) Undamaged, product release

7.2.3.1* Given examples of containers, including the DOT specification markings for non-bulk and bulk packaging, and associated reference guides, the hazardous materials technician shall identify the basic design and construction features of each container.

A.7.2.3.1 See Annex K for the appropriate reference guides.

7.2.3.1.1 The hazardous materials technician shall identify the basic design and construction features, including closures, of the following bulk containers:

(1) Cargo tanks
    (a) Compressed gas tube trailers
    (b) Corrosive liquid tanks
    (c) Cryogenic liquid tanks
    (d) Dry bulk cargo tanks
    (e) High-pressure tanks
    (f) Low-pressure chemical tanks
    (g) Nonpressure liquid tanks

Cargo tank specifications are found in 49 CFR Part 178 [4]. The DOT has established the following five classifications for cargo tanks:

1. MC-306 (DOT-406)
2. MC-307 (DOT-407)
3. MC-312 (DOT-412)
4. MC-331
5. MC-338

The classifications in parentheses became effective October 1, 1993. Some changes were made to the specifications, but both classifications of cargo tanks have the same general appearance. For illustrations of these various car types, see Section 5.2 of NFPA 472.

(2) Fixed facility tanks
    (a) Cryogenic liquid tanks
    (b) Nonpressure tanks
    (c) Pressure tanks

Anthony M. Ordile notes the following in the *Fire Protection Handbook*:

> The thickness of the metal used in tank construction is based not only on the strength required to hold the weight of the liquid, but also on an added allowance for corrosion. When intended for storing corrosive liquids, the specifications for the thickness of the tank shell are increased to provide additional metal and allow for the expected service life of the tank. In some cases, special tank linings are used to reduce corrosion. . . . All aboveground storage tanks should be built of steel or concrete, unless the character of the liquid necessitates the use of other materials. Both steel and concrete tanks resist heat from exposure fires. [2] p. 7-17

The required nominal thickness of shell plates must be the greater of the design shell thickness, including any corrosion allowance, or the hydrostatic test shell thickness, but in no case less than the values given in Commentary Table I.7.1. Concrete tanks require special engineering. Unlined concrete tanks should be used only for the storage of liquids with a specific gravity of 40 degrees API or heavier.

COMMENTARY TABLE I.7.1 *Shell Plate Thicknesses*

Nominal Tank Diameter (ft)	*Nominal Thickness (in.)*
Smaller than 50 ft (15.2 m)	$^{3}/_{16}$ in. (4.8 mm)
(15.2 m–36.4 m)	$^{1}/_{4}$ in. (6.6 mm)
120 ft to 200 ft inclusive (36.4 m–60.8 m)	$^{5}/_{16}$ in. (8 mm)
Over 200 ft (60.8 m)	$^{3}/_{8}$ in. (9.5 mm)

Source: *Fire Protection Handbook,* 20th edition [Table 7.2.1, p. 7-17]

For additional information, see NFPA 30, *Flammable and Combustible Liquids Code*, which provides specifications for fixed containers designed to store flammable and combustible liquids [10].

(3) Intermediate bulk containers (also known as tote tanks)
(4) Intermodal tanks
  (a) Nonpressure intermodal tanks
    i. IM-101 portable tank (IMO Type 1 internationally)
    ii. IM-102 portable tank (IMO Type 2 internationally)
  (b) Pressure intermodal tanks (DOT Specification 51; IMO Type 5 internationally)
  (c) Specialized intermodal tanks
    i. Cryogenic intermodal tanks (DOT Specification 51; IMO Type 7 internationally)
    ii. Tube modules

In the *Fire Protection Handbook*, Wright notes the following:

> Tank containers consist of a single, metal tank mounted inside a sturdy, metal supporting frame. This unique frame structure, built to international standards, makes tank containers multimodal (intermodal). . . . The tank container tank is generally built as a cylinder enclosed at the ends by heads. Other tank shapes – rectangular tanks – and configurations – tube modules – exist but they are rare. . . . Ninety percent or more of the tanks are built of stainless steel, and the rest are constructed of mild steel. . . . Aluminum and magnesium alloy tanks are available, but they cannot be used in water transport mode. Minimum head and shell thickness are measured in terms of "equivalent thickness in mild steel . . . after forming. . . . For regulated materials, the minimum thickness is $^{3}/_{8}$ in. (9.5 mm). For stainless steel tanks, the minimum thickness for nonregulated materials is slightly less than $^{1}/_{8}$ in. (3.2 mm). For regulated commodities, the minimum thickness is just under $^{3}/_{16}$ in. (4.8 mm). Most tanks are built according to the pressure-vessel standards of the ASME, and the welds are x-rayed." [2] p. 21-131

(5) One-ton containers (pressure drums)
(6) Pipelines
(7) Railroad cars
  (a) Cryogenic liquid tank cars
  (b) Nonpressure tank cars
  (c) Pneumatically unloaded hopper cars
  (d) Pressure tank cars

Wright notes the following in the *Fire Protection Handbook*:

Typically, tank cars are enclosed longitudinal cylinders (with or without compartments) with rounded ends called heads. Tank cars may or may not have underframes. If not, the tank must bear the stresses of train movement. Tank cars have railroad running gear and safety appliances. . . . Carbon steel is used in over 90 percent of tank car tanks, with aluminum making up most of the remainder. Stainless steel, referred to in the regulations as alloy steel, is used in a smaller number of cars. Nickel or nickel alloy is used for some tanks in acid or food services. Regulations specify the plate thickness of materials used to construct tank car tanks. [2] p. 21-123

See Commentary Table I.7.2.

COMMENTARY TABLE I.7.2 Minimum Tank and Jacket Plate Thickness

Minimum Plate Thickness After Forming	*Common Use of Plate Thickness*
Steel	
11 gauge (approximately $^1/_8$ in.), also aluminum	Jacket of insulated tank cars, or jacket for thermally protected cars
$^7/_{16}$ in.	Tank for nonpressure tank cars, outer tank for nonpressure tank within a tank; or shell portion of outer tank for cryogenic liquid tank cars
$^1/_2$ in.	Head puncture resistance (head shield); or head portion of outer tank for cryogenic liquid tank cars
$^9/_{16}$ in.	Tank for steel pressure tank cars with tank test pressures of 200 psi and below
$^{11}/_{16}$ in.	Tank for steel pressure tank cars with tank test pressure of 300 psi and greater
$^3/_4$ in.	Tank for steel pressure tank cars in chlorine service
Aluminum	
$^1/_2$ in.	Tank for nonpressure aluminum tank cars
$^5/_8$ in.	Tank for aluminum pressure tank cars

Notes:

1. If high-tensile strength steels are used, plate thickness for pressure tank car tanks may be reduced but in no case should that thickness be less than $^1/_2$ in.

2. Plate thickness for nonpressure steel tank cars with expansion domes is a function of where the plate is used in the tank and the diameter of the tank. Thickness ranges from $^1/_4$ to $^1/_2$ in. For tank cars built after 1969, minimum plate thickness is $^7/_{16}$ in., except for tank cars with a diameter of 112 in. to 122 in. where the thickness is $^1/_2$ in.

For SI units: 1 in. = 2.54 cm; 1 psi = 6.9 kPa.

7.2.3.1.2 The hazardous materials technician shall identify the basic design and construction features, including closures of the following nonbulk containers:

(1) Bags
(2) Carboys

Carboys are bottles of glass or plastic used to transport liquids, especially corrosives. The bottle could be protected in an outer box, wooden crate, or plywood drum. The carboy's capacity can range to more than 20 gal (75.7 L).

(3) Drums

A drum is a cylindrical container used to store and transport liquids and solids. Drums are constructed of metal, plastic, fiberboard, or other material and their capacity ranges up to 55 gal (208.2 L).

(4) Cylinders

Pressurized cylinders are steel cylinders of strong construction to hold compressed gas. The cylinders have covered fittings to allow connection to the material stored. The fittings must be protected by a cap to prevent damage and unintended release of the contents.

7.2.3.1.3 The hazardous materials technician shall identify the basic design features and testing requirements on the following radioactive materials packages:

(1) Excepted
(2) Industrial
(3) Type A
(4) Type B
(5) Type C

7.2.3.2 The hazardous materials technician shall describe how a liquid petroleum product pipeline can carry different products.

Many different products can be transported through a single pipeline. A "pig" isolates a product that could have just been transported from a different product that is to follow. The pig is a plug-like device that hugs the interior of the pipe but is able to move with the flow of a liquid product (see Exhibit I.7.13). The pig pushes one product in front of it through a pipeline while it is pushed by another product behind it. A monitoring device in the pig can transmit the pig's location and when it passes a specified location.

7.2.3.3 Given an example of a pipeline, the hazardous materials technician shall identify the following:

(1) Ownership of the line

The ownership of a line may be indicated with a pipeline marker. The responder can contact the company identified on the marker for information on the type of product in the line and

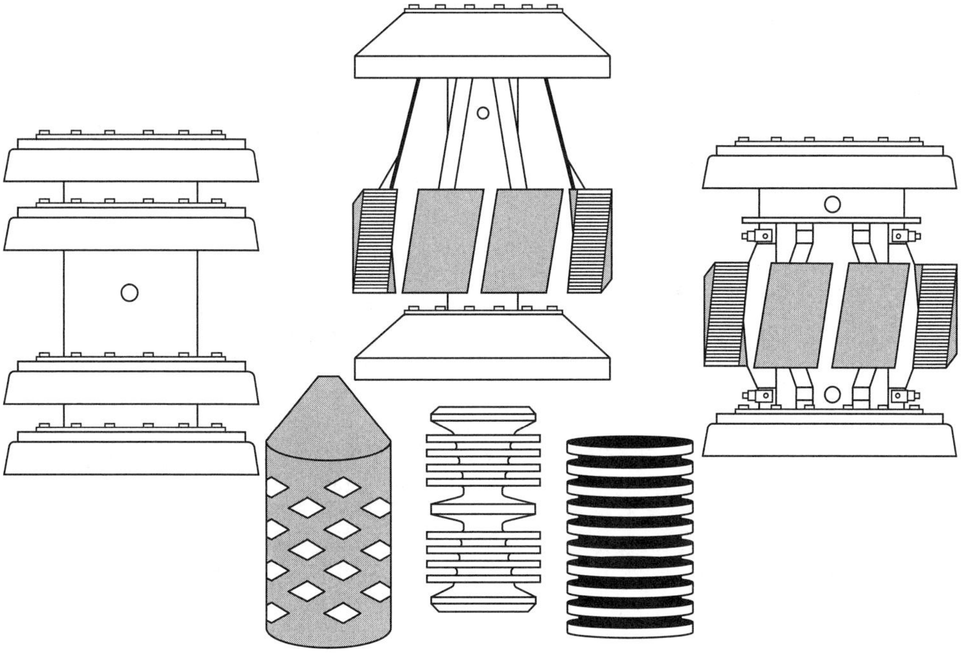

EXHIBIT I.7.13 *Pigs are plug-like devices that isolate products in a pipeline.*

for help in shutting down the line. Technician level responders should also consider acquiring pipeline maps for their area so they can pre-plan their response needs for pipeline incidents.

(2) Procedures for checking for gas migration
(3) Procedure for shutting down the line or controlling the leak
(4) Type of product in the line

7.2.3.4* Given examples of container stress or damage, the hazardous materials technician shall identify the type of damage in each example and assess the level of risk associated with the damage.

A.7.2.3.4 Some of the types of damage that containers can incur include the following:

(1) *Cracks.* A crack is a narrow split or break in the container metal that can penetrate through the metal of the container.

A cracked pressurized container should be considered to be critically damaged because determining the depth of a crack is difficult, which makes establishing the safest course of action even more problematic. Cracks in the base metal, no matter how small, warrant off-loading the container. If the crack is associated with a dent, the tank car should not be moved until it has been off-loaded.

(2) *Scores.* A score is a reduction in the thickness of the container shell. It is an indentation in the container made by a relatively blunt object. A score is characterized by the relocation of the container or weld metal in such a way that the metal is pushed aside along the track of contact with the blunt object.

Scores that are not accompanied by a dent are not critical. Nor is a score critical if it crosses a welded seam and does not cut into the heat-affected area of the weld. If a score does cut into the heat-affected area of a welded seam, however, the situation is critical and the product should be off-loaded.

(3) *Gouges.* A gouge is a reduction in the thickness of the container. It is an indentation in the shell made by a sharp, chisel-like object. A gouge is characterized by the cutting and complete removal of the container or weld metal along the track of contact.
(4) *Dents.* A dent is a deformation of the container metal. It is caused by impact with a relatively blunt object. With a sharp radius, there is the possibility of cracking.

Large dents are not serious unless gouges are also present. Longitudinal dents along the long axis of a tank are the most serious. Dents with a radius of less than 4 in. (10 cm) that occur in tank cars built before 1966 are critical, and tanks that are so damaged should be off-loaded without being moved. The situation is also critical if the tank car was built in 1966 or later and the radius of the dent is less than 2 in. (5 cm); in such cases, the tank car should be off-loaded without being moved.

EXHIBIT I.7.14 *The TI is usually equal to the maximum radiation level in mrem/hour. (Source: U.S. Department of Energy)*

7.2.3.5 Given a scenario involving radioactive materials, the hazardous materials technician, using available survey and monitoring equipment, shall determine if the integrity of any container has been breached.

Aside from performing a visual inspection of a package to assess package integrity, two methods can be used to help the responder determine if the integrity of a container has been breached in accordance with 7.2.3.5. One method involves surveying for radioactive contamination by taking smear or wipe samples on the ground surrounding a package. The wipe samples can then be surveyed in an area with low background radiation levels to determine if contamination was released from the package.

Another method involves using the transport index (TI), which is illustrated in Exhibit I.7.14, as a starting point for determining if the integrity of a package has been breached. The TI is used on Radioactive Yellow-II and Yellow-III labels. In addition to being shown on the shipping labels, the TI is also listed on the shipping papers. The TI is, in most cases, equal to the maximum radiation level in mrem/hour at 1 m from an undamaged package. The TI can be an indicator for determining the external radiation hazard of an undamaged package, and it can be a starting point for determining whether or not damage has occurred. For example, a package with a Radioactive Yellow-III label listing a TI of 2.5 should read approximately 2.5 mrem/hour at a distance of 1 m from the package. A reading of 15 mrem/hour 1 m from this type of package indicates potential damage.

7.2.4 Predicting Likely Behavior of Materials and Their Containers Where Multiple Materials Are Involved.

Given examples of hazardous materials/WMD incidents involving multiple hazardous materials or WMD, the hazardous materials technician shall predict the likely behavior of the material in each case and meet the requirements of 7.2.4.1 through 7.2.4.3.

7.2.4.1 The hazardous materials technician shall identify at least three resources available that indicate the effects of mixing various hazardous materials.

7.2.4.2 The hazardous materials technician shall identify the impact of the following fire and safety features on the behavior of the products during an incident at a bulk liquid facility and explain their significance in the analysis process:

(1) Fire protection systems

Fire protection systems allow responders to apply fire-extinguishing agents sooner and to control an incident in its initial stages, thus reducing the threat to adjoining containers.

(2) Monitoring and detection systems

Monitoring and detection systems permit early notification of potential problems and allow responders to initiate control actions while an incident is still relatively small, thereby limiting the threat to other containers.

(3) Pressure relief and vacuum relief protection

Pressure relief devices need to be capable of operating freely in order to keep the tank from failing violently. Vacuum relief devices also need to be operable in order to keep the tank from imploding.

(4) Product spillage and control (impoundment and diking)

Dikes and other impoundment features are designed to contain spilled product and minimize the exposure to adjoining tanks. See NFPA 30 for information about diking requirements. [10]

(5) Tank spacing

Adequate tank spacing minimizes the hazard to uninvolved tanks. See NFPA 30 for information about tank spacing requirements.

(6) Transfer operations

Transferring product from one tank to another minimizes the danger to surrounding containers.

7.2.4.3 The hazardous materials technician shall identify the impact of the following fire and safety features on the behavior of the products during an incident at a bulk gas facility and explain their significance in the analysis process:

(1) Fire protection systems
(2) Monitoring and detection systems

(3) Pressure relief protection
(4) Transfer operations

7.2.5 Estimating the Likely Size of an Endangered Area.

Given examples of hazardous materials/WMD incidents, the hazardous materials technician shall estimate the likely size, shape, and concentrations associated with the release of materials involved in an incident by using computer modeling, monitoring equipment, or specialists in this field and shall meet the requirements of 7.2.5.1 through 7.2.5.4.

7.2.5.1 Given the emergency response plan, the hazardous materials technician shall identify resources for dispersion pattern prediction and modeling, including computers, monitoring equipment, or specialists in the field.

The responder is required to be able to identify resources that help predict dispersion patterns. These resources include the weather service; computer models; industrial facilities; colleges or universities; county, state, or federal agencies, such as health departments; environmental protection agencies; and the U.S. Coast Guard, among others. Responders must be able to predict dispersion patterns to determine which areas are likely to become endangered by a spill.

7.2.5.2 Given the quantity, concentration, and release rate of a material, the hazardous materials technician shall identify the steps for determining the likely extent of the physical, safety, and health hazards within the endangered area of a hazardous materials/WMD incident.

Once the responder has identified the concentrations of the materials that have been released, the HMT is required to determine the acceptable exposure limits for those materials. This step can be done using resources that list acceptable exposure values. This competency is necessary when dealing with, for example, sulfuric acid, the chemical with the highest quantity production in the United States. Sulfuric acid comes in many different strengths, ranging from a mild acid to oleum or fuming sulfuric acid, which is a strong oxidizer.

7.2.5.2.1 The hazardous materials technician shall describe the following terms and exposure values and explain their significance in the analysis process:

The various exposure values are based on values established by the American Conference of Governmental Industrial Hygienists (ACGIH). OSHA, and the National Institute of Occupational Safety and Health (NIOSH). Responders should be familiar with all of these terms because different references can use different values, and responders must understand the differences and similarities so that they can make the appropriate comparisons.

(1) Counts per minute (cpm) and kilocounts per minute (kcpm)

Counts per minute and kilocounts per minute are measurements of radioactivity.

(2) Immediately dangerous to life and health (IDLH) value

The IDLH value is the maximum level of a hazardous material to which a healthy worker can be exposed without suffering irreversible health effects or impairment. If at all possible, exposure to this level should be avoided. If avoidance is not possible, responders should wear Level A or Level B protection with positive pressure self-contained breathing apparatus (SCBA) or a positive pressure supplied air respirator (SAR) with an auxiliary escape system. This limit is established by OSHA and NIOSH.

(3) Incubation period

Incubation period is the latency between exposure to a pathogen and onset of symptoms.

(4) Infectious dose

An infectious dose is the amount of a pathogen necessary to manifest its pathogenicity. It is dependent on pathogenic variables as well as host variables such as health, gender, predisposition, and several others.

(5) Lethal concentrations (LC_{50})

The LC_{50} is the median lethal concentration of a hazardous material. The term is defined as the concentration of a material in air that, on the basis of laboratory tests (inhalation route), is expected to kill 50 percent of a group of test animals when administered in a specific time period.

(6) Lethal dose (LD_{50})

The lethal dose of a substance is a single dose that causes the death of 50 percent of a group of test animals exposed to it by any route other than inhalation.

(7) Parts per billion (ppb)

Parts per billion (ppb) denotes one particle of a given substance for every 999,999,999 other particles. This is roughly equivalent to one drop of ink in a lane of a public swimming pool, one second per 32 years, or one part in 10_9.

(8) Parts per million (ppm)

The values used to establish exposure limits are quantified in parts per million or parts per billion. A good reference to remember is that 1 percent volume in air equals 10,000 ppm, 1 ppm equals 1000 ppb. So if the HMT obtains a reading from a sampling instrument of 0.5 percent volume in air, it is equivalent to 5000 ppm, or 5,000,000 ppb. If the threshold limit value (TLV) is determined to be 7500 ppm, the reading from the instrument can be related to determine the degree of hazard.

(9) Permissible exposure limit (PEL)

Permissible exposure limit (PEL) is a term OSHA uses in its health standards covering exposures to hazardous chemicals. This limit is similar to the threshold limit value time-weighted average (TLV/TWA) established by the ACGIH. PEL, which generally relates to legally enforceable TLV limits, is the maximum concentration, averaged over 8 hours, to which 95 percent of healthy adults can be repeatedly exposed for 8 hours per day, 40 hours per week.

(10) Radiation absorbed dose (rad)

The rad is a unit used to measure a quantity called absorbed dose. This relates to the amount of energy actually absorbed in some material, and is used for any type of radiation and any material. One rad is defined as the absorption of 100 ergs per gram of material. The unit rad can be used for any type of radiation, but it does not describe the biological effects of the different radiations. While the rad is still in use, the gray is the international standard unit for absorbed dose.

(11) Roentgen equivalent man (rem), millirem (mrem), microrem (μrem)

A roentgen is the international unit of the intensity of x-rays or gamma rays. It is the quantity of radiation that would produce, in air, ions carrying a positive or negative charge equal to one electrostatic unit in 0.001293 gram of air.

Rescue and recovery operations that involve radiological hazards can present very complex issues with regard to controlling personnel exposure. The type of response to these operations is generally left up to the officials in charge of the emergency situation. The official's judgment is guided by many variables, which include determining the risk versus the benefit of the action as well as how to involve other personnel in the operation. If the situation involves a substantial personal risk, volunteers should be used. The use of volunteers will be based on their age, training, experience, and previous exposure. The EPA has established guidelines for control of emergency exposures. These exposure guidelines are summarized in Commentary Table I.7.3.

***COMMENTARY TABLE I.7.3** Guidelines for Control of Emergency Exposures*

Dose Limit (rem)	*Activity Performed*	*Condition*
5	All	
10	Protection of major property	Where lower dose limit not practicable
25	Life-saving or protection of large populations	Where lower dose limit not practicable
>25	Life-saving or protection of large populations	Only on a voluntary basis to personnel fully aware of the risks involved

Source: U.S. Environmental Protection Agency, EPA 400-R-92-001, Manual of Protective Action Guides and Protective Actions for Nuclear Incidents, May 1992

(12) Threshold limit value ceiling (TLV-C)

The TLV-TWA is the maximum concentration, averaged over 8 hours, to which a healthy adult can be repeatedly exposed for 8 hours per day, 40 hours per week.

(13) Threshold limit value short-term exposure limit (TLV-STEL)

The TLV-STEL is the maximum average concentration, averaged over a 15-minute period, to which healthy adults can be safely exposed for up to 15 minutes continuously. Exposure should not occur more than 4 times a day with at least 1 hour between exposures.

(14) Threshold limit value time-weighted average (TLV-TWA)

The threshold limit value ceiling is the maximum concentration to which a healthy adult can be exposed without risk of injury. This limit is comparable to the IDLH, and exposures to higher concentrations should not occur.

7.2.5.2.2 The hazardous materials technician shall identify two methods for predicting the areas of potential harm within the endangered area of a hazardous materials/WMD incident.

Predicting the areas of potential harm in the endangered area involves determining the potential concentrations of the hazardous material that has been released. This determination includes the toxicity of the concentrations and the length of time that persons in the endangered area would be exposed.

7.2.5.3* The hazardous materials technician shall identify the steps for estimating the outcomes within an endangered area of a hazardous materials/WMD incident.

An estimate is a series of predictions that attempts to provide an overall picture of potential outcomes. Responders are required to assess the information gathered during analysis and predict the outcome based on that assessment.

According to the National Fire Academy's (NFA) *Initial Response to Hazardous Materials Incidents, Course II: Concept and Implementation,* it is necessary to break an incident into the following three components:

1. The product
2. The container
3. The environment [11]

Each of these components can then be broken into three subgroups of damage, hazard, and vulnerability risk. In addition, incidents could have the following three elements that can occur separately or in conjunction with one another:

1. A spill
2. A leak
3. A fire

The estimate identifies the relationships between the three components of an incident and the three elements of an incident.

The NFA course states, "An estimate is made by analyzing the physical, cognitive, and technical information that has been gathered. Then, by breaking the incident into the components dealing with product, container, environment and their respective sub-groups, a conclusion can be drawn . . . [a] conclusion with some measure of quantifiable accuracy that suggests what the full impact(s) of the relationships will be."

It is important for HMTs to understand that this analysis continues throughout an incident. As new information is gathered, old estimates should be verified or new estimates made. Predictions, which should be based on worst-case scenarios, allow the responder to develop an overall estimate of the incident's potential outcomes.

Incident commanders should keep in mind that the safety of both emergency personnel and the public is their primary objective. There might be times when the most prudent action is no action, and the establishment of an appropriate evacuation area is the best possible course to take.

A.7.2.5.3 The process for estimating the potential outcomes within an endangered area at a hazardous materials/WMD incident includes determining the dimensions of the endangered area; estimating the number of exposures within the endangered area; measuring or predicting concentrations of materials within the endangered area; estimating the physical, health, and safety hazards within the endangered area; identifying the areas of potential harm within the endangered area; and estimating the potential outcomes within the endangered area.

7.2.5.4 Given three examples involving a hazardous materials/WMD release and the corresponding instrument monitoring readings, the hazardous materials technician shall determine the applicable public protective response options and the areas to be protected.

7.3 Competencies — Planning the Response

7.3.1 Identifying Response Objectives.

7.3.1.1 Given scenarios involving hazardous materials/WMD incidents, the hazardous materials technician shall describe the response objectives for each problem.

7.3.1.2 Given an analysis of a hazardous materials/WMD incident, the hazardous materials technician shall be able to describe the steps for determining response objectives (defensive, offensive, and nonintervention).

In Chapter 5 and Chapter 6, the operations level responder is required to determine the appropriate response objectives for responders at that level based on training and equipment. The response objectives for the HMT are expanded to include additional control options and protective actions.

In the *Fire Protection Handbook*, Wright notes the following:

> The first task in planning the response is to determine the response objectives (strategy) based on the estimated outcomes. The response objectives, based on the stage of incident, are the strategic goals for stopping the event now occurring or keeping future events from occurring. Two basic principles apply to these decisions:
>
> 1. One cannot influence events that have already happened or change the outcomes of those events.
> 2. The earlier that the event sequence can be interrupted, the more acceptable the loss.

The following steps should be taken when determining response objectives:

1. Estimate the exposures that could be saved. The level of response and the acceptable risk associated with a response is based on the exposures that can be saved. The number of exposures that could be saved is based on the estimated outcomes minus the exposures already lost.
2. Determine the response objectives. The response objectives, based on the stage of the incident, are the strategic goals for stopping the event now occurring or keeping future events from occurring. Decisions should focus on changing the actions of the stressors, the containment system, and the hazardous material. [2] p. 13-137

The National Incident Management System (NIMS) specifies that the incident commander will develop response objectives that in turn drive response strategies and tactics. The HMT can recommend strategies to the incident commander based on an evaluation of the hazards and risk and vulnerability assessment. Incident objectives should reflect the actions to be taken to bring the incident to a successful conclusion, and may be included in the incident action plan and reflected in the site safety plan. See Exhibit I.7.15.

7.3.2 Identifying the Potential Response Options.

7.3.2.1 Given scenarios involving hazardous materials/WMD incidents, the hazardous materials technician shall identify the possible response options (defensive, offensive, and nonintervention) by response objective for each problem.

7.3.2.2 The hazardous materials technician shall be able to identify the possible response options to accomplish a given response objective.

Because they can conduct offensive operations, HMTs can take actions that are not available to the responder at the operations level without mission-specific training. Exhibit I.7.16 shows a sample Response Objective Analysis Form, which helps the HMT identify the options available to accomplish response objectives. The responder should note that many of the options available can be used either offensively or defensively. For example, the size of the endangered area may be changed defensively by placing barriers around the endangered area to contain a spill or offensively by entering the hot zone and plugging the leak. The HMT may also choose defensive operations as the most prudent method of dealing with an incident.

__EXHIBIT I.7.15__ This chart depicts the relationship between the sequence of events and potential losses.

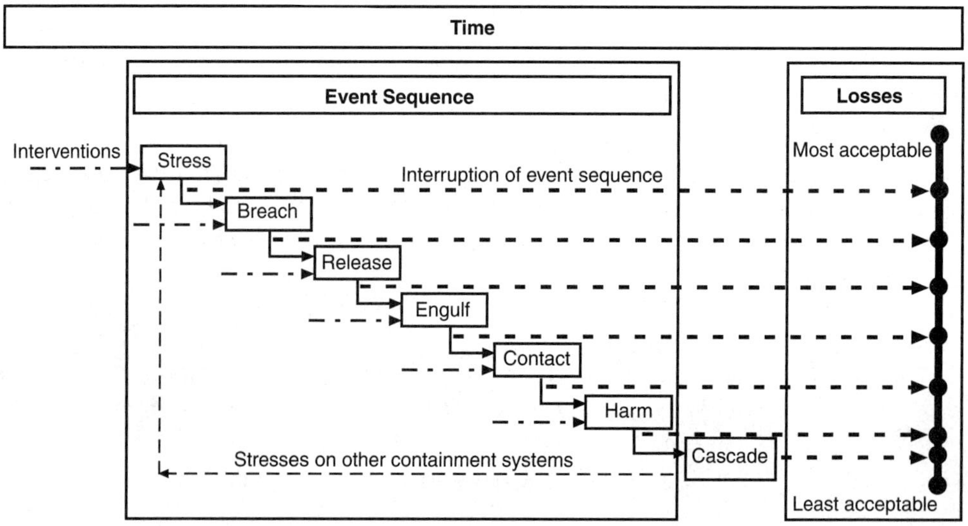

Response Objective Analysis Form	Containment System ID
	Material

Event Sequence

Stress	Breach	Release	Engulf	Contact	Harm

Response Objectives

Change Applied Stresses	Change Breach Size	Change Quantity Release	Change Size of Danger Zone	Change Exposures Contacted	Change Severity of Harm

Sample Response Options

Move stressor Move stressed system Shield stressed system	Chill contents Limit stress levels Activate venting devices Mechanical repair	Change container position Minimize pressure differential Cap off breach Remove contents	Barriers Dikes and dams Adsorbents Absorbents Diluents Reactants Overpack	Provide sheltering Begin evacuation Personal protective equipment	Rinse off contaminant Increase distance from source Provide shielding Provide prompt medical attention

EXHIBIT I.7.16 *This response objective analysis worksheet is used for identifying response options in a hazardous materials incident by response objective. (Source: Adapted from Ludwig Benner's Textbook for Use in the Study of Hazardous Materials Emergencies)*

7.3.3 Selecting Personal Protective Equipment.

Given scenarios of hazardous materials/WMD incidents with known and unknown hazardous materials/WMD, the hazardous materials technician shall determine the personal protective equipment for the response options specified in the incident action plan in each situation and shall meet the requirements of 7.3.3.1 through 7.3.3.4.7.

7.3.3.1 The hazardous materials technician shall identify and describe the four levels of personal protective equipment as specified by the Environmental Protection Agency (EPA) and the National Institute for Occupational Safety and Health (NIOSH).

Levels of protection are described in 29 CFR 1910.120, Appendix B, "General Description and Discussion of the Levels of Protection and Protective Gear" [1]. Technicians are also advised to read A.6.2.3.1(1) in NFPA 472, which identifies the need to use protective clothing from a performance-based application. Additionally, the performance-based requirements for hazardous materials/WMD response are identified in NFPA 1991, *Standard on Vapor-Protective Ensembles for Hazardous Materials Emergencies*, and NFPA 1994, *Standard on Protective Ensembles for First Responders to CBRN Terrorism Incidents* [12, 13].

Those responders who have not been adequately trained to select and use personal protective equipment should not be permitted to wear such equipment at a hazardous materials incident. Training in the selection and use of PPE should be thorough and frequent enough for the responder to become intimately familiar with the equipment's limitations. Responders should be trained to select, don, operate, test, clean, maintain, and care for the clothing that they may need to use at an incident.

During an incident, responders may have to change the level of protection they are wearing from a high level of protection to a lower one or vice versa. This decision is made by the incident commander, based on evaluation of the hazards to personnel present at the scene.

7.3.3.2 The hazardous materials technician shall identify and describe personal protective equipment options available for the following hazards:

(1) Thermal
(2) Radiological
(3) Asphyxiating
(4) Chemical (liquids and vapors)
(5) Etiological (biological)
(6) Mechanical (explosives)

7.3.3.3 The hazardous materials technician shall identify the process to be considered in selecting respiratory protection for a specified action option.

The type of respiratory protection required depends on many factors. A key factor is the level of protective clothing necessary to protect the HMT. Level A protection limits the user to positive pressure SCBA or a positive pressure SAR. The responder should always use the highest level of protection until the levels of concentration have been determined.

7.3.3.4 The hazardous materials technician shall identify the factors to be considered in selecting chemical-protective clothing for a specified action option.

One type of chemical-protective clothing cannot satisfy all protection needs for every hazardous materials incident. Based on a hazard analysis and risk assessment, the primary considerations required in selecting the proper PPE should focus on the known hazards, the specific tasks to be performed, the compliance with instructions and limitations as provided by the manufacturer, and the potential for exposure, especially through degradation, penetration, and permeation. An option now becoming more popular with many hazardous materials teams is the use of disposable suits.

7.3.3.4.1 The hazardous materials technician shall describe the following terms and explain their impact and significance on the selection of chemical-protective clothing:

(1) Degradation

Degradation of chemical-protective clothing (CPC) can be either chemical or physical. The result of degradation is an increased likelihood that a hazardous material will permeate and penetrate the garments, thus endangering the health of the responder.

Chemical degradation can be minimized by avoiding unnecessary contact with chemicals and by undergoing effective decontamination procedures. The choice of CPC is important. The garments a responder wears should be chosen based on their compatibility with the chemicals involved in an incident and breakthrough times consistent with their expected use.

Protective clothing can also degrade physically, such as might occur when the garment rubs against a rough surface. CPC wearers should recognize the physical limitations of their garments and make every effort to avoid circumstances that may cause the material to be damaged physically.

NFPA 1991 and NFPA 1994 contain criteria for abrasion or tear testing and for manufacturers' certification of CPC. When purchasing CPC, the responder should ascertain whether the garments are certified to the appropriate NFPA CPC standard.

(2) Penetration

Penetration is the movement of a material through a suit's closures, which include zippers, buttonholes, seams, flaps, and other design features of CPC. Hazardous materials can also penetrate CPC through cracks or tears in the suit's fabric.

Protection against penetration is vital. A regular and routine program of inspection can help uncover conditions that could lead to penetration. CPC must also be properly stored and regularly maintained and tested in order to ensure that it can still provide the proper level of protection. NFPA 1991 and NFPA 1992, *Standard on Liquid Splash-Protective Ensembles and Clothing for Hazardous Materials Emergencies,* provide criteria for testing CPC for penetration resistance and for manufacturer's certification of CPC [14].

(3) Permeation

Different fabrics have different resistance levels to chemical permeation and will absorb chemicals over a period of time. NFPA 1991 and NFPA 1994 provide guidelines for manufacturer permeation testing and certification. CPC technologies are constantly changing and it is essential that technicians stay abreast of these changes that will reduce inherent hazards of CPC and provide adequate protection.

NFPA 1991 requires the manufacturer to provide documentation on a garment's permeation resistance for 3 hours against at least the following chemicals:

- Acetone*
- Acetonitrile
- Anhydrous ammonia
- Carbon disulfide
- Chlorine
- Dichloromethane
- Diethyl amine*
- Dimethyl formamide
- Ethyl acetate
- Hexane*
- Nitrobenzene
- Sodium hydroxide*
- Sulfuric acid*
- Tetrachloroethylene*
- Tetrahydrofuran*
- Toluene*

Note that NFPA 1994 requires that the manufacturer provide documentation on a garment's permeation resistance for 1 hour for at least the chemicals noted with an asterisk in the preceding list.

Before buying CPC, the HMT should make sure that the garment meets, and has been certified as meeting, the appropriate standard. When choosing CPC for use at an incident, the HMT must be sure that the garment is compatible with the type of material to which it is going to be exposed. In any event, the wearer is advised to use extreme caution. Available data do not cover every situation the responder may encounter.

7.3.3.4.2 The hazardous materials technician shall identify at least three indications of material degradation of chemical-protective clothing.

Indications of material degradation that the HMT is required to know are as follows:

- Stiffness or excess pliability
- Tears, cuts, or abrasions
- Damage to zippers or other closures

This list is not all-inclusive, and users should check manufacturers' recommendations for the inspection of CPC.

7.3.3.4.3* The hazardous materials technician shall identify the different designs of vapor-protective and splash-protective clothing and describe the advantages and disadvantages of each type.

Type 1 protection is a fully encapsulating, airtight vapor-protective suit with SCBA worn on the inside. This protection offers responders the maximum level of protection. Challenges that might be encountered include potential heat exposure, communications, visibility, and mobility.

Type 2 protection is a nonencapsulating suit with SCBA worn on the outside. This protection provides more comfort and mobility and allows air bottles to be exchanged more easily. The facepiece serves as an effective barrier against chemical permeation for respiratory protection depending on compatibility.

Type 3 protection is an SAR with encapsulation suits. This protection provides positive pressure using an air line hose for extended operations. Limitations include the length of the hose line, up to 300 ft (91.4 m), which can become a potential tripping hazard, restrict maneuverability, and require an emergency air source, such as an SCBA or escape pack.

A.7.3.3.4.3 Refer to the American Chemistry Council and Association of American Railroads Hazardous Materials Technical Bulletin *Recommended Terms for Personal Protective Equipment,* issued in October 1985. Also refer to NFPA 1991, *Standard on Vapor-Protective Ensembles for Hazardous Materials Emergencies*; NFPA 1992, *Standard on Liquid Splash-Protective Ensembles and Clothing for Hazardous Materials Emergencies*; and NFPA 1994, *Standard on Protective Ensembles for First Responders to CBRN Terrorism Incidents.* It is important to remember that the EPA levels of protection are not "performance based," as are NFPA 1991, NFPA 1992, and NFPA 1994.

7.3.3.4.4 The hazardous materials technician shall identify the relative advantages and disadvantages of the following heat exchange units used for the cooling of personnel in personal protective equipment:

(1) Air cooled
(2) Ice cooled
(3) Water cooled
(4) Phase change cooling technology

Wearing CPC can cause wearers to suffer increased heat stress. Thus, it is important for responders to be monitored closely while working in CPC.

Some CPC garments have temperature control features. Some have air-cooling systems that require an air line and large quantities of breathable air. Others incorporate water-cooling systems that require an ice supply or refrigeration units and a pump. This adds additional weight and bulk to the suit. Users should conduct a thorough evaluation of such units to make sure they are appropriate for the intended use.

There are also vests that can hold coolant packs. This requires a supply of the frozen coolant packs or an ice source at the scene. The vests add additional weight to the CPC.

7.3.3.4.5 The hazardous materials technician shall identify the process for selecting protective clothing at hazardous materials/WMD incidents.

The HMT is required to determine the appropriate level of protection based on the criteria established by the EPA and OSHA and as shown in Commentary Table I.7.4. The CPC chosen must be compatible with the chemicals to which it will be exposed. The HMT must determine whether the breakthrough times of the chosen garment will allow him or her enough time to

COMMENTARY TABLE I.7.4 EPA/OSHA Protection Levels

Level of Protection	*Equipment*	*Protection Provided*	*Should Be Used When:*	*Limiting Criteria*
A	RECOMMENDED Pressure-demand, full-facepiece SCBA or pressure-demand, supplied-air respirator with escape SCBA Fully encapsulating chemical-resistant suit Inner chemical-resistant gloves Chemical-resistant safety boots/shoes Two-way radio communications OPTIONAL Cooling unit Coveralls Long cotton underwear Hard hat Disposable gloves and boot covers	The highest available level of respiratory, skin, and eye protection	The chemical substance has been identified and requires the highest level of protection for skin, eyes, and the respiratory system based on either: Measured (or potential for) high concentration of atmospheric vapors, gases, or particulates or Site operations and work functions involving a high potential for splash, immersion, or exposure to unexpected vapors, gases, or particulates of materials that are harmful to skin or capable of being absorbed through the intact skin Substances with a high degree of hazard to the skin are known or suspected to be present, and skin contact is possible Operations must be conducted in confined, poorly ventilated areas until the absence of conditions requiring Level A protection is determined	Fully encapsulating suit material must be compatible with the substances involved
B	RECOMMENDED Pressure-demand, full facepiece SCBA or pressure-demand supplied-air respirator with escape SCBA Chemical-resistant clothing (overalls and long-sleeved jacket; hooded, one- or two-piece chemical splash suit; disposable chemical-resistant one-piece suit) Inner and outer chemical-resistant gloves	The same level of respiratory protection but less skin protection than Level A. It is the minimum level recommended for initial site entries until the hazards have been further identified.	The type and atmospheric concentration of substances have been identified and require a high level of respiratory protection, but less skin protection. This involves atmospheres: With IDLH concentrations of specific substances that do not represent a severe skin hazard or	Use only when the vapor or gases present are not suspected of containing high concentrations of chemicals that are harmful to skin or capable of being absorbed through the intact skin

(continues)

COMMENTARY TABLE I.7.4 Continued

Level of Protection	*Equipment*	*Protection Provided*	*Should Be Used When:*	*Limiting Criteria*
	Chemical-resistant safety boots/shoes Hard hat Two-way radio communications OPTIONAL Coveralls Disposable boot covers Face shield Long cotton underwear		Containing less than 19.5 percent oxygen Presence of incompletely identified vapors or gases is indicated by direct-reading organic vapor detection instrument, but vapors and gases are not suspected of containing high levels of chemicals harmful to skin or capable of being absorbed through the intact skin. Use only when the vapor or gases present are not suspected of containing high concentrations of chemicals that are harmful to skin or capable of being absorbed through the intact skin	Use only when it is highly unlikely that the work being done generates either high concentrations of vapors, gases, or particulates or splashes of material that affect exposed skin
C	RECOMMENDED Full facepiece, air-purifying, canister-equipped respirator Chemical-resistant clothing (overalls and long-sleeved jacket; hooded, one- or two-piece chemical splash suit; disposable chemical-resistant one-piece suit) Inner and outer chemical-resistant gloves Chemical-resistant safety boots/shoes Hard hat Two-way radio communications OPTIONAL Coveralls Disposable boot covers Face shield Escape mask Long cotton underwear	The same level of skin protection as Level B, but a lower level of respiratory protection	The atmospheric contaminants, liquid splashes, or other direct contact does not adversely affect any exposed skin The types of air contaminants have been identified, concentrations measured, and a canister is available that can remove the contaminant All criteria for the use of air-purifying respirators are met	Atmospheric concentration of chemicals must not exceed IDLH levels
D	RECOMMENDED Coveralls Safety boots/shoes Safety glasses or chemical splash goggles Hard hat OPTIONAL Gloves Escape mask Face shield	No respiratory protection Minimal skin protection	The atmosphere contains no known hazard Work functions preclude splashes, immersion, or the potential for unexpected inhalation of or contact with hazardous levels of any chemicals	This level should not be worn in the hot zone and warm zone The atmosphere must contain at least 19.5 percent oxygen

enter the contaminated area safely, do the necessary work, leave the area, and undergo decontamination. In some cases, layering materials can provide increased protection. It is important to remember that no single garment protects a responder against all chemicals. The HMT should know the manufacturer's recommendations or how to find the information regarding the use of the CPC.

7.3.3.4.6 Given three examples of various hazardous materials, the hazardous materials technician shall determine the protective clothing construction materials for a given action option using chemical compatibility charts.

7.3.3.4.7 The hazardous materials technician shall identify the physiological and psychological stresses that can affect users of personal protective equipment.

Specialized protective clothing, particularly fully encapsulating garments, increases the stress a responder can feel when working at a hazardous materials incident. Persons wearing CPC usually experience a loss of dexterity and mobility. The higher the level of protection is, the greater this loss. The responder's visibility is also restricted, and his or her communication is affected. In addition, wearing CPC increases the likelihood of heat stress and heat exhaustion. Reductions in dexterity, mobility, visibility, and communication, in turn, create additional physical and mental stresses.

Familiarity with wearing and working in the garment reduces any mental anxiety associated with garment wear. Thus, frequent drills are essential. The HMT's awareness of these additional stresses is important, and he or she should receive adequate rest and rehabilitation before, during, and after suit wear. Drinking fluids such as water before donning CPC reduces some of the effects of excess heat.

7.3.4 Selecting Decontamination Procedures.

Given a scenario involving a hazardous materials/WMD incident, the hazardous materials technician shall select a decontamination procedure that will minimize the hazard, shall determine the equipment required to implement that procedure, and shall complete the following tasks:

(1) Describe the advantages and limitations of each of the following decontamination methods:

The two basic ways to decontaminate something are physical and chemical. Physical methods manually separate the chemical from the material being decontaminated by scrubbing or washing the material, or both. Physical decontamination is often easier than chemical decontamination, but it may not completely remove all the contaminants.

Chemical methods involve adding another chemical that changes the physical or chemical properties of one chemical into another or into a form that facilitates its removal. Unfortunately, the chemical process involved could introduce other hazards. Care must be taken to collect all the contamination that has been removed by either method and to dispose of it properly.

(a) Absorption

Absorption is the process by which materials hold liquids. Many types of commercial absorbents are available. Sand or soil can also be used for this purpose, although they are more suited for decontaminating equipment or the area surrounding a spill than they are for decontaminating personnel. Absorbents are often readily available, but they must be disposed of properly because the absorbent substance retains the properties of the material absorbed.

(b) Adsorption

Adsorption is a chemical method of decontamination involving the interaction of a hazardous liquid and a solid sorbent surface. Examples of adsorbents are activated charcoal, silica or aluminum gel, fuller's earth, and other clays. Adsorption produces heat and can cause spontaneous combustion. Adsorbents must be disposed of properly.

(c) Chemical degradation

Chemical degradation is the natural breakdown of the contaminants as they age. An example of chemical degradation is the evaporation of a flammable liquid spill. The decontamination of an oil spill on a beach because of manual (pressure washer) or natural (wave action) action is an example of physical degradation. Either of the two methods has limitations, depending on factors such as the location of the spill and the toxicity of the material. In some cases, however, these methods are the most practical.

(d) Dilution

Dilution, which simply reduces the concentration of a contaminant, is best used on materials that are soluble or miscible in water, such as chlorine and ammonia. An advantage of dilution is that solutes, especially water, are generally available in large quantities. A disadvantage is the necessary collection and disposal of runoff.

(e) Disinfecting
(f) Evaporation

In some cases, responders allow a hazardous material to evaporate, particularly if the vapors do not present a hazard. A small spill of gasoline, for example, can be allowed to evaporate as long as it does not present a vapor problem. Evaporation is an easy operation and requires minimal personnel. This method is not as effective on porous surfaces as it is on nonporous surfaces, however, and it could take quite a while, depending on the quantity of the chemical involved.

(g) Isolation and disposal

Disposal is the direct removal of a contaminant from a carrier. An example of this method is the removal of a contaminated object from a piece of equipment. This type of decontamination may not entirely remove all contamination.

(h) Neutralization

Neutralizers alter a contaminant chemically so that the resulting chemical is harmless. For example, the addition of soda ash to an acidic solution can increase the pH, making it a chemically harmless substance. Many neutralizing chemicals present hazards of their own, however, and should only be used by HMTs who are fully aware of the consequences. One advantage of neutralizers is that by rendering the remaining material harmless, they reduce the problem of disposal.

(i) Solidification

Commercial products are available that cause certain liquids to solidify. One advantage of solidification is that it allows responders to confine a small spill relatively quickly. As with other decontaminants, however, the resulting solid must be disposed of properly when the incident is over.

(j) Sterilization
(k) Vacuuming

Vacuuming allows for the collection of materials, either liquid or solid, into containers. The equipment being used must be appropriate for the material being vacuumed. If the material is corrosive or flammable, for example, specialized equipment is needed.

(l) Washing

A very effective decontamination process for many materials involves washing the contaminated person, building, or equipment. Materials that are not soluble in water, such as oil-based contaminants, can be washed with detergent solutions. Washing equipment, protective clothing, and personnel is one of the easiest methods of decontamination. However, collecting and properly disposing of any runoff is necessary.

(2) Identify three sources of information for determining the applicable decontamination procedure and identify how to access those resources in a hazardous materials/WMD incident.

Among the sources of technical information about decontamination an HMT is required to know are CHEMTREC, CANUTEC, SETIQ, MSDSs, product manufacturers, the National Response Center, and local or regional poison control centers.

7.3.5 Developing a Plan of Action.

Given scenarios involving hazardous materials/WMD incidents, the hazardous materials technician shall develop a plan of action, including site safety and a control plan, that is consistent with the emergency response plan and standard operating procedures and within the capability of available personnel, personal protective equipment, and control equipment for that incident, and shall meet the requirements of 7.3.5.1 through 7.3.5.5.

Wright notes the following in the *Fire Protection Handbook*:

> After selecting the response option for a hazardous material incident, a plan of action including safety and health considerations should be developed. This plan of action describes the response objectives and options and the personnel and equipment required to accomplish the objectives. The plan provides a permanent record of the decisions made at the incident. An organization's standard operating procedures provide the basis of this plan of action. Input from all segments of the response community is considered in developing the plan. Based on the specific incident conditions, the standard operating procedures are modified without having to write an entire plan for each incident.
>
> A plan of action also outlines the safety and health procedures to protect responders and the public from the potential hazards at an incident. These procedures should address incident management, communications protocol (both internal and external), control zones for incident security, personal protective equipment use, decontamination procedures, and documentation. They also include designation of a safety sector and a safety officer, emergency medical care procedures, environmental monitoring, emergency procedures, and personnel monitoring.
>
> Components for a typical plan of action would include the following:
>
> 1. Site description
> 2. Entry objectives
> 3. On-scene organization and coordination
> 4. On-scene control
> 5. Hazard evaluation
> 6. Personal protective equipment
> 7. On-scene work assignments
> 8. Communications procedures
> 9. Decontamination procedures
> 10. On-scene safety and health considerations including designation of the safety officer, emergency medical care procedures, environmental monitoring, emergency procedures, and personnel monitoring [2] p. 13-141

7.3.5.1 The hazardous materials technician shall describe the purpose of, procedures for, equipment required for, and safety precautions used with the following techniques for hazardous materials/WMD control:

(1) Absorption
(2) Adsorption

The technique of adsorption refers to the surface retention of certain solid, liquid, or gaseous molecules, which cause an agent to stick to or become chemically attached to the exterior of another material, such as soil.

(3) Blanketing
(4) Covering
(5) Damming
(6) Diking
(7) Dilution
(8) Dispersion
(9) Diversion
(10) Fire suppression
(11) Neutralization

The technique of neutralization is the process by which another chemical is applied to the original spill to form a less harmful by-product through an energetic exothermic reaction, which can produce toxic and flammable vapors. Advantages include the considerable reduction in the release of harmful vapors, and the by-product of the reaction can be disposed of at less cost and effort.

Pump sprayers work well in the mixing and application of neutralizing agents. Acid neutralization routinely involves the use of a weak base, such as soda ash/sodium carbonate. The most commonly used buffer is baking soda/sodium bicarbonate, which can react with both acids and bases, preventing extreme pH swings. Vinegar is the most common and least expensive weak acid.

(12) Overpacking

The technique of overpacking is a physical method of containment by placing a leaking or damaged container, drum, or vessel inside a larger, specially constructed container to confine any further release of product. Even though the container should be at least temporarily repaired before being placed inside of the overpack, it is important that the overpack containers be compatible with the released materials.

Common sizes of both steel and polyethylene liquid overpacks include 5, 15, 30, 55, and 85 gal (19, 57, 114, 208, and 322 L). Compressed gas cylinder coffins, such as those designed to handle 150 lb (68 kg) chlorine cylinders, have been recently introduced.

Several considerations must be taken into account as the leaking container is prepared to be overpacked. Not only could the container have become weakened from whatever impact caused the breach, but the actual physical act required to move it may involve mechanical equipment. A forklift or hoist may be used to prevent related injuries because of the sheer size and weight of the containers, which could range from 100 lb to 1000 lb (45.4 kg to 454 kg).

(13) Patching

Patches are used to repair leaks, holes, rips, gashes, or tears in container shells, piping systems, and valves by placing them over the breach and holding them in place to stop the flow.

Considering the specific conditions, container pressure, and chemical compatibility is important when determining the appropriate patching device. Whether homemade or commercial, patching devices include pipe sleeve patches, chain thumb screws, chlorine bonnets, inflatable drums, container sleeves, bags, toggle bolt compression patches, gasket patches, glued patches, and epoxy putties.

(14) Plugging

The process of plugging involves inserting, driving, or screwing a chemically compatible object into the breach of a container to reduce or temporarily stop the flow. Cracks around the plug should be filled to ensure a good seal, although the compounds used should only be considered a temporary repair, taking into account the strength of the material, the size of the hole, and the potential internal pressure.

Different plugs include those constructed of wood, rubber, metal drift pins (like those found in the Chlorine B and C kits), and solid metal pins. Even simple devices, which include boiler plugs, screws, golf tees, wooden wedges, and cones, can be most effective. Rubber, plastic, and wooden plugs, which are designed to be driven, must be constructed of softer material so that their shape can be modified to fill in a rough and jagged breach. Exhibit I.7.17 shows the techniques used in plugging and patching. As a safety precaution, the reason for the release must be identified prior to controlling the failure in order to prevent the potential for a more violent rupture from overpressurization, which can cause such devices to become lethal projectiles.

(15) Pressure isolation and reduction (flaring; venting; vent and burn; isolation of valves, pumps, or energy sources)
(16) Retention
(17) Solidification
(18) Transfer

Transfer operations are complex and can be hazardous. The HMT should be very cautious when considering this type of operation. Usually transfer operations are conducted by a technician with a tank car specialty, cargo tank specialty, or intermodal tank specialty (see Chapters 12, 13, and 14 of NFPA 472); by a specialty cleanup company; or by personnel from the shipper or manufacturer.

As Hildebrand and Noll point out in *Handling Gasoline Tank Truck Emergencies: Guidelines and Procedures*, the most common methods of product transfer for MC-306/DOT-406 cargo tanks are vacuum pumps, vehicles with power take-off (PTO) pumps, or air-driven portable pumps [15]. This statement is true for most nonpressure cargo tanks.

Transfer operations involving pressure tanks present additional hazards and should be undertaken only by personnel with the necessary technician specialty training and skills.

(19) Vapor control (dispersion, suppression)

7.3.5.2 Given a scenario involving a hazardous materials/WMD incident, the hazardous materials technician shall develop the site safety and control plan that must be included as part of the incident action plan.

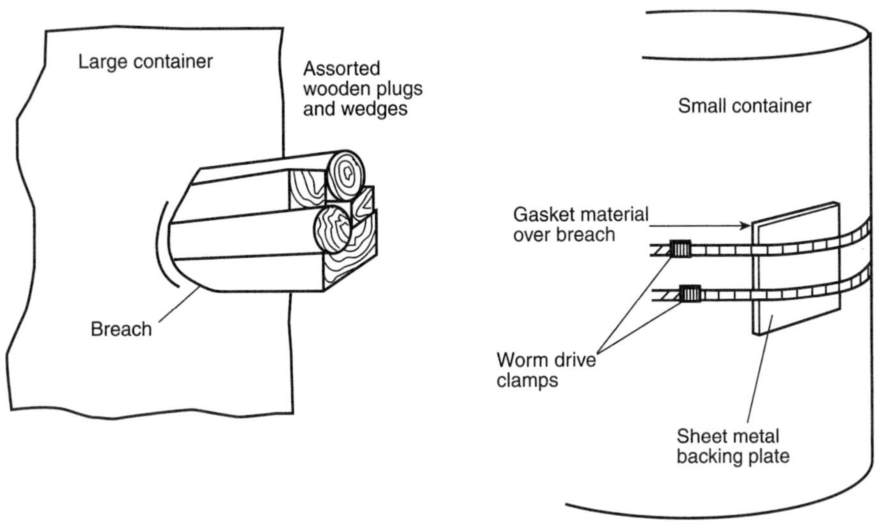

EXHIBIT I.7.17 *The use of plugging and patching materials is shown in these examples.*

According to 29 CFR 1910.1200, Appendix C, site safety and control plans should address the following:

- Analysis of hazards on the site and a risk analysis of those hazards
- Site map or sketch
- Site work (control) zones
- Use of buddy system
- Site communications
- Command post
- Standard operating procedures and safe work practices
- Medical assistance and triage area
- Other relevant topics

This plan should be a part of the employer's emergency response plan or an extension of it to the specific site.

7.3.5.2.1 The hazardous materials technician shall list and describe the safety considerations to be included.

7.3.5.2.2 The hazardous materials technician shall identify the points that should be made in a safety briefing prior to working at the scene.

7.3.5.3* The hazardous materials technician shall identify the atmospheric and physical safety hazards associated with hazardous materials/WMD incidents involving confined spaces.

This competency became necessary after the promulgation of the OSHA regulations on confined space operations. The HMT should become very familiar with the OSHA regulations. Failure to comply could lead to substantial fines.

A.7.3.5.3 Safety hazards associated with confined spaces could include the following:

(1) Atmospheric hazards
    (a) Oxygen-deficient atmosphere
    (b) Oxygen-enriched atmosphere
    (c) Flammable and explosive atmospheres
    (d) Toxic atmosphere
(2) Physical hazards
    (a) Engulfment hazards
    (b) Falls and slips
    (c) Electrical hazards
    (d) Structural hazards
    (e) Mechanical hazards

7.3.5.4 The hazardous materials technician shall identify the pre-entry activities to be performed.

Initially, activities for overall site safety must be performed. The HMT needs to determine the identification of the safety officer, the hazard control zones, an escape route, a designated withdrawal signal, and the identification of safe locations for personnel and equipment uphill and upwind.

Prior to entry, a safety briefing should be held for both the entry and backup teams to ensure that everyone understands the potential health and safety hazards, the objectives of the entry operations, and the specific tasks and procedures. This time should also be used to confirm designated radio channels, hand and verbal emergency signals, requirements for protective clothing, and the location and layout of the decontamination area. The entry team should not be permitted to enter the hot zone until the backup team is in place and the technical decontamination area is fully operational. All responders should remain alert and mindful of any unsafe practices or potentially dangerous conditions.

7.3.5.5 The hazardous materials technician shall identify the procedures, equipment, and safety precautions for preserving and collecting legal evidence at hazardous materials /WMD incidents.

Responders are required to follow organizational standard operating procedures for collecting evidence. A typical evidence collection kit includes marker pen, ties, ruler, camera, quick-slitter, flashlight, drop cloths, plastic bags, liquid/biological/air sampling packages, and log collection sheets. All evidence collected should be documented on the evidence collection form and formally turned over to law enforcement personnel using a chain of custody process. Proper protective clothing and equipment should be worn during the collection process, and hazardous materials should be stored in appropriate containers. For additional information on evidence collection and preservation, refer to the mission-specific competencies for Section 6.5.

7.4 Competencies — Implementing the Planned Response

7.4.1* Performing Incident Command Duties.

Given the emergency response plan or standard operating procedures and a scenario involving a hazardous materials/WMD incident, the hazardous materials technician shall demonstrate the duties of an assigned function in the hazardous materials branch or group within the incident command system and shall identify the role of the hazardous materials technician during hazardous materials/WMD incidents.

The emergency response plan is the link between the community's response plans and the operational personnel who are expected to implement those plans. Emergency response plans incorporate standard operating procedures, which should identify the type of response appropriate to a particular type of incident, as well as site-specific procedures. The emergency response plan should also address alerting procedures, response and coordination procedures, personnel, the command structure, communications, and training.

A.7.4.1 The functions within the hazardous materials group or branch can include the following:

(1) Hazardous materials branch director/group supervisor
(2) Assistant safety officer — Hazardous materials
(3) Site access control leader
(4) Decontamination leader
(5) Technical specialist — Hazardous materials leader
(6) Safe refuge area manager

7.4.2 Using Protective Clothing and Respiratory Protection.

The hazardous materials technician shall demonstrate the ability to don, work in, and doff liquid splash–protective, vapor-protective, and chemical-protective clothing and any other specialized personal protective equipment provided by the AHJ, including respiratory protection, and shall complete the following tasks:

In addition to the following tasks, the hazardous materials technician must also demonstrate the ability to record the use, repair, and testing of CPC according to the manufacturer's specifications and recommendations. The hazardous materials technician must also be able to describe the maintenance, testing, inspection, and storage procedures for PPE provided by the AHJ according to the manufacturer's specifications and recommendations.

(1) Describe three safety procedures for personnel working in chemical-protective clothing.

The safety procedures in 7.4.2(1) include the obvious considerations of keeping the individuals cool and protected from heat exposure, the prevention of dehydration, medical monitoring, and stringent accounting for time spent on air and in the suit. Three additional concerns that could affect the safety of individuals working a hot zone include visibility, mobility, and communications.

The lack of peripheral vision significantly alters the responder's visibility through not only the SCBA but also the window of the encapsulating Level A or Level B suits. If the suit is too large, the face shield might need to be held close to the mask for the individual to look around. The fogging up of the window is another concern, especially when the use of a hand towel from inside the suit to wipe it off or the application of antifog spray is not effective.

Mobility and dexterity are compromised each time the responder dons an additional layer of protection. Performing relatively simple tasks, such as operating detection and monitoring equipment, becomes more challenging when wearing nitrile, silver shield, or butyl rubber gloves, for example. Also, until the relief valve releases internal pressure, Level A suits inflate with exhaled air, creating a ballooning effect, which can compromise both mobility and visibility.

A responder can be difficult to understand when speaking through an SCBA to a partner or over hand-held radios. Communicating becomes even more challenging when a chemical-protective suit is worn. Throat or bone mikes are the most effective means of communication among responders as long as the background noise in the surrounding environment is conducive. Otherwise, standard hand signals and easily recognizable motions and gestures could be the extent of communications exchanged between two individuals in Level A or Level B suits.

(2)* Describe three emergency procedures for personnel working in chemical-protective clothing.

Because personnel wearing vapor-protective clothing might experience a loss of mobility, dexterity, vision, and communications capability, closely monitoring them is important. Backup personnel wearing the same level of protective clothing must be available, and hand signals should be established to aid in communications. Personnel must also be monitored for the effects of heat, and a proper rehabilitation program should be in place to replenish fluids and allow for rest and recovery.

A.7.4.2(2) Emergency procedures for personnel working in vapor-protective clothing should include procedures for the following:

(1) Loss of air supply
(2) Loss of suit integrity
(3) Loss of verbal communications
(4) Buddy down in hot zone

(3) Demonstrate the ability to don, work in, and doff self-contained breathing apparatus in addition to any other respiratory protection provided by the AHJ.
(4) Demonstrate the ability to don, work in, and doff liquid splash–protective, vapor-protective, and chemical-protective clothing in addition to any other specialized protective equipment provided by the AHJ.

HMTs are required to practice donning and doffing CPC to become proficient. One of the more effective ways to evaluate an HMT's ability to don and doff the protective clothing provided is to conduct training exercises that require putting on the personal protective equipment and to conduct simulated control activities, followed by simulated decontamination.

Because some types of CPC are very costly, using garments that are no longer adequate for emergency response solely for training purposes is advisable. Hazmat training suits that are less expensive are also available, and these allow cost-effective training in the use of totally encapsulating suits.

7.4.3 Performing Control Functions Identified in Incident Action Plan.

Given scenarios involving hazardous materials/WMD incidents, the hazardous materials technician shall select the tools, equipment, and materials for the control of hazardous materials/ WMD incidents and identify the precautions for controlling releases from the packaging/ containers and shall complete the following tasks:

(1)*Given a pressure vessel, select the material or equipment and demonstrate a method(s) to contain leaks from the following locations:

The purpose of the competency in 7.4.3(1) is to allow the HMT to demonstrate the ability to choose the appropriate equipment and methods to control releases in various situations. The equipment necessary to conduct the operations in 7.4.3(1) could vary, depending on the type of material that is leaking and the physical properties of the damaged area, valve, or plug. For example, nonsparking tools should be used when working with flammable liquids or gases. And in some cases, special tools may be needed for certain valves. Such situations should be identified during pre-incident planning.

A.7.4.3(1) Contact the Chlorine Institute for assistance in obtaining training on the use of the various chlorine kits (Chlorine Institute, 1300 Wilson Blvd., Arlington, VA 22209; www.chlorineinstitute.org).

(a) Fusible plug
(b) Fusible plug threads
(c) Side wall of cylinder

A patch placed over a leak in the side wall of a container and secured can control the leak.

(d) Valve blowout

If a valve has blown out, it could be possible to use a wooden plug to stop the flow temporarily until the line can be shut down or the container can be drained for replacement.

(e) Valve gland

Leaks from valve glands may be controlled by tightening the packing nut. Capping the outlet could also be possible.

(f) Valve inlet threads

If a leak is around the inlet threads of a valve, it could be possible to tighten the valve assembly.

(g) Valve seat

The valve should be tightened first to ensure that it is closed all the way. Opening and reclosing the valve to clear debris that is preventing the valve from seating properly might also be necessary. In addition, capping the outlet may be possible.

(h) Valve stem assembly blowout

If the valve assembly has blown out, driving a wooden plug into the hole to gain temporary control, depending on the pressure of the product in the container, could be possible.

(2)*Given the fittings on a pressure container, demonstrate the ability to perform the following:

A.7.4.3(2) See A.7.4.3(1).

(a) Close valves that are open
(b) Replace missing plugs
(c) Tighten loose plugs

This competency is intended to allow the HMT to demonstrate the ability to choose the appropriate equipment and methods to control the situations in 7.4.3(2). For the competency to be most effective, the HMT should demonstrate these skills while wearing CPC and equipment.

(3) Given a 55 gal (208 L) drum and applicable tools and materials, demonstrate the ability to contain the following types of leaks:

(a) Bung leak

Bung leaks can often be stopped by tightening the bung with a bung wrench.

(b) Chime leak

(c) Forklift puncture

Because a puncture made by a forklift could be large and irregularly shaped, a number of wooden plugs of different sizes or shapes can be used. While an appropriate plugging or patching device is being assembled, stopping the drum from leaking by turning it on its side, with the opening at the top, could be possible.

(d) Nail puncture

Punctures can often be stopped with a wooden plug. A sheet metal screw and a gasket can also be used.

(4) Given a 55 gal (208 L) drum and an overpack drum, demonstrate the ability to place the 55 gal (208 L) drum into the overpack drum using the following methods:

When conducting overpacking operations manually, responders must take care to lift and move the drum properly so as to avoid back strain and injuries to hands or feet.

(a) Rolling slide-in

The method in 7.4.3(4)(a) involves laying the drum on its side. Rollers are put underneath the drum, and the drum is rolled into the overpack drum. Items that can be used as rollers include lengths of pipe and other rounded materials.

(b) Slide-in

The slide-in method involves laying a leaking drum on its side and sliding it into an overpack drum.

(c) Slip-over

The slip-over method involves placing the overpack drum over the top of the leaking drum and manually rotating it upright. A device called a drum "up-ender" is available commercially to assist with this type of operation.

(5) Identify the maintenance and inspection procedures for the tools and equipment provided for the control of hazardous materials releases according to the manufacturer's specifications and recommendations.

The recommendations on maintenance, inspection, calibration, bump tests, cleaning, and warm-up periods are provided by the manufacturer and are required in 7.4.3(5) to be followed. External influences can also affect the quality of the readings given by instruments, such as extreme temperatures, altitude, and barometric pressure (oxygen monitors); radio frequencies, power lines, transformers, dust, and high humidity [photoionization detectors (PID)]; and limited shelf-life, response time from chemical to chemical, and the lack of tube interchangeability (colorimetric indicator or detector tubes). Weak batteries and electromagnetic fields can give false positive readings (radiation monitors). Not all flame ionization detectors (FID)/organic vapor analyzers (OVA) or halogen leak detectors are intrinsically safe. Prior to use, pH papers should be premoistened with water. All of these instruments are tools used to assist responders in determining the identification of hazardous substances, their approximate location and perimeter, and the estimated quantity of product. The results should always be confirmed with other equipment.

(6) Identify three considerations for assessing a leak or spill inside a confined space without entering the area.

Confined space operations are extremely dangerous, and the HMT must exercise the utmost caution during such operations. One of the most critical considerations is whether the confined space is oxygen-deficient. Another consideration is whether it contains a flammable or toxic atmosphere.

(7)*Identify three safety considerations for product transfer operations.

A.7.4.3(7) The safety considerations for product transfer operations should include the following:

(1) Bonding
(2) Grounding
(3) Elimination of ignition sources and shock hazards

A ground resistance tester and an ohmmeter should be utilized for grounding and bonding. The ground resistance tester measures the earth's resistance to a ground rod, and the ohmmeter measures the resistance of the connections to ensure electrical continuity. One ground rod might not be enough; more might have to be driven and connected to the first to ensure a good ground. In some cases, isolation would be a better option than bonding or grounding. In all cases involving vessels, the responder should consult appropriate vessel personnel who are familiar with the potential risks involved with electrical systems on marine tank vessels.

Bonding is the process of connecting two or more objects by means of a conductor. Grounding, a specific form of bonding, is the process of connecting one or more conductive objects to the ground.

A ground resistance tester and ohm meter should be utilized when grounding and bonding. The ground resistance tester measures the earth's resistance to a ground rod, and the ohm meter measures the resistance of the connections to ensure electrical continuity. One ground rod might not be enough; more might have to be driven and connected to the first to ensure a good ground. In some cases, isolation could be a better option than bonding or grounding. In all cases involving vessels, the responder should consult appropriate vessel personnel who are familiar with the potential risks involved with electrical systems on marine tank vessels.

Whenever flammables are transferred, responders must ensure that appropriate safety precautions in A.7.4.3(7) have been taken. Obviously, all ignition sources must be eliminated. In addition, proper bonding and grounding connections must be made.

(8) Given an MC-306/DOT-406 cargo tank and a dome cover clamp, demonstrate the ability to install the clamp on the dome.

This type of cargo tank is used frequently, so the HMT should become familiar with it and know how to secure the dome cover clamp, as shown in Exhibit I.7.18. Arrangements can sometimes be made with a local distributor to use a cargo tank for training purposes.

(9) Identify the methods and precautions used to control a fire involving an MC-306/DOT-406 aluminum shell cargo tank.

When exposed to fire, an MC-306/DOT-406 tank will melt, preventing the buildup of excessive pressure. Melting occurs only above the liquid level of the product.

Hildebrand and Noll note the following in *Handling Gasoline Tank Truck Emergencies: Guidelines and Procedures*:

> A fully involved MC306/DOT406 cargo tank truck fire will require a substantial amount of foam for final control and extinguishment. Attempting to apply 95 gpm (360 L/min) of foam from a single 5-gal (18.9-L) foam concentrate container onto a 9,000-gal (34,065-L) gasoline spill fire will ruin your day.

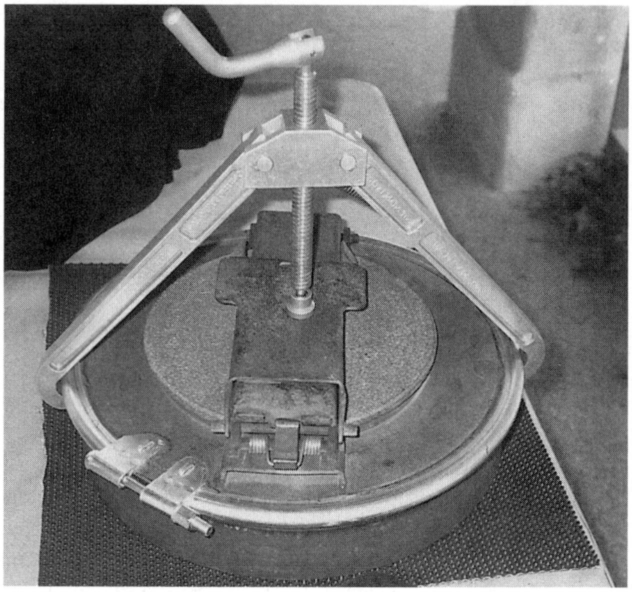

EXHIBIT I.7.18 *Clamping the dome cover on a cargo tanker is a necessary skill for the hazardous materials technician.*

> There may be situations where a controlled burn of the fire may be [appropriate] . . . gasoline burns at the rate of approximately one foot per hour. Although this may extend the duration of the incident, it will usually minimize both groundwater and surface water contamination. [15]

(10) Describe at least one method for containing each of the following types of leaks in MC-306/DOT-406, MC-307/DOT-407, and MC-312/DOT-412 cargo tanks:

(a) Dome cover leak
(b) Irregular-shaped hole
(c) Puncture
(d) Split or tear

The variety of methods to control leaks in cargo tanks are listed in 7.4.3(10). These methods might include the use of wooden plugs, patches, specially designed dome clamps, and others. The technician level responder should be aware of the type of product at hand and the compatible patching material types. The opportunity for the responder to train with the vehicles in 7.4.3(10) is important in order to have an understanding of the characteristics of each type of cargo tank and the special challenges each presents.

(11)*Describe three product removal and transfer considerations for overturned MC-306/DOT-406, MC-307/DOT-407, MC-312/DOT-412, MC-331, and MC-338 cargo tanks.

The cargo tanks in 7.4.3(11) present different problems if the tank overturns and the product inside must be transferred. In many cases this decision should be left up to the appropriate technician specialist. A cargo tank containing gasoline requires different procedures and equipment than a cargo tank of LP-gas. Again, it is important for the responder to be familiar with the various types of cargo tanks common in the response area and to work with shippers to gain an understanding of the considerations for conducting product transfer operations. In many cases, if not most, product transfer operations require the equipment and expertise of the shipper or a company involved in hazardous materials cleanup. The HMT needs to know where such assistance can be obtained and must have an understanding of the procedures and safety considerations involved.

A.7.4.3(11) Product removal and transfer considerations should include the following:

(1) Inherent risks associated with such operations
(2) Procedures and safety precautions
(3) Equipment required

7.4.4 Given MC-306/DOT-406, MC-307/DOT-407, MC-312/DOT-412, MC-331, and MC-338 cargo tanks, the hazardous materials technician shall identify the common methods for product transfer from each type of cargo tank.

7.4.5* Performing Decontamination Operations Identified in the Incident Action Plan.

The hazardous materials technician shall demonstrate the ability to set up and implement the following types of decontamination operations:

(1) Technical decontamination operations in support of entry operations
(2) Technical decontamination operations involving ambulatory and nonambulatory victims
(3) Mass decontamination operations involving ambulatory and nonambulatory victims

A.7.4.5 The decontamination processes identified in the incident action plan might be technical decontamination, mass decontamination, or both, depending on the circumstances of the incident. See 3.3.17.3 and 3.3.17.4.

7.5 Competencies — Evaluating Progress

All responders should understand why their efforts must be evaluated. If the HMTs are not making progress, the plan must be reevaluated to determine why progress is not being made. To decide whether the actions being taken at an incident are effective and the objectives are being achieved, the responder must determine whether the incident is stabilizing or increasing in intensity. The HMTs have no reason to remain in the immediate vicinity of an incident when nothing can be done to mitigate it and the situation might be about to deteriorate. If flames are impinging on an LP-gas vessel in the vapor space, for example, and the necessary volume of water to cool it is not available, prudent action would be to withdraw to a safe distance.

7.5.1 Evaluating the Effectiveness of the Control Functions.

Given scenarios involving hazardous materials/WMD incidents and the incident action plan, the hazardous materials technician shall evaluate the effectiveness of any control functions identified in the incident action plan.

7.5.2 Evaluating the Effectiveness of the Decontamination Process.

Given an incident action plan for a scenario involving a hazardous materials/WMD incident, the hazardous materials technician shall evaluate the effectiveness of any decontamination procedures identified in the incident action plan.

7.6 Competencies — Terminating the Incident

7.6.1 Assisting in the Debriefing.

Given a scenario involving a hazardous materials/WMD incident, the hazardous materials technician shall participate in the debriefing of the incident and shall meet the following requirements:

A debriefing is an opportunity to gather specific information from all operational personnel regarding positive, negative, and unique aspects of the response. The time should be used to determine whether strategic goals were achieved, which tasks were performed, by whom, when, and how, establishing an effective incident sequence.

(1) Describe three components of an effective debriefing.
(2) Describe the key topics of an effective debriefing.

Key topics should include specifics on HMTs involved in the response to the incident, what their objectives and tasks entailed, when the objectives were accomplished, the extent to which the objectives were successful, the injuries sustained, and the subsequent treatment provided.

(3) Describe when a debriefing should take place.

The debriefing focuses on information gathering, so the meeting should be conducted as soon after the incident as reasonably possible so that the details will be recalled simply and clearly. The debriefing does not have to be conducted at the same time with everyone in a specific location. The compiled information gathered during the discussion is the most beneficial to the modification of future responses.

(4) Describe who should be involved in a debriefing.

Hazardous materials incidents are fairly rare in most communities, with the exception of large industrial cities, so a debriefing is very helpful to document all the actions taken to control the incident. The debriefing should involve all participants and be done as soon as practical so that the responders still have a fresh recall of all the incident details.

7.6.2 Assisting in the Incident Critique.

Given a scenario involving a hazardous materials/WMD incident, the hazardous materials technician shall provide operational observations of the activities that were performed in the hot and warm zones during the incident and shall complete the following tasks:

(1) Describe three components of an effective critique.

An effective critique requires direction, participation, and solutions. The information acquired during a critique must remain focused, maintain forward momentum, and last only 1 to 2 hours to retain the appropriate quality and effectiveness. Participants must feel comfortable in openly sharing with the group observed deficiencies within the response system. The critique must be portrayed as a positive event and conducted in a way to get the most honest input from the responders. This opportunity is to identify shortcomings and deficiencies in operations, procedures, training, and site plans, while offering constructive recommendations and refraining from individual blame and criticism.

(2) Describe who should be involved in a critique.

Because a critique focuses on many levels of a response, representatives who were on the scene participating in the operations or command functions are the most beneficial to include. The individuals responsible for training, revising standard operating procedures and emergency response plans, and acquiring available resources should also be present to capture any necessary modifications. To make the critique session a positive, nonthreatening, learning experience, the critique leader must be a respected individual, preferably with counseling and/or arbitrator skills, who remains a neutral party, especially for the critiques of more sensitive incidents.

(3) Describe why an effective critique is necessary after a hazardous materials/WMD incident.

An effective critique provides an opportunity for those involved on the scene of a hazardous materials incident to identify and correct flaws and shortcomings. The critique also encourages suggestions and recommendations for improving future emergency response and preventing a recurrence. Valuable information can be derived from the discussions, including lessons learned, which have, in later cases, prevented related injuries and fatalities. By waiting several days to conduct the critique, the emotional stress from the incident should have subsided, creating a more relaxed and productive atmosphere as well as allowing time to gather relevant information from investigations and company reports.

A negative connotation is generally associated with the concept of a critique session, so a concerted effort must be made to address and emphasize those tasks that were performed well, including teamwork, safe operating procedures, and an effective response in which the incident action plan objectives were met.

(4) Describe which written documents should be prepared as a result of the critique.

Because of the infrequency of hazardous materials incidents, the timely completion and circulation of the critique among all the personnel of the organizations involved in the incident is essential. The critique report should try to focus on positive actions but should not ignore or discount areas where improvement is obviously needed. To acquire knowledge from a good critique report is better than to acquire it from a tragic experience at the next incident scene.

Some of this same information can be obtained by reading critiques from other hazardous materials teams' operations. All hazardous materials teams can benefit from publishing critiques in any of the national emergency services magazines.

7.6.3 Reporting and Documenting the Incident.

Given a scenario involving a hazardous materials/WMD incident, the hazardous materials technician shall complete the reporting and documentation requirements consistent with the emergency response plan or standard operating procedures and shall meet the following requirements:

(1) Identify the reports and supporting documentation required by the emergency response plan or standard operating procedures.
(2) Demonstrate completion of the reports required by the emergency response plan or standard operating procedures.
(3) Describe the importance of personnel exposure records.

Whether an exposure is the result of direct or cross-contamination, entry into unauthorized zones, donning improper PPE, or an actual PPE failure, the exposure must be reported to the medical director for evaluation and review to determine appropriate actions, treatments, testing, screening, and so forth. A copy of the incident report should be maintained in the individual's medical records because years may pass between the initial exposure and the development of related chronic effects. OSHA requires that exposure and medical records be maintained for at least 30 years after employees retire and that, upon request, these records must be made available to all affected employees and their representatives.

(4) Describe the importance of debriefing records.

Not only are the debriefing records important to assist in the chronological accounts of the incident and with the follow-up on the responders if problems arise at a later date, but careful recording of accurate information is subject to evidentiary summons, including all documentation, photographs, videotapes, audiotapes, and computer files.

(5) Describe the importance of critique records.

Critique records can be used to improve not only the safety of on-scene operations but also information related to the planning, training, identifying and prioritizing of hazards, and preventing negligent operations (for example, those in which procedures were not followed or equipment was not maintained). As critique records are revisited at a later date, they will assist in determining if the specific issues have been rectified and/or resolved.

(6) Identify the steps in keeping an activity log and exposure records.
(7) Identify the steps to be taken in compiling incident reports that meet federal, state, local, and organizational requirements.
(8) Identify the requirements for compiling hot zone entry and exit logs.
(9) Identify the requirements for compiling personal protective equipment logs.
(10) Identify the requirements for filing documents and maintaining records.

REFERENCES CITED IN COMMENTARY

1. Title 29, Code of Federal Regulations, Part 1910, U.S. Government Printing Office, Washington, DC.
2. Cote, A., ed., *Fire Protection Handbook®*, 20th ed., National Fire Protection Association, Quincy, MA, 2008.
3. Cote, A., ed., *Fire Protection Handbook®*, 19th ed., National Fire Protection Association, Quincy, MA, 2003.
4. Title 49, Code of Federal Regulations, Part 178, U.S. Government Printing Office, Washington, DC.
5. Title 49, Code of Federal Regulations, Part 173, U.S. Government Printing Office, Washington, DC.
6. *Emergency Response Guidebook,* U.S. Department of Transportation, Washington, DC 20590.
7. York, K., and Grey, G., *Hazardous Materials/Waste Handling for the Emergency Responder,* Fire Engineering, New York, 1989.
8. Title 40, Code of Federal Regulations, Part 261, U.S. Government Printing Office, Washington, DC.
9. Bronstein, A., *Emergency Care for Hazardous Materials Exposures,* 2nd ed., Mosby Yearbook, St. Louis, MO, 1994.
10. NFPA 30, *Flammable and Combustible Liquids Code,* National Fire Protection Association, Quincy, MA, 2008.
11. National Fire Academy, *Initial Response to Hazardous Materials Incidents, Course II: Concept and Implementation,* Emmitsburg, MD.
12. NFPA 1991, *Standard on Vapor-Protective Ensembles for Hazardous Materials Emergencies,* National Fire Protection Association, Quincy, MA, 2005.
13. NFPA 1994, *Standard on Protective Ensembles for First Responders to CBRN Terrorism Incidents,* National Fire Protection Association, Quincy, MA, 2007.
14. NFPA 1992, *Standard on Liquid Splash-Protective Ensembles and Clothing for Hazardous Materials Emergencies*, National Fire Protection Association, Quincy, MA, 2005.
15. Hildebrand, M., and G. Noll, *Handling Gasoline Tank Truck Emergencies: Guidelines and Procedures,* Fire Protection Publications, Stillwater, OK, 1991.

Additional References

Benner, L., *Textbook for Use in the Study of Hazardous Materials Emergencies,* 2nd ed., Lufred Industries Inc., Oakton, VA, 1978.

Bretherick, L., *Handbook of Reactive Chemical Hazards,* Butterworths, Boston, 1995.

Bowen, J., *Emergency Management of Hazardous Materials Incidents,* National Fire Protection Association, Quincy, MA, 1995.

Hawley's Condensed Chemical Dictionary, Van Nostrand Reinhold Co., New York, 1996.

Lewis, Sr., R., *Hazardous Chemicals Desk Reference,* Van Nostrand Reinhold Co., New York, 1996.

Sax, I., *Dangerous Properties of Industrial Materials,* Van Nostrand Reinhold Co., New York, 1996.

Spencer, A. B., and G. R. Colonna, *Fire Protection Guide to Hazardous Materials*, 13th ed., National Fire Protection Association, Quincy, MA, 2002.

Competencies for Incident Commanders

CHAPTER 8

The incident commander (IC) assumes control of an incident when skills beyond the capabilities of the awareness level responder are required. Because the IC is the supervisor of the entire incident, several competencies identical to those of the technician level that cover planning and management of the incident response must be met as part of IC training. In addition, ICs must have additional skills for managing the operation. Technician level competencies for mitigating the situation (e.g., plugging and diking) that are more "hands on" are not included in the competencies for an IC. See Exhibit I.8.1.

Regardless of the level of the incident or the personnel operating at the incident, the IC must be trained and competent, in accordance with 8.1.1, in the appropriate incident management system emergency operations plan, the risks of operating at the scene, state and federal resources, and the importance of decontamination.

The technical committee recognized that, in complex incidents involving the response of hazardous materials technicians (HMTs) or specialist employees, the IC would use their expertise in formulating response objectives, action options, and the plan of action. The IC relies on the hazardous materials officer (see Chapter 10 of NFPA 472) to directly supervise the team and provide appropriate technical information.

EXHIBIT I.8.1 *An incident command system enables the incident commander to manage the efforts of many specialists, including those trained in safety and decontamination. (Photo Courtesy of Henrik G. de Gyor)*

The IC is required to understand the types of resources that are available and the types of information each can provide. For example, the IC is not expected to have an in-depth knowledge of computer databases but should know the type of information that can be accessed in each of the major databases included in the hazardous materials response plan.

Similarly, the IC is not expected to be able to personally operate the various types of monitoring equipment used in the analysis of a hazardous materials incident, but he or she should be able to understand the functions and limitations of the various monitoring methods. The IC should be able to apply the results of air monitoring and detection in making and/or approving decisions pertaining to control zones, selection of personal protective equipment (PPE), public protection, and other related issues.

8.1 General

8.1.1 Introduction.

8.1.1.1 The incident commander (IC) shall be that person responsible for all incident activities, including the development of strategies and tactics and the ordering and release of resources.

8.1.1.2 The incident commander shall be trained to meet all competencies at the awareness level (Chapter 4), all core competencies at the operations level (Chapter 5), and all competencies in this chapter.

8.1.1.3 The incident commander shall receive any additional training necessary to meet applicable governmental occupational health and safety regulations.

8.1.1.4 The incident commander shall receive any additional training necessary to meet specific needs of the jurisdiction.

8.1.2 Goal.

8.1.2.1 The goal of the competencies at this level shall be to provide the incident commander with the knowledge and skills to perform the tasks in 8.1.2.2 safely.

8.1.2.2 In addition to being competent at the awareness and operations levels, the incident commander shall be able to perform the following tasks:

(1) Analyze a hazardous materials/WMD incident to determine the complexity of the problem and potential outcomes by completing the following tasks:
   (a) Collect and interpret hazard and response information from printed and technical resources, computer databases, and monitoring equipment.
   (b) Estimate the potential outcomes within the endangered area at a hazardous materials/WMD incident.
(2) Plan response operations within the capabilities and competencies of available personnel, personal protective equipment, and control equipment by completing the following tasks:
   (a) Identify the response objectives for hazardous materials/WMD incidents.
   (b) Identify the potential response options (defensive, offensive, and nonintervention) available by response objective.
   (c) Approve the level of personal protective equipment required for a given action option.
   (d)*Develop an incident action plan, including site safety and control plan, consistent with the emergency response plan or standard operating procedures and within the capability of available personnel, personal protective equipment, and control equipment.

A.8.1.2.2(2)(d) The following site safety and control plan considerations are from the EPA *Standard Operating Safety Guides*:

(1) Site description
(2) Entry objectives
(3) On-site organization
(4) On-site control
(5) Hazard evaluation
(6) Personal protective equipment
(7) On-site work plans
(8) Communication procedures
(9) Decontamination procedures
(10) Site safety and health plan

(3) Implement a response to favorably change the outcome consistent with the emergency response plan or standard operating procedures by completing the following tasks:
   (a) Implement an incident command system/unified command, including the specified procedures for notification and utilization of nonlocal resources (e.g., private, state, and federal government personnel).
   (b) Direct resources (private, governmental, and others) with task assignments and on-scene activities and provide management overview, technical review, and logistical support to those resources.
   (c) Provide a focal point for information transfer to media and local elected officials through the incident command system structure.
(4) Evaluate the progress of the planned response to ensure the response objectives are being met safely, effectively, and efficiently and adjust the incident action plan accordingly.
(5) Terminate the emergency phase of the incident by completing the following tasks:
   (a) Transfer command (control) when appropriate.
   (b) Conduct an incident debriefing.
   (c) Conduct a multiagency critique.
   (d) Report and document the hazardous materials/WMD incident and submit the report to the designated entity.

8.2 Competencies — Analyzing the Incident

8.2.1 Collecting and Interpreting Hazard and Response Information.

8.2.1.1 Given access to printed and technical resources, computer databases, and monitoring equipment, the incident commander shall collect and interpret hazard and response information not available from the current edition of the DOT *Emergency Response Guidebook* or an MSDS.

8.2.1.2 The incident commander shall be able to identify and interpret the types of hazard and response information available from each of the following resources and explain the advantages and disadvantages of each resource:

(1) Hazardous materials databases
(2) Monitoring equipment
(3) Reference manuals
(4) Technical information centers
(5) Technical information specialists

8.2.2 Estimating Potential Outcomes.

Given scenarios involving hazardous materials/WMD incidents, the surrounding conditions, and the predicted behavior of the container and its contents, the incident commander shall estimate the potential outcomes within the endangered area and shall complete the following tasks:

(1) Identify the steps for estimating the outcomes within an endangered area of a hazardous materials/WMD incident.

The IC must determine pertinent factors relating to the nature and type of the incident, what materials/containers are involved, and environmental influences in order to estimate the incident outcome. Types of incidents can involve spills, leaks, fires, odors, and other hazards. Product/container factors might include physical/chemical properties, hazards, quantities involved, type of container, stress or damage, safety features, among others. Environmental factors include what is at risk (i.e., people, environment, infrastructure, property), exterior versus interior operations, weather conditions, control/protective systems, air-handling systems, and terrain. All of these factors must be evaluated before the IC can make informed and valid decisions and develop an incident action plan.

(2) Describe the following toxicological terms and exposure values and explain their significance in the analysis process:

(a) Counts per minute (cpm) and kilocounts per minute (kcpm)
(b) Immediately dangerous to life and health (IDLH) value
(c) Infectious dose
(d) Lethal concentrations (LC_{50})
(e) Lethal dose (LD_{50})
(f) Parts per billion (ppb)
(g) Parts per million (ppm)
(h) Permissible exposure limit (PEL)
(i) Radiation absorbed dose (rad)
(j) Roentgen equivalent man (rem); millirem (mrem); microrem (μrem)
(k) Threshold limit value ceiling (TLV-C)
(l) Threshold limit value short-term exposure limit (TLV-STEL)
(m) Threshold limit value time-weighted average (TLV-TWA)

The IC must have a level of knowledge beyond that required at the operational level in order to clearly understand the potential hazards of the products that may be involved in a hazardous materials incident. See the commentary following 7.2.5.2.1 of NFPA 472 for an explanation of each preceding term.

Radiation exposure is measured in units of rad, roentgen, and rem. For all practical purposes, the three units are equal. The units roentgen and rem are most commonly used and often expressed in terms of milliroentgen or millirem. Milli means one one-thousandth (1 / 1000). Therefore, there are 1000 millirem (mrem) in one rem, or 1000 milliroentgens (mR) in one roentgen. The average radiation dose received by a person in the United States is approximately 360 mrem per year.

In the SI system (Systeme International d'Unites, or International System of Units), the unit for measuring radiation exposure is the sievert. One sievert is equal to 100 rem.

(3)* Identify two methods for predicting the areas of potential harm within the endangered area of a hazardous materials/WMD incident.

Predicting the areas of potential harm in the endangered area involves determining the potential concentrations of the hazardous material that has been released. This prediction includes the toxicity of the concentrations and the length of time that persons in the endangered area would be exposed.

A.8.2.2(3) Methods for predicting areas of potential harm can include use of the DOT *Emergency Response Guidebook* Table, Initial Isolation and Protective Action Distance, computer dispersion models, and portable and fixed air-monitoring systems.

(4) Identify the methods available to the organization for obtaining local weather conditions and predictions for short-term future weather changes.

It is important to determine specific meteorological conditions in order to effectively mitigate hazardous materials. Weather conditions, such as wind speed, wind direction, temperature, humidity, precipitation, dew point, and barometric pressure, impact the incident and plume modeling. Many radios already have channels dedicated specifically to weather radio (usually 162.55 MHz). Other sources of weather information include local dispatch, the weather channel on cable or satellite connections, local airport weather information, local weather phone recordings, and modem satellite technology on hazardous materials (hazmat) units.

(5) Explain the basic toxicological principles relative to assessment and treatment of personnel exposed to hazardous materials, including the following:

(a) Acute and delayed toxicity (chronic)

Acute toxicity refers to the sudden, severe onset of symptoms due to an exposure. The effects of delayed toxicity might not develop for hours or longer after an exposure. In some instances, such as biological exposures, symptoms might not appear until several days after the exposure.

(b) Dose response

The chemical, biological, or radiological dose relationship refers to the response produced in the human body. The relationship is one of cause and effect. The magnitude of the body's response depends on the concentration of the exposure at the site, the material, and the dose administered.

(c) Local and systemic effects

Local effects are those in which a toxic substance comes in direct contact with the skin or other tissue. Systemic effects are the effects a toxic product has either on the entire body or on a specific system or organ.

(d) Routes of exposure

The routes of exposure, sometimes referred to as routes of entry, include the following:

- *Inhalation:* The process by which irritants or toxins enter the body though the lungs as a result of the respiratory process
- *Ingestion:* The process of consuming contaminated food or water
- *Absorption:* The process by which hazardous materials are absorbed into the body through the skin or other external tissues
- *Injection:* The process by which a toxic substance is introduced directly into the blood by a needle, cannula, or some other mechanical means (Contaminants entering the bloodstream through an open wound can be considered to be injected.)

(e) Synergistic effects

Synergistic effects are when the combined effect of two or more chemicals is greater than the sum of the effect of each agent alone.

(6)* Describe the health risks associated with the following:

A.8.2.2(6) Some examples are shown in Table A.8.2.2(6)(a) and Table A.8.2.2(6)(b).

(a) Biological agents and biological toxins
(b) Blood agents
(c) Choking agents
(d) Irritants (riot control agents)
(e) Nerve agents
(f) Radiological materials
(g) Vesicants (blister agents)

TABLE A.8.2.2(6)(a) *Examples of Health Risks Associated with Chemical Agents*

Common Name of Chemical Agent	Military Abbreviation	NFPA 704* Ratings		
		H	F	R
Nerve agents				
Sarin	GB	4	1	1
Soman	GD	4	1	1
Tabun	GA	4	2	1
V agent	VX	4	1	1
Vesicants (blister agents)				
Mustard	H, HD	4	1	1
Lewisite	L	4	1	1
Blood agents				
Hydrogen cyanide	AC	4	4	2
Cyanogen chloride	CK	3	0	2
Choking agents				
Chlorine	CL	3	0	0
Phosgene	CG	4	0	0

H: health hazard, F: flammability hazard, R: reactivity hazard.

*NFPA 704, *Standard System for the Identification of the Hazards of Materials for Emergency Response.*

TABLE A.8.2.2(6)(b) *Examples of Health Risks Associated with Biological Agents and Toxins*

Common Name of Biological Agent or Toxin	Latency Period	Fatal?
Anthrax	1–5 days	Yes
Mycotoxin	2–4 hours	Often
Plague	1–3 days	Yes
Ricin	18–24 days	Yes
Viral hemorrhagic fevers	4–21 days	Yes
Smallpox	7–17 days	Yes

8.3 Competencies — Planning the Response

8.3.1 Identifying Response Objectives.

Given an analysis of a hazardous materials/WMD incident, the incident commander shall be able to describe the steps for determining response objectives (defensive, offensive, and nonintervention).

Response objectives can include modifying the stress being applied to the container, changing the size of the breach, changing the quantity being released, changing the size of the endangered area, reducing exposures by moving people away from the hot zone, and reducing the level of harm. These objectives can be met by implementing defensive, offensive, or nonintervention strategies. Applying risk-based principles, the potential loss and exposures might be reduced if actions can be safely taken to terminate the incident in a shorter period of time.

8.3.2 Identifying the Potential Response Options.

Given scenarios involving hazardous materials/WMD, the incident commander shall identify the possible response options (defensive, offensive, and nonintervention) by response objective for each problem and shall complete the following tasks:

(1) Identify the possible response options to accomplish a given response objective.

Many of the response options available can be used either offensively or defensively. For example, the size of the endangered area could be changed using a defensive strategy by placing barriers around the endangered area to contain a spill. An offensive strategy would be to enter the hot zone and plug the leak.

The IC might choose defensive operations as the most prudent method of dealing with an incident. For example, diking or blanketing a liquid spill of diesel fuel can often be accomplished easily. Transferring that same product from a damaged tank truck to another tank truck, however, would require specialized training and equipment that the technician level might not have. This task could be best completed by one of the new responder levels, such as technician with a tank car specialty. Other operations, such as vent and burn techniques, should be attempted only by technicians with a tank car specialty (see Chapter 12 in NFPA 472).

The competency described in this section is identical to the technician level competency described in 7.3.2 of NFPA 472.

(2) Identify the purpose of each of the following techniques for hazardous materials control:

- (a) Absorption
- (b) Adsorption
- (c) Blanketing
- (d) Covering
- (e) Damming
- (f) Diking
- (g) Dilution
- (h) Dispersion
- (i) Diversion
- (j) Fire suppression
- (k) Neutralization
- (l) Overpacking
- (m) Patching
- (n) Plugging
- (o) Pressure isolation and reduction (flaring; venting; vent and burn; isolation of valves, pumps, or energy sources)
- (p) Retention
- (q) Solidification
- (r) Transfer
- (s) Vapor control (dispersion, suppression)

Some of the methods listed here for controlling an incident require a high degree of specialized training and the use of sophisticated technical equipment. The IC must be familiar with the purposes of these techniques but does not necessarily have to be able to personally perform the more technical methods.

8.3.3 Approving the Level of Personal Protective Equipment.

Given scenarios involving hazardous materials/WMD with known and unknown hazardous materials/WMD, the incident commander shall approve the personal protective equipment for the response options specified in the incident action plan in each situation and shall complete the following tasks:

The difference between this competency for the IC and the similar competency for the HMT is important. At this level, the IC is expected to approve the PPE chosen. Based on their advanced training, the HMT and the hazardous materials officer are expected to be more knowledgeable in selecting the appropriate personal protective clothing, which the IC must then approve.

(1) Identify the four levels of chemical protection (EPA/OSHA) and describe the equipment required for each level and the conditions under which each level is used.

The four levels of chemical protection and the equipment required are as follows:

- ***Level A:*** The highest available level of respiratory, skin, and eye protection. This level requires a fully encapsulating suit constructed of material that is compatible with the substances involved and self-contained breathing apparatus.

- ***Level B:*** The same level of respiratory protection but less skin protection than Level A. This level is the minimum recommended for initial site entries where the hazards have not yet been identified. This level requires chemical-resistant clothing and self-contained breathing apparatus.
- ***Level C:*** The same level of skin protection as Level B but a lower level of respiratory protection. This level requires chemical-resistant clothing and air purifying respirator.
- ***Level D:*** No respiratory protection and minimal skin protection. This level requires normal work clothes.

(2) Describe the following terms and explain their impact and significance on the selection of chemical-protective clothing:

(a) Degradation

Degradation of chemical-protective clothing (CPC) can be either chemical or physical. The result of degradation is an increased likelihood that a hazardous material will permeate and penetrate the garments and endanger the health of the responder. Chemical degradation can be minimized by avoiding unnecessary contact with chemicals and by undergoing effective decontamination procedures. The garments a responder wears should be chosen based on their compatibility with the chemicals involved in an incident and with breakthrough times consistent with their expected use. Protective clothing can also degrade physically, such as damage that might happen from the garment rubbing against a rough surface.

(b) Penetration

Penetration is the movement of a hazardous material through a suit's closures. Closures include zippers, buttonholes, seams, flaps, and other design features of CPC. Hazardous materials can penetrate CPC through cracks or tears in the suit's fabric. Protection against such material penetration is vital. A regular and routine program of protective garment inspection can help uncover conditions that could lead to penetration. CPC must also be properly stored to avoid creating weak spots along seams or folds in the garment and other avenues for penetration.

(c) Permeation

Different fabrics have different resistance levels to chemical permeation, and all fabrics absorb chemicals over a period of time. NFPA 1991, *Standard on Vapor-Protective Ensembles for Hazardous Materials Emergencies,* and NFPA 1992, *Standard on Liquid Splash-Protective Ensembles and Clothing for Hazardous Materials Emergencies,* provide guidelines for manufacturer permeation testing and certification [1, 2].

(3) Describe three safety considerations for personnel working in vapor-protective, liquid splash–protective, and high temperature–protective clothing.

It takes practice to work efficiently while wearing protective clothing. Because personnel wearing vapor-protective clothing could experience a loss of dexterity, vision, and communications capability, closely monitoring these individuals is important. Backup personnel wearing the same level of protective clothing must be available to assist the entry team in an emergency, and hand signals should be established to aid in communications. All responding personnel must also be monitored for the effects of heat at the scene. A proper on-scene rehabilitation program should be in place to replenish fluids and allow for rest and recovery for all individuals responding to an incident, but especially for those wearing protective garments for hot zone and warm zone work.

(4) Identify the physiological and psychological stresses that can affect users of personal protective equipment.

Personal protective clothing, particularly fully encapsulating garments, increases the stress a responder may feel when working at a hazardous materials incident. Persons wearing CPC usually experience a loss of mobility and restricted visibility and communications. The higher the level of protection afforded by the garments, the greater these hindrances can be. Wearing CPC also increases the likelihood of heat stress and heat exhaustion in both fit and unfit individuals. Although fit individuals usually are able to work under conditions of extreme heat and physical exertion for long periods of time without adverse medical problems, a limit still exists to any person's endurance. Medical monitoring for all personnel in protective clothing at the scene is strongly recommended.

8.3.4 Developing an Incident Action Plan.

Given scenarios involving hazardous materials/WMD incidents, the incident commander shall develop an incident action plan, including site safety and control plan, consistent with the emergency response plan or standard operating procedures and within the capability of the available personnel, personal protective equipment, and control equipment, and shall complete the tasks in 8.3.4.1 through 8.3.4.5.5.

An incident action plan describes the response objectives and any options to achieving those objectives. The basis for the plan is the organization's standard operating procedures and the local emergency response plan. Safety and health considerations, necessary personnel, and control equipment should be listed in the plan for each objective. The plan provides a permanent record of the incident and can be the outline for the incident critique.

8.3.4.1 The incident commander shall identify the steps for developing an incident action plan.

In accordance with 8.3.4.1, the following components must be considered when developing an incident action plan:

- Site restrictions
- Entry objectives
- On-scene organization and control
- Selection of personal protective equipment
- Hazard evaluation
- Communications procedures
- Emergency procedures and personnel accountability
- Emergency medical care arrangements
- Rehabilitation plan
- Decontamination procedures
- On-scene work assignments (branches)
- Debriefing and critiquing of the incident once it is concluded

8.3.4.2 The incident commander shall identify the factors to be evaluated in selecting public protective actions, including evacuation and sheltering-in-place.

The evaluation is intended to reduce or prevent contamination of the public directly exposed to the hazardous material(s). If members of the public at the scene are safe in their present location, and the structure where they are located can be protected from contamination (by closing windows and doors, turning off ventilation systems that may draw in air from outside, and so forth), the preferable response is to leave these individuals in place until the incident can be controlled. Even after hazardous materials responders have notified the public at the scene that they are to stay in their present location, reassurances of their safety and updates on the progress of the incident from hazardous materials responders are necessary to reduce any anxiety.

8.3.4.3 Given the emergency response plan or standard operating procedures, the incident commander shall identify which agency will perform the following:

(1) Receive the initial notification.
(2) Provide secondary notification and activation of response agencies.

(3) Make ongoing assessments of the situation.
(4) Command on-scene personnel (incident management system).
(5) Coordinate support and mutual aid.
(6) Provide law enforcement and on-scene security (crowd control).
(7) Provide traffic control and rerouting.
(8) Provide resources for public safety protective action (evacuation or shelter in-place).
(9) Provide fire suppression services.
(10) Provide on-scene medical assistance (ambulance) and medical treatment (hospital).
(11) Provide public notification (warning).
(12) Provide public information (news media statements).
(13) Provide on-scene communications support.
(14) Provide emergency on-scene decontamination.
(15) Provide operations-level hazard control services.
(16) Provide technician-level hazard mitigation services.
(17) Provide environmental remedial action (cleanup) services.
(18) Provide environmental monitoring.
(19) Implement on-site accountability.
(20) Provide on-site responder identification.
(21) Provide incident command post security.
(22) Provide incident or crime scene investigation.
(23) Provide evidence collection and sampling.

The Emergency Planning and Community Right-to-Know Act requires local emergency planning committees (LEPCs) to develop local plans for emergency response to hazardous materials incidents [3]. The IC must understand the local emergency response plan, which should address the actions listed in 8.3.4.3 and those agencies responsible. Emergency response agencies should also develop their own standard operating procedures based on their roles in the local emergency plan. These procedures should address the actions listed in 8.3.4.3, indicate who is responsible for these actions, both inside and outside the agency, and relate how each task will be accomplished.

8.3.4.4 The incident commander shall identify the process for determining the effectiveness of a response option based on the potential outcomes.

Before a response option or combination of response options is selected, the potential outcome of the sequence of events should be reviewed by prioritizing the response options based on their effect on the outcomes. A worksheet, such as the one shown in Exhibit I.8.2, is used to estimate the outcomes in an emergency as well as the amount that could be saved. An al-

Estimating Loss and Salvageable Amount Worksheet

Exposures	Estimated Exposed	Estimated Type of Harm	Estimated Outcomes	Amount Already Lost	Amount That Could Be Saved
People	#	Deaths	#		
		Injuries	#		
Property	$	Damage	$		
Environment	$	Damage	$		

EXHIBIT I.8.2 *This worksheet is used for estimating loss and salvageable amounts. (Source: Fire Protection Handbook®, 20th edition, Figure 13.8.11)* ternative action plan should also be formulated in case the first action plan fails to achieve the desired outcomes. Constant evaluation is necessary during the course of an incident to prevent unsafe or ineffective operations and to assess subsequent options.

8.3.4.5 The incident commander shall identify the safe operating practices and procedures that are required to be followed at a hazardous materials/WMD incident.

The following practices should be considered to ensure safe operations at a hazardous materials incident scene:

- The IC and hazardous materials responders have met all of the appropriate level competencies in NFPA 472.
- Activities that present a significant risk to the safety of members are limited to situations where the potential exists to save endangered lives.
- No risk to the safety of members is acceptable when saving lives or property is not possible.
- All personnel working in the warm zone or hot zone are under the supervision of a hazardous materials branch officer.
- Personnel accountability procedures are utilized.
- A rest and rehabilitation area is completed and ready for first responders to finish their assignment.
- A hazardous materials branch safety officer is designated and operating.
- Communications are established on one simple radio channel that is not used by anyone close enough to interfere. Hand signals are available as a backup if the radios fail.
- Appropriate protective clothing and protective equipment are used whenever the responder is exposed or potentially exposed to hazardous materials.
- A rapid intervention crew consisting of at least two responders is available for rescue of a member or team if necessary. Responders are operating in the hot zone in teams of two or more.
- All responders are monitored before they can proceed to work in personal protective equipment in accordance with the guidelines found in Supplement 5 of this handbook.
- Hazardous materials responders are aware of clues indicating that the incident may be a chemical, biological, nuclear, or explosives incident. Efforts are made to notice secondary devices or attempts to disguise the true nature of the incident if terrorism is suspected.

8.3.4.5.1 The incident commander shall identify the importance of pre-incident planning relating to safety during responses to specific sites.

8.3.4.5.2 The incident commander shall identify the procedures for presenting a safety briefing prior to allowing personnel to work on a hazardous materials/WMD incident.

8.3.4.5.3* The incident commander shall identify at least three safety precautions associated with search and rescue missions at hazardous materials/WMD incidents.

A.8.3.4.5.3 Safety precautions should include the following:

(1) Buddy systems
(2) Backup team
(3) Personal protective equipment

8.3.4.5.4 The incident commander shall identify the advantages and limitations of the following and describe an example where each decontamination method would be used:

(1) Absorption
(2) Adsorption
(3) Chemical degradation
(4) Dilution
(5) Disinfection

(6) Evaporation
(7) Isolation and disposal
(8) Neutralization
(9) Solidification
(10) Sterilization
(11) Vacuuming
(12) Washing

8.3.4.5.5* The incident commander shall identify the atmospheric and physical safety hazards associated with hazardous materials/WMD incidents involving confined spaces.

Section 8.3.4.5.5 was added in response to the Occupational Health and Safety Administration (OSHA) regulations on confined space operations [4]. The IC should become very familiar with OSHA and other federal regulations. Failure to comply has led to several substantial fines.

A.8.3.4.5.5 Safety hazards associated with confined spaces could include the following:

(1) Atmospheric hazards
  (a) Oxygen-deficient atmosphere
  (b) Oxygen-enriched atmosphere
  (c) Flammable and explosive atmospheres
  (d) Toxic atmosphere
(2) Physical hazards
  (a) Engulfment hazards
  (b) Falls and slips
  (c) Electrical hazards
  (d) Structural hazards
  (e) Mechanical hazards

8.4 Competencies — Implementing the Planned Response

8.4.1 Implementing an Incident Command System.

Given a copy of the emergency response plan and annexes related to hazardous materials/WMD, the incident commander shall identify the requirements of the plan, including the procedures for notification and utilization of nonlocal resources (private, state, and federal government personnel), and shall meet the following requirements:

(1) Identify the role of the incident commander during a hazardous materials/WMD incident.

The role of the incident commander can change as the incident moves from the emergency phase to the post-emergency phase, where clean-up, recovery, and restoration services are required. One of the responsibilities of the IC is to ensure that the necessary notifications are made so that there is a seamless transfer of command from the emergency phase IC to those agencies responsible for the management and coordination of post-emergency cleanup and recovery operations.

(2) Describe the concept of unified command and its application and use at a hazardous materials/WMD incident.
(3) Identify the duties and responsibilities of the following hazardous materials branch/group functions within the incident command system:

The competency required by 8.4.1(3) is a good example of a competency required of the IC that is not required of the technician. According to this requirement, the IC must be familiar with and able to control and organize all phases of the complete emergency operation, not just mitigation procedures. This "big picture" strategic approach is necessary to ensure smooth and efficient interaction between different agencies, responder skill levels, and duties at the scene. Mitigation is only one vital part of that big picture.

(a) Decontamination
(b) Entry (backup)
(c) Hazardous materials branch director or group supervisor
(d) Hazardous materials safety
(e) Information and research

(4) Identify the steps for implementing the emergency response plans required under Title III Emergency Planning and Community Right-to-Know Act (EPCRA) of the Superfund Amendments and Reauthorization Act (SARA) Section 303, or other state and emergency response planning legislation.

Normally, emergency response plans are set in motion when someone notifies an emergency operations center that an incident has occurred. The emergency response plan then identifies the type of resources that are to be dispatched and determines whether there are any additional entities that should be notified. The IC should understand this process and be aware of which officials and/or agencies must be notified for each type or level of incident reported.

(5) Given the emergency response planning documents, identify the elements of each of the documents.

Title III of the Superfund Amendments and Reauthorization Act (SARA) requires that an emergency plan contain certain elements [3]. These issues must be addressed in the planning process. The emergency plan must address the following items if they are not covered elsewhere:

- Pre-emergency planning and coordination with outside parties
- Personnel roles, lines of authority, training, and communications
- Emergency recognition and prevention
- Safe distances and places of refuge
- Site security and control
- Evacuation routes and procedures
- Decontamination
- Emergency medical treatment and first aid
- Emergency alert and response procedures
- Critique response and follow-up
- Personal protective equipment and emergency equipment

If the standard operating procedures for the hazardous materials response team adequately cover these elements, as is sometimes the case, they do not need to be included in the emergency plan.

(6) Identify the elements of the incident management system necessary to coordinate response activities at hazardous materials/WMD incidents.

Several different models for hazardous materials incident command systems are available, some of which are presented in other sections of this handbook. NFPA 1561, *Standard on Emergency Services Incident Management System,* also contains additional information [5]. Each system generally has the same components, even though they may use different titles. For example, hazardous materials branch officers and hazardous materials group supervisors generally perform the same functions. Basically, they are responsible for implementing the incident action plan that deals with operations intended to control the hazardous materials portion of an incident.

Generally, the five primary functional areas within the incident management system are as follows:

1. Incident command
2. Operations
3. Planning
4. Logistics
5. Finance

A public information officer and an incident safety officer are usually included. For hazardous materials incidents, additional functions are identified, including the hazardous materials officer, hazardous materials safety officer, decontamination unit leader, rehabilitation unit leader, information research and resources unit leader, and entry/reconnaissance unit leader.

The IC's familiarity with the incident management system established in the local emergency response plan, and ability to implement it promptly and appropriately, is extremely important.

(7) Identify the primary government agencies and identify the scope of their regulatory authority (including the regulations) pertaining to the production, transportation, storage, and use of hazardous materials and the disposal of hazardous wastes.

The IC is required to know which agencies could become involved in a hazardous materials incident and must be able to identify their regulatory authority. The local emergency response plan should identify these agencies and delineate their roles, authority, and functions for each incident type and response level.

(8) Identify the governmental agencies and resources that can offer assistance during a hazardous materials/WMD incident and identify their role and the type of assistance or resources that might be available.

Some government agencies provide response or technical assistance to local authorities in specific cases. For example, the U.S. Coast Guard maintains a specialized team of regional responders trained to deal with hazardous materials incidents. The Coast Guard staffs three strategically located national strike teams that are trained and equipped to respond to major oil spills and chemical releases in U.S. waters. (See Exhibit I.8.3.) The U.S. Environmental Protection Agency also has a response component called the environmental response team (ERT). The ERT is a group of scientists and engineers who are trained in sampling and analysis, hazard assessment, cleanup techniques, and other technical support. In addition, many

EXHIBIT I.8.3 *The U.S. Coast Guard can provide assistance at hazardous materials incidents.*

state and local governments and private industries can provide technical assistance. The hazardous materials response teams from other municipalities or private industry can also be good resources for large incidents or for identifying specific hazardous material(s) that cannot be identified at the scene through other methods. The Federal Bureau of Investigation (FBI) has a WMD Coordinator in each field of their 56 field divisions. The WMD Coordinator is the liaison for the Hazmat Officer to the FBI and can also contact FBI Evidence Response Teams, FBI Hazardous Materials Response Teams, and other FBI assets that may be of assistance.

The private sector also has many resources available to emergency responders, and their services range from providing technical advice and assisting with on-scene monitoring to providing specialized equipment. In most areas, private sector companies that provide cleanup and disposal services are available. Some of these companies even stage equipment in an area and fly personnel to the scene when a hazardous materials incident occurs.

CHEMTREC/CANUTEC/SETIQ can put responders in touch with private sector resources if they are unfamiliar with any in their area.

The IC must know how to get specialized assistance when it is needed and must be familiar with the type of assistance available from each resource. These resources, both public and private, should be identified in the local emergency response plan. In situations involving services that must be contracted for in advance, arrangements should be addressed during the pre-incident planning phase to avoid complications during an actual emergency.

8.4.2* Directing Resources (Private and Governmental).

Given a scenario involving a hazardous materials/WMD incident and the necessary resources to implement the planned response, the incident commander shall demonstrate the ability to direct the resources in a safe and efficient manner consistent with the capabilities of those resources.

A.8.4.2 Criteria and factors should include the following:

(1) Task assignment (based on strategic and tactical options)
(2) Operational safety
(3) Operational effectiveness
(4) Planning support
(5) Logistical support
(6) Administrative support

8.4.3 Providing a Focal Point for Information Transfer to the Media and Elected Officials.

Given a scenario involving a hazardous materials/WMD incident, the incident commander shall identify information to be provided to the media and local, state, and federal officials and shall complete the following tasks:

(1) Identify the local policy for providing information to the media.

In accordance with 8.4.3(1), the emergency response plan and/or an agency's standard operating procedures should establish a procedure for providing information to the media, which can then inform the public. If the public is not to panic, people must receive accurate information. The media can also help responders alert the public to possible evacuations or any other protective actions that may be necessary. In addition, the media can announce the locations of any evacuation centers and phone numbers to check on the welfare of friends or family.

(2) Identify the responsibilities of the public information officer at a hazardous materials/WMD incident.

The public information officer (PIO) should function as a part of the IC's staff to serve as spokesperson at an incident. The PIO should have training and experience in public information and media relations. The PIO should establish a press area in a safe location and regularly provide the media with accurate information about the incident. Certain areas might be inaccessible to the press because of the hazards present. If the media are allowed to move about, however, the PIO should provide escorts for them or identify safe areas into which they can go unescorted.

(3) Describe the concept of a joint information center (JIC) and its application and use at a hazardous materials/WMD incident.

8.5 Competencies — Evaluating Progress

8.5.1 Evaluating Progress of the Incident Action Plan.

Given scenarios involving hazardous materials/WMD incidents, the incident commander shall evaluate the progress of the incident action plan to determine whether the efforts are accomplishing the response objectives and shall complete the following tasks:

(1) Identify the procedures for evaluating whether the response options are effective in accomplishing the objectives.

To determine whether the actions being taken at an incident are effective and the objectives are being met, responders must determine whether the incident is stabilizing or increasing in intensity. Feedback allows responders to modify either their strategic goals or the action options being implemented. This feedback should include information on the effectiveness of personnel, personal protective clothing and equipment, control zones, decontamination procedures, and the action options being implemented.

(2) Identify the steps for comparing actual behavior of the material and the container to that predicted in the analysis process.

When comparing actual behavior to predicted behavior, the IC should determine whether events at an incident are happening as predicted, are occurring out of sequence, or are different from expectations. The IC should also determine whether events expected as part of the mitigation, response, or overall plan are occurring as anticipated. This evaluation process should continue until the incident is terminated so that no "surprises" occur during the cleanup or overhaul phase.

(3) Determine the effectiveness of the following:

   (a) Control, containment, or confinement operations
   (b) Decontamination process
   (c) Established control zones
   (d) Personnel being used
   (e) Personal protective equipment

This competency is intended to allow the IC to show that he or she can analyze an incident to determine the effectiveness of the chosen action options in achieving the strategic goals. The competencies covered in 8.5.1(1) and 8.5.1(2) provide the basis for this analysis.

(4) Make modifications to the incident action plan as necessary.

8.6 Competencies — Terminating the Incident

The steps involved in terminating the emergency phase of an incident are transferring command, debriefing, critiquing the incident, and managing after-action activities. The incident debriefing involves gathering information from response personnel. This information should then be used to develop a chronological report of emergency activities. The critique is a review of the incident and is intended to identify and document lessons learned. It should allow the participants to review response activities and determine what worked well and what did not. The critique should be a positive process that allows responders to modify response procedures if problems are identified. After-action activities include analyzing the information gathered during the debriefing and the critique, documenting that analysis, and following up as necessary to ensure that any recommendations made to improve emergency operations are implemented.

8.6.1* Transferring Command and Control.

Given a scenario involving a hazardous materials/WMD incident, the emergency response plan, and standard operating procedures, the incident commander shall be able to identify the steps to be taken to transfer command and control of the incident and shall be able to demonstrate the transfer of command and control.

A.8.6.1 The appropriate steps to transfer command and control of the incident include the following:

(1) Command can be transferred only to an individual who is on-scene.

(2) Fully brief the incoming command and control person on the details of the incident, including response objectives and priorities, resources committed, unmet needs, and safety issues.

Transferring authority at an incident generally means transferring command, or the role of IC, from one person to another. This process should be identified in the incident management system's standard operating procedures. Authority might be transferred from one officer to another officer of higher rank or, in the case of some agencies, from one person to another person with a higher authority and more responsibility. Authority is also sometimes transferred when the emergency phase has ended and a nonemergency phase begins, as when cleanup and remediation activities must continue at the site. In such cases, some local, state, or federal agency is likely to be chosen to manage this phase. Whatever the case, procedures should be developed to identify who may be responsible for overseeing the operations in 8.6.1.

8.6.2 Conducting a Debriefing.

Given scenarios involving a hazardous materials/WMD incident, the incident commander shall conduct a debriefing of the incident and shall complete the following tasks:

The incident debriefing in 8.6.2 is not a critique; it is the gathering of information intended to provide an overall summary of the activities of each branch, sector, or division during an incident. The objectives of a debriefing are to identify who responded, what they did, when they did it, and how effective their operations were. The debriefing should also document any injuries suffered, note the type of treatment given, and indicate whether any medical follow-up is needed. Equipment that has been damaged should be reported, and any unsafe conditions should be noted. Responders need to be told what materials they were exposed to, warned of any symptoms those materials might produce, and advised on the decontamination procedures that they should undergo. These procedures likely include showering and washing or disposing of clothing.

(1) Describe three components of an effective debriefing.

See commentary for 7.6.1(1) of NFPA 472.

(2) Describe the key topics in an effective debriefing.

See commentary for 7.6.1(2) of NFPA 472.

(3) Describe when a debriefing should take place.

See commentary for 7.6.1(3) of NFPA 472.

(4) Describe who should be involved in a debriefing.

See commentary for 7.6.1(4) of NFPA 472.

(5) Identify the procedures for conducting incident debriefings at a hazardous materials/WMD incident.

8.6.3 Conducting a Critique.

Given details of a scenario involving a multiagency hazardous materials/WMD incident, the incident commander shall conduct a critique of the incident and shall complete the following tasks:

Ideally, the ICs of the various responding authorities should conduct an initial meeting to specify which personnel should be involved in the critique in order to ensure that all responding agencies or groups are properly represented.

(1) Describe three components of an effective critique.
(2) Describe who should be involved in a critique.
(3) Describe why an effective critique is necessary after a hazardous materials/WMD incident.
(4) Describe what written documents should be prepared as a result of the critique.
(5) Implement the procedure for conducting a critique of the incident.

Notes recorded throughout the critique session can become the basis for writing a post-critique report. This document should be clear and concise. It includes observations and conclusions presented by the participants during the critique and comments offered during the debriefing.

An after-action report is another document compiling relevant aspects of the critique. The first few pages need only cover a simple overview of the events, including the nature of the problem, the actions necessary to correct the problem, and the projected time frame to implement the necessary changes, plus designation of a responsible party to ensure that corrective actions are observed and implemented. Lessons learned should be listed and eventually incorporated into the existing emergency response plans as modifications and improvements. Recommendations for improvement should be listed at the end of the report. A semiannual review date should be selected to make sure that action items have been addressed.

8.6.4 Reporting and Documenting the Hazardous Materials/WMD Incident.

Given a scenario involving a hazardous materials/WMD incident, the incident commander shall demonstrate the ability to report and document the incident consistent with local, state, and federal requirements and shall complete the following tasks:

(1) Identify the reporting requirements of the federal, state, and local agencies.

The IC is required to know the reporting requirements necessary to deal with a hazardous materials incident. In many cases, the responsibility for reporting the incident to specific governmental agencies is outlined in an agency's, jurisdiction's, or organization's standard operating procedures. However, the IC must ensure that the proper agencies are notified and that the proper reports are completed.

(2) Identify the importance of the documentation for a hazardous materials/WMD incident, including training records, exposure records, incident reports, and critique reports.

Questions about an incident might not arise until someone files a claim some time after the incident is over. If information documenting the incident is not available, this lapse could have serious ramifications for all personnel involved. Thus, both documenting information about personnel training and exposure and keeping incident and critique reports on file are critical to ensuring that answers to questions that might arise about the handling of an incident can be efficiently, accurately, and appropriately answered. Typically, the IC is asked to explain the operations undertaken and the use of personnel during the incident, as well as why certain decisions were made or not made.

(3) Identify the steps in keeping an activity log and exposure records for hazardous materials/WMD incidents.

The IC should assign someone to maintain a record of incident events, which will be helpful in completing the incident analysis and conducting the critique. Personnel exposure records should also be maintained as required by federal and, in many cases, state law. The IC should assign someone to gather the necessary information about the type of exposure to which personnel were subjected, the exposure level and the length of the exposure, the type of personal protective clothing and equipment personnel were using, and the type of decontamination personnel underwent. Any on-scene medical assistance that personnel received should also be documented.

(4) Identify the requirements for compiling hazardous materials/WMD incident reports found in the emergency response plan or standard operating procedures.
(5) Identify the requirements for filing documents and maintaining records found in the emergency response plan or standard operating procedures.
(6) Identify the procedures required for legal documentation and chain of custody and continuity described in the standard operating procedures or the emergency response plan.

REFERENCES CITED IN COMMENTARY

1. NFPA 1991, *Standard on Vapor-Protective Ensembles for Hazardous Materials Emergencies,* National Fire Protection Association, Quincy, MA, 2005.
2. NFPA 1992, *Standard on Liquid Splash-Protective Ensembles and Clothing for Hazardous Materials Emergencies,* National Fire Protection Association, Quincy, MA, 2005.
3. Superfund Amendments and Reauthorization Act (SARA) of 1986, Title III, Emergency Planning and Community Right-to-know Act of 1986, (42 U.S.C. § 11001, et seq.), U.S. Government Printing Office, Washington, DC.
4. Title 29, Code of Federal Regulations, Part 1910.146, "Permit-Required Confined Spaces," U.S. Government Printing Office, Washington, DC.
5. NFPA 1561, *Standard on Emergency Services Incident Management System,* National Fire Protection Association, Quincy, MA, 2008.

Additional Reference

Cote, A., ed., *Fire Protection Handbook®,* 20th ed., National Fire Protection Association, Quincy, MA, 2008.

Competencies for Specialist Employees

CHAPTER 9

In the 1997 edition of NFPA 472, the title *off-site specialist employee* was changed to *private sector specialist employee*. The 2008 edition of the standard brings yet another change to *specialist employee*. The technical committee made these changes to more accurately describe the potential duties of these employees.

Specialist employees are not just assigned to respond to off-site hazardous materials incidents. They respond to incidents within their facility, in and outside their assigned work area, and outside their facility. In many cases, specialist employees are part of the emergency response team for the company.

As shown in Exhibit I.9.1, specialist employees must wear the appropriate protective clothing and breathing apparatus, just as those in the public sector must do when responding to a similar level of emergency.

EXHIBIT I.9.1 *As an employee of the facility, this specialist employee is most familiar with the breached pipe and its hazard.*

This chapter outlines the actions and conditions under which other employees are allowed to perform tasks without meeting any of the competency levels described in this chapter. For example, consider a chemical technician who is responsible for a piece of machinery that combines several chemicals (raw materials) into a new product for the company to market. If one of the chemicals is a hazardous material that spills or releases, the chemical technician is allowed to take some simple actions to mitigate the incident, such as turning off a valve. However, the employee must first inform the incident management structure of an emergency before taking any actions, have adequate personal protective equipment (PPE), have training in both use of the PPE and the procedures needed for mitigation, and break off all actions when the emergency response team arrives.

Specialist employees are also sent to off-site hazardous materials incidents to provide incident commanders with technical advice or assistance. Among the companies these individuals represent are chemical manufacturers, transportation companies, users of products, and container manufacturers.

All competencies for specialist employees at every level apply to the individuals' areas of specialization and include only those chemicals and containers to which the individual is expected to respond. The specialist employee must perform all activities in a manner consistent with the organization's emergency response plan and standard operating procedures and with the available resources.

9.1 General

9.1.1 Introduction.

9.1.1.1 This chapter shall address competencies for the following specialist employees:

(1) Specialist employee C
(2) Specialist employee B
(3) Specialist employee A

9.2 Specialist Employee C

9.2.1 General.

9.2.1.1 Introduction.

9.2.1.1.1 The specialist employee C shall be that person who responds to emergencies involving hazardous materials/WMD and/or containers in the organization's area of specialization, and the following:

(1) Consistent with the emergency response plan and/or standard operating procedures, the specialist employee C can be called on to gather and record information, provide technical advice, and arrange for technical assistance.
(2) The specialist employee C does not enter the hot or warm zone at an emergency.

9.2.1.1.2 The specialist employee C shall be trained to meet all competencies at the awareness level (Chapter 4) relative to the organization's area of specialization and all additional competencies in Section 9.2.

9.2.1.2 Goal.

9.2.1.2.1 The goal of the competencies at this level shall be to provide the specialist employee C with the knowledge and skills to perform the duties and responsibilities assigned in the emergency response plan and/or standard operating procedures and to perform the tasks in 9.2.1.2.2 safely and effectively.

9.2.1.2.2 When responding to hazardous materials/WMD incidents, the specialist employee C shall have the knowledge and skills to perform the following tasks safely:

(1) Assist the incident commander in analyzing the magnitude of an emergency involving hazardous materials/WMD or containers for hazardous materials/WMD by completing the following tasks:
    (a) Provide information on the hazards and harmful effects of specific hazardous materials/WMD.
    (b) Provide information on the characteristics of specific containers for hazardous materials/WMD.
(2) Assist the incident commander in planning a response to an emergency involving hazardous materials/WMD or containers for hazardous materials/WMD by providing information on the potential response options for hazardous materials/WMD or containers for hazardous materials/WMD.

9.2.2 Competencies — Analyzing the Incident.

9.2.2.1 Providing Information on the Hazards and Harmful Effects of Specific Hazardous Materials/WMD. Given a specific chemical(s) used in the organization's area of specialization and the corresponding MSDS or other applicable resource, the specialist employee C shall advise the incident commander of the chemical's hazards and harmful effects and shall complete the following tasks:

Specialist employees C are required by 9.2.2.1 to provide information about the chemical involved in an incident. The items listed in 9.2.2.1(1) represent the minimum requirements needed to determine proficiency in this area. Additional resources include the Emergency Response Guidebook (ERG) and *Emergency Handling of Hazardous Materials in Surface Transportation,* among others [1, 2].

(1) Identify the following hazard information from the MSDS or other resource:

- (a) Physical and chemical properties
- (b) Physical hazards of the chemical (including fire and explosion hazards)
- (c) Health hazards of the chemical
- (d) Signs and symptoms of exposure
- (e) Routes of entry
- (f) Permissible exposure limits
- (g) Reactivity hazards
- (h) Environmental concerns

The name of the material safety data sheet (MSDS) is changing to the global harmonization name, which is safety data sheet (SDS).

(2) Identify how to contact CHEMTREC/CANUTEC/SETIQ and local, state, and federal authorities.

The requirement in 9.2.2.1(2) can be met by determining whether a responder can find the correct telephone number listed in the ERG. The company to which the product belongs most likely includes the telephone number on an MSDS, on labels, on containers, and in shipping papers. The number must be included in the emergency response plan or standard operating procedures of the specialist employee's organization.

(3) Identify the resources available from CHEMTREC/CANUTEC/SETIQ and local, state, and federal authorities.

The Chemical Transportation Emergency Center (CHEMTREC), the Canadian Transport Emergency Center (CANUTEC), and SETIQ, which is the Emergency Transportation System for the Chemical Industry in Mexico, can provide information on products, fax SDSs, and contact manufacturers, mutual aid responders, and contractors for hire. CHEMTREC is a public service of the Chemical Manufacturers Association and is located in Arlington, Virginia. CHEMTREC provides the on-scene commander with immediate advice by telephone and contacts the involved shipper for detailed assistance and response follow-up. The organization can also notify the National Response Center (NRC) of significant incidents or bridge a caller to the NRC to report a spill. CHEMTREC operates 24 hours a day and can be contacted throughout the United States and Canada.

CHEMTREC can usually provide hazard information warnings and guidance when given a material's four-digit identification number, the name of the product, and the nature of the problem. If the product is unknown or more detailed information and assistance is needed, the caller should attempt to provide as much of the following information as possible:

- Caller's name and callback number
- Guide number being used
- Name of the shipper or manufacturer

- Railcar or truck number
- Carrier's name
- Consignee
- Local conditions

At an incident, a phone line should be kept open to CHEMTREC so that the center can provide guidance and assistance. CHEMTREC can also provide a teleconferencing bridge that connects technical experts to a caller's line as necessary.

CANUTEC is located in Ottawa, Canada, and operated by the Transport Dangerous Goods Directorate of Transport Canada. The organization provides technical assistance to emergency responders much the same as CHEMTREC. Personnel provide technical information regarding the physical, chemical, toxicological, and other properties of the products involved in an incident; recommend remedial actions for fires, spills, or leaks; provide advice on protective clothing and emergency first aid; and contact the shipper, manufacturer, or others who are deemed necessary.

SETIQ provides the same services as CHEMTREC and CANUTEC.

(4) Given the emergency response plan and/or standard operating procedures, identify additional resources of hazard information, including a method of contact.

Specialist employees C are required by 9.2.2.1(4) to be able to contact knowledgeable personnel, as prescribed in the organization's emergency response plan or standard operating procedures. They should also know where to find the organization's plan to locate the appropriate information.

9.2.2.2 Providing Information on the Characteristics of Specific Containers. Given examples of containers for hazardous materials/WMD in the organization's area of specialization, the specialist employee C shall advise the incident commander of the characteristics of the containers and shall complete the following tasks:

Specialist employees C are required to identify the containers, such as tank cars, cargo tanks, drums, and cylinders, routinely used in the manufacture, storage, and transport of hazardous materials, in accordance with 9.2.2.2. (See Exhibit I.9.2.) Specialist employees might already know about these containers, or they may learn about them from resources such as other company specialists and their own companies' packaging guides.

(1) Identify each container by name.
(2) Identify the markings that differentiate one container from another.

Specialist employees C are required by 9.2.2.2(2) to be able to identify various container markings used in their work area including tank car marks and numbers, cargo tank numbers, portable tank numbers, and fixed facility tank markings. Exhibit I.9.3 and Exhibit I.9.4 show the various warning placards and labels required by the U.S. Department of Transportation (DOT).

(3) Given the emergency response plan and/or standard operating procedures, identify the resources available that can provide information about the characteristics of the container.

Among the resources that can provide the responder with information about the container's characteristics are knowledgeable persons in the responder's organization, container manufacturers, the Association of American Railroads (AAR), CHEMTREC, and various carriers.

(4) Identify indicators of possible criminal or terrorist activity, including the following:
   (a) Intentional release of hazardous materials
   (b) Unexplained bomb- and munitions-like material

9.2.3 Competencies — Planning the Response.

9.2.3.1 Providing Information on Potential Response Options for Specific Hazardous Materials/WMD. Given a specific chemical used in the organization's area of specialization

EXHIBIT I.9.2 *Private sector specialist employees must be familiar with the containers used in the routine storage and transport of hazardous materials as well as the facility's emergency plans, piping cutoffs, and access points that public sector responders need to find during an emergency.*

and a corresponding MSDS or other resource, the specialist employee C shall advise the incident commander of the response information for that chemical by being able to complete the following tasks:

By virtue of their job duties, their knowledge of specific chemicals, or their access to appropriate resources within their organizations, specialist employees C are required by 9.2.3.1 to be able to provide the incident commander with response information. This information should include the physical and chemical properties of the hazardous material involved in an incident, health and environmental data, and containment and reactivity data.

(1) Obtain the following response information:
 (a) Precautions for safe handling, including industrial hygiene practices, protective measures, and procedures for cleanup of spills and leaks
 (b) Applicable emergency response control measures, including personal protective equipment
 (c) Emergency and first-aid procedures
(2) Relay any suspicions of criminal or terrorist activity to the incident commander.
(3) Identify additional resources for obtaining response information.

Additional resources the specialist employee might use include the ERG, the AAR's Emergency Action Guides (EAG), knowledgeable persons in his or her organization, and CHEMTREC/CANUTEC/SETIQ [3].

9.2.3.2 Providing Information on Potential Response Options for Specific Containers. Given a specific facility or transportation container used in the organization's area of specialization, the specialist employee C shall advise the incident commander of the response information for that chemical by being able to complete the following tasks:

(1) Identify safe operating procedures for that container, including acceptable pressures, temperatures, and materials of construction, and potential adverse outcomes resulting from those conditions.

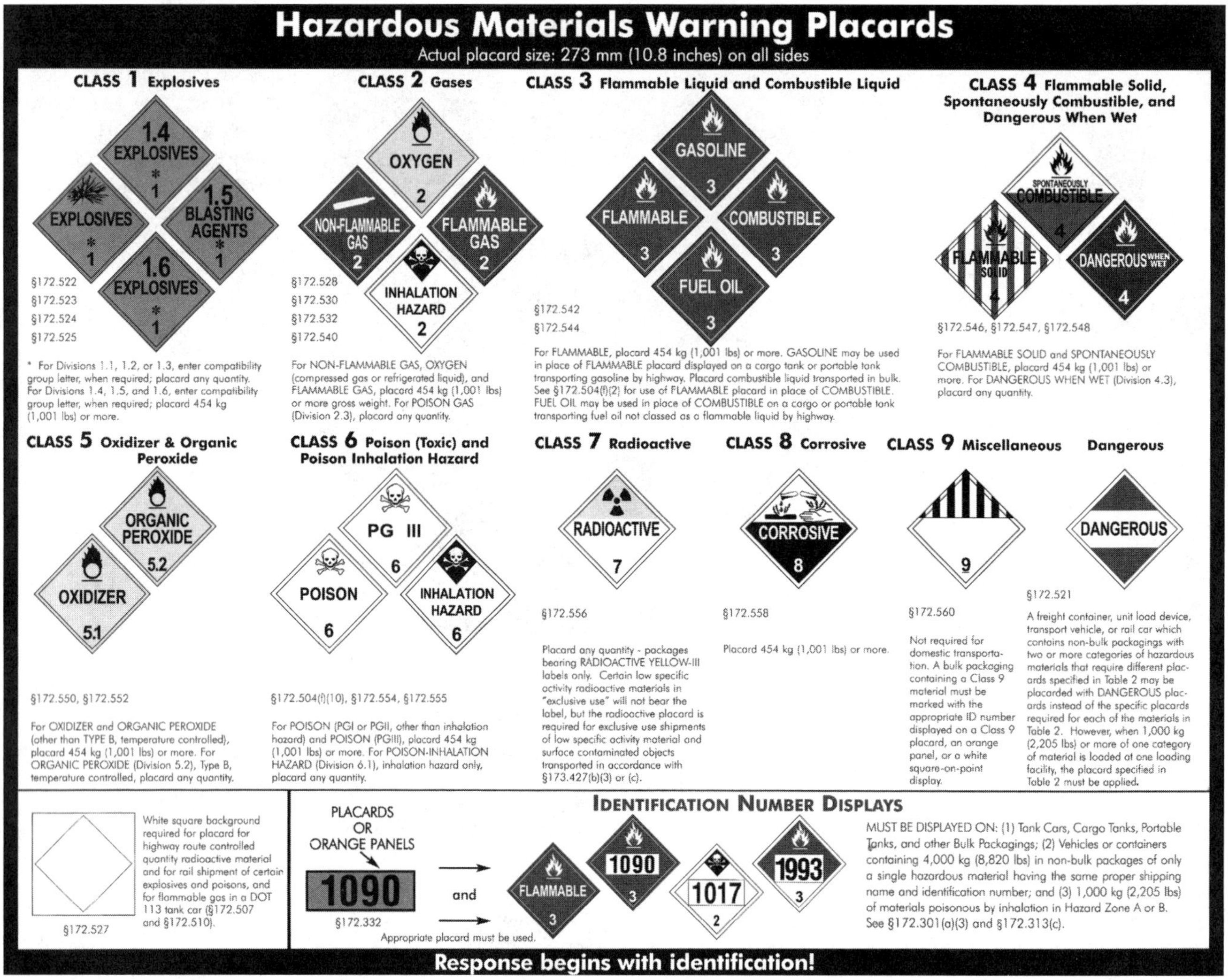

EXHIBIT I.9.3 *The U.S. Department of Transportation's Chart 12 shows the warning placards required for the transport of hazardous materials. Like their public sector counterparts, specialist employees need to be familiar with these placards. (Source: U.S. Department of Transportation, DOT Chart 12, Hazardous Materials Marking, Labeling, & Placarding Guide, DHM-50, p. 1)*

(2) Describe safety devices on the container, including emergency shutoff valves, pressure relief devices, and vacuum breakers.
(3) Identify early signs of container and safety device failure.
(4) Suggest emergency response procedures.

9.3 Specialist Employee B

9.3.1 General.

9.3.1.1 Introduction.

9.3.1.1.1 The specialist employee B shall be that person who, in the course of regular job duties, works with or is trained in the hazards of specific chemicals or containers in the individual's area of specialization and the following:

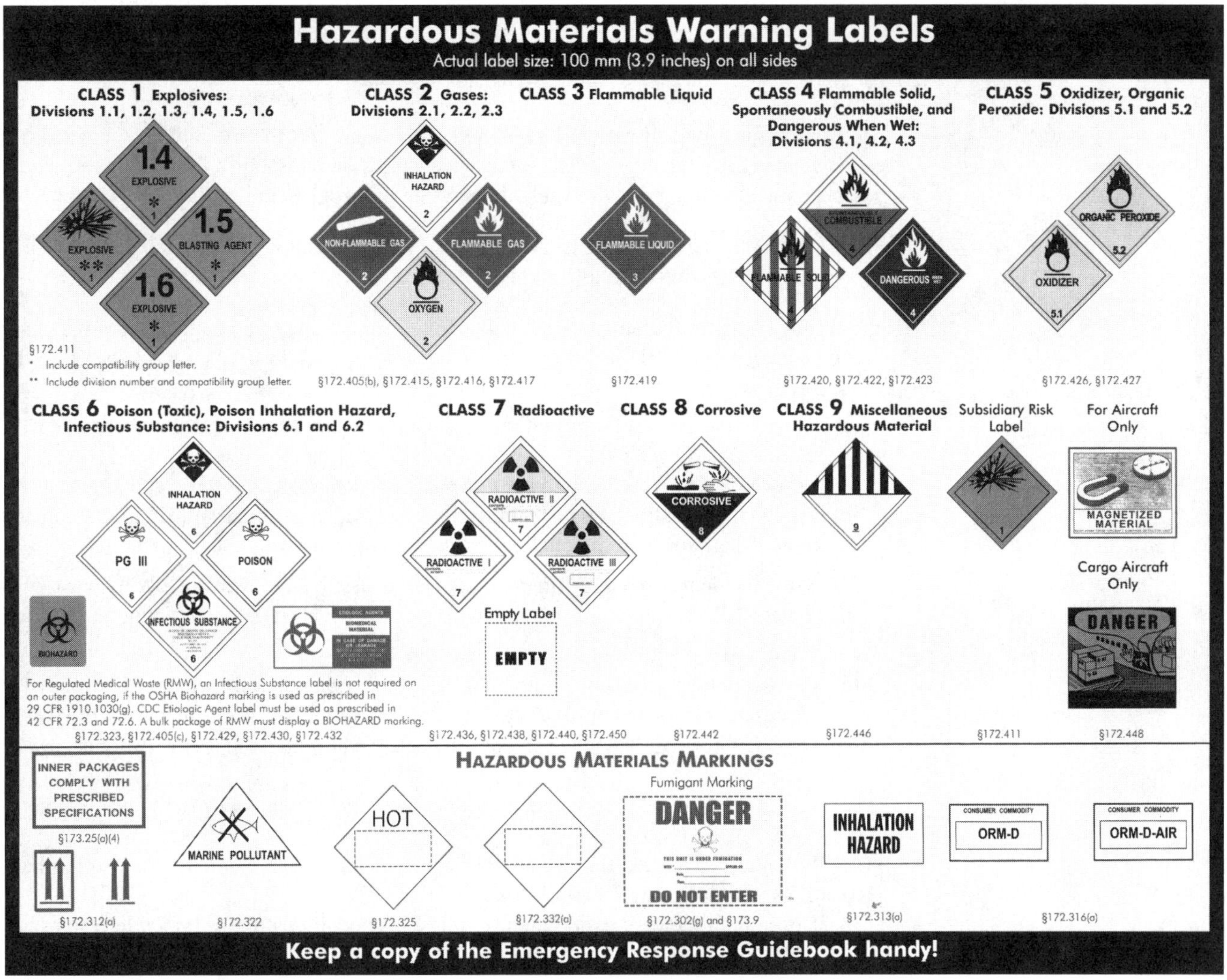

EXHIBIT I.9.4 *The U.S. Department of Transportation's Chart 12 shows the warning labels required for the transport of hazardous materials. (Source: U.S. Department of Transportation, DOT Chart 12, Hazardous Materials Marking, Labeling, & Placarding Guide, DHM-50, p. 2)*

(1) Because of the employee's education, training, or work experience, the specialist employee B can be called on to respond to incidents involving these chemicals or containers.
(2) The specialist employee B can be used to gather and record information, provide technical advice, and provide technical assistance (including work in the hot zone) at the incident, consistent with the emergency response plan and/or standard operating procedures.

9.3.1.1.2 The specialist employee B shall be trained to meet all competencies at the awareness level (Chapter 4) relative to the organization's area of specialization, all competencies at the specialist employee C level (Section 9.2), and all additional competencies in Section 9.3.

9.3.1.2* Goal.

A.9.3.1.2 An example of a specialist employee B is a person who regularly loads and unloads tank trucks of the specific chemical involved in the incident as part of his or her regular job.

At a hazardous materials/WMD incident, this person would be assigned the task of transferring the contents of the damaged tank truck into another container. The specialist employee B would not be involved with chemicals for which he or she has not been trained and would leave the hot or warm zone when this work is completed.

9.3.1.2.1 The goal of these competencies shall be to ensure that the specialist employee B has the knowledge and skills to safely perform the duties and responsibilities assigned in the emergency response plan and/or standard operating procedures and the tasks in 9.3.1.2.2.

9.3.1.2.2 Within the employee's individual area of specialization, the specialist employee B shall be able to perform the following tasks:

(1) Assist the incident commander in analyzing the magnitude of an incident involving hazardous materials/WMD or containers for hazardous materials/WMD by completing the following tasks:
   (a) Provide and interpret information on the hazards and harmful effects of specific hazardous materials/WMD.
   (b) Provide and interpret information on the characteristics of specific containers.
   (c) Provide information on concentrations of hazardous materials/WMD from exposure monitoring, dispersion modeling, or any other predictive method.
(2) Assist the incident commander in planning a response to an incident involving hazardous materials/WMD or containers for hazardous materials/WMD by completing the following tasks:
   (a) Provide information on the potential response options and their consequences for specific hazardous materials/WMD or containers for hazardous materials/WMD.
   (b) Provide information on the personal protective equipment requirements for a specific chemical.
   (c) Provide information on the technical decontamination methods for a specific chemical.
   (d) Provide information on the federal or provincial regulations that relate to the handling and disposal of a specific chemical.
   (e)* Develop an incident action plan (within the capabilities of the available resources), including site safety and control plan, for handling hazardous materials/WMD, or containers for hazardous materials/WMD, consistent with the emergency response plan and/or standard operating procedures.

A.9.3.1.2.2(2)(e) The following site safety plan considerations are from the EPA *Standard Operating Safety Guides:*

(1) Site description
(2) Entry objectives
(3) On-site organization
(4) On-site control
(5) Hazard evaluation
(6) Personal protective equipment
(7) On-site work plans
(8) Communication procedures
(9) Decontamination procedures
(10) Site safety and health plan

(3) Implement the planned response, as developed with the incident commander, for hazardous materials/WMD or containers for hazardous materials/WMD, consistent with the emergency response plan and/or standard operating procedures and within the capabilities of the available resources, by completing the following tasks:

(a) Perform response options specified in the incident action plan, as agreed upon with the incident commander and consistent with the emergency response plan and/or standard operating procedures.
(b) Don, work in, and doff personal protective equipment needed to implement the response options.

(4) Assist the incident commander to evaluate the results of implementing the planned response by completing the following tasks:
(a) Provide feedback on the effectiveness of the response options taken.
(b) Provide reporting and subsequent documentation of the incident involving hazardous materials/WMD as required.

9.3.2 Competencies — Analyzing the Incident.

9.3.2.1 Providing and Interpreting Information on Hazards of Specific Hazardous Materials/WMD. Given a specific chemical within the individual's area of specialization and a corresponding MSDS or other resource, the specialist employee B shall advise the incident commander of the chemical's hazards and harmful effects of specific hazardous materials/WMD and the potential consequences based on the incident and shall meet the following requirements:

(1) Given a specific chemical, identify and interpret the following hazard information:
(a) Physical and chemical properties
(b) Physical hazards of the chemical (including fire and explosion hazards)
(c) Health hazards of the chemical
(d) Signs and symptoms of exposure
(e) Routes of entry
(f) Permissible exposure limits
(g) Reactivity hazards
(h) Environmental concerns

All specialist employees B must be able to locate the hazard information in 9.3.2.1(1) in their organization's SDS and understand it. The product specialist described in Section 9.3 must be able to interpret this information and give advice to the incident commander. If specialist employee B responders cannot interpret this information, they must be able to contact the individuals in their organization who can interpret it. For an example of a typical SDS, see Chapter 4 of NFPA 472 in this handbook.

(2) Given examples of specific hazardous materials/WMD and the necessary resources, predict the potential behavior of the hazardous materials/WMD based on the damage found, including the consequences of that behavior.

Specialist employees B are required by 9.3.2.1(2) to be able to predict what will happen when a liquid, solid, or gas is released from its container. What will occur, for example, if the chemical is exposed to air? Will the chemical react, vaporize, or ignite?

(3) Identify the general types of hazard information available from the other resources identified in the emergency response plan and/or standard operating procedures.

Among the types of hazard information available to the specialist employee from sources identified in the organization's emergency response plan and standard operating procedures are containment techniques, medical treatment protocols, container design, reactivity data, and decontamination and mitigation procedures.

9.3.2.2 Providing Information on Characteristics of Specific Containers. Given a container for specific hazardous materials/WMD, the specialist employee B shall advise the incident commander of the characteristics and potential behavior of that container and shall meet the following requirements:

Specialist employees B are required by 9.3.2.2 to be able to identify the person or persons in their organizations they can contact for information on such activities as damage assessment, fitting arrangement, and probability of container failure.

(1) Given examples of containers for specific hazardous materials/WMD, identify the purpose and operation of the closures found on those containers.

Specialist employees must be able to identify the types of nonbulk and bulk containers used in their organization. Commentary Table I.9.1 displays sample types of containers and should

COMMENTARY TABLE I.9.1 *Types of Nonbulk and Bulk Containers*

Type of Container	*Example*	*Characteristics to be Identified*
Nonbulk containers	Drums	Type of bungs and seal caps Closed and open Head drums Closure rings Safety relief devices
	Cylinders	Valves Valve caps Relief devices Purge devices
Bulk containers	Tote bins	Valves Bungs, and safety and relief devices Spouts Secondary closures Purge devices Couplings
	Cargo tanks	Venting devices Valves Relief purge devices Connections Manways Emergency shutoff valves Excess flow valves Gauging devices Sampling devices Secondary closures Safety relief devices
	Tank cars	Valves Safety and relief devices Connections Sampling valves Vapor and liquid valves Heater coils Caps Bottom outlet Dome cover Dome gasket
	Hopper cars and hopper trucks	Gates Manways
	ISO containers	Venting devices Valves Relief purge devices Connections Manways Emergency shutoff valves Excess flow valves Gauging devices Sampling devices Secondary closures Safety and relief devices

also include any other types of containers, such as sample containers and supersacks, which the responder's organization may use. The specialist employee B should know how to contact the organization's container specialist, who can provide this type of information.

(2) Given a chemical container, list the types of damage that could occur.

The types of damage specified in 9.3.2.2(2) might include punctures, scores, gouges, blown rupture disks, damaged gaskets, corrosion, damaged "o-rings," liner failure, weld seam failure, cracked bungs, and frictional damage on drums, among other things. The specialist employee B should know how to contact the organization's container specialist, who can provide this type of information.

(3) Given examples of containers for specific hazardous materials/WMD and the necessary resources, predict the potential behavior of the containers and the consequences, based on the damage found.

Based on the types of damage listed in 9.3.2.2(3), all specialist employees B should be able to assess potential container failures. They should know, for example, that a scored pressurized container could lead to a boiling liquid expanding vapor explosion (BLEVE), that a sheared valve on a cylinder could cause the container to rocket, that a bulging drum could rupture, and that damaged or stripped threads could cause a product release. Exhibit I.9.5 shows an example of the warning signs that lead to detection of a leak.

EXHIBIT I.9.5 *This plugged drum has the warning signs of leakage, such as corrosion visible on the far left, dents and scratches visible along the seam side by the second plug, and seepage below the plugged hole, that the private sector specialist employee should be able to spot.*

(4) Given the emergency response plan and/or standard operating procedures, identify resources (including a method of contact) for knowledge of the design, construction, and damage assessment of containers for hazardous materials/WMD.

The resources in 9.3.2.2(4) could include, but are not limited to, chemical and mechanical engineers, packaging specialists, and container manufacturers. The carrier, especially the cargo tank carrier, could be another resource.

9.3.2.3 Providing Information on Concentrations of Hazardous Materials/WMD.

9.3.2.3.1 Given a chemical and the applicable monitoring equipment provided by the organization for that chemical or the available predictive capabilities (e.g., dispersion modeling, exposure modeling), the specialist employee B shall advise the incident commander of the concentrations of the released chemical and the implications of that information to the incident.

9.3.2.3.2 The specialist employee B shall meet the following additional requirements:

(1) Identify the applicable monitoring equipment.

Specialist employees B, who are trained to select and use monitoring equipment as part of their regular duties, are required by 9.3.2.3.2 (1) to be able to identify the equipment appropriate for monitoring a chemical used in their areas of specialization. The equipment should be operated in accordance with the manufacturer's instructions.

(2) Use the monitoring equipment provided by the organization to determine the actual concentrations of a specific chemical.
(3) Given information on the concentrations of a chemical, interpret the significance of that concentration information to the incident relative to the hazards and harmful effects of the chemical.

Specialist employees B should be able to interpret the results of monitoring in terms of known hazards. When unable to do so, the specialist employee should be able to contact the appropriate person in the organization who can.

(4) Demonstrate field calibration and testing procedures, as necessary, for the monitoring equipment provided by the organization.
(5) Given the emergency response plan and/or standard operating procedures, identify the resources (including a method of contact) capable of providing monitoring equipment, dispersion modeling, or monitoring services.

Specialist employees B are required by 9.3.2.3.2 (5) to be familiar with their organization's emergency response plan, standard operating procedures, and other resources so that they can identify the industrial hygienist, the site safety officer, the equipment supplier, and any other source who can provide this information.

9.3.3 Competencies — Planning the Response.

9.3.3.1 Providing Information on Potential Response Options and Consequences for Specific Hazardous Materials/WMD. Given specific hazardous materials/WMD or containers within the employee's individual area of specialization and the associated resources, the specialist employee B shall advise the incident commander of the potential response options and their consequences and shall complete the following tasks:

(1) Given a specific chemical and a corresponding MSDS, identify and interpret the following response information:
   (a) Precautions for safe handling, including industrial hygiene practices, protective measures, and procedures for cleanup of spills or leaks

Using their organizations' MSDSs, specialist employees B should be able to determine what PPE is needed to deal with spills and decontamination procedures, taking into consideration the class and state of the material, the external hazards, and the possible secondary hazards.

When dealing with flammables, for example, specialist employees B should eliminate ignition sources, use sparkproof tools, and consider using foam to suppress vapors. When dealing with corrosives, the specialist employee B should ensure that the equipment and tools specific to the materials involved in the incident are compatible with the material and type of neutralizing materials being used. For example, certain chemicals can react with some metal shovels, pumps, and hoses. When dealing with poison, the specialist employee B should consider evacuation or protection in-place; should not allow eating, smoking, or gum chewing in hazardous areas; and should ensure that any PPE and tools that have been used are properly decontaminated.

   (b) Applicable control measures, including personal protective equipment

Specialist employees B should be able to locate and identify the information on applicable control measures from an MSDS. Interpreting specific control and remediation procedures might require the expertise of a product specialist, an industrial hygienist, or another appropriate person in the organization.

(c) Emergency and first-aid procedures

The first-aid section of an MSDS is generally comprehensive and self-explanatory. The specialist employee B should pay particular attention to the use and availability of antidotes.

(2) Given the emergency response plan and/or standard operating procedures, identify additional resources for interpreting the hazards and applicable response information for a hazardous material/WMD.

Specialist employees B are required by 9.3.3.1(2) to be able to demonstrate how to access additional sources of assistance from other resources, such as CHEMTREC/CANUTEC/SETIQ.

(3) Describe the advantages and limitations of the potential response options for a specific chemical.

In accordance with 9.3.3.1(3), the specialist employee B should be able to select the appropriate response option to the particular chemical involved. For example, Commentary Table I.9.2 lists three event scenarios and possible response options for each event.

***COMMENTARY TABLE I.9.2** Examples of Response Options*

Event	*Response*	*Advantages*	*Disadvantages*
Pesticide fire	Allow to burn	Minimal runoff Minimal containment	Products of combustion Demand on fire departments Vapor release
Acid spill	Neutralize runoff	Reduces hazard	Exposure from process Lessens corrosivity
Poisonous liquid releasing vapor	Apply foam blanket	Suppresses vapor	Contributes to contamination, increases cleanup

(4) Given the emergency response plan and/or standard operating procedures, identify resources (including a method of contact) capable of the following:
   (a) Repairing containers for hazardous materials
   (b) Removing the contents of containers for hazardous materials
   (c) Cleaning and disposing of hazardous materials/WMD or containers for hazardous materials/WMD

The specialist employee B should be able to locate individuals or organizations, identified by organizational resources such as the emergency response plan and standard operating procedures that can accomplish the tasks in Commentary Table I.9.3, which are tasks they normally perform in their day-to-day activities.

***COMMENTARY TABLE I.9.3** Individuals Who Can Assist the Private Sector Specialist Employee at Level B*

Normal Job Function	*Task(s)*
Material handler Container specialist	Repair containers
Material handler Loader/unloader Container specialist	Removing container contents
Environmental specialist	Cleaning up site Disposing of chemicals

9.3.3.2 Providing Information on Personal Protective Equipment Requirements. Given specific hazardous materials/WMD or containers for hazardous materials/WMD within the employee's individual area of specialization and the associated resources, the specialist employee B shall advise the incident commander of the personal protective equipment necessary for various response options and shall meet the following requirements:

(1) Given a specific chemical and a corresponding MSDS or other chemical-specific resource, identify personal protective equipment, including the materials of construction, that is compatible with that chemical.
(2) Given the emergency response plan and/or standard operating procedures, identify other resources (including a method of contact) capable of identifying the personal protective equipment that is compatible with a specific chemical.

Some of the resources a specialist employee B can use to identify PPE compatible with a specific chemical are compatibility charts, databases, and PPE manufacturers' literature.

(3) Given an incident involving a specific chemical and the response options for that incident, determine whether the personal protective equipment is appropriate for the options presented.

All specialist employees B who are trained to select and/or use PPE are required by 9.3.3.2(3) to be able to identify, interpret, and apply compatibility data from compatibility charts and to consider the limitations associated with the use of such equipment, especially if the potential for fire is present.

9.3.3.3 Providing Information on Decontamination Methods. Given a specific chemical within the employee's individual area of specialization and the available resources, the specialist employee B shall identify the technical decontamination process for various response options and shall complete the following tasks:

(1) Given a specific chemical and a corresponding MSDS or other chemical-specific resource, identify the potential methods for removing or neutralizing that chemical.

In accordance with 9.3.3.3 (1), the specialist employee B should be able to point out the sections on an MSDS that include decontamination and neutralization information or be able to contact an individual in the organization who can provide this information.

(2) Given a specific chemical and a corresponding MSDS or other chemical specific resource, identify the circumstances under which disposal of contaminated equipment would be necessary.

Materials that cannot be decontaminated, such as porous materials like leather or wood; limited-use equipment, such as one-time-use protective clothing; or equipment for which decontamination procedures are unknown, might have to be disposed of through bagging. The specialist employee B should be able to clearly identify when disposal is necessary and the appropriate method to use, in accordance with 9.3.3.3(2).

(3) Given the emergency response plan and/or standard operating procedures, identify resources (including a method of contact) capable of identifying potential decontamination methods.

9.3.3.4 Providing Information on Handling and Disposal Regulations. Given a specific chemical within the employee's individual area of specialization and the available resources, the specialist employee B shall advise the incident commander of the federal or provincial regulations that relate to the handling, transportation, and disposal of that chemical and shall complete the following tasks:

(1) Given a specific chemical and a corresponding MSDS or other resource, identify federal or provincial regulations that apply to the handling, transportation, and disposal of that chemical.

Among the regulations that apply to the handling, transportation, and disposal of chemicals are DOT's Title 49 CFR; EPA's Title 40 CFR; and, in Canada, Transportation of Dangerous Goods Act (TDG) and/or the Ministry of the Environment (MOE) [4, 5, 6]. In the event of an incident, the agencies that handle these regulations should be contacted by someone identified in the organization's emergency response plan or standard operating procedures as the person responsible for making the necessary notifications according to regulations.

(2) Given a specific chemical and a corresponding MSDS or other resource, identify the agencies (including a method of contact) responsible for compliance with the federal or provincial regulations that apply to the handling, transportation, and disposal of a specific chemical.

In accordance with 9.3.3.4(2), the agencies responsible for compliance with regulations applying to the handling, transportation, and disposal of a specific chemical include the Occupational Safety and Health Administration (OSHA), which covers emergency response activities; the U.S. Department of Transportation (DOT), which deals with transportation; and the Environmental Protection Agency (EPA), which handles hazardous materials disposal in the United States. Transport Canada, Environment Canada, and Labor Canada cover the same issues for the provinces. In the event of an incident, these agencies should be contacted by someone identified in the organization's emergency response plan or standard operating procedures as the person responsible for making the necessary notifications according to regulations.

(3) Given the emergency response plan and/or standard operating procedures, identify resources for information pertaining to federal or provincial regulations relative to the handling and disposal of a specific chemical.

The specialist employee B is required by 9.3.3.4 (3) to be able to contact a knowledgeable person within the organization, such as an environmental engineer, chemist, or specialist, who knows where in the organization's emergency response plan or standard operating procedures the needed information can be located.

9.3.3.5 Developing an Incident Action Plan. Given a scenario involving hazardous materials/WMD or containers used in the employee's individual area of specialization, the specialist employee B shall (in conjunction with the incident commander) develop an incident action plan, consistent with the emergency response plan and/or standard operating procedures and within the capabilities of the available resources, for handling hazardous materials/WMD or containers in that incident and shall complete the following tasks:

(1) Given the emergency response plan and/or standard operating procedures, identify the process for development of an incident action plan, including roles and responsibilities under the Incident Command System site safety and control plan.

Each potential hazardous materials incident is required to have an overall plan of action. However, specialist employees B are expected to develop an action plan for only those tasks and procedures associated with their areas of expertise and regular job duties. They are also expected to define the processes needed to execute that plan. For example, loaders and unloaders who transload material as part of their normal duties may be called upon to develop an action plan for performing the same procedure at the scene of a hazardous materials incident and to define the steps and safety considerations required to execute that plan.

(2) Include a site safety and control plan in the incident action plan.

9.3.4 Competencies — Implementing the Planned Response.

9.3.4.1 Performing Response Options Specified in the Incident Action Plan. Given an assignment by the incident commander in the employee's individual area of specialization, the specialist employee B shall perform the assigned actions consistent with the emergency response plan and/or standard operating procedures and shall complete the following tasks:

(1) Perform assigned tasks consistent with the emergency response plan and/or standard operating procedures and the available personnel, tools, and equipment (including personal protective equipment), including the following:
   (a) Confinement activities
   (b) Containment activities
   (c) Product removal activities

(2)* Identify factors that can affect an individual's ability to perform the assigned tasks.

A.9.3.4.1(2) Such factors include heat, cold, working in a confined space, working in personal protective equipment, working in a flammable or toxic atmosphere, and pre-existing health conditions.

9.3.4.2 Using Personal Protective Equipment. Given an assignment within the employee's individual area of specialization that is consistent with the emergency response plan and/or standard operating procedures, the specialist employee B shall be able to complete the following tasks:

(1) Don, work in, and doff the correct respiratory protection and protective clothing for the assigned tasks.
(2) Identify the safety considerations for personnel working in personal protective equipment, including the following:
   (a) Buddy system

All responders at a hazardous materials incident must work in a minimum of two-person teams, as required by 9.3.4.2(2)(a). Each person must be within sight or sound of the other person at all times during the incident.

   (b) Backup personnel

Safety procedures in 9.3.4.2(2)(b) require that backup responders dress at the same level of protection as those on the entry team. Backup personnel may have to enter the same hazardous area as the primary entry team workers to perform a rescue and should be prepared to do so at any time.

   (c) Symptoms of heat and cold stress

Safety procedures in 9.3.4.2(2)(c) require that personnel be monitored for symptoms such as abnormal pulse, change in body temperature, respiratory difficulty, changes in skin color, and decreased mental alertness during or shortly after exposure to a hazardous material.

   (d) Limitations of personnel working in personal protective equipment

Safety procedures establish time limits for wearing personal protective clothing during an incident. Other limiting factors include the responder's level of physical fitness, endurance ability, and the psychological condition of the individual responder.

   (e) Indications of material degradation of chemical-protective clothing

Safety procedures in 9.3.4.2(2)(e) require that responders evaluate signs of material degradation, including discoloration, the loss of integrity and flexibility, the formation of blisters, and melting or stretching of material.

   (f) Physical and psychological stresses on the wearer

Safety procedures required to maintain the overall well-being of the responder include monitoring responders for signs of stress related to work in the hazardous materials incident environment. Such stress could cause a rise in body temperature, loss of body fluids, elevated pulse and respiration rates, vertigo, nausea, changes in skin color, disorientation, anxiety, and incoherence.

(g) Emergency procedures and hand signals

Safety procedures in 9.3.4.2(2)(g) require that responders maintain visual contact and that they demonstrate appropriate hand signals for loss of air and emergency escape assistance.

(3) Identify the procedures for cleaning, sanitizing, and inspecting personal protective equipment provided by the organization.

Cleaning, sanitizing, and inspecting PPE should be performed in accordance with the manufacturer's recommendations and the organization's standard operating procedures concerning the use and maintenance of PPE.

9.3.5 Competencies — Evaluating Progress.

9.3.5.1 Providing an Evaluation of the Effectiveness of Selected Response Options. Given an incident involving specific hazardous materials/WMD or containers for hazardous materials/WMD within the employee's individual area of specialization, the specialist employee B shall advise the incident commander of the effectiveness of the selected response options and shall complete the following tasks:

(1) Identify the criteria for evaluating whether the selected response options are effective in accomplishing the objectives.

The criteria in 9.3.5.1(1) must be based on the desired outcome. The responder must determine how the hazards to personnel, the environment, and property affect the outcome sought. Effective options result in diminished hazards, stabilization, or complete mitigation.

(2) Identify the circumstances under which it would be prudent to withdraw from a chemical incident.

In accordance with 9.3.5.1(2), the responder should withdraw from a chemical incident when the following conditions exist:

- Intervention will not or cannot produce a favorable outcome;
- The immediate hazard level is unacceptable;
- The incident is worsening as a result of the option selected.

9.3.5.2 Reporting and Documenting the Incident. Given a scenario involving hazardous materials/WMD or containers for hazardous materials/WMD used in the employee's individual area of specialization, the specialist employee B shall complete the reporting and subsequent documentation requirements consistent with the emergency response plan and/or standard operating procedures and shall complete the following tasks:

(1) Identify the importance of documentation (including training records, exposure records, incident reports, and critique reports) for an incident involving hazardous materials/WMD.

Organizations require documentation in order to help establish preventive and corrective actions. Documentation of hazardous materials incidents is also mandated by governmental regulation. Documentation is also required for emergency response training, medical monitoring, and PPE certification. Regulatory organizations require documentation for similar reasons. Finally, documentation is frequently used for trend analysis in hazardous materials incident prevention assessments.

(2) Identify the steps used in keeping an activity log and exposure records.

Most emergency response plans require that an employee's name and identifying code (usually a social security number) be recorded with each of the functions performed at a hazardous materials incident. In addition, the time and duration of each activity must be accurately documented. Exposure records are compiled in a similar manner, listing the materials to which the employee may be exposed, the duration of exposure, the manner in which the contaminant was identified, and the name of the person who made that determination.

(3) Identify the requirements for compiling incident reports.

The requirements in 9.3.5.2(3) for compiling an incident report are to provide a factual, objective format for defining the who, what, where, when, and how of an incident.

(4) Identify the requirements for compiling hot zone entry and exit logs.

Hot zone entry and exit logs are needed so that those in charge of an incident can monitor an employee's health and keep track of the work performed and the time spent in the hot zone. See 9.3.5.2 (2) for more information.

(5) Identify the requirements for compiling personal protective equipment logs.

In accordance with 9.3.5.2(5), PPE logs should include information about use time, inspections, testing, and the results of inspection and decontamination procedures for each garment or suit used. This documentation should also include a list of the contaminants to which the equipment has been exposed during its lifetime of use as well as the duration of each exposure.

(6) Identify the requirements for filing documents and maintaining records.

Most organizations keep records to protect themselves and their employees and provide a written account of each incident. Appropriate sources within the organization, such as regulatory specialists, should be consulted for both internal and external reporting requirements.

(7) Identify resources (including a method of contact) knowledgeable of the federal or provincial reporting requirements for hazardous materials/WMD incidents.

9.4 Specialist Employee A

9.4.1 General.

9.4.1.1 Introduction.

Specialist employees A can provide the incident commander with considerable assistance because their training deals with a range of products and containers specific to their employing organization and is usually equivalent to that of the hazardous materials technician for these specific products and containers. However, specialist employees A need only demonstrate those technician level competencies that apply to the materials and containers for which they are expected to respond at their company's property.

9.4.1.1.1 The specialist employee A shall be that person who is specifically trained to handle incidents involving chemicals or containers for chemicals used in the organization's area of specialization, and the following:

(1) Consistent with the emergency response plan and/or standard operating procedures, the specialist employee A is able to analyze an incident involving chemicals within his or her organization's area of specialization.
(2) The specialist employee A can then plan a response to that incident, implement the planned response within the capabilities of the resources available, and evaluate the progress of the planned response.

9.4.1.1.2 The specialist employee A shall be trained to meet all competencies at the awareness level (Chapter 4) relative to the organization's area of specialization, all competencies at the specialist employee C level (Section 9.2), and all competencies at the hazardous materials technician level (Chapter 7) relative to the hazardous materials/WMD and containers used in the organization's area of specialization.

9.4.1.2 Goal.

9.4.1.2.1 The goal of this level of competence shall be to ensure that the specialist employee A has the knowledge and skills to safely perform the duties and responsibilities assigned in the emergency response plan and/or standard operating procedures.

9.4.1.2.2 In addition to being competent at the specialist employee C and the hazardous materials technician levels, the specialist employee A shall be able to, in conjunction with the incident commander, perform the following tasks:

(1) Analyze an incident involving hazardous materials/WMD and containers for hazardous materials/WMD used in the organization's area of specialization to determine the magnitude of the incident by completing the following tasks:

  (a) Survey an incident involving hazardous materials/WMD and containers for hazardous materials/WMD, including the following:

    i. Identify the containers involved.

The specialist employee A is required by 9.4.1.2.2(1)(a)(i) to be able to identify the characteristics of any on-site containers that might indicate the presence of hazardous materials. These special containers include high-pressure containers (identified by their rounded ends), cryogenic cargo tanks or cylinders, and casks used for radioactive materials.

    ii. Identify or classify unknown materials.

The specialist employee A is required by 9.4.1.2.2(1)(a)(ii) to be able to identify unknown materials likely to be on hand in the employer's manufacturing or storage facilities in the event that a container's shipping papers, placards, MSDS, or other identifying items have been destroyed or are unavailable.

    iii. Verify the identity of the hazardous materials/WMD.

  (b) Collect and interpret hazard and response information from printed resources, technical resources, computer databases, and monitoring equipment for hazardous materials/WMD.

Specialist employees A must understand the importance of using multiple resources for gathering hazard and response information. In 9.4.1.2.2(1)(b), these responders are required to collect information from a variety of resources, be able to appropriately compare that information, and then make decisions based on this comparison and evaluation. Most responders generally give more weight to the more conservative information and base their actions on that information.

  (c) Determine the extent of damage to containers of hazardous materials/WMD.

The following terms allow the responder to use standard terminology when assessing containers to determine how badly they have been damaged. The condition of the container should be described using one of the following terms:

- Undamaged, no product release
- Damaged, no product release
- Damaged, product release
- Undamaged, product release

  (d) Predict the likely behavior of the hazardous materials/WMD and containers for hazardous materials/WMD.

After the chemical is identified, information about the chemical's hazards and behavior should be collected from appropriate reference sources. This information, which can be obtained simultaneously with determining the extent of container/packaging damage, is used to predict the behavior of the chemical and its container.

(e) Estimate the potential outcomes of an incident involving hazardous materials/WMD and containers for hazardous materials/WMD.

An estimate is a series of predictions that attempts to provide an overall picture of potential outcomes. Responders must assess the information gathered during analysis and predict the outcome based on that assessment.

(2) Plan a response (within the capabilities of available resources) to an incident involving hazardous materials/WMD and containers for hazardous materials/WMD used in the organization's area of specialization by completing the following tasks:

(a) Identify the response objectives for an incident involving hazardous materials/WMD and containers for hazardous materials/WMD.

The specialist employee A should note that many of the options available in 9.4.1.2.2(2)(a) can be used either offensively or defensively. For example, the size of the potential endangered area can be changed defensively, by diking to contain a spill, or offensively, by entering the hot zone and plugging the leak. Defensive operations might be the most prudent method of dealing with many hazardous materials.

(b) Identify the potential response options for each response objective for an incident involving hazardous materials/WMD and containers for hazardous materials/WMD.

Chapter 7 of this handbook provides information about the options available to the specialist employee A for accomplishing response objectives.

(c) Select the personal protective equipment required for a given response option for an incident involving hazardous materials/WMD and containers for hazardous materials/WMD.

As directed in 9.4.1.2.2(2)(c), the responder should always use the highest level of protection until the chemicals likely to be involved in an incident can be positively identified and the level of concentration can be determined. Determining the most appropriate type of respiratory protection required at each incident depends on many factors. A key factor is the level of protective clothing necessary to protect the specialist employee A. Level A protection limits the user to positive pressure self-contained breathing apparatus or positive pressure supplied-air respirators.

(d) Select the technical decontamination process for an incident involving hazardous materials/WMD and containers for hazardous materials/WMD.

The two methods of decontamination are physical and chemical. Physical methods manually separate the chemical from the material being decontaminated by scrubbing or washing the material, or both. Physical decontamination is often easier than chemical decontamination, but it may not completely remove all the contaminants. Chemical methods involve changing one chemical into another or into a form that facilitates its removal. Unfortunately, the chemical process involved could introduce other hazards. Care must be taken to collect all the contamination that has been removed by either method and to dispose of it properly.

(e) Develop an incident action plan (within the capabilities of the available resources), including site safety and control plan, for handling an incident involving hazardous materials/WMD and containers for hazardous materials/WMD consistent with the emergency response plan and/or standard operating procedures.

After selecting the response option for a hazardous materials incident, an incident action plan including safety and health considerations should be developed, in accordance with 9.4.1.2.2(2)(e). The incident action plan describes the response objectives and options and the personnel and equipment required to accomplish the objectives. The plan provides a permanent record of the decisions made at the incident. An organization's standard operating procedures provide the basis of the incident action plan. Components for a typical plan of action include the following:

- Site description
- Entry objectives
- On-scene organization and coordination
- On-scene control
- Hazard evaluation
- Personal protective equipment
- On-scene work assignments
- Communications procedures
- Decontamination procedures
- On-scene safety and health considerations including designation of the safety officer, emergency medical care procedures, environmental monitoring, emergency procedures, and personnel monitoring

Obviously, the complexity of an incident determines the detail identified in the incident action plan. Each of the items in the preceding list must be considered, however, in order to ensure that nothing is being overlooked.

(3) Operating under the Incident Command System, implement the planned response (as developed with the incident commander) to an incident involving hazardous materials/WMD and containers for hazardous materials/WMD used in the organization's area of specialization consistent with the emergency response plan and/or standard operating procedures by completing the following tasks:

(a) Don, work in, and doff correct personal protective equipment for use with hazardous materials/WMD.

Specialist employees A should practice donning and doffing chemical-protective clothing to become proficient in the competency required in 9.4.1.2.2(3)(a). One of the more effective ways to evaluate the ability to don and doff the protective clothing provided is to conduct training exercises that require putting on the PPE and to conduct simulated control activities, followed by simulated decontamination.

Because some types of chemical-protective clothing are very costly, using garments that are no longer adequate for emergency response might be advisable. Hazardous materials training suits that are less expensive are also available; these suits allow for cost-effective training in the use of totally encapsulating suits.

(b) Perform containment, control, and product transfer functions, as agreed upon with the incident commander, for hazardous materials/WMD and containers for hazardous materials/WMD.

The purpose of the competency required by 9.4.1.2.2(3)(b) is to allow the specialist employee A to demonstrate the ability to choose the appropriate equipment and methods to control a simulated incident. The equipment necessary to conduct the operations varies, depending on the type of material that is leaking and the physical properties of the damaged area, valve, or breach. For example, nonsparking tools should be used when working with flammable liquids or gases. And in some cases, special tools might be needed for certain valves. Such situations should be identified during pre-incident planning.

(4) Evaluate the results of implementing the planned response to an incident involving haz-

ardous materials/WMD and containers for hazardous materials/WMD used in the organization's area of specialization.

9.4.2 Competencies — Analyzing, Planning, Implementing, and Evaluating.

The specialist employee A shall demonstrate competencies at the specialist employee C level *(see Section 9.2)* and the hazardous materials technician level *(see Chapter 7)* relative to hazardous materials/WMD and containers used in the organization's area of specialization.

REFERENCES CITED IN COMMENTARY

1. *Emergency Response Guidebook,* U.S. Department of Transportation, Washington. DC, 2004.
2. *Emergency Handling of Hazardous Materials in Surface Transportation,* Association of American Railroads, Washington, DC.
3. *Emergency Action Guides,* Association of American Railroads, Washington, DC.
4. Title 49, Code of Federal Regulations, U.S. Government Printing Office, Washington, DC.
5. Title 40, Code of Federal Regulations, U.S. Government Printing Office, Washington, DC.
6. Transportation of Dangerous Goods Act 1992, Transport Canada, Ontario, Ottawa, Canada.

Additional References

Cote, A., ed., *Fire Protection Handbook®,* 20th ed., National Fire Protection Association, Quincy, MA, 2008.

Ministry of the Environment (MOE), Toronto, Ontario, Canada.

Competencies for Hazardous Materials Officers

CHAPTER 10

The term *hazardous materials branch officer* was added to the 1997 edition of NFPA 472 and has been changed to *hazardous materials officer* in the 2008 edition. The technical committee created the designation to meet the need for an individual who could function as the hazardous materials response team leader, the hazardous materials group supervisor, or as a hazardous materials technical specialist. Exhibit I.10.1 illustrates a typical incident command chart and shows where the hazardous materials officer fits into the team command.

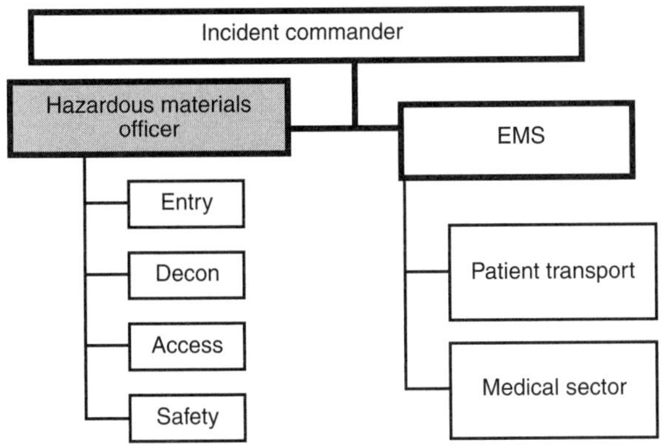

***EXHIBIT I.10.1** This incident command chart shows the role of the hazardous materials officer.*

10.1 General

10.1.1 Introduction.

10.1.1.1 The hazardous materials officer (NIMS: Hazardous Materials Branch Director/Group Supervisor) shall be that person who is responsible for directing and coordinating all operations involving hazardous materials/WMD as assigned by the incident commander.

10.1.1.2 The hazardous materials officer shall be trained to meet all competencies at the awareness level (Chapter 4), all core competencies at the operations level (Chapter 5), all competencies at the technician level (Chapter 7), and all competencies of this chapter.

10.1.1.3 The hazardous materials officer shall also receive training to meet governmental occupational health and safety regulations.

The hazardous materials officer has the same technical competencies as a technician and completes the chain of command needed at an incident, which, in turn, can increase the safety and efficiency at the hazardous materials/WMD incident site. In complex incidents involving numerous chemicals or more than one central location, the incident commander (IC) would use the hazardous materials officer's expertise in formulating response objectives, action options, and the plan of action. In addition, the IC would rely on the hazardous materials officer to provide direct supervision of the team to accomplish the objectives. More than one team can then work at an incident, and each has its own leader.

10.1.2 Goal.

10.1.2.1 The goal of the competencies at this level shall be to provide the hazardous materials officer with the knowledge and skills to perform the tasks in 10.1.2.2 safely.

10.1.2.2 When responding to hazardous materials/WMD incidents, the hazardous materials officer shall be able to perform the following tasks:

(1) Analyze a hazardous materials/WMD incident to determine the complexity of the problem by estimating the potential outcomes within the endangered area.
(2) Plan a response within the capabilities and competencies of available personnel, personal protective equipment, and control equipment by completing the following tasks:
    (a) Identify the response objectives (defensive, offensive, and nonintervention) for hazardous materials/WMD incidents.
    (b) Identify the potential response options (defensive, offensive, and nonintervention) available by response objective.
    (c) Determine the level of personal protective equipment required for a given action option.
    (d) Provide recommendations to the incident commander for the development of an incident action plan for the hazardous materials branch/group consistent with the emergency response plan and/or standard operating procedures and within the capability of available personnel, personal protective equipment, and control equipment.

The hazardous materials officer can perform many of the same planning functions as an IC, as outlined in 10.1.2.2(1) and 10.1.2.2(2). However, the officer can only recommend that various actions be taken. The IC must consider each incident in its entirety and make the final decision.

(3) Implement a response to favorably change the outcomes consistent with the emergency response plan and/or standard operating procedures by completing the following tasks:
    (a) Implement the functions within the incident command system as they directly relate to the specified procedures for hazardous materials branch/group operations.
    (b) Direct hazardous materials branch/group resources (private, governmental, and others) with task assignments and on-scene activities and provide management overviews, technical review, and logistical support to hazardous materials branch/group resources.
(4) Evaluate the progress of the planned response to ensure that the response objectives are effective, and adjust the incident action plan accordingly.
(5) Terminate the incident by completing the following:
    (a) Conduct a debriefing for hazardous materials branch/group personnel.
    (b) Conduct a critique for hazardous materials branch/group personnel.
    (c) Report and document the hazardous materials branch/group operations.

10.2 Competencies — Analyzing the Incident

Given scenarios involving hazardous materials/WMD incidents, including the surrounding conditions and the predicted behavior of the container and its contents, the hazardous materials officer shall estimate the potential outcomes within the endangered area.

Outcomes are the result of implementing the response objectives. These objectives can include the following:

- Modifying the stress being applied to the container
- Changing the size of the breach
- Changing the quantity being released
- Changing the size of the endangered area
- Reducing exposures
- Reducing the level of harm

Keep in mind that an acceptable alternative response that should always be considered is doing nothing and keeping everyone away.

The outcomes can be the result of defensive, offensive, or nonintervention actions. Several potential outcomes can be considered by the IC. Each outcome has its own risks, potential for success, and technical expertise required for successful completion. Each option must be thought through to its end or outcome. Only if the outcome can improve the situation should the option be considered.

10.3 Competencies — Planning the Response

10.3.1 Given a scenario involving a hazardous materials/WMD incident, the hazardous materials officer shall identify the response objectives (defensive, offensive, and nonintervention) for each incident.

10.3.2 Given a scenario involving hazardous materials/WMD incidents, the hazardous materials officer shall identify the potential response options (defensive, offensive and nonintervention) for each incident.

10.3.3 Selecting the Level of Personal Protective Equipment.

Given scenarios involving hazardous materials/WMD incidents with known and unknown hazardous materials/WMD, the hazardous materials officer shall select the personal protective equipment for the response options specified in the incident action plan in each situation.

In accordance with 10.3.3, the hazardous materials officer can select the level of personal protective equipment (PPE), but the IC must approve the selection. The type of respiratory protection required depends on many factors. A key factor is the level of protective clothing necessary for protection in the hot zone. A responder should always use the highest level of protection until the specific hazardous material and the levels of concentration have been determined.

10.3.4 Developing a Plan of Action

Given scenarios involving hazardous materials/WMD incidents, the hazardous materials officer shall develop a plan of action consistent with the emergency response plan and/or standard operating procedures that is within the capability of the available personnel, personal protective equipment, and control equipment and shall complete the following tasks:

A plan of action, as required by 10.3.4, describes the response objectives and any options available in achieving the objectives. Again, with each objective should be an estimate of the likely outcome. The basis for the plan is the estimate of the potential outcomes, the organization's standard operating procedures, and the local emergency response plan. In the plan, safety and health considerations, necessary personnel, and control equipment should be listed for each objective. This plan provides a permanent record of the incident and can be the outline for the incident critique.

(1) Identify the order of the steps for developing the plan of action.

In accordance with 10.3.4(1), the following steps need to be considered when developing a plan of action:

- Site restrictions
- Entry objectives
- On-scene organization and control
- PPE selection
- Hazard evaluation
- Communications procedures
- Emergency procedures and personnel accountability
- Emergency medical care arrangements
- Rehabilitation plan
- Decontamination procedures
- On-scene work assignments (branches)
- Debriefing and critiquing

(2) Identify the factors to be evaluated in selecting public protective actions, including evacuation and shelter-in-place.

The goal of the evaluation in 10.3.4(2) should be to reduce or prevent contamination of those members of the public who are or could be directly exposed to the hazardous materials. If these individuals are safe in their present location and the structure in which they are housed can be protected from contamination by methods such as closing windows and doors and turning off ventilation systems that might draw air in from the outside, leaving these individuals in place until the incident can be controlled might be the best option. Even after the public has been notified that they must remain in their present location, responders should provide these individuals with constant updates on the progress of the incident and reassurance of their safety.

(3) Given the emergency response plan and/or standard operating procedures, identify procedures to accomplish the following tasks:
   (a) Make ongoing assessments of the situation.
   (b) Command on-scene personnel assigned to the hazardous materials branch/group.
   (c) Coordinate hazardous materials/WMD support and mutual aid.
   (d) Coordinate public protective actions (evacuation or shelter-in-place).
   (e) Coordinate with fire suppression services as they relate to hazardous materials/WMD incidents.
   (f) Coordinate control, containment and confinement operations.
   (g) Coordinate with the medical branch to ensure medical assistance (ambulance) and medical treatment (hospital).
   (h) Coordinate on-scene decontamination.
   (i) Coordinate activities with those of the environmental remediation (cleanup) services.
   (j) Coordinate evidence preservation and sampling in a contaminated environment.

(4) Identify the process for determining the effectiveness of an action option on the potential outcomes.

Direct observations from the technicians and the hazardous materials officer should be compared against the expected outcomes as listed in the plan of action. If time is a factor, specific time frames should be determined before any actions are started. At the agreed-upon times, reports should be forwarded to the IC.

(5) Identify the procedures for presenting a safety briefing prior to allowing personnel to work on a hazardous materials/WMD incident.

Some of the areas that should be evaluated in the safety briefing, as required by 10.3.4(5), include the following:

- Incident commander, hazardous materials officers, and hazardous materials responders have met all the competencies for their appropriate levels in accordance with NFPA 472.

- Activities that present a significant risk to the safety of members are limited to situations that offer the potential to save endangered lives.
- No risk to the safety of members is acceptable when saving lives or property is not possible.
- All personnel working in the warm zone or hot zone are under the supervision of a hazardous materials officer.
- Personnel accountability procedures are being utilized.
- A rest and rehabilitation area that is ready and available for responders once their assignments have been completed is described and its location identified.
- A hazardous materials safety officer has been designated and is operational.
- Communications have been established on one simplex radio channel that is not used by anyone close enough to interfere. Hand signals are available as a backup if the radios fail.
- Appropriate protective clothing and protective equipment are used whenever the responder is exposed, or potentially exposed, to the hazardous materials.
- A rapid intervention crew consisting of at least two responders is available for rescue of a member or team if the need arises. Responders operating in the hot zone are operating in teams of two or more.
- Medical monitoring of all responders is completed before they can proceed to work in PPE.
- Responders are warned to be aware of any clues indicating that the incident could be a chemical, biological, nuclear, or explosives incident. If terrorism is suspected, responders are alert to the possibility of secondary devices or attempts to disguise the true nature of the incident.

10.4 Competencies — Implementing the Planned Response

10.4.1 Implementing the Functions in the Incident Management System.

Given a copy of the emergency response plan, the hazardous materials officer shall identify the requirements of the plan, including the required procedures for notification and utilization of nonlocal resources (private, state, and federal government personnel), and shall complete the following tasks:

Several different models for incident management systems are available. Each system generally has the same components, even though different titles for the various responders are sometimes used. For example, hazardous materials officers, hazardous materials branch directors, and hazardous materials group supervisors generally perform the same functions. Basically, these workers are responsible for implementing the incident action plan that deals with any operations intended to control the hazardous materials portion of an incident. Overall direction and approval of the action plan is the duty of the IC. Exhibit I.10.2 shows that there are generally five primary functional areas within the incident management system.

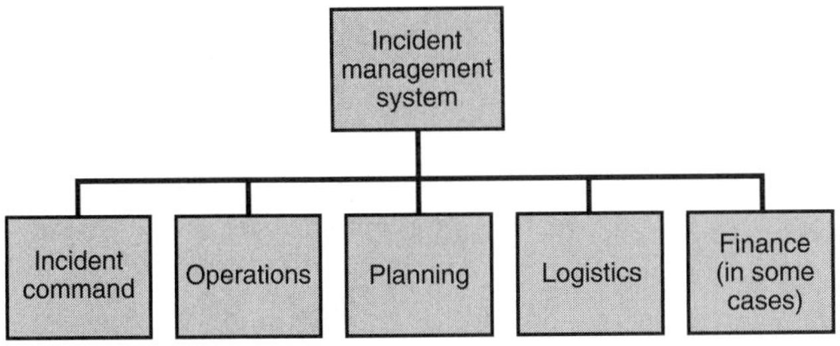

EXHIBIT I.10.2 *These are the five areas of an incident management system.*

A public information officer and an incident safety officer are usually included as part of the team as well. For hazardous materials incidents, additional functions are identified, including the hazardous materials officer, hazardous materials safety officer, decontamination unit leader, rehabilitation unit leader, information research and resources unit leader, and the entry/reconnaissance unit leader.

(1) Identify the process and procedures for obtaining cleanup and remediation services in the emergency response plan and/or standard operating procedures.
(2) Identify the steps for implementing the emergency response plans as required under SARA Title III Section 303 of the federal regulations or other emergency response planning legislation.
(3) Given the local emergency planning documents, identify the elements of each of the documents.
(4) Identify the elements of the local incident management system necessary to coordinate response activities at hazardous materials/WMD incidents.
(5) Identify the primary local, state, regional, and federal government agencies and identify the scope of their regulatory authority (including the regulations) pertaining to the production, transportation, storage, and use of hazardous materials/WMD and the disposal of hazardous wastes.

The hazardous materials officer is required by 10.4.1(5) to know which agencies could become involved and their regulatory authority in a hazardous materials incident. The local emergency response plan should identify these agencies and delineate their roles, authority, and functions.

(6) Identify the governmental agencies and resources offering assistance to the hazardous materials branch/group during a hazardous materials/WMD incident and identify their role and the type of assistance or resources available.

Some government agencies provide response or technical assistance to local hazardous materials response teams. For example, the U.S. Coast Guard staffs three strategically located National Strike Teams (Atlantic, Gulf, and Pacific) that are trained and equipped to respond to major oil spills and chemical releases. Strike teams are part of the National Strike Force, which consists of active duty, civilian, and support personnel. Strike teams have provided support and expertise to many recent events, including vessel groundings, hurricanes, air crashes, and oil spills. More information about the Coast Guard's Strike Force can be found at www.uscg.mil. The U.S. Environmental Protection Agency also has a response component called the Environmental Response Team (ERT). The ERT is a group of scientists and engineers who are trained in sampling and analysis, hazard assessment, cleanup techniques, and other avenues of technical support. More information on the ERT can be found at www.ert.org. In addition, many state and local governments and private industries can provide technical assistance. Hazardous materials response teams from other municipalities or private industry, for example, can be a great resource for large incidents and might have specialized monitoring equipment and knowledge to help identify complex or obscure hazardous materials.

The local emergency response plan should identify the law enforcement and related agencies and delineate their roles, authority, and functions. A key element will be the ability of local, state, and federal law enforcement personnel to collect evidence in a contaminated environment.

The Federal Bureau of Investigation (FBI) has a WMD Coordinator in each of their 56 field divisions. The WMD Coordinator is the liaison for the Hazmat Officer to the FBI and can also contact FBI Evidence Response Teams, FBI Hazardous Materials Response Teams, and other FBI assets that may be of assistance.

The private sector also has many resources available to emergency responders, from providing technical advice and assisting with on-scene monitoring to providing specialized equipment. In most areas, private sector companies that provide cleanup and disposal services are available. Some of these companies can even stage equipment in an area and fly personnel to the scene when a hazardous materials incident occurs. CHEMTREC/CANUTEC/SETIQ offers referrals to private sector resources.

The hazardous materials officer must know how to get specialized assistance when it is needed and must be familiar with various types of public and private sector assistance available. These resources and their contacts should be identified in the local emergency response plan.

(7) Identify the governmental agencies and resources offering assistance during a hazardous materials incident involving criminal or terrorist activities and identify their role and the type of assistance or resources available.

10.4.2* Directing Resources (Private and Governmental).

Given a scenario involving a hazardous materials/WMD incident and the necessary resources to implement the planned response, the hazardous materials officer shall demonstrate the ability to direct the hazardous materials branch/group resources in a safe and efficient manner consistent with the capabilities of those resources.

A.10.4.2 These abilities should include the following:

(1) Task assignment (based on strategic and tactical options)
(2) Operational safety
(3) Operational effectiveness
(4) Planning support
(5) Information and research
(6) Logistical support
(7) Administrative support

10.4.3 Providing a Focal Point for Information Transfer to Media and Elected Officials.

Given a scenario involving a hazardous materials/WMD incident, the hazardous materials officer shall demonstrate the ability to act as a resource to provide information to the incident commander or the public information officer for distribution to the media and local, state, and federal officials and shall complete the following tasks:

(1) Identify the local policy for providing information to the media.
(2) Identify the responsibilities of the public information officer at a hazardous materials/WMD incident.

10.5 Competencies — Evaluating Progress

Given scenarios involving hazardous materials/WMD incidents, the hazardous materials officer shall evaluate the progress of the incident action plan to determine whether the efforts are accomplishing the response objectives and shall complete the following tasks:

(1) Identify the procedures for evaluating whether the response options are effective in accomplishing the objectives.

To determine whether the actions being taken at an incident are effective and the objectives are being met, responders must frequently reevaluate whether the incident is stabilizing or increasing in intensity, in accordance with 10.5(1). By comparing the predicted outcomes in the action plan to what is unfolding in the actual incident, the hazardous materials officer has another indication of progress. Feedback from the entry teams and safety officer and other observations allow the hazardous materials officer to recommend modifying the action options to the IC. This feedback should include information on the effectiveness of personnel, on personal protective clothing and equipment, on control zones, on decontamination procedures, and on the action options being implemented.

(2) Identify the steps for comparing actual behavior of the material and the container to that predicted in the analysis process.

When comparing actual behavior to predicted behavior as required by 10.5(2), the hazardous materials officer should determine whether events at an incident are happening as predicted, are occurring out of sequence, or are different from expected. The hazardous materials officer should also determine whether events that were predicted to occur have actually occurred at the incident. This evaluation and re-evaluation process should continue until the incident has been terminated in order to minimize "surprises" during the cleanup or overhaul phase of the incident response.

(3) Determine the effectiveness of the following:
   (a) Personnel being used
   (b) Control zones
   (c) Personal protective equipment
   (d) Control, containment, and confinement operations
   (e) Decontamination
(4) Make appropriate modifications to the incident action plan.

10.6 Competencies — Terminating the Incident

10.6.1 Terminating the Emergency Phase of the Incident.

Given a scenario involving a hazardous materials/WMD incident, the hazardous materials officer shall demonstrate the ability to terminate the emergency phase of the incident consistent with the emergency response plan and/or standard operating procedures and shall complete the following tasks:

(1) Identify the steps required for terminating the emergency phase of a hazardous materials/WMD incident.

As required by 10.6.1(1), the steps involved in terminating the emergency phase of an incident include the following:

- Transferring command
- Debriefing
- Critiquing the incident
- Conducting after-action activities, such as restocking disposables or damaged equipment

(2) Identify the procedures for conducting incident debriefings at a hazardous materials/WMD incident.

10.6.2 Conducting a Debriefing.

Given a scenario involving a hazardous materials/WMD incident, the hazardous materials officer shall demonstrate the ability to conduct a debriefing of the incident for all units assigned to the hazardous materials branch/group and shall complete the following tasks:

(1) Describe three components of an effective debriefing.
(2) Describe the key topics in an effective debriefing.
(3) Describe when a debriefing should take place.

(4) Describe who should be involved in a debriefing.
(5) Identify the procedures for conducting incident debriefings at a hazardous materials/WMD incident.

The incident debriefing by 10.6.2(5) involves gathering information from response personnel as soon as possible after terminating the incident. This information should then be used to develop a chronological report of emergency activities.

10.6.3 Conducting a Critique.

Given the details of a scenario involving a hazardous materials/WMD incident, the hazardous materials officer shall demonstrate the ability to conduct a critique of the incident for all units assigned to the hazardous materials branch/group and shall complete the following tasks:

(1) Describe three components of an effective critique.
(2) Describe who should be involved in a critique.
(3) Describe why an effective critique is necessary after a hazardous materials/WMD incident.
(4) Describe what written documents should be prepared as a result of the critique.
(5) Identify the procedure for conducting a critique of the incident.
(6) Identify the requirements for conducting a post-incident analysis as defined in the emergency response plan, standard operating procedures, or local, state, and federal regulations.

The critique, as required in 10.6.3, is a review of the incident intended to identify and document the lessons learned from the actual incident response. This process should allow the participants to review response activities and determine what worked well and what did not. The critique should be conducted in a positive manner that allows responders to modify response procedures if problems are identified, without assigning blame. After analyzing the information gathered during the debriefing and the critique and documenting that analysis, the termination process should also include a plan to follow up as necessary to ensure that any recommendations made to improve emergency operations are implemented.

10.6.4 Reporting and Documenting the Incident.

Given an example of a hazardous materials/WMD incident, the hazardous materials officer shall demonstrate the ability to report and document the incident consistent with the local, state, and federal requirements and shall complete the following tasks:

In accordance with 10.6.4, the hazardous materials officer must be aware of the reporting requirements associated with a hazardous materials incident. In many cases, the responsibility for reporting to specific agencies is outlined in the responding agency or organization's standard operating procedures. However, the hazardous materials officer must ensure that the proper agencies have been notified and that the proper reports have been made.

(1) Identify the reporting requirements of federal, state, and local agencies.
(2) Identify the importance of documentation for a hazardous materials/WMD incident, including training records, exposure records, incident reports, and critique reports.

Throughout the incident, the hazardous materials officer should assign someone to maintain a record of incident events as they occur. This documentation will be helpful in completing the post-incident analysis and conducting the critique because the "live" report conveys information as it unfolded. Reconstructing an incident afterward can result in a flawed timeline and misinterpretation of events. Personnel exposure records must also be maintained at the scene in accordance with federal and state laws. The necessary information about the type of exposure to which personnel were subjected, the exposure level, the length of the exposure, the type of personal protective clothing and equipment personnel were using, and the type of decontamination that personnel had undergone should be included as part of each responder's record. Any on-scene medical assistance that personnel received should also be documented.

(3) Identify the steps in keeping an activity log and exposure records for hazardous materials/WMD incidents.
(4) Identify the requirements found in the emergency response plan and/or standard operating procedures for compiling hazardous materials/WMD incident reports.
(5) Identify the requirements for filing documents and maintaining records as defined in the emergency response plan and/or standard operating procedures.
(6) Identify the procedures required for legal documentation and chain of custody/continuity described in the emergency response plan and/or standard operating procedures.

Competencies for Hazardous Materials Safety Officers

CHAPTER 11

The hazardous materials safety officer must be knowledgeable in the tactical actions being undertaken to ensure responder safety. To perform this function, the hazardous materials safety officer must know and understand the risk associated with all the actions and interventions being undertaken and be able to identify situations that can pose immediate harm. The hazardous materials safety officer has the authority and responsibility to intervene and stop an unsafe activity. The activity is then reconciled and remedied with the hazardous materials group supervisor/branch director, the incident commander, and/or the incident safety officer (see Exhibit I.11.1). Under Occupational Safety and Health Administration (OSHA) Title 29 CFR Part 1910.120(q), the term *safety official* is used to describe the tasks for the safety officer or hazardous materials safety officer [1].

EXHIBIT I.11.1 *The hazardous materials safety officer advises the incident commander on the risks and potential health hazards involved in an incident. (Source: Henrik G. de Gyor)*

11.1 General

11.1.1* Introduction.

A.11.1.1 If the functions and responsibilities of the hazardous materials safety officer are performed by the overall incident safety officer or on-scene incident commander, that individual should meet the competencies of this chapter.

11.1.1.1 The hazardous materials safety officer (NIMS: Assistant Safety Officer — Hazardous Material) shall be that person who works within an incident management system (IMS) (specifically, the hazardous material branch/group) to ensure that recognized hazardous materials/WMD safe practices are followed at hazardous materials/ WMD incidents.

11.1.1.2 The hazardous materials safety officer shall be trained to meet all competencies at the awareness level (Chapter 4), all core competencies at the operations level (Chapter 5), all competencies at the technician level (Chapter 7), and all competencies of this chapter.

In addition to being competent as a hazardous materials technician, the hazardous materials safety officer needs to be knowledgeable in risk management techniques and have a thorough understanding of the safety of responders during hot and warm zone operations. While safety is everyone's responsibility, the hazardous materials safety officer is responsible for ensuring safe practices and identifying immediately dangerous to life or health (IDLH) conditions that have not been previously identified and /or protected against that may place the responder at risk. The hazardous materials safety officer must ensure that his or her activities are coordinated with the incident safety officer.

11.1.1.3 The hazardous materials safety officer shall receive additional training to meet applicable governmental occupational health and safety regulations.

11.1.2 Goal.

11.1.2.1* The goal of the competencies at this level shall be to provide the hazardous materials safety officer with the knowledge and skills to evaluate a hazardous materials/WMD incident for safety and ensure that recognized safe operational practices are followed and to perform the tasks in 11.1.2.2 safely.

A.11.1.2.1 Under this section, the hazardous materials safety officer is given specific responsibilities. It should be understood that even though these duties are to be carried out by the hazardous materials safety officer, the incident commander has overall responsibility for the implementation of these tasks.

The hazardous materials safety officer should meet all the competencies for the responder at the level of operations being performed. A hazardous materials safety officer directs the safety of operations in the hot and the warm zones. A hazardous materials safety officer should be designated specifically at all hazardous material incidents (29 CFR 1910.120) and is responsible for the following tasks:

(1) Obtain a briefing from the incident commander or incident safety officer.
(2) Participate in the preparation of and monitor the implementation of the incident site safety and control plan (including medical monitoring of entry team personnel before and after entry).
(3) Advise the incident commander/sector officer of deviations from the incident site safety and control plan and of any dangerous situations.
(4) Alter, suspend, or terminate any activity that is judged to be unsafe.

11.1.2.2 When responding to hazardous materials/WMD incidents, the hazardous materials safety officer shall be able to perform the following tasks safely and effectively:

(1) Analyze a hazardous materials/WMD incident to determine the complexity of the problem in terms of safety by observing a scene and reviewing and evaluating hazard and response information as it pertains to the safety of all persons in the hazardous materials branch/group.
(2) Assist in planning a safe response within the capabilities of available response personnel, personal protective equipment, and control equipment by completing the following tasks:
    (a) Identify the safety precautions for potential response options.
    (b) Provide recommendations regarding the site safety and control plan.
    (c) Assist in the development of an incident action plan.
    (d) Review the incident action plan and provide recommendations regarding safety.
    (e) Review the selection of personal protective equipment required for a given action option.
    (f) Review the decontamination plan and procedures.
    (g) Ensure that emergency medical services are provided.
(3) Ensure the implementation of a safe response consistent with the incident action plan, the emergency response plan, and/or standard operating procedures by completing the following tasks:
    (a) Perform the duties of the hazardous materials safety officer within the incident command system.
    (b) Identify safety considerations for personnel performing the control functions identified in the site safety and control plan.
    (c) Conduct safety briefings for personnel performing the control functions identified in the site safety and control plan.
    (d) Assist in the implementation and enforcement of the site safety and control plan.
    (e) Maintain communications within the incident command structure during the incident.
    (f) Monitor status reports of activities in the hot and the warm zones.
    (g) Ensure the implementation of exposure monitoring (personnel and environment).
(4) Evaluate the progress of the planned response to ensure that the response objectives are being met safely by completing the following tasks:
    (a) Identify deviations from the site safety and control plan or other dangerous situations.
    (b) Alter, suspend, or terminate any activity that can be judged to be unsafe.
(5) Assist in terminating the incident by completing the following tasks:
    (a) Perform the reporting, documentation, and follow-up required of the hazardous materials safety officer.
    (b) Assist in the debriefing of hazardous materials branch/group personnel.
    (c) Assist in the incident critique.

11.2 Competencies — Analyzing the Incident

11.2.1 Determining the Magnitude of the Problem in Terms of Safety.

Given scenarios involving hazardous materials/WMD incidents, the hazardous materials safety officer shall observe a scene, review and evaluate hazard and response information as it pertains to the safety of all persons within the hazardous materials branch/group, and meet the requirements of 11.2.1.1 through 11.2.1.6.

The hazardous materials safety officer must have a thorough understanding of risk-based response processes associated with the initial response to hazardous materials incidents. This response process is used when conditions can be rapidly assessed and interventions quickly initiated to lessen the overall impact and duration of the incident. The hazardous materials safety officer is responsible for implementing a more measured and procedural-driven risk management model for large-scale, unique, or long-duration incidents.

11.2.1.1 The hazardous materials safety officer shall explain the basic toxicological principles relative to the safety of personnel exposed to hazardous materials/WMD, including the following:

(1) Acute and chronic toxicity
(2) Dose response
(3) Local and systemic effects
(4) Routes of exposure to toxic materials
(5) Synergistic effects

11.2.1.2* The hazardous materials safety officer shall identify at least three conditions where the hazards from flammability would require chemical-protective clothing with thermal protection.

A.11.2.1.2 Conditions where protective clothing with thermal protection could be required if entry was made into an area where flammability was a concern can include the following:

(1) Unknown materials involved
(2) Oxygen-enriched atmosphere
(3) Detectable percentage of LEL on monitoring instruments
(4) Materials with a wide flammable range present
(5) Reactive materials present

11.2.1.3* The hazardous materials safety officer shall identify at least three conditions where personnel would not be allowed to enter the hot zone.

A.11.2.1.3 Conditions under which personnel would not be allowed in the hot zone include the following:

(1) Decontamination procedures not established or not in place
(2) Advanced first-aid and transportation not available
(3) Flammable or explosive atmosphere present
(4) Oxygen-enriched atmosphere of 23.5 percent or greater present
(5) Runaway reaction occurring
(6) Appropriate personal protective clothing not available
(7) No effective action to be taken
(8) Risk outweighing benefit
(9) Personnel not properly trained
(10) Insufficient personnel to perform tasks

11.2.1.4 Given the names of five hazardous materials/WMD and at least three reference sources, the hazardous materials safety officer shall identify the physical and chemical properties and their potential impact on the safety of personnel at an incident involving each of the materials or agents.

11.2.1.5 Given the names of five hazardous materials/WMD and at least three reference sources, the hazardous materials safety officer shall identify the health concerns and their potential impact on the safety and health of personnel at an incident involving each of the materials or agents.

11.2.1.6* Given the names of five hazardous materials and a description of their containers, the hazardous materials safety officer shall identify five hazards or physical conditions that would affect the safety of personnel at an incident involving each of the materials or agents.

A.11.2.1.6 Examples of scenarios that emergency responders might encounter in the field include the following:

(1) Ammonia leaking from a fitting or valve of a railroad tank car
(2) Chlorine leaking from the valve stem of a 150 lb (68 kg) cylinder
(3) Lacquer thinner leaking from a hole in a 55 gal (208 L) drum
(4) Gasoline leaking from a hole in the side of an aluminum tank truck
(5) Carbaryl, a powdered insecticide, found stored in a broken cardboard drum

11.3 Competencies — Planning the Response

The continued development of operations level, mission-specific personnel has blurred the traditional lines between offensive and defensive operations. The specific mission of all responders and support personnel needs to be understood and assessed based on risk, training, and equipment. All actions undertaken at an emergency scene have some inherent risk associated with them. It is the responsibility of the hazardous materials safety officer to be able to identify these risks, advise personnel, and provide recommendations to minimize or control identified risks. The hazardous materials safety officer must coordinate safety tasks and activities with the incident safety officer.

11.3.1* Identifying the Safety Precautions for Potential Response Options.

Given scenarios involving hazardous materials/WMD incidents, the hazardous materials safety officer shall assist the hazardous materials officer in developing a site safety and control plan to respond within the capabilities of available response personnel, personal protective equipment, and control equipment and shall complete the following tasks:

A.11.3.1 Potential response options are either defensive or offensive in nature. The site safety and control plan is integrated into the formal incident action plan.

(1)* Identify specific safety precautions to be observed during mitigation of each of the hazards or conditions identified in 11.2.1.6.

A.11.3.1(1) Safety precautions to be observed during mitigation of hazards or conditions can include the following:

(1) Elimination of ignition sources
(2) Use of monitoring instruments
(3) Stabilizing the container
(4) Establishing emergency evacuation procedures
(5) Ensuring availability of hose lines and foam, when appropriate
(6) Evacuating exposures
(7) Isolating the area
(8) Protecting in place
(9) Working in proper protective equipment

(2)* Identify safety precautions associated with search and rescue missions at hazardous materials/WMD incidents.

A.11.3.1(2) Safety precautions to be observed during search and rescue missions at hazardous materials/WMD incidents can include the following:

(1) Ensuring availability of appropriate personal protective equipment for all personnel
(2) Use of monitoring instruments
(3) Maintaining an escape path
(4) Knowledge of approved hand signals by all personnel

(5) Ensuring availability of communications equipment for each team
(6) Preplanning the search sequence prior to entry

11.3.2 Providing Recommendations Regarding Safety Considerations.

11.3.2.1 Given scenarios involving hazardous materials/WMD incidents, the hazardous materials safety officer shall provide the incident safety officer, hazardous materials officer, and incident commander with observation-based recommendations regarding considerations for the safety of on-site personnel.

The hazardous materials safety officer evaluates the situation, personnel, equipment, and information available in order to make specific recommendations to the hazardous materials officer, incident safety officer, or incident commander as required to ensure proper risk management for personnel operating on-site in the hot and warm zones of a hazardous materials incident. It must be understood that at incidents involving multiple discipline operations (law enforcement, fire, emergency medical services), there could be multiple assistant safety officers overseeing different activities. It is extremely important that these activities and the safety of responders performing them are coordinated through the incident command system.

11.3.2.2 The hazardous materials safety officer shall develop recommendations for the hazardous materials officer regarding safety considerations of the hazards and risks for each of the hazardous materials/WMD and containers identified in 11.2.1.6.

11.3.3 Assisting in the Development of a Site Safety and Control Plan for Inclusion in the Incident Action Plan.

Given scenarios involving hazardous materials/WMD incidents, the hazardous materials safety officer shall assist the incident safety officer and hazardous materials officer in the development of the site safety and control plan for inclusion in the incident action plan and shall complete the following tasks:

(1)* Identify the importance and list five benefits of pre-emergency planning relating to specific sites.

A.11.3.3(1) Benefits of pre-emergency planning include the following:

(1) Identification and mitigation of hazards during the planning process
(2) Familiarization of personnel with facility
(3) Identification of 24-hour responsible parties
(4) Identification of built-in containment systems
(5) Identification of the location of utility and other shutoff/shutdown valves and switches
(6) Identification of location of facility map
(7) Identification of location and quantities of hazardous materials/WMD
(8) Identification of vulnerable populations
(9) Identification of facility response capabilities

(2)* Identify and name five hazards and precautions to be observed when personnel approach a hazardous materials/WMD incident.

A.11.3.3(2) Hazards that should be observed when personnel approach a hazardous materials/WMD incident include the following:

(1) Inhalation hazards
(2) Dermal hazards
(3) Flammable hazards
(4) Reactive hazards
(5) Electrical hazards
(6) Mechanical hazards

The list in A.11.3.3(2) is not all inclusive. Additional hazards could exist that would not be observable. An example of a nonobservable hazard is radiation, which will need to be monitored with instrumentation.

(3)*List the elements of a site safety and control plan.

A.11.3.3(3) The following elements of a site safety plan are from the EPA *Standard Operating Safety Guides*:

(1) Site description
(2) Entry objectives
(3) On-site organization
(4) On-site control
(5) Hazard evaluation
(6) Personal protective equipment
(7) On-site work plans
(8) Communication procedures
(9) Decontamination procedures
(10) Site safety and health plan

(4) Given a pre-incident plan and a scenario involving one of the hazardous materials/WMD and containers described in 11.2.1.6, develop safety considerations for the incident.

11.3.4 Providing Recommendations Regarding Safety and Reviewing the Incident Action Plan.

Given a proposed incident action plan for an incident involving one of the hazardous materials/WMD and containers described in 11.2.1.6, the hazardous materials safety officer shall identify to the incident safety officer, the hazardous materials officer, and the incident commander the safety precautions for the incident action plan and shall complete the following tasks:

(1) Ensure that the site safety and control plan in the proposed incident action plan is consistent with the emergency response plan and/or standard operating procedures.
(2) Make recommendations to the incident commander on the safety considerations in the proposed incident action plan.

11.3.5 Reviewing Selection of Personal Protective Equipment.

Given scenarios involving hazardous materials/WMD incidents, the hazardous materials safety officer shall demonstrate the ability to review the selection of personal protective equipment required for a given action option and shall complete the following tasks:

(1) Identify five safety considerations for personnel working in personal protective equipment.
(2) Given the names of five different hazardous materials/WMD and a chemical compatibility chart for chemical-protective clothing, identify the chemical-protective clothing that would provide protection from the identified hazards to the wearer for each of the five substances.
(3)*Given the names of five different hazardous materials/WMD, identify the personal protective equipment options for specified response options.

A.11.3.5(3) Response options can include surveying the scene, sampling, monitoring, plugging, and patching.

(4) Identify the recommended methods for donning, doffing, and using all personal protective equipment provided by the AHJ for use in hazardous materials/WMD response activities.

11.3.6 Reviewing the Proposed Decontamination Procedures.

Given site-specific decontamination procedures by the hazardous materials officer or incident commander for a scenario involving a hazardous materials/WMD incident, the hazardous materials safety officer shall review the procedures to ensure that applicable safety considerations are included prior to implementation of the incident action plan.

11.3.7 Ensuring Provision of Emergency Medical Services.

Given a scenario involving a hazardous materials/WMD incident, the hazardous materials safety officer shall review the emergency medical services procedures to ensure that response personnel are provided medical care and shall complete the following tasks:

(1)* Identify the elements required in an emergency medical services plan.

Emergency medical services (EMS) activities occurring outside the warm zone and not involving hazardous materials responders could be outside the responsibility of the hazardous materials safety officer. During events that create mass casualities, it is often best to assign an assistant safety officer to oversee other activities taking place outside the hot and warm zones. The hazardous materials safety officer is responsible for ensuring that EMS are available for responders involved in the control and intervention of the hazardous materials release, and for other hazardous materials response activities conducted in the hot and warm zone

A.11.3.7(1) The elements of an emergency medical services plan according to NFPA 473, *Standard for Competencies for EMS Personnel Responding to Hazardous Materials/Weapons of Mass Destruction Incidents*, include the following:

(1) EMS control activities
(2) EMS component of an incident management system
(3) Medical monitoring of personnel utilizing chemical-protective and high temperature–protective clothing
(4) Triage of hazardous materials/WMD victims
(5) Medical treatment for chemically contaminated individuals
(6) Product and exposure information gathering and documentation

(2) Identify the importance of an on-site medical monitoring program.
(3) Identify the resources for the transportation and care of the injured personnel exposed to hazardous materials/WMD.

11.4 Competencies — Implementing the Planned Response

11.4.1 Performing the Duties of the Hazardous Materials Safety Officer.

Given a scenario involving hazardous materials/WMD incidents, the hazardous materials safety officer shall perform the duties of the position in a manner consistent with the emergency response plan and/or standard operating procedures and shall complete the following tasks:

(1) Identify the duties of the hazardous materials safety officer as defined in the emergency response plan and/or standard operating procedures.
(2) Demonstrate performance of the duties of the hazardous materials safety officer as defined in the emergency response plan and/or standard operating procedures.

11.4.2 Monitoring Safety of Response Personnel.

Given scenarios involving a hazardous materials/WMD incident, the hazardous materials safety officer shall ensure that personnel perform their tasks in a safe manner by identifying the safety considerations for the control functions identified in the site safety and control plan and shall complete the following tasks:

(1) Identify the safe operating practices that are required to be followed at a hazardous materials/WMD incident as stated in the emergency response plan and/or standard operating procedures.
(2) Identify how the following factors influence heat and cold stress for hazardous materials response personnel:
    (a) Activity levels
    (b) Duration of entry
    (c) Environmental factors
    (d) Hydration
    (e) Level of personal protective equipment
    (f) Physical fitness
(3) Identify the methods that minimize the potential harm from heat and cold stresses.
(4) Identify the safety considerations that minimize the psychological and physical stresses on personnel working in personal protective equipment.
(5) Describe five conditions in which it would be prudent to withdraw from a hazardous materials/WMD incident.

11.4.3 Conducting Safety Briefings.

11.4.3.1 Given a scenario involving a hazardous materials/WMD incident and site safety and control plan, the hazardous materials safety officer shall conduct safety briefings for personnel performing the functions identified in the incident action plan.

11.4.3.2 The hazardous materials safety officer shall be able to demonstrate the procedure for conducting a safety briefing to personnel for an incident involving one of the hazardous materials/WMD and its container identified in 11.2.1.6, as specified by the emergency response plan and/or standard operating procedures.

11.4.4 Implementing and Enforcing the Site Safety and Control Plan.

Given a scenario involving a hazardous materials/WMD incident and site safety and control plan, the hazardous materials safety officer shall assist the incident commander, the incident safety officer, and the hazardous materials officer in implementing and enforcing the safety considerations and shall complete the following tasks:

(1) Identify whether the boundaries of the established control zones are clearly marked, consistent with the site safety and control plan, and are being maintained.
(2) Identify whether the on-site medical monitoring required by the emergency response plan and/or standard operating procedures is being performed.
(3) Given an entry team, a backup team, and a decontamination team working in personal protective clothing and equipment, verify that each team is protected and prepared to safely perform its assigned tasks by completing the following:
    (a) Determine whether the selection of clothing and equipment is consistent with the site safety and control plan.
    (b) Determine whether each team has examined the clothing for barrier integrity and the equipment to ensure correct working order.
    (c) Determine whether protective clothing and equipment have been donned in accordance with the standard operating procedures and the manufacturer's recommendations.
(4) Determine whether each person entering the hot zone has a specific task assignment, understands the assignment, is trained to perform the assigned task(s), and is working with a designated partner at all times during the assignment.

(5) Determine whether a backup team is prepared at all times for immediate entry into the hot zone during entry team operations.
(6) Determine whether the decontamination procedures specified in the site safety and control plan are in place before any entry into the hot zone.
(7) Verify that each person exiting the hot zone and each tool or piece of equipment is decontaminated in accordance with the site safety and control plan and the degree of hazardous materials/WMD contamination.
(8) Demonstrate the procedure for recording the names of the individuals exiting the hot zone, as specified in the emergency response plan and/or standard operating procedures.
(9)* Identify three safety considerations that can minimize secondary contamination.

Once responders are operating in the warm and hot zones, it is the responsibility of the hazardous materials safety officer to use observation-based information and communications monitoring to ensure proper implementation of the safety and control plan. Deviations from the plan can be made after conferring with the hazardous materials officer and with the concurrence of the incident commander. Actions that pose an immediate threat to responders will be stopped by the hazardous materials safety officer.

A.11.4.4(9) Safety considerations that can minimize secondary contamination include the following:

(1) Control zones are established and enforced.
(2) All people and equipment exiting the hot zone are decontaminated.
(3) Personnel performing decontamination are properly trained.
(4) Personnel performing decontamination are properly protected.

See NFPA 473, *Standard for Competencies for EMS Personnel Responding to Hazardous Materials/Weapons of Mass Destruction Incidents*.

11.4.5 Maintaining Communications.

Given a scenario involving a hazardous materials/WMD incident and the site safety and control plan, the hazardous materials safety officer shall maintain routine and emergency communications within the incident command structure at all times during the incident and shall complete the following tasks:

(1)* Identify three types of communications systems used at hazardous materials/WMD incident sites.

A.11.4.5(1) Communications systems include in-suit radio communications, hand-held portable radios, emergency signaling devices, and hand signals.

(2) Verify that each person assigned to work in the hot zone understands the emergency alerting and response procedures specified in the safety considerations prior to entry into the hot zone.

The hazardous materials safety officer is responsible for performing the safety briefing for hazardous materials responders and support personnel working in the hot and warm zones (see Exhibit I.11.2). This briefing should ensure that responders understand the incident-specific emergency alerting procedures and any other procedures for other operations that might be taking place in the area. Redundant communications is always essential. Combinations of electronic and nonelectronic means might be necessary, especially in high-noise environments that could preclude verbal communications.

11.4.6 Monitoring Status Reports.

11.4.6.1 Given a scenario involving a hazardous materials/WMD incident and the site safety and control plan, the hazardous materials safety officer shall monitor routine and emergency communications within the incident command structure at all times during the incident.

EXHIBIT I.11.2 An incident command briefing at this FEMA Emergency Operations Center keeps everyone updated on progress and safety issues. Briefings are held twice daily with representatives of relief organizations and town, county, state, and federal governments. (Source: Greg Henshall/FEMA)

11.4.6.2 The hazardous materials safety officer shall ensure that entry team members regularly communicate the status of their work assignment to the hazardous materials officer.

11.4.7 Implementing Exposure Monitoring.

Given a scenario involving a hazardous materials/WMD incident and the site safety and control plan, the hazardous materials safety officer shall assist the incident commander, the incident safety officer, and the hazardous materials officer in implementing exposure monitoring.

11.4.8 Verifying Exposure Monitoring.

The hazardous materials safety officer shall identify that exposure monitoring (personnel and environment), as specified in the emergency response plan and/or standard operating procedures and site safety and control plan considerations, is performed.

11.5 Competencies — Evaluating Progress

11.5.1 Identifying Deviations from Safety Considerations or Other Dangerous Situations.

Given scenarios involving hazardous materials/WMD incidents and given deviations from the site safety and control plan for activities in both the hot and warm zones and dangerous conditions, the hazardous materials safety officer shall evaluate the progress of the planned response to ensure that the response objectives are being met safely and shall complete the following tasks:

(1) Identify those actions that deviate from the site safety and control plan or that otherwise violate accepted safe operating practices, organizational policies, or applicable occupational safety and health laws, regulations, codes, standards, or guidelines.
(2) Identify dangerous conditions that develop or are identified during work in the hot or warm zones that threaten the safety or health of persons in those zones.
(3) Identify the signs and symptoms of psychological and physical stresses on personnel wearing personal protective equipment.

11.5.2 Taking Corrective Actions.

Given scenarios involving hazardous materials/WMD incidents and given deviations from the site safety and control plan for activities in both the hot and warm zones and dangerous conditions, the hazardous materials safety officer shall take such corrective actions as are necessary to ensure the safety and health of persons in the hot and warm zones and shall complete the following tasks:

(1) Send emergency communications to and receive emergency communications from the incident safety officer, entry team personnel, the hazardous materials officer, and others regarding safe working practices and conditions:

(a)* Given a hazardous situation or condition that has developed or been identified following initial hot zone entry, demonstrate the application of the emergency alerting procedures specified in the site safety and control plan to communicate the hazard and emergency response information to the affected personnel.

Once the hazardous materials safety officer observes or becomes aware of a deviation from the site safety and control plan or a sudden change in the incident causes an IDLH situation, the hazardous materials safety officer will determine the appropriate course of action to ensure the safety of responders. This will include conferring with the incident safety officer, hazardous materials officer, and incident commander. If the situation poses an immediate threat to responders, the hazardous materials safety officer has the authority to alter or terminate activities until another course of action that mitigates the risk is decided upon. Termination of intervention activities should be a last resort.

A.11.5.2(1)(a) Examples of such situations or conditions can include, but are not limited to, the following:

(1) Fire or explosion
(2) Container failure
(3) Sudden change in weather conditions
(4) Failure of entry team's personal protective clothing and/or equipment
(5) Updated information on identification of hazardous material(s) involved that warrants reassessment of level of protective clothing and equipment being used

(b) Given a demonstrated emergency alert via hand signal by a member of the entry team operating within the hot zone, identify the meaning of that signal as specified in the site safety and control plan.

(2) Identify the procedures to alter, suspend, or terminate any activity that can be judged to be unsafe, as specified in the emergency response plan and/or standard operating procedures.
(3) Demonstrate the procedure for notifying the appropriate individual of the unsafe action and for directing alternative safe actions, in accordance with the site safety and control plan and standard operating procedures.
(4) Demonstrate the procedure for suspending or terminating an action that could result in an imminent hazard condition, in accordance with the emergency response plan and standard operating procedures.

11.6 Competencies — Terminating the Incident

11.6.1 Reporting and Documenting the Incident.

Given scenarios involving hazardous materials/WMD incidents, the hazardous materials safety officer shall complete and submit the reports, documentation, and follow-up required of the hazardous materials safety officer and shall complete the following tasks:

(1) Identify the safety reports and supporting documentation required by the emergency response plan and/or standard operating procedures.
(2) Demonstrate completion of the safety reports required by the emergency response plan and/or standard operating procedures.
(3) Describe the importance of personnel exposure records.

Reporting requirements for the incident are the responsibility of the incident commander and can be delegated to another individual, such as the documentation unit leader. It is incumbent upon the incident safety officer to assist the incident commander with any information or documentation that will assist in the incident documentation. The employer's emergency response plan and/or standard operating procedures should designate reporting requirements.

11.6.2 Debriefing of Hazardous Materials Branch/Group Personnel.

Given scenarios involving hazardous materials/WMD incidents, the hazardous materials safety officer shall debrief hazardous materials branch/group personnel regarding site-specific occupational safety and health issues.

11.6.2.1* The hazardous materials safety officer shall be able to identify five health and safety topics to be addressed in an incident debriefing.

A.11.6.2.1 Topics can include, but are not limited to, the following:

(1) Identity of the hazardous materials/WMD agent to which personnel have been or might have been exposed
(2) Signs and symptoms of exposure to the hazardous material(s) involved in the incident
(3) Signs and symptoms of critical incident stress
(4) Duration of recommended observation period for such signs and symptoms
(5) Procedures to follow in the event of delayed presentation of such signs or symptoms
(6) Name of the individual responsible for post-incident medical contact
(7) Safety and health hazards remaining at the site

11.6.2.2 The hazardous materials safety officer shall demonstrate the procedure for debriefing hazardous materials branch/group personnel regarding site-specific occupational safety and health areas of concern, as specified in the site safety and control plan, emergency response plan, and standard operating procedures.

The debriefing process is when the incident commander and others in the chain of command learn of the specific activities performed and pass along health and safety information to the responders. While the incident commander is responsible for ensuring that a debriefing takes place, this may be delegated in accordance with the emergency response plan and standard operating procedures.

11.6.3 Assisting in the Incident Critique.

Given scenarios involving hazardous materials/WMD incidents and the site safety and control plan, the hazardous materials safety officer shall provide safety and health-related critical observations of the activities that were performed in the hot and warm zones during the incident.

11.6.3.1 Information to be Presented. Given the site safety and control plan and the hazardous materials safety officer's report for a scenario involving a hazardous materials/WMD incident, the hazardous materials safety officer shall demonstrate the procedure for verbally presenting the following information in accordance with the emergency response plan and/or standard operating procedures:

(1) Safety and health-related critical observations of the activities that were performed in the hot and warm zones during the incident

(2) Recorded violations of the site safety and control plan or generally accepted safe operating practices, organizational policies, or applicable occupational safety and health laws, regulations, codes, standards, or guidelines
(3) Injuries or deaths that occurred as a result of reasonably unforeseen dangerous conditions that developed during the incident
(4) Injuries or deaths that occurred as a result of violations of the site safety and control plan, generally accepted safe operating practices, organizational policies, or applicable occupational safety and health laws, regulations, codes, standards, or guidelines
(5) The course of action(s) that likely would have prevented the injuries or deaths that occurred as a result of the safety violations identified in 11.6.3.1(4)
(6) Deficiencies or weaknesses in the site safety and control plan, emergency response plan, and standard operating procedures that were noted during or following the incident

It is important that the hazardous materials safety officer be a good communicator as well as an observer. The importance of a critique, after-action review, or post-incident analysis should not be lost due to poor communications skills. The critique should be as positive and upbeat as possible to allow involved personnel to be open and objective.

REFERENCES CITED IN COMMENTARY

1. Title 29, Code of Federal Regulations, Part 1910.120 (q), U.S. Government Printing Office, Washington, DC.

Additional References

NFPA 1521, *Standard for Fire Department Safety Officer*, National Fire Protection Association, Quincy, MA, 2008.

NFPA 1561, *Standard on Emergency Services Incident Management System*, National Fire Protection Association, Quincy, MA, 2008.

Title 49, Code of Federal Regulations, U.S. Government Printing Office, Washington, DC.

CHAPTER 12

Competencies for Hazardous Materials Technicians with a Tank Car Specialty

The use of tank cars began in the mid-1860s when wooden tubs were mounted on wooden flat cars. The tank cars were used to transport petroleum products in Pennsylvania. Since then, the tank car fleet has grown to more than 250,000, most of which are owned by non-railroad companies in the United States.

A tank car is an enclosed longitudinal cylinder (see Exhibit I.12.1) equipped with railroad running gear and safety appliances. It is equipped with fittings for loading/unloading, pressure relief, gauging, and other functions necessary for handling the product.

Tank cars transport a wide variety of nonhazardous and hazardous materials, including approximately 80 to 85 percent of the 1.5 million or more shipments of hazardous materials that travel annually by rail.

Tank cars are classified according to their construction, features, and fittings into three groups:

- Non-pressure tank cars
- Pressure tank cars
- Cryogenic liquid tank cars

While problems with tank cars occur infrequently, response personnel are called upon to handle those few problems. When faced with a tank car problem, the responder's ability to communicate an accurate and detailed description of the contents, condition of the tank, and

EXHIBIT I.12.1 This tank is an example of a nonpressure tank car. (Source: Union Pacific Railroad)

other circumstances is extremely important, as explained in *A General Guide to Tank Cars* from the Union Pacific Railroad [1].

12.1 General

12.1.1 Introduction.

12.1.1.1 The hazardous materials technician with a tank car specialty shall be that person who provides technical support pertaining to tank cars, provides oversight for product removal and movement of damaged tank cars, and acts as a liaison between technicians and outside resources.

12.1.1.2 The hazardous materials technician with a tank car specialty shall be trained to meet all competencies at the awareness level (Chapter 4), all core competencies at the operations level (Chapter 5), all competencies at the technician level (Chapter 7), and all competencies of this chapter.

In the 1987 edition of NFPA 472, one of the competencies required for the hazardous materials specialist (subsequently incorporated into the hazardous materials technician level in 1992) was merely to describe specific techniques. Specialty levels were introduced in the 1997 edition of NFPA 472 and covered extremely hazardous and highly skilled operations that had not had any standardized training or certification previously. The technician with a tank car specialty, the technician with a cargo tank specialty, and the technician with an intermodal tank specialty are levels that are above The Occupational Health and Safety Administration's (OSHA) technician and specialist levels and are not defined by the present OSHA regulations Title 29 CFR Part 1910.120 [2].

Commentary Table I.1.1 in Part I of this handbook offers a comparison of the various levels described by NFPA and OSHA. Responders with this specialty are now required by 12.1.1.2 to demonstrate competency in performing these tasks. See the competency requirements in 12.4.1(7) in the 2008 edition of NFPA 472 for an example.

12.1.1.3 The hazardous materials technician with a tank car specialty shall receive training to meet governmental occupational health and safety regulations.

12.1.2 Goal.

12.1.2.1 The goal of the competencies at this level shall be to provide the hazardous materials technician with a tank car specialty with the knowledge and skills to perform the tasks in 12.1.2.2 safely.

12.1.2.2 When responding to hazardous materials/WMD incidents, the hazardous materials technician with a tank car specialty shall be able to perform the following tasks:

(1) Analyze a hazardous materials/WMD incident involving tank cars to determine the complexity of the problem and potential outcomes by completing the following tasks:
    (a) Determine the type and extent of damage to tank cars.
    (b) Predict the likely behavior of tank cars and their contents in an emergency.
(2) Plan a response to an emergency involving tank cars within the capabilities and competencies of available personnel, personal protective equipment, and control equipment by determining the response options (offensive, defensive, and nonintervention) for a hazardous materials/WMD incident involving tank cars.
(3) Implement or oversee the implementation of the planned response to a hazardous materials/WMD incident involving tank cars.

12.1.3 Mandating of Competencies.

This standard shall not mandate that hazardous materials response teams performing offensive operations on tank cars have technicians with a tank car specialty.

12.1.3.1 Technicians operating within the bounds of their training as listed in Chapter 7 of this standard shall be able to intervene in railroad incidents.

12.1.3.2 If a hazardous materials response team decides to train some or all its technicians to have in-depth knowledge of tank cars, this chapter shall set out the required competencies.

The committee wanted to clearly state that existing and new responders at the hazardous materials technician level can perform the operations on tank cars for which they have been qualified. This specialty level covers those highly technical skills and knowledge rarely needed except at most major incidents. For example, the skills required by the competency in Section 12.4 are skills that the technician normally does not possess. A hazardous materials response might have a few members trained to this level in accordance with 12.1.3.2, or a request can be made to the railroad, shipper, or manufacturer to provide a qualified individual with this specialty at the incident scene. The committee also wants to clearly state that this specialty level is not mandated for any hazardous materials response team or its members. Arrangements to request assistance to perform these operations are perfectly appropriate and are encouraged in these situations.

12.2 Competencies — Analyzing the Incident

12.2.1 Determining the Type and Extent of Damage to Tank Cars.

Given examples of damaged tank cars, technicians with a tank car specialty shall describe the type and extent of damage to each tank car and its fittings and shall complete the following tasks:

In NFPA 472, tank cars are grouped into the following five types:

1. Nonpressure tank cars
2. Pressure tank cars
3. Cryogenic liquid tank cars
4. Pneumatically unloaded covered hopper cars

Nonpressure tank cars, also known as general service or low-pressure tank cars, transport hazardous and nonhazardous materials at vapor pressures below 25 psi (172 kPa) at 105°F to 115°F (46°C). Tank test pressures for nonpressure tank cars are 60 psi and 100 psi (414 kPa to 689 kPa). Capacities range from 4,000 gal to 45,000 gal (15,140 L to 170,325 L). Nonpressure tank cars are cylindrical with rounded heads. They have at least one manway for access to the tank's interior. Fittings for loading/unloading, pressure and/or vacuum relief, gauging, and other purposes are visible at the top and/or bottom of the car. Older nonpressure tank car tanks have at least one expansion dome with a manway. Nonpressure tank cars can be compartmented, with up to six compartments. Each compartment is constructed as a separate and distinct tank with its own set of fittings. Each compartment can have a different capacity and transport a different commodity. Nonpressure tank cars transport a variety of hazardous materials, including Class 3 (flammable liquids), Class 4 (flammable solids/reactive liquids and solids), Class 5 (oxidizers/organic peroxides), Class 6 (poisons and irritants), and Class 8 (corrosive) materials.

Pressure tank cars typically transport hazardous materials, including Class 2 (flammable, nonflammable, and poison gases) materials, at pressures greater than 40 psi (276 kPa) at (68°F). They may also transport flammable liquids. Tank test pressures for these tank cars are

100, 200, 300, 400, and 500 psi (689.5, 1379, 2068, 2344, 2758, 3447, and 4137 kPa) and range in capacity from 4,000 gal to 45,000 gal (15,140 L to 170,325 L). Pressure tank cars are cylindrical, noncompartmented steel or aluminum tanks with rounded heads. They typically are top-loading, with their fittings inside the protective housing mounted on the manway cover plate in the top center of the tank. Pressure tank cars, such as the one shown in Exhibit I.12.2, can be insulated and/or thermally protected. Those without insulation and without jacketed thermal protection have at least the top two-thirds of the tank painted white.

Cryogenic liquid tank cars carry low-pressure, usually 25 psi (172 kPa) or lower, refrigerated liquids, which are –130°F (–90°C) and below. Materials typically found in these types of tank cars include argon, ethylene, hydrogen, nitrogen, and oxygen. Cryogenic liquid tank cars, such as the one shown in Exhibit I.12.3, feature a "tank-within-a-tank" configuration, with a stainless steel inner tank supported within a strong carbon steel outer tank. The space between the inner tank and the outer tank is filled with insulation and kept under vacuum.

Pneumatically unloaded covered hopper cars are built to tank car specifications. Covered hopper cars are unloaded by pressure differential (pneumatic) through the application of air pressure. Although the pressure is used only during unloading, tank test pressures for the car range from 20 psi to 80 psi (138 kPa to 552 kPa). Dry caustic soda is one material frequently transported in this type of car.

(1) Given the specification mark for a tank car and the reference materials, describe the car's basic construction and features.

Tank cars are built to precisely defined standards established by the U.S. Department of Transportation (DOT) and the Association of American Railroads (AAR). A tank car's specification mark indicates the standards that were followed when the tank car was constructed (see Exhibit I.12.4). The specification mark is stenciled on both sides of the tank. As the responder faces the side of the car, the specification mark is to the right (at the opposite end from the reporting marks, or initials, and number). The specification mark is also stamped into the head of the tank where it is not readily visible. The specification mark may also be

EXHIBIT I.12.2 *This is an example of a typical pressure tank car. (Source: Union Pacific Railroad)*

EXHIBIT I.12.3 *This is an example of a typical cryogenic liquid tank car. (Source: Union Pacific Railroad)*

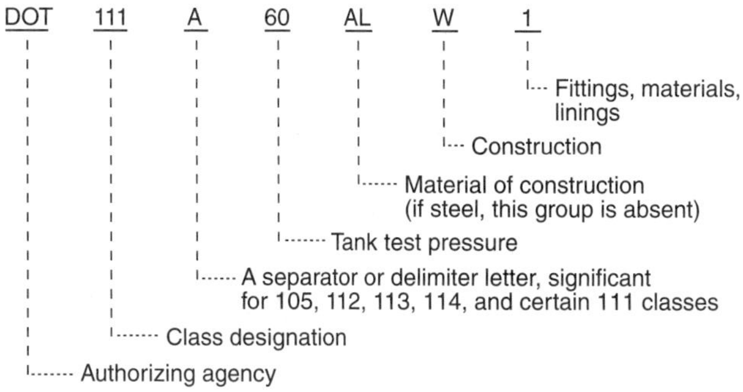

EXHIBIT I.12.4 *Specification markings on tank cars are explained in this diagram.*

found on the information plate. See Union Pacific's *A General Guide to Tank Cars* for more information [1].

Specification information can also be obtained from the railroad, shipper, car owner, or the AAR (from the car's certificate of construction) using the car's initials and number.

(2) Point out the "B" end of the car.

The "B" end of the car is the end where the hand brake is mounted, which the technician with a tank car specialty is required by 12.2.1(2) to know. The reference to various components of a tank car is based on facing the "B" end of the car. Additional information about tank cars and their markings, along with a diagram of marking locations, is found in Chapter 6 of NFPA 472.

(3) Given examples of various tank cars, point out and explain the design and purpose of each of the following tank car components, when present:

Technicians with a tank car specialty are required by 12.2.1(3) to be able to identify and explain the various components of a tank car so they can easily communicate the condition of a tank car at an incident.

(a) Body bolster

The body bolster is a structural cross member, mounted at right angles to the underframe, used to cradle the tank. Tank cars have two body bolsters, one at each end of the tank. Exhibit I.12.5 illustrates a typical body bolster.

(b) Head shield

Head shields, or head puncture resistance, are used to protect the heads of tank cars containing products such as flammable gases, ethylene oxide, and anhydrous ammonia from punctures. For jacketed tank cars, either "full" or "bottom half" head shields can be incorporated in the head jacket. For nonjacketed tank cars, a "half head" or "trapezoidal-shaped head" shield can be applied. Exhibit I.12.6 shows an example of a head shield.

(c) Heater coils — interior versus exterior

EXHIBIT I.12.5 *This is one of two body bolsters usually found on a tank car. (Source: Union Pacific Railroad)*

EXHIBIT I.12.6 *A head shield is commonly found on tank cars. (Source: Union Pacific Railroad)*

Some tank cars are equipped with a series of continuous parallel pipes or coils mounted internally (inside the tank) or externally (outside the tank). Steam, water, or hot oil from an external source is run through these coils to heat thick or solidified materials (i.e., asphalts, fused solids, heavy fuel oils, phenol, sulfur, metallic sodium, or petroleum waxes) to make them flow more easily when loading and unloading. Exterior heater coils are shown in Exhibit I.12.7.

EXHIBIT I.12.7 *This is an example of heater coils mounted on the exterior of a tank car. (Source: Union Pacific Railroad)*

(d) Jacket

The jacket is an outer covering, which is typically 11 gauge (1/8 in.) steel, used to hold both insulation and jacketed thermal protection in place as well as to protect the insulation or thermal protection from the weather. Wooden blocks or metal brackets hold the jacket away from the tank.

(e) Lining and cladding

The interiors of some tank cars are lined or clad with materials to protect the tank from the corrosive or reactive effects of the contents or to maintain the purity of the contents. A lining is a covering applied in strips or sections and fastened to the inside of the tank after the tank is constructed. Rubber is the most commonly used lining in tank cars used to transport hazardous materials. Glass, lead, nickel, polyurethane, and polyvinyl chloride are also used as linings. Claddings are coverings applied to the base metal before the plate is formed. Nickel and stainless steel are used as cladding materials.

(f) Shelf couplers

The top and bottom shelf coupler is a type of coupler with a vertical restraint mechanism on the top and bottom. The coupler reduces the potential for coupler override and possible head puncture. A typical coupler is shown in Exhibit I.12.8.

EXHIBIT I.12.8 *Shelf couplers such as these are often found on a tank car. (Source: Union Pacific Railroad)*

(g) Tank (including shell) and head

Each tank car tank, as listed in 12.2.1(3)(a), is made up of a shell enclosed at the ends by heads. The shell is constructed of two to seven metal plates formed into rings. Heads are made from plates pressed into an ellipsoidal shape. The components (rings and heads) are fusion-welded together. As they are built, representative samples of the tanks are x-rayed to identify possible flaws in metal or welds. Steel tank car tanks are heated to 1100°F (600°C) for 1 hour to relieve metal stresses caused by welding. Finally, the tanks are hydrostatically tested. Tank car tanks might have as many as six compartments, with each compartment constructed as a separate tank. Compartments within the same tank car may have different capacities and can transport different commodities. Tank car tanks with multiple compartments are identified by the multiple sets of fittings on top of the car, one set for each compartment.

(h) Trucks (pin and bowl)

The truck consists of wheels, axles, side frames, springs, truck bolster, and center bowl and pin. This component supports one end of the car and allows it to be moved on the rail.

(i) Underframe — continuous versus stub sill

Tank cars have either a continuous underframe or a stub sill underframe. The continuous underframe is a one-piece assembly, attached to the tank by the center anchor and tank bands at the body bolster, which bridges the trucks of a tank. This continuous underframe absorbs the draft and buff forces associated with train movement.

The stub sill type of frame is a short, longitudinal structural member welded to both ends of the tank to accommodate the coupler and draft gear. This part also attaches the tank to the truck. With this type of underframe, the tank transmits the draft and buff forces associated with train movement.

(4) Given examples of tank cars (some jacketed and some not jacketed), point out the jacketed tank cars.

Jacketed cars can be recognized by one or more of the following visual indicators:

- Flashing (shroud or cover) over the body bolster or tank bands
- Flat appearance of ends or flat sections on sides of tank car
- Rough appearance of visible welds, including lap welds, with welds generally thinner than the tank welds

(5) Describe the difference between insulation and thermal protection on tank cars.

There is a clear difference between insulation and thermal protection. Insulation as covered in 12.2.1(5) is used to safeguard the contents of the tank car from outside temperatures. Insulation can be used on both pressure and nonpressure tank cars and is always used on cryogenic liquid tank cars. Fiberglass and polyurethane foams are the most common types of insulating materials. Cork is used to insulate tanks used for hydrocyanic acid. Perlite is used in cryogenic liquid cars. The insulation is concealed by a jacket.

Thermal protection is used on certain tank cars [primarily tank cars carrying Class 2.1 materials such as liquefied petroleum gases (propane)] to protect them from flame impingement from either a pool or torch fire. Thermal protection is designed to keep tank metal temperatures below 800°F (427°C) for 100 minutes (pool fire exposure) or 30 minutes (torch fire exposure).

(6) Describe the difference between jacketed and sprayed-on thermal protection on tank cars.

The following two types of thermal protection are currently being used:

1. Jacketed thermal protection, which is mineral wool or various manmade ceramic fiber blankets held in place by a metal jacket
2. Sprayed-on thermal protection, which is a rough textured coating sprayed on the tank or the jacket that protects the tank by expanding when exposed to fire

Jacketed thermal protection is the most common type of thermal protection in use today.

(7) Describe the difference between interior and exterior heater coils on tank cars.

Interior heater coils are inside the tank. If these coils should become damaged, product can be released from the inlet or outlet pipes on the tank car. Exterior heater coils are welded to the outside of the tank shell and are not in contact with the product at all.

(8) Given examples of various fittings arrangements for pressure, nonpressure, cryogenic, and carbon dioxide tank cars (including examples of each of the following fittings), point out and explain the design, construction, and operation of each of the following fittings, when present:

(a) Fittings for loading and unloading tank cars, including the following:

i. Air valve

On nonpressure tank cars, the valve that controls the flow of vapor is called an air valve. Unloading is accomplished from pressure or gravity generated from the contents or by pressurizing the tank with air, nitrogen, or other gas. On nonpressure tank cars, the air valve is usually smaller than the liquid valve. On occasion, the vapor valve might be removed and replaced with a blind flange bolted to the cover plate. The blind flange reduces the possibility of tampering and subsequent release of the contents.

ii. Bottom outlet nozzle

The bottom outlet nozzle is the pipe or flange from the bottom of the tank to the bottom outlet.

iii. Bottom outlet valves (top operated with stuffing box, bottom operated — internal or external ball, wafersphere)

The bottom outlet valve is used to load or unload the tank from the bottom. This valve can be any of the following types:

- Plug-type valve that is operated from the top of the tank, called a top-operated bottom outlet valve
- Ball-type valve mounted inside or outside the tank, operated from ground level with some type of operating handle
- Wafersphere (butterfly valve) mounted outside the tank, operated from ground level with some type of operating handle

iv. Carbon dioxide tank car fittings

Fittings for carbon dioxide tank cars typically include two angle valves for loading and unloading, two pressure regulator valves, two sample valves (long pole/short pole used for gauging the amount of product), one safety valve, and one vent.

v. Cryogenic liquid tank car fittings
vi. Excess flow valve

Excess flow valves, such as the one shown in Exhibit I.12.9, are attached inside the tank between the manway cover plate and the eduction line. These valves almost completely cut off the flow of product when the valve is sheared off in an accident. They operate either by product flow or, when the car is overturned, by gravity. Under normal conditions, gravity keeps the excess flow valves open, but higher flow rates associated with a valve being sheared off or a hose break closes them. When the tank car is turned over, gravity closes the valves. An excess flow valve can be found on gauging devices and sample lines.

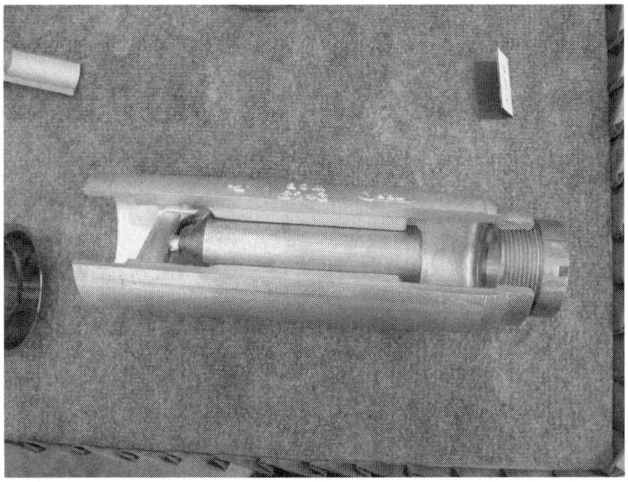

EXHIBIT I.12.9 *This is a typical excess flow valve. (Source: Union Pacific Railroad)*

vii. Flange for manway, valves, and so forth

The flange for manways, valves, and so forth is the rim used for strengthening the connection of fittings to the tank.

viii. Liquid valve and vapor valve (ball versus plug type)

Liquid valves are typically angle valves, with the primary closure being either plug type or ball type, and are mounted on the manway cover plate. The orifice of the liquid valve is closed by a secondary closure—a plug attached by a chain. Liquid valves are found in 1 in., 2 in., and 3 in. (25 mm, 51 mm, and 76 mm) sizes. Liquid valves are interchangeable with vapor valves, and 2 in. and 3 in. (51 mm and 76 mm) liquid valves can be used on the same car. A gasket is used to complete the seal between the liquid valve and the manway cover plate. An eduction line is a pipe that extends to within 1 in. to 2 in. (25 mm to 51 mm) of the bottom of the tank or into a sump. In many cases, an excess flow valve is attached between the liquid valve and the eduction line, just below the manway cover plate.

Vapor valves are typically angle valves, either plug type or ball type, and are mounted on the manway cover plate. The orifice of the vapor valve is closed by a secondary closure—a plug attached by a chain. Vapor valves are found in 1 in., 2 in., and 3 in. (25 mm, 51 mm, and 76 mm) sizes. Vapor valves are interchangeable with liquid valves. A gasket is used to complete the seal between the vapor valve and the manway cover plate. Exhibit I.12.10 shows an example of a typical angle type liquid and vapor valve.

EXHIBIT I.12.10 *This is an angle type liquid and vapor valve. (Source: Union Pacific Railroad)*

ix. Quick-fill hole cover

The fill hole on nonpressure acid cars is used for loading and unloading the tank.

(b) Fittings for pressure relief, including the following:

i. Pressure regulators on carbon dioxide cars and liquefied atmospheric gases in cryogenic liquid tank cars

The devices in 12.2.1(8)(b)(i) release vapors from the tank when the pressure of the contents reach the start-to-discharge pressure setting on the device.

ii. Pressure relief devices (pressure relief valve, safety vent, combination pressure relief valve)

Pressure relief devices are designed to reduce the buildup of excess internal pressure caused by commodity vaporization and expansion. These devices can be mounted on the manway cover plate or tank shell. Pressure relief devices include pressure relief valves, safety vents, and combination pressure relief valves. Pressure relief valves are pressure relief devices with the valve held closed by one or more springs. When pressure in the tank exceeds the start-to-discharge setting of the pressure relief valve setting, the valve opens. The valve recloses when the tank pressure is reduced below the pressure relief valve setting. The pressure relief valve on a pressure tank car is usually set at 75 percent of the tank test pressure but can be set as high as 82.5 percent of the tank test pressure. Safety vents are pressure relief devices that use a frangible disc, also called a rupture disc, to seal the vent opening. This disc is designed to rupture at a predetermined pressure to relieve internal pressure. Once ruptured, the safety vent does not reclose. Rupture discs are made of metal (lead is no longer authorized by the AAR), plastic, rubber, or a combination of metal, plastic, and rubber. Safety vents are used primarily on nonpressure tank cars but cannot be used with flammable liquids and poisons. Combination pressure relief valves incorporate rupture discs and breaking pins in series with spring-operated valves. This type of pressure relief valve is typically mounted on the manway cover plate and is usually found on tank cars transporting chlorine or sulfur dioxide. The spring is external to the tank and is protected from the effects of the lading by the rupture disc and breaking pin. (See Exhibit I.12.11.)

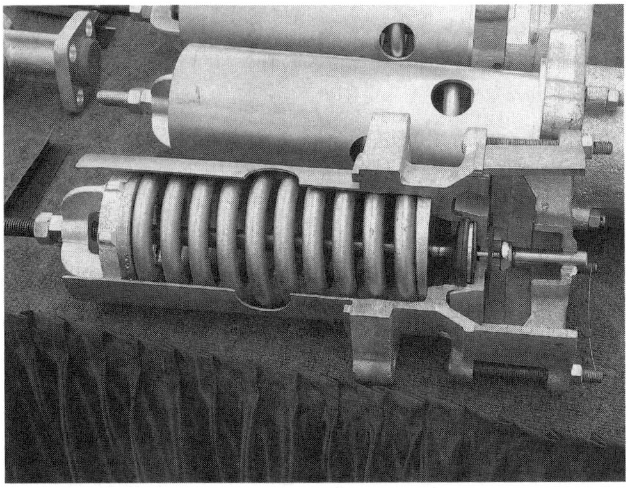

EXHIBIT I.12.11 *This is an example of a combination pressure relief valve. (Source: Union Pacific Railroad)*

iii. Staged pressure relief system for a carbon dioxide car

The staged pressure relief system accounts for the release of vapors in order to keep the liquid in the tank in autorefrigeration during transportation. The devices—regulator valves, pressure relief valve, and safety vent—have different settings to operate at different pressures.

iv. Vacuum relief valve (negative pressure or vacuum)

Vacuum relief valves, such as the ones shown in Exhibit I.12.12, are designed to prevent excessive internal negative pressure buildup (a vacuum greater than –0.75 psi (–5.2 kPa). The vacuum relief valve, used during unloading with a bottom outlet that is operable from ground level, opens to admit air when a vacuum occurs, then recloses after normal conditions have been restored.

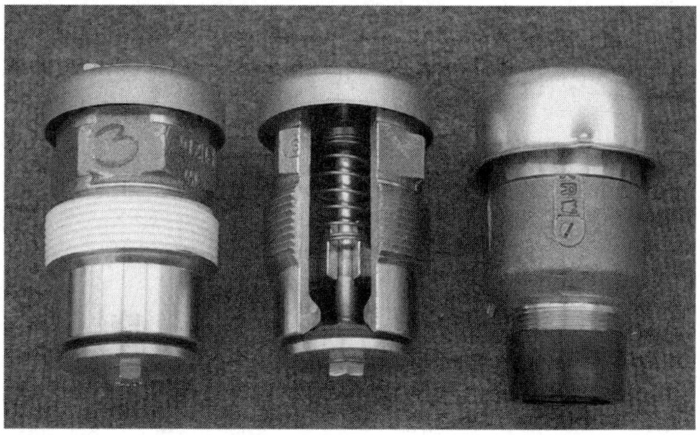

EXHIBIT I.12.12 *The illustration shows typical vacuum relief valves. (Source: Union Pacific Railroad)*

(c) Fittings for gauging, including the following:

Gauging devices are mounted on the manway cover plate and are used to measure the amount of commodity in a tank. These devices measure either inage, which is the amount of liquid in the tank, including liquefied or cryogenic gases, or outage, which is the amount of vapor space left in the tank.

The following two types of gauging devices are used on pressure tank cars:

1. Open-type gauging device, which is a slip tube gauging device with either a quick release cover or screw cover. Open-type gauging devices are being phased out. They are not used on new tank cars and are replaced during scheduled tests on older tank cars.
2. Closed-type gauging device, which is a rod-type magnetic or tape-type magnetic device.

i. Closed gauging devices (e.g., magnetic)

Closed-type gauging devices use a float coupled with a magnet on a measuring rod or a dial indicator to show the liquid level of the commodity (see Exhibit I.12.13).

ii. Open gauging devices (e.g., slip tube)

Open-type gauging devices release liquid once the liquid level in the tank reaches the bottom of a fixed- or adjustable-length tube. As mentioned previously, open-type gauging devices are being phased out. They are not used on new tank cars and are replaced during scheduled tests on older tank cars.

iii. Other gauging devices (T-bar, long pole, short pole)

The "T," or notch, of the gauge bar is mounted inside the manway of the tank. Liquid to the top of the notch or the bottom of the "T" indicates appropriate outage, normally 2 percent. The long pole/short pole type of gauging device resembles the sample line on pressure tank cars. This device consists of a valve and fixed-length tube that extends into the tank to a specified level for measuring outage.

(d) Miscellaneous fittings, including the following:

i. Manway, manway cover plate, hinged and bolted manway cover, protective housing

EXHIBIT I.12.13 *This photograph depicts a closed-type gauging device. (Source: Union Pacific Railroad)*

All pressure and nonpressure tank cars have manways. A manway is an opening on top of the tank car large enough to allow access to the tank's interior for cleaning, inspecting the interior, and making repairs. The manway on pressure tank cars is closed with a permanent 20-bolt manway cover plate that is used for attaching fittings. A gasket is used between the manway and its cover plate to form a seal. A protective housing is mounted on the manway. This housing protects the valves and fittings within from mechanical damage in a derailment. The protective housing has a hinged cover that must be opened to reach the valves and fittings. The protective housing is always found on pressure tank cars and can be found on some nonpressure tank cars.

ii. Sample line

A sample line can be mounted on the manway cover plate and is used to obtain a sample of the commodity without otherwise opening the tank. For flammable gases, the sample line has an excess flow valve on the end inside the tank.

iii. Sump

A sump is a closed depression in the bottom of the tank. This fitting allows the eduction line to extend slightly below the bottom of the tank to more completely unload the tank.

iv. Thermometer well

A thermometer well, mounted on the manway cover plate, consists of a closed tube that extends into the tank. This device is filled with a small amount of permanent antifreeze. By inserting a thermometer into the tube, the temperature of the commodity inside the tank can be measured.

v. Washout

A washout is a closed-off opening in the bottom of the tank car tank that is used to facilitate cleaning or purging the tank. Once the closure is removed, the commodity flow cannot be controlled.

(9) Given examples of various fitting arrangements on tank cars (including carbon dioxide and cryogenic liquid tank cars) with the following fittings included, point out the location(s) where each fitting is likely to leak and a reason for the leak:
- (a) Air valve
- (b) Bottom outlet nozzle
- (c) Bottom outlet valve and top operated bottom outlet valve (with stuffing box)
- (d) Closed gauging devices (e.g., magnetic)
- (e) Combination pressure relief valve
- (f) Flange for manway, valves, and so forth
- (g) Liquid valve and vapor valve (ball versus plug type)
- (h) Manway, manway cover plate, hinged and bolted manway cover, protective housing
- (i) Open gauging devices (e.g., slip tube)
- (j) Pressure regulators on carbon dioxide cars and liquefied atmospheric gases in cryogenic liquid tank cars
- (k) Quick-fill hole cover
- (l) Combination pressure relief valve
- (m) Pressure relief valve
- (n) Safety vent (with rupture or frangible) disk
- (o) Sample line
- (p) Thermometer well
- (q) Vacuum relief valve (negative pressure or vacuum)
- (r) Washout

The annual reports of the Association of American Railroads provide leak data that shows some of the types of problems found with tank cars transporting hazardous materials. Commentary Table I.12.1 offers guidance in assessing and repairing damaged car fittings.

(10) Given examples of each of the following types of tank car damage, identify the type of damage:

(a) Corrosion

Corrosion is pitting of the tank metal, thus reducing the thickness and, possibly, the strength of the tank metal.

(b) Crack

A crack is a narrow split or break in the container metal that can penetrate through the metal of the container. Cracks are typically associated with dents and rail burns. Some characteristics of cracks include the following:

- Typically occur in tension areas, not compression areas
- Cause failure because they can grow under stress
- Can grow at speeds approaching the speed of sound
- Grow very rapidly in brittle steels and relatively slowly in ductile steels

COMMENTARY TABLE I.12.1 Guidelines for Assessing and Repairing Damaged Tank Car Fittings

Location of Leaks	*Probable Cause*	*Basic Repair Methods*
Potential Locations of Leaks from Loading and Unloading Fittings		
Liquid or vapor valve (ball or plug type)		
Liquid or vapor leak from threaded orifice in valve	Valve not completely closed	Close valve
	Plug missing or loose	Replace and/or tighten plug
	Plug or seat worn	To be handled by a tank car specialist
Liquid or vapor leak from seat between valve and the manway cover plate	Loose flange nuts	Tighten flange nuts
	Bad gasket	To be handled by a tank car specialist
Liquid or vapor leak around valve stem	Packing retainer loose	Tighten packing retainer
	Missing split ring packing	To be handled by a tank car specialist
Fill hole cover		
Liquid or vapor leak around fill hole cover or missing	Loose cover nuts	Tighten loose cover nuts
	Fill hole gasket damaged	To be handled by a tank car specialist
Manway cover		
Liquid or vapor leak between manway nozzle and manway cover	Loose cover nuts	Tighten loose cover nuts
	Manway gasket damaged or missing	To be handled by a tank car specialist
Top operating mechanism (stuffing box) for bottom outlet valve		
Liquid or vapor leak from cover of stuffing box	Loose packing gland nut	Tighten packing gland nuts
	Defective packing material	To be handled by a tank car specialist
Bottom outlet		
Liquid leak from bottom outlet cap	Bottom outlet valve open	Close bottom outlet valve
	Bottom outlet cap/plug loose	Tighten bottom outlet cap/plug
	Bottom outlet cap gasket missing or defective	To be handled by a tank car specialist
Liquid line flange		
Liquid leak from flange	Loose flange nuts	Tighten flange nuts
	Missing or defective gasket	To be handled by a tank car specialist
Potential Locations of Leaks from Pressure/Vacuum Relief Fittings		
Safety relief valve (external, internal, or combination)		
Liquid or vapor leak from joint between base of valve and manway cover*	Loose flange nuts	Tighten flange nuts
Liquid or vapor leak from valve seat*	"O" ring or washer installed incorrectly or damaged from normal wear	To be handled by a tank car specialist; **do not remove or "gag" the safety relief valve**
	Caution: Spring or stem may be broken and not repairable in the field	
	Potentially overloaded tank	
Liquid or vapor leak from valve seat*	Valve stem bent or broken	To be handled by a tank car specialist
	Overloaded tank	To be handled by a tank car specialist
Safety vent		
Liquid or vapor leak from opening in center of safety vent	Ruptured frangible disc(rupture disc)	Replace frangible disc with new disc identical to the ruptured disc
	Liquid indicates overloaded or splash without overload	
Vacuum relief valve		
Liquid or vapor leak from under cap	"O" ring off seat or valve stem bent	To be handled by a tank car specialist
	Solidified product	To be handled by a tank car specialist

(continues)

***COMMENTARY TABLE I.12.1** Continued*

Location of Leaks	***Probable Cause***	***Basic Repair Methods***
Potential Locations of Leaks from Fittings for Gauging		
Open-type gauging device, slip tube with quick release or screw cover		
Liquid or vapor leak from gauging device control valve orifice plug	Gauging device control valve not closed; plug in valve is loose or missing	Close gauging device control valve Tighten plug in control valve
Liquid or vapor leak from joint between gauging device and manway cover plate	Loose flange nuts	Tighten loose flange nuts
Liquid or vapor leak from around cover at base of fitting	Loose flange nuts	Tighten loose flange nuts
Liquid or vapor leak around gauge rod packing gland	Packing gland nut loose Packing material defective or missing	Tighten gauging device cover To be handled by a tank car specialist
Closed-type gauging device, magnetic		
Liquid or vapor leak from base of gauging	Broken pipe	Tighten gauging device cover; **do not** device cover **remove cover**
Liquid or vapor leak from seal between gauging device and manway cover plate	Loose flange nuts	Tighten flange nuts
Closed-type gauging device, tape-type		
Liquid or vapor leak from seal between gauging device and manway cover plate	Loose flange nuts	Tighten flange nuts
Potential Locations of Leaks from Miscellaneous Fittings		
Sample line		
Liquid or vapor leak from sample line orifice or around plug	Sample line valve not closed Plug missing or loose	Close sample line valve Replace and/or tighten plug
Liquid or vapor leak from joint between base of sample line and manway cover	Damaged sample line pipe	To be handled by a tank car specialist
Thermometer well		
Liquid or vapor leak from thermometer well cap	Loose cap with damaged thermometer well pipe Missing or defective "O" ring in cap or on nipple with damaged thermometer well pipe	Tighten cap; **do not remove cap** To be handled by a tank car specialist; **do not remove cap**
Liquid leaking from between thermometer well nipple and manway cover	Damaged thermometer well pipe	To be handled by a tank car specialist; **do not tighten thermometer well nipple!!**
Thermometer well nipple broken off with no leak	Mechanical damage to thermometer well nipple	To be handled by a tank car specialist
Heater coil—internal		
Liquid leak from inlet or outlet pipes at bottom of tank	Condensation — residue from material used for heating contents Damage to internal heater coils	Tighten caps. If leak continues, contact a tank car specialist To be handled by a tank car specialist
Washout		
Liquid leaking from around seal between tank and washout plate	Flange nuts loose Gasket defective	Tighten flange nuts clockwise To be handled by a tank car specialist
Liquid leaking from plug in washout plate	Tell-tale plug loose	Tighten plug

Source: Union Pacific Railroad, *Participant's Manual: Tank Car Safety Course,* July 2007.

(c) Dent

A dent is a deformation of the container metal. This damage is caused by impact with a relatively blunt object. When a sharp radius of curvature is associated with the dent, the possibility of cracking exists. Long dents with cold work (gouges) at the bottom of the dent are called rail burns.

(d) Flame impingement

Flame impingement is fire striking the surface of the tank, either in the liquid space or in the vapor space.

(e) Puncture

A puncture is a hole in the tank.

(f) Score, gouge, wheel burn, rail burn

Scores and gouges, wheel burns, and rail burns equate to metal loss from the tank and thus a weakening of the tank structure. A score is a reduction in the thickness of the container shell. A score is an indentation in the container made by a relatively blunt object. A score is the relocation of the container or weld metal in such a way that the metal is pushed aside along the track of contact with the blunt object.

A gouge is a reduction in the thickness of the container caused by a gouge made by a sharp, chisel-like object. A gouge is the cutting and complete removal of the container or weld metal along the track of contact.

Wheel burns, sometimes called spot burns, are similar to a score in that the prolonged contact of the wheel with the tank reduces the thickness of the tank, pushing the metal aside at the point of contact. Long dents with cold work at the bottom of the dent are called rail burns. See 12.2.1(10)(e).

(11) Given examples (actual or simulated) of scores, gouges, wheel burns, and rail burns, perform each of the following tasks:

(a) Use a depth gauge to measure the depth of each score, gouge, wheel burn, and rail burn.
(b) Point out where each score, gouge, wheel burn, and rail burn crosses a weld, if that condition exists.
(c) Measure the depth of the weld metal removed at any point where the score, gouge, wheel burn, and rail burn crosses a weld.
(d)*Given examples (actual or simulated) of where a score, gouge, wheel burn, and rail burn crosses a weld, determine if the heat-affected zone has been damaged.

A.12.2.1(11)(d) The heat-affected zone is an area in the metal next to the actual weld. This zone is less ductile than either the weld or the metal due to the effect of the welding process. The heat-affected zone is vulnerable to cracks.

(12) Given examples (actual or simulated) of dents and rail burns, perform each of the following tasks:

The technician with a tank car specialty performs the tasks in 12.2.1(12) in a manner appropriate to the rail equipment used. These tasks should be performed using personal protective equipment appropriate for the material involved and the working conditions.

(a) Use a dent gauge to measure the radius of curvature for each dent or rail burn.

The tank car dent gauge is a "go/no-go" device used to compare the radius of curvature of a dent in a tank car tank to accepted standards in order to determine the severity of damage. This gauge is called a "go/no-go" because, depending on the reading, a tank car can be pulled out of service if the measurement is below accepted standards.

The technician with a tank car specialty might be required to use a dent gauge to determine whether the radius of curvature in a tank is critical or not critical in accordance with 12.2.1(12)(a).

(b) Identify those examples that include cracks at the point of minimum curvature.

The technician with the tank car specialty is required to identify cracks at the point of minimum curvature and explain the significance of this condition, in accordance with 12.2.1 (12)(b).

(13) Given examples of damaged tank car fittings, describe the extent of damage to those fittings.

The technician with a tank car specialty is required to describe the extent of damage to tank car fittings such as damaged and not leaking, damaged and leaking, repairable or not, and so forth, in accordance with 12.2.1(13).

(14) Given examples of tank car tank damage, describe the extent of damage to the tank car tank.

The technician with a tank car specialty is required to describe the extent of damage to a tank car tank such as damaged and not leaking, damaged and leaking, repairable or not, potential for release and so forth, in accordance with 12.2.1(14). The technician must also determine the depth of score, gouge, wheel burn or rail burn, radius of curvature of dents and rail burns, amount and location of corrosion, presence of cracks, and so forth, as appropriate for the example.

(15) Given a tank car and the applicable equipment and reference material, determine the pressure in the tank car, using either of the following methods:

(a) Pressure gauge

The technician with a tank car specialty is required to determine the pressure in the tank car using a pressure gauge attached to the tank car through one of the fittings (typically the sample line), in accordance with 12.2.1(15)(a).

(b) Temperature of the contents

Alternatively, the temperature of the material can be obtained and checked against temperature pressure charts to determine the pressure of the contents. This alternative method is unlikely to be used in an emergency.

(16)*Given a tank car, use the tank car's gauging device to determine the amount of liquid in it.

The technician with a tank car specialty is required to use various types of gauging devices to determine the inage (amount of liquid) or outage (amount of vapor space) in a tank car.

A.12.2.1(16) Other methods for determining the amount of liquid include shipping papers, the presence of frost line, the use of touch to feel for the colder liquid level, and the use of heat sensors.

12.2.2 Predicting the Likely Behavior of the Tank Car and Its Contents.

Technicians with a tank car specialty shall predict the likely behavior of the tank car and its contents and shall complete the following tasks:

The competencies in 12.2.2 also build on the competencies of the hazardous materials technician. The hazardous materials technician with a tank car specialty has expertise in predicting the likely behavior of the tank car and its contents. In some cases, a specialist employee A might also have these same skills and knowledge, but only for the specific tank car and chemical used by his or her company. For example, an employee of a petrochemical enterprise might be the resident expert on tank cars that carry gasoline. If the appropriate training or certification has been completed, this person can operate at the same level as the technician with a tank car specialty but only for tank cars that carry gasoline used by the petrochemical company.

(1) Given the following types of tank cars, describe the likely breach and release mechanisms associated with each type:
   (a) Cryogenic liquid tank cars
   (b) Nonpressure tank cars
   (c) Pneumatically unloaded covered hopper cars

The technician with a tank car specialty is required to be able to discuss the different types of breach and release mechanisms in 12.2.2(1) for various stress conditions on these tank cars.

   (d) Pressure tank cars

(2) Describe the difference in the following types of construction materials used in tank cars and their significance in assessing tank damage:
   (a) Alloy steel
   (b) Aluminum

Carbon steel is used in over 90 percent of the tank car tanks in use today, with aluminum making up most of the remainder. Stainless steel, referred to as alloy steel, is used in a smaller number of cars. Nickel or nickel alloy is used for some tanks that are used to transport acid or food. The minimum plate thickness of materials used to construct tank cars is specified by regulation.

   (c) Carbon steel

(3) Discuss the significance of selection of lading for compatibility with tank car construction material.

The technician with a tank car specialty is required to be able to explain the consequences of having an incompatible material in a tank car tank, in accordance with 12.2.2(3).

(4) Describe the significance of lining and cladding on tank cars in assessing tank damage.

The interiors of some tank cars are lined or clad with materials to protect the tank from the corrosive or reactive effects of the contents or to maintain the purity of the contents. During an emergency, responders must consider the potential effects of lining or cladding damage, as required in 12.2.2(4).

(5) Describe the significance of the jacket on tank cars in assessing tank damage.

An undamaged jacket can serve as a heat shield from radiated heat. Also, the jacket is not designed to hold the product, so mechanical damage to the jacket might not be indicative of the extent of damage to the tank itself.

(6) Describe the significance of insulation and thermal protection on tank cars in assessing tank damage.

Insulation is used to safeguard the contents of the tank car from outside temperatures. Thermal protection is used on certain tank cars to protect them from flame impingement from either a pool or torch fire. This protection is designed to keep tank metal temperatures below 800°F (427°C) for 100 minutes (pool fire exposure) or 30 minutes (torch fire exposure).

(7) Describe the significance of jacketed and sprayed-on thermal protection on tank cars in assessing tank damage.

Both types of thermal protection in 12.2.2(7) provide the same function, but like other forms of thermal protection, mechanical damage to the coating or jacket might not be indicative of the extent of damage to the tank itself.

(8) Describe the significance of interior and exterior heater coils on tank cars in assessing tank damage.

Exterior heater coils pose less of a problem than interior coils when damaged because they are outside of the tank and have no contact with the contents. However, heater coils inside the tank can allow the contents to get outside the tank because the coil is connected to a heater coil inlet on the outside of the tank. Without a cap, the flow cannot be controlled.

(9) Describe the significance of each of the following types of tank car damage on different types of tank cars in assessing tank damage:

(a) Corrosion
(b) Crack

It is difficult to determine when a crack in a tank is critical, so accurate and timely decisions about a tank's condition must be made. The following guidelines can be helpful:

- Any crack found in a tank, no matter how small, demands immediate action to relieve the stress in the tank by venting, flaring, or transferring the contents of the tank.
- Cracks in fillet welds (i.e., those used to attach brackets or reinforcement plates) are not critical unless a crack extends into tank metal.
- When a crack is in conjunction with a dent, score, or gouge, the tank should be unloaded as soon as possible before moving it.
- The pressure of the commodity should be considered, and the potential for a pressure rise should be evaluated.

(c) Dent

Sharp dents in the shell of the tank are the most serious, because these dents can reduce the strength of the tank significantly. In accordance with 12.2.2(9)(c), dents should be evaluated using the following guidelines:

1. Generally, for dents in the shell of tank cars built prior to 1967, the tank should be unloaded, without moving it, given the following conditions:

a. Minimum radius of curvature of 4 in. (102 mm) or less. Dents with a radius of curvature more than 4 in. (102 mm) are not problems by themselves.
b. Presence of a crack anywhere
c. Dent crossing a weld
d. Presence of a score or gouge
e. Evidence of cold work

2. Generally, for dents in the shell of tank cars built since 1967, the tank should be unloaded, without moving it, given the following conditions:

a. Minimum radius of curvature of 2 in. (51 mm) or less. [Dents with a radius of curvature more than 2 in. (51 mm) are not problems by themselves.]
b. Presence of a crack anywhere
c. Dent crossing a weld
d. Presence of a score or gouge
e. Evidence of cold work

3. Massive dents in heads of the tank are generally not serious unless gouges or cracks are present with the dents.
Note: Reduction in tank volume due to massive denting is not a major consideration unless it is suspected that atmospheric temperature might approach the "shell full temperature" of 115°F (46°C) during summer loading or 95°F (35°C) during winter loading. The loss of volume due to massive denting lowers the shell full temperature a range of 3°F to 4°F (1.7°C to 2.2°C). Massive denting could reduce tank shell capacity by as much as 5 percent. If massive denting causes the volume of the tank to equal the volume of the loading, the tank might undergo hydrostatic failure.

4. Small dents not exceeding 12 in. (305 mm) in diameter in heads in conjunction with cold work in the bottom of the dent are marginally safe if they show a radius of curvature less than 4 in. (102 mm) for tanks built prior to 1967 or less than 2 in. (51 mm) for tanks built since 1967. Such tanks should be unloaded in place. In any case, the tank should be moved as little as possible and should be unloaded promptly.

(d) Flame impingement
(e) Puncture
(f) Score, gouge, wheel burn, rail burn

Scores and gouges are evaluated in the following manner, using the depth of the indentation as a guideline:

- Longitudinal scores are of greatest concern; however, circumferential (hoop) scores must be evaluated as well.
- Scores or gouges crossing a weld with removal of the filler metal are of little concern.
- Longitudinal scores or gouges that cross a weld and damage the heat-affected zone are potentially critical; therefore, the contents of the tank should be transferred immediately by experienced personnel.
- When the internal pressure exceeds one-half of the allowable internal pressure listed for the tank, tanks having scores or gouges should be unloaded in place by experienced personnel.

Wheel burns do not induce a high probability of failure. For loaded tank cars with a wheel burn of less than $^1/_8$ in. (3.2 mm), the tank can be transported, not in ordinary train service, to the closest loading/unloading facility, and the tank emptied.

For loaded tank cars with a wheel burn of $^1/_8$ in. (3.2 mm) or greater, the tank should be unloaded as soon as possible without further transportation. When no liquid is behind the point of the wheel burn to cool the tank shell, the rubbing of the wheel tends to heat and deform the metal much more than on a loaded car. For a wheel burn on an empty tank car, the following steps should be taken:

- For a wheel bum less than $^1/_4$ in. (6.35 mm), the car should be sent to a shop for repair.
- For a wheel burn $^1/_4$ in. (6.35 mm) or greater, the internal pressure should be reduced by flaring or venting any remaining contents to not over 10 psi (69 kPa) before being sent to the shop.

All rail burns are serious and require that the contents be transferred to another tank car or cargo tank before the car is transported from the site. For a tank with a gouge less than 1/8 in. (3.2 mm) deep, the tank can be uprighted and transported a short distance prior to transfer. For a tank with a gouge 1/8 in. (3.2 mm) or greater, the tank should be uprighted and the contents transferred in place before being transported any distance.

(10) Describe the significance of the depth of scores, gouges, wheel burns, and rail burns on tank cars in assessing tank damage.

Scores, gouges, wheel burns, and rail burns reduce the thickness of the metal in the tank, thus reducing the strength of the tank metal.

(11) Describe the significance of scores, gouges, wheel burns, and rail burns crossing a weld on a pressure tank car in assessing tank damage.

Scores, gouges, wheel burns, and rail burns reduce the thickness of the metal in the tank, thus reducing the strength of the tank metal. This damage is also complicated by the reduction of strength associated with damage to the heat-affected zone of a welded seam.

(12) Describe the significance of damage to the heat-affected zone of a weld on a tank car in assessing tank damage.

Because of this change in composition of the tank metal, the ductility of the steel in the heat-affected zone is reduced. The heat-affected zone is a likely origin for cracks.

(13) Describe the significance of a condemning dent of a tank car in assessing tank damage.

The speed at which the car is to be offloaded becomes extremely important in relation to the competency in 12.2.2(13). There are times when the vapor in the tank must be released to reduce the pressure in the tank before transferring the contents.

(14) Given various types of tank cars, describe the significance of pressure increases in assessing tank damage.

Should the tank be weakened, increases in pressure could overstress the tank and cause failure.

(15) Given various types of tank cars, describe the significance of the amount of lading in the tank in assessing tank damage.

Overloaded tank cars have little or no room for expansion, so when the tank car is exposed to thermal stress, the risk for violent rupture is greater.

(16) Describe the significance of flame impingement on a tank car.

12.3 Competencies — Planning the Response

In Section 7.3.1 of NFPA 472, the technician level responder is required to identify the appropriate response objectives for responders at that level. The options available are offensive, defensive, or nonintervention. The response objectives for the technician with a tank car specialty are expanded in Section 12.3 to include highly technical, offensive operations. Response objectives now include those objectives that can lead to final mitigation. While the responder at the technician level can stabilize a situation until industrial experts arrive for final and complete mitigation, the technician with a tank car specialty can become trained or certified to assist in the following work:

- Plan the operation as follows:
  1. Develop a list of required equipment for the selected operation
  2. Prepare a plan for set up, implementation, and shutdown of the operation
  3. Prepare a site safety plan
- Set up the operation as follows:
  1. Hold a safety briefing
  2. Position the equipment
  3. Set up and activate the emergency shutoff system, if used
  4. Purge the liquid and vapor hoses and test for leaks
- Implement the operation
- Shut down the operation as follows:
  1. Purge the hoses and/or piping
  2. Disassemble and decontaminate the equipment
  3. Secure cars
- Determine disposition of cars and prepare them for transportation

12.3.1 Determining the Response Options.

Given the analysis of an emergency involving tank cars, technicians with a tank car specialty shall determine the response options for each tank car involved and shall complete the following tasks:

Product removal methods are outside the legitimate responsibility of the local emergency response personnel. However, oversight of their planning and implementation is within the responsibilities of local emergency response agencies.

(1) Describe the purpose of, potential risks associated with, procedures for, equipment required to implement, and safety precautions for the following product removal techniques for tank cars:

(a) Flaring liquids and vapors

Flaring is the controlled release and disposal of flammable materials by burning from the outlet of a flare pipe (horizontal or vertical). This method is used to reduce the pressure or dispose of the residual vapors in a damaged or overloaded tank car. Vapor flaring is the burning of vapors of a liquefied compressed gas at the outlet of a vertical flare pipe as the vapors exit the flare pipe. Liquid flaring is the vaporizing of liquid product and burning of the vapors at the end of a horizontal flare pipe. A pit is used to contain any product that is not completely burned. Flaring liquids and gases is a procedure that can be used to deplete the amount of hazardous material involved or to depressurize a vessel. This procedure should be performed only by persons who are trained in the technique. In some cases where the burning is already underway at a vent or pressure relief valve when responders arrive, a prescribed procedure is to allow the flaring process to continue. Determination must be made that the flaring does not present a source of ignition to other spilled flammable substances. In rail transportation, flaring is used for the following three basic purposes:

1. Reduce the pressure inside a tank car
2. Dispose of vapors remaining in a tank car during or after transfer of the liquid
3. Burn off liquid where transfer is impractical

Flaring can also be used to expedite recovery operations or as an interim method until a transfer can begin.

(b) Hot and cold tapping

Hot tapping describes a technique in which welding or cutting is done on a container or piping while it contains a flammable liquid or gas. This technique is an emergency procedure in which circumstances necessitate the work and the contents cannot be removed prior to performance of the work. The work must be done only by those who have been trained to do such procedures. The hot tap provides access to the contents of a tank car when damage to the valves and fittings precludes access to the contents. Once the hot tap is completed, transfer, flaring, or venting can take place. Hot tapping involves the welding of a threaded nozzle into an undamaged section of the tank that is in contact with the liquid. A liquid valve is attached to the nozzle. A hole is then drilled through the tank with a special drilling machine. This hot tapping (drilling) machine is equipped with seals that prevent loss of product during the drilling operation. Liquid hoses or pipe can be attached to the valve outlet.

Cold tapping is similar to hot tapping except that the attachment of the threaded nozzle to the tank is accomplished without welding and instead by strapping the nozzle plate onto the tank.

(c) Transferring liquids and vapors

The purpose of transferring liquids and gases is often associated with removal from a damaged or potentially damaged containment vessel to one that is undamaged. The equipment varies with the type of vessel involved and the hazardous material being transferred. Transfer procedures are generally established in advance by good industry practice. Only essential personnel, outlined, in most cases, in the overall incident management systems of the hazardous materials branch, should be involved in the process.

A transfer is the movement of the contents of a damaged or overloaded tank car into a receiving tank (e.g., a tank car, a cargo tank, an intermodal tank, or a fixed tank). In rail transportation, transfers can be used when the following conditions exist:

- The tank car tank itself is sound but, due to bolster or other mechanical damage, the car cannot be safely mounted on its trucks and rerailed.

- The site conditions prevent rerailing the damaged tank car (e.g., the terrain does not permit use of cranes or other rerailing equipment).
- The tank car tank is overloaded.
- The damage to leaking valves and fillings cannot be repaired.
- The tank car tank has been damaged to the extent that the tank cannot be safely rerailed and moved to an appropriate unloading point. Typically, transfers in the field are distinguished by the basic equipment used to move the contents from a damaged or overloaded tank car.

Gas or liquid transfers can be accomplished in the field using one of the following methods:

1. The vapor compressor product removal method uses a vapor compressor to move the contents of a damaged or overloaded tank car into a receiving tank (e.g., a tank car, cargo tank, or portable tank). The vapor compressor pulls the vapors from the receiving tank, compresses them, and forces them into the damaged tank car. The higher pressure in the damaged tank car pushes the liquefied gas into the receiving tank. The use of the vapor compressor results in a pressure increase in the damaged tank car. This transfer method should be used only when an increase in pressure in the damaged tank car is acceptable.
2. The vapor compressor and a liquid pump product removal method uses a vapor compressor and a liquid pump to move the contents of a damaged or overloaded tank car into a receiving tank (e.g., a tank car, cargo tank, or portable tank). The vapor compressor is used to accelerate the rate of transfer by withdrawing vapors from the receiving tank, compressing them, and forcing them into the damaged tank car. The higher pressure in the damaged tank car pushes the liquefied gas into the pump. The use of the vapor compressor results in a pressure increase in the damaged tank car. This transfer method should be used only when an increase in pressure in the damaged tank car is acceptable. This method is justified when the receiving tank is a greater distance from the damaged or overloaded tank car.
3. The compressed air or an inert gas product removal method uses a compressed gas (e.g., nitrogen or carbon dioxide) to move the contents of a damaged or overloaded tank car into a receiving tank (e.g., a tank car, cargo tank, or portable tank). The compressed gas creates a positive pressure differential in the damaged tank car that pushes the liquid into the receiving tank. Vapor from the receiving tank might have to be vented to the atmosphere or scrubbed. The use of the compressed gas results in a pressure increase in the damaged tank car. This transfer method should be used only when an increase in pressure in the damaged tank car is acceptable. Check with the shipper to determine the compatibility of the compressed gas to be used.
4. The liquid pump product removal method uses a liquid pump to move the contents of a damaged or overloaded tank car into a receiving tank (e.g., a tank car, cargo tank, or portable tank). The material in the damaged or overloaded tank car is then pumped into the receiving tank. The use of a liquid pump does not increase the pressure in the damaged tank car; however, if another means of creating positive pressure differential is used, a pressure increase may occur.
5. The vapor pressure (with or without flaring) product removal method uses the material's own vapor pressure to move the contents of a damaged or overloaded tank car into a receiving tank (e.g., a tank car, cargo tank, or portable tank). In addition, a vapor flare maintains the necessary positive pressure differential between the damaged or overloaded tank car and the receiving tank by burning off vapors in the receiving tank at the outlet of a flare pipe. The pressure in the receiving tank is kept as low as possible.

(d) Vent and burn

The vent and burn method is a highly specialized, and seldom practical, technique that involves the use of shaped charges that are designed to create an opening in a container and set fire to its contents. The purpose is to deplete the supply of contained hazardous flammable substances when other methods are unavailable. Vent and burn is a method of removing liquefied flammable compressed gases or flammable liquids from a tank car by creating controlled openings in the tank using explosives. Explosive charges are strategically placed on the tank, one at the highest point on the tank for venting vapor and the second at the lowest point on the tank for releasing liquid. The released contents are allowed to flow into a pit for evaporation or burnoff. Vent and burn is the last resort and is to be performed only by experienced personnel.

(e) Venting

Venting is the process of reducing the pressure in a tank by releasing liquefied compressed gas vapors into the atmosphere. This release can be direct or, in case of toxic products, indirect through an appropriate treatment system. Typically, venting is used with nonflammable gases. Extreme care must be exercised in a venting operation and only after consultation with the shipper, carrier, or other tank car specialist.

(2) Describe the inherent risks associated with, procedures for, equipment required to implement, and safety precautions for leak control techniques on various tank car fittings.
(3) Describe the effect flaring or venting gas or liquid has on the pressure in the tank (flammable gas or flammable liquid product).

Flaring and venting reduce the pressure or dispose of the residual vapors in a damaged or overloaded tank car.

(4) Describe the inherent risks associated with, procedures for, equipment required to implement, and safety precautions for lifting of tank cars.

The lifting of tank cars should be performed only by experienced personnel, such as railroad or shipper mechanical personnel or specially trained and equipped contractors.

(5) Describe the inherent risks associated with, procedures for, and safety precautions for the following operations:
   (a) Setting and releasing brakes on rail cars
   (b) Shutting off locomotives using the fuel shutoff and the battery disconnect
   (c) Uncoupling rail cars

The technician with a tank car specialty is required in 12.3.1(5) to perform these tasks in a manner appropriate to the rail equipment used. These tasks should be performed using appropriate personal protective equipment for the material involved and the working conditions.

(6) Describe the hazards associated with working on railroad property during emergencies.

The Union Pacific Railroad has provided the following summary of precautions to minimize hazards, and more information can be found on their website at www.up.com [3]:

> While working on railroad property, always take the following precautions:
>
> - Always wear a hard hat, safety glasses, and sturdy work boots, preferably steel-toed.
> - Tennis shoes and regular low-cut casual shoes are not acceptable.
> - Wear hearing protection in situations that call for its use.
> - Do not wear bright red or orange clothing on the right-of-way; train crews can confuse these colors with emergency situations.
> - Stay clear of the tracks whenever possible; trains can sneak up on you undetected due to atmospheric conditions and terrain.
> - When confronted with a passing train, always stand away from the track to prevent injury from flying debris or loose rigging. Also, observe the train as it passes so that you are prepared to take evasive action in the event of an emergency.
> - In double track territory, never stand between tracks when a train is passing. This could place you in a precarious situation if another train suddenly appears on the other track.

Never walk or stand on the track; rail surfaces can be extremely slippery and many rails in curves are lubricated. When you cross the rails, step over them.
- Try to maintain at least 20 ft from the rails when walking on the right-of-way. Never walk down the center of the track.
- Be prepared for the movement of cars at any time – in either direction. Stay at least 20 ft away from the ends of cars when crossing the track and never climb on, under, or between cars. Never rely solely on others to protect you from train movement – watch yourself.
- Stay away from remote-controlled switches. The switch points can move unexpectedly – with enough force to crush ballast rock!! Stay away from any other devices you are unsure of.
- Avoid pole lines within the right-of-way. These lines may carry from 500 volts to 2700 volts. Due to the terrain, sometimes these lines are located very close to the ground.
- Be careful when working around a derailment. Hazardous materials may be present, so check with the local railroad personnel before approaching.
- Equipment must never be moved across the tracks except at established road crossings and never moved across bridges or through tunnels.
- A 100-car train moving at 60 miles-per-hour can take more than one mile to stop in the event of an emergency. Never judge the distance or speed of a train by the headlight.
- Never lay metal objects across the rails. This is a safety hazard.

12.4 Competencies — Implementing the Planned Response

Most of the competencies in Section 12.4 require the actual demonstration of a task. These tasks require an in-depth knowledge of the tank car, its contents, and the specialized equipment needed to complete the task. Because these tasks must be done on actual tank cars, only a few training facilities are able to provide practical training and/or certification.

12.4.1 Implementing the Planned Response.

Given an analysis of an emergency involving tank cars and the planned response, technicians with a tank car specialty shall implement or oversee the implementation of the selected response options safely and effectively and shall complete the following tasks:

The equipment and resources necessary for demonstrating the tasks in 12.4.1 are not readily available. Requests for training for the performance of these tasks should be directed to the local shipper, carrier, or contractor personnel in the community, possibly through the local emergency planning committee.

(1) Given a leaking manway cover plate (loose bolts), control the leak.

The technician with a tank car specialty is required to demonstrate the ability to choose the appropriate tools and tighten the loose bolts on a leaking manway cover plate. This task should be performed using personal protective equipment appropriate for the material involved and for the working conditions.

(2) Given leaking packing on the following tank car fittings, control the leak:

(a) Gauging device packing nut

The technician with a tank car specialty is required in 12.4.l(2)(a) to choose the appropriate tools and tighten the gauging device packing nut, as appropriate for the specific make of valve. This task should be performed using personal protective equipment appropriate for the material involved and for the working conditions.

(b) Liquid or vapor valve packing nut

The technician with a tank car specialty is required by 12.4.l(2)(b) to choose the appropriate tools and tighten the liquid or vapor valve packing nuts or bolts, as appropriate for the specific make of valve. This task should be performed using personal protective equipment appropriate for the material involved and for the working conditions.

(c) Top operated bottom outlet valve packing gland

The technician with a tank car specialty is required by 12.4.1(2)(c) to choose the appropriate tools and tighten the packing gland on a top-operated bottom outlet valve, as appropriate for the specific make of valve. This task should be performed using personal protective equipment appropriate for the material involved and for the working conditions.

(3) Given an open bottom outlet valve with a defective gasket in the cap, control the leak.

The technician with a tank car specialty is required by 12.4.1(3) to choose the appropriate tools, ensure that the bottom outlet valve is closed, and tighten the cap on a bottom outlet valve, as appropriate for the specific make of valve. This task should be performed using personal protective equipment appropriate for the material involved and for the working conditions.

(4) Given a leaking top operated bottom outlet valve, close valve completely to control leak.

The technician with a tank car specialty is required by 12.4.1(4) to close the top-operated bottom outlet valve, as appropriate for the specific make of valve. This task should be performed using personal protective equipment appropriate for the material involved and for the working conditions.

(5) Given leaking fittings on a chlorine tank car, use the Chlorine C kit to control the leak

The technician with a tank car specialty is required by 12.4.1(5) to apply the C kit to control a leaking combination safety relief valve and leaking liquid and vapor valves on a chlorine tank car, following the instructions provided by the manufacturer of the C kit or local standard operating procedures. This task should be performed using personal protective equipment appropriate for the material involved and for the working conditions.

(6) Given the following types of leaks on various types of tank cars, plug or patch those leaks:

(a) Cracks, splits, or tears
(b) Irregular-shaped hole
(c) Puncture

The technician with a tank car specialty is required by 12.4.1(6) to select and apply various methods of plugging and patching to control or stop leakage of the tank car's contents. This task should be performed using personal protective equipment appropriate for the material involved and for the working conditions.

(7) Given the applicable equipment and resources, demonstrate the following:

(a) Flaring of liquids and vapors
(b) Transferring of liquids and vapors
(c) Venting

The technician with a tank car specialty is required by 12.4.1(7) to participate as a member of a team assigned to perform product removal. These tasks should be performed using personal protective equipment appropriate for the material involved and for the working conditions.

(8) Given the applicable resources, perform the following tasks:

(a) Set and release brakes on rail cars.
(b) Shut off locomotives using the fuel shutoff and the battery disconnect.
(c) Uncouple rail cars.

The technician with a tank car specialty is required by 12.4.1(8) to perform these tasks in a manner appropriate to the rail equipment used. These tasks should be performed using personal protective equipment appropriate for the material involved and for the working conditions. The equipment and resources necessary for demonstrating these tasks are not readily available. Requests for training for the performance of these tasks should be directed to the railroad or shippers in the community, possibly through the local emergency planning committee.

(9)* Demonstrate bonding and grounding procedures for the transfer of flammable and combustible products from tank cars or other products that can give off flammable gases or vapors when heated or contaminated, including the following:

(a) Selection of equipment
(b) Sequence of bonding and grounding connections
(c) Testing of bonding and grounding connections

A.12.4.1(9) When bonding and grounding are performed, a ground resistance tester and an ohmmeter should be used. The ground resistance tester measures the earth's resistance to a ground rod, and the ohmmeter measures the resistance of the connections to ensure electrical continuity. One ground rod might not be enough; more might have to be driven and connected to the first to ensure a good ground. Resistance varies with types of soils.

The technician with a tank car specialty should be able to verify (using the appropriate equipment) the following:

- Ground rod resistance to earth
- Connections between ground rod and tank car
- Connections between ground rod and main ground rod and ground cable

Local procedures for the tasks in 12.4.1(9) should be written and consistent with nationally accepted practices. The technician with a tank car specialty is required to participate as a member of a team assigned to set up bonding and grounding for the transfer of tank cars. These tasks should be performed using personal protective equipment appropriate for the material involved and for the working conditions.

(10) Given an example of a flammable liquid spill from a tank car, describe the procedures for site safety and fire control during cleanup and removal operations.

REFERENCES CITED IN COMMENTARY

1. *A General Guide to Tank Cars*, Hazardous Materials Management, Union Pacific Railroad, March 2007.
2. Title 29, Code of Federal Regulations, Part 1910.120, U.S. Government Printing Office, Washington, DC.
3. Union Pacific Railroad, *Participant's Manual: Tank Car Safety Course,* July 2007.

Additional References

"Damage Assessment of Tank Cars Involved in Accidents," Transportation Technology Center, Association of American Railroads for Federal Railroad Administration, Office of Research and Development, 1999.

"Field Product Removal Methods for Tank Cars," Transportation Test (Technology) Center, Association of American Railroads for the Federal Railroad Administration, Office of Research and Development, 1995.

Cote, Arthur E., Editor, *Fire Protection Handbook*, 20th ed., National Fire Protection Association, Quincy, MA, 2008.

Competencies for Hazardous Materials Technicians with a Cargo Tank Specialty

CHAPTER 13

The first truck to travel coast to coast across the United States was a Sauger, a Swiss-built truck that traveled from Los Angeles to New York in 1911. Less than 50 years later, the government began to invest heavily in a national network of interstates. The highways and trucking industry have become essential to the U.S. supply chain and have remade the country's social and economic landscape. Manufacturers now treat interstate highways as if they are part of the assembly line, clustering factories near the access ramps so that parts and raw materials can arrive at the right moment. It is in this complex world of pavement, interchanges, and traffic that the hazardous materials technician with a cargo tank specialty must learn to safely conduct operations.

13.1 General

13.1.1 Introduction.

13.1.1.1 The hazardous materials technician with a cargo tank specialty shall be that person who provides technical support pertaining to cargo tanks, provides oversight for product removal and movement of damaged cargo tanks, and acts as a liaison between technicians and outside resources.

13.1.1.2 The hazardous materials technician with a cargo tank specialty shall be trained to meet all competencies at the awareness level (Chapter 4), all core competencies at the operations level (Chapter 5), all competencies at the technician level (Chapter 7), and all competencies of this chapter.

The specialty levels of technician with a tank car specialty and technician with a cargo tank specialty cover extremely hazardous and highly skilled operations that have not had any standardized training or certification previously. The technician with a tank car specialty, the technician with a cargo tank specialty, and the technician with an intermodal tank specialty are levels that are above OSHA' s technician and specialist levels and are not defined by the present Occupational and Health Administration (OSHA) regulations. See Title 29 CFR Part 1910.120 for more details [1]. For a comparison of NFPA 472 and OSHA 29 CFR 1910.120, refer to Table I.1.1 in Part I of this handbook.

13.1.1.3 The hazardous materials technician with a cargo tank specialty shall also receive training to meet governmental occupational health and safety regulations.

13.1.2 Goal.

13.1.2.1 The goal of competencies at this level shall be to provide the technician with a cargo tank specialty with the knowledge and skills to perform the tasks in 13.1.2.2 safely.

13.1.2.2 When responding to hazardous materials/WMD incidents, the hazardous materials technician with a cargo tank specialty shall be able to perform the following tasks:

(1) Analyze a hazardous materials/WMD incident involving cargo tanks to determine the complexity of the problem and potential outcomes by completing the following tasks:
   (a) Determine the type and extent of damage to cargo tanks.
   (b) Predict the likely behavior of cargo tanks and their contents in an emergency.
(2) Plan a response for an emergency involving cargo tanks within the capabilities and competencies of available personnel, personal protective equipment, and control equipment by determining the response options (offensive, defensive, and nonintervention) for a hazardous materials emergency involving cargo tanks.
(3) Implement or oversee the implementation of the planned response to a hazardous materials/WMD incident involving cargo tanks.

13.1.3* Mandating of Competencies.

This standard shall not mandate that hazardous materials response teams performing offensive operations on cargo tanks have technicians with a cargo tank specialty.

A.13.1.3 Technicians operating within the bounds of their training as listed in Chapter 6 of this standard are able to intervene in cargo tank incidents. However, if a hazardous materials response team decides to train some or all of the technicians to have in-depth knowledge of cargo tanks, this chapter sets out the required competencies.

The committee wanted to clearly state that existing and new responders at the hazardous materials technician level can perform basic offensive operations on cargo tank trucks for which they have been qualified. This specialty level covers the highly technical skills and knowledge primarily needed for conducting product removal and transfer operations. For example, the skills required by the competency in 13.4.1(5) are skills that the technician normally does not possess. A hazardous materials response team might have a few members trained to this level, in accordance with 13.1.3, or can request that the shipper, carrier, or an environmental contractor bring someone with this specialty to the scene.

Programs covering these competencies are in place at a number of hazardous materials training centers, including the Association of American Railroads' Transportation Technology Center, the California Specialized Training Institute, and the New Jersey Office of Emergency Management—Hazardous Materials Emergency Response Program. The committee also wants to be clear that this specialty level is not mandated for any hazardous materials response team or its members. Arrangements to request assistance to perform these operations are perfectly appropriate in these situations.

13.1.3.1 Hazardous materials technicians operating within the scope of their training as listed in Chapter 7 of this standard shall be able to intervene in cargo tank incidents.

13.1.3.2 If a hazardous materials response team decides to train some or all of its hazardous materials technicians to have in-depth knowledge of cargo tanks, this chapter shall set out the required competencies.

13.2 Competencies — Analyzing the Incident

13.2.1 Determining the Type and Extent of Damage to Cargo Tanks.

Given examples of damaged cargo tanks, technicians with a cargo tank specialty shall describe the type and extent of damage to each cargo tank and its fittings and shall complete the following tasks:

(1) Given the specification mark for a cargo tank and the reference materials, describe the tank's basic construction and features.

In NFPA 472, cargo tank trucks are grouped into the following five categories:

1. MC-306/DOT-406 cargo tanks
2. MC-307/DOT-407 cargo tanks
3. MC-312/DOT-412 cargo tanks
4. MC-331 cargo tanks
5. MC-338 cargo tanks

MC-306/DOT-406 cargo tanks are commonly used to transport liquid petroleum products at atmospheric pressures. Since 1995, all cargo tanks in this service must be built to the DOT-406 specification. These are nonpressurized (atmospheric pressure) tanks, with maximum working pressures of 3 psi to 5 psi (21 kPa to 34 kPa). Most of these cargo tanks are not insulated and have a capacity between 7500 gal and 10,000 gal (28,387 L and 37,850 L). MC-306/DOT-406 is found in Title 49 CFR Part 178.346 [2].

MC-307/DOT-407 cargo tanks are commonly used to transport flammable and combustible liquids, mild corrosives, and chemicals with a vapor pressure of 18 psi (124 kPa) at 100°F (37.8°C) or greater but not more than 40 psi (276 kPa) at 170°F (76.7°C). Since 1995, all cargo tanks in this service must be built to the DOT-407 specification. Tank capacities range up to 7000 gal (26,495 L). MC-307/DOT-407 is found in Title 49 CFR 178.347.

MC-312/DOT-412 cargo tanks are commonly used to transport high-density liquids and strong corrosives, such as nitric and sulfuric acid. Since 1995, all cargo tanks in this service must be built to the DOT-412 specification. Tank design pressures range from 35 psi to 50 psi (241 kPa to 345 kPa), with maximum capacities of approximately 5000 gal to 6000 gal (18,925 L to 22710 L). MC-312/DOT-412 is found in Title 49 CFR 178.348.

MC-331 cargo tanks are pressurized containers commonly used for the transportation of liquefied and compressed gases (MC-331 is available in Title 49 CFR 178 337). Design pressures range from 100 psi to 500 psi (690 kPa to 3448 kPa) with capacities ranging from 2500 gal to 11,500 gal (9463 L to 43528 L).

MC-338 cargo tanks are commonly used to transport cryogenic liquids such as liquid nitrogen and liquid helium. These are a tank-within-a-tank design with a typical working pressure of 100 psig (690 kPa). Inner tank pressures can range from 235 to 500 psi (162 kPa to 3448 kPa), depending on the product being transported. Tank capacities range from 5000 to 14,000 gal (18925 L to 52990 L). MC-338 is found in Title 49 CFR 178.338.

Under federal law, a cargo tank constructed to Department of Transportation (DOT) specifications must have a certification plate mounted on the cargo tank. On truck and trailer units, the certification plate is mounted on the left front of the tank. Tanks constructed prior to 1985 have the specification plate mounted on the right front of the tank.

The certification plate provides the DOT container specification number, date of manufacture and test, shell material, container pressure ratings, number of compartments and their capacity, and maximum product load.

Some cargo tanks are designed to multiple container specifications, which allow them to transport more than one commodity. Common multipurpose configurations are the combination MC-306/DOT-406/MC-307/DOT-407 unit and the MC-307/DOT-407/MC-312/ DOT-412 unit.

In addition to the manufacturer's specification plate, these cargo tanks have a second multipurpose plate that identifies the specification under which the cargo tank is operated. These plates are color-coded, as are the fittings that are added to make the cargo tank meet the respective specifications. Color codes are as follows:

- MC-306/DOT-406 plate and fittings = red
- MC-307/DOT-407 plate and fittings = green
- MC-312/DOT-412 plate and fittings = yellow
- Nonspecification cargo tank = blue

(2) Given examples of cargo tanks (some jacketed and some not jacketed), point out the jacketed cargo tanks.

Jacketed cargo tanks can be recognized by one or more of the following visual indicators:

- Flat appearance of cargo tank ends
- MC-307/DOT-407 cargo tanks with oval, noncircular shape
- Lightweight, bright or shiny aluminum outer jacket on MC-307/DOT-407 or MC-312/DOT-412 cargo tanks
- Rough, relatively narrow welds

(3) Given examples of the following types of cargo tank damage, identify the type of damage in each example:

(a) Corrosion (internal and external)

Corrosion is pitting of the tank metal, thus reducing the thickness and possibly the strength of the tank metal.

(b) Crack

A crack is a narrow split or break in the container metal that could penetrate through the metal of the container. Cracks are typically associated with dents and rail burns. Cracks typically occur in tension areas, not compression areas, and can cause failure because they can grow under stress.

(c) Dent

A dent is a deformation of the container metal caused by impact with a relatively blunt object. When a sharp radius of curvature is associated with the dent, the possibility of cracking exists.

(d) Flame impingement

Flame impingement is flame striking on the surface of the tank, either in the liquid space or the vapor space.

(e) Puncture

A puncture is a hole in the tank.

(f) Scrape, score, gouge, or loss of metal

A score is a reduction in the thickness of the container shell. This damage is an indentation in the container made by a relatively blunt object. A score is characterized by the relocation of the container or weld metal in such a way that the metal is pushed aside along the track of contact with the blunt object.

A gouge is a reduction in the thickness of the container. This damage is an indentation in the shell made by a sharp, chisel-like object. A gouge is characterized by the cutting and complete removal of the container or weld metal along the track of contact.

A street burn is a deformation in the shell of a cargo tank. This damage is actually a long dent that is inherently flat. A street burn is generally caused by a container overturning and sliding some distance along a cement or asphalt road.

(4) Given examples of damage to an MC-331 cargo tank, determine the extent of damage to the heat-affected zone.

The heat-affected zone is an area in the metal next to the actual weld. This zone is less ductile than either the weld or the metal, due to the effect of the welding process. The heat-affected zone is vulnerable to cracks.

(5)* Given an MC-331 cargo tank containing a liquefied gas, determine the amount of liquid in the tank.

MC-331 measuring devices include the following:

- Magnetic float gauge with a rotary dial or roto-gauge that indicates quantity, by percentage, loaded in the tank
- Fixed-level gauge that indicates the specific quantity

A.13.2.1(5) See A.12.2.1(16).

(6) Given MC-306/DOT-406, MC-307/DOT-407, and MC-312/DOT-412 cargo tanks, point out and explain the design, construction, and operation of each of the following safety devices:

(a) Dome cover design

Dome cover and manhole assembly designs vary with the type of liquid cargo tank (e.g., MC-306/DOT-406 versus MC-312/DOT-412). Each compartment generally has one dome cover; however, in some instances a compartment can have more than one dome cover. Manhole assemblies for MC-306/DOT-406 cargo tanks usually incorporate other devices, such as the dome cover opening, fusible plug, and overfill sensor.

(b) Emergency remote shutoff device

Emergency remote shutoff devices, when actuated, automatically close all internal safety valves. The devices are primarily designed for use during product transfer operations when personnel cannot safely reach the discharge outlets and controls. These devices are always found at the left front of the cargo tank and can, in some instances, also be found at the right rear.

(c) Internal safety valve or external valve with accident protection, including method of activation (air, cable, hydraulic)

Internal safety valves sit inside the tank to protect the valve against mechanical stress and accident damage. The valves are commonly found on MC-306/DOT- 406 and MC-307/DOT-407 cargo tanks.

External valves with accident protection sit outside the tank and are surrounded with a metal framing to protect the valve against mechanical stress and accident damage. The valves are found on some MC-307/DOT-407 and MC-312/DOT-412 cargo tanks.

(d) Pressure and vacuum relief protection devices

Both normal and emergency relief protection is required on all cargo tanks. Common pressure relief devices include relief valves, breather vents, fusible plugs and caps, frangible discs, and pressure vents. Vacuum relief protection is designed to protect the integrity of the tank container during offloading operations.

(e) Shear-type breakaway piping

Internal safety valves have a "shear cut" section of piping, within 4 in. (102 mm) of the tank shell, designed to break under mechanical stress. This shear cut reduces the thickness of the piping by approximately 20 percent. If an accident causes stress at the shear point, the piping should fail at the shear point while the internal valve remains intact within the compartment.

(7) Given MC-331 and MC-338 cargo tanks, point out and explain the design, construction, and operation of each of the following safety devices:

(a) Emergency remote shutoff device

See the commentary following 13.2.1(6)(b).

(b) Excess flow valve

Excess flow valves are designed to almost completely shut off the flow of product when the discharge valve is sheared off and the flow rate increases.

(c) Fusible link and nut assemblies

In the event of a spill fire in or around a cargo tank, fusible links or nut assemblies will melt, releasing cable tension or air pressure and allowing the internal safety valve to automatically close. Fusible devices are required to actuate at temperatures not greater than 250°F (120°C).

(d) Internal safety valve or external valve with accident protection, including method of activation (air, cable, hydraulic)

See the commentary following 13.2.1(6)(c).

(e) Pressure relief protection devices

MC-331 cargo tanks are protected with spring-loaded pressure relief devices located along the top centerline of the container. MC-338 cargo tanks are a tank-within-a-tank design and have relief protection for both the inner tank and the outer tank. A combination of spring-loaded pressure relief devices and rupture discs are used.

(8) Given an MC-306/DOT-406 cargo tank, identify and describe the following normal methods of loading and unloading:

(a) Bottom loading

Bottom loading is the most prevalent method for the transfer of flammable liquids, including gasoline. Product is loaded into each compartment through its respective bottom discharge piping and internal safety valve. Bottom loading is a closed system in which no vapors are released into the environment.

(b) Top loading

Top loading is primarily used for the transfer of combustible liquids and low vapor pressure products (e.g., fuel oil), especially in rural areas. A fill stem tube is inserted into each compartment via the dome cover and the product is then transferred.

(c) Vapor recovery system

Vapor recovery systems are used to collect the product vapors generated as part of the transfer process. Although vapor recovery is an inherent part of bottom loading, it can also be found on top-loading operations. Each compartment has a vent connected to the vapor recovery piping system. These vents are connected either mechanically or pneumatically to each compartment's internal safety valve. When the internal valve is opened, the vent also opens automatically. This action provides both vacuum and pressure protection during transfer operations. As product is transferred, the vapors are collected into the vapor collection header on the cargo tank and then piped to either a vapor condensing unit or a burner. When the cargo tank is offloaded, vapors are returned to the cargo tank through a 3 in. (76 mm) vapor line. Vapor recovery piping and connections are often marked by an orange color code.

(9) Given the following types of cargo tank trucks and tube trailer, identify and describe the normal methods of loading and unloading:

(a) MC-307/DOT-407
(b) MC-312/DOT-412
(c) MC-331
(d) MC-338
(e) Compressed gas tube trailer

(10) Describe the normal and emergency methods of activation for the following types of cargo tank truck valve systems:

All internal safety valves rely on either air or hydraulic pressure or mechanical force to place the valve in the open position. If this pressure or force is removed, the internal safety valve automatically closes.

Emergency methods for activating the internal safety valve include use of the emergency remote shutoff device(s) and the activation of fusible devices in a fire situation. In a rollover situation, internal safety valves can also be manually opened; the method of activation varies with the type of valve and its method of actuation.

(a) Air

Air-actuated valves are found on all types of cargo tanks. On some cargo tanks, normal activation of the internal safety valve might be a two-step process comprised of the opening of a master air control valve and the opening of the respective compartment.

(b) Cable

Cable-actuated valves are found on all types of cargo tanks. Activation of a control handle moves the respective internal safety valve. If tension is released on the cable, such as when a fusible device is actuated, the internal safety valve automatically closes. A secondary safeguard found on MC-306/DOT-406 cargo tanks with cable-actuated systems is to close the compartment door on the right side of the tank where all internal valve controls are located. When the door is pushed closed, all internal safety control valve handles are forced to the vertical, or closed, position.

(c) Hydraulic

Hydraulic-actuated valves are found on MC-307/DOT-407 and MC-312/ DOT-412 cargo tanks. A hydraulic control unit is located near the discharge valves. To actuate the internal valve, the operator must increase the hydraulic system pressure through the use of a hand pump and open the respective internal safety valve. If hydraulic pressure is lost, the internal safety valve(s) automatically closes.

(11) Given a cargo tank involved in an emergency, identify the factors to be evaluated as part of the cargo tank damage assessment process, including the following:
  (a) Amount of product released and amount remaining in the cargo tank
  (b) Container stress applied to the cargo tank
  (c) Nature of the emergency (e.g., rollover, vehicle accident, struck by object)
  (d) Number of compartments
  (e) Pressurized or nonpressurized
  (f) Type and nature of tank damage (e.g., puncture, dome cover leak, valve failure)
  (g) Type of cargo tank (MC or DOT specification)
  (h) Type of tank metal (e.g., aluminum versus stainless steel)

13.2.2 Predicting the Likely Behavior of the Cargo Tank and Its Contents.

Technicians with a cargo tank specialty shall predict the likely behavior of the cargo tank and its contents and shall complete the following tasks:

The competencies of 13.2.2 build on the competencies of the hazardous materials technician. The hazardous materials technician with a cargo tank specialty has expertise in predicting the likely behavior of the cargo tank and its contents. In some cases, a private sector specialist employee A might also have these same skills and knowledge but only for the specific cargo tank truck and chemical used by his or her company. For example, an employee of an oil company might be the resident expert on cargo tank trucks that carry gasoline and fuel oils. If the appropriate training or certification has been completed, this person can operate at the same level as the technician with a cargo tank specialty but only for the cargo tanks used by the oil company. See also 7.2.3 of NFPA 472.

(1) Given the following types of cargo tanks (including a tube trailer), describe the likely breach and release mechanisms:

(a) MC-306/DOT-406 cargo tanks
(b) MC-307/DOT-407 cargo tanks
(c) MC-312/DOT-412 cargo tanks
(d) MC-331 cargo tanks
(e) MC-338 cargo tanks
(f) Compressed gas tube trailer

The technician with a cargo tank specialty is required to be able to discuss the different types of breach, release, dispersion characteristics, contact potential, and hazards for various stress conditions on these types of cargo tanks.

(2) Describe the difference in types of construction materials used in cargo tanks and their significance in assessing tank damage.

MC-306/DOT-406 cargo tanks are commonly constructed of aluminum, although older MC-305 steel and stainless steel tanks built prior to 1967 can occasionally be found. In some areas, MC-306/DOT-406 cargo tanks made from fiberglass-reinforced plastic (FRP) have also been tested.

MC-307/DOT-407 cargo tanks can be constructed of steel, aluminum, stainless steel, titanium, Hastaloy C, and related alloys. Most MC-307/DOT-407 cargo tanks are then insulated with fiberglass and covered by a stainless steel jacket.

MC-312/DOT-412 cargo tanks can be constructed of steel, aluminum, stainless steel, titanium, or Hastaloy C. The tanks can also be constructed of fiberglass-reinforced plastic under an exemption. In some instances, the tank can be lined with a material (e.g., rubber, plastic) that is suitable to protect the tank against chemical attack.

MC-331 cargo tanks are constructed of mild steel—nonquenched tempered (NQT) or high tensile steel—quenched and tempered (QT).

MC-338 cargo tanks are a tank-within-a-tank design similar to a thermos bottle. The inner tank is typically constructed of special steel alloys compatible with the product to be transported and capable of withstanding extremely cold temperatures. The outer container is typically steel. The space surrounding the entire inner tank is evacuated of atmosphere and insulated with a multilayered Mylar™ film.

Compressed gas tube trailers consist of a group of stainless steel cylinders, 9 in. to 48 in. (229 mm to 1219 mm) in diameter, permanently mounted on a semitrailer.

(3) Describe the significance of the jacket on cargo tanks in assessing tank damage.

An undamaged jacket can serve as a heat shield from radiated heat. Also, the jacket is not designed to hold the product, so mechanical damage to the jacket might not be indicative of the extent of damage to the tank itself.

(4) Describe the significance of each of the following types of damage on different types of cargo tanks in assessing tank damage:

(a) Corrosion (internal and external)
(b) Crack

It is difficult to determine when a crack in a tank is critical, so accurate and timely decisions about a tank's condition must be made. The following guidelines can be helpful:

- Any crack found in a tank, no matter how small, demands immediate action to relieve the stress in the cargo tank by venting, flaring, or transferring the contents of the tank.
- When a crack is in conjunction with a dent, score, or gouge, the cargo tank should be unloaded as soon as possible before moving it.
- The pressure of the cargo tank contents should be considered, and the potential for a pressure rise should be evaluated.

(c) Dent

Sharp dents in the shell of the cargo tank are the most serious because these dents can reduce the strength of the tank significantly. Dents should be evaluated using the following guidelines:

1. Dents with a sharp radius can develop cracks. Cracks usually develop on the convex side of a dent, usually inside the cargo tank where they cannot be readily detected.
2. Sharp dents in a cargo tank shell can significantly reduce the strength of a tank.
3. Cargo tanks with dents should be offloaded in place if the following conditions exist:
   - a. Presence of a crack anywhere within the dent
   - b. Evidence of the dent crossing a weld
   - c. Presence of a score or gouge
   - d. Evidence of cold work

(d) Flame impingement

The following points should be noted when evaluating flame impingement:

- Should the tank also be weakened due to mechanical stress, the increase of internal pressure created by the thermal stress of flame impingement could quickly overstress the tank and cause early failure of the cargo tank.
- Applying cooling water streams onto the outer shell of an MC-338 cryogenic liquid cargo tank is ineffective because of the tank's heavy insulation. In addition, water might freeze and block safety relief devices.

See also Hildebrand and Noll's *Handling Gasoline Tank Truck Emergencies: Guidelines and Procedures* [3].

(e) Puncture
(f) Scrape, score, gouge, or loss of metal

Using the depth of the indentation as a guideline, scores and gouges are evaluated in the following manner:

- Longitudinal scores are of greatest concern; however, circumferential (hoop) scores must be evaluated as well.
- Scores or gouges crossing a weld with removal of the filler metal are of little concern.
- Longitudinal scores or gouges that cross a weld and damage the heat-affected zone are potentially critical. Therefore, the contents of the cargo tank should be transferred immediately by experienced personnel.
- When the internal pressure exceeds one-half of the allowable internal pressure listed for the tank, cargo tanks having scores or gouges should be unloaded in place by experienced personnel.

(5) Given examples of damage to the heat-affected zone on an MC-331 cargo tank, describe the significance of the damage in assessing tank damage.

Because of this change in composition in the tank metal, the ductility of the steel in this heat-affected zone is reduced. The heat-affected zone is a likely origin for cracks.

13.3 Competencies — Planning the Response

In 7.3.1 of NFPA 472, the technician level responder is required to identify the appropriate response objectives for responders at that level. The options available are offensive, defensive, or nonintervention. The response objectives for the technician with a cargo tank specialty are expanded in Section 13.3 to include highly technical, offensive operations.

Response objectives now include those objectives that can lead to final transfer and recovery, uprighting, and mitigation. While the responder at the technician level can stabilize the situation until industrial experts arrive for final and complete mitigation, the technician with a cargo tank specialty can become trained or certified to do the following work:

1. Plan the operation as follows:
   a. Prepare a plan for setup, implementation, and termination of the operation.
   b. Develop and acquire the equipment necessary for the selected operation.
   c. Prepare a site safety plan.
2. Set up the operation as follows:
   a. Hold a safety briefing.
   b. Position the resources and equipment.
   c. Set up and activate the emergency shutoff system, if appropriate.
3. Purge the liquid/vapor transfer hoses and test for leaks.
4. Implement product transfer and recovery operations as follows:
   a. Terminate product transfer and recovery operations.
   b. Purge the hoses and/or piping.
   c. Disassemble and decontaminate the equipment.
   d. Secure the cargo tank.
5. Implement uprighting operations.
6. Determine disposition of the cargo tank and prepare for transportation.

13.3.1 Determining the Response Options.

Given the analysis of an emergency involving cargo tanks, technicians with a cargo tank specialty shall determine the response options for each cargo tank involved and shall complete the following tasks:

(1) Given an incident involving a cargo tank, describe the methods, procedures, risks, safety precautions, and equipment that are required to implement spill and leak control procedures.

The competency in 13.3.1(1) builds on the competencies of the hazardous materials technician.

Product removal methods might be outside the legitimate responsibility of the local emergency response personnel. However, oversight of the planning and implementation of spill and leak procedures are the responsibilities of local emergency response agencies.

(2) Given an overturned cargo tank, describe the factors to be evaluated for uprighting, including the following:

Tank truck manufacturers advise against attempts to upright an MC-306/DOT-406 cargo tank while it contains any product. For other types of cargo tanks, a detailed hazard and risk assessment must be conducted. Evaluation criteria are outlined in 13.3.1(2).

(a) Condition and weight of the cargo tank
(b) Lifting capabilities of wreckers and cranes
(c) Preferred lifting points
(d) Selection of lifting straps and air bags
(e) Site safety precautions
(f) Type and nature of stress applied to the cargo tank
(g) Type of cargo tank and material of construction

13.4 Competencies — Implementing the Planned Response

13.4.1 Implementing the Planned Response.

Given an analysis of an emergency involving a cargo tank and the planned response, technicians with a cargo tank specialty shall implement or oversee the implementation of the selected response options safely and effectively and shall complete the following tasks:

The equipment and resources necessary for demonstrating the tasks in 13.4.1 might not be readily available. Requests for training for the performance of these tasks should be directed to the local shipper, carrier, or contractor personnel in the community, possibly through the local emergency planning committee.

(1) Demonstrate the methods for containing the following leaks on liquid cargo tanks (e.g., MC-306/DOT-406, MC-307/ DOT-407, and MC-312/DOT-412):

(a) Dome cover leak
(b) Irregular-shaped hole
(c) Pressure relief devices (e.g., vents, burst disc)
(d) Puncture
(e) Split or tear
(f) Valves and piping

(2) Describe the methods for containing the following leaks in MC-331 and MC-338 cargo tanks:

The technician with a cargo tank specialty is required by 13.4.1(2) to evaluate tactical options and provide an appropriate method to control the release of contents. The technician with a cargo tank specialty should also describe the appropriate personal protective equipment required for the material involved.

(a) Crack
(b) Failure of pressure relief device (e.g., relief valve, burst disc)
(c) Piping failure

(3)* Demonstrate bonding and grounding procedures for the transfer of flammable and combustible products from cargo tanks, or other products that can give off flammable gases or vapors when heated or contaminated, including the following:

Local procedures for the tasks in 13.4.1(3) should be written out and consistent with NFPA 77, *Recommended Practice on Static Electricity* [4]. The technician with a cargo tank specialty is required to participate as a member of a team assigned to set up bonding and grounding for the transfer of product from a cargo tank. These tasks should be performed using appropriate personal protective equipment for the material involved and for the working conditions.

(a) Selection of equipment
(b) Sequence of bonding and grounding connections
(c) Testing of bonding and grounding connections

A ground resistance test meter and an ohm meter should be used when testing the bonding and grounding system. The ground resistance tester measures the earth's resistance to a ground rod, and the ohm meter measures the resistance of the connections to ensure electrical continuity. One ground rod might not be enough; more might have to be driven and connected to the first to ensure a good ground. Resistance varies with types of soils. The technician with a cargo tank truck specialty should be able to verify (using the appropriate equipment) the following:

- Ground rod resistance to earth
- Connections between tank car and ground rod
- Connection between highway cargo vehicle (damaged) and highway cargo vehicle (recovery) (bonding)

A.13.4.1(3) See A.12.4.1(9).

(4) Given the following product transfer and recovery equipment, demonstrate the safe application and use of each of the following:

The purpose of transferring liquids and gases is often associated with removal from a damaged or potentially damaged containment vessel to one that is undamaged. The equipment varies with the type of vessel involved and the hazardous material being transferred. Transfer procedures are generally established in advance by good industry practice. Only essential personnel—in most cases, those who are outlined in the overall incident management system of the hazardous materials branch—should be involved in the process. A transfer is the movement of the contents of a damaged or overloaded cargo tank into a receiving tank (e.g., a tank car, cargo tank, intermodal tank, or fixed tank).

In highway transportation, transfers can be used when any of the following conditions exist:

- Site conditions prevent uprighting the damaged cargo tank car (e.g., the terrain does not permit use of cranes).
- Damage to leaking valves and fittings cannot be repaired.
- The tank is sound but cannot be safely moved.
- The cargo tank has been damaged to the extent that it cannot be safely uprighted.

Generally, cargo tank transfers should be performed only when the following conditions exist:

- Site safety precautions are in place.
- Delayed container rupture is not likely.
- Required valves are accessible and functioning.
- The tank is in a position that allows for the transfer.
- An appropriate receiving tank is available.
- The transfer equipment is available.
- Experienced personnel are on-site to perform the transfer.
- All ignition sources are removed and/or controlled.
- Appropriate personal protective equipment is used.

(a) Portable pumps (air, electrical, gasoline, diesel)

The product removal method in 13.4.1(4)(a) uses a liquid pump to move the contents of a damaged or overloaded cargo tank into a receiving tank (e.g., a tank car, cargo tank, or portable tank). The material in the damaged or overloaded cargo tank is then pumped into the receiving tank. The use of a liquid pump does not increase the pressure in the damaged cargo tank. However, if another means of creating positive pressure differential is used, a pressure increase could occur.

In evaluating the use of portable and power-take-off (PTO) driven pumps, consideration should be given to the following factors:

- Materials of construction. Although not as critical with hydrocarbon products, the pump and associated hoses, valves, and gaskets must be compatible with the product(s) involved. Product contamination might also be a concern.
- Power rating and pressure capacity of the pump, including lift and flow capabilities.
- Energy source and sparking potential of the pump. Gasoline, diesel, and electrical pumps must be explosionproof and should be used only as a last resort.
- Grounding and bonding requirements.

(b) Pressure transfer

The several options for accomplishing a pressure transfer include the following:

- Compressed air or an inert gas product removal method
- Vapor pressure (with or without flaring) product removal method
- Vapor compressor product removal method
- Vapor compressor and a liquid pump product removal method

The compressed air or an inert gas product removal method uses a compressed gas (e.g., nitrogen or carbon dioxide) to move the contents of a damaged or overloaded cargo tank into a receiving tank (e.g., a tank car, cargo tank, or portable tank). The compressed gas creates a positive pressure differential in the damaged cargo tank that pushes the liquid into the receiving tank. Vapor from the receiving tank might have to be vented to the atmosphere or scrubbed. The use of the compressed gas results in a pressure increase in the damaged cargo tank. This transfer method should be used only when an increase in pressure in the damaged cargo tank is acceptable. The shipper should be consulted to determine the compatibility of the compressed gas to be used.

The vapor pressure (with or without flaring) product removal method uses the material's own vapor pressure to move the contents of a damaged or overloaded cargo tank into a receiving tank (e.g., a tank car, cargo tank, or portable tank). In addition, a vapor flare maintains the necessary positive pressure differential between the damaged or overloaded cargo tank and the receiving tank by burning off vapors in the receiving tank at the outlet of a flare pipe. The pressure in the receiving tank is kept as low as possible.

The vapor compressor product removal method uses a vapor compressor to move the contents of a damaged or overloaded pressurized cargo tank (e.g., MC-331, MC-338) into a receiving tank (e.g., a tank car, cargo tank, or portable tank). The vapor compressor pulls the vapors from the receiving tank, compresses them, and forces them into the damaged cargo tank. The higher pressure in the damaged cargo tank pushes the liquefied gas into the receiving tank. The use of the vapor compressor results in a pressure increase in the damaged cargo tank. The vapor compressor transfer method should be used only when an increase in pressure in the damaged cargo tank is acceptable.

The vapor compressor and a liquid pump product removal method uses a vapor compressor and a liquid pump to move the contents of a damaged or overloaded pressurized cargo tank (e.g., MC-331, MC-338) into a receiving tank (e.g., a tank car, cargo tank, or portable tank). The vapor compressor is used to accelerate the rate of transfer by withdrawing vapors from the receiving tank, compressing them and forcing them into the damaged cargo tank. The higher pressure in the damaged cargo tank pushes the liquefied gas into the pump. The use of the vapor compressor results in a pressure increase in the damaged cargo tank. This transfer method should be used only when an increase in pressure in the damaged cargo tank is acceptable. This method is justified when the receiving tank is a greater distance from the damaged or overloaded cargo tank.

(c) Vacuum trucks

Vacuum trucks are an efficient and expeditious tool for recovering flammable and combustible liquids. However, the vacuum truck is not inherently safe because the air surrounding the vacuum pump exhaust can become saturated with flammable vapors.

To minimize this flammability hazard, the following safety procedures should be used:

- The vacuum truck should be located upwind of any transfer point.
- An exhaust hose of sufficient length should be used to direct the flammable vapors to an area free from hazards and personnel, considering wind direction and velocity, terrain, exposures, and so forth.
- Continuous air monitoring of the area should be conducted.
- The materials being loaded should be confirmed to be compatible with materials previously loaded and to ensure that mixing does not present a hazard.
- Adequate venting must always be provided.

(d) Vehicles with power-takeoff (PTO) driven pumps

See the commentary for 13.4.1(4)(a).

(5) Given a scenario involving an overturned MC-306/DOT-406 cargo tank, demonstrate the safe procedures for the following methods of product removal and transfer:

The technician with a cargo tank specialty is required by 13.4.1(5) to participate as a member of a team assigned to perform product removal and transfer. These tasks should be performed using appropriate personal protective equipment for the material involved and for the working conditions. Additional information on these methods of product removal can be found in *Handling Gasoline Tank Truck Emergencies: Guidelines and Procedures* [3].

(a) Drilling
(b) Internal safety valve
(c) Unloading lines
(d) Vapor recovery lines

(6) Given a scenario involving an overturned MC-307/DOT-407 cargo tank, demonstrate the safe procedures for product removal and transfer.
(7) Given a scenario involving an overturned MC-331 cargo tank, demonstrate the safe procedures for product removal and transfer.

The technician with a cargo tank specialty is required by 13.4.1(7) to participate as a member of a team assigned to perform product removal and transfer. These tasks should be performed using appropriate personal protective equipment for the material involved and for the working conditions.

(8) Given the necessary resources, demonstrate the flaring of an MC-331 flammable gas cargo tank.

The technician with a cargo tank specialty is required by 13.4.1(8) to participate as a member of a team assigned to perform flaring operations. Flaring tasks should be performed using appropriate personal protective equipment for the material involved and for the working conditions. Flaring liquids and gases is a procedure that can be used to deplete the amount of hazardous material involved or to depressurize a vessel. Flaring should be performed only by persons who are trained in the technique. In some cases, where the burning is already under way at a vent or pressure relief valve when responders arrive, a prescribed procedure is to allow the flaring process to continue. Determination must be made that the flaring does not present a source of ignition to other spilled flammable substances.

Flaring is the controlled release and disposal of flammable materials by burning from the outlet of a flare pipe (horizontal or vertical). This method is used to reduce the pressure or dispose of the residual vapors in a damaged or overloaded tank car.

Flaring is used for the following three basic purposes:

1. Reduce the pressure inside a cargo tank
2. Dispose of vapors remaining in a pressurized cargo tank during or after transfer of the liquid
3. Burn off liquid where transfer is impractical

Flaring can also be used to expedite recovery operations or as an interim method until a transfer can begin. Vapor flaring is the burning of the vapors of a liquefied compressed gas at the outlet of a vertical flare pipe as the vapors exit the flare pipe. Liquid flaring is the vaporizing of a liquid product and burning of the vapors at the end of a horizontal flare pipe. A pit is used to contain any product that is not completely burned.

(9) Given a scenario involving a flammable liquid spill from a cargo tank, describe the procedures for site safety and fire control during cleanup and removal operations.

In their book, *Handling Gasoline Tank Truck Emergencies: Guidelines and Procedures*, Hildebrand and Noll state the following:

> Product removal operations cannot commence until after the incident site is stabilized. Again, stabilization means that all fires have been extinguished and ignition sources controlled, all spills have been confined, and the entire spill area has been foamed down and cleaned up as necessary. The incident scene should be continuously monitored. [3]

Specific site safety considerations that should be addressed include the following:

- Only a minimum number of properly trained and protected personnel should be allowed in the immediate hazard area.
- Air monitoring is continuously conducted throughout the incident.
- Backup crews with a minimum of $1\frac{1}{2}$ in. (38 mm) foam handlines and at least two 20 to 30 lb (9 to 14 kg) dry chemical fire extinguishers are in place to protect all personnel involved in the offloading and uprighting operation. Some organizations use a minimum of two foam handlines, placing one foam handline on each side of the cargo tank for maximum protection.
- A foam blanket should be applied, as necessary, throughout the incident. Just because a foam blanket is visible on the spill does not ensure adequate vapor suppression.
- Always have an escape signal and path for personnel working in the immediate hazard area.
- Ensure that all personnel remain alert.

REFERENCES CITED IN COMMENTARY

1. Title 29, Code of Federal Regulations, Part 1910.120, U.S. Government Printing Office, Washington, DC.
2. Title 49, Code of Federal Regulations, Part 178, U.S. Government Printing Office, Washington, DC.
3. Hildebrand, M., and Noll, G. *Handling Gasoline Tank Truck Emergencies: Guidelines and Procedures*, Fire Protection Publications, Stillwater, OK, 1991.
4. NFPA 77, *Recommended Practice on Static Electricity*, National Fire Protection Association, Quincy, MA, 2007.

Competencies for Technicians with an Intermodal Tank Specialty

CHAPTER 14

Intermodal containers (see Exhibit I.14.1) are a type of bulk container that can be placed in or on a transport vessel or vehicle. They are able to move in more than one mode of transportation, for example, by water to highway to rail. Common designs of intermodal containers include freight containers, which are also called box containers, and tank containers. This chapter specifies the competencies required of hazardous materials technicians when responding to incidents involving intermodal tanks.

14.1 General

14.1.1 Introduction.

14.1.1.1 The hazardous materials technician with an intermodal tank specialty shall be that person who provides technical support pertaining to intermodal tanks, provides oversight for product removal and movement of damaged intermodal tanks, and acts as a liaison between the hazardous materials technician and other outside resources.

14.1.1.2 The hazardous materials technician with an intermodal tank specialty shall be trained to meet all competencies at the awareness level (Chapter 4), all core competencies at the operations level (Chapter 5), all competencies at the technician level (Chapter 7), and all competencies of this chapter.

Specialty levels were introduced to the 1997 edition of NFPA 472 and addressed highly skilled operations that did not have any standardized training or certification previously. These levels are not specifically defined by Occupational Safety and Health Administration (OSHA) regulations. They are provided to develop necessary skills for those who are charged with responding to rail emergencies involving intermodal tanks.

Chapter 14 of NFPA 472 requires that competency in performing these tasks be demonstrated. For example, 14.4(6) states the following:

> Given the following product transfer and recovery equipment, demonstrate the safe and correct application and use of the following:
>
> **a.** Portable pumps (air, electrical, gasoline, diesel)
> **b.** Pressure transfer
> **c.** Vacuum trucks
> **d.** Vehicles with power-takeoff driven pumps

14.1.1.3 The hazardous materials technician with an intermodal tank specialty shall receive training to meet governmental occupational health and safety regulations.

14.1.2 Goal.

14.1.2.1 The goal of the competencies at this level shall be to provide the technician with an intermodal tank specialty with the knowledge and skills to perform the tasks in 14.1.2.2 safely.

14.1.2.2 When responding to a hazardous materials/WMD incident, the hazardous materials technician with an intermodal tank specialty shall be able to perform the following tasks:

EXHIBIT I.14.1 *These are two examples of intermodal containers: (top) one is being transported by rail and (bottom) one is being transported by an over-the-road vehicle. (Source: Union Pacific Railroad)*

(1) Analyze a hazardous materials/WMD incident involving an intermodal tank to determine the complexity of the problem and potential outcomes by completing the following tasks:

(a) Determine the type and extent of damage to an intermodal tank.

(b) Predict the likely behavior of an intermodal tank and its contents in an emergency.

(2) Plan a response for an emergency involving an intermodal tank within the capabilities and competencies of available personnel, personal protective equipment, and control equipment by determining the response options (offensive, defensive, and nonintervention) for a hazardous materials emergency involving intermodal tanks.

(3) Implement or oversee the implementation of the planned response to a hazardous materials/WMD incident involving intermodal tanks.

14.1.3 Mandating of Competencies.

This standard shall not mandate that hazardous materials response teams performing offensive operations on intermodal tanks have technicians with an intermodal tank specialty.

14.1.3.1 Hazardous materials technicians operating within the scope of their training as listed in Chapter 7 of this standard shall be able to intervene in intermodal tank incidents.

14.1.3.2 If a hazardous materials response team decides to train some or all its hazardous materials technicians to have in-depth knowledge of intermodal tanks, this chapter shall set out the minimum required competencies.

The committee wanted to make it clear that existing and new responders at the hazardous materials technician level can still perform operations on intermodal tanks for which they have always been qualified. This specialty level covers those highly technical skills and knowledge rarely needed in a major accident, skills that the technician normally does not possess. A hazardous materials response team might have a few members trained to this level, or a request can be made to the railroad, manufacturer, or shipper to provide someone with this specialty at the scene. Because the establishment of this level has been relatively recent, the expectation of seeing responders with either the training or certification necessary to achieve this level is still somewhat remote.

The committee also wanted to clarify that this specialty level is not mandated for any hazardous materials response team or its members, as stated in 14.1.3. Arrangements to request assistance to perform these operations are perfectly appropriate in these situations and are certainly encouraged.

14.2 Competencies — Analyzing the Incident

14.2.1 Determining the Type and Extent of Damage to Intermodal Tanks.

Given examples of damaged intermodal tanks, the hazardous materials technician with an intermodal tank specialty shall describe the type and extent of damage to each intermodal tank and its fittings and shall complete the following tasks:

The intermodal tank is generally built as a cylinder enclosed at the ends by heads. Other tank shapes (rectangular tanks) and configurations (tube modules) are rare. Tanks with multiple compartments are also rare. The intermodal tank's capacity ordinarily does not exceed 6340 gal (24,000 L) but can go as high as 7130 gal (27,000 L).

In NFPA 472, intermodal tanks are grouped into the following four types:

1. Nonpressure intermodal tanks, called intermodal (IM) portable tanks, comprise over 90 percent of the total number of tank containers. The tanks generally transport liquid and solid materials at maximum allowable working pressures (MAWP) of up to 100 psi (689 kPa). Tanks are tested to 1.5 times the MAWP. In the United States, the following two groups of nonpressure intermodal tanks are common:
   - **a.** IM-101 intermodal tanks are built to withstand MAWPs from 25.4 psi to 100 psi (175 kPa to 689 kPa). These tanks can transport both nonhazardous and hazardous materials, including toxic, corrosive, and flammable materials with flash points below 32°F (0°C). Internationally, an IM-101 intermodal tank is called an IMO Type I tank container.
   - **b.** IM-102 intermodal tanks are designed to withstand lower MAWPs from 14.5 psi to 25.4 psi (100 kPa to 175 kPa). These tanks transport materials like whiskey, alcohols,

some corrosives, pesticides, insecticides, resins, industrial solvents, and flammables with flash points between 32°F and 140°F (0°C and 60°C). (More commonly, they transport nonregulated materials like food-grade commodities.) Internationally, a JM-102 intermodal tank is basically equivalent to the IMO Type 2 tank container.

2. Pressure tank containers, known as DOT Specification 51 intermodal tanks, are less common in transport. These containers are designed to handle internal pressures ranging from 100 psi to 500 psi (689 kPa to 3,447 kPa) and generally transport gases liquefied under pressure, such as LP-Gas and anhydrous ammonia. They can also carry liquids such as motor fuel antiknock compound or aluminum alkyls. Internationally, pressure tank containers are called IMO Type S tank containers.
3. Cryogenic tank containers carry refrigerated liquid gases like argon, oxygen, and helium. Internationally, these containers are called IMO Type 7 tank containers.
4. Tube modules transport gases in high-pressure specification 3T cylinders tested to 3000 psi or 5000 psi (20,680 kPa to 34,470 kPa) and are permanently mounted within an ISO frame.

(1) Given the specification mark for an intermodal tank and the reference materials, describe the tank's basic construction and features.

The design, construction, and use of intermodal tanks must conform to the strict standards and regulations of many agencies, both in the United States and abroad. Markings on the intermodal tank indicate the standards to which the tank was built. Intermodal tanks must meet the design, construction, and safety standards set forth by the U.S. Department of Transportation (DOT). Tanks meeting these standards have the specification markings or an exemption number displayed on both sides, generally near the tank's reporting marks (initials) and number. Examples of specification markings found include the following:

- IM 101
- IM 102
- Specification 51

DOT authorizes special permits (previously called exemptions) for packages, containers, and the preparation and offering of hazardous materials for shipment, in accordance with Title 49 CFR Part 173 [1]. In these cases, the outside of the package must be plainly and durably marked in 2 in. (51 mm) letters and numerals with the letters DOT-SP (formerly DOT-E), followed by the number assigned, for example, DOT-SP 8623 (formerly DOT-E 8623).

For interchange purposes in rail transportation, tank containers should conform to the requirements of Section C, Part III, "Specifications for Tank Cars" found in the *Manual of Standards and Recommended Practices* from the Association of American Railroads (AAR) [2]. Tanks meeting these requirements display the AAR 600 marking in 2 in. (51 mm) letters on both sides near the tank's reporting marks (initials) and numbers. The European Agreement Concerning the International Carriage of Dangerous Goods by Road (ADR) and Regulations Concerning the International Carriage of Dangerous Goods by Rail (RID) are used in Europe for the movement of intermodal tanks. ADR/RID markings can remain when the tank is transported in the United States. See Exhibit I.14.2 for an example of specification markings.

(2) Given examples of intermodal tanks (some jacketed and some not jacketed), point out the jacketed intermodal tanks.

The technician with an intermodal tank specialty is required by 14.2.1(2) to be able to identify intermodal tanks with jackets.

(3) Given examples of various intermodal tanks, point out and explain the design and purpose of each of the following intermodal tank components, where present:

(a) Corner casting

EXHIBIT I.14.2 *This is an example of specification markings on a tank. (Source: Union Pacific Railroad)*

Supporting frames for all intermodal containers, including intermodal tanks, are built with standard corner fittings called corner castings. This component is used to secure the tank and to lift it with standard container handling equipment. See Exhibit I.14.3 for an illustration of a corner casting.

(b) Data plate

The corrosion-resistant data plate in 14.2.1(3)(b) provides additional technical, approval, and operational data. The data plate is permanently attached to the tank or frame.

(c) Heater coils (steam and electric)
(d) Insulation

EXHIBIT I.14.3 *A corner casting such as the one illustrated is used to secure the tank. (Source: Union Pacific Railroad)*

Insulation is used to moderate the effects of temperature on the commodity. Mineral wool, fiberglass, polyurethane foam, and polystyrene foam are used as insulating materials. Insulation is usually 3 in. to 4 in. (76 mm to 102 mm) thick.

(e) Jacket

Insulation is always covered with a jacket (called cladding) with flashing to make it weather-tight. This component in 14.2.1(3)(e) is made of metal, at least 1 mm (0.03937 in.) thick, or an equivalent thickness of plastic, reinforced with either glass or fiber.

(f) Refrigeration unit

A refrigeration unit on an intermodal tank, listed in 14.2.1(3)(f), is used to moderate the temperature of the commodity during transportation. Steam or electric heating can also be found on some intermodal tanks, especially those where the commodity must be heated during transportation or unloading.

(g) Supporting frame

The supporting frame of an intermodal tank, as listed in 14.2.1(3)(g), protects the tank and provides for stacking, lifting, and securing the tank. This component also supports the walkways and ladders. The most common supporting frame size for intermodal tanks is 20 ft by 8 or $8\frac{1}{2}$ ft, or 20 ft by 9 or $9\frac{1}{2}$ ft (6.1 m by 2.4 m or 2.6 m, or 6.1 m by 2.9 or 3.1 m) half-height intermodal tanks, which are 35 ft (10.7 m) long. Very few tanks longer than 20 ft (6.1 m) are used in the United States. The exception seems to be cryogenic intermodal tanks typically found in highway transportation. The two basic supporting frames are the box type, which encloses the tank in a cagelike framework, and the beam type, which uses frame structures only at the ends of the tank.

(4) Given examples of various fittings arrangements for pressure, nonpressure, and cryogenic intermodal tanks, point out and explain the design, construction, and operation of each of the following fittings, where present:

(a) Air line connection

An air line connection, located on the tank's top, can be used for pressure unloading, vapor return, and blanketing the contents with an inert gas. The air line connection may be flanged or have a valve attached.

(b) Bottom outlet valve

When the tank container is intended to carry hazardous materials, two externally operated bottom outlet valves are required. Bottom outlet valves, located on the discharge end of the tank, are connected in a series with a replacement gasket between them. A liquid-tight closure on the external bottom outlet valve is also required. This closure may be a blind flange, which is required on international shipments, a screw cap, or a cam-lock cap attached to the external valve.

(c) Gauging device

A dipstick can be inside the manhole. This component is used in conjunction with a calibration chart, known as a strapping chart, to measure the amount of commodity in the tank.

(d) Liquid or vapor valve

Intermodal tanks might have valves for loading and unloading liquids and vapors, depending on the characteristics of the commodity being transported.

(e) Thermometer

Some intermodal tanks have a built-in thermometer to measure the temperature of the commodity inside.

(f) Manhole cover

A manhole is located on top of the tank, at the center, to allow for access into the tank. The component in 14.2.1(4)(f) is secured by six or eight wing nuts on a hinged and bolted lid fitted with a replacement gasket.

(g) Pressure gauge

Some intermodal tanks have one or more pressure gauges to measure the pressure inside the tank.

(h) Sample valve

The sample valve provides a way to check the commodity.

(i) Spill box

On nonpressure intermodal tanks, the top fittings are surrounded by a spill box, which protects the tank's shell from spillage. Spilled material as well as rainwater in the spill box will empty through one or more small open drain pipes. See Exhibit I.14.4 for an illustration of a spill box.

(j) Thermometer well

The thermometer well is a fitting on a pressure intermodal tank where the temperature of the commodity can be determined when a thermometer is placed into the well.

(k) Top outlet

Top-loading valves in 14.2.1(4)(k) are attached to a removable eduction pipe (called a dip leg, dip tube, or siphon tube) running into the tank.

EXHIBIT I.14.4 *A spill box is used to protect the tank from spilled material. (Source: Union Pacific Railroad)*

(5) Given examples of various safety devices for pressure, nonpressure, and cryogenic intermodal tanks, point out and explain the design, construction, and operation of each of the following safety devices, where present:

(a) Emergency remote shutoff device

In an emergency, the internal valve (foot valve) can be shut off from a remote location. Facing the discharge end of the tank, the emergency shutoff device is on the right-hand side near the far end.

(b) Excess flow valve

Excess flow valves can be attached inside the tank between the manway cover plate and the eduction line. These devices almost completely cut off the flow of product when the valve is sheared off in an accident. They operate by either product flow or, when the car is overturned, by gravity. Under normal conditions, gravity keeps the excess flow valves open, but higher flow rates associated with the valve being sheared off or a break in the hose closes them. When the tank car is turned over, gravity closes the valves.

(c) Fusible link/nut assemblies

Fusible links or nuts can be found on cable-actuated remote control shutoff devices. Upon melting, the release of cable tension closes the valve.

(d) Regulator valve

Regulator valves are devices that control the release of vapors from the tank when the contents become heated.

(e) Rupture disc

Safety vents are pressure relief devices that use a frangible disc, also called a rupture disc, to seal the vent opening. The disc in 14.2.1(5)(e) is designed to rupture at a predetermined pressure to relieve internal pressure. Once ruptured, the safety vent does not reclose. Rupture discs are made of metal (lead is no longer authorized by the Association of American Railroads), plastic, rubber, or a combination of metal, plastic, and rubber. Safety vents are used primarily on nonpressure tank cars but cannot be used with flammable liquids and poisons.

(f) Pressure relief valve

Pressure relief valves are devices with the valve held closed by one or more springs. When pressure in the valve exceeds the safety setting, the valve opens. The valve recloses when the tank pressure is reduced below the safety setting. A combination pressure relief valve is generally found in pairs on nonpressure intermodal tanks. The valve protects the tank from both overpressure and vacuum. In many cases, the relief valve has a rupture disc located between the pressure relief valve spring and the commodity to protect the spring. The valve might also have a pressure gauge to determine if the disc is ruptured. See Exhibit I.14.5 for an illustration of a pressure relief valve.

(6) Given the following types of intermodal tank damage, identify the type of damage in each example and explain its significance:

(a) Corrosion (internal and external)

The effects of corrosion can reduce the thickness of the tank, thus weakening the tank.

(b) Crack

A crack is a narrow split or break in the container metal that can penetrate through the metal of the container. Cracks are typically associated with dents.

(c) Dent

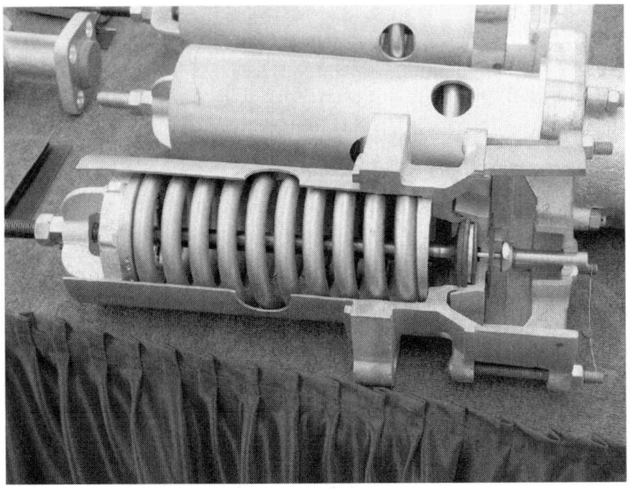

EXHIBIT I.14.5 A pressure relief valve such as this one is used to protect the tank from overpressure. (Source: Union Pacific Railroad)

A dent is a deformation of the container metal. This damage is caused by impact with a relatively blunt object. When a sharp radius of curvature is associated with the dent, the possibility of cracking exists.

(d) Flame impingement

Impingement is flame striking the surface of the tank, either in the liquid space or the vapor space.

(e) Metal loss (gouge and score)

Metal loss, including scores and gouges, is a problem with tank containers. A score is a reduction in the thickness of the container shell. A score is an indentation in the container made by a relatively blunt object. It is characterized by the relocation of the container or weld metal where the metal is pushed aside along the track of contact with the blunt object. Gouges are a reduction in the thickness of the container. This damage is an indentation in the shell made by a sharp, chisel-like object. They are characterized by the cutting and complete removal of the container or weld metal along the track of contact.

(f) Puncture

A puncture is when there is a hole in the tank.

(7) Given three examples of damage to the framework of intermodal tanks, describe the damage in each example and explain its significance in the analysis process.
(8) Given an intermodal tank involved in an emergency, identify the factors to be evaluated as part of the intermodal tank damage assessment process, including the following:

The technician with an intermodal tank specialty is required by 14.2.1(8) to be able to collect the information/data on a given intermodal tank involved in an emergency. Where necessary for safety reasons, this task should be performed using appropriate personal protective equipment for the material involved and for the working conditions.

(a) Amount of product released and amount remaining in the intermodal tank
(b) Container stress applied to the intermodal tank
(c) Nature of the emergency
(d) Number of compartments
(e) Pressurized or nonpressurized

(f) Type and nature of tank damage
(g) Type of intermodal tank
(h) Type of tank metal

(9)* Given a pressurized intermodal tank containing a liquefied gas, determine the amount of liquid in the tank.

The technician with an intermodal tank specialty uses various types of gauging devices to determine the innage (amount of liquid) or outage (amount of vapor space) in an intermodal tank, in accordance with 14.2.1(9).

A.14.2.1(9) Methods for determining the amount of liquid include the use of gauges, shipping papers, the presence of frost line, the use of touch or feel for the colder liquid level, and the use of heat sensors.

(10)*Given examples of damage to a pressurized intermodal tank, determine the extent of damage to the heat-affected zone.

Because of this change in composition in the tank metal, the ductility of the steel in this heat-affected zone is reduced. The heat-affected zone is a likely origin for cracks.

A.14.2.1(10) See A.12.2.1(11)(d).

14.2.2 Predicting the Likely Behavior of the Intermodal Tank and Its Contents.

Technicians with an intermodal tank specialty shall predict the likely behavior of the intermodal tank and its contents and shall complete the following tasks:

The competencies in 14.2.2 build on the competencies of the hazardous materials technician (HMT). The HMT with an intermodal tank specialty is an expert in predicting the likely behavior of the intermodal tank and its contents. In some cases, a private sector specialist employee A may also have these same skills and knowledge but only for a specific intermodal tank and chemical. For example, an employee of a petrochemical enterprise may be the resident expert on intermodal tanks that carry a flammable gas. If the appropriate training or certification has been completed, this person can operate at the same level as the technician with an intermodal tank specialty but only for the intermodal tanks that may be used by the petrochemical company.

(1) Given the following types of intermodal tanks, describe the likely breach/release mechanisms:

The technician with an intermodal tank specialty is required to describe the various types of breach/release mechanisms for the various types of intermodal tanks.

(a) IMO Type 1/IM-101
(b) IMO Type 2/IM-102
(c) IMO Type 5/DOT-51
(d) DOT-56
(e) DOT-57
(f) DOT-60
(g) Cryogenic (IMO Type 7)

(2) Describe the difference in types of construction materials used in intermodal tanks relative to assessing tank damage.

Ninety percent or more of intermodal tanks are built of stainless steel; almost all of the rest are built of mild steel. Stainless steel is used because of its excellent properties in cold temperatures. Aluminum and magnesium alloy tanks are built, but they cannot be used in water transportation. Minimum head and shell thickness is measured in terms of "equivalent thickness in mild steel" after forming. For stainless steel tanks, the minimum thickness is less than

$^{1}/_{8}$ in. (3.2 mm) for nonregulated materials and less than $3^{1}/_{2}$ in. (4.8 mm) for regulated materials. For steel tanks, the minimum thickness is $^{1}/_{4}$ in. (6.35 mm) for nonregulated materials and $^{3}/_{8}$ in. (9.5 mm) for regulated materials. Most tanks are built according to the pressure vessel standards of the American Society of Mechanical Engineers (ASME) with the welds x-rayed. Welds on carbon steel tanks are post-weld stress relieved. The shipper or the intermodal tank manufacturer is an excellent source of technical information on the stressing of the tank.

14.3 Competencies — Planning the Response

In 7.3.1 of NFPA 472, the technician level responder is required to identify the appropriate response objectives for responders at that level. The options available are offensive, defensive, or nonintervention. The response objectives for the technician with an intermodal tank specialty are expanded to include highly technical offensive operations. Response objectives now include those objectives that can lead to final mitigation. The HMT can only stabilize a situation until experts arrive for final and complete mitigation. The competencies outlined in Section 14.3 all carry with them an inherent danger. Because most of these operations are rarely done at an actual incident, the technician with an intermodal tank specialty should give a detailed briefing to the hazardous materials branch safety officer and branch officer as well as to the incident commander before staffing any operations.

Commentary Table I.14.1 describes the skills required for handling intermodal tank problems as outlined in the competencies in Section 14.3.

COMMENTARY TABLE I.14.1 *Intermodal Tank Problems and Response Options*

Problem	*Objectives*	*Actions*	*Methods*
Leaking fittings • Not secure • Wear • Damage	Stop release; forward to destination	Close Tighten Replace part Repair Replace	
	Stop release; forward for further action	Cap Repair after empty Replace after empty	
Overloaded tank	Reduce load; forward to destination	Product removal Fixed facility Field	Transfer
Tank damage or structure damage	Offload; forward for further action	Product removal Field	Transfer Flare Vent
	Reduce internal pressure	Product removal Field	Transfer Flare Vent
		Gain access to product	Hot tap Cold tap

Source: Union Pacific Railroad, *Participant's Manual: Tank Car Safety Course*, March 2007.

14.3.1 Determining the Response Options.

Given the analysis of an emergency involving intermodal tanks, technicians with an intermodal tank specialty shall determine the response options for each intermodal tank involved and shall complete the following tasks:

(1) Describe the purpose of, potential risks associated with, procedures for, equipment required to implement, and safety precautions for the following product removal techniques for intermodal tanks:

The product removal methods in 14.3.1(1) are outside the legitimate responsibility of the local emergency response personnel. However, oversight of their planning and implementation is within the responsibilities of local emergency response agencies. Local procedures for these tasks should be written out and consistent with NFPA 77, *Recommended Practice on Static Electricity* [3].

(a) Flaring liquids and vapors
(b) Hot tapping

Hot tapping is when welding or cutting is done on a container or piping while it contains a flammable liquid or gas. This is an emergency procedure that occurs when circumstances necessitate the work and where the contents cannot be removed prior to performance of the work. The work must be done only by those who have been trained to do such procedures. The hot tap is a technique for providing access to the contents of an intermodal tank when damage to the valves and fittings precludes access to the contents. Once the hot tap is completed, transfer, flaring, or venting can take place. Hot tapping involves the welding of a threaded nozzle into an undamaged section of the tank that is in contact with the liquid. A liquid valve is attached to the nozzle. A hole is then drilled through the tank with a special drilling machine. This hot tapping (drilling) machine is equipped with seals that prevent loss of product during the drilling operation. Liquid hoses or pipe can be attached to the valve outlet.

(c) Transferring liquids and vapors (pressure and pump)

The purpose of transferring liquids and gases is often associated with removal from a damaged or potentially damaged container to one that is undamaged. A transfer is the movement of contents of a damaged or overloaded intermodal tank into a receiving tank (for example, a tank car, a cargo tank, an intermodal tank or fixed car). The equipment varies with the type of container involved and the hazardous material being transferred. Transfer procedures are generally established in advance by good industry practice. Only essential personnel—outlined, in most cases, in the overall incident management system of the hazardous materials branch—should be involved in the process.

(2) Describe the purpose of, procedures for, and risks associated with controlling leaks from various fittings on intermodal tanks, including equipment needed and safety precautions.

The technician with an intermodal tank specialty is required to describe the items in 14.3.1(2) before attempting to make repairs within his or her level of training.

14.4 Competencies — Implementing the Planned Response

Given an analysis of an emergency involving intermodal tanks and the planned response, technicians with an intermodal tank specialty shall implement or oversee the implementation of the selected response options safely and effectively and shall complete the following tasks:

Most of the competencies in Section 14.4 require the actual demonstration of a task. These tasks require an in-depth knowledge of the intermodal tank, its contents, and the specialized equipment needed to complete the task. Because these tasks must be done on actual intermodal tanks, only a few training facilities are able to provide practical training and/or certification. Because some of the tasks in 14.4(1) require specialized resources that are not commonly available to local emergency response personnel, repairs beyond those indicated with each competency should be left to those with the specialized resources and training. Equipment and resources for these tasks are not readily available; therefore, requests for training on how to perform these tasks should be directed to the local shipper, carrier, or contractor personnel in the community, possibly through the local emergency planning committee.

(1) Given leaks from the following fittings on intermodal tanks, control the leaks using approved methods and procedures:

(a) Bottom outlet
(b) Liquid/vapor valve
(c) Manway cover
(d) Pressure relief device
(e) Tank

The technician with an intermodal tank specialty is required to select and apply appropriate procedures and/or various methods of plugging and patching to control or stop leakage of the commodity from the fittings listed in 14.4(1). These tasks should be performed using appropriate personal protective equipment for the commodity involved and for the working conditions.

(2) Demonstrate approved procedures for the following types of emergency product removal:

(a) Gas and liquid transfer (pressure and pump)
(b) Flaring

Flaring can be used to deplete the amount of hazardous material involved or to depressurize a vessel. The flaring of liquids and gases should be performed only by persons who are trained in the technique. In some cases, where the burning is already underway at a vent or pressure relief valve when responders arrive, a prescribed procedure is to allow the flaring process to continue. Determination must be made that the flaring does not present a source of ignition to other spilled flammable substances. Flaring is the controlled release and disposal of flammable materials by burning from the outlet of a flare pipe (horizontal or vertical). This procedure is used to reduce the pressure or dispose of the residual vapors in a damaged or overloaded intermodal tank.

(c) Venting

(3)* Demonstrate bonding and grounding procedures for the transfer of flammable and combustible products from an intermodal tank or other products that can give off flammable gases or vapors when heated or contaminated, including the following:

(a) Selection of equipment
(b) Sequence of bonding and grounding connections
(c) Testing of bonding and grounding connections

The technician with an intermodal tank specialty is required to be able to verify (using the appropriate equipment) the following:

- Ground rod resistance to earth
- Connections between ground rod and intermodal tank
- Connections between ground rod and main ground rod and ground cable

A.14.4(3) See A.12.4.1(9).

(4) Demonstrate the methods for containing the following leaks on liquid intermodal tanks (e.g., IM-101 and IM-102):

(a) Dome cover leak
(b) Irregular-shaped hole
(c) Pressure relief devices (e.g., vents, burst disc)
(d) Puncture
(e) Split or tear
(f) Valves and piping

(5) Describe the methods for containing the following leaks in pressure intermodal tanks:

The technician with an intermodal tank specialty is required to select and apply appropriate procedures and/or various methods of plugging and patching to control or stop leakage of the commodity. These tasks should be performed using appropriate personal protective equipment for the commodity involved and for the working conditions.

(a) Crack
(b) Failure of pressure relief device (e.g., relief valve, burst disc)
(c) Piping failure

(6) Given the following product transfer and recovery equipment, demonstrate the safe and correct application and use of the following:

The technician with an intermodal tank specialty is required to participate as a member of a team assigned to perform product removal. These tasks should be performed using appropriate personal protective equipment for the material involved and for the working conditions.

(a) Portable pumps (air, electrical, gasoline, diesel)
(b) Pressure transfer
(c) Vacuum trucks
(d) Vehicles with power-takeoff driven pumps

(7)* Given a scenario involving an overturned liquid intermodal tank, demonstrate the safe procedures for product removal and transfer.

A.14.4(7) See A.12.4.1(9).

(8)* Given a scenario involving an overturned pressure intermodal tank, demonstrate the safe procedures for product removal and transfer.

The key word in the competency in 14.4(8) is demonstrate. The competency listed in 7.2.4.2 for the hazardous materials technician requires only the ability to identify. The technician with an intermodal tank specialty will become competent in actually performing the operation. The technician with an intermodal tank specialty participates as a member of a team assigned to perform product removal. These tasks should be performed using appropriate personal protective equipment for the material involved and for the working conditions.

A.14.4(8) See A.12.4.1(9).

(9)* Given the necessary resources, demonstrate the flaring of a pressure flammable gas intermodal tank.

A.14.4(9) See A.12.4.1(9).

(10) Given a scenario involving a flammable liquid spill from an intermodal tank, describe the procedures for site safety and fire control during cleanup and removal operations.

REFERENCES CITED IN COMMENTARY

1. Title 49, Code of Federal Regulations, Part 173, U.S. Government Printing Office, Washington, DC.
2. *AAR Manual of Standards and Recommended Practices*, Association of American Railroads, Washington, DC, 2000.
3. NFPA 77, *Recommended Practice on Static Electricity*, National Fire Protection Association, Quincy, MA, 2007.

Additional References

A General Guide to Tank Containers, Hazardous Materials Management, Union Pacific Railroad, Omaha, NE, March 2007.

Title 29, Code of Federal Regulations, Part 1910.120, U.S. Government Printing Office, Washington, DC, July 1, 2001.

Competencies for Technicians with a Marine Tank Vessel Specialty

CHAPTER 15

Competencies for hazardous materials technicians with a marine tank vessel specialty is a new chapter in the 2008 edition of NFPA 472. Marine vessels, because of their structure and configuration, can present special challenges to the hazardous materials technician. The Technical Committee on Hazardous Materials Response Personnel extends its gratitude to the U.S. Coast Guard's Chemical Transportation Advisory Committee (CTAC) for its assistance in the preparation of this chapter and commentary.

15.1* General

A.15.1 Marine tank vessels are used to transport a wide range of different hazardous cargoes in bulk, including oils, chemicals, and liquefied gases. Many marine tank vessels are designed to carry a large number of segregated products simultaneously and can carry significantly greater volumes of cargo than other modes of transport. The operation of marine tank vessels differs from operation of other bulk cargo transportation. On a single voyage, a large number of cargoes with different properties, characteristics, and inherent hazards can be carried. Marine tank vessels are constructed in various types, sizes, and arrangements. Responders to hazardous material spills or releases from marine tank vessels face unique challenges. Marine tank vessels might be located at a dock, pier, or anchorage or might be underway, presenting special logistics issues. Marine tank vessels might be crewed with foreign nationals. Specialized equipment might be needed to properly respond to hazardous material spills and releases from marine tank vessels. In areas where hazardous materials are transported on waterways, responders to hazardous material incidents require a minimum level of specialized competency.

For the purposes of this chapter, a marine tank vessel is defined as a vessel that is constructed or adapted to carry or carries oil or hazardous material in bulk as cargo or cargo residue and operates on international navigable waters or that transfers oil or hazardous material in a port or place subject to international jurisdiction.

The term *tank ship* means a self-propelled tank vessel constructed or adapted primarily to carry oil or hazardous material in bulk in the cargo spaces.

The term *tank barge* means a non-self-propelled tank vessel.

The term *chemical carrier* means a tank ship or tank barge constructed or adapted and used for the carriage in bulk of any hazardous product listed in Chapter 17 of the *International Bulk Chemical Code*.

The term *liquefied gas carrier* means a tank ship or tank barge constructed or adapted and used for the carriage in bulk of any liquefied gas or other product listed in Chapter 19 of the *International Gas Carrier Code*.

Marine tank vessel responders should be familiar with the regulations that affect marine transportation, including, but not limited to, the following:

(1) 33 CFR, "Navigation and Navigatable Waters"
(2) 46 CFR, "Shipping"
(3) *MARPOL 73/78*

(4) *International Maritime Dangerous Goods Code* (IMDG Code)
(5) *Safety of Life at Sea (SOLAS)*
(6) *Code for the Construction and Equipment of Ships Carrying Dangerous Chemicals in Bulk* (BCH Code)
(7) *International Code for the Construction and Equipment of Ships Carrying Dangerous Chemicals in Bulk* (IBC Code)
(8) *International Code for the Construction and Equipment of Ships Carrying Dangerous Liquefied Gases in Bulk* (IGC Code)
(9) Additional maritime industry standards and codes of practice that provide useful information include, but are not limited to, the following:
   (a) *International Safety Guide for Oil Tankers and Terminals*
   (b) *International Chamber of Shipping Tanker Safety Guide* (chemicals)
   (c) *International Chamber of Shipping Tanker Safety Guide* (liquefied gases)
   (d) *OCIMF Ship to Ship Transfer Safety Guide* (petroleum) (liquefied gases)
   (e) *SIGTTO Liquefied Gas Handling Principles on Ships and in Terminals*
(10) Additional response reference material that provide useful information include, but are not limited to, the following:
   (a) *Emergency Response Guidebook*, DOT
   (b) *Chemical Data Guide for Bulk Shipment by Water*, U.S. Coast Guard
   (c) Chemical Hazards Response Information System (CHRIS), U.S. Coast Guard
   (d) Bulk Cargo Finding Aid Web Site, U.S. Coast Guard
   (e) Material Safety Data Sheets
   (f) CAMEO® (Computer-Aided Management of Emergency Operations), EPA and NOAA

15.1.1 Introduction.

15.1.1.1 The hazardous materials technician with a marine tank vessel specialty shall be that person who provides technical support pertaining to marine tank vessels, provides oversight for product removal and movement of damaged marine tank vessels, and acts as a liaison between technicians and outside resources.

For the purposes of this chapter, a marine tank vessel is defined as a vessel that is constructed or adapted to carry, or that carries, oil or hazardous material in bulk as cargo or cargo residue, and operates on the navigable waters of the United States; or transfers oil or hazardous material in a port or place subject to the jurisdiction of the United States.

The term *tank ship* refers to a self-propelled tank vessel constructed or adapted primarily to carry oil or hazardous material in bulk in the cargo spaces.

The term *tank barge* describes a non-self-propelled tank vessel.

The term *chemical tank ship* means a tank ship or tank barge constructed or adapted and used for the carriage in bulk of any hazardous material or hazardous product listed in Chapter 17 of the *International Bulk Chemical Code* [1], or as supplemented by the annual IMO Circular on the Provisional Categorization of Liquid Substances, MEPC.2/Circ.10 or later [2].

A *liquefied gas carrier* is a tank ship or tank barge constructed or adapted and used for the carriage in bulk of any liquefied gas or other product listed in Chapter 19 of the *International Gas Carrier Code* [3].

The transport of dangerous cargo in bulk in tanker ships on the ocean must follow the requirements of the International Maritime Organization (IMO). The ships are NOT required to have hazardous materials placards. The Master of the ship will have a list of dangerous cargoes. They are required to have knowledge of emergency response procedures, which can include Emergency Response to Dangerous Goods on Ships (EmS Guide), Medical First Aid for Dangerous Goods (MFAG), and safety data sheets [4, 5]. If the cargo is a flammable liquid or gas, the required posting is as follows:

NO SMOKING
NO VISITORS
NO OPEN LIGHTS

These ships are required to notify the U.S. Coast Guard prior to entering port. The ship must provide the Coast Guard with information about the commodity being transported.

A *dry bulk carrier*, which is shown in Exhibit I.15.1, is designed to carry dry bulk cargo, such as coal, iron and other ores, grains, fertilizers, animal feeds, scrap metal, or any other dry loose cargo. Dry bulk carriers typically have a flat deck with several waterproof hatches covering the cargo holds. Bulk carrier vessels vary greatly in size, depending on the intended cargo and regions served. This is an example of a non-tank vessel.

A *chemical tanker*, which is shown in Exhibit I.15.2, is designed to transport chemicals in bulk by a series of separate cargo tanks. These tanks are both integral (built into the hull form) and independent cylindrical tanks mounted on the main deck. A large-sized chemical tanker can have up to 30 integral cargo tanks. These cargo tanks are either stainless steel or coated, depending on the properties of the intended cargo. Typical chemical tanker cargoes include organic acids, vegetable oils, alcohols, and petrochemicals.

A *product tanker*, which is shown in Exhibit I.15.3, is designed to carry refined oil products, such as gasoline, diesel oil, and kerosene, in bulk. It is similar to a chemical tanker but has significantly less piping on deck, and the cargo tanks occupy more of the ship below deck. Product tankers typically have a large number of tanks to handle a variety of cargoes simultaneously and are usually double-hulled to help prevent any spillage in case of a collision or grounding.

An *LNG carrier*, which is shown in Exhibits I.15.4(a) and (b), is designed to transport liquefied natural gas (LNG), which is mostly methane that has been cooled below its boiling point to extremely low temperature of –260°F (–162°C) until it forms into liquid. Moss tanks are self-supporting tanks, spherical in shape and protrude above the ship's deck. Each sphere is housed in its own insulated double-hulled enclosure and is supported around its equator by a steel cylinder. Membrane tanks consist of a primary barrier composed of metal "waffles," a layer of insulation, another liquid-proof layer, referred to as the secondary barrier, and the outermost layer is more insulation. These layers are attached to the walls of the cargo hold, which is externally framed. Nitrogen gas is pumped between the primary and secondary barrier, to include the insulation spaces, as an inerting gas.

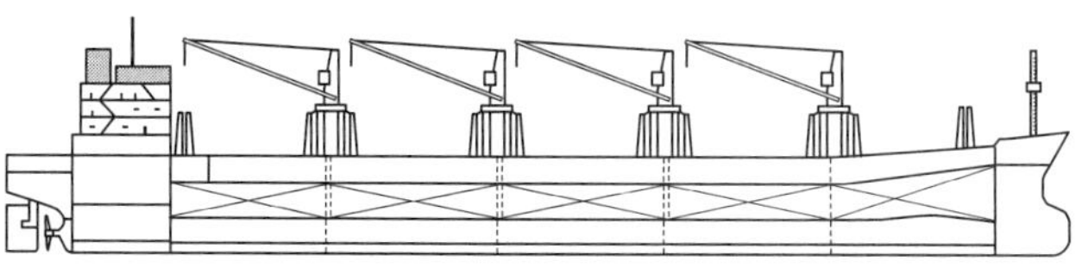

EXHIBIT I.15.1 *This is an example of a typical dry bulk carrier. (Courtesy of U. S. Coast Guard)*

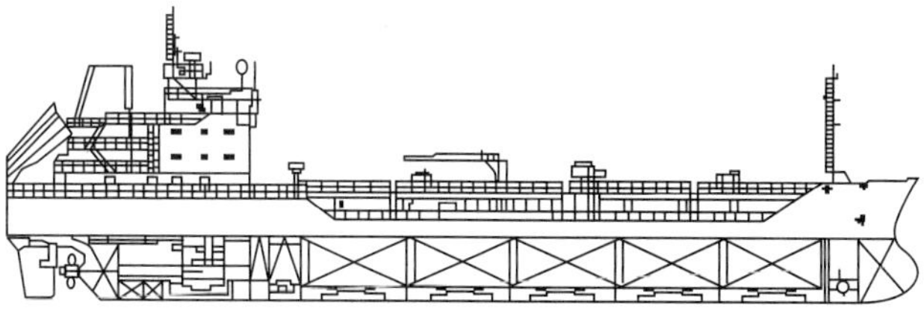

EXHIBIT I.15.2 *A typical chemical tanker transports chemicals in bulk. (Courtesy of U.S. Coast Guard)*

EXHIBIT I.15.3 *A product tanker carries refined oil products. (Courtesy of U. S. Coast Guard)*

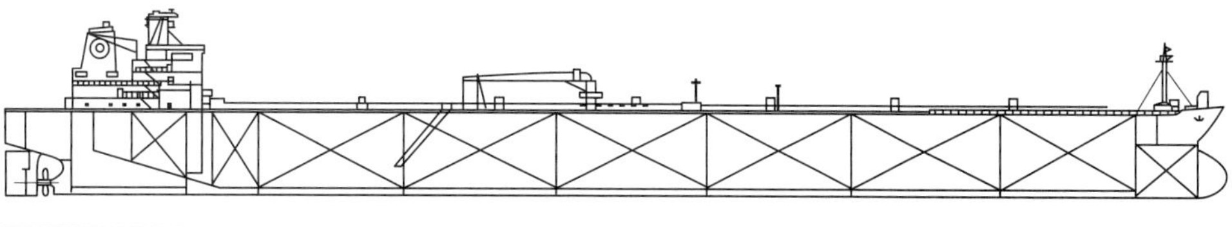

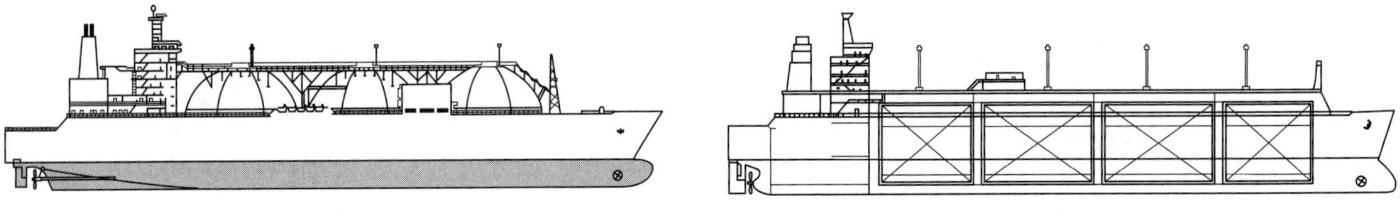

EXHIBIT I.15.4 *These are two types of LNG Carriers: (a) one with moss tanks and (b) one with membrane tanks. (Courtesy of U.S. Coast Guard)*

Materials transported on U.S. waterways are regulated by the Coast Guard. Barges transporting regulated materials are not placarded. Instead, barges transporting regulated cargo are required to have cargo signs facing outward on the port (left) and starboard (right) side of the barge. These signs must be rectangular in shape measuring 3 ft wide by 2 ft high. Lettering must be black on white 3 in. high with a 2 in. border. The following information must be printed on the signs:

WARNING	
DANGEROUS CARGO	(if the cargo is a listed material)
NO VISITORS	
NO SMOKING	(for flammable or combustible liquids)
NO OPEN LIGHTS	(for flammable or combustible liquids)

Additionally, barges with regulated cargo will be equipped with a red flag. The red flag is typically of metal construction and must be visible on all sides of the barge.

A cargo information card will be on the bridge or pilothouse of the power unit and near the cargo sign. This is a laminated card that is 7 in. by $9^1/_2$ in. The card will list the identity of the cargo and its characteristics, emergency procedures, fire-fighting procedures, and surveillance requirements. The cargo information card on the barge can be found in a document holder, which may be a mailbox or watertight tube.

A *barge* is designed mostly for the transport of heavy goods in bulk, whether it is in solid or liquid form. Two examples of a barge are shown in Exhibit I.15.5. They have a flat bottom and most are not self-propelled. For barges without self-propulsion, they can be moved by towboats either pulling them or pushing them by their flush stern. Typical bulk solid barges carry grains, sand, coal, ore, and scrap metal in holds that are either covered or uncovered. A tank barge carries liquid cargo in bulk, such as petroleum or chemical products, and are usually double-hulled to prevent spillage in case of a collision or grounding. Tank barges typically have a small structure on deck to house the cargo transfer pump.

A cargo information card will be on the bridge or pilothouse of the power unit and near the cargo sign. This is a laminated card 7 in. by $9^1/_2$ in. The card will list the identity of the

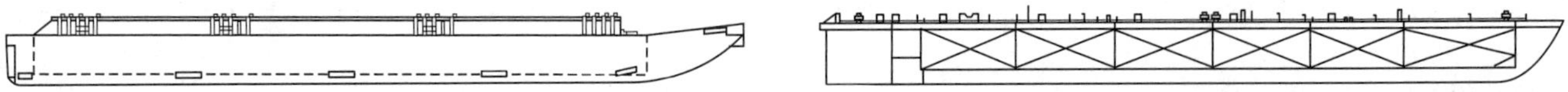

EXHIBIT I.15.5 *These are two tank barges, which typically transport heavy goods in bulk: (a) a single compartment tank barge and (b) a multi-compartment tank barge. (Courtesy of U.S. Coast Guard)*

cargo and its characteristics, emergency procedures, fire-fighting procedures, and surveillance requirements. The cargo information card on the barge can be found in a document holder, which may be a mailbox or watertight tube.

Note that the exhibits shown earlier depict only the most general shapes of marine vessels. Emergency response personnel must be aware that marine vessels vary widely in construction, fittings, purpose, and size. In attempting to determine the product carried by a particular vessel, its shape should be considered as the last resort if product cannot be identified by any other means.

15.1.1.2 The hazardous materials technician with a marine tank vessel specialty shall be trained to meet all competencies at the awareness level (Chapter 4), all core competencies at the operations level (Chapter 5), all competencies at the technician level (Chapter 7), and all competencies of this chapter.

A variety of situations and expertise could be needed in responding to an incident. In most situations, no single person will be able to provide all of the information to adequately respond to an incident. It is important to draw upon the expertise of various personnel who can provide that input and assistance in dealing with the incident. Personnel who may have expertise to support the response could include the following:

1. Vessel person-in-charge, such as a tankerman from a barge or a chief mate, master, or other officer from a tank ship.
2. Barge representative.
3. Towing vessel operator.
4. Facility person-in-charge.
5. Response organization.
6. Qualified individual (QI) and/or responsible party.
7. Marine chemist and/or certified industrial hygienist (CIH).

15.1.1.3 The hazardous materials technician with a marine tank vessel specialty shall receive training to meet governmental occupational health and safety regulations.

Marine tank vessel responders should be familiar with the following publications:

1. Title 33 CFR, U.S. Coast Guard – Navigation [6]
2. Title 46 CFR, U.S. Coast Guard – Shipping [7]
3. International Convention for Prevention of Pollution from Ships (MARPOL) [8]
4. International Convention for Safety of Life at Sea (SOLAS) [9]
5. Code for the Construction and Equipment of Ships Carrying Dangerous Chemicals in Bulk (IBC Code) [10]
6. International Code for the Construction and Equipment of ships carrying dangerous liquid gases in bulk (IGC Code) [11]
7. OSHA HAZWOPER Regulation (Title 29 CFR Part 1910.120) [12]

Additional maritime industry standards and codes of practice can also provide useful information, including the following resources:

1. *International Safety Guide for Oil Tankers and Terminals* (ISGOTT) [13]
2. International Society of Gas Tanker and Terminal Operators (ISGTTO)
3. International Chamber of Shipping Tanker Safety Guide for Chemicals (ICS TSG) [15]
4. "Chemical Hazard Response Information System (CHRIS)," which includes CHRIS+ and www.chrismanual.com [16]

15.1.2 Goal.

15.1.2.1 The goal of the competencies at this level shall be to provide the hazardous materials technician with a marine tank vessel specialty with the knowledge and skills to perform the tasks in 15.1.2.2 safely.

15.1.2.2 In addition to being competent at the technician level, the hazardous materials technician with a marine tank vessel specialty shall be able to perform the following tasks:

(1) Analyze a hazardous materials incident involving a marine tank vessel to determine the magnitude of the problem in terms of outcomes by completing the following tasks:

(a) Determine the type and extent of damage to a marine tank vessel and its cargo systems.

(b)*Predict the likely behavior of a marine tank vessel and its contents in an emergency.

A.15.1.2.2(1)(b) External parameters that could affect the incident, including, but not limited to, weather, currents, and tides, should be monitored.

(c)*Establish initial approved controls.

A.15.1.2.2(1)(c) Examples of appropriate controls in the marine environment include securing the vessel (i.e., anchoring or mooring), stabilizing the vessel, establishing exclusion zones, and precautions for public and personnel safety.

A hazardous materials response team responding to or preparing for an incident with a marine tank vessel should be aware of information that the local Coast Guard officer in charge of marine inspections (OCMI) and other local authorities will require. Before a vessel can continue on the intended route, move to another location, or complete temporary repairs, the OCMI will require specific information and plans in order to determine the feasibility and safety implications of the proposal. Some of the items that might need to be considered during an incident include the following:

1. Identity of the cargo or cargoes involved, identified by the trade name and proper shipping name under the bulk marine classification system—do not use local nickname of cargo, have actual chemical name and CHRIS Code
2. Nearest facility that could take the material and nearest location where the cargo could be trans-shipped (vessel to vessel)
3. Nearest location where the vessel could be repaired, including a shipyard
4. Agencies to be contacted, including the National Response Center and other local agencies
5. Requirement for Coast Guard approval prior to moving the vessel, and possibility that a classification society (for example, the American Bureau of Shipping) might also be required to assess damage and develop a plan for repairs
6. Logistical issues related to transferring cargo and/or moving the vessel
7. Stress and stability issues that could affect the safety of the vessel, including situations that could change those conditions
8. Ability to stop leaks, contain released material, or make temporary repairs that might be necessary before moving the vessel or transferring cargo
9. Development of transfer plans, transit plans, and repair proposals
10. Need for evacuation and ensuring the safety of the vessel crew, responders, and the public

(2) Plan a response for an emergency involving marine tank vessels within the capabilities and competencies of available personnel, personal protective equipment, and control equipment by completing the following tasks:

(a) Determine the response options (offensive, defensive, and nonintervention) for a hazardous materials emergency involving marine tank vessels.

(b) Ensure that the options are within the capabilities and competencies of available personnel, personal protective equipment, and control equipment.

(3) Implement the planned response to a hazardous materials incident involving marine tank vessels.

Other issues to be addressed include the following:

1. Develop a site safety plan, which would include the personal protective equipment (PPE) gear that would be required for each response level and sufficient training for personnel on vessel and maintenance of gear.
2. Identify other agencies/organizations that may be better equipped for responding to a high-level incident (i.e., USCG Strike Team, mutual aid organizations, or response organizations).
3. Conduct plume models (air, water, etc.) utilizing in-house or external modeling resources (i.e., scientific support coordinator).

The technician is a part of the team in responding to the incident, and will work in concert with other members of the response organization.

15.1.3* Mandating of Competencies.

This standard shall not mandate that hazardous materials response teams performing offensive operations on marine tank vessels have hazardous materials technicians with a marine tank vessel specialty.

Personnel responding to incidents with marine tank vessels should be aware of the particular issues and limitations associated with responding to incidents on this type of equipment and/or in the maritime environment. Where responders do not have specific knowledge of these issues and limitations, they should consult with other knowledgeable personnel who can provide that guidance (see 15.1.1.2).

A.15.1.3 Responders need to be trained in the competencies to address only the types of marine tank vessels to which they are expected to respond. For example, if a company ships cargo only by barges, their personnel need to be trained only to the competencies appropriate for barges and need not be trained to meet the competencies on other types of vessels. Competencies for responders are divided into barges and tankships.

15.1.3.1 Hazardous materials technicians operating within the scope of their training as listed in Chapter 7 of this standard shall be able to intervene in marine tank vessel incidents.

15.1.3.2 If a hazardous materials response team decides to train some or all its hazardous materials technicians to have in-depth knowledge of marine tank vessels, this chapter shall set out the minimum required competencies.

Responders only need to be trained in the competencies to address the types of marine tank vessels to which they are expected to respond. For example, if a company only ships cargo by barges, their personnel only need to be trained to the competencies appropriate for barges, and need not be trained to meet the competencies on other types of vessels. Competencies for responders are divided into tank barges and tank ships.

15.2 Competencies — Analyzing the Incident

15.2.1 Determining the Type and Extent of Damage to Marine Tank Vessels.

Given examples of damaged marine tank vessels, hazardous materials technicians with a marine tank vessel specialty shall describe the type and extent of damage to each marine tank vessel and its cargo systems and shall complete the following tasks:

Responders to hazardous material incidents involving marine tank vessels should acquire all available information related to the physical characteristics of the vessel. The OCMI and classification society might require the following information before any temporary repairs can be completed and the vessel can continue on the intended route. This information includes, but is not limited to, the following:

1. Vessel name and official numbers?
2. Is the vessel still aground? Is the vessel on ground or an object?
3. Is there ingress of water, what location?
4. Can you see any damage?
5. Are there other vessels in the area to assist?
6. If a barge, what is the tow configuration?
7. Stability of vessel?

Information regarding a particular vessel can be found in, but not limited to, the following sources on the vessel:

1. General Arrangement Plan
2. Capacity Plan
3. Procedures and Arrangement (P&A) Manual
4. Certificate of Fitness (Foreign flag Vessels) or Certificate of Inspection (U.S. flag vessels)
5. Cargo/Ballast Piping Plan
6. Vessel Response Plans and/or Shipboard Marine Pollution Emergency Plan
7. Fire and Emergency Plan

(1)* Given examples of marine tank vessels, describe a marine tank vessel's basic construction and arrangement features.

A.15.2.1(1) Examples of marine tank vessels include the following:

(1) Barges
 (a) Oil/chemical tank barges
 (b) Liquefied gas barges
(2) Tank ships
 (a) Oil/product ships
 (b) Chemical ships
(3) Product/chemical marine tank ships
(4) Sophisticated parcel chemical ships
(5) Specialized chemical ships
 (a) Liquefied gas ships
(6) Fully pressurized marine tank ships
(7) Semipressurized marine tank ships
(8) Fully refrigerated marine tank ships
 (a) Liquefied natural gas (LNG) ships

Examples of marine vessels include the following:

1. Tank ships
   - **a.** Oil/chemical tank ships
   - **b.** Sophisticated parcel chemical tank ships
   - **c.** Specialized chemical tank ships
2. Liquefied gas tank ships
   - **a.** Fully pressurized tank ships
   - **b.** Semi-pressurized tank ships
   - **c.** Ethylene (LPG and chemical gas) ships

d. Fully refrigerated tank ships
e. Liquefied natural gas (LNG) ships

3. Tank barges

a. Oil/chemical tank barges
b. Liquefied gas barges

4. Cargo vessels (Title 46 CFR Subchapter I Part 90-105) [7]

a. Dry cargo barge
b. Offshore supply vessel (OSV)

(2)* Given examples of various marine tank vessels, point out and explain the design and purpose of each of the various types of marine tank vessel cargo compartment design, structure, and components.

A.15.2.1(2) Types of marine tank vessel cargo compartments include the following:

(1) Barge cargo compartments
  (a) Oil/chemical tank barges
  (b) Liquefied gas barges
(2) Oil/product ship cargo compartments
(3) Chemical ship cargo compartments
  (a) Typical tank construction
  (b) Irregular shaped tank construction
  (c) Tank-within-a-tank construction
  (d) Baffled tank construction
(4) Liquefied gas ship cargo compartments
  (a) Independent type A
  (b) Independent type B
  (c) Independent type C
  (d) Membrane
  (e) Semimembrane
  (f) Internal insulation type 1
  (g) Internal insulation type 2
  (h) Integral
(5) Cargo compartment containment types (for barges and tank ships)
  (a) Coated, lined, uncoated, or cladded
  (b) Stainless steel or carbon steel
  (c) Insulated/thermal protection
(6) Other spaces (for barges and tank ships)
  (a) Cofferdams
  (b) Double bottoms and/or double sides
  (c) Pump rooms
  (d) Other void spaces adjacent to the cargo area

Responders to hazardous materials spills and releases from marine tank vessels should acquire all available information related to the physical characteristics of the vessel. In most cases, responders should work closely and consult with individuals who are experts in the construction of the vessel, its tanks, and other applicable details (the owner, operator, officers, crew, cargo owner, or other individuals as appropriate). Sources of information regarding a particular vessel include, but are not limited to, the following:

(1) General arrangement plan
(2) Procedures and arrangement (P&A) manual
(3) Fire and emergency plan

The following are all vessel container/compartment types:

1. Container cell
2. Cargo hold
   a. General cargo hold
   b. Bulk cargo hold
   c. Barge hopper
   d. RoRo deck
3. Weather decks
   a. Vehicle
   b. Railcar
   c. Container
   d. General
4. Other spaces
   a. Cofferdams
   b. Double bottoms and/or double sides
   c. Pump rooms
   d. Other void spaces adjacent to or within the cargo area
   e. Refrigeration spaces
   f. Ship stores
   g. Fuel tanks
   h. Deep tanks
   i. Pipe tunnel
   j. Duct keel
   k. Ballast tanks

Types of marine tank vessel cargo compartments include the following:

1. Tank barge cargo compartments
   a. Integral gravity tank
   b. Independent gravity tank
   c. Pressure vessels
2. Oil/Product tank ship cargo compartments (integral gravity tank)
3. Chemical tank ship cargo compartments
   a. Independent gravity deck tank
   b. Integral gravity tank
4. Liquefied gas ship cargo compartments
   a. Cylindrical
   b. Spherical
   c. Membrane/semi-membrane
5. Cargo Compartment Containment Types (for barges and tank ships):
   a. Coated, lined, uncoated, or cladded
   b. Stainless steel or carbon steel
   c. Insulation/thermal protection
6. Other spaces (for barges and tank ships)
   a. Cofferdams
   b. Double bottoms and/or double sides
   c. Pump rooms
   d. Other void spaces adjacent to or within the cargo area

(3)* Given examples of various fittings arrangements for marine tank vessels, point out and explain the design, construction, and operation of each.

A.15.2.1(3) Examples of fittings arrangements for tank vessels include the following:

(1) Cargo system valves
  (a) Gate valves
  (b) Globe valves
  (c) Butterfly valves
  (d) Ball valves
  (e) Check valves
  (f) Angle valves
  (g) Pneumatic, hydraulic, or electrically operated valves
(2) Cargo pipeline systems
  (a) Single loop (single line connected to all tanks)
  (b) Branch (multiple lines capable of operating in a segregated or common system of tanks)
  (c) Single tank (dedicated, fully segregated piping system)
(3) Cargo pumps
  (a) Centrifugal
  (b) Positive displacement
  (c) Screw drive
  (d) Deepwell
  (e) Portable emergency/backup pumps
  (f) Stripping systems
  (g) Systems for providing power to the pumps 378 hydraulic, electric, steam, direct diesel
(4) Cargo compartment fittings
  (a) Tank hatch/expansion trunk
  (b) Tank gauging/sampling points
  (c) Vents
  (d) Pressure gauges
  (e) Cleaning ports (Butterworth hatches)
  (f) Spill valves
(5) Emergency shutdown systems
  (a) Manual or automatic/integrated
  (b) Electrical
  (c) Pneumatic
  (d) Remote-actuated/radio
  (e) Thermal
(6) Pressure relief systems
  (a) Safety relief valves
  (b) Pressure relief valves
  (c) Vacuum relief valves
  (d) Regulator valves
  (e) Rupture discs
(7) Cargo temperature control systems
  (a) Steam/water
  (b) Thermal oil
  (c) Liquefaction systems (e.g., glycol)
  (d) Heat exchanger
(8) Cargo cooling (chemical ships) or refrigeration systems (liquefied gas ships)
(9) Cargo compressors (liquefied gas ships)
(10) Cargo vapor handling systems and piping

(11) Inert systems
   (a) Flue gas (tank ships only)
   (b) Inert gas generator (tank ships only)
   (c) Nitrogen generation/bottle supplied systems

(12) Cargo measurement systems
   (a) Open gauging systems
   (b) Closed gauging systems
   (c) Restricted gauging systems
   (d) Automatic gauging and high level alarm systems
   (e) Level indicating devices (slip tubes, sticks, etc.)

(13) Fire-fighting and fire protection equipment (see NFPA 1405, *Guide for Land-Based Fire Fighters Who Respond to Marine Vessel Fires*)

Examples of fittings on marine tank vessels include the following:

1. Valves:
   a. Gate valves
   b. Globe valves
   c. Butterfly valves
   d. Ball valves
   e. Check valves
   f. Angle valves
   g. Pneumatic, hydraulic, or electrically operated valves
   h. Sluice valves
2. Above-deck and below-deck pipeline systems:
   a. Single loop (single line connected to all tanks)
   b. Branch (multiple lines capable of operating in a segregated or common system of tanks)
   c. Single tank (dedicated, fully segregated piping system)
3. Pumps:
   a. Centrifugal
   b. Positive displacement
   c. Screw drive
   d. Deepwell
   e. Portable emergency/back-up pumps
   f. Stripping systems
   g. Pumping power systems (hydraulic, electric, steam, direct diesel)
4. Compartment Fittings
   a. Tank hatch/expansion trunk
   b. Hatch covers
   c. Tank gauging/sampling points/high-level alarms
   d. Vents
   e. Pressure gauges
   f. Cleaning ports (Butterworth hatches)
   g. Drop-line connections
   h. Spill valves
   i. Fixed tank cleaning machines
   j. Pontoons
   k. Doors, elevators, and ramps
   l. Sounding tubes
   m. Sight gauge
5. Emergency shut-down systems

   a. Manual or automatic/integrated
   b. Electrical
   c. Pneumatic
   d. Remote-actuated/radio
   e. Thermal
   f. Hydraulic

6. Pressure relief systems

   a. Safety relief valves
   b. Pressure relief valves
   c. Vacuum relief valves
   d. Regulator valves
   e. Rupture discs

7. Cargo temperature control systems

   a. Steam/water
   b. Thermal oil
   c. Cooling systems (i.e., glycol, ammonia, Freon™)
   d. Heat exchanger
   e. Electrical systems

8. Cargo cooling (chemical ships) or refrigeration systems (liquefied gas ships)
9. Cargo compressors (liquefied gas ships)
10. Cargo vapor handling systems and piping
11. Inert Systems

   a. Flue gas (tank ships only)
   b. Inert gas generator (tank ships only)
   c. Nitrogen generation/bottle-supplied systems
   d. CO_2 systems

12. Measurement and sampling systems

   a. Open gauging systems
   b. Closed gauging systems
   c. Restricted gauging systems
   d. Automatic gauging and high level alarm systems
   e. Level indicating devices (slip tubes, sticks, etc.)
   f. Closed sampling systems

13. Fire-fighting and fire protection equipment (See NFPA 1405, *Guide for Land-Based Fire Fighters Who Respond to Marine Vessel Fires*, for more details [16].)

(4) Given a barge or tank ship, identify and describe the normal methods of cargo transfer.

Examples of normal methods of cargo transfer include the following:

1. Deepwell pump
2. Pressuring cargo off
3. Over-the-top loading
4. Vapor return and/or vapor balancing

   Other examples of transfers include the following:

1. Vessel to shore
2. Vessel to vessel (i.e., barge to ship, barge to barge, ship to ship, etc.)

(5) Given a barge or tank ship, describe the following systems and processes used during cargo transfer:

   (a) Vapor recovery system
   (b) Vapor balancing

(c) Pressuring cargo
(d) Vacuum systems
(e) Padding tanks
(f) Inert gas system (tank ships only)

(6) Given the following types of cargo compartment damage on marine tank vessels, identify the type of damage in each example and explain its significance:

(a) Crack, puncture, slit, or tear
(b) Dent
(c) Flame impingement
(d) Over- or underpressurization
(e) Brittle fracture
(f) Pinhole or corrosion

The following is information that the responders, local agencies, Coast Guard OCMI, and classification society will need for the previously mentioned types of damage:

1. Vessel name and official numbers
2. Location of damage on the vessel
3. Size, vertical/horizontal, location with waterline
4. Whether there is ingress of water and what location
5. Whether any damage can be seen or whether it is located below water/cargo
6. Whether other vessels are in the area to assist
7. If a barge, the tow configuration
8. Stability of vessel

(7) Given examples of the types of emergency situations a marine tank vessel can experience that might result in damage to the vessel or its cargo transfer system, describe the following types of marine tank vessel emergencies and explain their significance related to the vessel's seaworthiness and cargo containment:

(a) Grounding
(b) Stranding

The following is information and items that vessel representatives and responders need to know when passing along information to USCG and/or classification society.

1. Are there other vessels in the area to assist, take you undertow, or push you in?
2. Can the vessel weigh anchor? What is company policy on requesting assistance and preventing the vessel from grounding or alliding/colliding with another vessel or object?

(c) Allision/collision

Information and items that vessel representatives and responders will need to know when passing along information to USCG and/or classification society include the following:

1. Does the space with the damage still have ingress of water, what is the location?
2. Can the vessel representative or responder see any damage? What is a description of the damage?
3. Are there other vessels in the area to assist?
4. If a barge, what is the tow configuration?
5. Stability?

(d) Foundering

Items that vessel representatives and responders need to know when passing along information to local agencies, USCG, and/or classification society include the following:

1. Are there other vessels in the area to assist the vessel in distress, take the vessel in tow, or push you in?

2. Will the company policy allow this?
3. Can you weigh anchor?

(e) Heavy weather damage

The responder should be aware of conditions that could affect the stress and stability of a vessel, which include the following:

1. Wind, waves, tides, and currents
2. Movement of nearby vessels
3. Shifting, adding, or removing weight
4. Reduction of reserve buoyancy
5. Free surface effects in ballast or cargo compartments
6. Free communication effects in a flooded compartment
7. Downflooding

(f) Fire

Items that vessel representatives and responders will need to know when passing along information to local agencies, USCG, and/or classification society include the following:

1. Is the fire out?
2. Does the crew have proper PPE and training to fight the fire?
3. Is the vessel able to maintain power or is it DIW?
4. What is the stability of the vessel after water was used to fight fire? Are voids and cargo spaces filled with water causing a free water effect?

(g) Explosion/boiling liquid expanding vapor explosion (BLEVE)

(8) Given a marine vessel involved in an emergency, identify the factors to be evaluated as part of the marine tank vessel damage assessment process, including the following:

(a) Type of marine tank vessel
(b) Pressurized or nonpressurized cargo system
(c) Number of cargo compartments
(d) Type of cargoes in the damaged cargo system
(e) Type of other cargoes on the marine tank vessel (outside the damaged area)
(f) Cargo compatibility (and incompatibility)
(g) Stability and stresses applied to the marine tank vessel
(h) Type and nature of cargo system damage
(i) Amount of product released and amount remaining in the cargo compartment

The local agencies, OCMI, and classification society will need to know the following:

1. Void spaces on the vessel/barge: Are the spaces dry or is there ingress of water? (What are the sights and sounds to be listening for to determine whether water is coming in?)
2. Cargo tank configuration; locations of cargo tanks and how they are arranged.
3. If there is water in the void/space, at what level is the water or cargo at? When was the last time the tanks were sounded?
4. Appropriate MSDS from actual facility that loaded the cargo named. Too many times variations do not match the actual cargo on board vessel.
5. Temporary repairs/permanent repairs that will need to be made to allow the vessel to continue on transit.
6. Transit plan.

(9) Given a cargo system containing a bulk liquid, determine the amount of liquid in the cargo tank.

15.2.2 Predicting the Likely Behavior of the Marine Tank Vessel and Its Contents.

Hazardous materials technicians with a marine tank vessel specialty shall understand the likely behavior of the marine tank vessel and its contents and shall complete the following tasks:

(1) Given the following types of marine tank vessels, provide examples of probable causes of releases:
   (a) Chemical ships
      i. Product/chemical tank ships
      ii. Sophisticated parcel chemical ships
      iii. Specialized chemical ships
   (b) Liquefied gas ships
      i. Fully pressurized tank ships
      ii. Semipressurized tank ships
      iii. Ethylene (LPG and chemical gas) ships
      iv. Fully refrigerated tank ships
      v. Liquefied natural gas (LNG) ships
   (c) Barges
(2) Describe the significance of lining and cladding on cargo compartments in assessing marine tank vessel damage.
(3) Describe the significance of coated and uncoated cargo compartments in assessing marine tank vessel damage.
(4) Describe the significance of insulation or thermal protection on cargo compartments in assessing marine tank vessel damage.
(5) Describe the significance of heating and refrigeration coils in cargo compartments in assessing marine tank vessel damage.
(6) Given the following examples of damage to the cargo compartments and cargo transfer systems on marine tank vessels, describe their significance in the risk analysis process:
   (a) Cargo spills or releases
   (b) Tank leakage within the vessel
   (c) Vacuum damage (liquefied gases highly soluble in water)
(7) Describe the significance of the following types of internal and external forces on a marine tank vessel's stability in assessing marine tank vessel damage:
   (a) Wind, waves, tides, and currents
   (b) Movement of nearby vessels
   (c) Shifting, adding, or removing weight
   (d) Reduction of reserve buoyancy
   (e) Free surface effects
   (f) Free communication effects in a flooded compartment

Responders and agencies should be aware of the possibility of incompatible cargoes aboard a vessel, along with the free surface effects. These problems can be magnified with reduction in liquid cargo volume and intensity of wave height.

Responders and agencies should also be aware that aboard an LNG/LPG carrier, any hull damage may result in loss of refrigeration of the cargo and rapid boil-off may occur resulting in a highly flammable area.

15.3 Competencies — Planning the Response

15.3.1 Determining the Response Options.

Given the analysis of an emergency involving marine tank vessels, hazardous materials technicians with a marine tank vessel specialty shall determine the response options for each marine tank vessel involved and shall complete the following tasks:

(1) Given an incident involving a marine tank vessel, describe the methods, procedures, risks, safety precautions, and equipment that are required to implement hazardous cargo spill, release, and leak control procedures.
(2) Describe the purpose of, potential risks associated with, procedures for, equipment required to implement, and safety precautions for the following product removal techniques for hazardous liquids and gases:
   (a) Vessel-to-shore transfer
   (b) Vessel-to-vessel transfer
   (c) Vessel-to-tank truck transfer
   (d) Vessel-to-rail car transfer
   (e) Internal transfer within the vessel
   (f) Other types of transfers [e.g., portable liquid storage tanks (frac tanks)]
(3) Describe the purpose of, procedures for, and risks associated with controlling leaks from various fittings on marine tank vessel cargo systems, including equipment needed and safety precautions.

Responding to an incident on a marine tank vessel can provide unique challenges with regards to personnel safety, including access/egress to the vessel, entry into confined spaces, and slipping/tripping hazards.

15.4 Competencies — Implementing the Planned Response

Given an analysis of an emergency involving marine tank vessels and the planned response, hazardous materials technicians with a marine tank vessel specialty shall implement or oversee the implementation of the selected response options safely and effectively and shall complete the following tasks:

(1) Given leaks from the following fittings on marine tank vessels, describe approved methods and procedures for controlling the leaks:
   (a) Tank hatch/expansion trunk
   (b) Liquid/vapor valve or fitting
   (c) Cargo compartment vent
   (d) Pressure relief device
   (e) Cargo system manifold or pipeline
(2) Describe approved procedures for the following types of emergency cargo removal:
   (a) Gas/liquid transfer (pressure/pump)
   (b) Flaring
   (c) Venting
(3)* Describe the importance of bonding and grounding procedures for the transfer of flammable and combustible cargoes from a marine tank vessel or other products that can give off flammable gases or vapors when heated or contaminated.

A.15.4(3) See A.12.4.1(9).

(4) Demonstrate the methods for containing the following leaks on marine tank vessels:
   (a) Puncture
   (b) Irregular-shaped hole
   (c) Split or tear
   (d) Dome cover leak
   (e) Valves and piping failure
   (f) Pressure relief devices (e.g., vents, burst disc)

When attempting to contain leaks on marine tank vessels, ensure that temporary repairs that are installed will be approved by the appropriate agency(s), such as the U.S. Coast Guard or classification society. Depending on the situation, (nonemergency) it is advisable to get temporary repair approval prior to installation. If the USCG or classification society do not approve of repairs, they will have to be removed and reinstalled to their satisfaction.

Also, ensure that permanent repair proposal is approved prior to barge moving, and ensure that the appropriate agency(s) clear the repairs and allow the barge to go back into service.

(5) Given the following product transfer and recovery equipment, describe the safe and correct application and use of the following:

(a) Portable pumps (air, electrical, hydraulic, gasoline, diesel)
(b) Vehicles with power-take-off-driven pumps
(c) Pressure transfer
(d) Vacuum trucks

(6)* Given the necessary resources, describe the flaring of a pressure flammable gas from a liquefied gas tank vessel (ship or barge, as applicable).

A.15.4(6) See A.12.4.1(9).

(7) Given a scenario involving flammable liquid spill from a marine tank vessel, describe the procedures for site safety and fire control during cleanup and removal operations.

REFERENCES CITED IN COMMENTARY

1. International Bulk Chemical Code, International Maritime Organization (IMO), London.
2. IMO Circular on the Provisional Categorization of Liquid Substances, International Maritime Organization (IMO), London.
3. *International Gas Carrier Code*, International Maritime Organization (IMO), London.
4. *Emergency Response Procedures for Ships Carrying Dangerous Goods (EmS Guide)* U.S. Environmental Protection Association, Washington, DC.
5. *Medical First Aid* Guide for Use in Accidents involving *Dangerous Goods (MFAG),* International Maritime Organization (IMO), London.
6. Title 33, Code of Federal Regulations, U.S. Coast Guard–Navigation and Navigable Waters, 2005, U.S. Government Printing Office, Washington, DC.
7. Title 46, Code of Federal Regulations, U.S. Coast Guard–Shipping, U.S. Government Printing Office, Washington, DC, 2006.
8. International Convention for Prevention of Pollution from Ships (MARPOL), International Maritime Organization (IMO), London, 1978.
9. International Convention for Safety of Life at Sea (SOLAS), International Maritime Organization, 1974.
10. International Code for the Construction and Equipment of Ships Carrying Dangerous Chemicals in Bulk (IBC Code), International Maritime Organization, London, 2005.
11. International Code for the Construction & Equipment of Ships Carrying Liquefied Gases in Bulk (IGC Code), International Maritime Organization (IMO), London, 1993.
12. Title 29, Code of Federal Regulations, Part 1910.120, OSHA HAZWOPER Regulation, U.S. Government Printing Office, Washington, DC, 2006.
13. *International Safety Guide for Oil Tankers and Terminals (ISGOTT),* International Chamber of Shipping, London, 2005.
14. *Tanker Safety Guide: Chemicals (TSG)*, International Chamber of Shipping (ICS), London, 2002.
15. "Chemical Hazard Response Information System (CHRIS)," CHRIS+ and www.chrismanual.com, U. S. Department of Transportation and U. S. Coast Guard, 2002.
16. NFPA 1405, *Guide for Land-Based Fire Fighters Who Respond to Marine Vessel Fires*, National Fire Protection Association, 2006.

Explanatory Material

ANNEX A

The material contained in Annex A is included in the text within this handbook and therefore is not repeated here.

Competencies for Responders Assigned Biological Agent–Specific Tasks

ANNEX B

This annex is not a part of the requirements of this NFPA document but is included for informational purposes only.

B.1 General

B.1.1 Introduction.

B.1.1.1 The responder assigned biological agent–specific tasks by the AHJ at hazardous materials/WMD incidents is that person, competent at the operations level, who, at hazardous materials/WMD incidents involving biological materials, is assigned to support the hazardous materials technician and other personnel, provides strategic and tactical recommendations to the on-scene incident commander, serves as a technical advisor to provide technical oversight for operations, and acts as a liaison between the hazardous material technician, response personnel, and outside resources regarding biological issues.

B.1.1.2 The responder assigned biological agent-specific tasks at hazardous materials/WMD incidents should be trained to meet all competencies at the awareness level (Chapter 4), all core competencies at the operations level (Chapter 5), all mission-specific competencies for personal protective equipment (Section 6.2), and all competencies in this annex.

B.1.1.3 The responder assigned biological agent–specific tasks at hazardous materials/WMD incidents should operate under the guidance of a hazardous materials technician, an allied professional, or standard operating procedures.

B.1.1.4 The responder assigned biological agent–specific tasks at hazardous materials/WMD incidents should receive the additional training necessary to meet specific needs of the jurisdiction.

B.1.2 Goal.

B.1.2.1 The goal of this section is to provide the responder assigned biological agent–specific tasks at hazardous materials/WMD incidents with the knowledge and skills to perform the tasks in paragraph B.1.2.2 safely and effectively.

B.1.2.2 When responding to hazardous materials/WMD incidents, the responder assigned biological agent–specific tasks should be able to perform the following tasks:

(1) Analyze an incident involving biological agents threat to determine the credibility and magnitude of the problem by completing the following tasks:
   (a) Understand biological-threat agents, methods of production, and potential harm from biological-threat agents involved in an incident.
   (b) Understand methods of threat agent dissemination, detection, laboratory testing, and surveillance systems.

(2) Plan a response for an incident involving biological threat agents within the capabilities and competencies of available personnel, personal protective equipment, and control equipment by completing the following tasks:

(a) Determine the response options (offensive, defensive, and nonintervention) for an incident involving biological threat agents.
(b) Ensure that the options are within the capabilities and competencies of available personnel, personal protective equipment, and control equipment.

(3) Implement the planned response to a hazardous materials incident involving biological threat agents

B.1.3 Mandating of Competencies.

This standard does not mandate that response organizations perform biological agent–specific tasks.

B.1.3.1 The responders assigned biological agent-specific tasks at hazardous materials/WMD incidents, operating within the scope of their training, should be able to perform their assigned biological agent–specific tasks.

B.1.3.2 If a response organization decides to train some or all its responders to perform biological agent–specific tasks at hazardous materials/WMD incidents, this annex sets out the minimum required competencies.

B.2 Competencies — Analyzing the Incident

B.2.1 The responder assigned biological agent–specific tasks should understand biological threat agents, methods of dissemination, and potential harm from biological threat agents involved in an incident.

B.2.1.1 Given examples of biological threat agents, the responder assigned biological agent–specific tasks should be able to perform the following tasks:

(1) Define the type of biological threat agent.
(2) Provide examples of each group.
(3) Identify potential sources of biological threat agents in industry and business.
(4) Describe potential methods of biological agent production.

B.2.1.2 The responder assigned biological agent–specific tasks should be able to perform the following tasks:

(1) Define the following terms germane to biological agents and biological incidents:
 (a) Infectious
 (b) Contagious
 (c) Pathogen
 (d) Endemic
 (e) Zoonotic
 (f) Morbidity
 (g) Mortality
 (h) Particle size
 (i) Spore
 (j) Infectious dose
 (k) Pandemic
 (l) Incubation period
 (m) Antibiotic
 (n) Prophylaxis
 (o) Syndromic surveillance
 (p) Index case

(2) Given the following types of biological threat agents, define each category and provide examples for each group:

(a) Bacteria
(b) Viruses
(c) Fungi
(d) Toxins

(3) Identify potential sources of microorganisms in the following:

(a) Business
(b) Industry
(c) Academia
(d) Government
(e) Criminal enterprises
(f) Natural reservoirs

(4) Provide examples of components used in biological threat agent production and describe the item and its potential use in agent production.
(5) Provide examples of items found in clandestine biological agent production laboratories that differ from items found in the production of illicit drugs and chemicals.
(6) Given the following types of biological pathogens, identify the potential harm associated with each agent as it relates to potential criminal use:

(a) Variola virus (smallpox)
(b) *Botulinum* toxin
(c) *E. coli*
(d) Ricin toxin
(e) *B. anthracis* (anthrax)
(f) Venezuelan equine encephalitis virus
(g) Rickettsia
(h) Q fever
(i) *Yersinia pestis* (plague)
(j) *Franciscella tularensis* (tularemia, rabbit fever)
(k) Viral hermoraic fever
(l) Any other CDC Category A, B, or C organisms

B.2.2 Identify Methods of Dissemination and Identification of Biological Threat Agents.

B.2.2.1 The responder assigned biological agent–specific tasks should be able to predict likely methods of dissemination of biological threat agents and methods for identification.

B.2.2.2 The responder assigned biological agent–specific tasks should be able to perform the following tasks:

(1) Given examples of the four types of exposure, identify the following potential routes of infection by biological agents:

(a) Inhalation
(b) Absorption
(c) Ingestion
(d) Injection

(2) Given examples of fixed surveillance, detection, or collection systems, define the method of operation, potential location for use, and detection technology utilized in each of the following specific systems:

(a) Particle size detector
(b) Automated biological agent detection system

(c) Dry filter units
(d) Liquid impinger
(e) Slit-to-agar air sampler

(3) Given examples of field detection systems, identify factors to be evaluated as part of the use of these systems, including system validation, capability, limitations, detection levels, operator training, interpretation of results, purity of sample, and destruction of evidence for confirmatory analysis for the following:
   (a) Hand-held assays
   (b) Fourier transform infrared spectroscopy
   (c) Screening tests kits
   (d) Protein assays
   (e) Field microscopy

(4) Explain the United States Laboratory Response Network (LRN) system and describe each of the following components as it relates to the network (for responders outside the United States, the applicable and equivalent laboratory network operating in their country is to be used wherever LRN references are made in this section):
   (a) Access to introduce samples into the laboratories in the network
   (b) Sampling procedures and required sampling equipment
   (c) Procedures for field screening items to be sent to network laboratories
   (d) Packaging requirements for items to be sent to network laboratories

(5) Given the following terms for analysis of biological threat agents, explain the methodology of agent identification:
   (a) Polymerase chain reaction
   (b) Culture tests
   (c) Gram stain
   (d) Morphology
   (e) Motility
   (f) Immunoassays (ELISA, Western blot, Southern blot, surface acoustic wave)
   (g) Time-resolved fluorescence

B.3 Competencies — Planning the Response

B.3.1 Determining the Response Options.

Given an analysis of an incident involving biological threat agents, the responder assigned biological agent–specific tasks should be able to determine the response options for the incident.

B.3.2 The responder assigned biological agent–specific tasks should be able to perform the following tasks:

(1) Given a release of biological agents, describe the considerations for establishing a hot zone for the following scenarios:
   (a) Biological agent release from a dissemination device or air-handling system
   (b) Biological agent release from an envelope or package
   (c) Biological agent spill or container breach of a liquid agent

(2) Describe the factors to be evaluated in selecting personal protective equipment for use at an incident involving biological threat agents.

(3) Given the following scenarios, describe the considerations for selecting personal protective clothing:

(a) Biological agent release from a dissemination device or air-handling system
(b) Biological agent release from an envelope or a package
(c) Biological agent spill or container breach of a liquid agent

(4) Describe the factors to be considered in selecting decontamination procedures for use at an incident involving biological threat agents.

(5) Given the following scenarios, describe the considerations for selecting decontamination procedures:

(a) Equipment exposed to the release of a dry or liquid biological agent
(b) Hard surfaces exposed to the release of a dry or liquid biological agent
(c) Victim exposed to a localized release, (e.g., hands or arms) of a dry or liquid biological agent
(d) Victim exposed to a significant release of a dry or liquid biological agent

(6) Describe the factors to be considered in identification of biological threat agents, including the following:

(a) Field screening and packaging consistent with LRN protocols
(b) Field test limitations, accuracy, and interpretation
(c) Preservation of forensic evidence
(d) Preservation of material for LRN testing
(e) Role of law enforcement agencies
(f) Role of the LRN
(g) Role of public health agencies
(h) Sampling of biological agents

B.4 Competencies — Implementing the Planned Response

B.4.1 Given an analysis involving the release or potential release of a WMD, the responder assigned biological agent–specific tasks should be able to determine the safety and effective response options.

B.4.2 The responder assigned biological agent–specific tasks should be able to perform the following tasks:

(1) Given a simulated incident involving a biological release from a dissemination device or air-handling system, describe the procedures for the following:

(a) Identification of hot zone
(b) Managing exposed victims
(c) Selection of protective clothing
(d) Decontamination
(e) Sampling, field screening, and packaging
(f) Laboratory analysis

(2) Given a simulated incident involving a biological release from an envelope or a package, describe the procedures for the following:

(a) Identification of hot zone
(b) Managing exposed victims
(c) Selection of protective clothing
(d) Decontamination
(e) Sampling, field screening, and packaging
(f) Laboratory analysis

(3) Given a simulated incident involving a biological agent spill or container breach of a liquid agent, describe the procedures for the following:

(a) Identification of hot zone
(b) Managing exposed victims
(c) Selection of protective clothing
(d) Decontamination
(e) Sampling, field screening, and packaging
(f) Laboratory analysis

B.5 Competencies — Evaluating Progress. (Reserved)

B.6 Competencies — Terminating the Incident. (Reserved)

Competencies for Responders Assigned Chemical Agent–Specific Tasks

ANNEX C

This annex is not a part of the requirements of this NFPA document but is included for informational purposes only.

C.1 General

C.1.1 Introduction.

C.1.1.1 The responder assigned chemical agent–specific tasks by the AHJ at hazardous materials/WMD incidents is that person, competent at the operations level, who, at hazardous materials/WMD incidents involving chemical materials, is assigned to support the hazardous materials technician and other personnel, provides strategic and tactical recommendations to the on-scene incident commander, serves as a technical advisor to provide technical oversight for operations, and acts as a liaison between the hazardous material technician, response personnel, and outside resources regarding chemical issues.

C.1.1.2 The responder assigned chemical agent–specific tasks at hazardous materials/WMD incidents should be trained to meet all competencies at the awareness level (Chapter 4), all core competencies at the operations level (Chapter 5), all mission-specific competencies for personal protective equipment (Section 6.2), and all competencies in this annex.

C.1.1.3 The responders assigned chemical agent–specific tasks at hazardous materials/WMD incidents should operate under the guidance of a hazardous materials technician, an allied professional, or standard operating procedures.

C.1.1.4 The responder assigned chemical agent–specific tasks at hazardous materials/WMD incidents should receive the additional training necessary to meet specific needs of the jurisdiction.

C.1.2 Goal.

C.1.2.1 The goal of the competencies in this annex is to provide the responder assigned chemical agent–specific tasks at hazardous materials/WMD incidents with the knowledge and skills to perform the tasks in C.1.2.2 safely and effectively.

C.1.2.2 When responding to hazardous materials/WMD incidents, the responder assigned chemical agent–specific tasks should be able to perform the following tasks:

(1) Analyze a hazardous materials/WMD incident involving potential release of WMD agents and determine the complexity of the problem and potential outcomes by completing the following tasks:
   (a) Determine if the incident is a potential dispersal of a WMD agent and identify the agent within the capabilities of the detection equipment available.
   (b) Identify unique aspects of a potential dispersal of a hazardous material/WMD agent incident.

(2) Within the capabilities and competencies of available personnel, personal protective equipment, and detection and monitoring equipment, plan a response for an incident where there is potential release of WMD agents by completing the following tasks:
   (a) Determine the response options necesssary to conduct detection and monitoring operations.
   (b) Ensure that the options are within the legal authorities, capabilities, and competencies of available personnel, personal protective equipment, and detection equipment.
(3) Implement the planned response to a WMD incident involving potential criminal intent.

C.1.3 Mandating of Competencies.

This standard does not mandate that response organizations perform chemical agent–specific tasks.

C.1.3.1 Responders assigned chemical agent–specific tasks at hazardous materials/WMD incidents, operating within the scope of their training in this annex, should be able to perform their assigned chemical agent–specific tasks.

C.1.3.2 If a response organization decides to train some or all its responders to perform chemical agent–specific tasks at hazardous materials/WMD incidents, this annex sets out the minimum required competencies.

C.2 Competencies — Analyzing the Incident

C.2.1 The responder assigned chemical agent–specific tasks should be able to determine if the incident has the potential for the release of a WMD and the type of detection devices to use based on the signs and symptoms of victims.

C.2.2 Given examples of WMD incidents involving potential release, the responder assigned chemical agent–specific tasks should be able to describe the type of detection devices to use based on the signs and symptoms of victims and chemical and physical properties observed.

C.2.3 The responder assigned chemical agent-specific tasks should be able to perform the following tasks:

(1) Given examples of various types of WMD chemicals, describe the products that might be encountered, chemical and physical properties of those chemicals, and the incident response considerations associated with each.
(2) Given examples of the following potential releases at WMD incidents, describe products potentially encountered and the incident response considerations associated with each situation.
   (a) WMD with no release but product present in container
   (b) WMD with release of visible vapor cloud, liquid pooling, solid dispersion
   (c) WMD with release of visible vapor cloud, liquid pooling, or solid dispersion with suspected victims (patients)
   (d) WMD with suspected victims (patients) but no apparent chemical release

C.2.4 The responder assigned chemical agent-specific tasks should be capable of identifying the unique aspects associated with chemical/WMD releases.

C.2.5 Given an incident involving the release or potential release of a WMD, the responder assigned chemical agent–specific tasks should be able to identify and implement the following tasks:

(1) Secure and isolate the scene
(2) Identify the correct detection device(s)
(3) Deploy the applicable detection device and interpret readings
(4) Notify appropriate explosive ordnance disposal (EOD) personnel if an explosive device has been used to disseminate product

C.3 Competencies — Planning the Response

C.3.1 Given an analysis of an incident involving release or potential release of a WMD, the responder assigned chemical agent–specific tasks should be able to determine possible response options.

C.3.2 The responder assigned chemical agent–specific tasks should be able to perform the following tasks:

(1) Describe the hazards, safety procedures, and tactical guidelines for responding to the following:
   (a) Environmental crime involving a hazardous materials/WMD incident
   (b) Illicit drug manufacturing
   (c) Release of or attack with a WMD agent
   (d) WMD clandestine laboratory
   (e) WMD suspicious package
   (f) WMD threatening communication
(2) Describe the factors to be evaluated in selecting the correct personal protective equipment, detection devices, and decontamination for the following types of incidents:
   (a) Environmental crime involving a hazardous materials/WMD incident
   (b) Illicit drug manufacturing
   (c) Release of or attack with a WMD agent
   (d) WMD clandestine laboratory
   (e) WMD suspicious package
   (f) WMD threatening communication
(3) Describe the detection options for gases, liquids, and solids found at the following types of incidents:
   (a) Environmental crime involving a hazardous materials/WMD incident
   (b) Illicit drug manufacturing
   (c) Release of or attack with a WMD agent
   (d) WMD clandestine laboratory
   (e) WMD suspicious package
   (f) WMD threatening communication
(4) Given examples of releases or potential releases involving a WMD, identify and describe the application, use, and limitations of the types of detection devices that can be utilized, including the following:
   (a) Combustible gas indicators
   (b) Electrochemical cells
   (c) Photoionization detector
   (d) Flame ionization detector
   (e) FT infrared spectrometer
   (f) Alpha, beta, gamma radiation detector
   (g) Colorimetric detection devices
   (h) Mass spectrometer, gas chromatograph
   (i) Any new technology or instrumentation utilized by the AHJ

(5) Describe the potential negative impact associated with detection devices that use destructive technologies.

C.4 Competencies — Implementing the Planned Response

C.4.1 Given an analysis involving the release or potential release of a WMD, the responder assigned chemical agent–specific tasks should determine the safety and effective response options.

C.4.2 The responder assigned chemical agent–specific tasks should be able to perform the following tasks:

(1) Given a simulated WMD incident involving a release or potential release, demonstrate the safe and effective methods for identifying the following:
   (a) Illicit drug manufacturing process
   (b) WMD threatening communication
   (c) WMD suspicious package
   (d) WMD clandestine laboratory
   (e) Release of or attack with a WMD agent
   (f) Environmental crime involving a hazardous material/WMD incident

(2) Given a simulated hazardous material/WMD incident involving release or potential release, demonstrate the methods for selecting the correct personal protective equipment, sampling equipment, detection devices, and decontamination for the following:
   (a) An illicit drug manufacturing process
   (b) A hazardous material/WMD threatening communication
   (c) A hazardous material/WMD suspicious package
   (d) A hazardous material/WMD clandestine laboratory
   (e) Release of or attack with a WMD agent
   (f) An environmental crime involving a hazardous material/WMD incident

(3) Given a simulated WMD incident involving a release or potential release, demonstrate the safe and effective methods for nondestructive detection of WMD products.

(4) Given a simulated WMD incident involving a release or potential release, demonstrate the safe and effective methods for detection of gas, liquid, and solid samples.

(5) Given an example of a WMD incident involving a release or potential release, demonstrate the different detection technologies that can be used with the following:
   (a) An illicit drug manufacturing process
   (b) A WMD threatening communication
   (c) A WMD suspicious package
   (d) A WMD clandestine laboratory
   (e) Release of or attack with a WMD agent
   (f) An environmental crime involving a hazardous material/WMD incident

(6) Given an example of a potential WMD incident, demonstrate the safe and effective methods for decontaminating detection instrumentation.

C.5 Competencies — Evaluating Progress. (Reserved)

C.6 Competencies — Terminating the Incident. (Reserved)

Competencies for Responders Assigned Radiological Agent–Specific Tasks

ANNEX D

This annex is not a part of the requirements of this NFPA document but is included for informational purposes only.

D.1 General

D.1.1 Introduction.

D.1.1.1 The responder assigned radiological agent–specific tasks by the AHJ at hazardous materials/WMD incidents is that person, competent at the operations level, who, at hazardous materials/WMD incidents involving radiological materials, is assigned to support the hazardous materials technician and other personnel, provides strategic and tactical recommendations to the on-scene incident commander, serves as a technical advisor to provide technical oversight for operations, and acts as a liaison between the hazardous material technician, response personnel, and outside resources regarding radiological issues.

D.1.1.2 The responder assigned radiological agent–specific tasks at hazardous materials/WMD incidents should be trained to meet all competencies at the awareness level (Chapter 4), all core competencies at the operations level (Chapter 5), all mission-specific competencies for personal protective equipment (Section 6.2), and all competencies in this annex.

D.1.1.3 The responder assigned radiological agent–specific tasks at hazardous materials/WMD incidents should operate under the guidance of a hazardous materials technician, an allied professional, or standard operating procedures.

D.1.1.4 The responder assigned radiological agent–specific tasks at hazardous materials/WMD incidents should receive additional training necessary to meet specific needs of the jurisdiction.

D.1.2 Goal.

D.1.2.1 The goal of the competencies in this annex is to provide the responder assigned radiological agent–specific tasks at hazardous materials/WMD incidents with the knowledge and skills to perform the tasks in D.1.2.2 safely and effectively.

D.1.2.2 When responding to hazardous materials/WMD incidents, the responder assigned radiological agent–specific tasks should be able to perform the following tasks:

(1) Analyze a hazardous materials/WMD incident involving radioactive material to determine the complexity of the problem and potential outcomes by completing the following tasks:
    (a) Understand types of radiation and potential harm of each type at an incident.
    (b) Predict the direct exposure pathways, including inhalation, ingestion, injection, and absorption.
(2) Plan a response for an emergency involving radioactive material within the capabilities and competencies of available personnel, personal protective equipment, and control

equipment by determining the response options for a hazardous materials/WMD emergency involving radioactive material.

(3) Implement or oversee the implementation of the planned response to a hazardous materials/WMD incident involving radioactive material.

D.1.3 Mandating of Competencies.

This standard does not mandate that response organizations perform radiological agent–specific responsibilities.

D.1.3.1 Responders assigned radiological agent–specific tasks at hazardous materials/WMD incidents, operating within the scope of their training in this chapter, should be able to perform their assigned radiological agent–specific tasks.

D.1.3.2 If a response organization decides to train some or all its responders to perform radiological agent–specific tasks at hazardous materials/WMD incidents, this annex sets out the minimum required competencies.

D.2 Competencies — Analyzing the Incident

D.2.1 Given examples of radiation, the responder assigned radiological agent–specific tasks should be able to define the types of radiation and provide examples of radiation sources, natural, manmade and other potential sources.

D.2.2 The responder assigned radiological agent–specific tasks should be able to the perform the following tasks:

(1) Define the following terms associated with radiological material:
   (a) Ionizing radiation
   (b) Nonionizing radiation
   (c) Radioactivity
   (d) Half-life
   (e) Dose, dose rate
   (f) Units of measure for radiation and radioactivity
   (g) Special nuclear material
   (h) Electromagnetic radiation, pulse
   (i) Radiological dispersion device (RDD)
   (j) Improvised nuclear device (IND)

(2) Identify the following types of radiation:
   (a) Alpha radiation
   (b) Beta radiation
   (c) Gamma radiation, X-ray
   (d) Neutron radiation

(3) Identify the following potential sources of radiation:
   (a) Naturally occurring
   (b) Manmade
   (c) Medical facilities
   (d) Research laboratories
   (e) Nuclear power plant
   (f) Uranium mines
   (g) Fuel processing plant
   (h) Radioactive material/waste shipments
   (i) Department of Defense facilities

(j) Department of Energy facilities
(k) Industrial applications

(4) Given the following types of radiation, identify the potential harm associated with each of the following:

(a) Alpha radiation
(b) Beta radiation
(c) Gamma radiation, X-ray
(d) Neutron radiation

(5) Identify the following terms related to a nuclear detonation from an IND:

(a) Blast and thermal effects
(b) Prompt radiation effects
(c) Fallout and ground shine

D.2.3 The responder assigned radiological agent–specific tasks should be able to identify the potential misuses of radioactive material, including radiological dispersal device, concealed source, improvised nuclear device, and nuclear bomb, and should be able to do the following:

(1) Given examples of the four exposure pathways for radioactive material, identify potential routes of exposure from the following:

(a) Inhalation
(b) Absorption
(c) Ingestion
(d) Injection

(2) Given examples of the classes of radiation detection systems, identify factors to be evaluated as part of the use of these systems, including system validation, capability, limitations, detection levels, operator training, and interpretation of results, for the following:

(a) Personal radiation detectors
(b) Radiation exposure survey meters
(c) Contamination survey meters
(d) Radioisotope identification meters
(e) Portal monitor systems
(f) Dosimetry devices

D.3 Competencies — Planning the Response

D.3.1 Given an analysis of an incident involving radiological material, the responder assigned radiological agent–specific tasks should be able to determine response options for the incident.

D.3.2 The responder assigned radiological agent–specific tasks should be able to perform the following tasks:

(1) Given the concealment of a radioactive material source in a public area, describe the considerations for the following:

(a) Identification of the source
(b) Determination of exposure rate and isolation distance
(c) Estimation of personnel exposure from the source

(2) Given a release of a radiological material, describe the considerations for establishing a hot zone for the following scenarios:

(a) Radioactive material release from a dissemination device or air-handling system
(b) Radioactive material release from an envelope or package

   (c) Radioactive material release or spill of a liquid agent
   (d) Radiological dispersion device (RDD), dirty bomb
   (e) Improvised nuclear device (IND)

(3) Describe the factors to be evaluated in selecting personal protective equipment for use at an incident involving radioactive material.

(4) Given the following scenarios, describe the considerations for selecting personal protective clothing:

   (a) Radioactive material release from a dissemination device or air-handling system
   (b) Radioactive material release from an envelope or package
   (c) Radioactive material release or spill of a liquid agent
   (d) Radiological dispersion device (RDD), dirty bomb
   (e) Improvised nuclear device (IND)

(5) Describe the factors to be considered for selecting decontamination procedures for use at an incident involving radiological materials

(6) Given the following scenarios, describe the considerations for selecting decontamination procedures:

   (a) Victim with localized external contamination (e.g., hands or feet)
   (b) Victim with significant or whole-body external contamination
   (c) Victim with internal contamination
   (d) Hard surfaces (e.g., floors and tables) contaminated with radioactive material
   (e) Porous surfaces or equipment with inaccessible areas contaminated with radioactive material

(7) Describe the factors to be considered in the identification of radioactive material, including the following:

   (a) Sampling techniques for radioactive material
   (b) Field test limitations, accuracy, and interpretation of results
   (c) Field screening and overpacking consistent with local protocols
   (d) Preservation of material for laboratory testing
   (e) Preservation of forensic evidence

(8) Identify the local, state, and federal resources available to assist the operations level responder identify a radioactive material and manage the incident.

D.4 Competencies — Implementing the Planned Response

D.4.1 Given an analysis of an incident involving radioactive material, the responder assigned radiological agent–specific tasks should implement or oversee the implementation of the selected response options safely and effectively.

D.4.2 The responder assigned radiological agent–specific tasks should be able to perform the following tasks:

(1) Given a simulated incident involving the concealment of a radioactive material source in a public area, describe the procedures for the following:

   (a) Locating the source
   (b) Identifying initial isolation zone
   (c) Identifying the source [i.e., isotope(s) involved]
   (d) Determining source exposure rate
   (e) Dose estimation for affected personnel

(2) Given a simulated incident involving a release of radioactive material from a dissemination device or air-handling system, describe the procedures for the following:

   (a) Identification of hot, warm, and cold zones
   (b) Managing exposed and contaminated victims
   (c) Selection of protective clothing
   (d) Decontamination
   (e) Sampling and identification of the material involved
   (f) Field screening and packaging the material involved
   (g) Laboratory analysis of the material involved

(3) Given a simulated incident involving a release of radioactive material from an envelope or a package, describe the procedures for the following:

   (a) Identification of hot, warm, and cold zones
   (b) Managing exposed and contaminated victims
   (c) Selection of protective clothing
   (d) Decontamination
   (e) Sampling and identification of the material involved
   (f) Field screening and packaging the material involved
   (g) Laboratory analysis of the material involved

(4) Given a simulated incident involving a release of radioactive material from a radiological dispersion device or a container breach, describe the procedures for the following:

   (a) Identification of hot, warm, and cold zones
   (b) Managing exposed and/or contaminated victims
   (c) Selection of protective clothing
   (d) Decontamination
   (e) Sampling and identification of the material involved
   (f) Field screening and packaging the material involved
   (g) Laboratory analysis of the material involved

(5) Given a simulated incident involving a release of radioactive material from a spill of a liquid agent, describe the procedures for the following:

   (a) Identification of hot, warm, and cold zones
   (b) Managing exposed and/or contaminated victims
   (c) Selection of protective clothing
   (d) Decontamination
   (e) Sampling and identifying of the material involved
   (f) Field screening and packaging the material involved
   (g) Laboratory analysis of the material involved

(6) Given a simulated incident involving a release of radioactive material from the detonation of an IND, describe the procedures for the following:

   (a) Identification of hot, warm, and cold zones
   (b) Managing exposed and contaminated victims
   (c) Selecting of protective clothing
   (d) Decontamination
   (e) Sampling and identifying of the material involved
   (f) Field screening and packaging the material involved
   (g) Laboratory analysis of the material involved

D.5 Competencies — Evaluating Progress. (Reserved)

D.6 Competencies — Terminating the Incident. (Reserved)

Competencies for Technicians with a Flammable Liquids Bulk Storage Specialty

ANNEX E

This annex is not a part of the requirements of this NFPA document but is included for informational purposes only.

E.1 General

E.1.1 Introduction.

Technicians with a flammable liquids bulk storage specialty should meet all requirements of the awareness, operations, and hazardous materials technician levels and the competencies of this annex. The technician with a flammable liquids bulk storage specialty also should receive additional training to meet applicable United States Environmental Protection Agency (EPA), Occupational Safety and Health Administration (OSHA), and other applicable state, local, or provincial occupational health and safety regulations.

E.1.2 The technician with a flammable liquids bulk storage specialty is that person who, in incidents involving bulk flammable liquid storage tanks, provides support to the hazardous materials technician and other personnel, provides strategic and tactical recommendations to the on-scene incident commander, provides oversight for fire control and product removal operations, and acts as a liaison between technicians, response personnel, and outside resources.

These technicians are expected to use appropriate personal protective clothing and specialized fire, leak, and spill control equipment.

E.1.3 Goal.

The goal of this annex is to provide the technicians with a flammable liquids bulk storage specialty with the knowledge and skills to perform the following tasks safely. In addition to being competent at the technician levels, the technician with a flammable liquids bulk storage specialty should be able to perform the following tasks:

(1) Analyze an incident involving a bulk flammable liquid storage tank to determine the magnitude of the problem by completing the following tasks:
   (a) Determine the type and extent of damage to the bulk liquid storage tank.
   (b) Predict the likely behavior of the bulk liquid storage tank and its contents in an incident.
(2) Plan a response for an incident involving a flammable liquid bulk storage tank within the capabilities and competencies of available personnel, personal protective equipment, and control equipment by completing the following tasks:
   (a) Determine the response options (offensive, defensive, and nonintervention) for a hazardous materials/WMD incident involving flammable liquid bulk storage tanks.
   (b) Ensure that the options are within the capabilities and competencies of available personnel, personal protective equipment, and control equipment.
(3) Implement the planned response to a hazardous materials/WMD incident involving a flammable liquid bulk storage tank.

E.1.4 Mandating of Competencies.

This standard does not mandate that hazardous materials response teams performing offensive operations on flammable liquids bulk storage tanks have technicians with a flammable liquids bulk storage specialty. Technicians operating within the bounds of their training as listed in Chapter 6 of this standard are able to intervene in flammable liquids bulk storage incidents. However, if a hazardous materials response team decides to train some or all its technicians to have in-depth knowledge of flammable liquids bulk storage facilities, this annex sets out the recommended competencies.

E.2 Competencies — Analyzing the Incident

E.2.1 Determining the Type and Extent of Damage to the Bulk Storage Tank.

Given examples of storage tank incidents, technicians with a flammable liquids bulk storage specialty should describe the type of storage tank and the type and extent of damage to the tank and its associated piping and fittings. The technician with a flammable liquids bulk storage specialty should be able to perform the tasks in E.2.1.1 through E.2.1.5.

E.2.1.1 Given examples of various flammable liquid bulk storage operations, the technician should be able to identify and describe the procedures for the normal movement and transfer of product(s) into and out of the facility and storage tanks. Examples should be based on local or regional facilities and could include marketing terminals, pipeline operations and terminals, refineries, and bulk storage facilities.

E.2.1.2 Given examples of the following atmospheric pressure bulk liquid storage tanks, describe each tank's design and construction features and types of products commonly found.

(1) Cone roof tank
(2) Open (external) floating roof tank
(3) Open floating roof tank with a geodesic dome external roof
(4) Covered (internal) floating roof tank

According to NFPA 30, *Flammable and Combustible Liquids Code*, atmospheric tanks are defined as storage tanks operating at pressures from atmospheric through a gauge pressure of 6.9 kPa (1.0 psi). The floating roof on an open floating roof tank can be a pan roof or a pontoon floating roof, while the floating roof on a covered floating roof tank can be constructed of aluminum, steel, or fiberglass.

E.2.1.3 Given examples of the following types of low pressure horizontal and vertical bulk liquid storage tanks, the technician should be able to describe the tank's uses and design and construction features.

(1) Horizontal tank
(2) Dome roof tank

According to NFPA 30, *Flammable and Combustible Liquids Code*, low pressure tanks are defined as storage tanks operating at internal pressure above a gauge pressure of 1.0 psi (6.9 kPa) but not more than 15 psi or 1 bar gauge (103.4 kPa).

E.2.1.4 Given examples of various atmospheric and low pressure bulk liquid storage tanks, describe the design and purpose of each of the following storage tank components, where present:

(1) Tank shell material of construction
(2) Type of roof and material of construction
(3) Primary and secondary roof seals (as applicable)
(4) Incident venting and pressure relief devices

(5) Tank valves
(6) Tank gauging devices
(7) Tank overfill device
(8) Secondary containment methods (as applicable)
(9) Tank piping and piping supports
(10) Fixed or semifixed fire protection system

E.2.1.5 Given three examples of primary and secondary spill confinement measures, describe the design, construction, and incident response considerations associated with each method provided.

E.2.2 Predicting the Likely Behavior of the Bulk Storage Tank and Contents.

Technicians with a flammable liquids bulk storage specialty should predict the likely behavior of the tank and its contents. The technician with a flammable liquids bulk storage specialty should be able to perform the tasks in E.2.2.1 through E.2.2.4.

E.2.2.1 Given examples of different types of flammable liquid bulk storage tank facilities, identify the impact of the following fire and safety features on the behavior of the products during an incident:

(1) Tank spacing
(2) Product spillage and control (impoundment and diking)
(3) Tank venting and flaring systems
(4) Transfer and product movement capabilities
(5) Monitoring and detection systems
(6) Fire protection systems

E.2.2.2 Given a flammable liquid bulk storage tank involved in a fire, identify the factors to be evaluated as part of the analysis process, including the following:

(1) Type of storage tank
(2) Product involved
(3) Amount of product within the storage tank
(4) Nature of the incident (e.g., seal fire, tank overfill, full-surface fire)
(5) Tank spacing and exposures
(6) Fixed or semifixed fire protection systems present

E.2.2.3 Given three types of incidents involving flammable liquid bulk storage tanks, describe the likely fire and spill behavior for each incident. Examples of fire and spill incidents include tank overfills, seal fires on floating roof tanks, floating roof with a sunken internal roof, tank or piping failures, and full-surface fire.

E.2.2.4 Describe the causes, hazards, and methods of handling the following conditions as they relate to fires involving flammable liquid bulk storage tanks:

(1) Frothover
(2) Slopover
(3) Boilover

For additional information, see NFPA 30, *Flammable and Combustible Liquids Code*, and API 2021, *Guide for Fighting Fires in and Around Flammable and Combustible Atmospheric Petroleum Storage Tanks*.

E.3 Competencies — Planning the Response

Given an analysis of an incident involving flammable liquid bulk storage tanks, technicians with a flammable liquids bulk storage specialty should determine response options for the storage tank involved. The technician with a flammable liquids bulk storage specialty should be able to perform the tasks in E.3.1 through E.3.11.

E.3.1 Describe the factors to be considered in evaluating and selecting Class B fire-fighting foam concentrates for use on flammable liquids.

E.3.2 Describe the factors to be considered for the portable application of Class B fire-fighting foam concentrates for the following types of incidents:

(1) Flammable liquid spill (no fire)
(2) Flammable liquid spill (with fire)
(3) Flammable liquid storage tank fire

E.3.3 Given examples of different types of flammable liquid bulk storage tanks, identify and describe the application, use, and limitations of the types of fixed and semifixed fire protection systems that can be used, including the following:

(1) Foam chambers
(2) Catenary systems
(3) Subsurface injection system
(4) Fixed foam monitors
(5) Foam and water sprinkler systems

E.3.4 Describe the hazards, safety procedures, and tactical guidelines for handling an accumulated (in-depth) flammable liquid-spill fire.

E.3.5 Describe the hazards, safety procedures, and tactical guidelines for handling the product and water drainage and runoff problems that can be created at a flammable liquid bulk storage tank fire.

E.3.6 Describe the hazards, safety procedures, and tactical guidelines for handling a flammable liquid bulk storage tank with a sunken floating roof.

E.3.7 Given a flammable liquid bulk storage tank fire, describe the methods and associated safety considerations for extinguishing the following types of fires by using portable application devices:

(1) Pressure vent fire
(2) Seal fire on an open floating roof tank
(3) Seal fire on an internal floating roof tank
(4) Full-surface fire on an internal floating roof tank
(5) Full-surface fire on an external floating roof tank
(6) Dike fire
(7) Pipeline manifold fire

E.3.8 Given the size, dimensions, and products involved for a flammable liquid spill fire, determine the following:

(1) Applicable extinguishing agent
(2) Approved application method (both portable and fixed system applications)
(3) Approved application rate and duration
(4) Required amount of Class B foam concentrate and required amount of water
(5) Volume and rate of application of water for cooling exposed tanks

For additional information, see NFPA 11, *Standard for Low-, Medium-, and High-Expansion Foam.*

E.3.9 Given the size, dimensions, and product involved for a flammable liquid bulk storage tank fire, determine the following:

(1) Applicable extinguishing agent
(2) Approved application method (both portable and fixed system applications)
(3) Approved application rate and duration
(4) Required amount of Class B foam concentrate and required amount of water
(5) Volume and rate of application of water for cooling involved and exposed tanks

For additional information, see NFPA 11, *Standard for Low-, Medium-, and High-Expansion Foam.*

E.3.10 Given the size, dimensions, and product involved for a fire involving a single flammable liquid bulk storage tank and its dike area, determine the following:

(1) Applicable extinguishing agent
(2) Approved application method (both portable and fixed system applications)
(3) Approved application rate and duration
(4) Required amount of Class B foam concentrate and required amount of water
(5) Volume and rate of application of water for cooling involved and exposed tanks

For additional information, see NFPA 11, *Standard for Low-, Medium-, and High-Expansion Foam.*

E.3.11 Given the size, dimensions, and product involved for multiple flammable liquid bulk storage tanks burning within a common dike area, determine the following:

(1) Applicable extinguishing agent
(2) Approved application method (both portable and fixed system applications)
(3) Approved application rate and duration
(4) Amount of Class B foam concentrate and water required
(5) Volume and rate of application of water for cooling involved and exposed tanks

For additional information, see NFPA 11, *Standard for Low-, Medium-, and High-Expansion Foam.*

E.4 Competencies — Implementing the Planned Response

Given an analysis of an incident involving flammable liquid bulk storage tanks, technicians with a flammable liquids bulk storage specialty should implement or oversee the implementation of the selected response options safely and effectively. The technician with a flammable liquids bulk storage specialty should be able to perform the tasks in E.4.1 through E.4.4.

E.4.1 Given a scenario involving a flammable liquid fire, demonstrate the safe and effective methods for extinguishing the following types of fires by using portable application devices:

(1) Valve and flange fires
(2) Pump fire (horizontal or vertical)
(3) Pressure vent fire
(4) Large spill fire
(5) Storage tank fire

E.4.2 Given a scenario involving a three-dimensional flammable liquid fire, demonstrate the safe and effective method for controlling the fire by using portable application devices.

E.4.3 Demonstrate bonding and grounding procedures for the transfer of flammable liquids, including the following:

(1) Selection of equipment
(2) Sequence of bonding and grounding connections
(3) Testing of bonding and grounding connections

E.4.4 Given a scenario involving a flammable liquid spill from a bulk storage tank or pipeline, describe the procedures for site safety and fire control during cleanup and removal operations.

Competencies for the Technician with a Flammable Gases Bulk Storage Specialty

ANNEX F

This annex is not a part of the requirements of this NFPA document but is included for informational purposes only.

F.1 General

F.1.1 Introduction.

Technicians with a flammable gases bulk storage specialty should meet all requirements of the first responder awareness, operations, and hazardous materials technician levels and the competencies of this annex. The technician with a flammable gases bulk storage specialty also should receive additional training to meet applicable United States Environmental Protection Agency (EPA), Occupational Safety and Health Administration (OSHA), and other appropriate state, local, or provincial occupational health and safety regulatory requirements.

F.1.2 Definition.

Technicians with a flammable gases bulk storage specialty are those persons who, in incidents involving flammable gas bulk storage tanks, provide support to the hazardous materials technician and other personnel, provide strategic and tactical recommendations to the on-scene incident commander, provide oversight for fire control and product removal operations, and act as a liaison between technicians, fire-fighting personnel, and other resources. These technicians are expected to use applicable personal protective clothing and specialized fire, leak, and spill control equipment.

F.1.3 Goal.

The goal of this annex is to provide the technicians with a flammable gases bulk storage specialty with the knowledge and skills to perform the following tasks safely:

(1) Analyze an incident involving a flammable gas bulk storage tank to determine the magnitude of the problem by completing the following tasks:
   (a) Determine the type and extent of damage to the bulk storage tank.
   (b) Predict the likely behavior of the bulk storage tank and its contents in an incident.

(2) Plan a response for an incident involving a flammable gas bulk storage tank within the capabilities and competencies of available personnel, personal protective equipment, and control equipment by completing the following tasks:
   (a) Determine the response options (offensive, defensive, and nonintervention) for a hazardous materials/WMD incident involving flammable gas bulk storage tanks.
   (b) Ensure that the options are within the capabilities and competencies of available personnel, personal protective equipment, and control equipment.

(3) Implement the planned response to a hazardous materials/WMD incident involving a flammable gas bulk storage tank.

F.1.4 Mandating of Competencies.

This standard does not mandate that hazardous materials response teams performing offensive operations on flammable gas bulk storage tanks have technicians with a flammable gases bulk storage specialty. Technicians operating within the bounds of their training as listed in Chapter 6 of this standard are able to intervene in flammable gas bulk storage incidents. However, if a hazardous materials response team decides to train some or all its technicians to have in-depth knowledge of flammable gas bulk storage facilities, this annex sets out the recommended competencies.

F.2 Competencies — Analyzing the Incident

F.2.1 Determining the Type and Extent of Damage to the Bulk Storage Tank.

Given examples of storage tank incidents, technicians with a flammable gases bulk storage specialty should describe the type of storage tank and extent of damage to the tank and its associated piping and fittings. The technician with a flammable gases bulk storage specialty should be able to perform the tasks in F.2.1.1 through F.2.1.3.

F.2.1.1 Given examples of various flammable gas bulk storage operations, identify and describe the procedures for the normal movement and transfer of product(s) into and out of the facility storage tanks. Examples should be based on local or regional facilities and could include marketing terminals, pipeline operations and terminals, refineries, bulk storage facilities, and underground storage caverns.

F.2.1.2 Given examples of the following types of high pressure bulk gas storage tanks, describe the tank's uses and design and construction features:

(1) Horizontal (bullet) tank
(2) Spherical tank

Additional information on the design and construction of high pressure bulk gas storage tanks can be referenced from NFPA 58, *Liquefied Petroleum Gas Code*, and API 2510-A, *Fire Protection Considerations for the Design and Operation of Liquefied Petroleum Gas (LPG) Storage Facilities*.

F.2.1.3 Given examples of various high pressure bulk gas storage tanks, point out and explain the design and purpose of each of the following storage tank components and fittings:

(1) Liquid valve and vapor valve
(2) Pressure relief valve
(3) Gauging device
(4) Tank piping and piping supports
(5) Fixed or semifixed fire protection system

F.2.2 Predicting the Likely Behavior of the Bulk Storage Tank and Contents.

Technicians with a flammable gases bulk storage specialty should predict the likely behavior of the tank and its contents. The technician with a flammable gases bulk storage specialty should be able to perform the tasks in F.2.2.1 through F.2.2.3.

F.2.2.1 Given examples of different types of bulk flammable gas storage tank facilities, identify the impact of the following fire and safety features on the behavior of the products during an incident:

(1) Tank spacing
(2) Product spillage and control (impoundment and diking)
(3) Tank venting and flaring systems
(4) Transfer and product movement capabilities
(5) Monitoring and detection systems
(6) Fire protection systems

F.2.2.2 Given examples of different types of flammable gas bulk storage tanks, identify and describe the application, use, and limitations of the types of fixed and semifixed fire protection systems that can be used, including the following:

(1) Water spray systems
(2) Fixed foam monitors
(3) Fixed hydrocarbon monitoring systems

F.2.2.3 Given a flammable gas bulk storage tank and its associated piping, describe the likely breach or release mechanisms and fire scenarios.

F.3 Competencies Planning the Response

Given an analysis of an emergency involving flammable gas storage tanks, technicians with a flammable gases bulk storage specialty should determine response options for the storage tank involved. The technician with a flammable gases bulk storage specialty should be able to perform the tasks in F.3.1 through F.3.6.

F.3.1 Describe the hazards, safety, and tactical considerations required for the following types of flammable gas incidents:

(1) Flammable vapor release (no fire)
(2) Flammable vapor release (with fire)
(3) Liquefied flammable gas release (no fire)
(4) Liquefied flammable gas release (with fire)

F.3.2 Given a flammable gas storage tank with a liquid leak from the pressure relief valve, describe the hazards, safety, and tactical considerations for controlling this type of leak.

F.3.3 Given a flammable gas fire from an elevated structure (e.g., tower or column), describe the hazards, safety, and tactical considerations for controlling this type of release.

F.3.4 Describe the purpose of, potential risks associated with, procedures for, equipment required to implement, and safety precautions for the following product removal techniques:

(1) Transfer of liquids and vapors
(2) Flaring of liquids and vapors
(3) Venting
(4) Hot and cold tapping

F.3.5 Describe the effect flaring or venting of gas or liquid has on the pressure in the tank (flammable gas or flammable liquid product).

F.3.6 Describe the hazards, safety procedures, and tactical guidelines for handling product and water drainage and runoff problems that can be created at a flammable gas bulk storage facility incident.

F.4 Competencies — Implementing the Planned Response

Given an analysis of an emergency involving flammable gas bulk storage tanks, technicians with a flammable gases bulk storage specialty should implement or oversee the implementation of the selected response options safely and effectively. The technician with a flammable gases bulk storage specialty should be able to perform the tasks in F.4.1 through F.4.4.

F.4.1 Given a scenario involving a flammable gas incident, demonstrate the safe and effective methods for controlling the following types of emergencies by using portable application devices:

(1) Unignited vapor release
(2) Valve and/or flange vapor release (no fire)
(3) Valve and/or flange fire
(4) Pump fire (horizontal or vertical)

F.4.2 Given a scenario involving the simultaneous release of both flammable liquids and flammable gases, demonstrate the safe and effective method for controlling the following types of emergencies by using portable application devices:

(1) Unignited vapor release
(2) Flange fire
(3) Pump seal fire

F.4.3 Demonstrate bonding and grounding procedures for the transfer of flammable gases, including the following:

(1) Selection of proper equipment
(2) Sequence of bonding and grounding connections
(3) Proper testing of bonding and grounding connections

F.4.4 Given a scenario involving a flammable gas incident from a bulk storage tank or pipeline, describe the procedures for site safety and fire control during cleanup and removal operations.

Competencies for the Technician with a Radioactive Material Specialty

ANNEX G

This annex is not a part of the requirements of this NFPA document but is included for informational purposes only.

G.1 General

G.1.1 Introduction.

Technicians with a radioactive material specialty should be trained to meet all competencies of the first responder awareness, operations, and hazardous materials technician levels and the competencies of this annex. The technician with a radioactive material specialty also should receive additional training to meet a United States Department of Transportation (DOT), United States Environmental Protection Agency (EPA), Occupational Safety and Health Administration (OSHA), and other applicable state, local, or provincial occupational health and safety regulatory requirements.

G.1.2 Definition.

Technicians with a radioactive material specialty are those persons who provide support to the hazardous materials technician on the use of radiation detection instruments and are expected to have the ability to manage the control of radiation exposure and conduct hazards assessment at an incident involving radioactive materials. These technicians are expected to use specialized protective clothing and survey instrumentation.

G.1.3 Goal.

The goal of this annex is to provide the technician with a radioactive material specialty with the knowledge and skills to perform the following tasks safely:

(1) Analyze a hazardous materials incident involving radioactive materials to determine the complexity of the problem and potential outcomes.
(2) Plan a response for an emergency involving radioactive material within the capabilities and competencies of available personnel, personal protective equipment, and control equipment based on an analysis of the radioactive material incident.
(3) Implement the planned response to a hazardous materials incident involving radioactive material.

G.1.4 Mandating of Competencies.

This standard does not mandate that hazardous materials response teams performing offensive operations on radioactive material incidents have technicians with a radioactive material specialty. Technicians operating within the bounds of their training as listed in this standard are able to intervene in radioactive material incidents. However, if a hazardous materials response team decides to train some or all of its technicians to have an in-depth knowledge and understanding of radioactive material, this annex sets out the required competencies.

G.2 Competencies — Analyzing the Incident

G.2.1 Understanding Nuclear Science and Radioactivity.

Technicians with a radioactive material specialty should have an understanding of nuclear science and radioactivity, including the units and terms used to describe radiation and radioactive material. The technician with a radioactive material specialty should be able to perform the following tasks:

(1) Define the following terms:
   - (a) Ionization
   - (b) Nucleon
   - (c) Nuclide
   - (d) Isotope
   - (e) Excitation
   - (f) Bremsstrahlung
   - (g) Fission
   - (h) Fusion
   - (i) Criticality

(2) Identify the basic principles of the mass-energy equivalence concept.

(3) Identify how the neutron-to-proton ratio is related to nuclear stability.

(4) Define the following terms related to nuclear stability:
   - (a) Radioactivity
   - (b) Radioactive decay

(5) Identify the characteristics of alpha, beta, and gamma radiations.

(6) Given simple equations, the technician with a radioactive material specialty should be able to identify the following radioactive decay modes:
   - (a) Alpha decay
   - (b) Beta decay
   - (c) Positron decay
   - (d) Electron capture

(7) Identify two aspects associated with the decay of a radioactive nuclide.

(8) Identify the differences between natural and artificial radioactivity.

(9) Explain why fission products are unstable.

(10) Given a nuclide, locate its block on the Chart of the Nuclides and identify the following for that nuclide:
   - (a) Atomic number
   - (b) Atomic mass
   - (c) Natural percent abundance
   - (d) Stability
   - (e) Half-life
   - (f) Types and energies of radioactive emissions

(11) Given the Chart of Nuclides, trace the decay of a radioactive nuclide and identify the stable end-product.

(12) Define the following units:
   - (a) Curie
   - (b) Becquerel

(13) Define *specific activity.*

(14) Define *half-life.*

(15) Calculate activity, time of decay, and radiological half-life using the formula for radioactive decay.

(16) Define the following terms:
   (a) Exposure
   (b) Absorbed dose
   (c) Dose equivalent
   (d) Quality factor
(17) Define the following units:
   (a) Roentgen
   (b) Rad, gray
   (c) Rem, sievert
(18) Identify the characteristics of materials best suited to shield from the following types of radiation:
   (a) Alpha
   (b) Beta
   (c) Gamma
   (d) Neutron

G.2.2 Understanding the Biological Effects of Ionizing Radiation.

Technicians with a radioactive material specialty should have an understanding of how ionizing radiation affects the human body. The technician with a radioactive material specialty should be able to perform the following tasks:

(1) Define the law of Bergonie and Tribondeau.
(2) Identify factors that affect the radiosensitivity of cells.
(3) Given a list of types of cells, identify which are the most and which are the least radiosensitive.
(4) Explain primary and secondary reactions on cells produced by ionizing radiation.
(5) Define the following terms and give examples of each:
   (a) Stochastic effect
   (b) Nonstochastic effect
(6) Identify the $LD_{50/30}$ value for humans.
(7) Identify the possible somatic effects of chronic exposure to radiation.
(8) Distinguish among the three types of acute radiation syndrome and identify the exposure levels and symptoms associated with each.
(9) Identify the risks of radiation exposure to the developing embryo and fetus.
(10) Distinguish between the terms *somatic* and *heritable* as they apply to biological effects.

G.2.3 Radiation Detector Theory.

Technicians with a radioactive material specialty should have an understanding of radiation detector theory in order to select the correct type of radiological survey instrument at an incident involving radioactive material. The technician with a radioactive material specialty should be able to perform the following tasks:

(1) Select the function of the detector and readout circuitry components in a radiation measurement system.
(2) Identify the parameters that affect the number of ion pairs collected in a gas-filled detector.
(3) Given a graph of the gas amplification curve, identify the regions of the curve.
(4) Identify the characteristics of a detector operated in each of the useful regions of the gas amplification curve.
(5) Explain the methods employed with gas-filled detectors to discriminate among various types of radiation and various radiation energies.

(6) Explain how a scintillation detector and associated components operate to detect and measure radiation.
(7) Explain how neutron detectors detect neutrons and provide an electrical signal.
(8) Explain the principles of detection and the advantages and disadvantages of a GeLi detector and an HPGe detector.

G.2.4 Radioactive Material Transportation.

Technicians with a radioactive material specialty should have an understanding of how radioactive material is transported and how to identify this material in an accident situation. The technician with a radioactive material specialty should be able to perform the following tasks:

(1) List the applicable agencies that have regulations governing the transport of radioactive material.
(2) Describe methods that can be used to determine the radionuclide contents of a package.
(3) Describe the radiation and contamination surveys that are performed on radioactive material packages and state the applicable limits.
(4) Describe the radiation and contamination surveys that are performed on exclusive-use vehicles and state the applicable limits.
(5) Identify the approved placement of placards on a transport vehicle.

G.3 Competencies — Planning the Response

G.3.1 External Exposure Control.

Given the analysis of an incident involving radioactive material, technicians with a radioactive material specialty should be able to determine the response options needed to minimize external exposure to radioactive material. The technician with a radioactive material specialty should be able to perform the following tasks:

(1) Using the equation Exposure Rate = 6CEN, calculate the gamma exposure rate for specific radionuclides.
(2) Using the stay time equation, calculate an individual's remaining allowable dose equivalent, or stay time.
(3) Identify "distance to radiation sources" techniques for minimizing personnel external exposures.
(4) Using the point source equation (inverse square law), calculate the exposure rate or distance for a point source of radiation.
(5) Using the line source equation, calculate the exposure rate or distance for a line source of radiation.
(6) Define *density thickness* and give its units.
(7) Calculate shielding thickness or exposure rates for gamma and X-ray radiation using the equations.

G.3.2 Internal Exposure Control.

Given the analysis of an incident involving radioactive material, technicians with a radioactive material specialty should determine the response options needed to minimize internal exposure to radioactive material. The technician with a radioactive material specialty should be able to perform the following tasks:

(1) Define and distinguish between the terms *annual limit on intake* (ALI) and *derived air concentration* (DAC).
(2) Identify the basis for determining annual limit on intake (ALI).
(3) Define the term *reference man*.

(4) Identify three factors that govern the behavior of radioactive materials in the body.
(5) Identify the two natural mechanisms that reduce the quantity of a radionuclide in the body.
(6) Identify the relationship of physical, biological, and effective half-lives.
(7) Given the physical and biological half-lives, calculate the effective half-life.
(8) Given a method used by medical personnel to increase the elimination rate of radioactive materials from the body, identify how and why that method works.

G.3.3 Radiation Survey Instrumentation.

Given the analysis of an incident involving radioactive material, technicians with a radioactive material specialty should be able to determine the correct instrument to use for radiation and contamination monitoring. The technician with a radioactive material specialty should be able to perform the following tasks:

(1) List the factors that affect the selection of a portable radiation survey instrument and identify appropriate instruments for external radiation surveys.
(2) Identify the following features of and specifications for ion chamber instruments:
    (a) Detector type
    (b) Instrument operating range
    (c) Detector shielding
    (d) Detector window
    (e) Types of radiation detected and measured
    (f) Operator-adjustable controls
    (g) Markings for detector effective center
    (h) Specific limitations and characteristics
(3) List the factors that affect the selection of a portable contamination monitoring instrument.
(4) Describe the following features of and specifications for commonly used count rate meter probes used for beta/gamma and alpha surveys:
    (a) Detector type
    (b) Detector shielding and window
    (c) Types of radiation detected and measured
    (d) Energy response for measured radiation
    (e) Specific limitations and characteristics.
(5) Describe the following features of and specifications for commonly used count rate instruments.
    (a) Types of detectors available
    (b) Operator-adjustable controls
    (c) Specific limitations and characteristics

G.4 Competencies — Implementing the Planned Response

G.4.1 Radiological Incidents.

Given an analysis of an incident involving radioactive material and the planned response, technicians with a radioactive material specialty should implement or oversee the response to a given radiological emergency. The technician with a radioactive material specialty should be able to perform the following tasks:

(1) Describe the general response and responsibilities of a specialist during any radiological incident.

(2) Identify the emergency equipment and facilities that are available, including the location and contents of emergency equipment kits.
(3) Describe the specialist's response to personnel contamination.
(4) Describe the specialist's response to off-scale or lost dosimetry.
(5) Describe the specialist's response to rapidly increasing, unanticipated radiation levels in an area.
(6) Describe the specialist's response to a dry or liquid radioactive material spill.
(7) Describe the specialist's response to a fire in a radiological area or involving radioactive materials.
(8) Describe specific procedures for documenting radiological incidents.
(9) Identify the available federal responder resources and explain the assistance that each group can provide.

G.4.2 Contamination Control.

Given an analysis of an incident involving radioactive material and the planned response, technicians with a radioactive material specialty should be able to implement or oversee contamination control techniques to minimize the spread of radiological contamination. The technician with a radioactive material specialty should be able to perform the following tasks:

(1) Define the terms *removable* and *fixed surface contamination*, state the difference between them, and list the common methods used to measure each.
(2) State the basic principles of contamination control and list examples of implementation methods.
(3) State the purpose of using protective clothing in radiologically contaminated areas.
(4) List the basic factors that determine protective clothing requirements for personnel protection.

G.4.3 Personnel Decontamination.

Given an analysis of an incident involving radioactive material and the planned response, technicians with a radioactive material specialty should be able to implement or oversee decontamination techniques for equipment and personnel. The technician with a radioactive material specialty should be able to perform the following tasks:

(1) Describe how personnel, personal protective equipment, apparatus, and tools become contaminated with radioactive material.
(2) State the purpose of radioactive material decontamination.
(3) Describe field decontamination techniques for equipment.
(4) List the three factors that determine the actions taken in decontamination of personnel.
(5) Identify methods and techniques for performing personnel decontamination.

Overview of Responder Levels and Tasks at Hazardous Materials/WMD Incidents

ANNEX H

This annex is not a part of the requirements of this NFPA document but is included for informational purposes only.

H.1 Responder Levels

H.1.1 Awareness Level.

Awareness level personnel are those persons who, in the course of their normal duties, can be the first on the scene of an emergency involving hazardous materials. Awareness level personnel are expected to recognize the presence of hazardous materials/WMD, protect themselves, call for trained personnel, and secure the area.

H.1.2 Operations Level.

Operations level responders are those persons who respond to hazardous materials/WMD incidents for the purpose of protecting nearby persons, the environment, or property from the effects of the release. They should be trained to respond in a defensive fashion to control the release from a safe distance and keep it from spreading.

Operations level responders can have additional competencies that are specific to their response mission, expected tasks, and equipment and training as determined by the AHJ.

H.1.3 Technician Level.

Hazardous materials technicians are those persons who respond to releases or potential releases of hazardous materials for the purpose of controlling the release. Hazardous materials technicians are expected to use specialized chemical protective clothing and specialized control equipment.

Hazardous materials technicians respond to hazardous materials/WMD incidents using a risk-based response process *[see 7.1.2.2(1)]* with the ability to analyze a problem involving hazardous materials/WMD, select appropriate decontamination procedures, and control a release using specialized protective clothing and control equipment. Hazardous materials technicians can have additional competencies that are specific to their response mission, expected tasks, and equipment and training as determined by the AHJ.

H.1.4 Command Level.

The incident commander is that person who is responsible for all decisions relating to the management of the incident. The incident commander is in charge of the incident site.

H.2 Responder Tasks

H.2.1 Analysis Tasks.

The list of analysis tasks by responder level is as follows:

(1) *Awareness Level.* Awareness level personnel analyze an incident to determine both the hazardous materials/WMD present and the basic hazard and response information for each hazardous materials/WMD by completing the following tasks:
  (a) Detect the presence of hazardous materials/WMD.
  (b) Survey a hazardous materials/WMD incident from a safe location to identify the name, UN/NA identification number, or type placard applied for any hazardous materials/WMD involved.
  (c) Collect hazard and response information from the current edition of the DOT *Emergency Response Guidebook.*
(2) *Operations Level.* Operations level responders must be competent at the awareness level and be able to analyze a hazardous materials/WMD incident to determine the scope of the problem and potential outcomes by completing the following tasks:
  (a) Survey the hazardous materials/WMD incident to identify the containers and materials involved, determine whether hazardous materials/WMD have been released, and evaluate the surrounding conditions.
  (b) Collect hazard and response information from material safety data sheets (MSDS), CHEMTREC/CANUTEC/ SETIQ, and shipper and manufacturer contacts.
  (c) Predict the likely behavior of a hazardous materials/WMD agent as well as its container.
  (d) Estimate the potential harm at a hazardous materials/WMD incident.
(3) *Technician Level.* Hazardous materials technicians must be competent at the awareness and operations levels and be able to analyze a hazardous materials/WMD incident to determine the complexity of the problem and potential outcomes by completing the following tasks:
  (a) Survey the hazardous materials/WMD incident to identify special containers involved, identify or classify unknown materials, and verify the presence and concentrations of hazardous materials/WMD through the use of monitoring equipment.
  (b) Collect and interpret hazard and response information from printed and technical resources, computer databases, and monitoring equipment.
  (c) Determine the type and extent of damage to containers.
  (d) Where multiple materials are involved, predict the likely behavior of released materials and their containers.
  (e) Estimate the size of an endangered area using computer modeling, monitoring equipment, or specialists in this field.
(4) *Command Level.* The incident commander analyzes a hazardous materials/WMD incident to determine the complexity of the problem and potential outcomes by completing the following tasks:
  (a) Collect and interpret hazard and response information from printed and technical resources, computer databases, and monitoring equipment.
  (b) Estimate the potential outcomes within the endangered area at a hazardous materials/WMD incident.

H.2.2 Planning Tasks.

The list of planning tasks by responder level is as follows:

(1) *Awareness Level.* No requirements.
(2) *Operations Level.* The operations level responder must be competent at the first responder awareness level and be able to plan an initial response within the capabilities and

competencies of available personnel, personal protective equipment, and control equipment by completing the following tasks:

(a) Describe the response objectives for hazardous materials/WMD incidents.
(b) Describe the defensive options available by response objective.
(c) Determine whether the personal protective equipment provided is appropriate for implementing each action option.
(d) Identify the emergency decontamination process.

(3) *Technician Level.* The hazardous materials technician must be competent at both the first responder awareness and operations levels and be able to plan a response within the capabilities of available personnel, personal protective equipment, and control equipment by completing the following tasks:

(a) Identify the response objectives for hazardous materials/WMD incidents.
(b) Identify the potential response options available by response objective.
(c) Select the personal protective equipment required for a given action option.
(d) Select the applicable technical decontamination process.
(e) Develop an incident action plan, including site safety and control plan, consistent with the emergency response plan and/or standard operating procedures and within the capability of the available personnel, personal protective equipment, and control equipment.

(4) *Command Level.* The incident commander plans response operations within the capabilities and competencies of available personnel, personal protective equipment, and control equipment by completing the following tasks:

(a) Identify the response objectives for hazardous materials/WMD incidents.
(b) Identify the potential response options (defensive, offensive, and nonintervention) available by response objective.
(c) Approve the level of personal protective equipment required for a given action option.
(d) Develop an incident action plan, including site safety and control plan, consistent with the emergency response plan and/or standard operating procedures and within the capability of available personnel, personal protective equipment, and control equipment.

H.2.3 Implementation Tasks.

The list of implementation tasks by responder level is as follows:

(1) *Awareness Level.* The awareness level personnel must be able to implement actions consistent with the emergency response plan, standard operating procedures, and the current edition of the DOT *Emergency Response Guidebook* by completing the following tasks:

(a) Initiate protective actions.
(b) Initiate the notification process.

(2) *Operations Level.* The operations level responder must be competent at the awareness level and be able to implement the planned response to favorably change the outcomes consistent with the emergency response plan and/or standard operating procedures by completing the following tasks:

(a) Establish and enforce scene control procedures, including control zones, decontamination, and communications.
(b) Establish a means of evidence preservation where criminal or terrorist acts are suspected.
(c) Initiate an incident management system (IMS).
(d) Don, work in, and doff personal protective equipment provided by the authority having jurisdiction.

(e) Perform the defensive control actions identified in the incident action plan.
(f) Perform mass decontamination as required.

(3) *Technician Level.* The hazardous materials technician must be competent at both the first responder awareness and operations levels and be able to implement the planned response to favorably change the outcomes consistent with the standard operating procedures or site safety and control plan by completing the following tasks:

(a) Perform the duties of an assigned position within the local IMS.
(b) Don, work in, and doff appropriate personal protective clothing, including, but not limited to, liquid splash– and vapor-protective clothing with approved respiratory protection.
(c) Perform the control functions identified in the incident action plan.

(4) *Command Level.* The incident commander must be competent at the operations level and be able to implement a response to favorably change the outcomes consistent with the emergency response plan and/or standard operating procedures by completing the following tasks:

(a) Implement the IMS including the specified procedures for notification and utilization of nonlocal resources (including private, state, and federal government personnel).
(b) Direct resources (private, governmental, and others) with expected task assignments and on-scene activities and provide management overview, technical review, and logistical support to private and governmental sector personnel.
(c) Provide a focal point for information transfer to media and local elected officials through the IMS structures.

H.2.4 Evaluation Tasks.

The list of evaluation tasks by responder level is as follows:

(1) *Awareness Level.* No requirements.
(2) *Operations Level.* The operations level responder must be competent at the awareness level and be able to evaluate the progress of the actions taken to ensure that the response objectives are being met safely, effectively, and efficiently by completing the following tasks:

(a) Evaluate the status of the defensive actions taken in accomplishing the response objectives.
(b) Communicate the status of the planned response.

(3) *Technician Level.* The hazardous materials technician must be competent in evaluating the progress of the planned response by completing the following tasks:

(a) Evaluate the effectiveness of the control functions.
(b) Evaluate the effectiveness of the decontamination process.

(4) *Command Level.* The incident commander must be competent at the operations level and be able to evaluate the progress of the planned response to ensure the response objectives are being met safely, effectively, and efficiently and adjust the incident action plan accordingly by evaluating the effectiveness of the control functions.

H.2.5 Termination Tasks.

The list of termination tasks by responder level is as follows:

(1) *Awareness Level.* No requirements.
(2) *Operations Level.* No requirements.
(3) *Technician Level.* The hazardous materials technician must be competent to terminate an incident by completing the following tasks:

(a) Assist in the incident debriefing.
(b) Assist in the incident critique.
(c) Provide reports and documentation of the incident.

(4) *Command Level.* The incident commander must be competent to terminate an incident by completing the following tasks:

(a) Transfer command (control) when appropriate.
(b) Conduct an incident debriefing.
(c) Conduct a multi-agency critique.
(d) Report and document the hazardous materials/WMD incident and submit the reports to the proper entity.

Definitions of Hazardous Materials

ANNEX I

This annex is not a part of the requirements of this NFPA document but is included for informational purposes only.

I.1 General

Many definitions and descriptive names are used for the term *hazardous material*, each of which depends on the nature of the problem being addressed. Unfortunately, no one list or definition covers everything. U.S. government agencies, as well as state and local governments, have different purposes for regulating hazardous materials that, under certain circumstances, pose a risk to the public or the environment.

I.2 Hazardous Materials Terms

The following hazardous materials terms, as used by the indicated government agencies, show the variety of definitions that can be applied.

I.2.1 Hazardous Materials.

The U.S. Department of Transportation (DOT) uses the term *hazardous materials* to cover 11 hazard classes, some of which have subcategories called divisions. DOT includes in its regulations hazardous substances and hazardous wastes as Class 9 (Miscellaneous Hazardous Materials), both of which are regulated by the U.S. Environmental Protection Agency (EPA), if their inherent properties would not otherwise be covered.

I.2.2 Hazardous Substances.

EPA uses the term *hazardous substances* for chemicals that if released into the environment above a certain amount must be reported, and, depending on the threat to the environment, federal involvement in handling the incident can be authorized. A list of the hazardous substances is published in Table 302.4 of 40 CFR 302. The U.S. Occupational Safety and Health Administration (OSHA) uses the term *hazardous substances* in 29 CFR 1910.120, which resulted from Title I of the Superfund Amendments and Reauthorization Act (SARA) (40 CFR 355) and covers emergency response. Unlike EPA, OSHA uses the term *hazardous substances* to cover every chemical regulated by both DOT and EPA.

I.2.3 Extremely Hazardous Substances.

EPA uses the term *extremely hazardous substances* for chemicals that must be reported to the appropriate authorities if released above the threshold reporting quantity. Each substance has a threshold reporting quantity. The list of extremely hazardous substances is identified in Title III of SARA (40 CFR 355).

I.2.4 Toxic Chemicals.

EPA uses the term *toxic chemicals* for chemicals whose total emissions or releases must be reported annually by owners and operators of certain facilities that manufacture, process, or otherwise use a listed toxic chemical. The toxic chemicals are listed in Title III of SARA (40 CFR 355).

I.2.5 Hazardous Wastes.

EPA uses the term *hazardous wastes* for chemicals that are regulated under the Resource, Conservation, and Recovery Act (40 CFR 261.33). Hazardous wastes in transportation are regulated by DOT (49 CFR 170–180).

I.2.6 Hazardous Chemicals.

OSHA uses the term *hazardous chemicals* for any chemical that would be a risk to employees if they were exposed in the workplace. The term *hazardous chemicals* covers a broader group of chemicals than the other chemical terms.

I.2.7 Dangerous Goods.

In United Nations model codes and regulations, hazardous materials are called *dangerous goods*.

I.2.8 Highly Hazardous Chemicals.

OSHA uses the term *highly hazardous chemicals* for those chemicals that fall under the requirements of 29 CFR 1910.119, "Process Safety Management of Highly Hazardous Chemicals." Highly hazardous chemicals are those chemicals that possess toxic, reactive, flammable, or explosive properties. A list of covered substances is published in Annex A of 29 CFR 1910.119.

UN/DOT Hazard Classes and Divisions

ANNEX J

This annex is not a part of the requirements of this NFPAdocument but is included for informational purposes only.

J.1 General

The definitions of UN/DOT hazard classes and divisions (49 CFR 170–180) are as follows.

J.2 Class 1 — Explosives

An explosive is any substance or article, including a device, that is designed to function by explosion (i.e., an extremely rapid release of gas and heat) or that, by chemical reaction within itself, is able to function in a similar manner even if not designed to function by explosion. Explosives in Class 1 are divided into six divisions. Each division has a letter designation.

J.2.1 Division 1.1.

Division 1.1 consists of explosives that have a mass explosion hazard. A mass explosion is one that affects almost the entire load instantaneously. Examples of Division 1.1 explosives include black powder trinitrotoluene, dynamite, and trinitrofoluene (TNT).

J.2.2 Division 1.2.

Division 1.2 consists of explosives that have a projection hazard but not a mass explosion hazard. Examples of Division 1.2 explosives include aerial flares, detonating cord, and power device cartridges.

J.2.3 Division 1.3.

Division 1.3 consists of explosives that have a fire hazard and a minor blast hazard, a minor projection hazard, or both, but not a mass explosion hazard. Examples of Division 1.3 explosives include liquid-fueled rocket motors and propellant explosives.

J.2.4 Division 1.4.

Division 1.4 consists of explosive devices that present a minor explosion hazard. No device in the division can contain more than 0.9 oz (25 g) of a detonating material. The explosive effects are largely confined to the package, and no projection of fragments of appreciable size or range are expected. An external fire must not cause virtually instantaneous explosion of almost the entire contents of the package. Examples of Division 1.4 explosives include line-throwing rockets, practice ammunition, and signal cartridges.

J.2.5 Division 1.5.

Division 1.5 consists of very insensitive explosives. This division comprises substances that have a mass explosion hazard but are so insensitive that there is very little probability of initiation or of transition from burning to detonation under normal conditions of transport. Examples of Division 1.5 explosives include pilled ammonium nitrate fertilizer–fuel oil mixtures (blasting agents).

J.2.6 Division 1.6.

Division 1.6 consists of extremely insensitive articles that do not have a mass explosive hazard. This division comprises articles that contain only extremely insensitive detonating substances and that demonstrate a negligible probability of accidental initiation or propagation.

J.3 Class 2 — Gases

J.3.1 Division 2.1.

Division 2.1 (flammable gas) consists of materials that are a gas at 68°F (20°C) or less and 14.7 psi (101.3 kPa) of pressure, have a boiling point of 68°F (20°C) or less at 14.7 psi (101.3 kPa), and have the following properties:

(1) Are ignitable at 14.7 psi (101.3 kPa) when in a mixture of 13 percent or less by volume with air
(2) Have a flammable range at 14.7 psi (101.3 kPa) with air of at least 12 percent regardless of the lower limit

Examples of Division 2.1 gases include inhibited butadienes, methyl chloride, and propane.

J.3.2 Division 2.2.

Division 2.2 (nonflammable, nonpoisonous compressed gas, including compressed gas, liquefied gas, pressurized cryogenic gas, and compressed gas in solution, asphyxiant gas, and oxidizing gas) consists of materials (or mixtures) that exert in the packaging an absolute pressure of 41 psi (280 kPa) at 68°F (20°C). A cryogenic liquid is a refrigerated liquefied gas having a boiling point colder than –130°F (–90°C) at 14.7 psi (101.3 kPa).

Examples of Division 2.2 gases include anhydrous ammonia, cryogenic argon, carbon dioxide, and compressed nitrogen.

J.3.3 Division 2.3.

Division 2.3 (gas poisonous by inhalation) consists of materials that are a gas at 68°F (20°C) or less and a pressure of 14.7 psi, or 1 atm (101.3 kPa), have a boiling point of 68°F (20°C) or less at 14.7 psi (101.3 kPa), and have the following properties:

(1) Are known to be so toxic to humans as to pose a hazard to health during transportation
(2) In the absence of adequate data on human toxicity, are presumed to be toxic to humans because, when tested on laboratory animals, they have an LC_{50} value of not more than 5000 ppm. Examples of Division 2.3 gases include anhydrous hydrogen fluoride, arsine, chlorine, and methyl bromide.

Hazard zones associated with Division 2.3 materials are the following:

(1) Hazard zone A — LC_{50} less than or equal to 200 ppm
(2) Hazard zone B — LC_{50} greater than 200 ppm and less than or equal to 1000 ppm

(3) Hazard zone C — LC_{50} greater than 1000 ppm and less than or equal to 3000 ppm
(4) Hazard zone D — LC_{50} greater than 3000 ppm and less than or equal to 5000 ppm

J.4 Class 3 — Flammable Liquids

Flammable liquids are liquids having a flash point of not more than 140°F (60°C) or materials in a liquid phase with a flash point at or above 100°F (37.8°C) that are intentionally heated and offered for transportation or transported at or above their flash point in a bulk packaging.

Examples of Class 3 liquids include acetone, amyl acetate, gasoline, methyl alcohol, and toluene.

J.4.1 Combustible Liquids.

Combustible liquids are liquids that do not meet the definition of any other hazard class and that have a flash point above 140°F (60°C) and below 200°F (93°C). Flammable liquids with a flash point above 100°F (38°C) can be reclassified as combustible liquids.

Examples of combustible liquids include mineral oil, peanut oil, and No. 6 fuel oil.

J.5 Class 4 — Flammable Solids

J.5.1 Division 4.1.

Division 4.1 (flammable solids) comprises the following three types of materials:

(1) Desensitized explosives — explosives wetted with sufficient water, alcohol, or plasticizers to suppress explosive properties
(2) Self-reactive materials — materials that are thermally unstable and that can undergo a strongly exothermic decomposition even with participation of oxygen (air)
(3) Readily combustible solids — solids that can cause a fire through friction and any metal powders that can be ignited.

Examples of Division 4.1 materials include magnesium (pellets, turnings, or ribbons) and nitrocellulose.

J.5.2 Division 4.2.

Division 4.2 (spontaneously combustible material) comprises the following materials:

(1) Pyrophoric materials — liquids or solids that, even in small quantities and without an external ignition source, can ignite within 5 minutes after coming in contact with air
(2) Self-heating materials — materials that, when in contact with air and without an energy supply, are liable to self-heat

Examples of Division 4.2 materials include aluminum alkyls, charcoal briquettes, magnesium alkyls, and phosphorus.

J.5.3 Division 4.3.

Division 4.3 (dangerous-when-wet materials) comprises of materials that, by contact with water, are liable to become spontaneously flammable or to give off flammable or toxic gas at a rate greater than 1 L/kg of the material per hour. Examples of Division 4.3 materials include calcium carbide, magnesium powder, potassium metal alloys, and sodium hydride.

J.6 Class 5 — Oxidizers and Organic Peroxides

J.6.1 Division 5.1.

Division 5.1 (oxidizers) comprises materials that can, generally by yielding oxygen, cause or enhance the combustion of other materials. Examples of Division 5.1 materials include ammonium nitrate, bromine trifluoride, and calcium hypochlorite.

J.6.2 Division 5.2.

Division 5.2 (organic peroxides) comprises organic compounds that contain oxygen (O) in the bivalent -O-O- structure that can be considered a derivative of hydrogen peroxide, where one or more of the hydrogen atoms have been replaced by organic radicals. Examples of Division 5.2 materials include dibenzoyl peroxide, methyl ethyl ketone peroxide, and peroxyacetic acid. Division 5.2 (organic peroxide) materials are assigned to one of the following seven types:

(1) Type A — organic peroxides that can detonate or deflagrate rapidly as packaged for transport. Transportation of Type A organic peroxides is forbidden.
(2) Type B — organic peroxides that neither detonate nor deflagrate rapidly but that can undergo a thermal explosion.
(3) Type C — organic peroxides that neither detonate nor deflagrate rapidly and that cannot undergo a thermal explosion.
(4) Type D — organic peroxides that detonate only partially or deflagrate slowly, with medium to no effect when heated under confinement.
(5) Type E — organic peroxide that neither detonate nor deflagrate and that show low or no effect when heated under confinement.
(6) Type F — organic peroxides that will not detonate, do not deflagrate, show only a low or no effect if heated when confined, and have low or no explosive power.
(7) Type G — organic peroxides that will not detonate, do not deflagrate, show no effect if heated when confined, have no explosive power, are thermally stable, and are desensitized

J.7 Class 6 — Poisonous Materials

J.7.1 Division 6.1.

Division 6.1 (poisonous materials) comprises materials other than gases that either are known to be so toxic to humans as to afford a hazard to health during transportation or in the absence of adequate data on human toxicity are presumed to be toxic to humans, including materials that cause irritation. Examples of Division 6.1 materials include aniline, arsenic compounds, carbon tetrachloride, hydrocyanic acid, and tear gas.

J.7.2 Division 6.2.

Division 6.2 (infectious substances) comprises materials known to contain or suspected of containing a pathogen. A pathogen is a micro-organism (including viruses, plasmids, and other genetic elements) or a proteinaceous infectious particle (prion) that has the potential to cause disease in humans or animals. The terms *infectious substance* and *etiologic agent* are synonymous. Examples of Division 6.2 materials include anthrax, botulism, rabies, and tetanus. Hazard zones associated with Class 6 materials are as follows:

(1) Hazard zone A — LC_{50} less than or equal to 200 ppm
(2) Hazard zone B — LC_{50} greater than 200 ppm and less than or equal to 1000 ppm

J.8 Class 7 — Radioactive Materials

Radioactive material is any material containing radionuclides where both the activity concentration and the total activity in the consignment exceed specified values. Examples of Class 7 materials include cobalt, uranium hexafluoride, and "yellow cake."

J.9 Class 8 — Corrosive Materials

Corrosive materials are liquids or solids that cause full-thickness destruction of skin at the site of contact within a specified period of time. A liquid that has a severe corrosion rate on steel or aluminum is also a corrosive material. Examples of Class 8 materials include nitric acid, phosphorus trichloride, sodium hydroxide, and sulfuric acid.

J.10 Class 9 — Miscellaneous Hazardous Materials

Miscellaneous hazardous materials are materials that present a hazard during transport but that do not meet the definition of any other hazard class. Miscellaneous hazardous materials, include the following:

(1) Any material that has an anesthetic, noxious, or other similar property that could cause extreme annoyance or discomfort to a flight crew member so as to prevent the correct performance of assigned duties
(2) Any material that is not included in any other hazard class but that is subject to DOT requirements (e.g. elevated-temperature material, hazardous substance, hazardous waste, marine pollutant). Examples of Class 9 materials include adipic acid, hazardous substances (e.g., PCBs), and molten sulfur.

J.11 ORM-D Material

ORM-D materials are materials that present a limited hazard during transportation due to their form, quantity, and packaging. Examples of ORM-D materials include consumer commodities and small arms ammunition.

J.12 Forbidden

Forbidden means prohibited from being offered or accepted for transportation. Prohibition does not apply if these materials are diluted, stabilized, or incorporated into devices.

J.13 Marine Pollutant

A marine pollutant is a material that has an adverse effect on aquatic life.

J.14 Elevated-Temperature Material

Elevated temperature materials are materials that, when offered for transportation in a bulk packaging, meet one of the following conditions:

(1) Are liquid at or above 212°F (100°C)
(2) Are liquid with a flash point at or above 100°F (37.8°C) and are intentionally heated and transported at or above their flash point
(3) Are solid at or above 464°F (240°C)

Informational References

ANNEX K

K.1 Referenced Publications

The documents or portions thereof listed in this annex are referenced within the informational sections of this standard and are not part of the requirements of this document unless also listed in Chapter 2 for other reasons.

K.1.1 NFPA Publications.

National Fire Protection Association, 1 Batterymarch Park, Quincy, MA 02169-7471.

NFPA 11, *Standard for Low-, Medium-, and High-Expansion Foam*, 2005 edition.

NFPA 30, *Flammable and Combustible Liquids Code*, 2008 edition.

NFPA 58, *Liquefied Petroleum Gas Code*, 2008 edition.

NFPA 473, *Standard for Competencies for EMS Personnel Responding to Hazardous Materials/Weapons of Mass Destruction Incidents*, 2008 edition.

NFPA 704, *Standard System for the Identification of the Hazards of Materials for Emergency Response*, 2007 edition.

NFPA 1405, *Guide for Land-Based Fire Fighters Who Respond to Marine Vessel Fires*, 2006 edition.

NFPA 1971, *Standard on Protective Ensembles for Structural Fire Fighting and Proximity Fire Fighting*, 2007 edition.

NFPA 1991, *Standard on Vapor-Protective Ensembles for Hazardous Materials Emergencies*, 2005 edition.

NFPA 1992, *Standard on Liquid Splash-Protective Ensembles and Clothing for Hazardous Materials Emergencies*, 2005 edition.

NFPA 1994, *Standard on Protective Ensembles for First Responders to CBRN Terrorism Incidents*, 2007 edition.

Hazardous Materials/Weapons of Mass Destruction Response Handbook, 2008.

Wright, Charles J., "Managing the Hazardous Materials Incident," Section 7, Chapter 9 in *Fire Protection Handbook*, 19th edition, 2003.

K.1.2 Other Publications.

K.1.2.1 American Chemistry Council (formerly Chemical Manufacturers Association) Publications. American Chemistry Council, 1300 Wilson Blvd., Arlington, VA 22209.

Recommended Terms for Personal Protective Equipment, 1985.

K.1.2.2 API Publications. American Petroleum Institute, 1220 L Street, N.W., Washington, DC 20005-4070.

API 2021, *Guide for Fighting Fires in and Around Flammable and Combustible Liquid Atmospheric Petroleum Storage Tanks*, 2001.

API 2510-A, *Fire Protection Considerations for the Design and Operation of Liquefied Petroleum Gas (LPG) Storage Facilities*, 1996.

K.1.2.3 IMO Publications. International Maritime Organization, 4 Albert Embankment, London SEI 7SR, UK..

Code for the Construction and Equipment of Ships Carrying Dangerous Chemicals in Bulk, (BCH Code).

International Code for the Construction and Equipment of Ships Carrying Dangerous Chemicals in Bulk (IBC Code).

International Code for the Construction and Equipment of Ships Carrying Dangerous Liquefied Gases in Bulk (IGC Code).

International Maritime Dangerous Goods Code (IMDG Code).

MARPOL 73/78.

Safety of Life at Sea (SOLAS).

K.1.2.4 NRT Publications. U.S. National Response Team, Washington, DC 20593, www.nrt.org.

NRT-1, *Hazardous Materials Emergency Planning Guide*, 2001.

K.1.2.5 U.S. Government Publications. U.S. Government Printing Office, Washington, DC 20402.

Department of Homeland Security (DHS), Responder Knowledge Base. http://www.rkb.mipt.org

Environmental Protection Agency, *Standard Operating Safety Guides*, June 1992.

National Incident Management System (NIMS), *Site Safety and Control Plan (form ICS 208 HM).*

National Toxicology Program, U.S. Department of Health and Human Services, *9th Report on Carcinogens*, Washington, DC, 2001.

National Incident Management System (NIMS), March 2004, http://www.fema.gov/nims/nims_compliance.shtm#nimsdocument.

National Preparedness Goal, March 2005, https://www.llis.dhs.gov.

National Preparedness Guidance, April 2005, https://www.llis.dhs.gov.

National Response Plan, December 2004, http://www.dhs.gov/Xprepresp/committees/editorial_0566.shtm.

NIOSH/OSHA/USCG/EPA, *Occupational Safety and Health Guidance Manual for Hazardous Waste Site Activities*, October 1985.

NIOSH Pocket Guide to Chemical Hazards, DHHS (NIOSH) Publication No. 2005-149, September 2005: http://www.cdc.gov/niosh.npg.

Target Capabilities List, May, 2005, https://www.llis.dhs.gov.

Title 18, U.S. Code, Section 232a, "Use of Weapons of Mass Destruction."

Title 29, Code of Federal Regulations, Parts 1910.119–1910.120.

Title 29, Code of Federal Regulations, Part 1910.134.

Title 33, Code of Federal Regulations, "Navigation and Navigable Waters."

Title 40, Code of Federal Regulations, Part 261.33.

Title 40, Code of Federal Regulations, Part 302.

Title 40, Code of Federal Regulations, Part 355.

Title 46, Code of federal Regulations, "Shipping."

Title 49, Code of Federal Regulations, Parts 170–180.

Title 49, Code of Federal Regulations, Part 173.431.

Universal Task List, May 2005, https://www.llis.dhs.gov.

U.S. Army Research, Development, and Engineering Command (RDECOM), Edgewood Chemical Biological Center, Emergency Response, Command, and Planning Guidelines (various documents) for terrorist incidents involving chemical and biological agents. http://www.ecbc.army.mil/hld.

U.S. Department of Transportation, *Emergency Response Guidebook*, 2004 edition.

K.1.2.6 Additional Publications.

Grey, G. L., et al., *Hazardous Materials/Waste Handling for the Emergency Responder*, Fire Engineering Publications, New York, 1989.

Maslansky, C. J., and Stephen P. Maslansky, *Air Monitoring Instrumentation*, New York, Van Nostrand Reinhold, 1993.

Noll, G., and M. Hildebrand, *Hazardous Materials: Managing the Incident*, 3rd edition, Fire Protection Publications, Stillwater, OK, 2005.

K.2 Informational References

The following documents or portions thereof are listed here as informational resources only. They are not a part of the requirements of this document.

NFPA 25, *Standard for the Inspection, Testing, and Maintenance of Water-Based Fire Protection Systems*, 2008 edition.

NFPA 306, *Standard for the Control of Gas Hazards on Vessels*, 2003 edition.

NFPA 424, *Guide for Airport/Community Emergency Planning*, 2008 edition.

NFPA 600, *Standard on Industrial Fire Brigades*, 2005 edition.

NFPA 1404, *Standard for Fire Service Respiratory Protection Training*, 2006 edition.

NFPA 1500, *Standard on Fire Department Occupational Safety and Health Program*, 2007 edition.

NFPA 1561, *Standard on Emergency Services Incident Management System*, 2005 edition.

NFPA 1581, *Standard on Fire Department Infection Control Program*, 2005 edition.

NFPA 1951, *Standard on Protective Ensembles for Technical Rescue Incidents*, 2007 edition.

K.3 References for Extracts in Informational Sections. (Reserved)

PART II

NFPA® 473, *Standard for Competencies for EMS Personnel Responding to Hazardous Materials/Weapons of Mass Destruction Incidents*, 2008 Edition, with Commentary

Part II of this handbook presents the full text of NFPA 473, *Standard for Competencies for EMS Personnel Responding to Hazardous Materials Incidents,* and explanatory commentary to guide the reader through the code. As in Part I of this handbook, the text, figures, and tables of NFPA 473 appear in black. The commentary and its exhibits and tables are printed in brown.

The goal of the 2008 edition of NFPA 473 is to provide emergency medical services personnel with guidance on how to deliver basic life support (BLS) and advanced life support (ALS) at hazardous materials/weapons of mass destruction incidents. Essentially, the objective is to define BLS and ALS competencies for EMS responders, both mission-specific and agent-specific, for incidents involving chemical, biological, nuclear, radiological, and explosive events, when human injury or illness is suspected or confirmed.

A diverse task group of EMS industry experts (both within and outside of the technical committee structure) was assembled to compile and refine comments, achieve consensus, and prepare a draft for technical committee review. The task group worked under the following basic assumptions:

1. Responders would be protected first, and then the patients would be managed.
2. No ALS would be performed in the hot zone.
3. Hazardous materials response training would be left to NFPA 472, *Standard for Competence of Responders to Hazardous Materials/Weapons of Mass Destruction Incidents*, to address, with NFPA 473 concentrating on the EMS element.
4. EMS responders at hazardous materials incidents essentially would be the "first receivers of the first receivers." NFPA 473 would function as the missing link in

hazardous materials/EMS response, taking patient care from the scene to definitive hospital care.

5. Fundamentally, the level of hazardous materials training would be the difference between a Level I and Level II responder as described in NFPA 473, not the level of medical care they would be expected to provide. An EMS responder would deliver either BLS or ALS – the authority having jurisdiction (AHJ) would decide their responsibilities in terms of hazardous materials response.

Level I EMS Responders are required to receive awareness level training in addition to the EMS competencies outlined in NFPA 473. Level I EMS Responders could be called to the scene of a hazardous materials incident but would not be responsible for rescue, incident mitigation, or any other duties that would require personal protective equipment (PPE) beyond Level D of the U.S. Occupational Safety and Health Administration and the U.S. Environmental Protection Agency's levels of protection. Conceptually, Level I EMS Responders could be expected to render BLS/ALS care in the cold zone only and/or provide transportation after a patient has been fully decontaminated by other responders.

Level II EMS Responders are required to receive operations-level training, including all mission- and agent-specific training, in addition to the EMS competencies outlined in NFPA 473. Level II EMS Responders could be called to the scene of a hazardous materials incident and would be responsible for victim rescue/recovery, operations-level incident mitigation efforts, or any other duties that require PPE beyond the awareness level. Conceptually, Level II EMS responders could be expected to render BLS/ALS care prior to decontamination or later in the cold zone after a patient has been fully decontaminated by other responders.

Administration

CHAPTER 1

The goal of the 2008 edition of NFPA 473, *Standard for Competencies for EMS Personnel Responding to Hazardous Materials/Weapons of Mass Destruction Incidents,* is to provide emergency medical services (EMS) personnel with guidance on how to deliver basic life support (BLS) and advanced life support (ALS) at hazardous materials/weapons of mass destruction (WMD) incidents.

Essentially, the objective is to define BLS and ALS competencies for EMS responders, both mission specific and agent specific, for incidents involving chemical, biological, nuclear, radiological, and explosive events when human injury and illness is suspected or confirmed.

A diverse task group of EMS industry experts (both within and outside of the technical committee structure) was assembled to compile and refine comments, achieve consensus, and prepare a draft for the technical committee to review.

The task group worked under the following basic assumptions:

1. Protect the responders first, and then manage the patient
2. No ALS would be performed in the hot zone
3. Hazardous materials response training would be left to NFPA 472 to address, with NFPA 473 concentrating on the EMS element
4. EMS responders at hazardous materials incidents are essentially the "first receivers of the first receivers." NFPA 473 would function as the missing link in hazardous materials/ EMS response, taking patient care from the scene to definitive hospital care.
5. Fundamentally, it is the level of hazardous materials training that is the difference between a Level I and Level II responder as described in earlier editions of NFPA 473, not the level of medical care they are expected to provide. An EMS responder either delivers BLS or ALS—it is up to the authority having jurisdiction (AHJ) to decide their responsibilities in terms of hazardous materials response.

1.1 Scope

This standard identifies the levels of competence required of emergency medical services (EMS) personnel who respond to incidents involving hazardous materials or weapons of mass destruction (WMD). It specifically covers the requirements for basic life support and advanced life support personnel in the pre-hospital setting.

1.1.1 This standard is based on the premise that all EMS responders are trained to meet at least the core competencies of the operations level responders as defined in Chapter 5 of NFPA 472, *Standard for Competence of Responders to Hazardous Materials/Weapons of Mass Destruction Incidents.*

1.2 Purpose

The purpose of this standard is to specify minimum requirements of competence and to enhance the safety and protection of response personnel and all components of the emergency medical services system. It is not the intent of this standard to restrict any jurisdiction from exceeding these minimum requirements.

NFPA 473 recognizes the inherent differences in pre-hospital care systems nationwide and the variety of ways in which EMS responders can be called upon to render care to victims of hazardous materials/WMD incidents. To that end, the document is not protocol driven. It is intended to provide a set of broad competencies that an AHJ can use to better train BLS and ALS responders to render care at a hazardous materials/WMD incident. The goal of NFPA 473 is to create a template for informed and effective decision making when identifying and treating exposed patients.

The 2008 edition of NFPA 473 is hopefully more reflective of the "real world" and of the challenges that may be encountered at a hazardous materials/WMD incident.

This standard might require additional training for some EMS providers, but it is believed that this additional training is prudent and essential for the safety of EMS personnel responding to a hazardous materials/WMD incident. Although NFPA 473 is designed to promote thoughtful and informed medical care, it is unrealistic to expect that any standard can address all possible situations.

NFPA 472, which is covered in Part I of this handbook, is designed to provide guidance for a wide variety of first responder disciplines such as fire, law enforcement, EMS, public works, and others. NFPA 473 is designed to build upon that training with EMS-specific competencies.

1.3* CDC Categories A, B, and C

This standard uses the U.S. Centers for Disease Control and Prevention (CDC) categories of diseases and agents.

A.1.3 The CDC categories of bioterrorism diseases and agents are as follows (for more information, see the CDC website www.bt.cdc.gov):

(1) Category A
  (a) Anthrax (*Bacillus anthracis*)
  (b) Botulism (*Clostridium botulinum* toxin)
  (c) Plague (*Yersinia pestis*)
  (d) Smallpox (variola major)
  (e) Tularemia (*Francisella tularensis*)
  (f) Viral hemorrhagic fevers (filoviruses [e.g., Ebola, Marburg] and arenaviruses [e.g., Lassa, Machupo])
(2) Category B
  (a) Brucellosis (*Brucella* species)
  (b) Epsilon toxin of *Clostridium perfringens*
  (c) Food safety threats (e.g., *Salmonella* species, *Escherichia coli* O157:H7, *Shigella*)
  (d) Glanders (*Burkholderia mallei*)
  (e) Melioidosis (*Burkholderia pseudomallei*)
  (f) Psittacosis (*Chlamydia psittaci*)
  (g) Q fever (*Coxiella burnetii*)
  (h) Ricin toxin from *Ricinus communis* (castor beans)
  (i) Staphylococcal enterotoxin B
  (j) Typhus fever (*Rickettsia prowazekii*)
  (k) Viral encephalitis [alphaviruses (e.g., Venezuelan equine encephalitis, eastern equine encephalitis, western equine encephalitis)]
  (l) Water safety threats (e.g., *Vibrio cholerae*, *Cryptosporidium parvum*)
(3) Category C — emerging infectious diseases, such as Nipah virus and hantavirus

Category A Diseases/Agents. The U.S. public health system and primary healthcare providers must be prepared to address various biological agents, including pathogens that are rarely seen in the United States. These high-priority agents include organisms that pose a risk to national security because of the following:

(1) They can be easily disseminated or transmitted from person to person.
(2) They result in high mortality rates and have the potential for major public health impact.
(3) They might cause public panic and social disruption.
(4) They require special action for public health preparedness.

Category B Diseases/Agents. These second-highest priority agents have the following characteristics:

(1) They are moderately easy to disseminate.
(2) They result in moderate morbidity rates and low mortality rates.
(3) They require specific enhancements of CDC's diagnostic capacity and enhanced disease surveillance.

Category C Diseases/Agents. These third-highest priority agents include emerging pathogens that could be engineered for mass dissemination in the future because of the following characteristics:

(1) Availability
(2) Ease of production and dissemination
(3) Potential for high morbidity and mortality rates and major health impact

Referenced Publications

CHAPTER 2

This chapter lists the publications that are referenced within the mandatory chapters of NFPA 473. These mandatory referenced publications are needed for effective use of and compliance with NFPA 473. The requirements contained within these references constitute part of the requirements of NFPA 473. Annex B lists nonmandatory publications that are referenced within the nonmandatory annexes of NFPA 473.

2.1 General

The documents or portions thereof listed in this chapter are referenced within this standard and shall be considered part of the requirements of this document.

2.2 NFPA Publications

National Fire Protection Association, 1 Batterymarch Park, Quincy, MA 02169-7471.

NFPA 472, *Standard for Competence of Responders to Hazardous Materials/Weapons of Mass Destruction Incidents,* 2008 edition.

NFPA 704, *Standard System for the Identification of the Hazards of Materials for Emergency Response,* 2007 edition.

2.3 Other Publications

Emergency Response Guidebook, Washington, D.C.: U.S. Department of Transportation, 2004.

Merriam-Webster's Collegiate Dictionary, 11th edition, Merriam-Webster, Inc., Springfield, MA, 2003.

Title 18 U.S. Code Section 2332a, "Use of weapons of mass destruction," Washington, D.C.: Government Printing Office.

2.4 References for Extracts in Mandatory Sections (Reserved)

Definitions

CHAPTER 3

The definitions contained in this chapter have been updated from the previous edition of NFPA 473 to more closely align with current emergency medical services (EMS) terminology.

3.1 General

The definitions contained in this chapter shall apply to the terms used in this standard. Where terms are not defined in this chapter or within another chapter, they shall be defined using their ordinarily accepted meanings within the context in which they are used. *Merriam-Webster's Collegiate Dictionary,* 11th edition, shall be the source for the ordinarily accepted meaning.

3.2 NFPA Official Definitions

3.2.1* Approved. Acceptable to the authority having jurisdiction.

The authority having jurisdiction (AHJ) can also seek the expertise of other allied professionals, including certified safety professional (CSP), certified health physicist (CHP), certified industrial hygienist (CIH), certified hazardous materials (CHM), or similar credentialed or competent individuals. These individuals might also be referred to as a subject matter expert (SME) in a mission-specific area.

A.3.2.1 Approved. The National Fire Protection Association does not approve, inspect, or certify any installations, procedures, equipment, or materials; nor does it approve or evaluate testing laboratories. In determining the acceptability of installations, procedures, equipment, or materials, the authority having jurisdiction may base acceptance on compliance with NFPA or other appropriate standards. In the absence of such standards, said authority may require evidence of proper installation, procedure, or use. The authority having jurisdiction may also refer to the listings or labeling practices of an organization that is concerned with product evaluations and is thus in a position to determine compliance with appropriate standards for the current production of listed items.

3.2.2* Authority Having Jurisdiction (AHJ). An organization, office, or individual responsible for enforcing the requirements of a code or standard, or for approving equipment, materials, an installation, or a procedure.

A.3.2.2 Authority Having Jurisdiction (AHJ). The phrase "authority having jurisdiction," or its acronym AHJ, is used in NFPA documents in a broad manner, since jurisdictions and approval agencies vary, as do their responsibilities. Where public safety is primary, the authority having jurisdiction may be a federal, state, local, or other regional department or individual such as a fire chief; fire marshal; chief of a fire prevention bureau, labor department, or health department; building official; electrical inspector; or others having statutory authority. For insurance purposes, an insurance inspection department, rating bureau, or other insurance company representative may be the authority having jurisdiction. In many circumstances, the property owner or his or her designated agent assumes the role of the authority having jurisdiction; at government installations, the commanding officer or departmental official may be the authority having jurisdiction.

3.2.3* Listed. Equipment, materials, or services included in a list published by an organization that is acceptable to the authority having jurisdiction and concerned with evaluation of products or services, that maintains periodic inspection of production of listed equipment or materials or periodic evaluation of services, and whose listing states that either the equipment, material, or service meets appropriate designated standards or has been tested and found suitable for a specified purpose.

A.3.2.3 Listed. The means for identifying listed equipment may vary for each organization concerned with product evaluation; some organizations do not recognize equipment as listed unless it is also labeled. The authority having jurisdiction should utilize the system employed by the listing organization to identify a listed product.

3.2.4 Shall. Indicates a mandatory requirement.

3.2.5 Should. Indicates a recommendation or that which is advised but not required.

3.2.6 Standard. A document, the main text of which contains only mandatory provisions using the word "shall" to indicate requirements and which is in a form generally suitable for mandatory reference by another standard or code or for adoption into law. Nonmandatory provisions shall be located in an appendix or annex, footnote, or fine-print note and are not to be considered a part of the requirements of a standard.

3.3 General Definitions

3.3.1 Advanced Life Support (ALS). Emergency medical treatment beyond basic life support level as defined by the medical authority having jurisdiction in conjunction with the American Heart Association guidelines.

Because advanced life support (ALS) can be broadly defined, the intent of this definition is to give the AHJ enough latitude to determine which category most closely reflects the EMS responders within their system.

> *3.3.1.1* ***Emergency Medical Technician — Intermediate (EMT-I).*** An individual who has completed a course of instruction that includes selected modules of the U.S. Department of Transportation National Standard EMT — Paramedic curriculum and who holds an intermediate level EMT-I or EMT-C certification from the authority having jurisdiction.

The Emergency Medical Technician – Intermediate (EMT-I) is an individual who has successfully completed a course of instruction based on the U.S. Department of Transportation (DOT) EMT-Intermediate: National Standard Curriculum and who holds a license to practice at the EMT-I level from the AHJ.

Although the *National EMS Scope of Practice Model* has not yet been widely adopted by the EMS community, it has defined an advanced emergency medical technician (AEMT) as follows:

> Individual who has successfully completed a course based on the U.S. DOT AEMT Education Standards and provides basic and limited advanced emergency medical care and transportation for critical and emergent patients. Performs interventions with the basic and advanced equipment typically found on an ambulance. Licensed to practice by their State and credentialed by the authority having jurisdiction (AHJ). [1]

3.3.1.2 Emergency Medical Technician — Paramedic (EMT-P). An individual who has successfully completed a course of instruction that meets or exceeds the requirements of the U.S. Department of Transportation National Standard EMT — Paramedic curriculum and who holds an EMT-P certification from the authority having jurisdiction.

The Emergency Medical Technician – Paramedic (EMT-P) is an individual who has successfully completed a course o\f instruction based on the DOT EMT-Paramedic: National Standard Curriculum and who holds a license to practice at the EMT-P level from the AHJ.

Although the *National EMS Scope of Practice Model* has not yet been widely adopted by the EMS community, it has defined a paramedic as follows:

> Individual who has successfully completed a course based on the U.S. DOT Paramedic Education Standards and provides advanced emergency medical care for critical and emergent patients. Performs interventions with the basic and advanced equipment typically found on an ambulance. Licensed to practice by their State and credentialed by the authority having jurisdiction (AHJ). [1]

3.3.1.3 Medical Director. Plans and directs all aspects of an organizations or systems medical policies and programs, including operations and offline (protocol) and online medical direction (direct communication consultation); is responsible for strategic clinical relationships with other physicians; oversees the development of the clinical content in materials; ensures all clinical programs are in compliance; writes and reviews research publications appropriate to support clinical service offerings; requires an active degree in medicine with specialty experience or training in emergency mitigation, administration, and management; relies on experience and judgment to plan and accomplish goals; and typically coordinates with the incident command.

3.3.1.4 Medical Team Specialist. Any healthcare provider or medically trained specialist acting under the authority of the medical director and within the context of the National Incident Management System authorized to act as the medical point of contact for an incident. This can include, but is not exclusive to, nurses, nurse practitioners, EMTs, ECAs, physician assistants, and in some cases a health and safety officer.

3.3.2 Basic Life Support (BLS). Emergency medical treatment at a level as defined by the medical authority having jurisdiction in conjunction with American Heart Association guidelines.

Because basic life support (BLS) can be broadly defined, the intent of this definition is to give the AHJ enough latitude to determine which category most closely reflects the EMS responders within their system.

3.3.2.1* Emergency Care First Responder (ECFR). An individual who has successfully completed the specified emergency care first responder course developed by the U.S. Department of Transportation and who holds an ECFR certification from the authority having jurisdiction.

The Emergency Care First Responder (ECFR) is an individual who has successfully completed a course based on the DOT First Responder: National Standard Curriculum and who holds a license to practice at the FR level from the authority having jurisdiction.

Although the *National EMS Scope of Practice Model* has not yet been widely adopted by the EMS community, it has defined emergency medical responder (EMR) as follows:

> Individual who has successfully completed a course based on the U.S. DOT EMR Education Standards and performs basic interventions with minimal equipment while awaiting additional EMS response. Licensed to practice as an EMR by their State and credentialed by the authority having jurisdiction (AHJ). [1]

A.3.3.2.1 Emergency Care First Responder (ECFR). In Canada, the terminology used is

Emergency Medical Assistant-1 (EMA-1), Emergency Medical Assistant-2 (EMA-2), and Emergency Medical Assistant-3 (EMA-3).

3.3.2.2 ***Emergency Medical Technician — Ambulance/Basic (EMT- A/B).*** An individual who has successfully completed an EMT-A or EMT-B curriculum developed by the U.S. Department of Transportation or equivalent, who holds an EMT-A/B certification from the authority having jurisdiction.

The Emergency Medical Technician – Ambulance/Basic (EMTA/B) is an individual who has successfully completed a course of instruction based on the DOT EMT A or EMT B - Basic: National Standard Curriculum and who holds a license to practice at the EMTA/B level from the authority having jurisdiction.

Although the *National EMS Scope of Practice Model* has not yet been widely adopted by the EMS community, it has defined emergency medical technician (EMT) as follows:

> Individual who has successfully completed a course based on the U.S. DOT EMT Education Standards and performs interventions with the basic equipment typically found on an ambulance on the scene and during transport. Licensed to practice as an EMT by their State and credentialed by the authority having jurisdiction (AHJ). [1]

3.3.3 Competence. The possession of knowledge, skills, and judgment needed to perform indicated objectives satisfactorily.

3.3.4* Components of Emergency Medical Service (EMS) System. The parts of a comprehensive plan to treat an individual in need of emergency medical care following an illness or injury.

A.3.3.4 Components of Emergency Medical Service (EMS) System. These components include the following:

(1) First responders
(2) Emergency dispatching
(3) EMS agency response
(4) Hospital emergency departments
(5) Specialized care facilities

3.3.5 Contaminant. A hazardous material, or the hazardous component of a weapon of mass destruction (WMD), that physically remains on or in people, animals, the environment, or equipment, thereby creating a continuing risk of direct injury or a risk of exposure.

3.3.6 Core Competencies. The knowledge, skills, and judgment needed by operations level responders who can respond to releases or potential releases of hazardous materials/WMD.

3.3.7* Demonstrate. To show by actual performance.

A.3.3.7 Demonstrate. This performance can be supplemented by simulation, explanation, illustration, or a combination of these.

3.3.8 Describe. To explain verbally or in writing using standard terms recognized in the hazardous materials response community.

3.3.9 Emergency Medical Services (EMS). The provision of treatment, such as first aid, cardiopulmonary resuscitation, basic life support, advanced life support, and other prehospital procedures, including ambulance transportation, to patients.

3.3.10 EMS Hazardous Materials (EMS/Hazardous Materials/WMD) Responder.

3.3.10.1 ***Emergency Medical Services Responders to Hazardous Materials/Weapon of Mass Destruction at the BLS Level (BLS Level Responder).*** In addition to their BLS certification, shall be trained to meet at least the core competencies of the operations level responders as defined in NFPA 472, *Standard for Competence of Responders to Hazardous Materials/Weapons of Mass Destruction Incidents*, and all competencies of Chapter 4 of this standard.

3.3.10.2 Emergency Medical Services Responders to Hazardous Materials/Weapon of Mass Destruction at the ALS Level (ALS Level Responder). In addition to their ALS certification, shall be trained to meet at least the core competencies of the operations level responders as defined in NFPA 472, *Standard for Competence of Responders to Hazardous Materials/Weapons of Mass Destruction Incidents*, and all competencies of Chapter 5 of this standard.

3.3.11 Exposure. The act or condition whereby responders or civilians come into contact with hazardous materials/WMD that results in any level of physical injury or acute/delayed health effect.

3.3.12* Hazardous Material. A substance (matter — solid, liquid, or gas — or energy) that when released is capable of creating harm to people, the environment, and property, including weapons of mass destruction (WMD) as defined in 18 U.S. Code, Section 2332a, as well as any other criminal use of hazardous materials, such as illicit laboratories, environmental crimes, or industrial sabotage. Hazardous materials/WMD shall be used throughout this document to represent hazardous materials/weapons of mass destruction.

A.3.3.12 Hazardous Material. Other criminal use of hazardous materials includes CBRNE, or chemical, biological, radiological, nuclear, and high yield explosives.

3.3.13 Identify. To select or indicate verbally or in writing using standard terms to establish the identity of; the fact of being the same as the one described.

3.3.14 Incident. An emergency involving the release or potential release of hazardous materials/WMD.

3.3.15* Incident Commander (IC). The individual responsible for all incident activities, including the development of strategies and tactics and the ordering and the release of resources.

A.3.3.15 Incident Commander (IC). This position is equivalent to the on-scene incident commander. The IC has overall authority and responsibility for conducting incident operations and is responsible for the management of all incident operations at the incident site.

3.3.16 Incident Command System (ICS). A management system designed to enable effective and efficient on-scene incident management by integrating a combination of facilities, equipment, personnel, procedures, and communications operating within a common organizational structure.

3.3.17* Incident Management System (IMS). A plan that defines the roles and responsibilities to be assumed by personnel and the operating procedures to be used in the management and direction of emergency operations to include the incident command system, multiagency coordination system, training, and management of resources.

A.3.3.17 Incident Management System (IMS). The IMS provides a consistent approach for all levels of government, private sector, and volunteer organizations to work together effectively and efficiently to prepare for, respond to, and recover from domestic incidents, regardless of cause, size, or complexity. An IMS provides for interoperability and compatibility among all levels of government, private sector, and volunteer organization capabilities. The IMS includes a core set of concepts, principles, terminology, and technologies covering the ICS, multiagency coordination systems, training, and identification and management of resources.

3.3.18 Medical Control. The physician providing direction for patient care activities in the prehospital setting.

3.3.19 Medical Surveillance. The ongoing process of medical evaluation of hazardous materials response team members and public safety personnel who respond to a hazardous materials incident.

3.3.20 Mission-Specific Competencies. The knowledge, skills, and judgment needed by operations level responders who have completed the requisite core competencies and who are designated by the authority having jurisdiction to perform mission-specific tasks, such as decontamination, victim/hostage rescue and recovery, evidence preservation and sampling, etc.

In the 2008 edition of NFPA 472, Chapter 5 specifies the competencies required of all emergency responders at the operations level. In Chapter 4 of the 2002 edition, operations level responders were also required to have product control competencies, which included personal protective equipment.

The competencies specified in Chapter 6 of the 2008 edition of NFPA 472 are optional and are provided so that the AHJ can match the expected tasks and duties of its personnel with the required competencies to perform those tasks.

Mission-specific competencies are available for operations level responders who are assigned to perform the following tasks:

- Use personal protective equipment, as provided by the AHJ
- Perform technical decontamination
- Perform mass decontamination
- Perform product control
- Perform air monitoring and sampling
- Perform victim rescue and recovery operations
- Evidence preservation and sampling
- Respond to illicit laboratory incidents

Operations-level mission-specific tasks must be performed under the guidance of a hazardous materials technician, allied professional, or standard operating procedure.

3.3.21 Patient. Any person or persons requiring or requesting a BLS/ALS evaluation or intervention at the scene of a hazardous materials/WMD incident.

3.3.22 Protocol. A guideline for a series of sequential steps describing the precise patient treatment.

3.3.23 Region. A geographic area that includes the local and neighboring jurisdiction for an EMS agency.

REFERENCE CITED IN COMMENTARY

1. *The National EMS Scope of Practice Model,* U.S. Department of Transportation/National Highway Traffic Safety Administration, Washington, DC, 2007, www.nhtsa.gov.

Competencies for Hazardous Materials/WMD Basic Life Support (BLS) Responder

CHAPTER 4

Hazardous materials/WMD incidents differ from other situations in which EMS personnel provide patient care. These incidents require greater coordination with other response personnel, and usually it is not the basic life support (BLS) responder that effects a rescue. These kinds of incidents are time-consuming from the BLS responder standpoint, as they involve decontamination, protection of the transport vehicle, and protection from cross contamination.

4.1 General

4.1.1 Introduction.

All EMS personnel at the hazardous materials/WMD BLS responder level, in addition to their BLS certification, shall be trained to meet at least the core competencies of the operations level responders as defined in NFPA 472, *Standard for Competence of Responders to Hazardous Materials/Weapons of Mass Destruction Incidents*, and all competencies of this chapter.

4.1.2 Goal.

The goal of the competencies at the BLS responder level shall be to provide the individual with the knowledge and skills necessary to safely deliver BLS at hazardous materials/WMD incidents, function within the established incident management system, and perform the following duties:

(1) Analyze a hazardous materials/WMD incident to determine the potential health hazards encountered by the BLS level responder, other responders, and anticipated and actual patients by completing the following tasks:

It is imperative that BLS responders maintain good situational awareness when responding to hazardous materials/WMD incidents. In some circumstances, the local ambulance company might arrive first at a hazardous materials/WMD incident. In those cases, the BLS responders should take time to size up the scene and understand the hazards present. When the BLS providers arrive after an organized hazmat response has begun, it is equally important to again size up the incident to understand the potential threats to the health and safety of the responders and civilian population.

(a) Survey an incident where hazardous materials/WMD have been released and evaluate suspected and identified patients for signs and symptoms of exposure.

BLS responders must be aware of the subtle signs and symptoms that can accompany a chemical exposure. Not all patients present with an overt or clear-cut collection of signs and symptoms. When multiple patients are generated by an event, the BLS responders should look for consistencies (and inconsistencies) in the patient presentations. It may be possible to determine the presence of a nerve agent, for example, by the signs and symptoms exhibited by exposed civilians.

(b) Collect hazard and response information from available technical resources to determine the nature of the problem and potential health effects of the substances involved.

BLS responders must take the necessary steps to correlate the physical and chemical properties of a released substance to the potential health effects and treatment(s) provided to exposed patients.

If BLS responders are called to the scene to "stand by" during a working hazmat incident, they should actively seek information regarding the chemical and physical properties of the released substance. In some cases, alerting the anticipated receiving hospital would be prudent if specific antidotes or other therapies might be needed. It is wise to aggressively plan to render care — it will be too late to formulate a good plan after the exposure happens.

(2) Plan to deliver BLS to any exposed patient within the scope of practice by completing the following tasks:

(a) Identify preplans of high-risk areas and occupancies to identify potential locations where significant human exposures can occur.

It is important for BLS care providers to identify fixed facilities or other locations where hazardous substances are routinely used. The idea is to prepare for the low-frequency-high-impact situations before they occur. As an example, it would be good to know that a facility within a response area uses hydrofluoric acid (HF) and also keeps available (or identifies readily available sources of) topical calcium gluconate gel to treat skin exposures. Additionally, it would be beneficial to contact the appropriate receiving hospital to discuss procedures for handling a patient exposed to HF.

(b) Identify the capabilities of the hospital network to accept exposed patients and perform emergency decontamination if required.

Just as with other specialty requirements, such as trauma, pediatrics, and burns, hospitals vary in their ability to handle contaminated patients. Pre-hospital care providers should be familiar with local hospitals and their ability to perform emergency decontamination. Ideally, all patients are fully decontaminated prior to transport.

(c) Identify the medical components of the communication plan.

All BLS responders should be familiar with the radio communications procedures for their jurisdictions.

(d) Describe the role of the BLS level responder as it relates to the local emergency response plan and established incident management system.

All BLS level responders should understand their role in the incident command system (ICS). The ICS is a reliable and effective method of organizing and managing a hazardous materials/WMD incident. Responders of any discipline who act outside their accepted role at an incident run the risk of undermining the goals and objectives set by the incident commander (IC) and put themselves and fellow responders at greater risk. It may be tempting to arrive at a large incident, for example, and start working without direction. Unfortunately, this is counterproductive to the overall goal and objectives of the incident.

(3) Implement a prehospital treatment plan within the scope of practice by completing the following tasks:

(a) Determine the nature of the hazardous materials/WMD incident as it relates to anticipated or actual patient exposures and subsequent medical treatment.

BLS responders should learn as much as possible about the physical and chemical properties of the released substance in order to develop an informed treatment plan for exposed individuals.

(b) Identify the need for and the effectiveness of decontamination efforts.

Prior to transport, it is incumbent on the medical personnel accepting responsibility for a patient (or patients) to ensure that patient has been fully decontaminated. It is unacceptable to spread contamination from the scene to the hospital.

(c) Determine if the available medical resources will meet or exceed patient care needs.

All BLS response personnel should have the ability to estimate the medical resources necessary to deliver optimal care to victims of a hazardous materials/WMD event. The BLS responder should be capable of advising the incident commander regarding quality and quantity of necessary resources, the availability of those resources, and special conditions affecting access to the scene and delivery of patient care. In general, the BLS responder should be able to look at the incident from a bird's eye perspective and understand the needs of the incident as a whole. It is easy to get tunnel vision and focus on the needs of a single patient while neglecting the needs of the patient care system.

(d) Describe evidence preservation issues associated with patient care.

Evidence preservation is not solely the responsibility of law enforcement personnel. Each responder involved with the incident is responsible for identifying and preserving potential evidence. This includes contaminated clothing, personal articles belonging to the exposed individuals, and bodies of victims.

(e) Develop and implement a medical monitoring plan for responders.

Pre-entry medical monitoring, as it relates to the potential for rendering treatment for an accidentally exposed responder, is an important facet of providing EMS at hazardous materials/WMD incidents. EMS responders must think beyond the basics of vital signs when doing pre-entry medical monitoring. They must aggressively plan to render care (for an unplanned exposure) based on the nature of the substance involved.

Medical support personnel should also address the medical history and baseline physical status of responders immediately prior to and immediately following an operational rotation. Medical history should include any present illness or condition, past medical history, allergies, and medications. Physical evaluation should include such routine measures as pulse, blood pressure, body temperature, and respiratory rate.

Other measures that should be evaluated include pre- and post-entry body weight. These statistics will provide invaluable information in discerning between dehydration and free water toxicity or overhydration. It has been established that prolonged excessive ingestion of water can lead to electrolyte derangement and hyponatremia, and the physical signs and symptoms are often similar to the manifestations of dehydration. It is important that comparison weights be recorded so that an incorrect treatment course can be avoided.

(f) Report and document the actions taken by the BLS level responder at the incident scene.

BLS providers infrequently encounter patients suffering significant illness as a result of an exposure. When those situations occur, however, it is critical that the BLS provider understand the nature of the exposure and treat it accordingly. Documenting and/or reporting those findings is a valuable tool for the continuum of care. A good report at the transfer of care is important.

4.2 Competencies — Analyzing the Incident

Medical support personnel should be capable of understanding the common overt presentations of human and animal exposures. This capacity is necessary so that the information can be communicated to the incident commander and to the next level of pre-hospital care providers (if present) to provide valuable intelligence for strategic determinations.

4.2.1 Surveying Hazardous Materials/WMD Incidents.

Given scenarios of hazardous materials/WMD incidents, the BLS level responder shall assess the nature and severity of the incident as it relates to anticipated or actual EMS responsibilities at the scene.

4.2.1.1 Given examples of the following types of containers, the BLS level responder shall identify the potential mechanisms of injury/harm and possible treatment modalities:

(1) Pressure
(2) Nonpressure
(3) Cryogenic
(4) Radioactive

Good situational awareness enables the BLS responder to see and interpret the "big picture" at a hazardous materials/WMD incident. Without knowing anything about the nature of the product inside, the BLS responder should, simply by identifying the type of container, link together some basic physical characteristics of the contained substance that could have an impact on an exposed patient. The same applies for other types of containers. Again, the BLS responder should develop the ability to pick out the important visual clues present and link them to the delivery of patient care.

4.2.1.2 Given examples of the nine U.S. Department of Transportation (DOT) hazard classes, the BLS level responder shall identify possible treatment modalities associated with each hazard class.

Even when the exact chemical name is unknown, it is still possible to formulate a treatment plan based on the broad hazards of a particular classification of a chemical. When no other information is known about the hazardous material/WMD, routine medical care could be the most appropriate care provided.

4.2.1.3 Given examples of various hazardous materials/WMD incidents at fixed facilities, the BLS level responder shall identify the following available health-related resource personnel:

(1) Environmental heath and safety representatives

Typically, corporate environmental health and safety representatives are knowledgeable about the chemical inventories, processes, and other important health- and safety-related issues at the site. These employees are typically trained in safety and industrial hygiene and are familiar with their workplace and the hazards that could be present, whether stored chemicals or materials used in manufacturing processes.

(2) Radiation safety officers

The requirements for a fixed facility to have a radiation safety officer (RSO) vary with the type of license and/or the radioactive substances used. If BLS providers respond to a radiation exposure at a fixed facility, attempts should be made to contact the RSO. That person could provide useful guidance on the potential health effects of an exposure to radiation.

(3) Occupational physicians and nurses

These individuals may provide medical information pertaining to the victims (plant or site workers) to the medical support personnel.

(4) Site emergency response teams

Site-specific emergency response teams (ERTs) can be a useful asset to BLS responders called to the scene of a hazardous materials/WMD incident. In many cases, formalized fire brigades and/or ERTs are able to provide site-specific information — from an emergency response perspective — to the arriving responders.

(5) Product or container specialists

4.2.1.4 Given various scenarios of hazardous materials/WMD incidents, the BLS level responder, working within an incident command system, shall evaluate the off-site consequences of the release based on the physical and chemical nature of the released substance and the prevailing environmental factors, to determine the need to evacuate or to shelter in place affected persons.

4.2.1.5 Given the following biological agents, the BLS level responder shall define the signs and symptoms of exposure and the likely means of dissemination:

(1) Variola virus (smallpox)
(2) *Botulinum* toxin
(3) *E. coli* O157:H7
(4) Ricin toxin
(5) *B. anthracis* (anthrax)
(6) Venezuelan equine encephalitis virus
(7) *Rickettsia*
(8) *Yersinia pestis* (plague)
(9) Tularemia
(10) Viral hemorrhagic fever
(11) Other CDC Category A, B, or C–listed organism

The U.S. Centers for Disease Control and Prevention (CDC)'s web site for bioterrorism agents and diseases states the following:

> The U.S. public health system and primary healthcare providers must be prepared to address various biological agents, including pathogens that are rarely seen in the United States.
>
> **Category A Diseases/Agents.** High-priority agents that pose a risk to national security because they
>
> - Can be easily disseminated or transmitted from person to person;
> - Result in high mortality rates and have the potential for major public health impact;
> - Might cause public panic and social disruption; and
> - Require special action for public health preparedness.
>
> **Category B Diseases/Agents.** Second highest priority agents include those that
>
> - Are moderately easy to disseminate;
> - Result in moderate morbidity rates and low mortality rates; and
> - Require specific enhancements of CDC's diagnostic capacity and enhanced disease surveillance.
>
> **Category C Diseases/Agents.** Third highest priority agents include emerging pathogens that could be engineered for mass dissemination in the future because of
>
> - Availability;
> - Ease of production and dissemination; and
> - Potential for high morbidity and mortality rates and major health impact.
>
> Source: U.S. Centers for Disease Control and Prevention, www.bt.cdc.gov/agent/agentlist-category.asp

4.2.1.6 Given examples of various types of hazardous materials/WMD incidents involving toxic industrial chemicals (TICs) and toxic industrial materials (TIMs) (e.g., corrosives, reproductive hazards, carcinogens, nerve agents, flammable and/or explosive hazards, blister agents, blood agents, choking agents, and irritants), the BLS level responder shall determine the general health risks to patients exposed to those substances in the case of any release with the following:

(1) A visible cloud

There are countless substances that can harm civilians and responders alike. It is not the intention of 4.2.1.6 to require BLS responders to be familiar with every potential substance. The BLS responder should, however, be familiar with those substances that are present in their jurisdiction, as well as the most common WMDs, toxic industrial chemicals (TICs), and toxic industrial materials (TIMs).

The U.S. Occupational Safety and Health Administration (OSHA) states that TICs are industrial chemicals that are manufactured, stored, transported, and used throughout the world. The physical state of TICs can be gas, liquid, or solid. They can be chemical hazards (e.g., carcinogens, reproductive hazards, corrosives, or agents that affect the lungs or blood) or physical hazards (e.g., flammable, combustible, explosive, or reactive). See www.osha.gov for a table of TICs, and see www.osha.gov/SLTC/emergencypreparedness/guides/chemical.html for more information on chemicals.

The National Institute of Justice (NIJ) states that TIMs are similar to TICs but are chemicals other than chemical warfare agents that have harmful effects on humans [1]. They are used in a variety of settings such as manufacturing facilities, maintenance areas, and general storage areas.

Training programs for BLS responders should include a section on the basic physical and chemical properties of the broad classifications of hazardous materials/WMD they might encounter. Typically, the scope of practice for BLS level responders is limited when it comes to handling exposures resulting from hazardous materials/WMD incidents (see Exhibit II.4.1). Therefore, training in this area should focus on recognizing the signs and symptoms of typical hazardous materials/WMD substances, basic treatment modalities, and the best practices to avoid becoming exposed.

(2) Liquid pooling
(3) Solid dispersion

4.2.1.7 Determining If a Hazardous Materials/WMD Incident Is an Illicit Laboratory Operation. Given examples of hazardous materials/WMD incidents involving illicit laboratory operations, BLS level responders assigned to respond to illicit laboratory incidents shall identify the potential drugs/WMD being manufactured and shall meet the following related requirements:

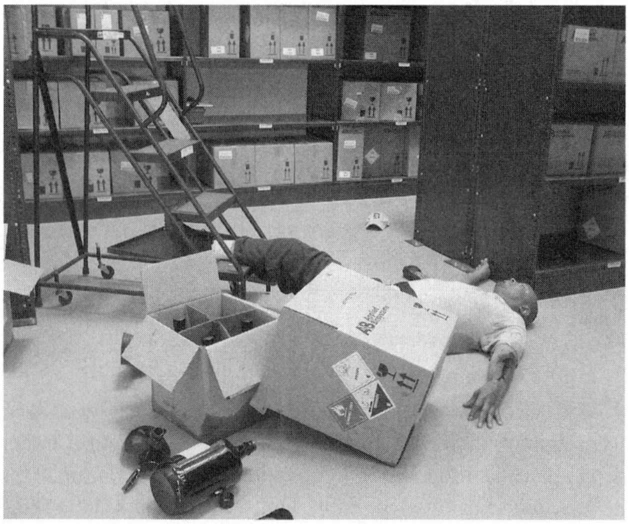

EXHIBIT II.4.1 *The BLS responder has to determine the health risks to individuals exposed to toxic materials. (Source: Rob Schnepp)*

A complete site risk assessment must be done prior to entry to identify potential hazards to the responders, civilian population, and the community as a whole. Each lab process will have distinct hazards that must be preplanned before the event. This preplanning of resources identifies the existing hazards and the potential hazards the responder can come across. In all cases the appropriate level of personal protective equipment must be worn. Some common examples of general hazards include the following:

1. Confined spaces
2. Electrical hazards
3. Pathogen hazards
4. Damaged or unstable containers
5. Flammable, corrosive, oxygen deficient or enriched atmospheres

(1)*Given examples of illicit drug manufacturing methods, describe the operational considerations, hazards, and products involved in the illicit process.

As with any emergency event, realizing the potential hazards and health effects are a part of the risk assessment. In any incident, especially those where an illicit lab might be present, a complete hazard-risk assessment must be accomplished. Some items to consider when responding to an illicit lab include the following:

1. Large amount of waste product(s) can be generated from the synthesis process. For example, for every 1 lb of production within a methamphetamine lab, 5 lb to 6 lb of waste can be generated.
2. The level of concern is based on the precursors used within these processes. Each lab type will create a different level of apprehension. A variety of acids, bases, and solvents are among the most common chemicals used.
3. Some processes cause reduced oxygen levels within enclosed environments that are sometimes associated with flammable gases and with asphyxiating gases with ignition sources.

Common chemicals used in an illicit drug lab include the following:

- Acetic acid (glacial)
- Acetone
- Anhydrous ammonia
- Benzene
- Bromine
- Calcium hypochlorite
- Chlorine gas
- Chloroform
- Cyclohexane
- Dimethylamine
- Dioxane
- Ethyl alcohol
- Formic acid
- Hexane
- Hydrochloric acid
- Hydrogen
- Hydrogen bromide
- Hydrogen sulfide gas
- Isopropyl alcohol
- Methylene chloride
- Nitromethane
- Picric acid
- Potassium fluoride
- Red phosphorus
- Sulfuric acid
- Thionly chloride
- Toluene
- Xylene

Potential health hazards can fall into one or more of the following categories:

1. Flammable
2. Corrosive
3. Oxygen deficient/enriched
4. Asphyxiant

A.4.2.1.7(1) Examples of products involved in illicit drug manufacturing include the following:

(1) Ammonia
(2) Ephedrine and pseudoephedrine
(3) Flammable solvents such as ether compounds and methanol
(4) Fluorinated/chlorinated hydrocarbons (Freon)
(5) Hydrogen chloride
(6) Aluminum chloride
(7) Iodine
(8) Lithium or sodium metal
(9) Phosphine gas
(10) Red phosphorus
(11) Sodium hydroxide or other caustic substances

(2) Given examples of illicit chemical WMD methods, describe the operational considerations, hazards, and products involved in the illicit process.

The production of chemicals used for illegal purposes presents a myriad of challenges for the BLS responder. Management of these labs can, in certain circumstances, involve the handling of chemicals, including acids, bases, solvents, reagents, and intermediates.

These chemicals with their associated reagents sometimes produce intermediates. This reaction process can further require heating or cooling reactants and/or products. Cooling a substance that requires heat or heating a substance that requires cooling could cause further release of product. An understanding of the process is necessary before interacting with the lab.

Common chemicals that might be used in an illicit chemical lab include the following:

- Acetylene
- Acetone
- Ammonium fluoride
- Arsenic
- Arsenic trichloride
- Benzyl chloride
- Bromine
- Carbon dioxide
- Carbon tetrachloride
- Chlorine gas
- Cyclohexanol
- Dimethylamine
- Ethylene
- Ethylene dichloride
- Hydrofluoric acid
- Malononitrile
- Methyl chloride
- Nitromethane
- Phosphorus trichloride
- Potassium cyanide
- Sodium cyanide
- Sodium fluoride
- Sulfuric acid
- Toluene
- Vinyl chloride

Potential health hazards fall into one or more of the following categories:

1. Neurotoxins (nerve agents)
2. Chemical asphyxiants (blood agents)
3. Irritating (lachrymators)
4. Strong bases (mustards)
5. Psychomotor (disabling) toxins (psychogenic)

(3) Given examples of illicit biological WMD methods, describe the operational considerations, hazards, and products involved in the illicit process.

The biological laboratory can have both chemical hazards and the biological agent that is being produced. As with the drug, chemical, and explosives labs, the biological lab has chemicals in the precursor form as reagents before the desired product is formed.

The goal of a biological process is to grow or extract biological materials. These can include bacteria, viruses, or toxins extracted from the microorganism.

A growth medium called agar is used in the growing stage of the process. Agar can be protein, blood, or chemical based. These samples are then incubated for potential growth production in which the culture is produced. Some biological agents can be concentrated and dried for easy dissemination.

Biological weapons can be placed into three general categories based on potential outcome: lethal, potentially lethal, or incapacitating. Commentary Table II.4.1 provides some examples of these categories.

COMMENTARY TABLE II.4.1 *Three General Categories of Biological Weapons*

Lethal	*Potentially Lethal*	*Incapacitating*
Anthrax	Diphtheria	West Nile encephalitis
Glanders	Brucellosis	Q-Fever
Plague	Monkey pox infection	Influenza
Smallpox	Tularemia	Dengue fever
Ebola	Psittacosis	VEE, EEE

As with any illicit lab, waste production has components of the desired end product along with process waste. These components can be just as hazardous as the initial production products.

(4) Given examples of illicit laboratory operations, describe the potential booby traps that have been encountered by response personnel.

As with any illegal activity, the perpetrator does not want to be found or recognized. These labs are producing product for financial return. The end manufactured goods have a street value several times the cost of production. In many cases, these operations are guarded by booby traps, which can be one or more of the following:

- Trip wires
- Pipe bombs (IEDs)
- Explosives in a variety of common household appliances
- IEDs in light sockets
- Light bulbs filled with gasoline
- Containers of acids/bases/cyanides over door openings or behind doors
- Weapons
- Guard dogs

(5) Given examples of illicit laboratory operations, describe the agencies that have investigative authority and operational responsibility to support the response.

The resources within any community must be planned before the event. Some examples of resources, agencies, and disciplines that might be required include the following:

- Field hospitals
- Emergency department loads, support staff, physicians, and so on
- Medical antidote and/or prophylaxis

Community resources include the following:

- EMS
- Law enforcement

- Fire department
- Local health department
- Mutual aid/automatic response from surrounding communities
- Local interhospital agreements
- Medical reserve corps

State-level resources include the following:

- Statewide mutual aid for catastrophic events (multiple disciplinary)
- Pharmaceutical stockpiles
- Time-phased resources for long-term events (state planning or federal resources)
- Civil support teams
- State health department
- State law enforcement
- National Guard

Federal-level resources include the following:

- U.S. Disaster Medical Assistance Teams (DMAT)
- Pharmaceutical stockpiles
- Federal Bureau of Investigation (FBI)
- Centers for Disease Control and Prevention (CDC)
- Military assets (National Guard)
- National Medical Response Team (NMRT)
- Metropolitan Medical Response System (MMRS)

4.2.1.8 Given examples of a hazardous materials/WMD incident involving radioactive materials, including radiological dispersion devices, the BLS level responder shall determine the probable health risks and potential patient outcomes by completing the following tasks:

(1) Determine the most likely exposure pathways for a given radiation exposure, including inhalation, ingestion, and direct skin exposure.
(2) Identify the difference between radiation exposure and radioactive contamination and the health concerns associated with each.

Radiation exposure occurs when a person is near a source of radiation. The person might receive a radiation exposure but might not become radioactive. As an example, a person having an X-ray receives a radiation exposure but does not become radioactive. Persons suffering a radiation exposure do not require decontamination prior to medical treatment.

Radioactive contamination occurs when loose particles, contaminated with a radioactive material, settle on clothing, personal protective equipment, or skin. A person with radioactive contamination is radioactive. Radiation illness can result from inhaled or ingested particles. Persons with radioactive contamination must be decontaminated prior to medical treatment.

4.2.1.9 Given three examples of pesticide labels and labeling, the BLS level responder shall use the following information to determine the associated health risks:

(1) Hazard statement
(2) Precautionary statement

Precautionary statements guide the applicator in taking proper precautions to protect humans or animals that could be exposed.

(3) Signal word

Each pesticide label must contain a signal word. The three signal words, in order of increasing toxicity, are *caution*, *warning*, and *danger*.

The following are the toxicity categories for pesticides, which are established by the U.S. Environmental Protection Agency (EPA) according to their acute (short-term) toxicity [7]:

- **Toxicity Category I:** All pesticide products meeting the criteria of Toxicity Category I must bear on the front panel the signal word "Danger." In addition, if the product was assigned to Toxicity Category I on the basis of its oral, inhalation, or dermal toxicity (as distinct from skin and eye local effects), the word "Poison" must appear in red on a background of distinctly contrasting color and the skull and crossbones must appear in immediate proximity to the word "poison."
- **Toxicity Category II:** All pesticide products meeting the criteria of Toxicity Category II must bear on the front panel the signal word "Warning."
- **Toxicity Category III:** All pesticide products meeting the criteria of Toxicity Category III must bear on the front panel the signal word "Caution."
- **Toxicity Category IV:** All pesticide products meeting the criteria of Toxicity Category IV must bear on the front panel the signal word "Caution."
- **Child Hazard Warning Statement:** The Child Hazard Warning Statement "Keep Out of Reach of Children" is required on all product labels, unless the requirement is waived. The warning statement requirement may be waived when the registrant (the individual or entity registering the pesticide with EPA) adequately demonstrates that the likelihood of contact with children during distribution, storage, or use is extremely remote or if the pesticide is approved for use on infants or small children.

The EPA uses the criteria shown in Commentary Table II.4.2 to determine the toxicity category of pesticides. These criteria are based on the results of animal tests done in support of registration of the pesticide.

COMMENTARY TABLE II.4.2 Pesticide Toxicity Category Criteria

Hazard Indicators	*I*	*II*	*III*	*IV*
Oral LD_{50}	Up to and including 50 mg/kg	50 thru 500 mg/kg	500 thru 5,000 mg/kg	5,000 mg/kg
Dermal LD_{50}	Up to and including 200 mg/kg	200 thru 2000 mg/kg	2000 thru 20,000 mg/kg	20,000 mg/kg
Inhalation LC_{50}	Up to and including 0.2 mg/liter	0.2 thru 2 mg/liter	2 thru 20 mg/liter	20 mg/liter
Eye irritation	Corrosive; corneal opacity not reversible within 7 days	Corneal opacity reversible within 7 days; irritation persisting for 7 days	No corneal opacity; irritation reversible within 7 days	No irritation
Skin irritation	Corrosive	Severe irritation at 72 hours	Moderate irritation at 72 hours	Mild or slight irritation at 72 hours

Source: Title 40 CFR Part 156.62, Toxicity Category.

(4) Pesticide name

A specific name, usually registered as a trademark, will identify a pesticide as being produced by a particular manufacturer.

4.2.2 Collecting and Interpreting Hazard and Response Information.

The BLS level responder shall obtain information from the following sources to determine the nature of the medical problem and potential health effects:

(1) Hazardous materials databases
(2) Clinical monitoring
(3) Reference materials
(4)*Technical information centers (e.g., CHEMTREC, CANUTEC, and SETIQ) and local state and federal authorities

A.4.2.2(4) CHEMTREC, the Chemical Transportation Emergency Center, is a round-the-clock resource for obtaining immediate emergency response information for accidental chemical releases. CANUTEC, the Canadian Transport Emergency Centre, is operated by Transport Canada to assist emergency response personnel in handling dangerous goods emergencies. SETIQ is the Mexican Emergency Transportation System for the Chemical Industry.

(5) Technical information specialists
(6) Regional poison control centers

4.2.3 Establishing and Enforcing Scene Control Procedures.

Given two scenarios involving hazardous materials/WMD incidents, the BLS level responder shall identify how to establish and enforce scene control, including control zones and emergency decontamination, and communications between responders and to the public and shall meet the following requirements:

(1) Identify the procedures for establishing scene control through control zones.
(2) Identify the criteria for determining the locations of the control zones at hazardous materials/WMD incidents.
(3) Identify the basic techniques for the following protective actions at hazardous materials/WMD incidents:
  (a) Evacuation
  (b) Sheltering-in-place protection
(4) Demonstrate the ability to perform emergency decontamination.
(5) Identify the items to be considered in a safety briefing prior to allowing personnel to work at the following:
  (a) Hazardous materials incidents
  (b) Hazardous materials/WMD incidents involving criminal activities
(6) Identify the procedures for ensuring coordinated communication between responders and to the public.

Even though it might not be a primary responsibility for EMS personnel to establish control zones, it is still an important concept to understand. There could be times when BLS care providers, especially those assigned to transport ambulances, arrive first at the scene of a hazardous materials/WMD incident. In those instances, it might be necessary to control the scene instead of stopping to treat the first victim encountered.

In the event that control zones are not in place prior to the arrival of BLS level personnel, control zones should be quickly established. BLS care providers should draw on their hazmat training to understand the nature of the release and set zones accordingly. In some cases, it is more beneficial to begin to manage the incident instead of finding and treating the first patient encountered.

All BLS responders should be familiar with the tools, equipment, and standard operating procedures of the AHJ when it comes to emergency decontamination. It might be necessary to improvise on the scene — be flexible. If the concept of emergency decontamination is understood, the procedures can be worked out on the scene.

It is vital to provide the public with adequate information when it comes to illness and injuries as a result of a hazardous materials/WMD incident. The emergency response community as a whole should be cognizant of the need to provide good public information. This may involve incorporation in a joint information center or JIC.

4.3 Competencies — Planning the Response

Planning the response might involve the medical response/support personnel as subject matter contributors to the IC, safety officer, or other responders as appropriate. The strategic considerations that are directly addressed by medical specialists include hospital capability, hospital considerations, caregiver-to-victims ratios, special resources, medical logistics, and emergent, immediate, mid- and long-range clinical considerations.

4.3.1 Identifying High Risk Areas for Potential Exposures.

The goal of 4.3.1.1 is to remind BLS level responders that preplanning is essential. Almost every jurisdiction has an area of concern when it comes to hazardous materials/WMD events. It is unwise not to preplan at these locations. A sound preplan could be the difference between effectively handling an event and being completely overwhelmed.

4.3.1.1 The BLS level responder, given an events calendar and pre-incident plans, which can include the local emergency planning committee plan, as well as the agency's emergency response plan and standard operating procedures (SOPs), shall identify the venues for mass gatherings, industrial facilities, potential targets for terrorism, and any other location where an accidental or intentional release of a harmful substance can pose an unreasonable health risk to any person in the local geographical area as determined by the AHJ and shall identify the following:

(1) Locations where hazardous materials/WMD are used, stored, or transported
(2) Areas and locations that present a potential for a high loss of life or rate of injury in the event of an accidental or intentional release of hazardous materials/WMD
(3)* External factors that may complicate a hazardous materials/WMD incident

A.4.3.1.1(3) External factors can include geographic, environmental, mechanical, and transportation factors such as prevailing winds, water supply, vehicle and pedestrian traffic flow, ventilation systems, and other natural or man-made influences, including air and rail corridors.

4.3.2 Determining the Capabilities of the Local Hospital Network.

4.3.2.1 The BLS level responder shall identify the following methods and vehicles available to transport hazardous materials patients and shall determine the location and potential routes of travel to the medically appropriate local and regional hospitals, based on the patients' needs:

BLS responders, especially those charged with patient transport, should be familiar with the local and regional healthcare systems when it comes to the location, type, and capabilities of local hospitals. During the preplanning process, it is necessary to identify those facilities that are capable of providing care to exposed patients (see Exhibit II.4.2). BLS responders should be aware of the level of preparedness of the local hospitals.

(1) Adult trauma centers
(2) Pediatric trauma centers
(3) Adult burn centers
(4) Pediatric burn centers
(5) Hyperbaric chambers
(6) Established field hospitals
(7) Dialysis centers
(8) Supportive care facilities
(9) Forward deployable assets
(10) Other specialty hospitals or medical centers

4.3.2.2 Given a list of receiving hospitals in the region, the BLS level responder shall describe the location, availability, and capability of hospital-based decontamination facilities.

The methods used for decontamination will vary from hospital to hospital. Responders need to be familiar with the capabilities of local and regional primary care facilities as they relate

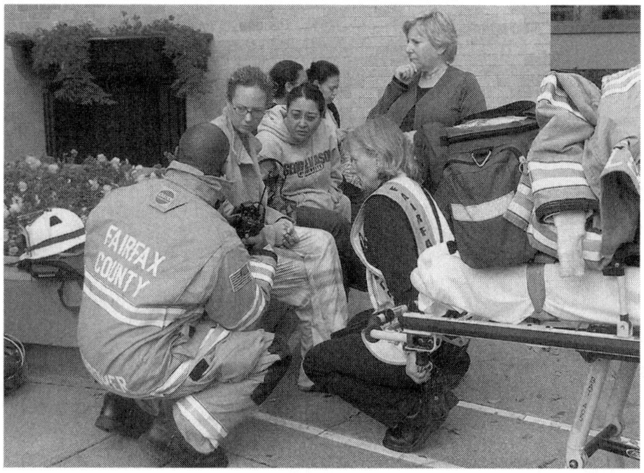

EXHIBIT II.4.2 *The photograph shows exposed patients at a hazardous materials incident. (Courtesy of Fairfax County Fire and Rescue Department)*

to guarding against cross contamination and treatment of hazardous materials exposures. This will aid in ensuring that the most appropriate receiving facility is chosen to ensure the best care for the patient and safety of attending personnel.

4.3.2.3 The BLS level responder shall describe the BLS protocols and SOPs at hazardous materials/WMD incidents as developed by the AHJ and the prescribed role of medical control and poison control centers, as follows:

The local emergency response plan, protocols, and procedures must define the actions that the BLS level responder should follow when confronted with a hazardous material–related mass causality incident. The action plan should provide direction for times when normal communications are disrupted. An alternative means of contacting medical control or poison control centers is required.

(1) During mass casualty incidents
(2) Where exposures have occurred
(3) In the event of disrupted radio communications

4.3.2.4 The BLS level responder shall identify the formal and informal mutual aid resources (hospital- and nonhospital-based) for the field management of multicasualty incidents, as follows:

(1) Mass-casualty trailers with medical supplies
(2) Mass-decedent capabilities
(3) Regional decontamination units
(4) Replenishment of medical supplies during long-term incidents
(5) Rehabilitation units for the EMS responders
(6) Replacement transport units for vehicles lost to mechanical trouble, collision, theft, and contamination

4.3.2.5 The BLS level responder shall identify the special hazards associated with inbound and outbound air transportation of patients exposed to hazardous materials/WMD.

The threat of a contaminant spread by inbound aircraft and the potential contamination of the aircraft by the release can pose a significant risk to responders, the public, and possibly the aircraft itself. The threat to the outbound aircraft is the risk that exposed patients might off-gas a substance that would affect the flight crew while the aircraft is in flight. Additionally, the threat of contaminating the aircraft itself is another risk. The proper management of air operations at hazardous materials events is extremely important to both the aircraft flight crews and those on the ground.

4.3.3 Identifying Incident Communications.

4.3.3.1 Given an incident communications plan, the BLS level responder shall identify the following:

(1) Medical components of the communications plan
(2) Ability to communicate with other responders, transport units, and receiving facilities

4.3.3.2 Given examples of various patient exposure scenarios, the BLS level responder shall describe the following information to be transmitted to the medical or poison control center or the receiving hospital prior to arrival:

BLS responders should understand that an accurate and complete report to the receiving hospital is important. This provides the necessary lead time for hospital staff to prepare for accepting the exposed patient.

(1) The name of the substance(s) involved
(2) Physical and chemical properties of the substance(s) involved
(3) Number of victims being transported
(4) Age and sex of transported patient
(5) Patient condition and chief complaint
(6) Medical history
(7) Circumstances and history of the exposure, such as duration of exposure and primary route of exposure
(8) Vital signs, initial and current
(9) Symptoms described by the patient, initial and current
(10) Presence of associated injuries, such as burns and trauma
(11) Decontamination status
(12) Treatment rendered or in progress
(13) Patient response to treatment(s)
(14) Estimated time of arrival

4.3.4 Identifying the Role of the BLS Level Responder.

The agency's hazardous materials emergency response plan should outline the role of the basic life support level responder. Primarily, that role includes responding to an emergency, assessing the nature of the incident, implementing protective measures, notifying other agencies, asking for additional assistance, establishing or working within an incident management system, and performing basic life support medical treatment, triage, and transport in accordance with local protocols and procedures.

4.3.4.1 Given scenarios involving hazardous materials/WMD, the BLS level responder shall identify his or her role during hazardous materials/WMD incidents as specified in the emergency response plan and SOPs developed by the AHJ, as follows:

(1) Describe the purpose, benefits, and elements of the incident command system as it relates to the BLS level responder.

The ICS is an organized structure of roles, responsibilities, and procedures for the command and control of emergency operations. ICS is modular and can expand or contract based on the need, the size, and nature of an incident. It enables multiple disciplines and multiple jurisdictions to work together safely and effectively.

The ICS utilizes the following three management concepts: unity of command, span of control, and functional positions.

Unity of command stipulates that only one incident command or unified command is ultimately responsible for the entire incident. The command structure encompasses clearly defined lines of authority in which everyone is responsible to, and directed by, one person.

Span of control is established so that only three to seven individuals report to one position so that no one position becomes overloaded, with optimum span of control at five.

The functional positions concept means that all resources assigned to one functional position (for example, fire fighter, BLS responder, hazardous materials officer, hazardous materials technician) are to remain in that position until reassigned or released from the incident.

(2) Describe the typical incident command structure, for the emergency medical component of a hazardous materials/WMD incident as specified in the emergency response plan and SOPs, as developed by the AHJ.

The medical component of the command structure will normally be a functional group either directly under the incident commander or the operations sections, if established. The function can be expanded to a medical branch with a medical group performing treatment and triage and a patient transportation group. Further expansion of the medical branch might include dividing the medical group into the triage group and treatment group, depending on the scope of the incident and available resources. Some additional functions of the medical group could be extrication and air operations. Additionally, BLS level responders could be assigned to the medical unit under the logistic section, if established. The medical unit is responsible for responder medical treatment while the medical group/branch under the operations section would be responsible for public medical treatment. The size and complexity of the medical function will expand and collapse, depending on the needs of the incident.

(3) Demonstrate the ability of the BLS level responder to function within the incident command system.

The BLS level responder must be able to function within the ICS. The BLS responder first reports to the incident command post or staging area to check in. Upon receiving an assignment, the BLS level responder reports to the assignment (i.e., incident, response area) and reports to the branch director/group supervisor at the incident or response area and performs the task assigned. Upon completion of the task, the BLS level responder reports back to the supervisor for release or reassignment.

(4) Demonstrate the ability to implement an incident command system for a hazardous materials/WMD incident where an ICS does not currently exist.

The BLS level responder might be the first on the scene of an emergency. When this happens, it is imperative that the BLS level responder initiate the ICS and assume command until relieved by a more senior authority. The importance of this action is to ensure responder safety and coordination of resources to best effect the desired outcome of the incident.

(5) Identify the procedures for requesting additional resources at a hazardous materials/WMD incident.

Responders at every level are required to know what types of resources are available and how to request them. The employer's emergency response plan and procedures should identify the processes by which the BLS responder will request additional resources both within and from outside the organization.

4.3.4.2 The hazardous materials/WMD BLS responder shall describe his or her role within the hazardous materials response plan developed by the AHJ or identified in the local emergency response plan, as follows:

(1) Determine the toxic effect of hazardous materials/WMD.

As early as possible in the incident, the identity or classification of the product should be obtained so that the BLS level responder can take appropriate actions to minimize the impact of toxic effect on the patient(s). Once the material is identified or classified, the BLS level responder can call medical control for orders on treatment if no standing protocols exist. Information can also be obtained from a poison control center. Consideration must also be given to personal protection and decontamination of victims prior to patient treatment.

(2) Estimate the number of patients.

As with any multi-casualty incident, the estimate of the number of potential patients needs to be ascertained to determine the number and types of resources necessary to triage, treat, and transport the injured.

(3) Recognize and assess the presence and severity of symptoms.

The BLS level responder should be able to associate symptoms consistent with the product exposure as confirming the determination of the toxic effects of exposure to a material. Severity of symptoms might not be immediately known, depending on the material, since symptoms are the result of a dose and the material's effect on the body and our ability to intervene.

(4) Take and record vital signs.

The proper taking and recording of vital signs such as respiration, pulse, and blood pressure on regular intervals will assist in evaluating the severity of the exposure and monitoring the body's vital functions in response to the toxic effects.

(5) Determine resource maximization and assessment.

A vital part of initial triage is the identification of additional resource needs as early as possible and to utilize them to their maximum efficiency and effectiveness within the ICS established.

(6) Assess the impact on the health care system.

Once the number of patients and the identity or classification of the materials have been determined, the local heath care system should be alerted and medical control notified. The number of decontaminated patients being transported and the number of self-transporting victims enroute to a facility should be communicated to medical control as early as possible. Every effort should be given to minimize the impact to any facility, both in numbers of transported injured and potential for cross contamination of transport vehicles and emergency facilities.

(7) Perform appropriate patient monitoring.

While treating and transporting patients, it is imperative that the patient be closely monitored to identify any changes in condition, extent of symptoms, or impacts of underlying medical problems.

(8) Communicate pertinent information.

Information related to the incident as well as the patients' numbers, types, and severity of exposure or injuries should be communicated and updated regularly throughout the event to receiving facilities and medical control as well as the incident command structure.

4.4 Competencies — Implementing the Planned Response

Because the BLS responder will play a key role in the EMS part of the response plan, on which the outcome for both patients and responders depends, the BLS responder is required by Section 4.4 to know enough about the hazardous materials to analyze the incident and effectively determine the risks and medical care necessary for his or her part of the response.

4.4.1 Determining the Nature of the Incident/Providing Medical Care.

The BLS level responder shall demonstrate the ability to identify the mechanisms of injury or harm and the clinical implications and provide emergency medical care to those patients exposed to hazardous materials/WMD agent by completing the following tasks:

The BLS responder is required by 4.4.1 to be able to determine from clues presented during dispatch, response, and approach whether a hazardous material is present at the scene and whether the released material poses a risk to the patient and, in turn, to the responders at the scene. Typical indicators of the presence of a released hazardous material include the following: operators or witnesses, placards, the normal occupancy of buildings at the scene (such as chemical storage buildings), the type of containers involved, and the presence of fires or explosions.

It is critical for BLS responders to have a good working understanding of the many physical states of potentially released substances that could be present at a hazardous materials/WMD incident as they perform their emergency medical care duties. Understanding these states and the associated mechanisms of injury and health implications associated with the released substance in these various states is critical to their care of patients as well as their personal safety while performing their medical care duties. It is also important for BLS responders to be aware that the released substances could be in more than one state at the incident, as well as more than one type of released substance.

(1) Determine the physical state of the released substance, in addition to the environmental influences surrounding the release, as follows:

(a) Solid

A solid is a material in the state of matter characterized by resistance to deformation and changes of volume. At the microscopic scale, a solid has the properties of the atoms or molecules being packed closely together, its constituent elements have fixed positions in space relative to each other, and because any solid has some thermal energy, its atoms vibrate. However, this movement is very small and very rapid and cannot be observed under ordinary conditions. It is important to recognize that solid materials can rapidly change state under certain environmental conditions (of particular concern are those solids that rapidly volatilize under certain environmental conditions).

(b) Liquid

A liquid is a state of matter whose shape is usually determined by the container it fills. Liquid particles (normally molecules or clusters of molecules) are free to move within the liquid volume, but their mutual attraction limits ability of particles to leave the volume. The volume of a quantity of liquid is fixed by its temperature and pressure. It is also important to recognize that liquids can rapidly change state under certain environmental conditions (of particular concern are those liquids that rapidly volatilize under certain environmental conditions).

(c) Gas

A gas is a state of matter that has atoms or molecules basically moving independently, with no forces keeping them together or pushing them apart. Hazardous substances in the gaseous state are of particular concern for the inhalation exposure route.

(d) Vapor

A vapor refers to a gas-phase material that normally exists as a liquid or solid under a given set of environmental conditions. Vapors are composed of single gas-phase molecules. Many, but not all, vapors are colorless and therefore invisible. Vapors do not wet objects with which they come in contact.

(e) Dust

A dust is fine (small) particles of dry matter. Handling, crushing, grinding, rapid impact, detonation, and breakdown of certain organic or inorganic materials, such as rocks, ore, metal, coal, wood, grains, or other such material, can generate dusts. Particles ranging in size from 0.1 μm (micrometer, or micron) in diameter to about 30 μm in diameter and are referred to as total suspended particulate matter (TSP). Particles in the size range between 0.1 μm and 10 μm are of particular concern for inhalation exposures.

(f) Mist

A mist or fog is a microscopic suspension of liquid droplets in a gas. Do not confuse a mist with a vapor. Mists can generally be seen and reduce visibility. Mists generally wet objects with which they come in contact.

(g) Aerosol

The BLS responder should also be aware of the environmental conditions at the incident and how these environmental conditions can alter the health impact and physical states of the released substances. For example, on very hot days or in situations where the event results in elevated temperature at the scene, a substance normally not very volatile can become quite volatile with increased gaseous releases. Windy conditions can create lower concentrations of the released substance in the working area. If aware of these conditions, BLS responders can position themselves upwind of the source and away from higher exposures. By being knowledgeable of situations like this, BLS responders are better able to perform their medical care duties safely and more efficiently.

(2) Identify potential routes of exposure and correlate those routes of exposure to the physical state of the released substance, to determine the origin of the illness or injury, as follows:

The BLS responders need to be aware of how the associated released substances can enter the patient, and themselves, if they are not careful. Also, the BLS responders should continually evaluate all routes of exposure as they render medical care to patient(s). For example, care must be taken not to introduce the released substance into a patient through ingestion or injection as they administer medications to the patient. Likewise, they should be aware that while caring for patients who have received inhalation doses of the released substance, they could be subjecting themselves as well as other patients to the substance from the exhalation of that patient. Again, it is important to consider all routes of exposure while proceeding to administer to patients at a hazardous materials/WMD incident.

(a) Inhalation

Inhalation is the means by which contaminants enter the body through the normal respiratory process (i.e., uptake through normal breathing process, with contaminants deposited along the respiratory tract into the lungs).

(b) Absorption

Absorption is the process by which contaminants are absorbed into the body through the skin and other exposed tissue (often referred to as dermal absorption or dermal uptake).

(c) Ingestion

Ingestion is the process of consuming contaminants through the normal ingestion process (i.e., usually through the process of consuming food and water).

(d) Injection

Injection is the process by which contaminants are introduced directly into the bloodstream by means of a needle, cannula, or other mechanical process. Contaminants entering the bloodstream through an open wound are considered to be introduced by injection.

(3)* Describe the potential routes of entry into the body, the common signs and symptoms of exposure, and the BLS treatment options approved by the AHJ for exposure(s) to the following classification of substances:

The BLS responder should maintain a strong working knowledge of the signs and symptoms of exposure from each of these various classes of substances. These signs and symptoms will provide the BLS responder with key insights to integrate with their awareness of the material(s) release, their associated routes of exposure, and their associated health effects and implications. These signs and symptoms are important checkpoints for a BLS responder to use in providing medical care to the patient(s). It is also important for the BLS responders to understand the health implications of these different classes of substances in order to protect themselves at the scene, as well as provide better medical care for their patients.

Patients exposed to these different classes of hazardous materials could pose a risk of cross contamination to others who come in contact with them, including the BLS responder. A BLS responder's knowledge of toxic exposure, patent assessment, and decontamination procedures is essential for the responder to determine what actions are necessary to prepare patients to be treated and transported safely. In some cases, treatment might need to wait until the hazardous materials technicians at the scene decontaminate and transfer a patient to the cold zone.

A.4.4.1(3) Examples of classified substances include the following:

(1) Acids, alkalis, and corrosives
(2) Fumigants and pesticides: organophosphates, carbamates, zinc or aluminum phosphide, strychnine, sulfuryl fluoride
(3) Chemical asphyxiants: cyanide, carbon monoxide, hydrogen sulfide
(4) Simple asphyxiants: nitrogen, helium
(5) Organic solvents: xylene, benzene, methylene chloride
(6) Nerve agents: Tabun, Sarin, Soman, V agent
(7) Vesicants and blister agents: mustard, Lewisite
(8) Blood agents: cyanide, cyanogen chloride, arsine
(9) Choking agents: ammonia, chlorine, diphosgene, phosgene
(10) Pepper spray, irritants, and riot-control agents: CS (orthochlorobenzalmalononitrile), CN (chloroacetophenone), CR (dibenzoxazepine), MACE (phenylchloromethylketone), OC (oleoresin capsicum)
(11) Biological agents and toxins: anthrax, mycotoxin, plague, viral hemorraghic fevers, smallpox, ricin
(12) Incapacitating agents: BZ, LSD
(13) Radiological materials: plutonium, cesium, iridium, technetium
(14) Nitrogen-containing compounds: aniline, nitrates
(15) Opiate compounds: fentanyl, morphine
(16) Fluorine compounds: hydrogen fluoride, hydrofluoric acid
(17) Phenolic compounds: carbolic acid, cresylic acid

(a) Corrosives

Corrosives are chemicals that cause visible destruction of, or irreversible alterations in, living tissue by chemical action at the site of contact. A chemical is considered to be corrosive if, when tested on the intact skin of albino rabbits by the method described in Appendix A of Title 49 CFR Part 173, it destroys or changes irreversibly the structure of the tissue at the site of contact following an exposure period of four hours [2]. For purposes of this standard, the term does not refer to action on inanimate surfaces.

(b) Pesticides

A pesticide is any substance or mixture of substances intended for preventing, destroying, repelling, or mitigating any pest. A pesticide might be a chemical substance or biological agent (such as a virus or bacteria) used against pests, including insects, plant pathogens, weeds, mollusks, birds, mammals, fish, nematodes (roundworms), and microbes that compete with humans for food, destroy property, spread disease, or are a nuisance. Many pesticides are poisonous to humans.

(c) Chemical asphyxiants

Chemical asphyxiants reduce the body's ability to absorb, transport, or utilize inhaled oxygen. They are often active at very low concentrations — a few parts per million (ppm).

(d) Simple asphyxiants

An asphyxiant is a substance that can cause unconsciousness or death by suffocation (asphyxiation). Asphyxiants, which have no other health effects, are referred to as simple asphyxiants. Asphyxiation is an extreme hazard when working in enclosed spaces. Responders must be trained in confined space entry before working in sewers, storage tanks, and so on, where gases such as methane can displace oxygen from the atmosphere.

(e) Organic solvents

Organic solvents are a chemical class of compounds that are used routinely in commercial industries. They share a common structure (at least 1 carbon atom and 1 hydrogen atom), low molecular weight, lipophilicity, and volatility, and they exist in liquid form at room temperature. They may be grouped further into aliphatic-chain compounds, such as *n*-hexane, and as aromatic compounds with a 6-carbon ring, such as benzene or xylene. Aliphatics and aromatics can contain a substituted halogen element and might be referred to as halogenated hydrocarbons, such as perchloroethylene (PCE or PER), trichloroethylene (TCE), and carbon tetrachloride. Alcohols, ketones, glycols, esters, ethers, aldehydes, and pyridines are substitutions for a hydrogen group. Organic solvents can dissolve oils, fats, resins, rubber, and plastics.

(f) Nerve agents

The following information on nerve agents is extracted from the Federation of American Scientists' web site:

> Nerve agents are a group of particularly toxic chemical warfare agents. They were developed just before and during World War II and are related chemically to the organophosphorus insecticides. The principle agents in this group are:
>
> - GA - tabun
> - GB - sarin
> - GD - soman
> - GF - cyclosarin
> - VX - methylphosphonothioic acid
>
> The "G" agents tend to be non-persistent whereas the "V" agents are persistent. Some "G" agents may be thickened with various substances in order to increase their persistence, and therefore the total amount penetrating intact skin. At room temperature GB is a comparatively volatile liquid and therefore non-persistent. GD is also significantly volatile, as is GA though to a lesser extent. VX is a relatively non-volatile liquid and therefore persistent. It is regarded as presenting little vapor hazard to people exposed to it. In the pure state nerve agents are colorless and mobile liquids. In an impure state nerve agents may be encountered as yellowish to brown liquids. Some nerve agents have a faint fruity odor.
>
> - GB and VX doses which are potentially life threatening may be only slightly larger than those producing least effects. Death usually occurs within 15 minutes after absorption of a fatal VX dosage.
> - Although only about half as toxic as GB by inhalation, GA in low concentrations is more irritating to the eyes than GB. Symptoms appear much more slowly from a skin dosage

(continues)

than from a respiratory dosage. Although skin absorption great enough to cause death may occur in 1 to 2 minutes, death may be delayed for 1 to 2 hours. Respiratory lethal dosages kill in 1 to 10 minutes, and liquid in the eye kills almost as rapidly.

Toxicological Data

Route	*Form*	*Effect*	*Type*	*GA*	*GB*	*GD*	*VX*	*Dosage*
Ocular	Vapor	Miosis	ECt_{50}	—	<2	<2	<0.09	mg•min/m^3
Inhalation at RMV = 15 l/min	Vapor	Runny Nose	ECt_{50}	—	<2	<2	<0.09	mg•min/m^3
Inhalation at RMV = 15 liters/min	Vapor	Incapacitation	ICt_{50}	—	35	35	25	mg•min/m^3
Inhalation at RMV = 15 liters/min	Vapor	Death	LCt_{50}	135	70	70	30	mg•min/m^3
Percutaneous	Liquid	Death	LD_{50}	4,000	1,700	350	10	mg

Ct (Concentration time; mg•min/m^3) - A measure of exposure to a gas, the effective vapor exposure, determined by the concentration of the gas (mg/m^3) and the length of exposure (min).

ECt_{50} (Effective Concentration Time; mg•min/m^3) - The Ct at which a gas debilitates 50% of the exposed population in a specific way.

ICt_{50} (Incapacitating Concentration Time; mg•min/m^3) - The Ct at which a gas incapacitates 50% of the exposed population.

LCt_{50} (Lethal concentration time; mg•min/m^3) - The Ct at which a gas kills 50% of the exposed population.

LD_{50} (Lethal dose; mg) - The dose or amount at which a substance kills 50% of the exposed population.

RMV (Respiratory minute volume; liters/min) - Volume of air inhaled per minute.

The values are estimates of the doses, which have lethal effects on a 70kg man. Effective dosages of vapor are estimated for exposure durations of 2-10 minutes. The effects of the nerve agents are mainly due to their ability to inhibit acetylcholinesterase throughout the body. Since the normal function of this enzyme is to hydrolyse acetylcholine wherever it is released, such inhibition results in the accumulation of excessive concentrations of acetylcholine at its various sites of action. These sites include the endings of the parasympathetic nerves to the smooth muscle of the iris, ciliary body, bronchial tree, gastrointestinal tract, bladder and blood vessels; to the salivary glands and secretory glands of the gastrointestinal tract and respiratory tract; and to the cardiac muscle and endings of sympathetic nerves to the sweat glands.

The sequence of symptoms varies with the route of exposure. While respiratory symptoms are generally the first to appear after inhalation of nerve agent vapor, gastrointestinal symptoms are usually the first after ingestion. Tightness in the chest is an early local symptom of respiratory exposure. This symptom progressively increases, as the nerve agent is absorbed into the systemic circulation, whatever the route of exposure. Following comparable degrees of exposure, respiratory manifestations are most severe after inhalation, and gastrointestinal symptoms may be most severe after ingestion.

The lungs and the eyes absorb nerve agents rapidly. In high vapor concentrations, the nerve agent is carried from the lungs throughout the circulatory system; widespread systemic effects may appear in less than 1 minute.

- The earliest ocular effect, which follows minimal symptomatic exposure to vapor, is miosis. The pupillary constriction may be different in each eye. Within a few minutes after the onset of exposure, there also occurs redness of the eyes. Following minimal exposure, the earliest effects on the respiratory tract are a watery nasal discharge, nasal hyperaemia, sensation of tightness in the chest and occasionally prolonged wheezing.
- Exposure to a level of a nerve agent vapor slightly above the minimal symptomatic dose results in miosis, pain in and behind the eyes and frontal headache. Some twitching of the eyelids may occur. Occasionally there is nausea and vomiting.

- In mild exposures, the systemic manifestations of nerve agent poisoning usually include tension, anxiety, jitteriness, restlessness, emotional lability, and giddiness. There may be insomnia or excessive dreaming, occasionally with nightmares.
- If the exposure is more marked, the following symptoms may be evident: headache, tremor, drowsiness, difficulty in concentration, impairment of memory with slow recall of recent events, and slowing of reactions. In some casualties there is apathy, withdrawal and depression.
- With the appearance of moderate systemic effects, the casualty begins to have increased fatigability and mild generalized weakness, which is increased by exertion. This is followed by involuntary muscular twitching, scattered muscular fasciculation's and occasional muscle cramps. The skin may be pale due to vasoconstriction and blood pressure moderately elevated.
- If the exposure has been severe, the cardiovascular symptoms will dominate and twitching (which usually appear first in the eyelids and in the facial and calf muscles) becomes generalized. Many rippling movements are seen under the skin and twitching movements appear in all parts of the body. This is followed by severe generalized muscular weakness, including the muscles of respiration. The respiratory movements become more labored, shallow and rapid; then they become slow and finally intermittent.
- After moderate or severe exposure, excessive bronchial and upper airway secretions occur and may become very profuse, causing coughing, airway obstruction and respiratory distress. Bronchial secretion and salivation may be so profuse that watery secretions run out of the sides of the mouth. The secretions may be thick and tenacious. If the exposure is not so overwhelming as to cause death within a few minutes, other effects appear. These include sweating, anorexia, nausea and heartburn. If absorption of nerve agent has been great enough, there may follow abdominal cramps, vomiting, diarrhea, and urinary frequency. The casualty perspires profusely, may have involuntary defecation and urination and may go into cardio respiratory arrest followed by death.
- If absorption of nerve agent has been great enough, the casualty becomes confused and ataxic. The casualty may have changes in speech, consisting of slurring, difficulty in forming words, and multiple repetition of the last syllable. The casualty may then become comatose, reflexes may disappear and generalized convulsions may ensue. With the appearance of severe central nervous system symptoms, central respiratory depression will occur and may progress to respiratory arrest.
- After severe exposure the casualty may lose consciousness and convulse within a minute without other obvious symptoms. Death is usually due to respiratory arrest requires prompt initiation of assisted ventilation to prevent death. If assisted ventilation is initiated, the individual may survive several lethal doses of a nerve agent.
- If the exposure has been overwhelming, amounting to many times the lethal dose, death may occur despite treatment as a result of respiratory arrest and cardiac arrhythmia. When overwhelming doses of the agent are absorbed quickly, death occurs rapidly without orderly progression of symptoms.

Nerve agent poisoning may be identified from the characteristic signs and symptoms. If exposure to vapor has occurred, the pupils will be very small, usually pin-pointed. If exposure has been cutaneous or has followed ingestion of a nerve agent in contaminated food or water, the pupils may be normal or, in the presence of severe systemic symptoms, slightly to moderately reduced in size. In this event, the other manifestations of nerve agent poisoning must be relied on to establish the diagnosis. No other known chemical agent produces muscular twitching and fasciculation's, rapidly developing pinpoint pupils, or the characteristic train of muscarinic, nicotinic and central nervous system manifestations.

The rapid action of nerve agents call for immediate self-treatment. Unexplained nasal secretion, salivation, tightness of the chest, shortness of breath, constriction of pupils, muscular twitching, or nausea and abdominal cramps call for the immediate intramuscular injection of 2 mg of atropine, combined if possible with oxime.

Source: Federation of American Scientists, www.fas.org/cw/cwagents.htm.

(g) Vesicants and blister agents

The following information on vesicant agents is extracted from the Federation of American Scientists' web site:

> Blister or vesicant agents are likely to be used both to produce casualties and to force opposing troops to wear full protective equipment thus degrading fighting efficiency, rather than to kill, although exposure to such agents can be fatal. Blister agents can be thickened in order to contaminate terrain, ships, aircraft, vehicles or equipment with a persistent hazard.
>
> Vesicants burn and blister the skin or any other part of the body they contact. They act on the eyes, mucous membranes, lungs, skin and blood-forming organs. They damage the respiratory tract when inhaled and cause vomiting and diarrhea when ingested. The vesicant agents include:
>
> - HD - sulfur mustard
> - HN - nitrogen mustard
> - L - lewisite (arsenical vesicants may be used in a mixture with HD)
> - CX - phosgene (properties and effects are very different from other vesicants)
>
> HD and HN are the most feared vesicants historically, because of their chemical stability, their persistency in the field, the insidious character of their effects by attacking skin as well as eyes and respiratory tract, and because no effective therapy is yet available for countering their effects. Since 1917, mustard has continued to worry military personnel with the many problems it poses in the fields of protection, decontamination and treatment. It should be noted that the ease with which mustard can be manufactured and its great possibilities for acting as a vapor would suggest that in a possible future chemical war HD would be preferred to HN.
>
> Due to their physical properties, mustards are very persistent in cold and temperate climates. It is possible to increase the persistency by dissolving them in non-volatile solvents. In this way thickened mustards are obtained that are very difficult to remove by decontaminating processes.
>
> Exposure to mustard is not always noticed immediately because of the latent and sign-free period that may occur after skin exposure. This may result in delayed decontamination or failure to decontaminate at all. Whatever means is used has to be efficient and quick acting. Within 2 minutes contact time, a drop of mustard on the skin can cause serious damage. Chemical inactivation using chlorination is effective against mustard and lewisite, less so against HN, and is ineffective against phosgene oxime.
>
> - In a single exposure the eyes are more susceptible to mustard than either the respiratory tract or the skin. The effects of mustard on the eyes are very painful. Conjunctivitis follows exposure of about 1 hour to concentrations barely perceptible by odor. This exposure does not affect the respiratory tract significantly. A latent period of 4 to 12 hours follows mild exposure, after which there is lachrymation and a sensation of grit in the eyes. The conjunctival and the lids become red. Heavy exposure irritates the eyes after 1 to 3 hours and produces severe lesions.
> - The hallmark of sulfur mustard exposure is the occurrence of a latent symptom and sign free period of some hours post exposure. The duration of this period and the severity of the lesions are dependent upon the mode of exposure, environmental temperature and probably on the individual himself. High temperature and wet skin are associated with more severe lesions and shorter latent periods.
> - If only a small dose is applied to the skin, the skin turns red and itches intensely. At higher doses blister formation starts, generally between 4 and 24 hours after contact, and this blistering can go on for several days before reaching its maximum. The blisters are fragile and usually rupture spontaneously giving way to a suppurating and necrotic wound. The necrosis of the epidermal cells is extended to the underlying tissues, especially to the dermis. The damaged tissues are covered with slough and are extremely susceptible to infection. The regeneration of these tissues is very slow, taking from several weeks to several months.

- Mustard attacks all the mucous membranes of the respiratory tract. After a latent period of 4 to 6 hours, it irritates and congests the mucous membranes of the nasal cavity and the throat, as well as the trachea and large bronchi. Symptoms start with burning pain in the throat and hoarseness of the voice. A dry cough gives way to copious expectoration. Airway secretions and fragments of necrotic epitheliums may obstruct the lungs. The damaged lower airways become infected easily, predisposing to pneumonia after approximately 48 hours. If the inhaled dose has been sufficiently high the victim dies in a few days, either from pulmonary edema or mechanical asphyxia due to fragments of necrotic tissue obstructing the trachea or bronchi, or from superimposed bacterial infection, facilitated by an impaired immune response.

The great majority of mustard gas casualties survive. There is no practical drug treatment available for preventing the effects of mustard. Infection is the most important complicating factor in the healing of mustard burns. There is no consensus on the optimum form of treatment.

A full protective ensemble can only achieve protection against these agents. The respirator alone protects against eye and lung damage and gives some protection against systemic effects. No drug is available for the prevention of the effects of mustard on the skin and the mucous membranes caused by mustards. It is possible to protect the skin against very low doses of mustard by covering it with a paste containing a chlorinating agent, e.g., chloramine. The only practical prophylactic method is physical protection such as is given by the protective respirator and special clothing.

In a pure form lewisite is a colorless and odorless liquid, but usually contains small amounts of impurities that give it a brownish color and an odor resembling geranium oil. It is heavier than mustard, poorly soluble in water but soluble in organic solvents. L is a vesicant (blister agent); also, it acts as a systemic poison, causing pulmonary edema, diarrhea, restlessness, weakness, subnormal temperature, and low blood pressure. In order of severity and appearance of symptoms, it is: a blister agent, a toxic lung irritant, absorbed in tissues, and a systemic poison. When inhaled in high concentrations, may be fatal in as short a time as 10 minutes.

- Liquid arsenical vesicants cause severe damage to the eye. On contact, pain and blepharospasm occur instantly. Edema of the conjunctival and lids follow rapidly and close the eye within an hour. Inflammation of the iris usually is evident by this time. After a few hours, the edema of the lids begins to subside, while haziness of the cornea develops.
- Liquid arsenical vesicants produce more severe lesions of the skin than liquid mustard. Stinging pain is felt usually in 10 to 20 seconds after contact with liquid arsenical vesicants. The pain increases in severity with penetration and in a few minutes becomes a deep, aching pain. Contamination of the skin is followed shortly by erythema, then by vesication, which tends to cover the entire area of erythema. There is deeper injury to the connective tissue and muscle, greater vascular damage, and more severe inflammatory reaction than are exhibited in mustard burns. In large, deep, arsenical vesicant burns, there may be considerable necrosis of tissue, gangrene and slough.
- The vapors of arsenical vesicants are so irritating to the respiratory tract that conscious casualties will immediately put on a mask to avoid the vapor. No severe respiratory injuries are likely to occur except among the wounded that cannot put on masks or the careless, caught without masks. Lewisite is irritating to nasal passages and produces a burning sensation followed by profuse nasal secretion and violent sneezing. Prolonged exposure causes coughing and production of large quantities of froth mucus. Injury to respiratory tracts, due to vapor exposure is similar to mustard's; however, edema of the lung is more marked and frequently accompanied by pleural fluid.

An antidote for lewisite is dimercaprol (British anti-lewisite (BAL). This ointment may be applied to skin exposed to lewisite before actual vesication has begun. Some blistering is inevitable in most arsenical vesicant cases. The treatment of the erythema, blisters and denuded areas is identical with that for similar mustard lesions. Burns severe enough to cause

(continues)

> shock and systemic poisoning are life threatening. Even if the patient survives the acute effects, the prognosis must be guarded for several weeks.
>
> Phosgene oxime (CX) is a white crystalline powder. It melts between 39–40°C, and boils at 129°C. By the addition of certain compounds it is possible to liquefy phosgene oxime at room temperature. It is fairly soluble in water and in organic solvents. In aqueous solution phosgene oxime is hydrolyses fairly rapidly, especially in the presence of alkali. It has a high vapor pressure and its odor is very unpleasant and irritating. Even as a dry solid, phosgene oxime decomposes spontaneously and has to be stored at low temperatures.
>
> In low concentrations, phosgene oxime severely irritates the eyes and respiratory organs. In high concentrations, it also attacks the skin. A few milligrams applied to the skin cause severe irritation, intense pain, and subsequently a necrotizing wound. Very few compounds are as painful and destructive to the tissues.
>
> Phosgene oxime also affects the eyes, causing corneal lesions and blindness and may affect the respiratory tract causing pulmonary edema. The action on the skin is immediate: phosgene oxime provokes irritation resembling that caused by a stinging nettle. A few milligrams cause intense pain, which radiates from the point of application, within a minute the affected area turns white and is surrounded by a zone of erythema (skin reddening), which resembles a wagon wheel in appearance. In 1 hour the area becomes swollen, and within 24 hours, the lesion turns yellow and blisters appear. Recovery takes 1 to 3 months.
>
> Source: Federation of American Scientists, www.fas.org/cw/cwagents.htm.

(h) Blood agents

A blood agent or cyanogen agent is a chemical compound that contains the cyanide group, which prevents the body from using oxygen. The term *blood agent* is a misnomer, however, because these agents do not actually affect the blood in any way. Rather, they exert their toxic effect at the cellular level, by interrupting the electron transport chain in the inner membranes of mitochondria.

(i) Choking agents

The following information on choking agents is extracted from the Federation of American Scientists' web site:

> Choking agents are chemical agents which attack lung tissue, primarily causing pulmonary edema, are classed as lung damaging agents. To this group belong:
>
> - CG - phosgene
> - DP - diphosgene
> - Cl - chlorine
> - PS - chloropicrin
>
> The toxic action of phosgene is typical of a certain group of lung damaging agents. Phosgene is the most dangerous member of this group and the only one considered likely to be used in the future. Phosgene was used for the first time in 1915, and it accounted for 80% of all chemical fatalities during World War I.
>
> Phosgene is a colorless gas under ordinary conditions of temperature and pressure. Its boiling point is 8.2°C, making it an extremely volatile and non-persistent agent. Its vapor density is 3.4 times that of air. It may therefore remain for long periods of time in trenches and other low-lying areas. In low concentrations it has a smell resembling new mown hay.
>
> The outstanding feature of phosgene poisoning is massive pulmonary edema. With exposure to very high concentrations death may occur within several hours; in most fatal cases pulmonary edema reaches a maximum in 12 hours followed by death in 24–48 hours. If the casualty survives, resolution commences within 48 hours and, in the absence of complicating infection, there may be little or no residual damage.
>
> During and immediately after exposure, there is likely to be coughing, choking, a feeling of tightness in the chest, nausea, and occasionally vomiting, headache and lachrymation.

The presence or absence of these symptoms is of little value in immediate prognosis. Some patients with severe coughs fail to develop serious lung injury, while others with little sign of early respiratory tract irritation develop fatal pulmonary edema. A period follows during which abnormal chest signs are absent and the patient may be symptom-free. This interval commonly lasts 2 to 24 hours but may be shorter. The signs and symptoms of pulmonary edema terminate it. These begin with cough (occasionally substernally painful), dyspnea, rapid shallow breathing and cyanosis. Nausea and vomiting may appear. As the edema progresses, discomfort, apprehension and dyspnea increase and frothy sputum develops. The patient may develop shock-like symptoms, with pale, clammy skin, low blood pressure and feeble, rapid heartbeat. During the acute phase, casualties may have minimal signs and symptoms and the prognosis should be guarded. Casualties may very rapidly develop severe pulmonary edema. If casualties survive more than 48 hours they usually recover.

Source: Federation of American Scientists, www.fas.org/cw/cwagents.htm.

(j) Irritants

An irritant is a chemical that is not corrosive, but which causes a reversible inflammatory effect on living tissue by chemical action at the site of contact. A chemical is a skin irritant if, when tested on the intact skin of albino rabbits by the methods of 16 CFR Part 1500.41 for four hours exposure or by other appropriate techniques, it results in an empirical score of five or more [3]. A chemical is an eye irritant if so determined under the procedure listed in 16 CFR Part 1500.42 or other appropriate techniques [4].

(k) Biological agents and toxins

Biological warfare (BW) agents, also known as germ warfare, is the use of any pathogen (bacterium, virus, or other disease-causing organism) or toxin found in nature as a weapon of war. BW agents can be intended to kill, incapacitate, or seriously impede an adversary. Ideal characteristics of biological agents and toxins are high infectivity, high potency, availability of vaccines, and delivery as an aerosol. Diseases most likely to be considered for use as biological weapons are contenders because of their lethality (if delivered efficiently) and robustness (making aerosol delivery feasible).

The biological agents used in biological weapons can often be manufactured quickly and easily. The primary difficulty is not the production of the biological agent but delivery in an infective form to a vulnerable target. For example, anthrax is considered an effective agent for several reasons. First, it forms hardy spores, perfect for dispersal aerosols. Second, pneumonic (lung) infections of anthrax usually do not cause secondary infections in other people. Thus, the effect of the agent is usually confined to the target. A pneumonic anthrax infection starts with ordinary "cold" symptoms and quickly becomes lethal, with a fatality rate that is 80 percent or higher. Finally, friendly personnel can be protected with suitable antibiotics.

A mass attack using anthrax would require the creation of aerosol particles of 1.5 micrometers to 5 micrometers. Too large and the aerosol would be filtered out by the respiratory system. Too small and the aerosol would be inhaled and exhaled. Also, at this size, nonconductive powders tend to clump and cling because of electrostatic charges. This hinders dispersion. So, the material must be treated with silica to insulate and discharge the charges. The aerosol must be delivered so that rain and sun do not rot it, and yet the human lung can be infected. There are other technological difficulties as well. Diseases considered for weaponization or known to be weaponized include anthrax, Ebola, Bubonic Plague, Cholera, Tularemia, Brucellosis, Q fever, Machupo, Coccidioides mycosis, Glanders, Melioidosis, Shigella, Rocky Mountain Spotted Fever, Typhus, Psittacosis, Yellow Fever, Japanese B Encephalitis, Rift Valley Fever, and Smallpox. Naturally occurring toxins that can be used as weapons include Ricin, SEB, Botulism toxin, Saxitoxin, and many Mycotoxins. The organisms causing these diseases are known as select agents. Their possession, use, and transfer are regulated by the Centers for Disease Control and Prevention's Select Agent Program.

(l) Incapacitating agents

The term *incapacitating agent* is defined by the U.S. Department of Defense as follows:

> An agent that produces temporary physiological or mental effects, or both, which will render individuals incapable of concerted effort in the performance of their assigned duties. [5]

Incapacitating agents are not primarily intended to kill, but supposedly nonlethal incapacitating agents can kill many of those exposed to them. The term *incapacitation*, when used in a general sense, is roughly equivalent to the term *disability* as used in occupational medicine and denotes the inability to perform a task because of a quantifiable physical or mental impairment. In this sense, any of the chemical warfare agents can incapacitate a victim; however, again by the military definition of this type of agent, incapacitation refers to impairments that are temporary and nonlethal. Thus, riot-control agents are incapacitating because they cause temporary loss of vision due to blepharospasm, but they are not considered military incapacitants because the loss of vision does not last long. Although incapacitation can result from physiological changes such as mucous membrane irritation, diarrhea, or hyperthermia, the term *incapacitating agent* as militarily defined refers to a compound that produces temporary and nonlethal impairment of military performance by virtue of its psychobehavioral or central nervous system (CNS) effect.

(m) Radiological materials

The following information is extracted from the EPA's web site:

Three basic concepts apply to all types of ionizing radiation. When developing regulations or standards that limit how much radiation a person can receive in a particular situation, one considers how these concepts affect a person's exposure.

BASIC CONCEPTS OF RADIATION PROTECTION

Time. The amount of radiation exposure increases and decreases with the time people spend near the source of radiation.

In general, think of the exposure time as how long a person is near radioactive material. It's easy to understand how to minimize the time for external (direct) exposure. Gamma and X-rays are the primary concern for external exposure.

However, if radioactive material gets inside the body, one can't move away from it. The only options once internal uptake occurs are to wait until it decays or until the body can eliminate it. When this happens, the biological half-life of the radionuclide controls the time of exposure. Biological half-life is the amount of time it takes the body to eliminate one half of the radionuclide initially present. Alpha and beta particles are the main concern for internal exposure.

When establishing a radiation standard that assumes an exposure over a certain period, the concept of time is applied. For example, exposures are often expressed in terms of a committed dose. A committed dose is one that accounts for continuing exposures over long periods of time (such as 30, 50, or 70 years). It refers to the exposure received from radioactive material that enters and remains in the body for many years.

When assessing the potential for exposure in a situation, consider the amount of time a person is likely to spend in the area of contamination. For example, in assessing the potential exposure from radon in a home, estimate how much time people are likely to spend in the basement.

Distance. The farther away a person is from a radiation source, the less their exposure. How close to a source of radiation one can get without getting a high exposure depends on the energy of the radiation and the size (or activity) of the source. Distance is a prime concern when dealing with gamma rays, because they can travel long distances. Alpha and beta particles don't have enough energy to travel very far.

As a rule, if you double the distance, you reduce the exposure by a factor of four (i.e., halving the distance increases the exposure by a factor of four). The area of the circle depends on the distance from the center to the edge of the circle (radius). It is proportional to the square of the radius. As a result, if the radius doubles, the area increases four times. Using the light bulb analogy, think of the radiation source as a bare light bulb. The bulb gives off light equally in every direction, in a circle. The energy from the light is distributed evenly over the whole area of the circle. When the radius doubles, the radiation is spread out over four times as much area, so the dose is only one fourth as much. (In addition, as the distance from the source increases so does the likelihood that some gamma rays will lose their energy.

The exposure of an individual sitting 4 feet from a radiation source will be $^1/_4$ the exposure of an individual sitting 2 feet from the same source.

Shielding. The greater the shielding around a radiation source, the smaller the exposure. Shielding simply means having something that will absorb radiation between you and the source of the radiation. The amount of shielding required to protect against different kinds of radiation depends on how much energy they have.

α (Alpha) A thin piece of light material, such as paper, or even the dead cells in the outer layer of human skin provides adequate shielding because alpha particles can't penetrate it. However, living tissue inside body, offers no protection against inhaled or ingested alpha emitters.

β (Beta) Additional covering, for example heavy clothing, is necessary to protect against beta-emitters. Some beta particles can penetrate and burn the skin.

γ (Gamma) Thick, dense shielding, such as lead, is necessary to protect against gamma rays. The higher the energy of the gamma ray, the thicker the lead must be. X-rays pose a similar challenge, so X-ray technicians often give patients receiving medical or dental X-rays a lead apron to cover other parts of their body.

DIRTY BOMBS/RADIOACTIVE DISPERSAL DEVICES (RDDS)

Although "dirty bombs," or radioactive dispersal devices (RDDs), are not weapons of mass destruction, in the past few years terrorists have indicated their interest in acquiring such weapons. RDDs disperse radioactive material by using conventional explosives or other means. There are only a few radioactive sources that can be used effectively in an RDD. The greatest security risk is posed by Cobalt-60, Cesium-137, Iridium-192, Strontium-90, Americium-241, Californium-252, and Plutonium-238.

Source: U.S. Environmental Protection Agency, www.epa.gov/radiation.

(n) Nitrogen compounds

Nitrogen oxides are produced during most combustion processes. About 80 percent of the immediately released nitrogen oxide is in the form nitric oxide (NO). Small amounts of nitrous oxide (N_2O) are also produced. Nitric oxide reacts with oxygen in the air to produce nitrogen dioxide (NO_2). Further oxidation during the day causes the nitrogen dioxide to form nitric acid and nitrate particles. In the dark, nitrogen dioxide can react with ozone and form a very reactive free radical. The free radical then can react with organic compounds in the air to form nitrogenated organic compounds, some of which have been shown to be mutagenic and carcinogenic.

Nitrogen dioxide is the most important nitrogen oxide compound with respect to acute adverse health effects. Under most chemical conditions, it is an oxidant. However, it takes about 10 times more nitrogen dioxide than ozone to cause significant lung irritation and inflammation.

Nitrates and nitrites are known to cause several health effects. In general, the following are the most common effects:

- Reactions with hemoglobin in blood, causing the oxygen-carrying capacity of the blood to decrease (nitrite)
- Decreased functioning of the thyroid gland (nitrate)
- Vitamin A shortages (nitrate)

- Formation of nitro amines, which are known as one of the most common causes of cancer (nitrates and nitrites)

(o) Opiate compounds

The main opiates derived from opium are morphine, codeine, thebaine, and diacetylmorphine (heroin). Papaverine and noscapine are also present, but have essentially no effect on the central nervous system, and are not usually placed in the same category as the others. Papaveretum is a standardized preparation of mixed opium alkaloids used on cardiac patients. Heroin is not generally used therapeutically and is illegal to produce, sell, or possess in many parts of the world because of its high potential for abuse.

Opioids work on specific receptors in the brain to decrease the sensation of pain, which is their primary pharmacological use. The following are possible side effects of opioids:

- Nausea. Nausea is the most common side effect. Opioids slow down gastrointestinal activity, which can make many patients queasy. Dehydration or constipation can also cause nausea or make it worse.
- Constipation. Constipation depresses gut activity, causing stool to stay in the body longer. It can accumulate and harden, which makes bowel movements difficult or impossible.
- Central nervous system (CNS) effects. Opioids can have a wide range of side effects on CNS functions. Opioids can either inhibit or excite the CNS, although inhibition is more common. Patients with depressed CNS functions might feel varying levels of drowsiness, lightheadedness, euphoria or dysphoria, or confusion. If opioids excite CNS functions, it can result in side effects like hyperalgesia (extreme sensitivity to pain), myoclonus (involuntary jerking of muscles), or seizures, in rare cases.

Less common side effects of opioids include the following:

- Urinary retention
- Respiratory depression, particularly in elderly or debilitated patients
- Pruritus (itching)
- Miosis (constriction of the pupil, a common effect of opioids)

(p) Fluorine compounds

Pure fluorine (F_2) is a corrosive pale yellow or brown gas that is a powerful oxidizing agent. It is the most reactive of all the elements and readily forms compounds with most other elements. Fluorine even combines with the noble gases krypton, xenon, and radon. Even in dark cool conditions, fluorine reacts explosively with hydrogen. It is so reactive that glass, metals, and even water, as well as other substances, burn with a bright flame in a jet of fluorine gas. It is far too reactive to be found in elemental form and has such an affinity for most elements, including silicon, that it can neither be prepared nor be kept in ordinary glass vessels. Instead, fluorine must be kept in specialized quartz tubes lined with a very thin layer of fluorocarbons. In moist air it reacts with water to form also-dangerous hydrofluoric acid.

Fluorides are compounds that combine fluorine with some positively charged counterpart. They often consist of crystalline ionic salts. Fluorine compounds with metals are among the most stable of the salts. Both elemental fluorine and fluoride ions are highly toxic and must be handled with great care. Contact with skin and eyes should be strictly avoided. In its free element state, fluorine has a characteristic pungent odor that is detectable in concentrations as low as 20 nL/L.

Hydrofluoric acid (HF) contact with exposed skin poses one of the most extreme and dangerous hazards. These effects are intensified by the fact that HF damages nerves in such a way as to make such burns initially painless. The HF molecule is capable of rapidly migrating through lipid layers of cells that would ordinarily stop an ionized acid, and the burns are typically deep. HF can react with calcium, permanently damaging the bone. More seriously, reaction with the body's calcium can cause cardiac arrhythmias, followed by cardiac arrest brought on by sudden chemical changes within the body. These reactions cannot always be prevented by local or intravenous injection of calcium salts. HF spills as small as 2.5 percent of the body's surface area (an area of about 9 in^2 or 23 cm^2), despite copious immediate washing, have been fatal. If the patient survives, HF burns typically produce open wounds of an especially slow-healing nature.

(q) Phenolic compounds

Phenols, sometimes called phenolics, are a class of chemical compounds consisting of a hydroxyl group (-OH) attached to an aromatic hydrocarbon group. The simplest of the class is phenol (C_6H_5OH). Although similar to alcohols, phenols have unique properties and are not classified as alcohols (because the hydroxyl group is not bonded to a *saturated* carbon atom). They have relatively higher acidities due to the aromatic ring tightly coupling with the oxygen and a relatively loose bond between the oxygen and hydrogen.

A number of health effects from breathing phenol in air have been reported. Short-term effects include respiratory irritation, headaches, and burning eyes. Chronic effects of high exposures include weakness, muscle pain, anorexia, weight loss, and fatigue. Effects of long-term low-level exposures included increases in respiratory cancer and heart disease and effects on the immune system. In animal laboratory studies, exposure to high concentrations of phenol in air for a few minutes irritates the lungs, and repeated exposure for several days produces muscle tremors and loss of coordination. Exposure to high concentrations of phenol in the air for several weeks results in paralysis and severe injury to the heart, kidneys, liver, and lungs, followed by death in some cases. When exposures involve the skin (dermal uptake), the size of the total surface area of exposed skin can influence the severity of the toxic effects. Ingestion of very high concentrations of phenol has resulted in death.

Effects reported in humans following dermal exposure to phenol include liver damage, diarrhea, dark urine, and red blood cell destruction. Skin exposure in humans to a relatively small amount of concentrated phenol has resulted in death. Small amounts of phenol applied to the skin of laboratory animals for brief periods can produce blisters and burns on the exposed surface, and spilling dilute phenol solutions on large portions of the body (greater than 25 percent of the body surface) can result in death.

(4) Describe the basic toxicological principles relative to assessment and treatment of persons exposed to hazardous materials, including the following:

An understanding of the toxicological principles of the various classes of compounds as coupled with the clinical signs and symptoms associated with exposure to these compounds is critical for the BLS responder to provide medical care to their patients, as well as protect themselves at an incident. While managing acute effects is often critical to the very survival of the patient, the BLS responder should also be aware of the delayed (chronic) effects. An understanding of both acute and chronic effects of the released materials will help the BLS responder give the most prudent care to patients. It is the dose–response relationship of the release material that can provide the BLS responder with the best insight as to the health effects associated with the different internal concentration levels of an exposed patient.

(a) Acute and delayed effects

Acute toxicity refers to the sudden, severe onset of symptoms due to an exposure to the contaminant(s) of concern. Delayed toxicity might not develop for hours, days, or even years following an exposure to the contaminant(s) of concern. In some cases, such as exposure to biological agents, symptoms might not appear until three or more days following an exposure to such agents.

(b) Local and systemic effects

Local effects are those in which a toxic substance comes in direct contact with the skin or other sensitive tissue. Systemic effects are the effects of a toxic substance on either the entire body or a specific organ or organ system.

(c) Dose–response relationship

The chemical, biological, or radiological dose–response relationship refers to the response a specific dose produces in the human body. The magnitude of the body's response depends on the on-scene concentration (as can be measured by monitoring systems) of the hazardous substance, the patient exposure concentration and duration, and the actual dose (considering uptake rate for each applicable exposure route) received by the patient. The maximum ambient concentration at the scene determines the maximum concentration available for exposure. The exposure concentration is the concentration available to the pertinent routes of exposure, and the duration is the amount of time the patient is exposed to this available concentration. The actual dose is that amount taken up by the patient through the applicable uptake mechanisms. The dose will be the total amount of a patient's uptake during the exposure time considering all routes of uptake.

(5) Given examples of various hazardous materials/WMD, define the basic toxicological terms as applied to patient care:

(a) Threshold limit value — time-weighted average (TLV-TWA)

The threshold limit value — time-weighted average (TLV-TWA) is the time-weighted average concentration for a conventional 8-hour workday and 40-hour workweek, to which it is believed that nearly all workers might be repeatedly exposed, day after day, without adverse health effect.

(b) Permissible exposure limit (PEL)

Permissible exposure limit (PEL) is a term OSHA uses in its health standards covering exposures to hazardous chemicals. It is similar to the TLV-TWA established by the American Conference of Governmental Industrial Hygienists (ACGIH). The PEL, which generally relates to the legally enforceable TLV limits, is the maximum concentration, averaged over 8 hours, to which 95 percent of healthy adults can be repeatedly exposed for 8 hours per day, 40 hours per week.

(c) Threshold limit value — short-term exposure limit (TLV-STEL)

Threshold limit value — short-term exposure limit (TLV-STEL) is the maximum average concentration, averaged over a 15-minute period, to which healthy adults can be safely exposed for up to 15 minutes continuously. Exposure should not occur more than 4 times a day with at least 1 hour between exposures.

(d) Immediately dangerous to life and health (IDLH)

IDLH is the maximum level to which a healthy worker can be exposed for 30 minutes and escape without suffering irreversible health effects or impairment. If at all possible, exposure to this level should be avoided. If that is not possible, responders must wear positive pressure self-contained breathing apparatus or a positive pressure supplied-air respirator with an auxiliary escape system. OSHA and NIOSH establish this limit.

(e) Threshold limit value — ceiling (TLV-C)

Threshold limit value — ceiling (TLV-C) is the maximum concentration to which a healthy adult can be exposed without risk of injury. TLV-C is comparable to IDLH, and exposures to higher concentrations should not occur.

(f) Parts per million/parts per billion/parts per trillion (ppm/ppb/ppt)

The values used to establish the exposure limits are quantified in parts per million, parts per billion, or parts per trillion. A good reference to remember is that 1 percent equals 10,000 ppm, 1 ppm equals 1,000 ppb, and 1ppb equals 1,000 ppt. Thus, for a reading from a sampling instrument of 0.5 percent, that is equivalent to 5,000 ppm, 5,000,000 ppb, or 5,000,000,000 ppt. If the TLV is determined to be 7,500 ppm, the reading from the instrument can be related to determine the degree of hazard that exposure concentration represents.

(6) Given examples of hazardous materials/WMD incidents with exposed patients, evaluate the progress and effectiveness of the medical care provided at a hazardous materials/WMD incident to ensure that the overall incident response objectives, along with patient care goals, are being met by completing the following tasks:
   (a) Locate and track all exposed patients at a hazardous materials/WMD incident, from triage and treatment to transport to a medically appropriate facility.
   (b) Review the incident objectives at periodic intervals to ensure that patient care is being carried out within the overall incident action plan.
   (c) Ensure that the required incident command system forms are completed, along with the patient care forms, during the course of the incident.
   (d) Evaluate the need for trained and qualified EMS personnel, medical equipment, transport units, and other supplies based on the scope and duration of the incident.

Understanding the critical factors in 4.4.1 for a particular hazardous material released at an incident is essential for a responder to determine the nature of the hazards to both the patients and the responders themselves. The *Emergency Response Guidebook* is an excellent reference to use in finding the needed information [6].

It is essential for responders to know which references and/or on-line databases provide the information necessary to determine the health effects of the hazardous materials involved in an incident in order to establish the medical treatment to combat them, in consultation with medical control. They must also understand which properties have an impact on a patient's reactions and on the medical care the patient must receive. These references become critical resources to perform these tasks. Sources of this information include EMS reference books, the poison control center, EMS/HM data systems, and the Agency for Toxic Substances and Disease Registry (ATSDR), which is part of the CDC. Telephone numbers for voice and data communications for these resources should be part of the EMS response teams' resources.

The BLS responder must be able to understand and apply the information from these reference sources to the patient presentation. For example, if a patient states that he or she was completely covered by a particular hazardous material and reference sources indicate that the material is very difficult to remove once it has contaminated the skin, the chance of secondary exposure is slight because the material does not dislodge easily. However, if the responder must touch the patient to provide medical care, the contaminant could be transferred by their physical contact.

4.4.2 Decontamination.

Given the emergency response plan and SOPs developed by the AHJ, the BLS level responder shall do the following:

(1) Determine if patient decontamination activities were performed prior to accepting responsibility and transferring care of exposed patients.

To anticipate and understand patient decontamination requirements, the BLS responder must make every effort to obtain information about the released substance. This information will help the BLS responder make an informed decision on performing decontamination or more fully evaluate the effectiveness of the decontamination performed. These are important decisions, especially for transporting agencies. It is unwise to accept a contaminated patient into a transport unit or to be unsure of the level of decontamination performed. A poor decision in the field can have significant ramifications at the door of the hospital. If hospital staff are unconvinced that proper decontamination was performed in field, they may require further decontamination prior to accepting the patient. This could delay care and have an adverse effect on the patient.

There can also be a need to balance decontamination with providing emergency medical care. In some cases, it might be wise to provide airway management, bleeding control, basic trauma care, or CPR prior to or concurrent with decontamination. There are no clear-cut guidelines for these situations. The BLS responder must weigh the benefits and risks of the need to provide BLS with the need to decontaminate.

As with all other sections in this handbook, the NFPA 473 section is designed to promote thoughtful and informed medical care. It is impossible to expect a standard to address all possible situations. To that end, the BLS responder, at the moment he or she must make a decision regarding patient care, should be well trained and confident to make that decision.

BLS responders must weigh all options and make reasonable decisions in the field. Again, it is important to understand the nature and level of contamination in order to make the best decision regarding decontamination and patient care (see Exhibit II.4.3).

EXHIBIT II.4.3 *This is an example of patient decontamination. (Source: Rob Schnepp)*

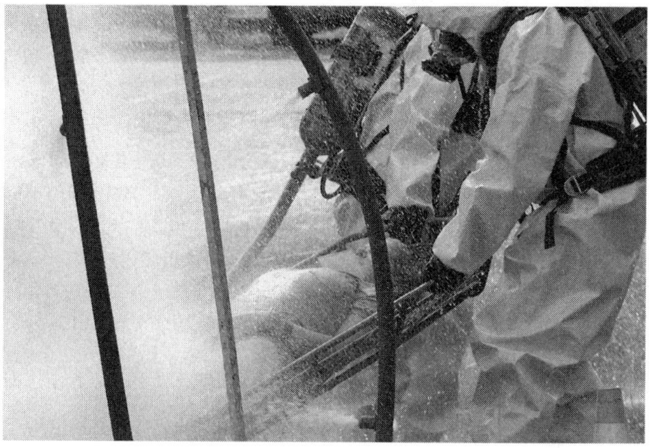

(2) Determine the need and location for patient decontamination, including mass casualty decontamination, in the event none has been performed prior to arrival of EMS personnel and complete the following tasks:

(a) Given the emergency response plan and SOPs developed by the AHJ, identify sources of information for determining the appropriate decontamination procedure and identify how to access those resources in a hazardous materials/WMD incident.

(b) Given the emergency response plan and SOPs developed by the AHJ, identify (within the plan) the supplies and equipment required to set up and implement the following:

The BLS responder should access and apply information from reliable reference sources in order to determine appropriate decontamination procedures. These reference sources include MSDSs, CHEMTREC/CANUTEC/SETIQ, Regional Poison Control Centers, the U.S. Department of Transportation (DOT)'s *Emergency Response Guidebook*, Hazardous Materials Information System (HMIS), Shipper/manufacturer contacts, Agency for Toxic Substances and Disease Registry (ATSDR) medical management guidelines, medical toxicologists, and electronic databases.

i. Emergency decontamination operations for ambulatory and nonambulatory patients

ii. Mass decontamination operations for ambulatory and nonambulatory patients

Decontamination efforts are influenced by the number of ambulatory and nonambulatory patients and must be addressed accordingly. It is more difficult, and requires more personnel, to decontaminate nonambulatory patients. The BLS responder should be familiar enough with the concepts and procedures for decontamination to adapt to the situation. Flexibility is the key to handling decontamination issues when it comes to life safety. Keep in mind that patient decontamination should be rapid, complete, and geared at getting the patient clean enough to treat and transport.

(c) Identify procedures, equipment, and safety precautions for the treatment and handling of emergency service animals brought to the decontamination corridor at hazardous materials/WMD incidents.
(d) Identify procedures, equipment, and safety precautions for communicating with critical, urgent, and potentially exposed patients and identify population prioritization as it relates to decontamination purposes.

All responders are reminded to address the importance of communicating with large groups of people who might be ill or injured. Panic can be contagious and the lack of crowd control can create significant problems for the responders. BLS responders should expect panic and unexpected behaviors from a large group of people exposed to a hazardous material/WMD.

(e) Identify procedures, equipment, and safety precautions for preventing cross contamination.

4.4.3 Determining the Ongoing Need for Medical Supplies.

4.4.3.1 Given examples of single-patient and multicasualty hazardous materials/WMD incidents, the BLS level responder shall determine the following:

(1) If the available medical equipment will meet or exceed patient care needs throughout the duration of the incident
(2) If the available transport units will meet or exceed patient care needs throughout the duration of the incident

Multicasualty incidents present different challenges than a single patient medical response (see Exhibit II.4.4). Multicasualty incidents require more personnel and equipment and are usually more chaotic. In these cases, the BLS responder should look at the entire incident and determine the need for medical equipment, qualified EMS personnel, and transport units. These and other resources should be called to the scene as soon as possible. Additionally, the ICS should be implemented as soon as possible to ensure responder accountability and the most efficient use of on-scene resources.

4.4.4 Preserving Evidence.

Given examples of hazardous materials/WMD incidents where criminal acts are suspected, the BLS level responder shall make every attempt to preserve evidence during the course of delivering patient care by completing the following tasks:

(1) Determine if the incident is potentially criminal in nature and cooperate with the law enforcement agency having investigative jurisdiction.
(2) Identify the unique aspects of criminal hazardous materials/WMD incidents, including crime scene preservation and evidence preservation, to avoid the destruction of potential evidence on medical patients during the decontamination process.
(3) Identify within the emergency response plan and SOPs developed by the AHJ procedures, equipment, and safety precautions for securing evidence during decontamination operations at hazardous materials/WMD incidents.
(4) Ensure that any information regarding suspects, sequence of events during a potentially criminal act, and observations made based on patient presentation or during patient

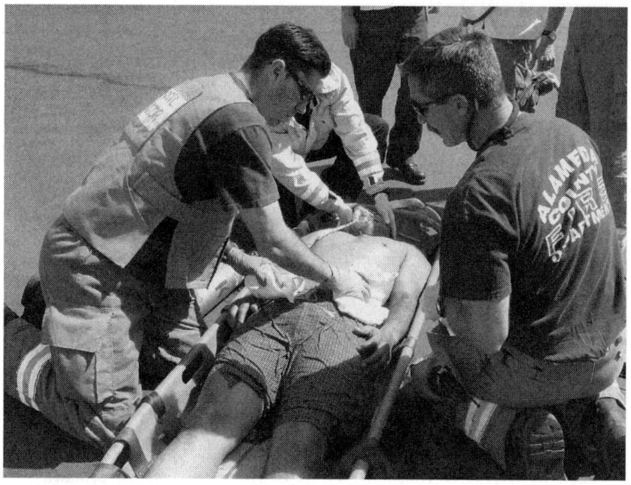

EXHIBIT II.4.4 *This shows an example of single-patient medical response. Multicasualty hazardous materials/WMD incidents will pose significantly more challenges to the BLS responder. (Source: Rob Schnepp)*

assessment are documented and communicated to the law enforcement agency having investigative jurisdiction.

The rationale behind 4.4.4 is to remind BLS level responders that they have a role in identifying and preserving evidence at a crime scene. If there is a suspicion that the event might be criminal in nature, it is incumbent on the BLS responder to be observant; pay attention to what is touched or moved; understand patient presentations relative to the suspected hazardous materials/WMD; observe and note the locations of patients, and other precautions. Furthermore, complete and thorough documentation is important and can be used in the prosecution of any suspects. These and other actions taken by BLS responders assist law enforcement with criminal investigation.

4.4.5 Medical Support at Hazardous Materials/WMD Incidents.

Given examples of hazardous materials/WMD incident, the BLS level responder shall describe the procedures of the AHJ for performing medical monitoring and support of hazardous materials incident response personnel and shall complete the following tasks:

The use of a systematic approach toward medical monitoring is required of the hazmat responder, especially when entry into a hazardous environment is prolonged, affected by meteorological factors, or difficult to mitigate. When the primary objective is potential rescue, on-scene medical personnel might default back to the annual "fit to work" physical. In either case, at an absolute minimum, the responder should be evaluated for heat stress, which includes hydration, and potential exposure.

(1) Given examples of various hazardous materials/WMD incidents requiring the use chemical protective ensembles, the BLS level responder shall complete the following tasks:

   (a) Demonstrate the ability to set up and operate a medical monitoring station.

The medical monitoring station can serve many purposes at the scene of a hazardous materials event. The first and primary role of the medical monitoring station is to evaluate the core vitals of the entry team. Evaluation can include the back-up team and the decontamination line personnel. Each individual assigned to work at the medical monitoring station should observe all responders for any physiological response that becomes detrimental toward the responder and the task of that responder.

The basic setup of the monitoring station should facilitate taking vital signs along with pulse oximetry, the responder's weight, skin status, body temperature, and hydration. Medical monitoring stations should be positioned to protect responders from adverse weather conditions.

In larger incidents, a medical manager within the incident command structure will have to organize his or her resources to compile the basic physiological information from each responder, but also must anticipate future medical needs of the hazmat technician. This may include, but is not limited to, specialized transport capability, on-scene advanced life support, and analysis of the capacity of the health care system.

(b) Demonstrate the ability to recognize the signs and symptoms of heat stress, cold stress, heat exhaustion, and heat stroke.

BLS responders should be trained to recognize and treat various types of heat/cold stress. Heat stroke is the most dangerous off all the heat-related illnesses. BLS care providers should understand the need for prompt recognition of the early signs of heat stroke.

(c) Determine the BLS needs for responders exhibiting the effects of heat stress, cold stress, and heat exhaustion.

Aggressive monitoring of individuals will become the priority for the BLS care providers. Once heat or cold stress has been identified, the treatment might only include monitoring of the individual and oral hydration. All medical providers that are used at a hazardous materials/WMD incident must always look for the signs and symptoms of heat stress especially when environmental factors present challenges and encapsulating protective equipment is in use.

(d) Describe the medical significance of heat stroke and the importance of rapid transport to an appropriate medical receiving facility.

(e) Given a simulated hazardous materials incident, demonstrate the appropriate documentation of medical monitoring activities.

(2) The BLS level responder responsible for pre-entry medical monitoring shall obtain hazard and toxicity information on the hazardous materials/WMD from the designated hazardous materials technical reference resource or other sources of information at the scene.

(3) The following information shall be conveyed to the entry team, incident safety officer, hazardous materials officer, other EMS personnel at the scene, and any other responders responsible for the health and well-being of those personnel operating at the scene:

(a) Chemical name

Looking up synonyms and trade names can give the responder additional information when databases are referenced.

(b) Hazard class

Although direct information about medical application is limited, this may give clues to the type of injury the individual may have. Example, Class 2 gases (pressurized or liquefied), gives us the highest potential for a respiratory hazard if PPE is not worn, not worn properly, or if a breach has occurred.

(c) Multiple hazards and toxicity information

Because many chemicals have multiple hazards, it is up to the medical personnel to fully reference the suspected/confirmed substance. Understanding how these factors affect individuals allows the medical staff to have a proactive treatment plan.

(d) Applicable decontamination methods and procedures

BLS responders should be familiar with the local procedures used for victim decontamination. In some jurisdiction, decontamination is considered to be a form of medical treatment. The BLS responder should look at the overall scene and balance the decontamination efforts with the need to provide patient care.

(e) Potential for cross contamination

Liquids and particular solids provide the highest level of cross contamination; gases rarely have any cross contamination issues. The organophosphates as a chemical family have the highest level of cross contamination due to the mechanism of injury. Again one must understand the state of matter, hazards associated with the substance at different concentrations, and the multiple hazards that a substance might have.

(f) Procedure for transfer of patients from the constraints of the incident to the EMS
(g) Prehospital management of medical emergencies and exposures

As with any medical emergency that occurs outside of the hospital, rapid evaluation, BLS measures, and transportation with information transfer are the keys toward a positive outcome. This includes but is not limited to the type of substance involved, the action(s) at time of incident, referenced information of the substance, and the potential degree of exposure.

(4) The BLS level responder shall evaluate the pre-entry health status of responders to hazardous materials/WMD incidents prior to their donning personal protective equipment (PPE) by performing the following tasks (consideration shall be given to excluding responders if they do not meet criteria specified by the AHJ prior to working in chemical protective clothing):

The medical authority having jurisdiction must decide before an event what situations or criteria will exclude a responder from donning PPE and working in a contaminated area. These considerations must be in accordance with prevailing federal, state, and local criteria.

(a) A full set of vital signs

Vital signs should include, but are not limited to, body temperature, pulse, respiration (rate and character), and blood pressure.

(b) Body weight measurements to address hydration considerations

Fluid loss is best measured in the field by measuring the responder's pre- and post-entry body weight. By achieving a percent loss of body weight, the BLS care provider can estimate the level of hydration. Hydration procedures should have input from the medical control officer (medical director). Fluid loss will become the most important factor to monitor and observe for working effectively within protective ensembles.

(c) General health observations

An idea of the responder's general health status can be obtained by observing his or her physical appearance and asking specific questions about their well being. Anyone who has had signs and symptoms of general common illness such as a cold or the flu, or has just been released to work after having a general illness should be further evaluated and should consult with the medical director.

(d) Core body temperature: hypothermia/hyperthermia

Extreme heat or cold will make it difficult for personnel to function in protective ensembles. Response personnel must be evaluated in either extreme for signs and symptoms that may indicate hypothermia/hyperthermia. The wearing of protective ensembles can exacerbate these conditions especially when working in high level PPE.

(e) Blood pressure: hypotension/hypertension

It is important to monitor the diastolic blood pressure pre- and post-entry. Pre-entry exam diastolic in relation to the systolic should be evaluated and exclusion criteria planned for by the medical director. Look for a 3- to 5-minute recovery back to baseline after exiting PPE during post entry exam.

(f) Pulse rate: bradycardia/tachycardia as defined

Each jurisdiction must adhere to its own definitions of bradycardia and tachycardia. The intent here is not to set hard limits, but to include these vital signs as part of a pre entry examination. As with any vital sign, bradycardia and tachycardia must be viewed in the context of an overall patient presentation. As an example, a pulse rate over 100 beats per minute might not be out of line for a responder preparing to don PPE on a hot and humid day. The NFPA 473 committee encourages a reasonable and well thought out approach to medical monitoring.

(g) Respiratory rate: bradypnea/tachypnea

Both the rate and character of respiration should be evaluated along with these conditions, which should have a planned response from the medical control officer and this vital sign related to the blood pressure, historical conditions, and annual physical findings.

(5) The BLS level responder shall determine how the following factors influence heat stress on hazardous materials/WMD response personnel:

(a) Baseline level of hydration
(b) Underlying physical fitness
(c) Environmental factors
(d) Activity levels during the entry
(e) Level of PPE worn
(f) Duration of entry
(g) Cold stress

(6) The BLS level responder shall medically evaluate all team members after decontamination and PPE removal, using the following criteria:

Each organization should develop and have preplanned medical monitoring protocols that identify conditions that are the basic evaluative concerns for the first responder. Within 3 to 5 minutes of exit from the protective ensemble, all vitals should be close to the pre-entry physical. Temperature should be monitored longer, and evaluated. This is especially important when performing entry functions within the two extreme conditions of heat and cold.

(a) Pulse rate determined within the first minute
(b) Pulse rate determined 3 minutes after initial evaluation
(c) Temperature
(d) Body weight
(e) Blood pressure
(f) Respiratory rate

(7) The BLS level responder shall recommend that any hazardous materials team member be prohibited from redonning chemical protective clothing if any of the following criteria is exhibited:

The use of exclusion criteria is important to establish guidelines for the post exam and for team members re-entering a hazardous environment. Again, the NFPA 473 committee encourages a reasonable approach to medical monitoring. The standard is designed to provide a framework for monitoring, but the limits and exclusion points should be determined on a case-by-case basis, adhering to the standard of care of the AHJ.

(a) Signs or symptoms of heat stress or heat exhaustion

Heat stress can affect the performance of any responder. It is up to the BLS care provider to recognize, evaluate and treat the responder who may be suffering from a heat stress event. This type of event should have input from the local medical control officer for guidance within the systems response manual. This guidance is a blueprint for the medical authority to provide the necessary evaluation and treatment for a responder that may be subject to heat stress.

(b) Pulse rate: tachycardia/bradycardia

(c) Core body temperature: hyperthermia/hypothermia
(d) Recovery heart rate with a trend toward normal rate and rhythm
(e) Blood pressure: hypertension/hypotension
(f)* Weight loss of >5 percent

A.4.4.5(7)(f) Regarding the issue of weighing individuals, recent medical research has focused on the concerns relating to water consumption and the difficulty in managing oral fluid intake. Often the distinction of water intoxication and resulting hyponatremia versus dehydration from insufficient water consumption, especially during sustained and prolonged operations, cannot be determined by vital sign measurements alone in the prehospital setting. One invaluable measure in making this distinction is a comparison weight of the individual prior to and following entry and re-entry to the operational theater. It is for this reason that comparison weighing is an included recommendation for evaluation of fitness.

(8) Any team member exhibiting the signs or symptoms of extreme heat exhaustion or heat stroke shall be transported to the medical facility.
(9) The BLS level responder responsible for medical monitoring and support shall immediately notify the persons designated by the incident action plan that a team member required significant medical treatment or transport. Transportation shall be arranged through the designee identified in the emergency response plan.

4.5 Reporting and Documenting the Incident

Given a scenario involving a hazardous materials/WMD incident, the responder assigned to use PPE shall complete the reporting and documentation requirements consistent with the emergency response plan or SOPs and identify the reports and supporting documentation required by the emergency response plan or SOPs.

4.6 Compiling Incident Reports

The BLS responder shall describe his or her role in compiling incident reports that meet federal, state, local, and organizational requirements, as follows:

(1) List the information to be gathered regarding the exposure of all patient(s) and describe the reporting procedures, including the following:
  (a) Detailed information on the substances released
  (b) Pertinent information on each patient treated and transported
  (c) Routes, extent, and duration of exposures
  (d) Actions taken to limit exposure
  (e) Decontamination activities
(2) At the conclusion of the hazardous materials/WMD incident, identify the methods used by the AHJ to evaluate transport units that might have been contaminated and the process and locations available to decontaminate those units.

The intent of Section 4.6 is to offer basic guidance in the area of documentation. Each jurisdiction has unique characteristics and requirements for patient care reports and incident documentation. The BLS responder should, in all cases, write timely documentation that is accurate and reflective of his/her actions on the scene. When it comes to hazardous materials/WMD incidents, patient care reports may be used during a prosecution. To that end, the report should be written in a professional manner.

REFERENCES CITED IN COMMENTARY

1. "Guide 100-00: Guide for Selection of Chemical Agent and Toxic Industrial Material Detection Equipment for Emergency Responders," U.S. Department of Justice, National Institute of Justice, Office of Science and Technology, Washington, DC, June, 2000. Available online at www.ncjrs.gov/pdffiles1/nij/184449.pdf.
2. Title 49, Code of Federal Regulations, Part 173, Shippers — General Requirements for Shipments and Packagings, U.S. Government Printing Office, Washington, DC.
3. Title 16, Code of Federal Regulations, Part 1500.41, Hazardous Substances and Articles; Administration and Enforcement Regulations: Method of testing primary irritant substances, U.S. Government Printing Office, Washington, DC.
4. Title 16, Code of Federal Regulations, Part 1500.42, Hazardous Substances and Articles; Administration and Enforcement Regulations: Test for eye irritants, U.S. Government Printing Office, Washington, DC.
5. Publication 1-02, *Dictionary of Military and Associated Terms*, U.S. Department of Defense, Washington, DC, 2007. Available online at www.dtic.mil/doctrine/jel/doddict/
6. *Emergency Response Guidebook*, Department of Transportation, U.S. Government Printing Office, Washington, DC.
7. Title 40, Code of Federal Regulations, Part 156, Labeling Requirements for Pesticides and Devices, U.S. Government Printing Office, Washington, DC.

Additional Reference

Agency for Toxic Substances and Disease Registry (ATSDR), U.S. Centers for Disease Control. See www.atsdr.cdc.gov.

Competencies for Hazardous Materials/WMD Advanced Life Support (ALS) Responder

CHAPTER 5

There are many similarities between the advanced life support (ALS) and basic life support (BLS) competencies. The standard was written and structured on the premise that good BLS precedes ALS level care. All EMS responders, regardless of their scope of practice, should share a basic set of hazmat response skills in order to work safely on the scene and deliver effective patient care. The ALS competencies are not protocol driven and not tied to antidotal therapies. The rationale behind the approach is that there are relatively few antidotes that can be administered in the field and the patient presentations requiring those antidotes are rare. Therefore, the intent of the standard is to raise the skill level of the EMS responder by understanding the link between EMS and hazardous materials response, the need to look at a scene with a critical eye to pick up the clues that might assist with understanding the nature of the exposure, and the need to look at the EMS system as a whole when it comes to treating exposed victims.

5.1 General

5.1.1 Introduction.

All EMS personnel at the hazardous materials/WMD ALS responder level, in addition to their ALS certification, shall be trained to meet at least the core competencies of the operations level responders as defined in Chapter 5 of NFPA 472, *Standard for Competence of Responders to Hazardous Materials/Weapons of Mass Destruction Incidents*, and all competencies of this chapter.

5.1.2 Goal.

The goal of the competencies at the ALS responder level shall be to provide the individual with the knowledge and skills necessary to safely deliver ALS at hazardous materials/WMD incidents and to function within the established incident command system, as follows:

(1) Analyze a hazardous materials/WMD incident to determine the potential health hazards encountered by the ALS level provider, other responders, and anticipated/actual patients by completing the following tasks:

It is imperative that ALS responders maintain good situational awareness when responding to hazardous materials/WMD incidents. In some circumstances, it might be the local ambulance company that arrives first at a hazardous materials/WMD incident. In those cases, the ALS responders should take time to size up the scene and understand the hazards present. When the ALS providers arrive after an organized hazmat response has begun, it is equally important to again size up the incident to understand the potential threats to the health and safety of the responders and civilian population.

(a) Survey a hazardous materials/WMD incident to determine whether harmful substances have been released and to evaluate suspected and identified patients for telltale signs of exposure.

ALS responders must be aware of the subtle signs and symptoms that can accompany a chemical exposure. Not all patients present with an obvious or clear-cut collection of signs and symptoms. When multiple patients are generated by an event, the ALS responders should look for consistencies (and inconsistencies) in the patient presentations. It might be possible to determine the presence of a nerve agent, for example, by the signs and symptoms exhibited by an exposed population of civilians.

(b) Collect hazard and response information from reference sources and technical experts on the scene, to determine the nature of the problem and potential health effects of the substances involved. *(See Annex B for a list of informational references.)*

ALS responders must take the necessary steps to correlate the physical and chemical properties of a released substance to the potential health effects and treatment(s) provided to exposed patients.

If EMS personnel are called to the scene to "stand-by" during a working hazmat incident, they should actively seek information regarding the chemical and physical properties of the released substance. It might be prudent, in some cases, to alert the anticipated receiving hospital if specific antidotes or other therapies could be needed. It is wise to aggressively plan to render care — it will be too late to formulate a good plan after the exposure happens.

(c) Survey the hazardous materials/WMD scene for the presence of secondary devices and other potential hazards.

The intent of discussing secondary devices is to raise responder awareness to the need for acknowledging such hazards. Not all scenes are tied to terrorism or some other criminal act intended to harm responders. It is prudent, however, to maintain good situational awareness and stay alert on all scenes.

(2) Plan to deliver ALS to exposed patients, within the scope of practice and training competencies established by the AHJ, by completing the following tasks:

(a) Evaluate preplans of high-risk areas/occupancies within the AHJ to identify potential locations where significant human exposures can occur.

It is important for ALS care providers to identify fixed facilities or other locations where hazardous substances are routinely used. The idea is to prepare for the low-frequency–high-impact situations before they occur. As an example, it would be good to know that a facility within a response area uses hydrofluoric acid (HF) and has a topical calcium gluconate gel available to treat skin exposures. Additionally, it would be beneficial to contact the appropriate receiving hospital to discuss procedures for handling a patient exposed to HF. Generally speaking, it is possible to preplan potential exposure prior to an event.

(b) Identify the capabilities of the hospital network within the AHJ to accept exposed patients and to perform emergency decontamination if required.

Just as with other specialty requirements, such as trauma, pediatrics, and burns, hospitals vary in their ability to handle contaminated patients. Pre-hospital care providers should be familiar with local hospitals and their ability to perform emergency decontamination. Ideally, all patients will be fully decontaminated prior to transport.

(c) Evaluate the components of the incident communication plan within the AHJ.

All ALS responders should be familiar with the radio communications procedures for their jurisdictions. This is a problem that is common to responders of all disciplines. Good communications are vital to an organized response.

(d) Describe the role of the ALS level responder as it relates to the local emergency response plan and established incident management system.

All ALS level responders should understand their role in the incident command system (ICS). The ICS is a reliable and effective method of organizing and managing a hazardous materials/WMD incident. Responders of any discipline who act outside their accepted role at an incident run the risk of undermining the goals and objectives set by the incident commander. It might be tempting to arrive at a large incident, for example, and start working without direction. Unfortunately, this is counter productive to the overall goal and objectives of the incident.

(e) Identify supplemental regional and national medical resources, including assets of the strategic national stockpile (SNS) and the metropolitan medical response system (MMRS).

(3) Implement a prehospital treatment plan for exposed patients, within the scope of practice and training competencies established by the AHJ, by completing the following tasks:

(a) Determine the nature of the hazardous materials/WMD incident as it relates to anticipated or actual patient exposures and subsequent medical treatment.

ALS responders should learn as much as possible about the physical and chemical properties of the released substance in order to develop an informed treatment plan for exposed individuals. The foundation of effective patient care is the ability to understand the nature of the exposure.

(b) Determine the need or effectiveness of decontamination prior to accepting an exposed patient.

Prior to providing advanced life support and/or transport, it is incumbent on the medical personnel accepting responsibility for a patient (or patients) to ensure that patient has been fully decontaminated. It is unacceptable to spread contamination from the scene to the hospital.

(c) Determine if the available medical equipment, transport units, and other supplies, including antidotes and therapeutic drugs, will meet or exceed patient care needs.

All ALS response personnel should have the ability to estimate the amounts of medical resources necessary to deliver optimal care to victims of a hazardous materials/WMD event (see Exhibit II.5.1). The ALS responder should be capable of advising the incident

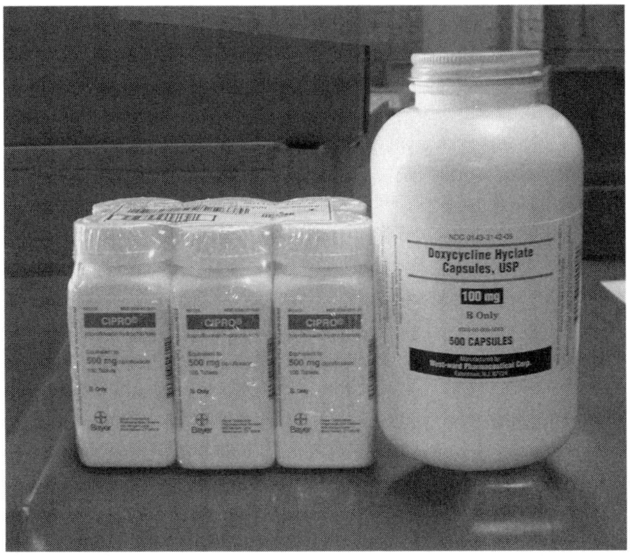

EXHIBIT II.5.1 *Available medical equipment has to meet or exceed patient care needs. (Source: Rob Schnepp)*

commander (IC) regarding quality and quantity of necessary resources, availability, and special conditions affecting access to the scene and delivery of patient care. In general, the ALS responder should be able to look at the incident from a bird's eye perspective and to understand the needs of the incident as a whole. It is easy to get tunnel vision and focus on the needs of a single patient while neglecting the needs of the patient care "system."

(d) Describe the process of evidence preservation where criminal or terrorist acts are suspected or confirmed.

Evidence preservation is not solely the responsibility of law enforcement personnel. Each responder involved with the incident is responsible for identifying and preserving potential evidence. This includes contaminated clothing, personal articles belonging to the exposed individuals, and bodies of victims.

(e) Develop and implement a medical monitoring plan for those responders operating in chemical protective clothing at a hazardous materials/WMD incident.

Pre-entry medical monitoring, as it relates to the potential for rendering treatment for an exposed responder, is an important facet of providing care at hazardous materials/WMD incidents. ALS responders, however, must think beyond the basics of vital signs when doing pre-entry medical monitoring. Aggressively plan to render care (for an unplanned exposure) based on the nature of the substance involved.

Medical support personnel should address the medical history and baseline physical status of responders prior to, and immediately following, an operational rotation. Medical history should include any present illness or condition, past medical history, allergies, and medications. Physical evaluation should include such routine measures as pulse, blood pressure, body temperature, and respiratory rate.

Other measures that should also be evaluated include pre- and post-entry body weight. This will provide invaluable information in discerning between dehydration and free water toxicity or overhydration. It has been established that prolonged excessive ingestion of water can lead to electrolyte derangement and hyponatremia. The physical signs and symptoms of this illness are often similar to the manifestations of dehydration. To that end, it is important that comparison weights be recorded so that an incorrect treatment course can be avoided.

(f) Evaluate the need to administer antidotes to reverse the effects of exposure in affected patients.

ALS providers infrequently encounter patients suffering significant illness as a result of a hazardous materials/WMD exposure and so are rarely required to administer antidotes to counter these exposures (see Exhibit II.5.2). When those situations occur, however, it is critical that the ALS provider understand the nature of the exposure and treat accordingly. Documenting and/or reporting those findings is a valuable tool for the continuum of care. A good report at the transfer of care is important.

(4) Participate in the termination of the incident by completing the following tasks:
  (a) Participate in an incident debriefing.
  (b) Participate in an incident critique with the appropriate agencies.
  (c) Report and document the actions taken by the ALS level responder at the scene of the incident.

5.2 Competencies — Analyzing the Hazardous Materials Incident

Medical support personnel should be capable of understanding the common overt presentations of human and animal exposures. This capacity is necessary so that the information can

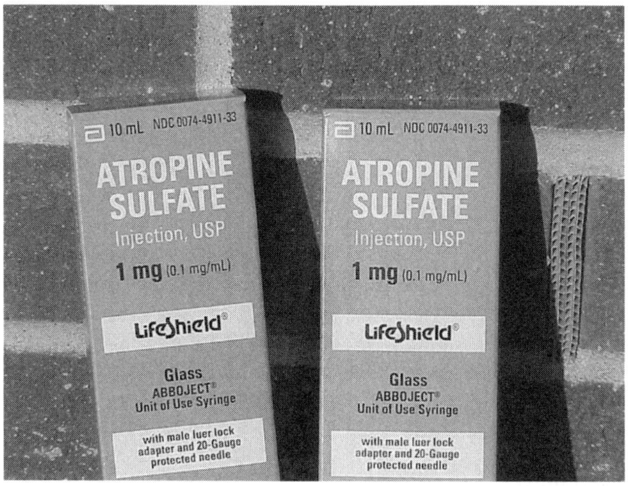

EXHIBIT II.5.2 *Atropine may have to be administered to reverse the effects of exposure to nerve agents. (Source: Rob Schnepp)*

be communicated to the incident commander, and to the next level of pre-hospital care providers (if present), to provide valuable intelligence for strategic determinations.

5.2.1 Surveying Hazardous Materials/WMD Incidents.

Given scenarios of hazardous materials/WMD incidents, the ALS level responder shall assess the nature and severity of the incident as it relates to anticipated or actual EMS responsibilities at the scene.

5.2.1.1 Given examples of the following marked transport vehicles (and their corresponding shipping papers or identification systems) that can be involved in hazardous materials/WMD incidents, the ALS level responder shall evaluate the general health risks based on the physical and chemical properties of the anticipated contents:

(1) Highway transport vehicles, including cargo tanks
(2) Intermodal equipment, including tank containers
(3) Rail transport vehicles, including tank cars

5.2.1.2 Given examples of various hazardous materials/WMD incidents at fixed facilities, the ALS level responder shall demonstrate the ability to perform the following tasks:

Good situational awareness enables the ALS responder to see and interpret the "big picture" at a hazardous materials/WMD incident. Without knowing anything about the nature of the product inside, the ALS responder should, simply by identifying the type of container, link some basic physical characteristics of the contained substance that could have an impact on an exposed patient. The same applies for other types of containers. Again, the ALS responder should hone the ability to pick out the important visual clues present and link them to the delivery of patient care.

Even when the exact chemical name is unknown, it is still possible to formulate a treatment plan based on the broad hazards of a particular classification of a chemical. When no other information is known about the hazardous materials/WMD, routine medical care can be the most appropriate care provided.

(1) Identify a variety of containers and their markings, including bulk and nonbulk packages and containers, drums, underground and aboveground storage tanks, specialized storage tanks, or any other specialized containers found in the AHJ's geographic area, and evaluate the general health risks based on the physical and chemical properties of the anticipated contents.

(2) Identify the following job functions of health-related resource personnel available at fixed facility hazardous materials/WMD incidents:

(a) Environmental heath and safety representatives

Typically, environmental health and safety representatives are knowledgeable about the chemical inventories, processes, and other important health- and safety-related issues at the site. These employees are typically trained in safety and industrial hygiene and are familiar with their workplace and the hazards that might be present, whether they be stored chemicals or materials used in manufacturing processes.

(b) Radiation safety officers

The requirements for a fixed facility to have a radiation safety officer (RSO) vary with the type of license and/or the radioactive substances used. If ALS providers respond to a radiation exposure at a fixed facility, attempts should be made to contact the RSO. That person can provide useful guidance on the potential health effects of an exposure to radiation.

(c) Occupational physicians and nurses

These individuals can provide medical information pertaining to the victims (plant or site workers) to the medical support personnel.

(d) Site emergency response teams

Site-specific emergency response teams (ERTs) can be a useful asset to ALS responders called to the scene of a hazardous materials/WMD incident. In many cases, formalized fire brigades and/or ERTs are able to provide site-specific information — from an emergency response perspective — to the arriving responders.

(e) Specialized experts

These individuals might be industry experts who have the most complete and up-to-date understanding of the hazardous materials/WMD agents and their health effects. They might be public health experts or other such personnel who could assist with determining the most appropriate treatment of the exposed patients.

5.2.1.3 The ALS level responder shall identify two ways to obtain a material safety data sheet (MSDS) at a hazardous materials/WMD incident and shall demonstrate the ability to identify the following health-related information:

(1) Proper chemical name or synonyms
(2) Physical and chemical properties
(3) Health hazards of the material
(4) Signs and symptoms of exposure
(5) Routes of entry
(6) Permissible exposure limits
(7) Emergency medical procedures or recommendations
(8) Responsible party contact

When treating the victim of a chemical exposure, it is important for ALS responders to obtain a material safety data sheet (MSDS). Having this sheet enables the care provider to identify the physical and chemical properties of the substance as well as the expected signs and symptoms of exposure. An MSDS, however, is not designed to provide detailed information on treatment options. The MSDS should accompany the patient to the hospital and be turned over to the treating physician. This transfer of information can speed up further treatment in the hospital.

5.2.1.4 Given scenarios at various fixed facilities, transportation incidents, pipeline release scenarios, maritime incidents, or any other unexpected hazardous materials/WMD incident, the ALS level responder, working within an incident command system must evaluate the off-site consequences of the release, based on the physical and chemical nature of the released substance, and the prevailing environmental factors to determine the need to evacuate or shelter in place affected persons.

5.2.1.5 Given examples of the following biological threat agents, the ALS level responder shall define the various types of biological threat agents, including the signs and symptoms of exposure, mechanism of toxicity, incubation periods, possible disease patterns, and likely means of dissemination:

(1) Variola virus (smallpox)
(2) *Botulinum* toxin
(3) *E. coli* O157:H7
(4) Ricin toxin
(5) *B. anthracis* (anthrax)
(6) Venezuelan equine encephalitis virus
(7) *Rickettsia*
(8) *Yersinia pestis* (plague)
(9) Tularemia
(10) Viral hemorrhagic fever
(11) Other CDC Category A listed organism or threat

The web site for the U.S. Centers for Disease Control and Prevention (CDC) for bioterrorism agents and diseases says the following:

> The U.S. public health system and primary healthcare providers must be prepared to address various biological agents, including pathogens that are rarely seen in the United States.
>
> **Category A Diseases/Agents**
>
> High-priority agents that pose a risk to national security because they
>
> - Can be easily disseminated or transmitted from person to person;
> - Result in high mortality rates and have the potential for major public health impact;
> - Might cause public panic and social disruption; and
> - Require special action for public health preparedness.
>
> **Category B Diseases/Agents**
>
> Second highest priority agents include those that
>
> - Are moderately easy to disseminate;
> - Result in moderate morbidity rates and low mortality rates; and
> - Require specific enhancements of CDC's diagnostic capacity and enhanced disease surveillance.
>
> **Category C Diseases/Agents**
>
> Third highest priority agents include emerging pathogens that could be engineered for mass dissemination in the future because of
>
> - Availability;
> - Ease of production and dissemination; and
> - Potential for high morbidity and mortality rates and major health impact.
>
> Source: U.S. Centers for Disease Control and Prevention, www.bt.cdc.gov/agent/agentlist-category.asp

5.2.1.6* Given examples of various types of hazardous materials/WMD incidents involving toxic industrial chemicals (TICs), toxic industrial materials (TIMs), blister agents, blood agents, nerve agents, choking agents and irritants, the ALS level responder shall determine the general health risks to patients exposed to those substances and identify those patients who may be candidates for antidotes.

There are countless substances that may harm civilians and responders alike. It is not the intention of 5.2.1.6 to require ALS responders to be familiar with every potential substance. The ALS responder should, however, be familiar with those substances that are present in their jurisdiction, as well as the most common WMDs, toxic industrial chemicals (TICs), and toxic industrial materials (TIMs).

The U.S. Occupational Safety and Health Administration (OSHA) regulations state that TICs are industrial chemicals that are manufactured, stored, transported, and used throughout the world. The physical state of TICs can be gas, liquid, or solid. They can be chemical hazards (e.g., carcinogens, reproductive hazards, corrosives, or agents that affect the lungs or blood) or physical hazards (e.g., flammable, combustible, explosive, or reactive). See www.osha.gov for a table of TICs, and see www.osha.gov/SLTC/emergencypreparedness/guides/chemical.html for more information on chemicals.

The National Institute of Justice (NIJ) states that TIMs are similar to TICs but are chemicals other than chemical warfare agents that have harmful effects on humans [1]. They are used in a variety of settings such as manufacturing facilities, maintenance areas, and general storage areas.

Training programs for ALS responders should include a section on the basic physical and chemical properties of the broad classifications of hazardous materials/WMD they might encounter. Typically, the scope of practice for ALS level responders is limited when it comes to handling exposures resulting from hazardous materials/WMD incidents. Therefore, training in this area should focus on recognizing the signs and symptoms of typical hazardous materials/WMD substances, basic treatment modalities, and the best practices to avoid becoming exposed.

A.5.2.1.6 Examples of toxic industrial materials are corrosives, reproductive hazards, carcinogens, flammable hazards, and explosive hazards.

5.2.1.7* Given examples of hazardous materials/WMD found at illicit laboratories, the ALS level responder shall identify general health hazards associated with the chemical substances that are expected to be encountered.

A complete site risk assessment must be done prior to entry to identify potential hazards to the responders, the civilian population, and the community as a whole. Each lab process will have distinct hazards that must be preplanned before the event. This preplanning of resources identifies the existing hazards and the potential hazards the responder can come across. In all cases the appropriate level of personal protective equipment (PPE) must be worn. Common examples of general hazards include the following:

1. Confined spaces
2. Electrical hazards
3. Pathogen hazards
4. Damaged or unstable containers
5. Flammable, corrosive, oxygen-deficient, or enriched atmospheres

As with any emergency event, realizing the potential hazards and health effects are critical parts of the risk assessment. In any incident, especially those where an illicit lab might be present, a complete hazard-risk assessment must be accomplished. Concerns to consider when responding to an illicit lab include the following:

1. Large amounts of waste product(s) from the synthesis process might be present. For example, for every 1 pound of production within a methamphetamine lab, 5 to 6 pounds of waste can be generated.
2. The level of concern based on the precursors that are used within these processes must be established. Each lab type will create a different level of apprehension. A variety of acids, bases, and solvents are among the most common chemicals used.
3. Some processes can cause reduced oxygen levels within enclosed environments, which is sometimes associated with flammable gases, as well as asphyxiating gases with ignition sources.

Potential health hazards fall into one or more of the following categories:

1. Flammable
2. Corrosive
3. Oxygen deficient/enriched
4. Asphyxiant

A.5.2.1.7 Examples of hazardous materials at illicit laboratories are as follows:

(1) Ammonia
(2) Ephedrine and pseudoephedrine
(3) Flammable solvents such as ether compounds and methanol
(4) Fluorinated/chlorinated hydrocarbons (Freon)
(5) Hydrogen chloride
(6) Iodine
(7) Lithium or sodium metal
(8) Phosphine gas
(9) Red phosphorus
(10) Sodium hydroxide or other caustic substances

5.2.1.8 Given examples of a hazardous materials/WMD incident involving radioactive materials, including radiological dispersion devices, the ALS level responder shall determine the probable health risks and potential patient outcomes by completing the following tasks:

(1) Determine the types of radiation (alpha, beta, gamma, and neutron) and potential health effects of each.

Radiation exposure occurs when a person is near a source of radiation. The person might receive a radiation exposure but not become radioactive. As an example, a person having an X-ray receives a radiation exposure but does not become radioactive. Persons suffering a radiation exposure do not require decontamination prior to medical treatment.

Radioactive contamination occurs when loose particles, contaminated with radiation, settle on clothing, personal protective equipment, or skin. Testing is conducted on individuals to determine whether a person with radioactive contamination is radioactive (see Exhibit II.5.3). Radiation illness can result from inhaled or ingested particles. Persons with radioactive contamination must be decontaminated prior to medical treatment.

(2) Determine the most likely exposure pathways for a given radiation exposure, including inhalation, ingestion, and direct skin exposure.

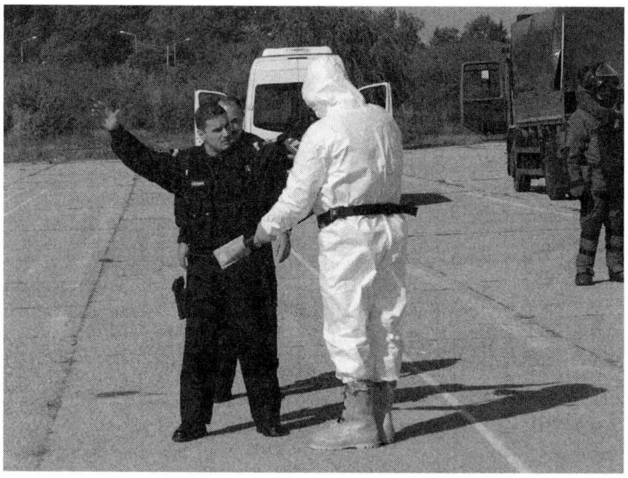

EXHIBIT II.5.3 *Individuals are tested for radioactive contamination to determine decontamination priorities. (Source: Rob Schnepp)*

(3) Describe how the potential for cross contamination differs for electromagnetic waves compared to radioactive solids, liquids, or vapors.
(4) Identify priorities for decontamination in scenarios involving radioactive materials.
(5) Describe the manner in which acute medical illness or traumatic injury can influence decisions about decontamination and patient transport.

5.2.1.9 Given examples of typical labels found on pesticide containers, the ALS level responder shall define the following terms:

(1) Pesticide name
(2) Pesticide classification (e.g., insecticide, rodenticide, organophosphate, carbamate, organochlorine)
(3) Environmental Protection Agency (EPA) registration number
(4) Manufacturer name
(5) Ingredients broken down by percentage
(6) Cautionary statement (e.g., Danger, Warning, Caution, Keep from Waterways)
(7) Strength and concentration
(8) Treatment information

Each pesticide label must contain a signal word. The three signal words, in order of increasing toxicity, are *caution*, *warning*, and *danger*.

The following are the toxicity categories for pesticides, which are established by the U.S. Environmental Protection Agency (EPA) according to their acute (short-term) toxicity [7]:

- **Toxicity Category I:** All pesticide products meeting the criteria of Toxicity Category I must bear on the front panel the signal word "Danger." In addition, if the product was assigned to Toxicity Category I on the basis of its oral, inhalation, or dermal toxicity (as distinct from skin and eye local effects), the word "Poison" must appear in red on a background of distinctly contrasting color and the skull and crossbones must appear in immediate proximity to the word "poison."
- **Toxicity Category II:** All pesticide products meeting the criteria of Toxicity Category II must bear on the front panel the signal word "Warning."
- **Toxicity Category III:** All pesticide products meeting the criteria of Toxicity Category III must bear on the front panel the signal word "Caution."
- **Toxicity Category IV:** All pesticide products meeting the criteria of Toxicity Category IV must bear on the front panel the signal word "Caution."
- **Child Hazard Warning Statement:** The Child Hazard Warning Statement "Keep Out of Reach of Children" is required on all product labels, unless the requirement is waived. The warning statement requirement may be waived when the registrant (the individual or entity registering the pesticide with EPA) adequately demonstrates that the likelihood of contact with children during distribution, storage, or use is extremely remote or if the pesticide is approved for use on infants or small children.

The EPA uses the criteria shown in Commentary Table II.5.1 to determine the toxicity category of pesticides. These criteria are based on the results of animal tests done in support of registration of the pesticide.

5.2.2 Collecting and Interpreting Hazard and Response Information.

The ALS level responder shall demonstrate the ability to utilize various reference sources at a hazardous materials/WMD incident, including the following:

(1) MSDS
(2) CHEMTREC/CANUTEC/SETIQ
(3) Regional poison control centers
(4) DOT *Emergency Response Guidebook*

COMMENTARY TABLE II.5.1 *Pesticide Toxicity Category Criteria*

Hazard Indicators	I	II	III	IV
Oral LD_{50}	Up to and including 50 mg/kg	50 thru 500 mg/kg	500 thru 5,000 mg/kg	5,000 mg/kg
Dermal LD_{50}	Up to and including 200 mg/kg	200 thru 2000 mg/kg	2000 thru 20,000 mg/kg	20,000 mg/kg
Inhalation LC_{50}	Up to and including 0.2 mg/liter	0.2 thru 2 mg/liter	2 thru 20 mg/liter	20 mg/liter
Eye irritation	Corrosive; corneal opacity not reversible within 7 days	Corneal opacity reversible within 7 days; irritation persisting for 7 days	No corneal opacity; irritation reversible within 7 days	No irritation
Skin irritation	Corrosive	Severe irritation at 72 hours	Moderate irritation at 72 hours	Mild or slight irritation at 72 hours

Source: Title 40 CFR Part 156.62, Toxicity Category.

(5) NFPA 704, *Standard System for the Identification of the Hazards of Materials for Emergency Response* identification system
(6) Hazardous Materials Information System (HMIS)
(7) Local, state, federal, and provincial authorities
(8) Shipper/manufacturer contacts
(9) Agency for Toxic Substances and Disease Registry (ATSDR) medical management guidelines
(10) Medical toxicologists
(11) Electronic databases

5.2.2.1 Identifying Secondary Devices. Given scenarios involving hazardous materials/WMD, the ALS level responders shall describe the importance of evaluating the scene for secondary devices prior to rendering patient care, including the following safety points:

(1) Evaluate the scene for likely areas where secondary devices can be placed.
(2) Visually scan operating areas for a secondary device before providing patient care.
(3) Avoid touching or moving anything that can conceal an explosive device.
(4) Designate and enforce scene control zones.
(5) Evacuate victims, other responders, and nonessential personnel as quickly and safely as possible.

Secondary devices present a considerable challenge to the responder. The incident scene must be carefully screened for these devices, and extreme caution must be taken if any devices are discovered (see Exhibit II.5.4).

Planning the response can involve the medical response/support personnel as subject matter contributors to the IC, safety officer, or other responders, as appropriate. The strategic considerations that are directly addressed by medical specialists include hospital capability, hospital considerations, caregiver-to-victims ratios, special resources, medical logistics, and emergent, immediate, mid- and long-range clinical considerations.

5.3 Competencies — Planning the Response

5.3.1 Identifying High-Risk Areas for Potential Exposures.

5.3.1.1 The ALS level responder, given an events calendar and pre-incident plans, which can include the local emergency planning committee plan as well as the agency's emergency response plan and SOPs, shall identify the venues for mass gatherings, industrial facilities, potential targets for terrorism, or any other locations where an accidental or intentional release

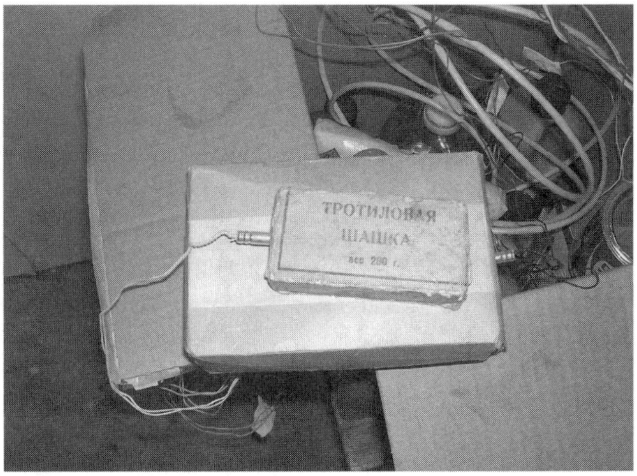

EXHIBIT II.5.4 *Secondary devices can be found at hazmat/WMD incident scenes. This illustration shows an explosive device hidden in a cardboard storage box. (Source: Rob Schnepp)*

of a harmful substance can pose an unreasonable health risk to any person within the local geographical area as determined by the AHJ and shall do the following:

(1) Identify locations where hazardous materials/WMD are used, stored, or transported.
(2) Identify areas and locations presenting a potential for a high loss of life or rate of injury in the event of an accidental/intentional release of a hazardous materials/WMD substance.
(3) Evaluate the geographic and environmental factors that can complicate a hazardous materials/WMD incident, including prevailing winds, water supply, vehicle and pedestrian traffic flow, ventilation systems, and other natural or man-made influences, including air and rail corridors.

The goal of 5.3.1.1 is to remind ALS level responders that preplanning is important. Almost every jurisdiction has an area of concern when it comes to hazardous materials/WMD events. It is unwise not to preplan at these locations. A sound preplan can be the difference between handling an event and being completely overwhelmed.

5.3.2 Determining the Capabilities of the Local Hospital Network.

5.3.2.1 The ALS level responder shall identify the methods and vehicles available to transport hazardous materials patients and shall determine the location and potential routes of travel to the following appropriate local and regional hospitals, based on patient need:

ALS responders should be familiar with the local and regional healthcare systems when it comes to the location, type, and capabilities of local hospitals. During the preplanning process, it is necessary to identify those facilities capable of providing care to exposed patients. ALS responders should be aware of the level of preparedness of the local hospitals.

(1) Adult trauma centers
(2) Pediatric trauma centers
(3) Adult burn centers
(4) Pediatric burn centers
(5) Hyperbaric chambers
(6) Established field hospitals
(7) Other specialty hospitals or medical centers

5.3.2.2 Given a list of local receiving hospitals in the AHJ's geographic area, the ALS level responder shall describe the location and availability of hospital-based decontamination facilities.

The methods used for decontamination vary from hospital to hospital. Responders need to be familiar with the capabilities of local and regional primary care facilities as they relate to guarding against cross-contamination and treatment of hazardous materials exposures. This will aid in ensuring that the most appropriate receiving facility is chosen to ensure the best care for the patient and safety of attending personnel.

5.3.2.3 The ALS level responder shall describe the ALS protocols and SOPs developed by the AHJ and the prescribed role of medical control and poison control centers during mass casualty incidents, at hazardous materials/WMD incidents where exposures have occurred, and in the event of disrupted radio communications.

The local emergency response plan, protocols, and procedures must define the actions that the ALS level responder should follow when confronted with a hazardous material related mass causality incident. The action plan should provide direction for times when normal communications are disrupted and alternative means of contacting medical control or poison control centers is required.

5.3.2.4 The ALS level responder shall identify the following mutual aid resources (hospital and nonhospital based) identified by the AHJ for the field management of multicasualty incidents.

(1) Mass-casualty trailers with medical supplies
(2) Mass-decedent capability
(3) Regional decontamination units
(4) Replenishment of medical supplies during long-term incidents
(5) Locations and availability of mass-casualty antidotes for selected exposures, including but not limited to the following:
  (a) Nerve agents and organophosphate pesticides
  (b) Biological agents and other toxins
  (c) Blood agents
  (d) Opiate exposures
  (e) Selected radiological exposures
(6) Rehabilitation units for the EMS responders
(7) Replacement transport units for those vehicles lost to mechanical trouble, collision, theft, and contamination

5.3.2.5 The ALS level responder shall identify the special hazards associated with inbound and outbound air transportation of patients exposed to hazardous materials/WMD.

The threat of a contaminant spread by inbound aircraft and the potential contamination of the aircraft by the release can pose a significant risk to responders, the public, and possibly the aircraft itself. The threat to the outbound aircraft is the risk that exposed patients could off-gas a substance while the aircraft is in flight that would affect the flight crew. Additionally, the threat of contaminating the aircraft itself is another risk. The proper management of air operations at hazardous materials events is extremely important to both the aircraft flight crews and those on the ground.

5.3.2.6 The ALS level responder shall describe the available medical information resources concerning hazardous materials toxicology and response.

5.3.3 Identifying Incident Communications.

5.3.3.1 The ALS level responder shall identify the components of the communication plan within the AHJ geographic area and determine that the EMS providers have the ability to communicate with other responders on the scene, with transport units, and with local hospitals.

5.3.3.2 Given examples of various patient exposure scenarios, the ALS level responder shall describe the following information to be transmitted to the medical control or poison control center or the receiving hospital prior to arrival:

(1) The exact name of the substance(s) involved
(2) The physical and chemical properties of the substance(s) involved
(3) Number of victims being transported
(4) Age and sex of transported patients
(5) Patient condition and chief complaint
(6) Medical history
(7) Circumstances and history of the exposure, such as duration of exposure and primary route of exposure
(8) Vital signs, initial and current
(9) Symptoms described by the patient, initial and current
(10) Presence of associated injuries, such as burns and trauma
(11) Decontamination status
(12) Treatment rendered or in progress, including the effectiveness of antidotes administered
(13) Estimated time of arrival

ALS responders should understand that an accurate and complete report to the receiving hospital is important. This would provide the necessary lead time for hospital staff to prepare for accepting the exposed patient.

5.3.4 Identifying the Role of the ALS Level Responder.

5.3.4.1 Given scenarios involving hazardous materials/WMD, the ALS level responder shall identify his or her role during hazardous materials/WMD incidents as specified in the emergency response plan and SOPs developed by the AHJ, as follows:

The agency's hazardous materials emergency response plan should outline the role of the ALS level responder. Primarily, that role includes responding to an emergency, assessing the nature of the incident, implementing protective measures, notifying other agencies, asking for additional assistance, establishing or working within an incident management system, and performing advanced life support medical triage, treatment, and transport in accordance with local protocols and procedures.

(1) Describe the purpose, benefits, and elements of the incident command system at it relates to the ALS level responder.

The ICS is an organized structure of roles, responsibilities, and procedures for the command and control of emergency operations. ICS is modular and can expand or contract based on the need, size, and nature of an incident. It enables multiple disciplines and multiple jurisdictions to work together safely and effectively.

ICS utilizes the following three management concepts: unity of command, span of control, and functional positions.

Unity of command stipulates that only one incident command or unified command is ultimately responsible for the entire incident. The command structure encompasses clearly defined lines of authority in which everyone is responsible to, and directed by, one person.

Span of control is established so that only three to seven individuals report to one position so that no one position becomes overloaded, with optimum span of control at five.

The functional positions concept means that all resources assigned to one functional position (for example, fire fighter, BLS responder, hazardous materials officer, hazardous materials technician) are to remain in that position until reassigned or released from the incident.

(2) Describe the typical incident command structure for the emergency medical component of a hazardous materials/WMD incident as specified in the emergency response plan and SOPs developed by the AHJ.

The medical component of the command structure will normally be a functional group directly under either the incident commander or the operations sections, if established. The function can be expanded to a medical branch with a medical group performing treatment and triage and to a patient transportation group. Further expansion of the medical branch can include dividing the medical group into the triage group and treatment group, depending on the scope of the incident and available resources. Some additional functions of the medical group could be extrication and air operations. Additionally, ALS level responders could be assigned to the medical unit under the logistic section if established. The medical unit would be responsible for responder medical treatment while the medical group/branch under the operations section would be responsible for public medical treatment. The size and complexity of the medical function will expand and collapse, depending on the needs of the incident.

(3) Demonstrate the ability of the ALS level responder to function within the incident command system.

The ALS level responder must be able to function within the ICS. The ALS responder first reports to the incident command post or staging area to check in. Upon receiving an assignment, the ALS level responder reports to the assignment (i.e., incident, response area) and reports to the branch director/group supervisor at the incident or response area and performs the task assigned. Upon completion of the task, the ALS level responder reports back to the supervisor for release or reassignment.

(4) Demonstrate the ability to implement an incident command system for a hazardous materials/WMD incident where an ICS does not currently exist.

The ALS level responder might be the first on the scene of an emergency. When this happens, it is imperative that the ALS level responder initiate the ICS and assume command until relieved by a more qualified or senior authority. The importance of this action is to ensure responder safety and coordination of resources to best effect the desired outcome of the incident.

(5) Identify the procedures for requesting additional resources at a hazardous materials/WMD incident.

Responders at every level are required to know what types of resources are available and how to request them. The employer's emergency response plan and procedures should identify the processes by which the ALS level responders will request additional resources both within and outside the organization.

5.3.4.2 Describe the hazardous materials/WMD ALS responder's role in the hazardous materials/WMD response plan developed by the AHJ or identified in the local emergency response plan as follows:

(1) Determine the toxic effect of hazardous materials/WMD.

As early as possible in the incident, the identity or classification of the product should be obtained so that the ALS level responder can take appropriate actions to minimize the impact of toxic effect on the patient(s). Once the material is identified or classified, the ALS level responder can communicate with medical control for orders on treatment if no standing protocols exist. Information can also be obtained from a poison control center. Consideration must also be given to personal protection and decontamination of victims prior to patient treatment.

(2) Estimate the number of patients.

As with any multicasualty incident, the estimate of the number of potential patients needs to be ascertained to determine the number and types of resources necessary to triage, treat, and transport the injured.

(3) Recognize and assess the presence and severity of symptoms.

The ALS level responder should be able to recognize symptoms consistent with the material exposure as confirming the determination of the toxic effects. Severity of symptoms might not be immediately known because some materials have delayed toxic effects. Depending on the material, symptoms are the result of a dose and the dose will dictate the affect on the body and our ability to intervene.

(4) Assess the impact on the health care system.

Once the number of patients and the identity or classification of the materials have been determined, the local heath care system should be alerted and medical control notified. The number of decontaminated patients being transported and the number of self-transporting victims that may be enroute to a facility should be communicated to medical control as early as possible. Every effort should be made to minimize the impact to a facility by notifying the facility of the numbers of injured being transported.

(5) Perform appropriate patient monitoring as follows:

   (a) Pulse oximetry
   (b) Cardiac monitor
   (c) End tidal CO_2

The proper taking and recording of vital signs such as respirations, pulse, and blood pressure on regular intervals assists in evaluating the severity of the exposure and monitoring the body's vital functions in response to toxic effects. ALS level responders have more diagnostic tools available to them than the BLS level responder. The use and interpretation of these tools can provide more certainty in defining a patient's true condition.

(6) Communicate pertinent information.

Information related to the incident, including the number of patients and the type and severity of exposure or injuries, should be communicated and updated regularly throughout the event to receiving facilities and medical control as well as to the incident command structure.

(7) Estimate pharmacological need.

Based on the knowledge of the materials, the signs and symptoms of the patient, and the estimation of the dose, the ALS level responder should be able to estimate the type and amount of pharmaceuticals necessary for adequate intervention based on standing protocols or medical control.

(8) Address threat potential for clinical latency.

Many materials demonstrate delayed effects related to the exposure to the material and the dose involved. It is usually necessary to watch patients closely for delayed symptomology or clinical latency. With known products, medical control or poison controls centers may be able to alert the ALS level responder to the threat of delayed onset of effects.

(9) Estimate dosage — exposure.

The ALS level responder should be able, given appropriate information, to estimate the amount of time the patient was exposed and the amount of the product introduced into the body to determine an exposure estimate.

(10) Estimate dosage — treatment.
(11) Train in appropriate monitoring.

While treating and transporting patients it is imperative that the patient be closely monitored to identify any changes in condition, extent of symptoms, or impacts of underlying medical problems.

5.3.5 Supplemental Medical Resources.

Given scenarios of various hazardous materials/WMD mass casualty incidents, the ALS level responder shall identify the supplemental medical resources available to the AHJ, including the following:

(1) Describe the strategic national stockpile (SNS) program, including the following components:

(a) Intent and goals of the SNS program

The mission of the Strategic National Stockpile (SNS), which is maintained by the CDC, is to provide a re-supply of large quantities of essential medical material to states and communities during an emergency within 12 hours of the federal decision to deploy.

The SNS is a complex program involving such programs as the Cities Readiness Initiative (CRI), Chempak, the push package, the vendor-managed inventory, and the Pandemic Influenza Plan.

The CRI is a program designed for rapid deployment of antibiotic prophylaxis within 48 hours of identification of an anthrax attack. The Chempak program is exclusive to nerve agents. It provides the positioning of EMS and hospital-tailored caches of antidotes.

The vendor-managed inventory is a sustainable, event-tailored inventory that can be initially delivered within 36 hours of the decision to deploy.

The Pandemic Influenza Plan is designed to meet the requirements of a large population of people threatened by pandemic influenza, by monitoring disease outbreaks, maintaining stockpiles of antivirals and vaccines, providing public education and communication, and preparing a network of federal, state, and local preparedness.

(b) Procedures and requirements for deploying the SNS to a local jurisdiction

Refer to your authority having jurisdiction (AHJ) for the specific procedures that are required for deployment.

(c) Typical supplies contained in 12-hour push package

A push package, which is supplied by the CDC SNS, is a general cache of medical supplies configured to restock hospitals and public health during times of large-scale emergencies. Generally, a state governor's office initiates the request for a push package. The push package can be on site within 12 hours of the decision to deploy

(d) Role of the technical advisory response unit (TARU)

(2) Describe the metropolitan medical response system (MMRS) including the following components:

(a) Scope, intent, and goals of the MMRS

The metropolitan medical response system (MMRS) program is an initiative of the U.S. Department of Homeland Security, which assists highly populated regions of the country to develop plans, conduct training and exercises, acquire pharmaceuticals and personal protective equipment, and to achieve the enhanced capability to respond to a mass casualty event caused by a WMD terrorist act. This assistance supports the jurisdictions' activities to increase their response capabilities during the initial crucial hours of lifesaving and population protection.

(b) Capabilities and resources of the MMRS

(c) Eight capability focus areas of the MMRS

5.4 Competencies — Implementing the Planned Response

The ALS responder will play a key role in the EMS part of the response plan for which the outcome for both patients and responders depends. Therefore, the ALS responder is required by Section 4.5 to know enough about the hazardous materials to analyze the incident and effectively determine the risks and medical care necessary for their part of the response.

5.4.1 Determining the Nature of the Incident and Providing Medical Care.

The ALS level responder shall demonstrate the ability to provide emergency medical care to those patients exposed to hazardous materials/WMD by completing the following tasks:

The ALS responder is required by 5.4.1 to be able to determine from clues presented during dispatch, response, and approach whether a hazardous material is present at the scene and whether the released material poses a risk to the patient and, in turn, to the responders at the scene. Typical indicators of the presence of a released hazardous material include the following: operators or witnesses, placards, the normal occupancy of buildings at the scene (such as chemical storage buildings), the type of containers involved, and the presence of fires or explosions.

It is critical that ALS responders have a good understanding of the many physical states of potentially released substances that may be present at a hazardous materials/WMD incident as they perform their emergency medical care duties. Understanding these states and the associated mechanisms of injury and health implications associated with the released substance in these various states is critical to their care of patients as well as their personal safety while performing their medical care duties. It is also important for the ALS responder to be aware that the released substances might be in more than one state at the incident, as well as more than one released substance.

(1) The ALS level responder shall determine the physical state of the released substance and the environmental influences surrounding the release, as follows:

(a) Solid

A solid is a material in the state of matter characterized by resistance to deformation and changes of volume. At the microscopic scale, a solid has the properties of the atoms or molecules being packed closely together, its constituent elements have fixed positions in space relative to each other, and because any solid has some thermal energy, its atoms vibrate. However, this movement is very small and very rapid and cannot be observed under ordinary conditions. It is important to recognize that solid materials can rapidly change state under certain environmental conditions (of particular concern are those solids that rapidly volatilize under certain environmental conditions).

(b) Liquid

A liquid is a state of matter whose shape is usually determined by the container it fills. Liquid particles (normally molecules or clusters of molecules) are free to move within the liquid volume, but their mutual attraction limits ability of particles to leave the volume. The volume of a quantity of liquid is fixed by its temperature and pressure. It is also important to recognize that liquids can rapidly change state under certain environmental conditions (of particular concern are those liquids that rapidly volatilize under certain environmental conditions).

(c) Gas, vapor, dust, mist, aerosol

A gas is a state of matter that has atoms or molecules basically moving independently, with no forces keeping them together or pushing them apart. Hazardous substances in the gaseous state are of particular concern for the inhalation exposure route.

A vapor refers to a gas-phase material that normally exists as a liquid or solid under a given set of environmental conditions. Vapors are composed of single gas-phase molecules. Many, but not all, vapors are colorless and therefore invisible. Vapors do not wet objects with which they come in contact.

A dust is fine (small) particles of dry matter. Dusts can be generated by handling, crushing, grinding, rapid impact, detonation, and breakdown of certain organic or inorganic materials, such as rocks, ore, metal, coal, wood, grains, or other such material. Particles ranging in size from 0.1 μm (micron or micrometer) in diameter to about 30 μm in diameter and are referred to as total suspended particulate matter (TSP). Particles in the size range between 0.1μm and 10 μm are of particular concern for inhalation exposures.

A mist or fog is a microscopic suspension of liquid droplets in a gas. Do not confuse a mist with a vapor. Mists can generally be seen and reduce visibility. Mists generally wet objects with which they come in contact.

The ALS responder should also be aware of the environmental conditions at the incident and how these environmental conditions can alter the health impacts and physical state of the released substances. For example, on very hot days or in situations where the event results in elevated temperature at the scene, a substance normally not very volatile can become quite volatile with increased gaseous releases.

Windy conditions can create lower concentrations of the released substance in the working area. If aware of these conditions, ALS responders can position themselves upwind of the source and away from higher exposures. By being knowledgeable of situations like this, ALS responders are better able perform their medical care duties safely and more efficiently.

(2)* The ALS level responder shall identify potential routes of exposure, and correlate those routes of exposure to the physical state of the released substance, to determine the origin of the illness or injury, as follows:

The ALS responder needs to be very aware of how the associated released substances can enter the patient and themselves if they are not careful. Also, the ALS responder should continually evaluate all routes of exposure as they prepare to give medical care to patients. For example, care must be taken to not introduce the released substance into a patient through ingestion or injection as they administer medications to the patient. Likewise, they should be aware that while administering to patients who have received inhalation doses of the released substance, they can be subjecting themselves as well as other patients to the substance from the exhalation of that patient. Again, it is important to consider all routes of exposure while proceeding to administer to patients at a hazardous materials/WMD incident.

A.5.4.1(2) See A.4.4.1(3)(3).

(a) Inhalation

Inhalation is the means by which contaminants enter the body through the normal respiratory process (i.e., uptake through normal breathing process, with contaminants deposited along the respiratory tract into the lungs).

(b) Absorption

Absorption is the process by which contaminants are absorbed into the body through the skin and other exposed tissue, and it is often referred to as dermal absorption or dermal uptake.

(c) Ingestion

Ingestion is the process of consuming contaminants through the normal ingestion process (i.e., usually through the process of consuming food and water).

(d) Injection

Injection is the process by which contaminants are introduced directly into the bloodstream by means of a needle, cannula, or other mechanical process. Contaminants entering the bloodstream through an open wound are considered to be introduced by injection.

(3) The ALS level responder shall describe the potential routes of entry into the body, the common signs and symptoms of exposure, and the ALS treatment options approved by the AHJ (e.g., advanced airway management, drug therapy), including antidote administration where appropriate, for exposure(s) to the following classification of substances:

The ALS responder should maintain a strong working knowledge of signs and symptoms of exposure from each of these various classes of substances. These signs and symptoms will provide the ALS responder with key insights to integrate with their awareness of the material(s) release, their associated routes of exposure, their associated health effects and implications. These signs and symptoms are important checkpoints for an ALS responder to use to provide medical care to the patient(s). It is also important for the ALS responder to understand the health implications of these different classes of substances to protect themselves at the scene, as well as provide better medical care for their patients.

Patients exposed to these different classes of hazardous materials can pose a risk of secondary contamination to others who come in contact with them, including the ALS responder. An ALS responder's knowledge of toxic exposure, patent assessment, and decontamination procedures is essential for the responder to determine what actions are necessary to prepare patients to be treated and transported safely. In some cases, treatment might need to wait until the hazardous materials technicians at the scene decontaminate and transfer a patient to the cold zone.

(a) Corrosives

Corrosives are chemicals that cause visible destruction of, or irreversible alterations in, living tissue by chemical action at the site of contact. A chemical is considered to be corrosive if, when tested on the intact skin of albino rabbits by the method described in Appendix A of Title 49 CFR Part 173, it destroys or changes irreversibly the structure of the tissue at the site of contact following an exposure period of 4 hours [2]. For purposes of this standard, the term does not refer to action on inanimate surfaces.

(b) Pesticides

A pesticide is any substance or mixture of substances intended for preventing, destroying, repelling, or mitigating any pest. A pesticide might be a chemical substance or biological agent (such as a virus or bacteria) used against pests including insects, plant pathogens, weeds, mollusks, birds, mammals, fish, nematodes (roundworms), and microbes that compete with humans for food, destroy property, spread disease, or are a nuisance. Many pesticides are poisonous to humans.

(c) Chemical asphyxiants

Chemical asphyxiants reduce the body's ability to absorb, transport, or utilize inhaled oxygen. They are often active at very low concentrations — a few parts per million (ppm).

(d) Simple asphyxiants

An asphyxiant is a substance that can cause unconsciousness or death by suffocation (asphyxiation). Asphyxiants that have no other health effects are referred to as simple asphyxiants. Asphyxiation is an extreme hazard when working in enclosed spaces. Responders must be trained in confined space entry before working in sewers, storage tanks, and the like, where gases such as methane can displace oxygen from the atmosphere.

(e) Organic solvents

Organic solvents are a chemical class of compounds that are used routinely in commercial industries. They share a common structure (at least 1 carbon atom and 1 hydrogen atom), low molecular weight, lipophilicity, and volatility, and they exist in liquid form at room temperature. They can be grouped further into aliphatic-chain compounds, such as *n*-hexane, and as aromatic compounds with a 6-carbon ring, such as benzene or xylene. Aliphatics and aromatics can contain a substituted halogen element and might be referred to as halogenated hydrocarbons, such as perchloroethylene (PCE or PER), trichloroethylene (TCE), and carbon tetrachloride. Alcohols, ketones, glycols, esters, ethers, aldehydes, and pyridines are substitutions for a hydrogen group. Organic solvents can dissolve oils, fats, resins, rubber, and plastics.

(f) Nerve agents

The following information on nerve agents is extracted from the Federation of American Scientists' web site:

Nerve agents are a group of particularly toxic chemical warfare agents. They were developed just before and during World War II and are related chemically to the organophosphorus insecticides. The principle agents in this group are:

- GA - tabun
- GB - sarin
- GD - soman
- GF - cyclosarin
- VX - methylphosphonothioic acid

The "G" agents tend to be non-persistent whereas the "V" agents are persistent. Some "G" agents may be thickened with various substances in order to increase their persistence, and therefore the total amount penetrating intact skin. At room temperature GB is a comparatively volatile liquid and therefore non-persistent. GD is also significantly volatile, as is GA though to a lesser extent. VX is a relatively non-volatile liquid and therefore persistent. It is regarded as presenting little vapor hazard to people exposed to it. In the pure state nerve agents are colorless and mobile liquids. In an impure state nerve agents may be encountered as yellowish to brown liquids. Some nerve agents have a faint fruity odor.

- GB and VX doses which are potentially life threatening may be only slightly larger than those producing least effects. Death usually occurs within 15 minutes after absorption of a fatal VX dosage.
- Although only about half as toxic as GB by inhalation, GA in low concentrations is more irritating to the eyes than GB. Symptoms appear much more slowly from a skin dosage than from a respiratory dosage. Although skin absorption great enough to cause death may occur in 1 to 2 minutes, death may be delayed for 1 to 2 hours. Respiratory lethal dosages kill in 1 to 10 minutes, and liquid in the eye kills almost as rapidly.

Toxicological Data

Route	*Form*	*Effect*	*Type*	*GA*	*GB*	*GD*	*VX*	*Dosage*
Ocular	Vapor	Miosis	ECt_{50}	—	<2	<2	<0.09	mg•min/m^3
Inhalation at RMV = 15 l/min	Vapor	Runny Nose	ECt_{50}	—	<2	<2	<0.09	mg•min/m^3
Inhalation at RMV = 15 liters/min	Vapor	Incapacitation	ICt_{50}	—	35	35	25	mg•min/m^3
Inhalation at RMV = 15 liters/min	Vapor	Death	LCt_{50}	135	70	70	30	mg•min/m^3
Percutaneous	Liquid	Death	LD_{50}	4,000	1,700	350	10	mg

Ct (Concentration time; mg•min/m^3) - A measure of exposure to a gas, the effective vapor exposure, determined by the concentration of the gas (mg/m^3) and the length of exposure (min).

ECt_{50} (Effective Concentration Time; mg•min/m^3) - The Ct at which a gas debilitates 50% of the exposed population in a specific way.

ICt_{50} (Incapacitating Concentration Time; mg•min/m^3) - The Ct at which a gas incapacitates 50% of the exposed population.

LCt_{50} (Lethal concentration time; mg•min/m^3) - The Ct at which a gas kills 50% of the exposed population.

LD_{50} (Lethal dose; mg) - The dose or amount at which a substance kills 50% of the exposed population.

RMV (Respiratory minute volume; liters/min) - Volume of air inhaled per minute.

The values are estimates of the doses, which have lethal effects on a 70kg man. Effective dosages of vapor are estimated for exposure durations of 2-10 minutes. The effects of the

(continues)

nerve agents are mainly due to their ability to inhibit acetylcholinesterase throughout the body. Since the normal function of this enzyme is to hydrolyse acetylcholine wherever it is released, such inhibition results in the accumulation of excessive concentrations of acetylcholine at its various sites of action. These sites include the endings of the parasympathetic nerves to the smooth muscle of the iris, ciliary body, bronchial tree, gastrointestinal tract, bladder and blood vessels; to the salivary glands and secretory glands of the gastrointestinal tract and respiratory tract; and to the cardiac muscle and endings of sympathetic nerves to the sweat glands.

The sequence of symptoms varies with the route of exposure. While respiratory symptoms are generally the first to appear after inhalation of nerve agent vapor, gastrointestinal symptoms are usually the first after ingestion. Tightness in the chest is an early local symptom of respiratory exposure. This symptom progressively increases, as the nerve agent is absorbed into the systemic circulation, whatever the route of exposure. Following comparable degrees of exposure, respiratory manifestations are most severe after inhalation, and gastrointestinal symptoms may be most severe after ingestion.

The lungs and the eyes absorb nerve agents rapidly. In high vapor concentrations, the nerve agent is carried from the lungs throughout the circulatory system; widespread systemic effects may appear in less than 1 minute.

- The earliest ocular effect, which follows minimal symptomatic exposure to vapor, is miosis. The pupillary constriction may be different in each eye. Within a few minutes after the onset of exposure, there also occurs redness of the eyes. Following minimal exposure, the earliest effects on the respiratory tract are a watery nasal discharge, nasal hyperaemia, sensation of tightness in the chest and occasionally prolonged wheezing
- Exposure to a level of a nerve agent vapor slightly above the minimal symptomatic dose results in miosis, pain in and behind the eyes and frontal headache. Some twitching of the eyelids may occur. Occasionally there is nausea and vomiting.
- In mild exposures, the systemic manifestations of nerve agent poisoning usually include tension, anxiety, jitteriness, restlessness, emotional lability, and giddiness. There may be insomnia or excessive dreaming, occasionally with nightmares.
- If the exposure is more marked, the following symptoms may be evident: headache, tremor, drowsiness, difficulty in concentration, impairment of memory with slow recall of recent events, and slowing of reactions. In some casualties there is apathy, withdrawal and depression.
- With the appearance of moderate systemic effects, the casualty begins to have increased fatigability and mild generalized weakness, which is increased by exertion. This is followed by involuntary muscular twitching, scattered muscular fasciculation's and occasional muscle cramps. The skin may be pale due to vasoconstriction and blood pressure moderately elevated.
- If the exposure has been severe, the cardiovascular symptoms will dominate and twitching (which usually appear first in the eyelids and in the facial and calf muscles) becomes generalized. Many rippling movements are seen under the skin and twitching movements appear in all parts of the body. This is followed by severe generalized muscular weakness, including the muscles of respiration. The respiratory movements become more labored, shallow and rapid; then they become slow and finally intermittent.
- After moderate or severe exposure, excessive bronchial and upper airway secretions occur and may become very profuse, causing coughing, airway obstruction and respiratory distress. Bronchial secretion and salivation may be so profuse that watery secretions run out of the sides of the mouth. The secretions may be thick and tenacious. If the exposure is not so overwhelming as to cause death within a few minutes, other effects appear. These include sweating, anorexia, nausea and heartburn. If absorption of nerve agent has been great enough, there may follow abdominal cramps, vomiting, diarrhea, and urinary frequency. The casualty perspires profusely, may have involuntary defecation and urination and may go into cardio respiratory arrest followed by death.
- If absorption of nerve agent has been great enough, the casualty becomes confused and ataxic. The casualty may have changes in speech, consisting of slurring, difficulty in forming words, and multiple repetition of the last syllable. The casualty may then be-

come comatose, reflexes may disappear and generalized convulsions may ensue. With the appearance of severe central nervous system symptoms, central respiratory depression will occur and may progress to respiratory arrest.
- After severe exposure the casualty may lose consciousness and convulse within a minute without other obvious symptoms. Death is usually due to respiratory arrest requires prompt initiation of assisted ventilation to prevent death. If assisted ventilation is initiated, the individual may survive several lethal doses of a nerve agent.
- If the exposure has been overwhelming, amounting to many times the lethal dose, death may occur despite treatment as a result of respiratory arrest and cardiac arrhythmia. When overwhelming doses of the agent are absorbed quickly, death occurs rapidly without orderly progression of symptoms.

Nerve agent poisoning may be identified from the characteristic signs and symptoms. If exposure to vapor has occurred, the pupils will be very small, usually pin-pointed. If exposure has been cutaneous or has followed ingestion of a nerve agent in contaminated food or water, the pupils may be normal or, in the presence of severe systemic symptoms, slightly to moderately reduced in size. In this event, the other manifestations of nerve agent poisoning must be relied on to establish the diagnosis. No other known chemical agent produces muscular twitching and fasciculation's, rapidly developing pinpoint pupils, or the characteristic train of muscarinic, nicotinic and central nervous system manifestations.

The rapid action of nerve agents call for immediate self-treatment. Unexplained nasal secretion, salivation, tightness of the chest, shortness of breath, constriction of pupils, muscular twitching, or nausea and abdominal cramps call for the immediate intramuscular injection of 2 mg of atropine, combined if possible with oxime.

Source: Federation of American Scientists, www.fas.org/cw/cwagents.htm.

(g) Vesicants

The following information on vesicant agents is extracted from the Federation of American Scientists' web site:

Blister or vesicant agents are likely to be used both to produce casualties and to force opposing troops to wear full protective equipment thus degrading fighting efficiency, rather than to kill, although exposure to such agents can be fatal. Blister agents can be thickened in order to contaminate terrain, ships, aircraft, vehicles or equipment with a persistent hazard.

Vesicants burn and blister the skin or any other part of the body they contact. They act on the eyes, mucous membranes, lungs, skin and blood-forming organs. They damage the respiratory tract when inhaled and cause vomiting and diarrhea when ingested. The vesicant agents include:

- HD - sulfur mustard
- HN - nitrogen mustard
- L - lewisite (arsenical vesicants may be used in a mixture with HD)
- CX - phosgene (properties and effects are very different from other vesicants)

HD and HN are the most feared vesicants historically, because of their chemical stability, their persistency in the field, the insidious character of their effects by attacking skin as well as eyes and respiratory tract, and because no effective therapy is yet available for countering their effects. Since 1917, mustard has continued to worry military personnel with the many problems it poses in the fields of protection, decontamination and treatment. It should be noted that the ease with which mustard can be manufactured and its great possibilities for acting as a vapor would suggest that in a possible future chemical war HD would be preferred to HN.

Due to their physical properties, mustards are very persistent in cold and temperate climates. It is possible to increase the persistency by dissolving them in non-volatile solvents. In this way thickened mustards are obtained that are very difficult to remove by decontaminating processes.

(continues)

Exposure to mustard is not always noticed immediately because of the latent and sign-free period that may occur after skin exposure. This may result in delayed decontamination or failure to decontaminate at all. Whatever means is used has to be efficient and quick acting. Within 2 minutes contact time, a drop of mustard on the skin can cause serious damage. Chemical inactivation using chlorination is effective against mustard and lewisite, less so against HN, and is ineffective against phosgene oxime.

- In a single exposure the eyes are more susceptible to mustard than either the respiratory tract or the skin. The effects of mustard on the eyes are very painful. Conjunctivitis follows exposure of about 1 hour to concentrations barely perceptible by odor. This exposure does not affect the respiratory tract significantly. A latent period of 4 to 12 hours follows mild exposure, after which there is lachrymation and a sensation of grit in the eyes. The conjunctival and the lids become red. Heavy exposure irritates the eyes after 1 to 3 hours and produces severe lesions.
- The hallmark of sulfur mustard exposure is the occurrence of a latent symptom and sign free period of some hours post exposure. The duration of this period and the severity of the lesions are dependent upon the mode of exposure, environmental temperature and probably on the individual himself. High temperature and wet skin are associated with more severe lesions and shorter latent periods.
- If only a small dose is applied to the skin, the skin turns red and itches intensely. At higher doses blister formation starts, generally between 4 and 24 hours after contact, and this blistering can go on for several days before reaching its maximum. The blisters are fragile and usually rupture spontaneously giving way to a suppurating and necrotic wound. The necrosis of the epidermal cells is extended to the underlying tissues, especially to the dermis. The damaged tissues are covered with slough and are extremely susceptible to infection. The regeneration of these tissues is very slow, taking from several weeks to several months.
- Mustard attacks all the mucous membranes of the respiratory tract. After a latent period of 4 to 6 hours, it irritates and congests the mucous membranes of the nasal cavity and the throat, as well as the trachea and large bronchi. Symptoms start with burning pain in the throat and hoarseness of the voice. A dry cough gives way to copious expectoration. Airway secretions and fragments of necrotic epitheliums may obstruct the lungs. The damaged lower airways become infected easily, predisposing to pneumonia after approximately 48 hours. If the inhaled dose has been sufficiently high the victim dies in a few days, either from pulmonary edema or mechanical asphyxia due to fragments of necrotic tissue obstructing the trachea or bronchi, or from superimposed bacterial infection, facilitated by an impaired immune response.

The great majority of mustard gas casualties survive. There is no practical drug treatment available for preventing the effects of mustard. Infection is the most important complicating factor in the healing of mustard burns. There is no consensus on the optimum form of treatment.

A full protective ensemble can only achieve protection against these agents. The respirator alone protects against eye and lung damage and gives some protection against systemic effects. No drug is available for the prevention of the effects of mustard on the skin and the mucous membranes caused by mustards. It is possible to protect the skin against very low doses of mustard by covering it with a paste containing a chlorinating agent, e.g., chloramine. The only practical prophylactic method is physical protection such as is given by the protective respirator and special clothing.

In a pure form lewisite is a colorless and odorless liquid, but usually contains small amounts of impurities that give it a brownish color and an odor resembling geranium oil. It is heavier than mustard, poorly soluble in water but soluble in organic solvents. L is a vesicant (blister agent); also, it acts as a systemic poison, causing pulmonary edema, diarrhea, restlessness, weakness, subnormal temperature, and low blood pressure. In order of severity and appearance of symptoms, it is: a blister agent, a toxic lung irritant, absorbed in tissues, and a systemic poison. When inhaled in high concentrations, may be fatal in as short a time as 10 minutes.

- Liquid arsenical vesicants cause severe damage to the eye. On contact, pain and blepharospasm occur instantly. Edema of the conjunctival and lids follow rapidly and close the eye within an hour. Inflammation of the iris usually is evident by this time. After a few hours, the edema of the lids begins to subside, while haziness of the cornea develops.
- Liquid arsenical vesicants produce more severe lesions of the skin than liquid mustard. Stinging pain is felt usually in 10 to 20 seconds after contact with liquid arsenical vesicants. The pain increases in severity with penetration and in a few minutes becomes a deep, aching pain. Contamination of the skin is followed shortly by erythema, then by vesication, which tends to cover the entire area of erythema. There is deeper injury to the connective tissue and muscle, greater vascular damage, and more severe inflammatory reaction than are exhibited in mustard burns. In large, deep, arsenical vesicant burns, there may be considerable necrosis of tissue, gangrene and slough.
- The vapors of arsenical vesicants are so irritating to the respiratory tract that conscious casualties will immediately put on a mask to avoid the vapor. No severe respiratory injuries are likely to occur except among the wounded that cannot put on masks or the careless, caught without masks. Lewisite is irritating to nasal passages and produces a burning sensation followed by profuse nasal secretion and violent sneezing. Prolonged exposure causes coughing and production of large quantities of froth mucus. Injury to respiratory tracts, due to vapor exposure is similar to mustard's; however, edema of the lung is more marked and frequently accompanied by pleural fluid.

An antidote for lewisite is dimercaprol (British anti-lewisite (BAL)). This ointment may be applied to skin exposed to lewisite before actual vesication has begun. Some blistering is inevitable in most arsenical vesicant cases. The treatment of the erythema, blisters and denuded areas is identical with that for similar mustard lesions. Burns severe enough to cause shock and systemic poisoning are life threatening. Even if the patient survives the acute effects, the prognosis must be guarded for several weeks.

Phosgene oxime (CX) is a white crystalline powder. It melts between 39-40°C, and boils at 129°C. By the addition of certain compounds it is possible to liquefy phosgene oxime at room temperature. It is fairly soluble in water and in organic solvents. In aqueous solution phosgene oxime is hydrolyses fairly rapidly, especially in the presence of alkali. It has a high vapor pressure and its odor is very unpleasant and irritating. Even as a dry solid, phosgene oxime decomposes spontaneously and has to be stored at low temperatures.

In low concentrations, phosgene oxime severely irritates the eyes and respiratory organs. In high concentrations, it also attacks the skin. A few milligrams applied to the skin cause severe irritation, intense pain, and subsequently a necrotizing wound. Very few compounds are as painful and destructive to the tissues.

Phosgene oxime also affects the eyes, causing corneal lesions and blindness and may affect the respiratory tract causing pulmonary edema. The action on the skin is immediate: phosgene oxime provokes irritation resembling that caused by a stinging nettle. A few milligrams cause intense pain, which radiates from the point of application, within a minute the affected area turns white and is surrounded by a zone of erythema (skin reddening), which resembles a wagon wheel in appearance. In 1 hour the area becomes swollen, and within 24 hours, the lesion turns yellow and blisters appear. Recovery takes 1 to 3 months.

Source: Federation of American Scientists, www.fas.org/cw/cwagents.htm.

(h) Blood agents

A blood agent prevents the body from utilizing oxygen. A cyanogen agent is a chemical compound, containing the cyanide group, which acts as a chemical asphyxiant at the cellular level. The term *blood agent* is a misnomer, however, because these agents do not actually affect the blood in any way. Rather, they exert their toxic effect at the cellular level, by interrupting the electron transport chain in the inner membranes of mitochondria.

(i) Choking agents

The following information on choking agents is extracted from the Federation of American Scientists' web site:

> Choking agents are chemical agents which attack lung tissue, primarily causing pulmonary edema, are classed as lung damaging agents. To this group belong:
>
> - CG - phosgene
> - DP - diphosgene
> - Cl - chlorine
> - PS - chloropicrin
>
> The toxic action of phosgene is typical of a certain group of lung damaging agents. Phosgene is the most dangerous member of this group and the only one considered likely to be used in the future. Phosgene was used for the first time in 1915, and it accounted for 80% of all chemical fatalities during World War I.
>
> Phosgene is a colorless gas under ordinary conditions of temperature and pressure. Its boiling point is 8.2°C, making it an extremely volatile and non-persistent agent. Its vapor density is 3.4 times that of air. It may therefore remain for long periods of time in trenches and other low-lying areas. In low concentrations it has a smell resembling new mown hay.
>
> The outstanding feature of phosgene poisoning is massive pulmonary edema. With exposure to very high concentrations death may occur within several hours; in most fatal cases pulmonary edema reaches a maximum in 12 hours followed by death in 24–48 hours. If the casualty survives, resolution commences within 48 hours and, in the absence of complicating infection, there may be little or no residual damage.
>
> During and immediately after exposure, there is likely to be coughing, choking, a feeling of tightness in the chest, nausea, and occasionally vomiting, headache and lachrymation. The presence or absence of these symptoms is of little value in immediate prognosis. Some patients with severe coughs fail to develop serious lung injury, while others with little sign of early respiratory tract irritation develop fatal pulmonary edema. A period follows during which abnormal chest signs are absent and the patient may be symptom-free. This interval commonly lasts 2 to 24 hours but may be shorter. The signs and symptoms of pulmonary edema terminate it. These begin with cough (occasionally substernally painful), dyspnea, rapid shallow breathing and cyanosis. Nausea and vomiting may appear. As the edema progresses, discomfort, apprehension and dyspnea increase and frothy sputum develops. The patient may develop shock-like symptoms, with pale, clammy skin, low blood pressure and feeble, rapid heartbeat. During the acute phase, casualties may have minimal signs and symptoms and the prognosis should be guarded. Casualties may very rapidly develop severe pulmonary edema. If casualties survive more than 48 hours they usually recover.
>
> Source: Federation of American Scientists, www.fas.org/cw/cwagents.htm.

(j) Irritants (riot control agents)

An irritant is a chemical that is not corrosive but causes a reversible inflammatory effect on living tissue by chemical action at the site of contact. A chemical is a skin irritant if, when tested on the intact skin of albino rabbits by the methods of Title 16 CFR Part 1500.41 for 4 hours exposure or by other appropriate techniques, it results in an empirical score of five or more [3]. A chemical is an eye irritant if so determined under the procedure listed in Title 16 CFR Part 1500.42 or other appropriate techniques [4].

(k) Biological agents and toxins

Biological warfare (BW) agents, also known as germ warfare, is the use of any pathogen (bacterium, virus or other disease-causing organism) or toxin found in nature as a weapon of war. BW agents might be intended to kill, incapacitate, or seriously impede an adversary. Ideal characteristics of biological agents and toxins are high infectivity, high potency, availability of vaccines, and delivery as an aerosol. Diseases most likely to be considered for use as biological weapons are contenders because of their lethality (if delivered efficiently) and robustness (making aerosol delivery feasible).

The biological agents used in biological weapons can often be manufactured quickly and easily. The primary difficulty is not the production of the biological agent but delivery in an infective form to a vulnerable target. For example, anthrax is considered an effective agent for several reasons. First, it forms hardy spores, perfect for dispersal aerosols. Second, pneumonic (lung) infections of anthrax usually do not cause secondary infections in other people. Thus, the effect of the agent is usually confined to the target. A pneumonic anthrax infection starts with ordinary "cold" symptoms and quickly becomes lethal, with a fatality rate that is 80 percent or higher. Finally, friendly personnel can be protected with suitable antibiotics.

A mass attack using anthrax would require the creation of aerosol particles of 1.5 to 5 micrometers. Too large and the aerosol would be filtered out by the respiratory system. Too small and the aerosol would be inhaled and exhaled. Also, at this size, nonconductive powders tend to clump and cling because of electrostatic charges. This hinders dispersion. So, the material must be treated with silica to insulate and discharge the charges. The aerosol must be delivered so that rain and sun do not rot it, and yet the human lung can be infected. There are other technological difficulties as well. Diseases considered for weaponization or known to be weaponized include anthrax, Ebola, Bubonic Plague, Cholera, Tularemia, Brucellosis, Q fever, Machupo, Coccidioides mycosis, Glanders, Melioidosis, Shigella, Rocky Mountain Spotted Fever, Typhus, Psittacosis, Yellow Fever, Japanese B Encephalitis, Rift Valley Fever, and Smallpox. Naturally occurring toxins that can be used as weapons include Ricin, SEB, Botulism toxin, Saxitoxin, and many Mycotoxins. The organisms causing these diseases are known as select agents. Their possession, use, and transfer are regulated by the Centers for Disease Control and Prevention's Select Agent Program.

(l) Incapacitating agents

The term *incapacitating agent* is defined by the U.S. Department of Defense as follows:

> An agent that produces temporary physiological or mental effects, or both, which will render individuals incapable of concerted effort in the performance of their assigned duties. [5]

Incapacitating agents are not primarily intended to kill, but supposedly nonlethal incapacitating agents can kill many of those exposed to them. The term *incapacitation*, when used in a general sense, is roughly equivalent to the term *disability* as used in occupational medicine and denotes the inability to perform a task because of a quantifiable physical or mental impairment. In this sense, any of the chemical warfare agents can incapacitate a victim; however, again by the military definition of this type of agent, incapacitation refers to impairments that are temporary and nonlethal. Thus, riot-control agents are incapacitating because they cause temporary loss of vision due to blepharospasm, but they are not considered military incapacitants because the loss of vision does not last long. Although incapacitation can result from physiological changes such as mucous membrane irritation, diarrhea, or hyperthermia, the term *incapacitating agent* as militarily defined refers to a compound that produces temporary and nonlethal impairment of military performance by virtue of its psychobehavioral or central nervous system (CNS) effect.

(m) Radiological materials

The following information is extracted from the EPA's web site:

> Three basic concepts apply to all types of ionizing radiation. When developing regulations or standards that limit how much radiation a person can receive in a particular situation, one considers how these concepts affect a person's exposure.
>
> **BASIC CONCEPTS OF RADIATION PROTECTION**
>
> **Time.** The amount of radiation exposure increases and decreases with the time people spend near the source of radiation.
>
> In general, think of the exposure time as how long a person is near radioactive material. It's easy to understand how to minimize the time for external (direct) exposure. Gamma and X-rays are the primary concern for external exposure.

(continues)

However, if radioactive material gets inside the body, one can't move away from it. The only options once internal uptake occurs are to wait until it decays or until the body can eliminate it. When this happens, the biological half-life of the radionuclide controls the time of exposure. Biological half-life is the amount of time it takes the body to eliminate one half of the radionuclide initially present. Alpha and beta particles are the main concern for internal exposure.

When establishing a radiation standard that assumes an exposure over a certain period, the concept of time is applied. For example, exposures are often expressed in terms of a committed dose. A committed dose is one that accounts for continuing exposures over long periods of time (such as 30, 50, or 70 years). It refers to the exposure received from radioactive material that enters and remains in the body for many years.

When assessing the potential for exposure in a situation, consider the amount of time a person is likely to spend in the area of contamination. For example, in assessing the potential exposure from radon in a home, estimate how much time people are likely to spend in the basement.

Distance. The farther away a person is from a radiation source, the less their exposure. How close to a source of radiation one can get without getting a high exposure depends on the energy of the radiation and the size (or activity) of the source. Distance is a prime concern when dealing with gamma rays, because they can travel long distances. Alpha and beta particles don't have enough energy to travel very far.

As a rule, if you double the distance, you reduce the exposure by a factor of four (i.e., halving the distance increases the exposure by a factor of four). The area of the circle depends on the distance from the center to the edge of the circle (radius). It is proportional to the square of the radius. As a result, if the radius doubles, the area increases four times. Using the light bulb analogy, think of the radiation source as a bare light bulb. The bulb gives off light equally in every direction, in a circle. The energy from the light is distributed evenly over the whole area of the circle. When the radius doubles, the radiation is spread out over four times as much area, so the dose is only one fourth as much. (In addition, as the distance from the source increases so does the likelihood that some gamma rays will lose their energy.

The exposure of an individual sitting 4 feet from a radiation source will be 1/4 the exposure of an individual sitting 2 feet from the same source.

Shielding. The greater the shielding around a radiation source, the smaller the exposure. Shielding simply means having something that will absorb radiation between you and the source of the radiation. The amount of shielding required to protect against different kinds of radiation depends on how much energy they have.

α (Alpha)	A thin piece of light material, such as paper, or even the dead cells in the outer layer of human skin provides adequate shielding because alpha particles can't penetrate it. However, living tissue inside body, offers no protection against inhaled or ingested alpha emitters.
β (Beta)	Additional covering, for example heavy clothing, is necessary to protect against beta-emitters. Some beta particles can penetrate and burn the skin.
γ (Gamma)	Thick, dense shielding, such as lead, is necessary to protect against gamma rays. The higher the energy of the gamma ray, the thicker the lead must be. X-rays pose a similar challenge, so X-ray technicians often give patients receiving medical or dental X-rays a lead apron to cover other parts of their body.

DIRTY BOMBS/RADIOACTIVE DISPERSAL DEVICES (RDDS)

Although "dirty bombs," or radioactive dispersal devices (RDDs), are not weapons of mass destruction, in the past few years terrorists have indicated their interest in acquiring such weapons. RDDs disperse radioactive material by using conventional explosives or other means. There are only a few radioactive sources that can be used effectively in an RDD. The greatest security risk is posed by Cobalt-60, Cesium-137, Iridium-192, Strontium-90, Americium-241, Californium-252, and Plutonium-238.

Source: U.S. Environmental Protection Agency, www.epa.gov/radiation.

(n) Nitrogen compounds

Nitrogen oxides are produced during most combustion processes. About 80 percent of the immediately released nitrogen oxide is in the form nitric oxide (NO). Small amounts of nitrous oxide (N_2O) are also produced. Nitric oxide reacts with oxygen in the air to produce nitrogen dioxide (NO_2). Further oxidation during the day causes the nitrogen dioxide to form nitric acid and nitrate particles. In the dark, nitrogen dioxide can react with ozone and form a very reactive free radical. The free radical then can react with organic compounds in the air to form nitrogenated organic compounds, some of which have been shown to be mutagenic and carcinogenic.

Nitrogen dioxide is the most important nitrogen oxide compound with respect to acute adverse health effects. Under most chemical conditions, it is an oxidant. However, it takes about 10 times more nitrogen dioxide than ozone to cause significant lung irritation and inflammation.

Nitrates and nitrites are known to cause several health effects. In general, the following are the most common effects:

- Reactions with hemoglobin in blood, causing the oxygen-carrying capacity of the blood to decrease (nitrite)
- Decreased functioning of the thyroid gland (nitrate)
- Vitamin A shortages (nitrate)
- Formation of nitro amines, which are known as one of the most common causes of cancer (nitrates and nitrites)

(o) Opiate compounds

The main opiates derived from opium are morphine, codeine, thebaine, and diacetylmorphine (heroin). Papaverine and noscapine are also present but have essentially no effect on the central nervous system and are not usually placed in the same category as the others. Papaveretum is a standardized preparation of mixed opium alkaloids used on cardiac patients. Heroin is not generally used therapeutically and is illegal to produce, sell, or possess in many parts of the world because of its high potential for abuse.

Opioids work on specific receptors in the brain to decrease the sensation of pain, which is their primary pharmacological use. The following are possible side effects of opioids:

- Nausea. Nausea is the most common side effect. Opioids slow down gastrointestinal activity, which can make many patients queasy. Dehydration or constipation can also cause nausea or make it worse.
- Constipation. Constipation depresses gut activity, causing stool to stay in the body longer. It can accumulate and harden, which makes bowel movements difficult or impossible.
- Central nervous system (CNS) effects. Opioids can have a wide range of side effects on CNS functions. Opioids can either inhibit or excite the CNS, although inhibition is more common. Patients with depressed CNS functions may feel varying levels of drowsiness, lightheadedness, euphoria or dysphoria, or confusion. If opioids excite CNS functions, it can result in side effects like hyperalgesia (extreme sensitivity to pain), myoclonus (involuntary jerking of muscles), or seizures, in rare cases.

Less common side effects of opioids include the following:

- Urinary retention
- Respiratory depression, particularly in elderly or debilitated patients
- Pruritus (itching)
- Miosis (constriction of the pupil, a common effect of opioids)

(p) Fluorine compounds

Pure fluorine (F_2) is a corrosive pale yellow or brown gas that is a powerful oxidizing agent. It is the most reactive of all the elements and readily forms compounds with most other elements. Fluorine even combines with the noble gases, krypton, xenon, and radon. Even in dark cool conditions, fluorine reacts explosively with hydrogen. It is so reactive that glass, metals, and even water, as well as other substances, burn with a bright flame in a jet of fluorine gas. It is far too reactive to be found in elemental form and has such an affinity for most elements, including silicon, that it can neither be prepared nor be kept in ordinary glass vessels. Instead, it must be kept in specialized quartz tubes lined with a very thin layer of fluorocarbons. In moist air it reacts with water to form also-dangerous hydrofluoric acid.

Fluorides are compounds that combine fluorine with some positively charged counterpart. They often consist of crystalline ionic salts. Fluorine compounds with metals are among the most stable of the salts. Both elemental fluorine and fluoride ions are highly toxic and must be handled with great care. Contact with skin and eyes should be strictly avoided. In its free element state, fluorine has a characteristic pungent odor that is detectable in concentrations as low as 20 nL/L.

Hydrofluoric acid (HF) contact with exposed skin posses one of the most extreme and dangerous hazards. These effects are intensified by the fact that HF damages nerves in such a way as to make such burns initially painless. The HF molecule is capable of rapidly migrating through lipid layers of cells that would ordinarily stop an ionized acid, and the burns are typically deep. HF can react with calcium, permanently damaging the bone. More seriously, reaction with the body's calcium can cause cardiac arrhythmias, followed by cardiac arrest brought on by sudden chemical changes within the body. These cannot always be prevented with local or intravenous injection of calcium salts. HF spills over just 2.5 percent of the body's surface area, despite copious immediate washing, have been fatal (this corresponds with an area of about 9 in^2 or 23 cm^2). If the patient survives, HF burns typically produce open wounds of an especially slow-healing nature.

(q) Phenolic compounds

Phenols, sometimes called phenolics, are a class of chemical compounds consisting of a hydroxyl group (-OH) attached to an aromatic hydrocarbon group. The simplest of the class is phenol (C_6H_5OH). Although similar to alcohols, phenols have unique properties and are not classified as alcohols (because the hydroxyl group is not bonded to a *saturated* carbon atom). They have relatively higher acidities due to the aromatic ring tightly coupling with the oxygen and a relatively loose bond between the oxygen and hydrogen.

A number of health effects from breathing phenol in air have been reported. Short-term effects include respiratory irritation, headaches, and burning eyes. Chronic effects of high exposures include weakness, muscle pain, anorexia, weight loss, and fatigue. Effects of long-term low-level exposures include increases in respiratory cancer, heart disease, and effects on the immune system. In animal laboratory studies, exposure to high concentrations of phenol in air for a few minutes irritates the lungs, and repeated exposure for several days produces muscle tremors and loss of coordination. Exposure to high concentrations of phenol in the air for several weeks results in paralysis and severe injury to the heart, kidneys, liver, and lungs, followed by death in some cases. When exposures involve the skin (dermal uptake), the size of the total surface area of exposed skin can influence the severity of the toxic effects. Ingestion of very high concentrations of phenol has resulted in death.

Effects reported in humans following dermal exposure to phenol include liver damage, diarrhea, dark urine, and red blood cell destruction. Skin exposure to a relatively small amount of concentrated phenol has resulted in the death of humans. Small amounts of phenol applied to the skin of laboratory animals for brief periods can produce blisters and burns on the exposed surface, and spilling dilute phenol solutions on large portions of the body (greater than 25 percent of the body surface) can result in death.

(4) The ALS level responder shall describe the basic toxicological principles relative to assessment and treatment of persons exposed to hazardous materials, including the following:

An understanding of the toxicological principles of the various classes of compounds as coupled with the clinical signs and symptoms associated with exposure to these compounds is critical for the ALS responder to provide medical care to their patients, as well as protect themselves at an incident. While managing acute effects is critical often to the very survival of the patient, the ALS responder should also be aware of the delayed (chronic) effects. An understanding of both acute and chronic effects of the released materials will help the ALS responder give the most prudent care to the patients. It is the dose–response relationship of the release material that can provide the ALS responder with the best insight as to the health effects associated with the different internal concentration levels of an exposed patient.

(a) Acute and delayed toxicological effects

Acute toxicity refers to the sudden, severe onset of symptoms due to an exposure to the contaminant(s) of concern. Delayed toxicity might not develop for hours, days, or even years following an exposure to the contaminant(s) of concern. In some cases, such as exposure to biological agents, symptoms might not appear until three or more days following an exposure to such agents.

(b) Local and systemic effects

Local effects are those in which a toxic substance comes in direct contact with the skin or other sensitive tissue. Systemic effects are the effects of a toxic substance on either the entire body or a specific organ or organ system.

(c) Dose-response relationship

The chemical, biological, or radiological dose–response relationship refers to the response a specific dose produces in the human body. The magnitude of the body's response depends on the on-scene concentration (as can be measured by monitoring systems) of the hazardous substance, the patient exposure concentration and duration, and the actual dose (considering uptake rate for each applicable exposure route) received by the patient. The maximum ambient concentration at the scene determines the maximum concentration available for exposure. The exposure concentration is the concentration available to the pertinent routes of exposure, and the duration is the amount of time the patient is exposed to this available concentration. The actual dose is that amount taken up by the patient through the applicable uptake mechanisms. The dose will be the total amount of a patient's uptake during the exposure time, considering all routes of uptake.

(5) Given examples of various hazardous substances, the ALS level responder shall define the basic toxicological terms as they relate to the treatment of an exposed patient, as follows:

(a) Threshold limit value — time weighted average (TLV-TWA)

The threshold limit value — time-weighted average (TLV-TWA) is the time-weighted average concentration for a conventional 8-hour workday and 40-hour workweek, to which it is believed that nearly all workers might be repeatedly exposed, day after day, without adverse health effect.

(b) Lethal doses and concentrations, as follows:

The lethal dose (LD) of a material is a single dose that causes the death of a specified number of the group of test animals exposed by any route other than inhalation. The lethal concentration (LC) is the median lethal concentration of a hazardous material. The LC is defined as the concentration of a material in air that, on the basis of laboratory tests (inhalation route), is expected to kill a specified number of the group of test animals when administered over a specified period of time. The following are the various types of LD and LC:

1. LD_{lo} is the is lowest dosage per unit of bodyweight (typically stated in milligrams per kilogram) of a substance known to have resulted in fatality in a particular animal species. This is also called the lowest dosage causing death, lowest detected lethal dose, and lethal dose low.
2. LD_{50} or median lethal dose of a toxic material is the dose required to kill half (50 percent) of the members of a tested population. LD_{50} figures are frequently used as a general indicator of a substance's toxicity.
3. LD_{hi} or LD_{100} is the absolute dose of a toxic material required to kill all (100 percent) of the members of a tested population.
4. LC_{lo} is the lowest lethal concentration of a material reported to cause death in a particular animal species, when administered via the inhalation route. It is the lowest lethal concentration for gases, dusts, vapors, mists.
5. LC_{50} is the lethal concentration of a material in air that is expected to kill 50 percent of the group of a particular animal species when administered via the inhalation route.
6. LC_{hi} or LC_{100} is the absolute lethal concentration of a toxic material required to kill all (100 percent) of the members of a tested population, administered via inhalation route.

    i. LD_{lo}
    ii. LD_{50}
    iii. LD_{hi}
    iv. LC_{lo}
    v. LC_{50}
    vi. LC_{hi}

(c) Parts per million/parts per billion/parts per trillion (ppm/ppb/ppt)

The values used to establish the exposure limits are quantified in parts per million, parts per billion, or parts per trillion. A good reference to remember is that 1 percent equals 10,000 ppm, 1 ppm equals 1,000 ppb, and 1 ppb equals 1,000 ppt. Thus, for a reading from a sampling instrument of 0.5 percent, that is equivalent to 5,000 ppm, 5,000,000 ppb, or 5,000,000,000 ppt. If the TLV is determined to be 7,500 ppm, the reading from the instrument can be related to determine the degree of hazard that exposure concentration represents.

(d) Immediately dangerous to life and health (IDLH)

Immediate danger to life and health (IDLH) is the maximum level to which a healthy worker can be exposed for 30 minutes and escape without suffering irreversible health effects or impairment. If at all possible, exposure to this level should be avoided. If that is not possible, responders must wear positive pressure self-contained breathing apparatus or a positive pressure supplied-air respirator with an auxiliary escape system. This limit is established by OSHA and NIOSH.

(e) Permissible exposure limit (PEL)

Permissible exposure limit (PEL) is a term OSHA uses in its health standards covering exposures to hazardous chemicals. It is similar to the TLV-TWA established by the American Conference of Governmental Industrial Hygienists (ACGIH). The PEL, which generally relates to the legally enforceable TLV limits, is the maximum concentration, averaged over 8 hours, to which 95 percent of healthy adults can be repeatedly exposed for 8 hours per day, 40 hours per week.

(f) Threshold limit value — short-term exposure limit (TLV-STEL)

Threshold limit value — short-term exposure limit (TLV-STEL) is the maximum average concentration, averaged over a 15-minute period, to which healthy adults can be safely exposed for up to 15 minutes continuously. Exposure should not occur more than 4 times a day with at least 1 hour between exposures.

(g) Threshold limit value — ceiling (TLV-C)

Threshold limit — value ceiling (TLV-C) is the maximum concentration to which a healthy adult can be exposed without risk of injury. TLV-C is comparable to the IDLH, and exposures to higher concentrations should not occur.

(h) Solubility

Solubility, or the degree to which a substance is soluble in water, is useful in determining effective extinguishing agents and methods. Solubility should be considered along with specific gravity.

(i) Poison — a substance that causes injury, illness, or death

Poisons are substances that can cause injury, illness, or death to organisms, usually by chemical reaction or other activity on the molecular scale, when a sufficient quantity is absorbed by an organism.

(j) Toxic — harmful nature related to amount and concentration

Toxicity is a measure of the degree to which something is toxic or poisonous. The study of poisons is known as toxicology. Toxicity can refer to the effect on a whole organism, such as a human or a bacterium or a plant, or to a substructure, such as a cell (cytotoxicity) or the liver. Toxicity addresses the harmful nature of a hazardous material related to amount and concentration.

(6) Given examples of hazardous materials/WMD incidents with exposed patients, the ALS level responder shall evaluate the progress and effectiveness of the medical care provided at a hazardous materials/WMD incident, to ensure that the overall incident response objectives, along with patient care goals, are being met by completing the following tasks:
   (a) Locate and track all exposed patients at a hazardous materials/WMD incident, from triage and treatment to transport to the appropriate hospital.
   (b) Review the incident objectives at periodic intervals to ensure that patient care is being carried out within the overall incident response plan.
   (c) Ensure that the incident command system forms are completed, along with the patient care forms required by the AHJ, during the course of the incident.
   (d) Evaluate the need for trained and qualified EMS personnel, medical equipment, transport units, and other supplies, including antidotes based on the scope and duration of the incident.

Understanding all of these critical factors under 5.4.1 for a particular hazardous material released at an incident is essential for a responder to determine the nature of the hazards to both the patients and the responders themselves. The *Emergency Response Guidebook* is an excellent reference to use in finding the needed information [6].

It is essential for responders to know which references and/or on-line databases provide the information necessary to determine the health effects of the hazardous materials involved in an incident to establish the medical treatment that can be used to combat them, in consultation with medical control. They must also understand which properties have an impact on a patient's reactions and on the medical care the patient must receive. These references become critical resources to perform these tasks. Sources of this information include EMS reference books, the poison control center, EMS/HM data systems, and the Agency for Toxic Substances and Disease Registry (ATSDR), which is part of the CDC. ATSDR can be contacted using the following contact: www.atsdr.cdc.gov. Telephone numbers for voice and data communications for these resources should be part of the EMS responders' resources.

The ALS responder must be able to apply the information from these reference sources on the released material to the information obtained while assessing the patient to determine the risk of secondary contamination to others. For example, if a patient states that he or she was completely covered by a particular hazardous material and reference sources indicate that the material is a very difficult material to remove once it has contaminated the skin, the chance of secondary exposure is slight because the material does not dislodge easily. However, if the responder must touch the patient to provide medical care, the contaminant may be transferred by their physical contact.

5.4.2* Decontaminating Exposed Patients.

Given the emergency response plan and SOPs developed by the AHJ and given examples of hazardous materials/WMD incidents with exposed patients, the ALS level responder shall do as follows:

(1) Given the emergency response plan and SOPs developed by the AHJ, identify and evaluate the patient decontamination activities performed prior to accepting responsibility for and transferring care of exposed patients.

It is important for the ALS responder to determine the degree of decontamination conducted on exposed patients prior to accepting responsibility for exposed patients (see Exhibit II.5.5).

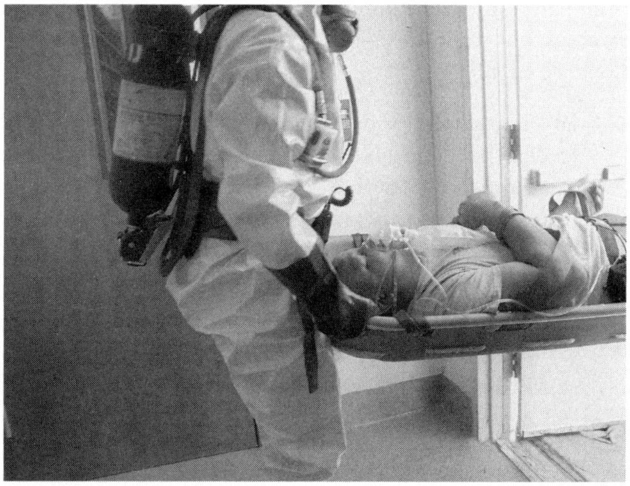

EXHIBIT II.5.5 *Before accepting responsibility for exposed patients, the ALS responder must first determine the degree of decontamination conducted. (Source: Rob Schnepp)*

(2) Determine the need and location for patient decontamination, including mass-casualty decontamination, in the event none has been performed prior to arrival of EMS personnel, and complete the following tasks:

   (a) Given the emergency response plan and SOPs developed by the AHJ, identify and evaluate the patient decontamination activities performed prior to accepting responsibility for and transferring care of exposed patients; identify sources of information for determining the appropriate decontamination procedure and how to access those resources in a hazardous materials/WMD incident.

To anticipate and understand patient decontamination requirements, the ALS responder must make every effort to obtain information about the released substance. This information will help the ALS responder make an informed decision on performing decontamination or more fully evaluate the effectiveness of the decontamination performed. These are important decisions, especially for transporting agencies. It is unwise to accept a contaminated patient into a transport unit or to be unsure of the level of decontamination performed. A poor decision in the field can have significant ramifications at the door of the hospital. If hospital staff are unconvinced that proper decontamination was performed in field, they might require further decontamination prior to accepting the patient. This may delay care and have an adverse effect on the patient.

There may be a need to balance decontamination with providing emergency medical care. In some cases, it might be wise to provide airway management, bleeding control, basic trauma care, or CPR prior to or concurrent with decontamination. There are no clear-cut guidelines for these situations. The ALS responder must weigh the benefits and risks of the need to provide ALS with the need to decontaminate.

As with all other sections in the NFPA 473 part of the handbook, this section is designed to promote thoughtful and informed medical care. It is impossible to expect a standard to address all possible situations. To that end, the ALS responder, at the moment he or she must make a decision regarding patient care, should be well trained and confident to make that decision.

ALS responders must weigh all options and make reasonable decisions in the field. Again, it is important to understand the nature and level of contamination in order to make the best decision regarding decontamination and patient care.

(b) Given the emergency response plan and SOPs developed by the AHJ, identify and evaluate the patient decontamination activities performed prior to accepting responsibility for and transferring care of exposed patients.

The ALS responder should access and apply information from reliable reference sources in order to determine appropriate decontamination procedures. These reference sources include MSDSs, CHEMTREC/CANUTEC/SETIQ, Regional Poison Control Centers, U.S. Department of Transportation (DOT) Emergency Response Guidebook, Hazardous Materials Information System (HMIS), shipper/manufacturer contacts, Agency for Toxic Substances and Disease Registry (ATSDR) medical management guidelines, medical toxicologists, and electronic databases.

Decontamination efforts are influenced by the number of ambulatory and nonambulatory patients and must be addressed accordingly. It will be more difficult, and require more personnel, to decontaminate nonambulatory patients. The ALS responder should be familiar enough with the concepts and procedures for decontamination to adapt to the situation. Flexibility is the key to handling decontamination issues when it comes to life safety. Keep in mind that patient decontamination should be rapid, complete, and geared at getting the patient clean enough to treat and transport.

(c) Given the emergency response plan and SOPs provided by the AHJ, identify the supplies and equipment required to set up and implement technical or mass-casualty decontamination operations for ambulatory and nonambulatory patients.
(d) Given the emergency response plan and SOPs developed by the AHJ, identify the procedures, equipment, and safety precautions for securing evidence during decontamination operations at hazardous materials/WMD incidents.
(e) Identify procedures, equipment, and safety precautions for handling tools, equipment, weapons, and law enforcement and K-9 search dogs brought to the decontamination corridor at hazardous materials/WMD incidents.
(f) Identify procedures, equipment, and safety precautions for communicating with critically, urgently, and potentially exposed patients, and population prioritization and management techniques.

The intent of 5.4.2(2)(f) is to remind responders to address the importance of communicating with large groups of people who might be ill or injured. Panic can be contagious and the lack of crowd control can create significant problems for the responders. ALS responders should expect panic and unexpected behaviors from a large group of people exposed to a hazardous materials/WMD.

(g) Determine the threat of cross contamination to all responders and patients by completing the following tasks:
   i. Identify hazardous materials/WMD with a high risk of cross contamination.
   ii. Identify hazardous materials/WMD agents with a low risk of cross contamination.
   iii. Describe how the physical state of the hazardous materials/WMD provides clues to its potential for secondary contamination, when the exact identity of the hazardous materials/WMD is not known.

It is extremely important to understand the potential for cross contamination. All patients involved in an incident will not be contaminated equally.

A.5.4.2 Most ALS medical treatment at hazardous materials/WMD incidents will be delivered in the cold zone, after decontamination. In some cases, ALS level skills need to be delivered in the warm or hot zone prior to or concurrent with decontamination. In those situations, ALS level providers need to balance the need for performing life-saving interventions with decontamination, taking into consideration the nature and severity of the incident; the medical needs of the patient; and the need to perform decontamination prior to rendering care.

Life safety of the responder is paramount. ALS level responders who anticipate functioning under these conditions should receive training and meet the mission-specific personal protective equipment competencies as defined in Section 6.2 of NFPA 472, *Standard for Competence of Responders to Hazardous Materials/Weapons of Mass Destruction Incidents*.

It is critical that EMS providers review their responsibilities within their local emergency response plan before an incident occurs to ensure that EMS responders are adequately trained for their expected roles within the Incident Management System at the hazardous materials/WMD incident. The priorities for triage, treatment, or decontamination in the setting of other significant injuries should be based on the following requirements:

(1) *Priority I — Medical Care First.* Medical care outweighs immediate decontamination, and patients should be grossly decontaminated only as priority to transport. Contaminated patients with serious or critical illness, trauma, or burns should be decontaminated while their life-threatening injuries are being addressed.
(2) *Priority II — Combined Priorities.* Medical care needs are balanced with a priority to decontaminate. These patients present with a serious illness other than from the chemical exposure, have trauma or burn injuries, and have not been decontaminated but might have a high level of contamination. There might be a risk to the EMS provider from an ongoing exposure to the hazardous substance. In this situation, it might not be safe to render medical care without the appropriate personal protective equipment. The ABCs (airway/breathing/circulation) and threats to life should be managed along with rapid decontamination.
(3) *Priority III — Decontaminate First.* Decontamination should be performed prior to providing medical care. In this situation, it might not be safe to render medical care without the appropriate personal protective equipment.

Patient conditions are categorized as follows:

(1) A = Critical condition: airway compromised, serious signs or symptoms of shock, cardiac arrest, life-threatening trauma or burns
(2) B = Unstable condition: shortness of breath, unstable vital signs, altered level of consciousness after the exposure, significant trauma or burns
(3) C = Stable condition: stable vital signs, no altered level of consciousness, no significant trauma or burns

See Table A.5.4.2.

TABLE A.5.4.2 Patient Priority Levels

	Priority Based on Condition		
Level of Contamination	***Medically Critical (A)***	***Medically Unstable (B)***	***Medically Stable (C)***
Heavily contaminated with highly toxic substance	II	III	III
Heavily contaminated with low-toxicity substance	I	II	II
Low-level contamination with highly toxic substance	II	III	III
Low-level contamination with low-toxicity substance	I	I	II
Chemical in eyes: Decontaminate eyes immediately and thoroughly.			

5.4.3 Evaluating the Need for Medical Supplies.

Given examples of single-patient and multicasualty hazardous materials/WMD incidents, the ALS level responder shall determine if the available medical equipment, transport units, and other supplies, including antidotes, will meet or exceed expected patient care needs throughout the duration of the incident.

Multicasualty incidents present different challenges than a single-patient medical response. Multicasualty incidents require more personnel and equipment and are usually more chaotic. In these cases, the ALS responder should look at the entire incident and determine the need for medical equipment, qualified EMS personnel, and transport units. These and other resources should be called to the scene as soon as possible. Additionally, the ICS should be implemented as soon as possible to ensure responder accountability and the most efficient use of on-scene resources.

5.4.4 Evidence Preservation.

Given examples of hazardous materials/WMD incidents where criminal acts are suspected, the ALS level responder shall make every attempt to preserve evidence during the course of delivering patient care by completing the following tasks:

(1) Determine if the incident is potentially criminal in nature and cooperate with the law enforcement agency having investigative jurisdiction.
(2) Identify the unique aspects of criminal hazardous materials/WMD incidents, including crime scene preservation, evidence preservation, and destruction of potential evidence found on medical patients, and/or the destruction of evidence during the decontamination process.
(3) Ensure that any information regarding suspects, sequence of events during a potential criminal act, or observations made based on patient presentation or during patient assessment are documented and communicated and passed on to the law enforcement agency having investigative jurisdiction.

The rationale behind 5.4.4 is to remind ALS level responders that they have a role in identifying and preserving evidence at a crime scene. If there is a suspicion that the event might be criminal in nature, it is incumbent on the ALS responder to be observant; pay attention to what is touched or moved; understand patient presentations relative to the suspected hazardous materials/WMD; observe and note the locations of patients, and so forth. Furthermore, complete and thorough documentation is important and could be used in the prosecution of any suspects. These and other actions taken by ALS responders will assist law enforcement with the criminal investigation.

5.4.5 Medical Support at Hazardous Materials/WMD Incidents.

Given the emergency response plan and SOPs developed by the AHJ and examples of various hazardous materials/WMD incidents, the ALS level responder shall describe the procedures for performing medical support of hazardous materials/WMD incident response personnel, and shall complete the following tasks:

The use of a systematic approach toward medical monitoring is required of the hazmat responder, especially when entry into a hazardous environment is prolonged, affected by meteorological factors, or difficult to mitigate. In conditions when the primary objective is potential rescue, on-scene medical personnel might default back to the annual "fit to work" physical. In either case, at an absolute minimum, the responder should be evaluated for heat stress, which includes hydration and potential exposure.

(1) The ALS level responder responsible for pre-entry medical monitoring shall obtain hazard and toxicity information on the released substance from the designated hazardous materials technical reference resource or other reliable sources of information at the scene. The following information shall be conveyed to the entry team, incident safety officer, hazardous materials officer, other EMS personnel at the scene, and any other responders responsible for the health and well-being of those personnel operating at the scene:

The medical monitoring station can serve many purposes at the scene of a hazardous materials event. The first and primary role of the medical monitoring station is to evaluate the core vitals of the entry team. This may include the backup team and the decontamination line personnel. Each individual assigned to work at the medical monitoring station should observe all responders for any physiological response that could become detrimental toward the responder and the task of that responder.

The basic setup of the monitoring station should facilitate taking vital signs along with pulse oximetry, weight, skin status, body temperature, and hydration. Medical monitoring stations should be positioned to protect responders from adverse weather conditions.

In larger incidents, a medical manager within the incident command structure must organize his or her resources to compile the basic physiological information from each responder, as well as anticipate future medical needs of the hazmat technician. This could include, but is not limited to, specialized transport capability, on-scene advanced life support, and analysis of the capacity of the health care system.

ALS responders should be trained to recognize and treat various types of heat/cold stress. Heat stroke is the most dangerous off all the heat-related illnesses. ALS care providers should understand the need for prompt recognition of the early signs of heat stroke.

Aggressive monitoring of individuals will become the priority for the ALS care providers. Once heat or cold stress has been identified, treatment might only include monitoring of the individual and oral hydration. All medical providers that are used at a hazardous materials/WMD incident must always look for the signs and symptoms of heat stress, especially when the environmental factors are present along with the use of encapsulating protective equipment.

(a) Chemical name

Looking up synonyms and trade names can give the responder additional information when databases are referenced.

(b) Hazard class

Although direct information about medical application is limited, this may give clues to the type of injury the individual may have. For example, Class 2 gases (pressurized or liquefied) give us the highest potential for a respiratory hazard if PPE is not worn or not worn properly, or if a breach has occurred.

(c) Hazard and toxicity information

Many chemicals have multiple hazards, and it is up to the medical personnel to fully reference the suspected/confirmed substance. Understanding how these factors affect individuals allows the medical staff to have a proactive treatment plan.

(d) Applicable decontamination methods and procedures

ALS responders should be familiar with the local procedures used for victim decontamination. In some jurisdiction, decontamination is considered to be a form of medical treatment. The ALS responder should look at the overall scene and balance the decontamination efforts with the need to provide patient care.

(e) Potential for secondary contamination

Liquids and particular solids provide the highest level of cross contamination; gases rarely have any cross contamination issues. The organophosphates as a chemical family have the highest level of cross contamination due to the mechanism of injury. Again one must understand the state of matter, hazards associated with the substance at different concentrations, and the multiple hazards that a substance might have.

(f) Procedure for transfer of patients from the constraints of the incident to the emergency medical system
(g) Prehospital management of medical emergencies and exposures, including antidote administration

As with any medical emergency that occurs outside of the hospital, rapid evaluation, BLS measures, and transportation with information transfer are the keys toward a positive outcome. This includes but is not limited to the type of substance involved, the action(s) at time of incident, referenced information of the substance, and the potential degree of exposure.

(2) The ALS level responder shall evaluate the pre-entry health status of hazardous materials/WMD responders prior to donning PPE by performing the following tasks:

(a) Record a full set of vital signs

Vital signs should include but are not limited to body temperature, pulse, respiration (rate and character), and blood pressure.

(b) Record body weight measurements

Fluid loss is best measured in the field by measuring the responder's pre- and post-entry body weight. By achieving a percent loss of body weight, the ALS care provider can estimate the level of hydration. Hydration procedures should have input from the medical control officer (medical director). Fluid loss will become the most important factor to monitor and observe for working effectively within protective ensembles.

(c) Record general health observations

An idea of the responder's general health status can be obtained by observing his or her physical appearance and asking specific questions about their well-being. Anyone who has had signs and symptoms of general common illness, such as a cold or the flu, or has just been released to work after having a general illness should be further evaluated and should consult with the medical director.

(3) The ALS level responder shall determine the medical fitness of those personnel charged with donning chemical protective clothing, using the criteria set forth in the emergency action plan (EAP) and the SOP developed by the AHJ. Consideration shall be given to excluding responders if they do not meet the following criteria prior to working in chemical protective clothing:

(a) Core body temperature: hypothermia/hyperthermia

Extreme heat or cold will make it difficult for personnel to function in protective ensembles. Response personnel must be evaluated in either extreme for signs and symptoms that could indicate hypothermia/hyperthermia. The wearing of protective ensembles can exacerbate these conditions, especially when working in high-level PPE.

(b) Blood pressure: hypotension/hypertension

It is important to monitor the diastolic blood pressure pre- and post-entry. Pre-entry exam diastolic in relation to the systolic should be evaluated and exclusion criteria planned for by the medical director. Look for a 3- to 5-minute recovery back to baseline after exiting PPE during post-entry exam.

(c) Heart rate: bradycardia/tachycardia

Each jurisdiction must adhere to its own definitions of bradycardia and tachycardia. The intent here is not to set hard limits but to include these vital signs as part of a pre-entry examination. As with any vital sign, bradycardia and tachycardia must be viewed in the context of an overall patient presentation. As an example, a pulse rate over 100 beats per minute might not be out of line for a responder preparing to don PPE on a hot and humid day. The NFPA 473 committee encourages a reasonable and well thought-out approach to medical monitoring.

(d) Respiratory rate: bradypnea/tachypnea

Both the rate and character of respiration should be evaluated along with these conditions, which should have a planned response from the medical control officer, the vital sign related to the blood pressure, historical conditions, and annual physical findings.

(4) The ALS level responder shall determine how the following factors influence heat stress on hazardous materials/WMD response personnel:

Heat stress can affect the performance of any responder. It is up to the ALS care provider to recognize, evaluate, and treat the responder who may be suffering from a heat stress event. This type of event should have input from the local medical control officer for guidance within the systems response manual. This guidance is a blueprint for the medical authority to provide the necessary evaluation and treatment for a responder who may be subject to heat stress.

(a) Baseline level of hydration
(b) Underlying physical fitness
(c) Environmental factors
(d) Activity levels during the entry
(e) Level of PPE worn
(f) Duration of entry
(g) Cold stress

(5) Given examples of various hazardous materials/WMD incidents requiring the use of chemical protective ensembles, the ALS level responder shall complete the following tasks:

(a) Demonstrate the ability to set up and operate a medical monitoring station.
(b) Demonstrate the ability to recognize the signs and symptoms of heat stress, heat exhaustion, and heat stroke.
(c) Determine the ALS needs for responders exhibiting the effects of heat stress, cold stress, and heat exhaustion.
(d) Describe the medical significance of heat stroke and the importance of rapid transport to an appropriate medical receiving facility.

(6) Given a simulated hazardous materials/WMD incident, the ALS level responder shall demonstrate documentation of medical monitoring activities.

(7) The ALS level responder shall evaluate all team members after decontamination and PPE removal, using the following criteria:

Each organization should develop and have preplanned medical monitoring protocols that identify conditions that are the basic evaluative concerns for the first responder. Within 3 to 5 minutes of exit from the protective ensemble, all vitals should be close to the pre-entry physical. Temperature should be monitored longer and evaluated. This is especially important when performing entry functions within the two extreme conditions of heat and cold.

(a) Pulse rate done within the first minute
(b) Pulse rate 3 minutes after initial evaluation
(c) Temperature
(d) Body weight
(e) Blood pressure
(f) Respiratory rate

(8) The ALS level responder shall recommend that any hazardous materials team member exhibiting any of the following signs be prohibited from redonning chemical protective clothing:

The use of exclusion criteria is important to establish guidelines for the post exam and for team members re-entering a hazardous environment. Again, the NFPA 473 committee encourages a reasonable approach to medical monitoring. The standard is designed to provide a framework for monitoring, but the limits and exclusion points should be determined on a case-by-case basis, adhering to the standard of care of the AHJ.

(a) Heat stress or heat exhaustion
(b) Pulse rate: tachycardia/bradycardia
(c) Core body temperature: hyperthermia/hypothermia
(d) Recovery heart rate with a trend toward normal rate and rhythm
(e) Blood pressure: hypertension/hypotension
(f) Weight loss of >5 percent
(g) Signs or symptoms of extreme heat exhaustion or heat stroke, which requires transport by ALS ambulance to the appropriate hospital

(9) The ALS level responder shall notify immediately the appropriate persons designated by the emergency response plan if a team member requires significant medical treatment or transport (arranged through the appropriate designee identified by the emergency response plan).

5.5 Competencies — Terminating the Incident

Upon termination of the hazardous materials/WMD incident, the ALS level responder shall complete the reporting, documentation, and EMS termination activities as required by the local emergency response plan or the organization's SOPs and shall meet the following requirements:

(1) Identify the reports and supporting documentation required by the emergency response plan or SOPs.
(2) Demonstrate completion of the reports required by the emergency response plan or SOPs.
(3) Describe the importance of personnel exposure records.
(4) Describe the importance of debriefing records.
(5) Describe the importance of critique records.

(6) Identify the steps in keeping an activity log and exposure records.
(7) Identify the steps to be taken in compiling incident reports that meet federal, state, local, and organizational requirements.
(8) Identify the requirements for compiling personal protective equipment logs.
(9) Identify the requirements for filing documents and maintaining records, as follows:
   (a) List the information to be gathered regarding the exposure of all patient(s) and describe the reporting procedures, including the following:
      i. Detailed information on the substances released
      ii. Pertinent information on each patient treated or transported
      iii. Routes, extent, and duration of exposures
      iv. Actions taken to limit exposure
      v. Decontamination activities
   (b) Identify the methods used by the AHJ to evaluate transport units for potential contamination and the process and locations available to decontaminate those units.

The intent of Section 5.5 is to offer basic guidance in the area of documentation. Each jurisdiction has unique characteristics and requirements for patient care reports and incident documentation. The ALS responder should, in all cases, write timely documentation that is accurate and reflective of his or her actions on the scene. When it comes to hazardous materials/WMD incidents, patient care reports might be used during a prosecution. To that end, the report should be written in a professional manner.

REFERENCES CITED IN COMMENTARY

1. "Guide 100-00: Guide for Selection of Chemical Agent and Toxic Industrial Material Detection Equipment for Emergency Responders," U.S. Department of Justice, National Institute of Justice, Office of Science and Technology, Washington, DC, June, 2000. Available online at www.ncjrs.gov/pdffiles1/nij/184449.pdf.
2. Title 49, Code of Federal Regulations, Part 173, Shippers — General Requirements for Shipments and Packagings, U.S. Government Printing Office, Washington, DC.
3. Title 16, Code of Federal Regulations, Part 1500.41, Hazardous Substances and Articles; Administration and Enforcement Regulations: Method of testing primary irritant substances, U.S. Government Printing Office, Washington, DC.
4. Title 16, Code of Federal Regulations, Part 1500.42, Hazardous Substances and Articles; Administration and Enforcement Regulations: Test for eye irritants, U.S. Government Printing Office, Washington, DC.
5. Publication 1-02, *Dictionary of Military and Associated Terms*, U.S. Department of Defense, Washington, DC, 2007. Available online at www.dtic.mil/doctrine/jel/doddict/
6. *Emergency Response Guidebook*, Department of Transportation, U.S. Government Printing Office, Washington, DC.
7. Title 40, Code of Federal Regulations, Part 156, Labeling Requirements for Pesticides and Devices, U.S. Government Printing Office, Washington, DC.

Additional Reference

Agency for Toxic Substances and Disease Registry (ATSDR), U.S. Centers for Disease Control. See www.atsdr.cdc.gov.

Explanatory Material

ANNEX A

The material contained in Annex A is included in the text within this handbook and therefore is not repeated here.

Informational References

ANNEX B

B.1 Referenced Publications

The documents or portions thereof listed in this annex are referenced within the informational sections of this standard and are not part of the requirements of this document unless also listed in Chapter 2 for other reasons.

B.1.1 NFPA Publications.

National Fire Protection Association, 1 Batterymarch Park, Quincy, MA 02169-7471.

NFPA 472, *Standard for Competence of Responders to Hazardous Materials/Weapons of Mass Destruction Incidents*, 2008 edition.

B.1.2 Other Publications. (Reserved)

B.2 Informational References

The following documents or portions thereof are listed here as informational resources only. They are not a part of the requirements of this document.

B.2.1 NFPA Publications.

National Fire Protection Association, 1 Batterymarch Park, Quincy, MA 02169-7471.

NFPA 11, *Standard for Low-, Medium-, and High-Expansion Foam*, 2005 edition.
NFPA 30, *Flammable and Combustible Liquids Code*, 2008 edition.
NFPA 58, *Liquefied Petroleum Gas Code*, 2008 edition.
NFPA 1561, *Standard on Emergency Services Incident Management System*, 2005 edition.
NFPA 1991, *Standard on Vapor-Protective Ensembles for Hazardous Materials Emergencies*, 2005 edition.
NFPA 1992, *Standard on Liquid Splash-Protective Ensembles and Clothing for Hazardous Materials Emergencies*, 2005 edition.
Hazardous Materials Response Handbook, 2008.

B.2.2 Other Publications.

B.2.2.1 ACC Publications. American Chemistry Council (formerly Chemical Manufacturers Association), 1300 Wilson Blvd., Arlington, VA 22209.

Recommended Terms for Personal Protective Equipment, 1985.

B.2.2.2 API Publications. American Petroleum Institute, 1220 L Street, N.W., Washington, DC 20005-4070.

API 2021, *Guide for Fighting Fires in and Around Flammable and Combustible Liquid Atmospheric Petroleum Storage Tanks*, 2001.

API 2510-A, *Fire Protection Considerations for the Design and Operation of Liquefied Petroleum Gas (LPG) Storage Facilities*, 1996.

B.2.2.3 NFA Publications. National Fire Academy, Federal Emergency Management Agency, Emmitsburg, MD 21727.

Hazardous Materials Incident Analysis, 1984.

B.2.2.4 NRT Publications. National Response Team, National Oil and Hazardous Substances Contingency Plan, Washington, DC 20593.

NRT-1, *Hazardous Materials Emergency Planning Guide*, 2001.

B.2.2.5 U.S. Government Publications. U.S. Government Printing Office, Superintendent of Documents, Washington, DC 20402.

Title 29, Code of Federal Regulations, Parts 1910.119–1910.120.
Title 40, Code of Federal Regulations, Part 261.33.
Title 40, Code of Federal Regulations, Part 302.
Title 40, Code of Federal Regulations, Part 355.
Title 49, Code of Federal Regulations, Parts 170–180.
Emergency Response Guidebook, U.S. Department of Transportation, 2004 edition.
Emergency Response, Command, and Planning Guidelines (various documents) for terrorist incidents involving chemical and biological agents. U. S. Army Research, Development, and Engineering Command (RDECOM), available through the Edgewood Chemical Biological Center, website http://www.ecbc.army.mil/hld.

B.2.2.6 Additional Publications. EPA, Emergency Response Program publications, Washington, DC: Environmental Protection Agency, www.epa.gov.

Grey, G. L., et al., *Hazardous Materials/Waste Handling for the Emergency Responder*, New York: Fire Engineering Publications, 1989.
Maslansky, C. J., and S. P. Maslansky., *Air Monitoring Instrumentation*, New York: Van Nostrand Reinhold, 1993.
National Toxicology Program, *Report on Carcinogens*, 9th edition, Washington, DC: U.S. Department of Health and Human Services, 2001.
NIOSH/OSHA/USCG/EPA, *Occupational Safety and Health Guidance Manual for Hazardous Waste Site Activities*, October 1985.
Noll, G. G., et al., *Hazardous Materials, Managing the Incident*, 2nd edition, Stillwater, OK: Fire Protection Publications, 1995.
Wright, C. J., "Managing the Hazardous Materials Incident," *Fire Protection Handbook*, 18th edition, Quincy, MA: National Fire Protection Association, 1997.

B.3 References for Extracts in Informational Sections. (Reserved)

PART III

Supplements

The seven supplements included in this part of the *Hazardous Materials/Weapons of Mass Destruction Response Handbook* provide additional information about key areas of concern for hazardous materials responders. The following supplements are not part of the official NFPA documents or the commentary but present additional information for the reader:

1. Excerpts from *Protecting Emergency Responders: Lessons Learned from Terrorist Attacks*
2. Recognizing and Identifying Hazardous Environments
3. Fire Fighter Fatalities, West Helena, Arkansas, May 8, 1997
4. Propane Tank Explosion Results in the Death of Two Volunteer Fire Fighters, Hospitalization of Six Other Volunteer Fire Fighters and a Deputy Sheriff — Iowa
5. Selection of Chemical-Protective Clothing Using NFPA Standards
6. Response Levels
7. Incident Mitigation

SUPPLEMENT 1

Excerpts from *Protecting Emergency Responders: Lessons Learned from Terrorist Attacks*

Editor's Note: *The following excerpts are reprinted with permission from conference proceedings "Protecting Emergency Responders: Lessons Learned from Terrorist Attacks," by Brian A. Jackson, D.J .Peterson, James T. Bartis, Tom LaTourrette, Irene Brahmakulam, Ari Houser, and Jerry Sollinger. The conference was organized and the proceedings published by the Science and Technology Institute. The Institute is a federally funded research and development center sponsored by the National Science Foundation and managed by RAND. For additional information, see www.rand.org.*

On December 9–11, 2001, a conference was held in New York City that brought together individuals with experience in responding to acts of terrorism. The purpose of the conference was to hear and document the firsthand experiences of emergency responders regarding the performance, availability, and appropriateness of their personal protective equipment as they responded to these incidents. The meeting considered the responses to the September 11, 2001 attacks at the World Trade Center and the Pentagon; the 1995 attack at the Alfred P. Murrah Federal Building in Oklahoma City, Oklahoma; and the emergency responses to the anthrax incidents that occurred in several locations through autumn 2001. The conference was sponsored by the National Institute for Occupational Safety and Health of the U.S. Centers for Disease Control and Prevention, which also arranged for RAND to organize and conduct the conference and prepare this report.

This report presents a synthesis of the discussions held at the December meeting. It is intended to help federal managers and decision makers

- Understand the unique working and safety environment associated with terrorist incidents.
- Develop a comprehensive personal protective technology research agenda.
- Improve federal education and training programs and activities directed at the health and safety of emergency responders.

The report should also help state and municipal officials, trade union leaders, industry executives, and researchers obtain a better understanding of equipment and training needs for protecting emergency workers.

EXECUTIVE SUMMARY

Just as it has for the nation as a whole, the world in which emergency responders work has changed in fundamental ways since September 11, 2001. Members of professions already defined by their high levels of risk now face new, often unknown threats on the job. At a basic level, the September 11 terrorist events have forced emergency responders to see the incidents they are asked to respond to in a new light. At the World Trade Center, 450 emergency responders perished while responding to the terrorist attacks — about one-sixth of the total number of victims. Hundreds more were seriously injured. In this light, the terrorist events are also forcing emergency responders to reconsider the equipment and practices they use to protect themselves in the line of duty.

Preparation is key to protecting the health and safety of emergency responders, and valuable lessons can be learned from previous responses. To this end, the National Institute for Occupational Safety and Health (NIOSH) sponsored and asked the RAND Science and Technology Policy Institute to organize a conference of individuals with firsthand knowledge of emergency response to terrorist attacks. The purpose of the conference was to review the adequacy of personal protective equipment (PPE) and practices, such as training, and to make recommendations on how the equipment and practices worked and how they might be improved. Attendees included persons who responded to the 1995 attack on the Alfred P. Murrah Federal Building in Oklahoma City, the September 11 attacks on the World Trade Center and the Pentagon, and the anthrax incidents that occurred during autumn 2001. They represented a wide range of occupations and skills: firefighters, police, emergency medical technicians, construction workers, union officials, and government representatives from local, state, and federal agencies. The conference was held December 9–11, 2001, in New York City, and this report synthesizes the discussions that took place there.

NEW RISKS, NEW ROLES FOR EMERGENCY RESPONDERS

Although the terrorist incidents shared some characteristics with large natural disasters, the NIOSH/RAND conference participants highlighted ways in which those incidents posed unique challenges. They were large in scale, long in duration, and complex in terms of the range of hazards presented. As a result of these characteristics, these events thrust responders into new roles for which they may not have been properly prepared or equipped. The themes of scale, duration, and range of hazards were repeated frequently during the discussions at the conference because they were seen as having critical implications for protecting the health and safety of emergency responders — during both the immediate, urgent phase and the sustained campaign phase of the responses.

The September 11 terrorist incidents were notable for their large scale — in terms of both the damage incurred and the human and material resources needed to respond. Conference participants spoke extensively about the difficulty of conducting search and rescue, fire suppression, and shoring and stabilization operations, as well as hazard monitoring. Responses were hampered by collateral developments, in particular the grounding of commercial air transport, which slowed the implementation of command and logistical support infrastructures.

The responses to the terrorist attacks involved days and weeks of constant work. At the World Trade Center, an initial urgent phase persisted for several days and then gradually transitioned into a sustained campaign that lasted for several months. An important message of the conference was that PPE generally worked well for its designed purpose in the initial response. However, such equipment typically was not designed for the continuous use associated with a sustained response campaign. Firefighter turnout gear, for example, is constructed to be worn for, at most, hours. Accordingly, responders spoke of being hampered by basic problems such as wet garments and blistered feet.

Furthermore, at major terrorist-attack sites, emergency workers face a staggering range of hazards. Not only do they confront the usual hazards associated with building fires — flames, heat, combustion by-products, smoke — they also must be prepared to deal with rubble and debris, air choked with fine particles, human remains, hazardous materials (anhydrous ammonia, freon, battery acids), and the potential risk of secondary devices or a follow-on attack. Conference participants indicated that many currently available PPE ensembles and training practices were not designed to protect responders from this range of hazards or were not supplied in sufficient quantity at the attack sites to meet the scale of the problem.

The scale of the terrorist events, their duration, and the range of hazards required that many emergency responders take on atypical tasks for which they were insufficiently equipped and trained. The nature of the destruction at the World Trade Center and the Pentagon reduced opportunities for primary reconnaissance and rescue — important tasks for firefighters in large structural fires. Conversely, firefighters became engaged in activities they usually do not do: "busting up and hauling concrete," scrambling over a rubble pile, and removing victims and decayed bodies and body parts.

Construction workers were also deployed at the scenes and placed in hazardous environments early on. In all of the terrorist-incident responses, emergency medical personnel were on-scene, performing rescue operations, for example, in the rubble pile at the World Trade Center. Complicating activity at these already chaotic, hazardous, and demanding attack sites was the fact that the sites are also crime scenes. In addition, there were massive influxes of skilled and unskilled volunteers that created a significant challenge in managing the incident sites and assuring that all were properly protected.

In sum, the definition and roles of an *emergency responder* expanded greatly in the wake of the terrorist attacks, but few of the responders had adequate PPE, training or information for such circumstances.

PERSONAL PROTECTIVE EQUIPMENT PERFORMANCE AND AVAILABILITY

From the experiences at these attack sites, it is clear that there were significant shortfalls in the way responders were protected. Many responders suggested that the PPE even impeded their ability to accomplish their missions.

Within the overall PPE ensemble used by responders at these sites, some equipment performed better than others. While head protection and high-visibility vests functioned relatively well for most responders, protective clothing and respirators exhibited serious shortcoming. Conference participants reported that the available garments did not provide sufficient protection against biological and infectious disease hazards, the heat of fires at the sites, and the demanding physical environment of unstable rubble piles, nor were they light and flexible enough to allow workers to move debris and enter confined spaces. Attendees also indicated that the available eye protection, while protecting well against direct impact injury, provided almost no protection against the persistent dust at the World Trade Center site.

Of all personal protective equipment, respiratory protection elicited the most extended discussion across all of the professional panels. Attendees indicated that under most circumstances, the self-contained breathing apparatus (SCBA) was grossly limited by both the weight of the systems and the short lengths of time (about 15 to 30 minutes) they can be used before their air bottles must be refilled. Most participants complained that respirators reduced their field of vision at best, and their facepieces fogged up at worst. Filters for air-purifying respirators (APRs) often did not match available facepieces, and many responders questioned the level of protection they provided, especially during anthrax responses.

For almost all protective technologies, responders indicated serious problems with equipment not being comfortable enough to allow extended wear during demanding physical labor. It was frequently observed that current technologies require a tradeoff between the amount of protection they provide and the extent to which they are light enough, practical enough, and wearable enough to allow responders to do their jobs. While conference attendees were concerned about having adequate protection, many were even more concerned about equipment hindering them from accomplishing their rescue and recovery missions in an arduous and sustained campaign. Respirators available at the sites were uncomfortable, causing many wearers to use them only intermittently (one participant dubbed them "neck protectors") or to discard them after a short period.

For many firefighters at the conference, PPE availability was as important a concern as PPE performance. Some health-and-safety panelists expressed a similar view. There was an acute shortage of respirators early in the response at the World Trade Center, for example. Subsequently, providing appropriate equipment to the large numbers of workers at these sites was made even more difficult because of the many types and brands of equipment that were being used by the various responder organizations or were being supplied from various sources. The problem was further exacerbated by a lack of interoperability among different types of equipment. These issues, coupled with the very large volume of equipment sent to the World Trade Center site, in particular, made it very difficult to match responders with appropriate equipment and supplies.

PERSONAL PROTECTIVE EQUIPMENT TRAINING AND INFORMATION

The responses to the terrorist attacks uncovered a range of PPE training and information needs. Before an incident occurs, those who are likely to be involved in a response should be trained on the proper selection and operation of personal protective equipment. Emergency medical technicians who were themselves treating casualties in the heart of the disaster site should have been wearing PPE but frequently were not, in large part because this equipment was not part of their standard training regimen.

The experiences in these incidents also showed that there is a need for significant on-site training to protect the health and safety of workers. The attack sites involved large numbers of workers, particularly construction workers and volunteers, many of whom were not familiar with most PPE. They needed to be trained in the proper selection and fitting of respirators, how to maintain them, and when to change filters. The situation with anthrax was more severe. Health and safety panel members felt that training support during the anthrax attacks was inadequate on all fronts: The response protocols were being developed *during* the actual response.

Emergency responders repeatedly stressed the importance of having timely and reliable health and safety information. "What kills rescue responders is the unknown," commented an emergency medical services (EMS) panel member. Several shortcomings were noted by conference participants. Special-operations and law-enforcement responders reported problems caused by different information sources telling them different things. Such information conflicts were often attributed to differences in risk assessment and PPE standards among reporting parties. Especially in the case of anthrax incidents, keeping up with changing information being provided by numerous agencies was a serious challenge for front-line responder organizations. For many conference participants, the problem was not a lack of information on hazards. Rather, they spoke of difficulties trying to manage and make sense of a surplus of information. Finally, conference attendees suggested that better and more consistent information provision could motivate responders to wear PPE and could decrease the tendency to modify it or take it off when it becomes uncomfortable.

SITE MANAGEMENT

One message that emerged clearly from virtually all panel discussions is that proper site management had a decisive effect on whether personal protective equipment was available, appropriately prescribed, used, and maintained.

The most critical need for site management is a coherent command authority. An effective command structure is essential to begin solving three critical issues affecting PPE: information provision, equipment logistics, and enforcement. Due to logistical problems early in the response, for example, supplies of PPE were misplaced, the stocks of equipment that were available were largely unknown, and responders often did not receive or could not find the equipment they needed.

Conference attendees also emphasized the need for immediate and effective perimeter or scene control. Initially, this entailed responders personally "holding people back" and isolating the scene. As the response evolved, it was necessary to erect a "hard perimeter," such as a chain link fence to make sure only essential personnel operating under the direction of the scene commander were on-site.

Conference attendees also indicated that enforcement of PPE use is very important. Although panelists acknowledged that there is a period early in a chaotic response when it is not practical to rigorously enforce the use of protective equipment, they indicated that strict enforcement must eventually begin in order to protect the health of the responders. Other factors that complicated enforcement of PPE use were the large number of organizations (with different PPE standards) operating on-site, the lack of a unified command, and shortcomings in scene control. Because of the difficulty of defining when it is appropriate to begin enforcing PPE use — and removing workers from the site if they do not comply with use requirements — attendees indicated that this role might be best played by an organization not directly involved in or affected by the incident.

RECOMMENDATIONS

After having discussed PPE performance, information and training, and site-management issues, NIOSH/RAND conference participants were asked to put forward concrete recommendations about technologies and procedures that could help protect the health and safety of emergency workers as they respond to acts of terrorism. The following points represent a brief sample of the themes that emerged and the solutions put forth by conference discussions.

Personal Protective Equipment Performance

- Develop guidelines for the appropriate PPE ensembles for long-duration disaster responses involving rubble, human remains, and a range of respiratory threats. If appropriate equipment is not currently available, address any roadblocks to its development. Such equipment could be applicable to other major disasters, such as earthquakes or tornadoes, as well as to terrorist attacks.
- Define the appropriate ensembles of PPE needed to safely and efficiently respond to biological incidents, threats, and false alarms. Key considerations include providing comparable levels of protection for all responders and addressing the logistical and decontamination issues associated with large numbers of responders in short time periods.

Personal Protective Equipment Availability

- Explore mechanisms to effectively outfit all responders at large incident sites with appropriate personal protective equipment as rapidly as possible.
- Examine any barriers to equipment standardization or interoperability among emergency-responder organizations. Strategies could include coordination of equipment procurement among organizations or work with equipment manufacturers to promote broader interoperability within classes of equipment.

Training and Information

- Define mechanisms to rapidly and effectively provide responders at incident sites with useful information about the hazards they face and the equipment they need for protection. Approaches could include more-effective coordination among relevant organizations and development of technologies that provide responders with individual, real-time information about their enforcement.
- Explore ways to ensure that responders at large-scale disaster sites are appropriately trained to use the protective equipment they are provided. All types of responders must be addressed, and mechanisms that provide training and experience with the equipment before a disaster occurs should be investigated.

- Consider logistical requirements of extended response activities during disaster drills and training. Such activities provide response commanders with information on the logistical constraints that could restrict response capabilities.

Management

- Provide guidelines and define organizational responsibilities for enforcing protective-equipment use at major disaster sites. While such guidelines must address the risks responders are willing to take when the potential exists to save lives, they must also consider that during long-term responses, the health and safety of responders should be a principal concern.
- Develop mechanisms to allow rapid and efficient control at disaster sites as early as possible during a response.

CONCLUDING REMARKS

The emergency workers and managers who attended the NIOSH/RAND conference provided a wealth of information on availability, use, performance, and management of personal protective equipment. Throughout the conference, a number of important issues were explicitly addressed during the meeting; others were implicit consequences of the lessons learned. This concluding chapter draws out several of these strategic policy issues for further reflection.

Guidelines

One of the clear messages of the conference was that most emergency workers do not believe that they are prepared with the necessary information, training, and equipment to cope with many of the challenges associated with the response to a major disaster such as the World Trade Center attack or for threats associated with anthrax and similar agents. These challenges include the large scale of the operations, the long duration of the response, the broad range of known and potential hazards encountered, and the assumption of nonstandard tasks by emergency responders.

Lessons learned from the response to the terrorist attacks suggest that near-term efforts to develop and upgrade equipment and operating guidelines could significantly improve the safety of emergency workers.

- Guidelines are needed for designing personal protective equipment ensembles appropriate for long-term responses to a range of major disasters.[1] An obvious case would be a disaster involving the collapse of one or more large buildings and the consequent need to work on rubble in the presence of a variety of hazards, including human remains, smoldering fires, and airborne contaminants derived from the building and its collapse.
- Recognizing that different responders have different personal protection requirements, these guidelines could also address the various professional groups working at a disaster site. Moreover, the guidelines should take into account the reality that individual responders may fulfill various tasks entailing different hazards and that hazards vary within the inner and outer perimeters of a disaster site.
- Protective equipment and safety guidelines could lead to better responses to biological incidents, not only for anthrax but for other potential biological threats.
- Well-designed guidelines and protocols could significantly improve real-time on-site hazard assessments. Essential elements include sensing equipment, measurement sites, organizational responsibilities and authorities, and data interpretation consistent with operational requirements.
- Discussions about the management of the terrorist attack sites often touched (sometimes indirectly) on sensitive and debated topics such as the appropriate time to declare an end to rescue efforts, the way off-duty and volunteer assistance should be managed, and the accommodation of VIPs and other concerned parties. Given the understandable difficulty of making such decisions in the midst of a response effort, site commanders could greatly benefit from guidelines developed in advance of an incident.
- To be useful, guidelines must be practical in the sense that they consider the capabilities of emergency-response organizations, are easy to use in the field, and do not unduly impair the ability of emergency responders to perform critical lifesaving missions.

Cost

The conference participants identified many new technologies for personal protection that would be desirable, based on the lessons learned from the terrorist attacks. Some argued that many desired technologies already exist and progress may simply be a question of procuring the appropriate equipment. Participants highlighted, however, that in the case of both existing and new technology, cost can be a very serious barrier to adoption of equipment by state and local response organizations. Powered-air respirators, for example, can cost ten times as much as the simpler nonpowered variety. Providing each emergency worker

[1] By an ensemble, we mean the entire list of PPE responders should carry, including respirators, clothing, eye protection, sensors, etc.

with his or her own ensemble of equipment specific to a range of hazards could be prohibitively expensive for most local emergency-response organizations.

Efforts could be directed toward making these technologies more affordable or, alternatively, developing efficient ways to deliver the appropriate equipment to incident sites. In instances where a desired technology is commercially available, expanding the number of prepositioned caches of such equipment that could be moved to response sites could be a good compromise solution. The know-how in supply logistics resident in the U.S. military could be helpful for developing supply strategies for the domestic emergency-response community. Another option would be preplanned equipment-sharing with non-neighboring emergency-response units.[2] For smaller departments, it may be appropriate to examine alternative approaches to increasing purchasing power, such as banding together and conducting coordinated procurements.

Research, Development, and Technology Transfer

Several panels put forth recommendations for new equipment and technologies, most of which were for modest and incremental improvements to existing technologies. Research and development (R&D), however, may yield significant benefits to the emergency-responder community. For example, a major theme that ran through many of the panels was the apparent tradeoff between the level of protection provided by equipment and the discomfort and physical burden the equipment placed upon those using it. Directing R&D toward advanced respirators, clothing sensors, and other safety gear may be able to reduce that tradeoff. Other areas suggested by the conference discussions include applications of information technology and communications systems for better management of worker safety at disaster sites and continued emphasis on technologies for locating responders buried or trapped under rubble.

As previously discussed, a theme that arose in several panel discussions was that the purchasing power of the emergency-response community was limited, given its relatively small size and tight budgets, especially at the local level. These factors constrain the community's ability to drive R&D on new technologies. However, much of the safety-related technology that is in use came through technology transfer from other industries, and in some cases, the military. Technology transfer is expected to continue to play an important role in providing emergency responders with improved safety equipment, for example, equipment using information technology, telecommunications, and advanced sensor systems originally developed for purposes other than emergency response.

Technology transfer can help reduce personal protective equipment costs by spreading R&D outlays across a larger user community. It can also speed the introduction of new technologies to the emergency-response community. But the emergency-response community also has special safety needs that may not be adequately met through technology transfer alone. Many at the meeting suggested that publicly supported R&D would be appropriate for addressing the safety needs of emergency responders. The recent terrorist attacks have raised awareness of this issue.

Equipment Standardization and Interoperability

Equipment standardization and interoperability, as well as the development of more uniform training, maintenance, and use protocols, were mentioned as important needs throughout the conference discussions. Although these are not new issues, the scale and complexity of the terrorist attacks and the problems encountered in the responses appear to have drawn greater attention to them and have increased their importance as policy matters for all members of the emergency-response community. The recommendations put forth by conference participants indicate that these issues may be addressed from the top down (through promulgation of uniform safety standards) or from the bottom up (through greater interagency cooperation).

Safety Management

One of the most important lessons learned from the responders at the terrorist-attack sites is the importance of on-site safety management. Effective safety management is unlikely to be achieved if the overall site is not under a defined management structure, with clear lines of authority and responsibility. The operational side of safety management involved hazard monitoring and assessment, safety-equipment logistics and maintenance, site access control, health and safety monitoring, and medical treatment of emergency workers.

Given the magnitude of these tasks, conference participants argued that the safety officer at a disaster site should be an independent official whose sole responsibility is safety enforcement. In cases where incident sites are managed through a unified command structure, those responsible for responder safety could be part of that command.

From the federal perspective, an important issue is reassessing and clearly defining the roles and relationships of various federal agencies with health and safety responsibilities at a major disaster site.

[2] In the event of a major disaster, neighboring emergency-response organizations are likely to be part of the response team and unavailable to share equipment.

SUPPLEMENT 2

Recognizing and Identifying Hazardous Environments

Editor's Note: *This supplement contains Chapter 3, "Recognizing and Identifying Hazardous Environments," from Cocciardi, J.A., Worksafe Series: Operating Safely in Hazardous Environments (Jones and Bartlett, Boston, 2001). This chapter provides detailed information on thermal, radiological, asphyxiant, chemical, etiological, and mechanical environmental hazards.*

Hazardous environments may occur within buildings, outside structures, or may exist in the air, water, and soil. In general, six environmentally hazardous situations may be encountered. See Exhibit S2.1.

1. Thermal hazards
2. Radiological hazards
3. Asphyxiant hazards
4. Chemical hazards
5. Etiological hazards
6. Mechanical hazards

The mnemonic TRACEM can be used to quickly recall these six primary hazards for individuals identifying potentially hazardous environments.

EXHIBIT S2.1 *Hazardous environments include thermal, radiological, asphyxiant, chemical, etiological, or mechanical hazards.*

THERMAL HAZARDS

The inner body is designed to work efficiently under ordinary temperatures (i.e., temperatures at which the body can heat or cool itself to maintain its core body temperature of 98.6°F [37°C]). Any situations that may prevent maintenance of body temperature should be professionally evaluated and appropriate actions taken to assist with body cooling or warming. In addition to ambient temperatures, the addition or lack of protective clothing and equipment, or the increase or decrease in work load may substantially affect the body's ability to heat or cool itself. Severe thermal hazards result in immediate physical problems, such as burns (the destruction of skin tissue). Older individuals or those with circulatory system problems are particularly susceptible to thermal-related problems.

Heat-Related Disorders

Many individuals spend time in hot environments. The human body maintains a fairly constant internal temperature even though it is exposed to varying degrees of environmental heat.

To maintain body safety, the body must rid itself of excess heat, primarily through varying the rate and amount of blood circulation through the skin (as heart rate increases and blood flows closer to skin surfaces, excessive heat is lost) and releasing fluid onto the skin through sweating and subsequent evaporation (cooling). These are automatic responses. As environmental temperatures approach normal skin temperatures, cooling the body becomes more difficult. Blood brought to the body surface cannot lose its heat, and sweating takes over as the primary body cooling mechanism. During conditions of high humidity (or when sweat cannot evaporate adequately through body coverings) the body's ability to maintain acceptable temperatures may become impaired. An individual's alertness and ability to work may also be affected.

Excessive exposure to hot work environments may induce a variety of heat-related disorders, such as heat stroke. Heat stroke is the most serious health problem associated with working in hot environments and is a medical emergency. Heat exhaustion, heat cramps, fainting, or transient heat fatigue may also occur. There are four environmental stresses in a hot environment.

1. Temperature
2. Humidity
3. Radiant heat
4. Air velocity (movement of air over the body surface)

When these factors increase (or decrease, in the case of air velocity) at a rate that cannot be handled by the body's natural mechanisms for shedding heat, physical heat disorders may occur.

Various organizations such as the National Institute for Occupational Safety and Health (NIOSH), the Occupational Safety and Health Administration (OSHA), and the American Conference of Governmental and Industrial Hygienists (ACGIH) have published recommendations for minimizing stress in hot environments. General recommendations for individuals who must work in hot environments can be found in Table S2.1. Typically, safety protocol for hot environments should be considered when temperatures reach 72°F to 77°F. Symptoms of heat strain should never be ignored. Factors such as work regimen, area ventilation, ambient sunlight, and worker health and fitness affect the initiation of hot environment monitoring. As referenced environmental conditions change, protocol may change as well.

In addition, time of year, time of day, or other environmental factors specific to sites may affect the need for a heat stress monitoring program, as well as the acclimatization level of workers (i.e., how their bodies have been trained to compensate for heat). The recognition and identification of heat-related physiological events is found in Table S2.2.

In extreme situations, personal protective equipment, such as layered thermal protective clothing (e.g., firefighters' gear) or reflective layered protective clothing (e.g., aluminized glass suits used for work within the proximity of hot areas), is recommended. Further information concerning thermal protective equipment is found in Chapter 7 [of Cocciardi, J.A., *Worksafe Series: Operating Safely in Hazardous Environments*, Jones and Bartlett, Boston, 2001].

TABLE S2.1 Key Programmatic Components: Workers Exposed to Hot Environments

NIOSH
Workplace limits and surveillance
Medical surveillance for workers in hot environments
Surveillance of heat-induced sentinel health events
Posting of hazardous areas
Protective clothing and equipment programs
Worker information and training
Control of heat stress
Recordkeeping and data collection for evaluation of acclimatization
NIOSH — Additional Considerations
Cool-off period provision
Heart rate monitoring
Body temperature monitoring
Weight loss monitoring
Environmental surveillance monitoring initiation and recommendations
OSHA
Provision of cool rest areas
Sampling of the environment
Environmental controls (ventilation, air cooling)
Fluid replacement
Engineering controls
Administrative controls and work practices (training)
Worker monitoring programs
Personal protective equipment (reflective clothing, auxiliary body cooling devices, wetted clothing, water cooled garments, circulating air)
ACGIH
Measurement of the environment
Development and characterization of work load categories
Development of work rest regimen
Water and salt supplementation
Other considerations
Clothing
Acclimatization and fitness
Adverse health effects

Cold-Related Disorders

Events that may cause the body's core temperature to drop below 96.8°F (36°C) should be avoided (i.e., long periods of time in cold, nonprotected environments, such as air or water). Hypothermia can be caused by routine exposure to water, wind, and moisture in air, or adjacent solid materials whose temperature is below human body temperature. The combination or synergistic effects of these materials may cause cooling with increased rapidity.

In general, hypothermia is recognized by a dull tone to the skin, loss of mental alertness, a feeling of numbness in the extremities, and eventually the loss of consciousness. This is caused by the body's internal reaction to cold. In its

TABLE S2.2 Recognition of Heat-Related Physiological Events

Event	Symptoms
Heat stress, heat exhaustion, transient heat fatigue, heat rash	Profuse sweating, weakness, nausea, and headache, possibly associated with heat syncope (fainting), due to loss of fluids
Heat cramps	Painful muscle spasms due to failure to replace salts; cramps may occur after work
Heat stroke	Mental confusion, loss of consciousness, coma; body temperature above 106°F; hot, red/blue dry skin; medical emergency

TABLE S2.3 Recognition and Identified Actions for Cold-Induced Events

60.8°F (16°C)	• Initiate and record temperature measurements.
39.2°F (4°C)	• Provide gloves for stationary workers. • Provide gloves for light work and appropriate total body protection. • Provide additional protection to individuals exposed to evaporating liquids.
35.6°F (2°C)	• Treat wet workers and remove any wet clothing. • Provide eye protection, including UV protection for work in snow- or ice-covered terrain.
30.2°F (–1°C)	• Initiate dry bulb temperature measurements every four hours; record wind speed if in excess of 5 mph. These readings are necessary to calculate equivalent chill temperature (ECT). At these temperatures, a safety officer should be designated to determine ECT and applicable actions. • Review/exclude workers with body temperature regulation concerns. • Prevent contact with cold surfaces.
19.4°F (–7°C) ECT	• Provide gloves for moderate work. • Provide warming shelters and warming fluids; record measurements every four hours.
10.4°F (–12°C) ECT	• Use buddy system and acclimatization; provide safety/awareness training to workers.
4°F	• Provide mittens for hand warming.
0°F (–18°C)	• Ensure medical approval for workers at these temperatures (this may be reduced to –11.2°F if wind speeds are below 5 mph).
25.6°F (–32°C) ECT	• At this ECT, prohibit continuous skin exposure. ECT includes the cooling wind power on exposed flesh relative to the actual temperature. It was developed by the U.S. Army Research Institute of Medicine (Natick, MA).

attempt to warm the heart, the body shuts down the flow of blood to the extremities. When severe shivering becomes evident, all work should be terminated and body warming initiated.

Actions to warm individuals suffering from hypothermia include removing them from the cold environment and raising the body's core temperature. The recognition and identification of cold-related events are found in Table S2.3. Layered clothing, in addition to acclimatization to cold events, is recommended for cold-related recreational or work environments. As with hot environments, remaining outside the environment where cold temperature extremes are present is always advisable.

RADIOLOGICAL HAZARDS

Radiological hazards can be classified as ionizing and non-ionizing radiation. Although radioactive decay continuously occurs around us, levels above normal background levels may cause personal harm. See Exhibit S2.2.

Ionizing Radiation

Ionizing radiation has sufficient energy to break the chemical bonds of atoms. This potentially strips electrons from atomic structures and subsequently may damage cellular materials. Items emitting ionizing radiation (e.g., a pair of positively charged protons and neutrons spontaneously emitted from the nucleus of an atom) may deposit this energy in their path of travel, thus forcing changes (ionizing the chemical make-up) within the body's system, which they affect or pass through.

The characteristics of various types of ionizing radiation — alpha, beta, and gamma radiation (radioactive materials) — are found in Table S2.4. The effects of ionizing radiation are found in Table S2.5. Additional types of ionizing radiation (X-ray, photon radiation) may produce hazardous environments as well. Worker exposure to radiation is generally regulated by the U.S. Nuclear Regulatory Commission (NRC). See Exhibit S2.3.

The amount or strength of the radioactivity is related to both the amount of material and the half life (the time it

EXHIBIT S2.2 *In some cases, physiological monitoring might be required for workers in hot environments.*

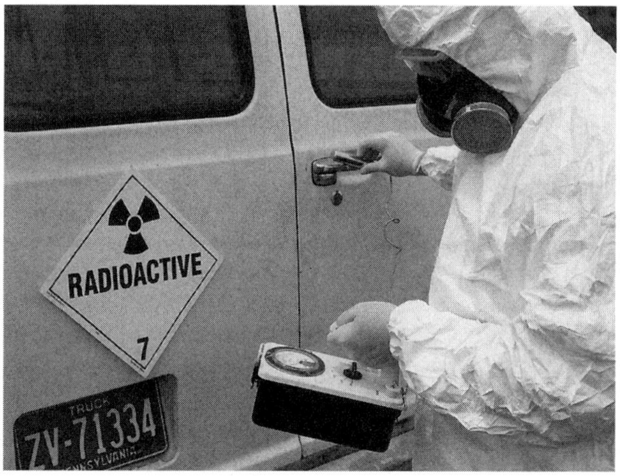

EXHIBIT S2.3 *Radioactive materials can be found in many environments.*

TABLE 2.4 *Characteristics of Ionizing Radiation*

Alpha	Two protons/two neutrons charged (+2); heavy, short range in air (4″); 0.01 mm in tissue (stopped by dead skin) (<4 MeV–8 MeV energy)
Beta	Electron emitted upon creation from the nucleus; usually a (–1) charge (positions are +1 change); mid range, (0–20′ in air); .2–0.5 cm in tissue, less energy (1 KeV–1 MeV)
Gamma	No charge/mass (1 KeV–10 MeV energy) large range, small energy deposition
X-Ray	10 eV–120 KeV energy, originates in the electron field of the atom

TABLE S2.5 *Effects of Ionizing Radiation*

50 R	Decrease in white blood cells
200 R	Nausea, vomiting, dizziness, hair loss, diarrhea, and infection
450 R	$LD_{50/30}$: 50% of those exposed are expected to die within 30 days without medical intervention
600 R	LD_{100}: 100% of those exposed are expected to die without medical intervention
10,000 R	Death results from central nervous system disorder
Note 1:	Biological half life decreases total body burden
Note 2:	Individuals exposed to 5 rem are considered irradiated (individuals evidencing 0.5 mR/Hr [surface] or 0.1 mR/Hr [thyroid] are considered contaminated). The OSHA/EPA Worker Protection Guideline is 5 rem/year. The NRC Public Exposure Limit is 0.5 rem/year. Background levels of ionizing radiation are generally 0.01–0.02 mR/Hr, and as high as 0.05 mR/Hr in some jurisdictions.

takes for the radioactive material to reduce [decay] to half its original strength).

Nonionizing Radiation

Nonionizing radiation does not have sufficient energy to break chemical bonds and is therefore considered nonionizing. The primary effect of nonionizing radiation (e.g., radio frequency radiation, microwave emissions, radar) is heating caused by the increased movement and collisions of molecules.

Various recommended exposure limits are published by the National Council of Radiation Protection and Measurements and have been adopted by the U.S. Federal Communication Commission. The American National Standards Institute, the American Conference of Governmental Industrial Hygienists, the U.S. Department of Defense, and the Institute of Electrical and Electronics Engineers have also published recommended exposure limits.

ASPHYXIANT HAZARDS

Asphyxiant hazards are those hazards associated with the lack of oxygen in the atmosphere, the introduction of particulate matter into air that may obstruct respiration, or the introduction of certain chemicals into breathable air that may prohibit the up-take of oxygen by the blood. See Exhibit S2.4.

The first situation described (low oxygen) is considered immediately dangerous and may be found in many occupational environments. Standard breathing air contains approximately 20.9% oxygen and 79% nitrogen by volume. Nitrogen, the inert component, is not used by the body. The oxygen is circulated and used by the body's car-

EXHIBIT S2.4 *U.S. Nuclear Regulatory Commission logo.*

diopulmonary system and used in cellular respiration, where it is converted to carbon dioxide, transported back to the lungs, and expelled. Table S2.6 lists typical air composition.

Human beings must breathe oxygen in order to survive, and they will begin to suffer adverse health effects when the oxygen level of their breathing air drops below the normal atmospheric level. Table S2.7 lists some physical effects of oxygen volume/pressure changes. This is a difficult situation to detect because observable changes are generally not noted in oxygen deficient atmospheres (i.e., there is no color, odor, or visible indicator). Below 19.5% oxygen by volume, air is considered oxygen-deficient. At concentrations of 16% to 19.5%, workers engaged in any form of exertion can rapidly become symptomatic as their tissues fail to obtain the oxygen necessary to function properly. Increased breathing rates, accelerated heartbeat, and impaired thinking or coordination occur more quickly in an oxygen-deficient environment. Even a momentary loss of coordination may be devastating to a worker if it occurs while the worker is performing a potentially dangerous activity, such as climbing a ladder. Concentrations of 12% to 16% oxygen cause increased breathing rates; accelerated heartbeat; and impaired attention, thinking, and coordination even in people who are resting.

TABLE S2.6 *Typical Air Composition*

Nitrogen	78.1%
Oxygen	20.9%
Argon	0.9%
Carbon dioxide	0.03%
Neon	0.002%
Helium	0.0005%
Krypton	0.0001%
Xenon	0.000009%
Radon	0.000000000000007%

TABLE S2.7 *Physical Effects of Oxygen Volume Changes*

Oxygen Level	Effects
23%	Fire hazard
21%	Normal volume
19.5%	Immediately dangerous to life and health (IDLH) level
16%	Nausea, headaches, and symptoms occur; self-rescue may not be possible
12%	Intermittent respiration and exhaustion
10%	Intermittent respiration, lethargic movement, and unconsciousness
6%	Convulsions and cardiac arrest

At oxygen levels of 10% to 14%, faulty judgment, intermittent respiration, and exhaustion can be expected even with minimal exertion. Breathing air containing 6% to 10% oxygen results in nausea, vomiting, lethargic movements, and perhaps unconsciousness. Breathing air containing less than 6% oxygen produces convulsions, then cessation of breathing, followed by cardiac arrest. These symptoms occur immediately. Even if a worker survives the low oxygen event, organs may show evidence of hypoxic damage (injury to the human anatomy due to lack of oxygen), which may be irreversible.

A number of workplace conditions can lead to oxygen deficiency. Simple asphyxiants, or gases that are physiologically inert, can cause asphyxiation when present in high enough concentrations to lower the oxygen content in the air. See Exhibit S2.5.

At nonstandard atmospheric pressures, the quantity of oxygen that the body may absorb is reduced (i.e., at altitudes above 8,000 feet, the partial pressure of oxygen may be below 100 mm/hg). In addition, at increased pressures such as conditions underwater, nitrogen may be forced into the bloodstream, producing situations such as nitrogen narcosis (divers' euphoria) in which adequate oxygen is not found in the bloodstream. Once introduced, bubbles of nitrogen cannot easily be removed from the bloodstream, producing painful pressure (the "bends"). Table S2.8 identifies situations that may produce low oxygen levels.

Conversely, high levels of oxygen may be considered immediately hazardous. High oxygen concentrations may increase the burning rate and temperature of some materi-

EXHIBIT S2.5 *Fire events typically produce low oxygen levels.*

als or reduce the temperature necessary to ignite substances. Consequently, high oxygen levels (higher than 23%) are generally considered to be a fire or explosion hazard. In addition to the hazards of displaced oxygen, particles such as dust may block the respiratory system, or certain chemicals can poison hemoglobin or enzyme systems that permit the transfer and transport of oxygen from the lungs to the blood.

CHEMICAL HAZARDS

Hazardous environments may be produced by chemicals. Several classification systems are used to recognize and identify potentially hazardous atmospheres caused by chemicals. These include the United States Department of Transportation, Categories of Hazardous Materials, the National Fire Protection Association (Code #704: The Identification of Hazards of Materials for Emergency Response), and the Hazardous Materials Information System (Hazardous Materials Information Guides, which are produced by a variety of manufacturers). See Exhibit S2.6. Further information concerning the recognition and identification of hazardous chemicals is found in Chapter 4 [of Cocciardi, J.A., *Worksafe Series: Operating Safely in Hazardous Environments*, Jones and Bartlett, Boston, 2001].

EXHIBIT S2.6 *Sanitary sewer systems may be oxygen deficient.*

ETIOLOGICAL HAZARDS

Etiological materials are considered to be hazardous because of their potential to cause harm when introduced into the human body (toxicological effects). See Exhibit S2.7.

Etiological agents may be found in a variety of pack-

TABLE S2.8 *Situations Producing Low Oxygen Pressures or Volumes*

	Situation	Cause
Inert gases or gases heavier than air	Oxygen displacement	Releases of nitrogen, argon, carbon dioxide
Oxygen below 19.5%	Oxygen deficient	Fires, sewer work, confined space activities
Chemical asphyxiation	Oxygen present but not absorbed into the bloodstream	CO poisoning, chemical release of cyanides
Dusts, mists, aerosols in high concentrations	Mechanical blockage of oxygen uptake	High concentration of dusts or aerosols

EXHIBIT S2.7 *Signs and labels are used to identify chemical hazards. Container shapes and sizes also give clues to chemical content.*

EXHIBIT S2.8 *Microbial growth in air distribution systems presents a unique etiological hazard.*

ages, many times identified by location. Although many etiological agents coexist naturally, inappropriate introduction of these biological agents into the human body may cause illness or disease. Etiological agents include those of plant, animal, and microbial origin, as well as aerosols (i.e., fungi, bacteria, molds, viruses, allergens). The body's natural defenses to certain etiological agents can be increased through vaccinations. Over time, the Public Health System has eliminated the etiological hazard of these agents through the vaccination system and through combined sanitation and medical management programs. Individuals with compromised immune systems are at particular risk from etiological agents.

MECHANICAL HAZARDS

Hazardous mechanical environments are generally recognized by site inspections. Sharp edges, fragmented equipment, or the chance to be struck by or against items should be reviewed when mechanical hazards may present themselves. Guarding is generally used as a protective mechanism against mechanical hazards. Excess pressures and overpressures may cause mechanical problems as well. Excessive pressure on the eardrums from loud noise may cause hearing damage.

The U.S. Department of Labor and OSHA require guards for the protection of employees from hazardous mechanical energy, such as moving parts or cylinders that may release pressure. See Exhibit S2.8.

In addition, the OSHA Hazardous Energy Standard requires lockout of newer energy storage or distribution equipment (e.g., electrical sources) and tagout of older equipment incapable of accepting lockouts prior to servicing or maintenance operations. Employee training in these procedures is required. Specific written lockout procedures are required for equipment with multiple energy sources or stored or residual energy. A variety of regulations apply to individuals who work at heights, on scaffolds, and in trenches that may produce mechanical hazards. Further information concerning protection from these hazards is found in Chapter 7 [of Cocciardi, J.A., *Worksafe Series: Operating Safely in Hazardous Environments*, Jones and Bartlett, Boston, 2001].

Noisy Environments

Noisy work environments are regulated by OSHA or other regulatory agencies with appropriate jurisdiction (e.g., the Mine Safety and Health Administration). Noisy environments must be evaluated using a calibratable sound level meter. Many sound levels may cause hearing losses. However, those above 85 decibels on the "A" weighted scale are generally considered capable of temporary or permanent hearing impairment. The U.S. Department of Labor's hearing conservation regulations apply to most environments with sound pressure levels above 85 decibels (see Exhibit S2.9). This is also the recommendation of NIOSH, although their measurement calculations differ from those of OSHA. At 90 decibels, OSHA requires mandatory hearing protection. General requirements for noisy environments are found in Table S2.9. [Typical noisy environments (see Exhibit S2.10) are listed in Table S2.10.]

Community noise protection, a function of public health, is designed to protect individuals from both nuisance and hazardous noise pressures. Most community noise standards analyze noise by octave band, prohibiting excessive pressures in each octave of the sound environment. Community noise standards are typically

***TABLE S2.9** Health Recommendations for Noisy Environments*

Noisy Environment	Assess exposures and individual sound pressure doses.
Environments measuring greater than 85 dBA TWA_8 (decibels on the "A" weighted scale, taken as an 8 hour time weighted average)	Hearing Conservation Program required by OSHA for general industry settings and recommended by NIOSH. Hearing Conservation Program required by MSHA for mining sites. This Hearing Conservation Program includes area monitoring, baseline audiograms, optional use of hearing protectors, and annual training. The NIOSH-recommended standard requires noise monitoring, engineering, or administrative controls of noise, mandatory hearing protectors, medical surveillance, hazard communication, program evaluation, and recordkeeping.
Environmental noise greater than 90 dBA TWA	Mandatory OSHA and MSHA Noise Control Program, which, in addition, includes mandatory engineering and administrative controls and use of hearing protection when the engineering or administrative controls will not reduce employee doses. Impact noise above 140 dB is prohibited.

Adapted from: U.S. Department of Labor, OSHA: Occupational Noise Exposure, 29 CFR 1910.95, 1996; and U.S. Department of Health and Human Services, Centers for Disease Control and Prevention, National Institute for Occupational Safety and Health, "Occupational Noise Exposure, Revised Standard," Cincinnati, OH 1998.

***EXHIBIT S2.9** Guarding of areas with potential mechanical energy is required by the Occupational Safety and Health Administration.*

***EXHIBIT S2.10** Noise level engineering and administrative controls are required to be used prior to the use of hearing protection devices.*

***TABLE S2.10** Typical Noisy Environments*

Sound Pressure Level (in decibels)	
30	Very soft whisper
55	Conversation, at 3 feet
60	Air conditioner (window unit)
65	Passenger car; 55 mph at 35 feet
70	Vacuum cleaner
80	Garbage disposal at 3 feet
85	Diesel truck, 40 mph at 35 feet
90	Fire alarm system
95	Power lawn mower; some manufacturing and printing equipment
115	Pneumatic chipper; wood-working equipment
140	Jet engine

applied at airports, transportation corridors, and mining operations.

CHAPTER SUMMARY

Workers and the public must recognize and identify hazardous environments to which they may be exposed. Thermal, radiological, asphyxiant, chemical, etiological, and mechanical hazards present hazardous environments. Quantification of these environments allows specific selection of engineering, administrative, or protective equipment controls. The initial recognition and identification of the environment allows individuals the opportunity to avoid, replace, or remove the hazard; remain distant; or review health and safety requirements prior to entry into the hazardous environment.

TERMS

Asphyxiant Hazards: Problems occurring due to the lack of oxygen, the constituent in air required for human respiration. In addition to displacing oxygen in air, asphyxiant hazards may be caused by particles or substances in air that either block the human respiratory system (e.g., dust particles small enough to inhale) or prohibit the blood from acquiring and distributing oxygen at standard temperatures and pressures (e.g., cyanide).

Chemical Hazards: Exposure to hazardous chemical substances intentionally, unintentionally, or in emergency situations.

Etiological Hazards: Hazards posed by bacteria, viruses, or other human or zoonotic pathogenic agents.

Mechanical Hazards: These hazards include the absorption of pressure from sudden events such as falls, explosions, or incidents in which individuals are struck by or against objects. In addition, mechanical hazards include sharp or hard surfaces, uneven or improperly balanced work environments, and the excess pressures of agents such as noise.

Radioactivity: The process by which unstable atoms "try" to become stable and, as a result, emit radiation (energy). This process is called disintegration or decay.

Radiological Hazards: Exposure above background levels of ionizing or nonionizing radiation.

Thermal Hazards: Exposure to temperature extremes that may cause hypothermic or heat-related disorders.

REFERENCES

American Conference of Governmental Industrial Hygienists, *Threshold Limit Values for Chemical Substances and Physical Agents and Biological Exposure Indices*. Cincinnati: ACGIH Worldwide, 1999.

American Industrial Hygiene Association, *Field Guide for the Determination of Biological Containments in Environmental Samples*, Fairfax, VA: ACGIH Worldwide, 1996.

Barns, J.R., "Cellular Phones: Are They Safe?" *Professional Safety, Journal of the American Society of Safety Engineers* 44, no. 12 (1998).

Cocciardi, J.A., "The Development and Validation of Procedures for Working in Hot Environments." *American Society of Safety Engineers, Industrial Hygiene Division Newsletter*, June 2000.

———, "The Development and Validation of Procedures for Working in Hot Environments in an Industrial Cleanup Situation During the Summer of 1997." Paper presented at the American Industrial Hygiene Conference and Exhibition, Atlanta, May 1998.

———, *Emergency Planning: Medical Services-1: A Manual for Medical Service Personnel Addressing Contaminated or Irradiated and Otherwise Physically Injured Persons and Equipment Following a Nuclear Power Plant Accident.* Commonwealth of Pennsylvania, Emergency Management Agency: Mechanicsburg, PA, 1996.

Department of Defense, Instruction Number 6055.11. Washington, DC: Government Printing Office, 1995.

Federal Communications Commission, "Evaluating Compliance with FCC Guidelines for Human Exposure to Radiofrequency Electromagnetic Fields." Office of Engineering & Technology Bulletin 65. Washington, DC: Government Printing Office, 1997.

———, "Guidelines for Evaluating the Environmental Effects of Radiofrequency Radiation." Washington DC: Government Printing Office, 1997, 96–326.

Federal Communications Commission, Office of Engineering & Technology, *Information on Human Exposure to Radiofrequency Fields*. Washington, DC: Government Printing Office, 1998.

Federal Register, "Hazardous Waste Operations and Emergency Response." U.S. Department of Labor, Occupational Safety and Health Administration, March 8, 1989.

Federal Register, "Respiratory Protection, Final Rule," Department of Labor, Occupational Safety and Health Administration, January 8, 1998.

Hazardous Materials Information Guides, *Laboratory Safety: Industrial Supplies*, Janesville, WI: Lab Safety Supply Company, 1998.

"Heat Related Mortality, U.S., 1997." MMWR 47(1998): 23, 473.

"IEEE-USA Position Statement," *Bioelectromagnetics Society Newsletter*, Jan./Feb. 1993.

Molder, J.E., "Cellular Phone Antennas and Human Health." In *Electromagnetic Fields and Human Health.* Milwaukee: Medical College of Wisconsin, 1998.

National Council on Radiation Protection and Measurements, "Biological Effects and Exposure Criteria for Radiofrequency Electromagnetic Fields," NCRP Report No. 86. Bethesda, MD: NCRP, 1986.

National Fire Academy, National Emergency Training Center, *Hazardous Materials Incident Analysis*. Emmittsburg, MD: NFA-SM-HMIA/TTT, February 1, 1985.

NFPA 704, *Identification of the Hazards of Materials*, National Fire Protection Association, Quincy, MA, 1996.

National Institute of Occupational Safety and Health, U.S. Public Health Service, Centers for Disease Control and Prevention, *Chemical Control Corporation, Health Hazard Evaluation Report #TA80 77-853*. Elizabeth, NJ: NIOSH, U.S. Public Health Service, CDC, April 1981.

Occupational Safety and Health Administration, "Heat Stress." In OSHA Technical Manual. Washington, DC: Government Printing Office, 1986.

"The Phone Tree." *Bioelectromagnetics Society Newsletter*. July/August 1995.

Rom, W., *Environmental Occupational Medicine*, 2nd ed. Boston: Little, Brown, 1992.

U.S. Department of Health and Human Services, Centers for Disease Control and Prevention, National Institute for Occupational Safety and Health, "Criteria for a Recommended Standard...Occupational Exposure to Hot Environments." [DHHS-No. 86-113], Washington, DC: Government Printing Office, 1986.

———, "Occupational Noise Exposure, Revised Standard." Cincinnati, OH: NIOSH, 1998.

U.S. Department of Labor, Occupational Safety and Health Administration, "Hazardous Energy," 29 CFR 1910.147. Washington, DC: Government Printing Office, 1994.

———, "Occupational Noise Exposure," 29 CFR 1910.95. Washington, DC: Government Printing Office, 1996.

U.S. Department of Transportation, "Hazardous Materials Table," 49 CFR 171.101. Washington, DC: Government Printing Office.

SUPPLEMENT 3

Fire Fighter Fatalities, West Helena, Arkansas, May 8, 1997 NFPA Fire Investigation Summary

Editor's Note: *This supplement reprints the summary of a fire investigation conducted by NFPA. An explosion at a pesticides repackaging facility in West Helena, Arkansas, on May 8, 1997 claimed the lives of three fire fighters and injured 16 more. The lessons learned from this tragedy reinforced the need to conduct a comprehensive size-up of any potential hazardous materials incident before personnel are placed in potentially hazardous situations.*

On Thursday, May 8, 1997, the West Helena Fire Department responded to a reported fire at a pesticides repackaging facility. An explosion occurred as fireground operations were beginning. As a result, four fire fighters were struck and buried by debris. One of the fire fighters was rescued but seriously injured, and the other three died before they could be rescued. The building was destroyed by the fire and explosion. See Exhibit S3.1.

The building involved was approximately two years old and of unprotected, noncombustible construction. Most of the building's area was used for storage of product. However, in one small production area where pesticides were repackaged there were several offices in the building. The building was served by a wet-pipe sprinkler system.

Facility personnel discovered a smoking sack of commodity in the facility's receiving area and attempted to extinguish the smoldering fire before calling the fire department at 1:02 p.m. In response, the West Helena Fire Department sent two engines, and several fire fighters drove to the scene in their own vehicles. The West Helena fire chief reported smoke showing upon arrival and requested a full response from the West Helena Fire Department and mutual aid assistance from the Helena Department.

The Helena fire chief and his driver were just down the street when they heard the request for mutual aid, and they responded immediately. Upon arrival, both the chief and his driver observed yellow smoke coming from the facility. The Helena chief approached the West Helena chief, who was meeting with the facility personnel, for assignment. The West Helena chief handed him the MSDS sheets and asked him to evaluate the hazards being presented by the products.

The Helena chief reviewed the MSDS sheets, and based on his evaluation felt that it would be appropriate to pull back and develop a plan of attack prior to approaching the building. He was approaching the West Helena chief to relay this information to him when the explosion occurred.

The explosion occurred as the West Helena fire fighters approached the building to investigate the source of the smoke. The four fire fighters were on the outside of the building and were struck and buried by debris. Immediate efforts were made to extricate the trapped fire fighters by others on the scene. West Helena and Helena fire fighters were able to rescue only one fire fighter because of the severe fire. The other three were buried under debris that could not be removed quickly. The fire was rapidly growing, and the incident commander believed it involved

EXHIBIT S3.1 *A pesticides repackaging facility is destroyed by an explosion during fireground operations.*

chemicals that posed a high risk to all fire fighters in the area. As a result, the incident commander ordered everyone withdrawn before the last fire fighters could be removed and he kept all personnel at a safe distance until a hazardous materials response team from West Memphis, Arkansas, arrived.

Since fire fighters could not attack the fire and smoke was considered to be extremely toxic, the focus of the fire department turned toward protecting the community from exposure. City, county, and state law enforcement and emergency management agencies were notified. Evacuation of areas that could be exposed to the smoke was initiated. The local hospital was one of the many facilities in the evacuation zone.

EXHIBIT S3.2 *A hazmat team conducts a comprehensive evaluation of the scene.*

When the West Memphis hazardous materials response team arrived, they assessed the situation and planned a fire attack to determine whether they could extinguish the fire. Their attack had no effect on the fire so the team decided that they could not extinguish the fire. Instead, they concentrated on recovering the three victims. This was successfully completed.

The EPA dispatched a team to the incident, and they assumed command of the scene. Over the following days an incident command structure was slowly created, incorporating the many agencies involved in the suppression and recovery operations.

Several days into the incident another private hazardous materials team arrived on the scene and began evaluating the situation. They conducted a more comprehensive evaluation of the scene. Based on this evaluation and their airborne monitoring, they established a new hot zone that was larger then the one that had originally been established. See Exhibit S3.2.

The fire gradually decreased as it consumed the fuel, and by noon on Sunday, May 11, 1997, only smoldering spot fires remained.

The building where the incident occurred was reportedly fully sprinklered. However, due to the damage, NFPA's fire investigators were unable to approach the building to verify the details. Furthermore, the plans to the building were destroyed in the explosion. There were 50 employees in the building at the time of the incident.

The exact cause of the fire and explosion are unknown at the time of this report.

Based on the NFPA's investigation and analysis of this fire, the following significant factors were considered as having contributed to the loss of life and property in this incident:

- Inadequate size-up
- Delayed alarm
- Ignition of material
- Proximity of fire personnel to building containing identified hazardous materials
- Lack of a rapid intervention crew

SUPPLEMENT 4

Propane Tank Explosion Results in the Death of Two Volunteer Fire Fighters, Hospitalization of Six Other Volunteer Fire Fighters and a Deputy Sheriff — Iowa

Editor's Note: *This supplement reprints the summary of a National Institute for Occupational Safety and Health (NIOSH) fire fighter fatality investigation (Fire Fighter Investigation Report 98 F-14, NIOSH, Atlanta, GA, September 11, 1998). The Fire Fighter Fatality Investigation and Prevention Program is conducted by NIOSH. The purpose of the program is to determine factors that cause or contribute to fire fighter deaths suffered in the line of duty. Identification of causal and contributing factors enable researchers and safety specialists to develop strategies for preventing future similar incidents. To request additional copies of this report (specify the case number), other fatality investigation reports, or further information, visit the program website at: www.cdc.gov/niosh/firehome.html. Note that some information in this supplement has been updated to reflect the most current information available.*

SUMMARY

On April 9, 1998, 20 fire fighters from a volunteer fire department responded to a propane tank fire located at a turkey farm about 2.5 miles from the fire department. Upon arrival at the fire scene a decision was made to water down the buildings adjacent to the propane tank and allow the tank to burn itself out since the tank was venting. Some of the fire fighters positioned themselves between the burning propane tank and the turkey sheds and were watering down the buildings as the remaining fire fighters performed other tasks, e.g., pulling hose and operating pumps.

About 8 minutes after the fire fighters arrived on scene, the tank exploded (see Exhibit S4.1). When the tank exploded it separated into four parts and traveled in four different directions. Two fire fighters about 105 feet from the tank were struck by one piece of the exploding tank and killed instantly. Six other fire fighters and a deputy sheriff, who had arrived on scene just before the explosion, were also injured.

NIOSH investigators concluded that, to prevent similar incidents, fire departments should

- Follow guidelines as outlined in published literature and guidebooks for controlling fire involving tanks containing propane
- Adhere to emergency response procedures contained in 29 CFR 1910.120(q) — emergency response to hazardous substance release procedures
- Educate fire fighters to the many dangers associated with a propane tank explosion, which is also known

EXHIBIT S4.1 *An 18,000-Gallon Propane Tank with Protective Fencing (Similar to Tank That Exploded).*

as a Boiling Liquid Expanding Vapor Explosion (BLEVE).

Additionally, owners and users of propane tanks should

- Protect aboveground external piping against physical damage via fencing or some other means of protection
- Equip propane tank piping with excess-flow valves and/or emergency shutoff valves, where applicable.

INTRODUCTION

On April 9, 1998, two male volunteer fire fighters aged 45 and 46 years old (the victims) were killed when an 18,000-gallon propane tank exploded. Additionally, six other volunteer fire fighters and a deputy sheriff were seriously injured. The fire fighters were part of a volunteer fire company that arrived on the fire scene about 2320 hours. The fire fighters were watering down buildings adjacent to the tank when it exploded at 2328 hours. The two victims were killed instantly by a flying tank part, and six other fire fighters and a deputy sheriff received varying injuries including burns and fractures.

On April 10, 1998, the United States Fire Administration notified NIOSH of the deaths. On April 15, 1998, two safety and occupational health specialists, Richard Braddee and Frank Washenitz, traveled to Iowa to conduct an investigation of this incident. Meetings were held with the Iowa State Fire Marshal, the 1st and 2nd assistant chiefs of the fire department involved in the incident, the State OSHA inspectors, representatives from the U.S. Chemical Safety and Hazard Investigation Board, and the local police department. Copies of photographs and measurements of the incident site were obtained along with the medical examiner and police department reports, and a site visit was conducted.

The volunteer fire department involved in the incident serves a population of 850 in a geographic area of 100 square miles, and is comprised of approximately 25 volunteer fire fighters. The fire department provides all new fire fighters with 24 hours of mandatory training. The training is designed to cover personal safety, forcible entry, ventilation, fire apparatus, ladders, self-contained breathing apparatus, hose loads, streams, and special hazards. The victims had 15 and 16 years of fire fighting experience, respectively.

INVESTIGATION

On April 9, 1998, at 2311 hours, a call came into the fire department regarding a turkey barn fire at a turkey farm about 2.5 miles from the department. The Chief in the command car #1190, and 19 fire fighters in Pumpers #1193 and #1194, Tanker #1195, Emergency Van #1196, and two private vehicles were dispatched to the scene. En route to the scene, the Chief advised the command center that the fire involved a large propane tank and not a turkey barn. All the fire fighters arrived on the fire scene between 2314 and 2317 hours. At about 2305 hours two people riding a 4-wheel, off-road vehicle struck one of the two fixed metal pipes between the propane tank and the two vaporizers (a device other than a container that receives LP-Gas [Liquefied Petroleum-Gas] in liquid form and adds sufficient heat to convert the liquid to a gaseous state), breaking one pipe off completely. As the liquid propane spewed from the pipe the operator of the 4-wheel, off-road vehicle drove away to call 911. The heavier-than-air propane vapors, which have a vapor density of 1.45 to 2.0, spread along the ground and were eventually ignited by the pilot flame at the vaporizers. Burning propane vapors spread throughout the area and began to impinge on the tank, causing the pressure relief valve to activate and send burning propane flames high into the air.

Upon arrival at the fire scene an assessment by the Chief was made of the burning tank. Fire had engulfed the propane tank and it was venting burning propane vapors via two pressure relief vent pipes located on top of the tank. Also, an extremely loud noise similar to a jet engine was being emitted by the tank's pressure relief vent pipes. The tank was manufactured in 1964 and the tank's shell was constructed of $^{3}/_{4}$-inch carbon steel. The tank was 42 feet, 2 inches long, with an inside diameter of 106 inches and an 18,000-gallon capacity, and it rested on two concrete supports/saddles. The tank was cylindrical and contained about 10,000 gallons of liquid propane. The tank had been fitted with two internal spring-type pressure relief valves which would vent to the atmosphere via two 2-inch-diameter pressure relief vent pipes when the internal tank pressure reached a set limit. The tank also had a fixed metal piping system between the tank and two vaporizers located on the ground about 35 feet away. The piping was $^{3}/_{4}$ inches in diameter, positioned about 36 inches above the ground, was unprotected and had not been fitted with an excess flow valve.

After seeing the flames and hearing the high-pitched shrill being emitted by pressure relief vent pipes located on the top west section of the tank, a decision was made to allow the tank to burn itself out and to try to save the adjacent buildings by watering them down. The fire fighters positioned themselves in various areas in a semicircle north, northeast and northwest of the tank (see Exhibit S4.2). The two victims, whose location was about 105 feet away from the tank and on the northwest side of a building used to house turkeys, began to water down the building. At 2328 a Boiling Liquid Expanding Vapor Explosion (BLEVE) occurred. The BLEVE ripped the tank into four parts, each flying in a different direction. One part of the tank traveled in a northwest direction toward the two victims, striking them and killing them instantly. Six other fire fighters and a deputy sheriff received varying degrees of burns and assorted injuries.

CAUSE OF DEATH

The cause of death was listed by the medical examiner as massive trauma to all systems.

Key:

P	Position of propane tank
1&2	Position of victims
F	Position of other fire fighters
A	Position of apparatus

EXHIBIT S4.2 *Aerial View of Incident Scene.*

RECOMMENDATIONS/DISCUSSION

Recommendation No. 1: Fire departments should follow guidelines as outlined in published literature and guidebooks for controlling fire involving tanks containing propane.

Discussion: Information contained from the *Emergency Response Guidebook* (ERG) [1], the National Propane Gas Association (NPGA) [2], the International Fire Service Training Association (IFSTA) [3], and the National Fire Protection Association (NFPA) [4], contain guidelines for controlling Fire Involving Tanks, LP-Gas Fire Control, Suppressing Class "B" Fires, and *Recommended Practice for Responding to Hazardous Materials Incidents*, respectively. As these guidelines are comprehensive and expansive, they are not reflected here in their entirety. An example of the ERG guidelines for controlling fire involving propane tanks includes, but is not limited to the following:

1. Fight fire from maximum distance or use unmanned hose holders or monitor nozzles,
2. Cool containers with flooding quantities of water until well after fire is out,

3. Do not direct water at source of leak or safety devices; icing may occur,
4. Withdraw immediately in case of rising sound from venting safety devices or discoloration of tank,
5. Always stay away from the ends of tanks, and
6. For massive fire, use unmanned hose holders or monitor nozzles; if this is impossible, withdraw from area and let fire burn.

For additional information, see the References section at the end of this report.

Recommendation No. 2: Emergency response personnel should adhere to the procedures outlined in 29 CFR 1910.120(q) - Emergency response to hazardous substance releases.

Discussion: 29 CFR 1910.120(q) contains procedures for dealing with emergency responses [5]. These procedures include the following:

1. Emergency response plan,
2. Elements of an emergency response plan,
3. Procedures for handling emergency response,
4. Skilled support personnel,
5. Specialist employees,
6. Training,
7. Trainers,
8. Refresher training,
9. Medical surveillance and consultation,
10. Chemical protective clothing, and
11. Post-emergency response operations.

Recommendation No. 3: All fire fighters should be educated to the many dangers associated with a BLEVE.

Discussion: Dangers associated with a BLEVE include, but are not limited to the following:

1. The fire ball can engulf and burn fire fighters operating near a burning tank when a BLEVE occurs,
2. Metal parts of the tank can travel hundreds of feet and strike and critically injure fire fighters after a BLEVE,
3. A trail of burning liquid propane can douse and burn fire fighters, and
4. The shock wave, air blast, or flying metal tank parts created by a BLEVE can collapse buildings onto fire fighters or blow fire fighters out of windows or off roofs. [6]

Additionally, owners and users of propane tanks should:

Recommendation No. 4: Ensure that aboveground external piping is protected against physical damage via fencing or some other means of protection.

Discussion: Unprotected aboveground piping is subject to damage from a variety of sources (e.g., machinery, vehicles, people, animals, etc.). NFPA 58, *Liquefied Petroleum Gas Code*, 6.9.3.10, states the following:

> Aboveground piping shall be supported and protected against physical damage by vehicles [7].

Property or tank owners should ensure that bulk propane tank piping is protected against damage by fencing or some other means to prevent accidental contact and physical damage.

Recommendation No. 5: Equip propane tank piping with excess-flow valves and/or emergency shutoff valves, where applicable.

Discussion: See NFPA 58, *Liquefied Petroleum Gas Code,* for specific information related to excess flow valves and emergency shutoff valves.

REFERENCES

1. U.S. Department of Transportation, Transport Canada, and the Secretariat of Communications and Transportation of Mexico, *Emergency Response Guidebook.* U.S. Department of Transportation, Research and Special Programs Administration, Office of Hazardous Materials Initiatives and Training, Washington, DC, p. 179. 2004.
2. *LP-Gas Fire Control and Hazmat Training Guide*, Safety Bulletin No. 211, National Propane Gas Association, Lisle, IL, 1992.
3. *Essentials of Fire Fighting*, 4th ed., International Fire Service Training Association, Stillwater, OK, 1998.
4. NFPA 471, *Recommended Practice for Responding to Hazardous Materials Incidents,* is now incorporated into NFPA 472 and NFPA 473, National Fire Protection Association, Quincy, MA, 2008.
5. Title 29, Code of Federal Regulations, Part 1910.120, U.S. Government Printing Office, Washington, DC.
6. Dunn V., *Safety and Survival on the Fireground.* Penn Well, Tulsa, OK, 1992.
7. NFPA 58, *Liquefied Petroleum Gas Code*, National Fire Protection Association, Quincy, MA, 2008.

SUPPLEMENT 5

Selection of Chemical-Protective Clothing Using NFPA Standards

Jeffrey O. Stull, International Personnel Protection, Inc.

Editor's Note: *One of the most important concerns for a hazardous materials response team is the selection of chemical-protective clothing (CPC). A team must purchase clothing that is appropriate to the potential exposures they are likely to encounter at incidents in their response area. In this supplement, Jeff Stull, an expert on CPC design, construction, and selection criteria, describes the key considerations in choosing the right garments, with information drawn from his experience and from several NFPA standards dealing with the performance of CPC.*

INTRODUCTION

Chemical-protective clothing (CPC) comes in a variety of shapes, colors, and sizes. Manufacturers of CPC are constantly developing and offering new products or new features on existing products. Although product documentation has improved, claims for increased performance are not always readily apparent. End users are faced with a myriad of choices of encapsulating and splash suits with a wide range of performance and cost. With all of these choices, response organizations sometimes have a difficult time sorting out products in the marketplace. Choosing the appropriate protective ensemble requires a general understanding of these products, knowledge of factors important in their selection, and knowledge of the standards governing their performance.

Standards have a lot to do with the way that those protective ensembles are designed and how they perform. The NFPA now offers three different standards addressing the needs of the emergency responder. NFPA 1991, *Standard on Vapor-Protective Ensembles for Hazardous Materials Emergencies* [1], and NFPA 1992, *Standard on Liquid Splash-Protective Ensembles and Clothing for Hazardous Materials Emergencies* [2], establish requirements for vapor-protective ensembles (replaces Level A) and liquid splash-protective ensembles (replaces Level B), respectively. NFPA 1994, *Standard on Protective Ensembles for First Responders to CBRN Terrorism Incidents* [3], sets criteria for three classes of ensembles for protecting first responders during terrorism incidents involving chemical agents, biological agents, or radiological particulates. Having a fundamental understanding of these standards enables easier CPC selection with a higher level of confidence that the protective ensembles provide the needed protection.

GENERAL UNDERSTANDING OF PRODUCTS

Chemical-Protective Ensembles

To provide protection to the entire body, chemical-protective ensembles must be part of an overall ensemble of personnel protective equipment (PPE). As defined by the U.S.

Environmental Protection Agency (EPA) and in the Occupational Safety and Health Administration (OSHA) regulations for hazardous waste site remediation and emergency response (OSHA Title 29 CFR 1910.120) [4], four different levels of protection are established based on different ensembles. These ensembles consist of CPC, a respirator, gloves and boots, hard hats, communications equipment, cooling devices, and various types of undergarments. See Table S5.1 for a detailed list of the EPA levels of PPE protection. Exhibits S5.1 through S5.3 show examples of various types of PPE.

First responders generally use Level A and Level B

TABLE S5.1 *EPA Levels of Protection*

Level	*Equipment*	*Protection Provided*	*Should Be Used When*	*Limiting Criteria*
A	Recommended: Pressure- demand, full-facepiece SCBA or pressure-demand, supplied-air respirator with escape SCBA Fully encapsulating chemical-resistant suit Inner chemical-resistant gloves Chemical-resistant safety boots/shoes Two-way radio communications Optional: Cooling unit Coveralls Long cotton underwear Hard hat Disposable gloves and boot covers	The highest available level of respiratory, skin, and eye protection	The chemical substance has been identified and requires the highest level of protection for skin, eyes, and the respiratory system based on either: • measured (or potential for) high concentration of atmospheric vapors, gases, or particulates, or • site operations and work functions involving a high potential for splash, immersion, or exposure to unexpected vapors, gases, or particulates of materials that are harmful to skin or capable of being absorbed through the intact skin Substances with a high degree of hazard to the skin are known or suspected to be present, and skin contact is possible. Operations must be conducted in confined, poorly ventilated areas until the absence of conditions requiring Level A protection is determined.	Fully encapsulating suit material must be compatible with the substances involved.
B	Recommended: Pressure-demand, full-facepiece SCBA or pressure-demand supplied-air respirator with escape SCBA Chemical-resistant clothing (overalls and long-sleeved jacket; hooded, one- or two-piece chemical splash suit; disposable chemical-resistant one-piece suit) Inner and outer chemical-resistant gloves Chemical-resistant safety boots/shoes	The same level of respiratory protection but less skin protection than Level A It is the minimum level recommended for initial site entries until the hazards have been further identified	The type and atmospheric concentration of substances have been identified and require a high level of respiratory protection but less skin protection. This involves the following atmospheres: • With IDLH concentrations of specific substances that do not represent a severe skin hazard; or • That do not meet the criteria for use of air-purifying respirators Atmosphere that contain less than 19.5% oxygen	Use only when the vapor or gases present are not suspected of containing high concentrations of chemicals that are harmful to skin or capable of being absorbed through the intact skin. Use only when it is highly unlikely that the work being

(continues)

TABLE S5.1 Continued

Level	*Equipment*	*Protection Provided*	*Should Be Used When*	*Limiting Criteria*
	Hard hat Two-way radio communications Optional: Coveralls Disposable boot covers Face shield Long cotton underwear		Presence of incompletely identified vapors or gases is indicated by direct-reading organic vapor detection instrument, but vapors and gases are not suspected of containing high levels of chemicals harmful to skin or capable of being absorbed through the intact skin.	done will generate either high concentrations of vapors, gases, or particulates or splashes of material that will affect exposed skin.
C	Recommended: Full-facepiece, air- purifying, canister-equipped- respirator Chemical-resistant clothing (overalls and long-sleeved jacket; hooded, one- or two-piece chemical splash suit; disposable chemical-resistant one-piece suit) Inner and outer chemical-resistant gloves Chemical-resistant safety boots/shoes Hard hat Two-way radio communications Optional: Coveralls Disposable boot covers Face shield Escape mask Long cotton underwear	The same level of skin protection as Level B but a lower level of respiratory protection	Atmospheric contaminants, liquid splashes, or other direct contact will not adversely affect any exposed skin. The types of air contaminants have been identified, concentrations have been measured, and a canister is available that can remove the contaminant. All criteria for the use of air-purifying respirators are met.	Atmospheric concentration of chemicals must not exceed IDLH levels. Atmosphere must contain at least 19.5% oxygen.
D	Recommended: Coveralls Safety boots/shoes Safety glasses or chemical splash goggles Hard hat Optional: Gloves Escape mask Face shield	No respiratory protection Minimal skin protection	Atmosphere contains no known hazard. Work functions preclude splashes, immersion, or the potential for unexpected inhalation of, or contact with, hazardous levels of any chemicals.	This level should not be worn in the hot and warm zones. Atmosphere must contain at least 19.5% oxygen.

protective ensembles. Level A ensembles provide the highest level of protection and consist of a totally encapsulating suit, a self-contained breathing apparatus (SCBA) or combination SCBA and supplied-air respirator, chemically resistant gloves and footwear, and a communications system. Level A ensembles are for use in situations where the highest level of respiratory, skin, and eye protection is needed. Level B ensembles employ the same respiratory protection but pertain to situations where hazards to the skin and eyes are not as significant as those encountered in Level A situations. Consequently, for Level B ensembles, one- or multipiece chemical splash suits replace the totally encapsulating suits used in Level A ensembles. Although used to a much lesser extent in the emergency response community, Level C ensembles use the identical clothing systems found in Level B ensembles but replace SCBA or

EXHIBIT S5.1 *This is an example of Level A chemical-protective ensemble. (Source: Jeffrey Stull)*

EXHIBIT S5.2 *A Level B chemical-protective ensemble is illustrated. (Source: Jeffrey Stull)*

combination SCBA/supplied-air respirators with air-purifying respirators for situations where lower levels of respiratory hazards are perceived.

Although the EPA levels of protection describe what the ensemble should look like, little guidance is offered for how the ensemble should perform. It is vital that the ensemble elements work together to provide the intended level of protection, which means that ensemble items should fit together (provide good interfaces) and offer consistent performance for the wearer's entire body. Unfortunately, duct tape is often used in an attempt to correct ill-fitting suits and poorly designed interfaces.

Chemical-protective suits should be considered as a system consisting of the base material, seams, closures, and the overall suit design. Often, attention is paid only to the base material, neglecting other parts of the suit that have a significant impact on the suit's effectiveness. For example, the barrier characteristics of the best material in the market can be diminished when put into a design that is awkward to wear and has poor quality seams, which readily permit the permeation or penetration of chemicals. Therefore, all aspects of suit design are important in the choice of a chemical-protective suit, not simply the base material and its chemical resistance properties as generally promoted in the marketplace for industrial protection. The following are examples of other desirable attributes:

- Seams should offer chemical resistance consistent with base material.
- Closures, the weak link of the system, should be protected in a manner that keeps liquid (splashes) from reaching the inside of the clothing.
- Exhaust valves on totally encapsulating chemical-protective suits should be protected from direct splashes and include means to prevent their blockage if pressed in a confined space.
- Fittings installed on the suit body for attachment of air lines, cooling devices, and other equipment should be reinforced to prevent ruptures in the wearer's protective envelope.
- Suits should be designed to be easily donned and to minimize impairment of wearer movement, field of vision, dexterity, and tactility.

EXHIBIT S5.3 An example of Level C chemical-protective ensemble is depicted. (Source: Jeffrey Stull)

For the most part, this sort of information can come only from direct inspections or trials with potential products and from asking manufacturers for data they normally do not provide. In addition, contacting other purchasers and users to get their observations and experiences is helpful.

Risk Assessment

The choice of CPC must be based on first completing a risk assessment. Two types of risk assessments aid in selecting PPE for purchase or for use: those performed on the general, expected situations that response teams encounter and assessments that are performed for a specific hazard. In each case, the risk assessment should consist of the following steps:

- Identifying the hazards present or likely to be present
- Estimating the likelihood of exposure
- Understanding the consequences of exposure
- Determining the risk

A simple model for conducting a risk assessment is shown in Table S5.2. In this model, the likelihood of exposure and the consequences of exposure are estimated for each expected hazard. Risk is determined by multiplying the exposure likelihood by the exposure consequences. In this way, risks associated with specific hazards can be determined and ranked to ascertain protection and clothing performance needs.

The hazards alone need not be the determining factor for choosing PPE; rather the potential for exposure should govern selection of CPC. For example, the risk is different when dealing with 1 gal (3.8 L) of toluene versus dealing with a tank car full of the same chemical. Underprotection should be avoided to prevent exposure, but overprotection can be just as dangerous. Overprotection can lead to injury through heat stress and hinder the wearer from safely performing the needed tasks. Contingencies must be planned for, but a sense of realism should prevail when it comes to suit selection.

SELECTION FACTORS

The following factors should be collectively considered when selecting CPC:

- Overall suit integrity
- Material chemical resistance
- Material strength and physical properties
- Overall suit design
- Service life
- Cost

Overall Suit Integrity

Although material permeation resistance is important, other material and clothing characteristics should not be overlooked. Suit integrity, for example, should be carefully examined. Encapsulating suits are easily tested for gastight integrity using pressure (inflation) testing. The pressure or inflation test evaluates the encapsulating suit for pinholes and other leaks, such as at faulty closures. While this test is simple to perform, it does not evaluate the entire suit because exhaust valves are closed off during the test procedure. Pressure testing is recommended when receiving new suits and as a periodic test of suit condition. (See Exhibit S5.4.) More sophisticated testing of an encapsulating suit's performance in providing vapor protection is accomplished by dynamic leak testing. In this testing, a test subject wears the suit in a closed chamber filled with an innocuous gas (sulfur hexafluoride or SF6 is used because it is relatively nontoxic and, is easily detected; thus SF6 can be used at relatively low challenge levels without any harm to the test subject). The test subject then proceeds

TABLE S5.2 Risk Assessment Areas for Selection of CPC

	Body Area Affected					
Hazard Area	*Full Body*	*Respiratory System*	*Head Area*	*Torso*	*Arms/Hands*	*Legs/Feet*
Chemical vapor inhalation						
Chemical vapor skin absorption						
Chemical liquid skin contact						
Chemical ingestion						
Falling objects						
Flying debris						
Sharp objects						
Rough surfaces						
Slippery surfaces						
Extreme cold						
High heat						
Flame exposure						
Chemical flash fire						
Static discharge						
Electrical shock						
Poor visibility						
Falling from height						
Falling into water						
Cold stress						
Heat stress						
Mobility restriction						
Dexterity restriction						
Vision restriction						
Hearing restriction						

Note: Some boxes are shaded because these hazards would not apply to that part of the body.

through a series of exercises (e.g., overhead reaching, deep knee bends) while the interior of the suit is sampled for any SF6 that may leak into the suit. In this fashion, the entire suit is evaluated and an "intrusion coefficient" can be measured that shows the percentage of gas that could enter the suit during use.

More recently, a new dynamic vapor test has been introduced for the evaluation of encapsulating and non-encapsulating full-body chemical-protective suits. The test referred to as *MIST* (man-in-simulant test) involves similar principles as the aforementioned SF6 test, but using a different stimulant (methyl salicylate) and different approach to evaluating penetration of vapor into the clothing. Methyl salicylate (oil of wintergreen) serves as a surrogate for the chemical warfare agent, distilled mustard. Detection of methyl salicylate penetration into the suit is accomplished by the test subject wearing small patches that contain a special absorbent that captures any penetrating test chemical. These patches are placed over different parts of the test subject's body, particularly at interface areas, and therefore can yield body area specific information as to where most penetration takes place. Exhibit S5.5 shows the locations for the

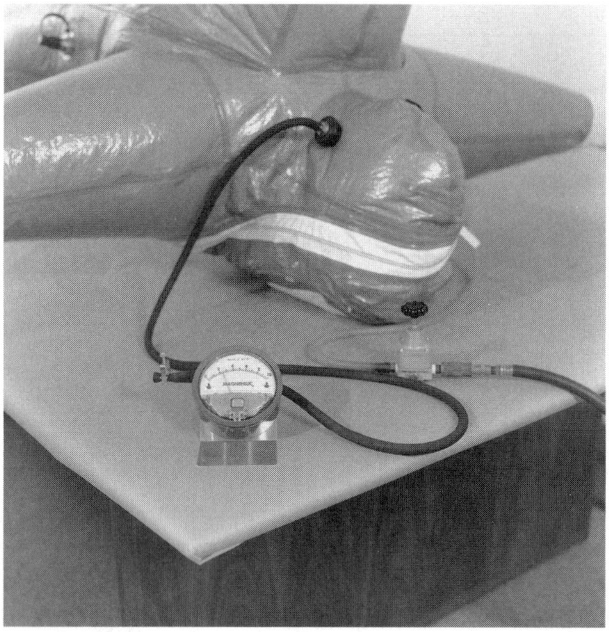

EXHIBIT S5.4 *In this photograph, a totally encapsulating suit is being pressure tested. (Source: Jeffrey Stull)*

patches or pads. The information is used to calculate protection factors (ratio of outside challenge concentration to concentration measured in the patches) both for each local body area and for the overall clothing system as a whole.

The integrity of splash suits is rarely given the same attention as encapsulating suits. These suits should provide liquidtight integrity by keeping liquids off the wearer's skin or clothing. Even though most coated or laminated materials successfully offer this performance, seams, closures, and interface areas might not. These areas must be protected so that the entire suit and ensemble will function properly under liquid exposure situations. Liquidtight integrity can be evaluated by using simple techniques that challenge critical areas with liquid spray and then examining the suit's interior or inner garment for evidence of penetrating liquid. This form of testing is useful to evaluate how different parts of the liquid-protective ensemble work together.

Material Chemical Resistance

The most commonly used factor for selecting CPC is permeation resistance data on the base clothing material. However, CPC resistance can be evaluated in the following three ways:

1. Degradation resistance
2. Penetration resistance
3. Permeation resistance

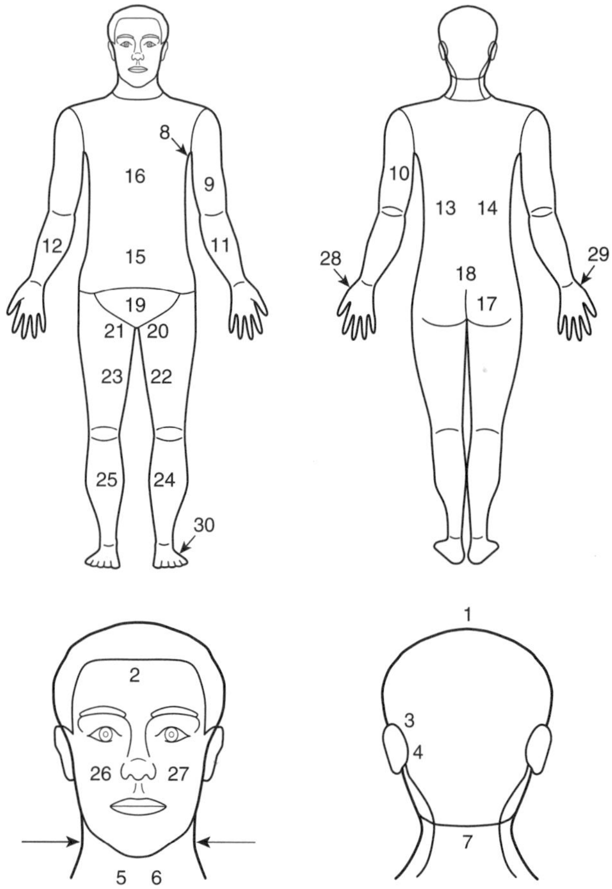

1: scalp (SCA)	16: chest (C)
2: forehead (F)	17: right buttock (RB)
3: behind left ear upper (LED)	18: lower back (LB)
4: behind left ear (LE)	19: groin (GR)
5: neck right (NED)	20: crotch (LCR)
6: neck left (NE)	21: crotch (RCR)
7: nape (NA)	22: left inner thigh (LIT)
8: left armpit (LA)	23: right inner thigh (RIT)
9: left inner upper arm (LIU)	24: left inner shin (LIS)
10: left outer upper arm (LOU)	25: right inner shin (RIS)
11: left forearm (LFA)	26: cheek (RM)
12: right forearm (RFA)	27: cheek (LM)
13: middle back (MB)	28: left hand (G)
14: middle back dup. (MBD)	29: right hand (GD)
15: abdomen (AB)	30: foot (B)

EXHIBIT S5.5 *This diagram shows the placement location of adsorbent pads on the test subject body.*

This information can also be applied to all materials used in a suit's construction.

Degradation Resistance. This evaluation involves a determination of chemical effects on specific clothing material properties. In degradation resistance testing, specimens of clothing material are typically placed in contact with the test chemical for a relatively long period. Material degradation is determined in terms of visible changes (e.g., discoloration or swelling), weight gain or loss, or changes in one or more material properties (e.g., strength). Most often, chemical degradation resistance information takes the form of qualitative ratings, which ranks material performance on a five or more part scale. Many manufacturers refer to this information as "chemical compatibility recommendations." These ratings most frequently are provided for gloves, footwear, and splash suits but usually do not appear for encapsulating suits. For many products, this chemical resistance information would be the only information provided. Unfortunately, degradation ratings are only good for excluding a product from consideration because materials may show no obvious degradation but yet allow chemicals to penetrate or permeate.

Penetration Resistance. This evaluation determines how well protective clothing materials prevent the passage of liquid. Test results are reported as either pass or fail. Materials pass the test when they show no visible penetration after a given test exposure period (usually 1 hour) with some pressure behind the liquid. Any liquid that penetrates through the material and becomes visible on the interior of the clothing material constitutes a failure. Most coated or laminated clothing easily passes this test unless the chemical degrades the material and causes a hole. As a consequence, penetration resistance testing is most often used for breathable materials, seams, and closures.

Permeation Resistance. Permeation occurs on a submicroscopic level where chemical molecules pass through the protective clothing material. Permeation can occur when a gas, liquid, or solid is against the outer surface of the CPC. Permeating chemical is then released as a vapor on the clothing interior. Permeation resistance testing involves measuring "breakthrough time" and the rate the chemical passes through the material ("permeation rate"). Breakthrough time represents the time it takes for a chemical to move through a material until it is first detected on the other side. The permeation rate serves as a measure for how quickly the chemical passes through a given exposed surface area of the material. Because permeation resistance testing involves a very sensitive determination of clothing material barrier effectiveness, it is the preferred measurement for chemical resistance among protective clothing manufacturers.

In practice, end users typically choose clothing materials that have breakthrough times longer than the expected maximum exposure time. However, changes are being made in the CPC industry to move toward the reporting of "cumulative" permeation. Cumulative permeation is the total mass of chemical that permeates through the material in a given time period. This information is useful because it can be compared to a permissible exposure limit. In fact, cumulative permeation data are used to judge the effectiveness of clothing materials against chemical warfare agents where exposure limits have been established. Unfortunately, skin exposure limits have not been established for all chemicals and the industry is still working on models for how to use cumulative permeation in the future determination of CPC material chemical permeation resistance.

Material Physical Properties

Many manufacturers present physical property data in their marketing literature as indications of material strength and durability. Commonly reported physical properties include the following:

- Tensile strength, which measures force to break material apart when pulled in one direction
- Burst strength, which measures force to rupture material when the force is applied to the face of the material
- Tear resistance, which measures the force to tear a material apart
- Cut resistance (gloves and footwear), which measures force for a sharp blade to cut through a material
- Puncture resistance (gloves and footwear), which measures force for a pointed nail-like probe to push through a material
- Flex durability, which assesses how well a material stays intact (with cracks) when repeatedly flexed
- Abrasion resistance, which assesses damage to material when rubbed on a rough surface
- Flame resistance, which assesses the ease for igniting a material when contacted by a flame and the burning characteristics of the material once the flame is removed

Unfortunately, unless end users are familiar with this type of data, the information might not be easily understood.

Physical properties are best used for comparing the relative strength or durability of different materials, provided the same test is used on all the materials compared. To understand physical properties, the end user must be able to compare the result on a candidate material with the result on a material with which he or she has had experience. Physical failure of suits in the field might not always be attributable to material strength but in fact might be the result of a poor design or improper fit. Therefore, examining a physical property that has relevance to conditions for using CPC is important. The selected clothing should include a material that is strong enough to avoid tearing during use. Nevertheless, appropriate suit sizing plays a key role in maintaining the integrity of clothing. See Exhibit S5.6.

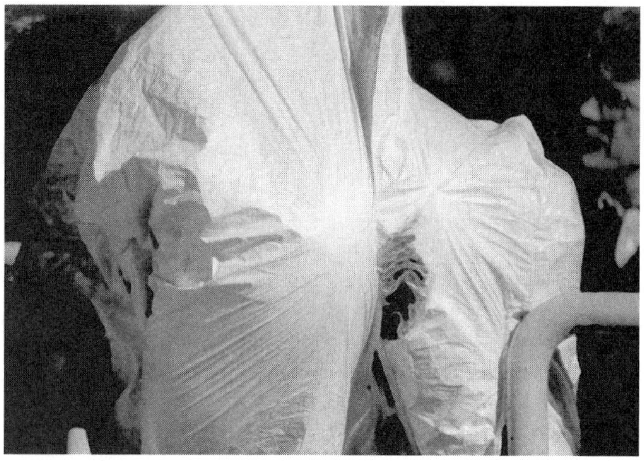

EXHIBIT S5.6 *This is an illustration of a tear that is mostly due to the material not being strong enough to hold up while performing the task; although had the suit been the proper size, the tear could have been avoided. (Source: Jeffrey Stull)*

Design Features

Design features affecting wearer function, fit, and comfort are difficult to measure and are most often subjective, but they are still an important part of the selection process. The best way to evaluate suit design is through trial wearing of CPC. These trials need to include tasks that replicate the same types of movements and stresses that would be placed on suits during actual use. Through this type of evaluation, end users can determine how the suit impacts their ability to perform work. The following are examples of relevant design features:

- Location and length of the closure (affects ease of putting on and taking off suit)
- Position, size, and type of the visor (affects user ability to see outside the suit)
- The bulk of the suit materials, in terms of the number of layers and the relative stiffness (affects user movement, ability to perform tasks)
- The type of glove system (the number of gloves that must be used and the relative bulk and stiffness)
- The type of footwear system (combination of all footwear articles needed for foot protection)
- Interfaces between the suits and gloves and footwear (affects system integrity)
- The volume inside the suit hood area for accommodating the wearing of a respirator facepiece, head protection, and other equipment
- The overall volume inside the suit for those suits where the respirator is worn inside
- Available suit, glove, and footwear sizes (for accommodating different sized individual responders)

These features dramatically affect user functioning. In particular, glove systems have been found to decrease dexterity and cause hand function problems. (See Exhibit S5.7.) Gloves are a problem for most hazardous materials responders. Currently, most responders use double or triple gloving techniques to compensate for a limited size range and material selection. However, using this multiglove technique is not without its tradeoffs, including limited dexterity (e.g., difficulty using meter button controls). Some responders carry a small pencil stub or other disposable tool taped to their suit sleeves to help press small meter buttons with accuracy. Sizing is important because manufacturers offer these suits in a number of sizes. Accommodating particularly large or small people can be difficult with a limited number of sizes. Ill-fitting clothing is particularly apparent for persons not being able to clearly see out of suits having visors. CPC does not have to fit poorly.

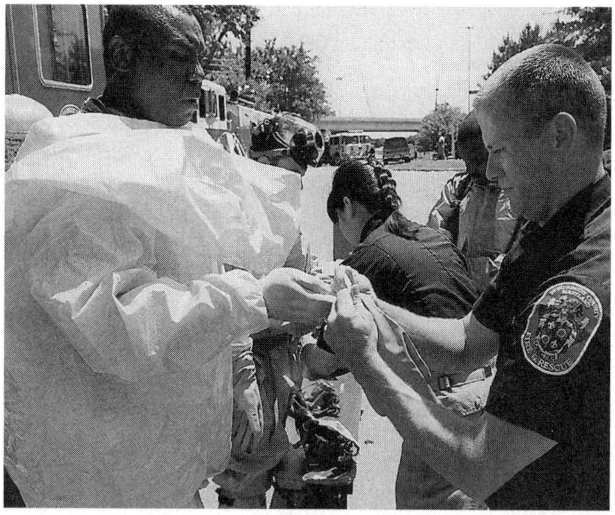

EXHIBIT S5.7 *Gloves can present a problem for hazardous materials responders. (Source: Henrik G. de Gyor)*

Service Life

Service life is a complicated issue. In general, most users perceive inexpensive, lightweight plastic-based products as less durable and disposable, and relatively more expensive, heavy, rubber-based products as reusable. The service life of a product is actually based on its life cycle cost, durability, and ease of decontamination.

Life cycle cost includes all costs associated with the use of the product including the initial purchase, maintenance, decontamination, storage, and disposal costs. There are also costs for putting clothing back into use and ensuring that it is safe. While purchase costs might be the principal cost for product use, disposal costs are taking on greater significance.

Determining how well the suit maintains its original condition for providing protection to the wearer best assesses durability. This factor can be evaluated by measuring product chemical resistance for representative products following simulated use (information generally not provided in manufacturer literature). If the clothing loses its chemical resistance from abrasion, repeated flexing, or other forms of wear, this suggests suits might not maintain their barrier performance during use. The ruggedness of the CPC is also a factor for how durable the suit might be.

Ease of decontamination is important too. A product that cannot be easily decontaminated becomes disposable regardless of the initial purchase cost. Products that are reused must be decontaminated to a level whereby the user is safe from exposure of residual levels of chemical contaminants.

Cost

The issue of cost cannot be dismissed. In an ideal world, the "best" suit in the marketplace would be purchased. But the fact is that organization resources are limited. While some organizations have been able to set up programs to recoup PPE costs from those responsible for the incident, this form of chemical-protective suit reimbursement is not always reliable. Response organizations want the optimum number and types of CPC in their inventory to minimize selection decisions and obtain the best protection for their team members. Therefore, issues of product service life are important to the cost.

UNDERSTANDING AND USING COMPREHENSIVE PERFORMANCE STANDARDS

Creation of NFPA Standards on CPC

In 1985, the National Transportation Safety Board recommended that government agencies support the development of protective standards for chemical protection after several first responders were exposed to a hazardous chemical from a leaking railcar, even though the manufacturer recommended the use of their suits for the chemical involved. Shortly thereafter, a subcommittee was formed by the National Fire Protection Association to develop standards on hazardous CPC. The subcommittee chose to develop the following two standards to meet emergency responder needs:

1. NFPA 1991, *Standard on Vapor-Protective Ensembles for Hazardous Materials Emergencies*
2. NFPA 1992, *Standard on Liquid Splash-Protective Ensembles and Clothing for Hazardous Materials Emergencies*

Each standard provides performance-oriented requirements for specific protective suit ensembles, including gloves for hand protection and boots for foot protection. NFPA 1991 covers protective suits that offer protection from all forms of chemicals including vapors, gases, liquids, and solids. NFPA 1992 pertains to ensembles and clothing that prevents wearer exposure to liquid splashes or solid chemicals. NFPA 1991 and 1992 were prepared specifically for emergency response applications with the following characteristics:

1. The hazards range in severity.
2. The physical environment varies.
3. Aspects of the response remain uncharacterized.

The current NFPA standards define a hierarchy of hazard protection as shown in Table S5.3. Using this hierarchy, the committee defined protective clothing types on the basis of needed performance as demonstrated by test

TABLE S5.3 *Definitions of Protective Clothing Types by Performance Tests in NFPA Standards*

Type of Clothing	*Material Performance*	*Overall Clothing Performance*	*Replaces*
Vapor-Protective (NFPA 1991)	Permeation resistance	Integrity against inward vapor leakage (gastight)	EPA Level A
Liquid Splash-Protective (NFPA 1992)	Penetration resistance	Integrity against inward liquid leakage	EPA Level B

methods designed to measure the type of protection provided. This approach associates clothing gastight integrity and material permeation resistance with vapor protection and clothing liquidtight integrity, and material penetration resistance with liquid-splash protection. These performance-based definitions are intended to replace the historical EPA levels of protection that define what clothing should look like but not how it should perform.

In addition to defining a performance hierarchy among different types of CPC, the NFPA standards define consistent requirements for chemical resistance for all parts of the clothing ensemble to ensure consistent protective performance. For example, the same chemical resistance requirements are applied to the garment material; visor, glove, and footwear materials; and the garment and visor seams. Furthermore, specific chemical batteries have been developed that are relevant for either vapor or liquid splash protection. The chemical lists, based on an ASTM International standard, provide a broad-based assessment of clothing material performance against a range of chemicals as shown in Table S5.4.

Flame resistance is a mandatory requirement in NFPA 1991 but is optional in NFPA 1992. Originally, a flame resistance requirement was included in both standards, but the subcommittee believed that some minimum requirement should be added to NFPA 1991 because it represents the highest level of protection involving entries into environments where the hazards might not be completely characterized. The requirement exists not for fire protection, but rather so that the clothing does not contribute to the

***TABLE S5.4** Standard Chemical Batteries for Evaluating CPC*

Chemical	*ASTM F 1001*		*NFPA 1991*	*NFPA 1992*
	Liquids	*Gases*		
Acetone	♦		♦	♦
Acetonitrile	♦		♦	
Ammonia		♦	♦	
1,3-Butadiene		♦	♦	
Carbon Disulfide	♦		♦	
Chlorine		♦	♦	
Dichloromethane	♦		♦	
Diethylamine	♦		♦	
Dimethylformamide	♦		♦	♦
Ethyl Acetate	♦		♦	♦
Ethylene Oxide		♦	♦	
Hexane	♦		♦	
Hydrogen Chloride		♦	♦	
Methanol	♦		♦	
Methyl Chloride		♦	♦	
Nitrobenzene	♦		♦	♦
Sodium Hydroxide (50%)	♦		♦	♦
Sulfuric Acid (93.1%)	♦		♦	♦
Tetrachloroethylene	♦		♦	
Tetrahydrofuran	♦		♦	♦
Toluene	♦		♦	
Carbonyl Chloride			♦	
Cyanogen Chloride			♦	
Distilled Sulfur Mustard (HD)			♦	
Dimethyl Sulfate			♦	
Hydrogen Cyanide			♦	
Sarin (GB)			♦	

injury of the wearer in the event of accidental flame or high heat exposure.

All parts of the ensemble are addressed in terms of physical protection, durability, and functional performance in addition to chemical resistance requirements as follows:

- Visors are tested for clarity and strength.
- Gloves are evaluated for cut, puncture, abrasion, flex fatigue resistance, and dexterity.
- Footwear is tested for requirements that address sole slippage, abrasion, puncture, toe impact, and compression resistance as well as upper footwear cut and puncture resistance. Because many suits are configured with soft booties, manufacturers are responsible for supplying or specifying outer boots to provide consistent levels of protection.
- Seams and closures are evaluated for breaking strength.
- Exhaust values are evaluated for inward leakage.

The overall suit is also evaluated for impact on the wearer's mobility and function in tests that simulate emergency response tasks. Other tests are taken to evaluate important parts of the suit's functional performance.

In the 2005 edition of NFPA 1991, chemical/biological terrorism agent protection was changed from an option to a mandatory set of requirements. These requirements dictate testing chemical-protective suits to the SF6 test and against specific chemical warfare agents and toxic industrial chemicals that are considered likely threats during chemical terrorism. There are still two optional areas of product certification included in the 2005 edition of the standard:

1. Chemical flash fire escape protection
2. Liquefied gas protection

For NFPA 1992, additional product certification to chemical flash fire escape protection is the only option. These supplemental requirements were established to allow manufacturers to demonstrate additional protection against these hazards.

Chemical flash fires were considered to represent a serious hazard to the emergency responder. By their nature, many organic chemicals pose the threat of creating flammable atmospheres, particularly in confined spaces. The sudden ignition of a flammable atmosphere creates an intense fireball that releases significant amounts of heat capable of easily igniting many forms of clothing, including many of the CPC materials used in industrial applications. To address this level of protection, the subcommittee established requirements for clothing materials in terms of the following criteria:

- Increased flame resistance
- Some minimum amount of thermal insulation

In addition, an overall test was created to simulate the environment of a flash fire. The test involves placing suited mannequins in a closed chamber, filling the chamber with propane in its flammable range, and igniting the chamber contents to produce a flash fire lasting 6 to 8 seconds in duration. The requirements were established only to aid escape, not entry into a flammable environment, and do not provide for the suit to survive the rescue and be reusable.

In assessing clothing performance against liquefied gases, requirements were added to NFPA 1991 that addressed concerns for embrittlement and subsequent cracking of suit materials that allowed chemicals to pass through.

Products Compliant with NFPA 1991

NFPA 1991 imposes tough requirements. The standard requires that materials provide combined chemical and flame resistance. Permeation resistance to a 21 chemical battery with breakthrough times greater than 1 hour is required for all "primary" materials in the standard. (Primary materials are defined as those used in the garment, visor, gloves, and boots.) This same level of chemical resistance performance must also be demonstrated following repeated material flexing and abrasion for garment, glove, and footwear materials. As of the 2005 edition, primary materials must also be tested for four toxic industrial chemicals and two warfare agents. The flame resistance test requires that primary materials resist ignition when exposed to a flame for 3 seconds and show self-extinguishing characteristics after a subsequent 12-second flame exposure.

Chemical-protective suit manufacturers have used two different approaches for meeting the requirements for NFPA 1991. Some manufacturers have compliant products that meet all of the requirements using a single layer. These suits, such as the one shown in Exhibit S5.8, are typically classified as reusable because of their high initial purchase costs and relatively rugged construction. Another approach has been to use two-layer systems where an inner suit is constructed of a plastic laminate and an overcover is provided that is made of an aluminized flame-resistant textile material. This type of suit is shown in Exhibit S5.9.

The plastic laminate–based products must include overcovers to achieve both required flame and abrasion resistance performance. This feature alone has caused criticism of the standards by users who question the need for the overcover because they feel that the majority of situations where chemical-protective suits are worn do not

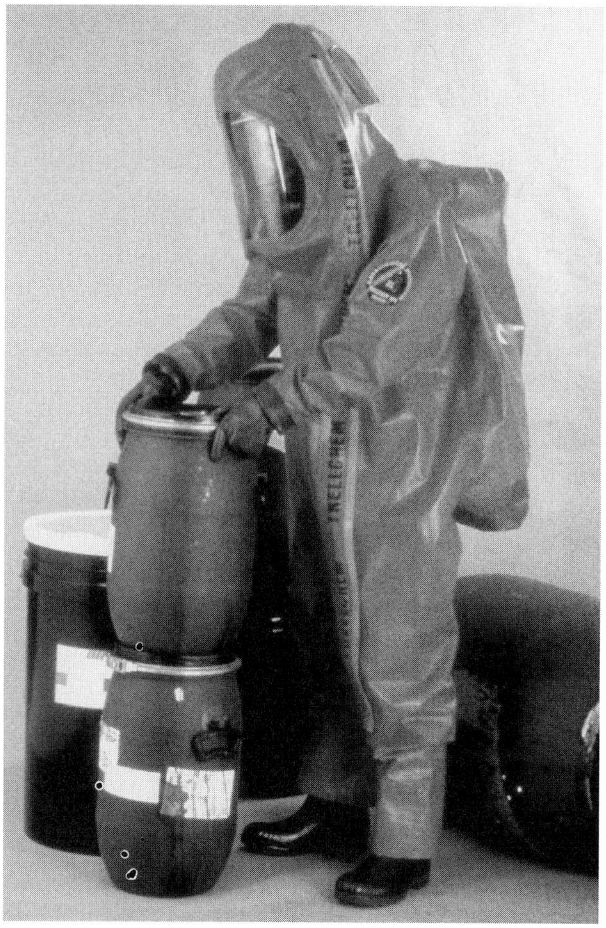

EXHIBIT S5.8 *A single-layer, totally encapsulating suit is shown in this example. (Source: Jeffrey Stull)*

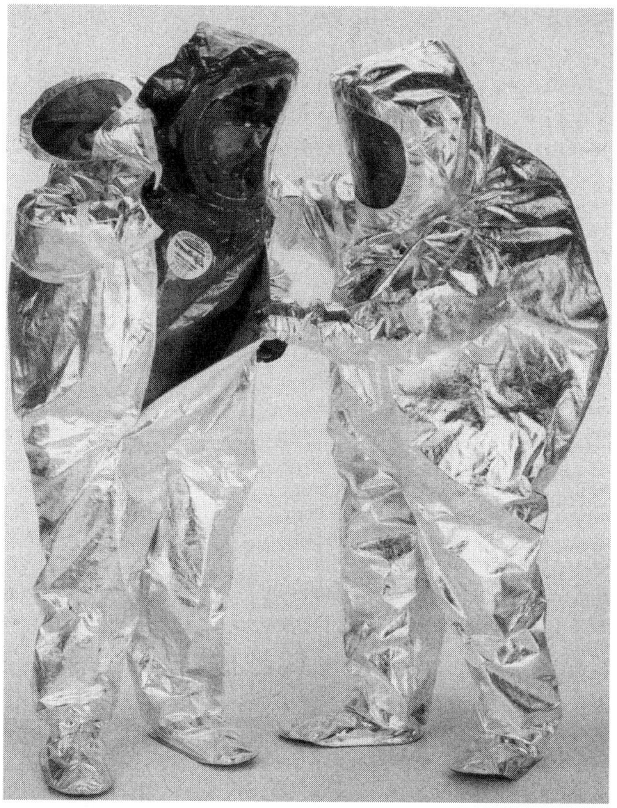

EXHIBIT S5.9 *An example of a multilayer totally encapsulating suit is pictured. (Source: Jeffrey Stull)*

involve the potential for flame contact or chemical flash fires. They consider the overcover to be burdensome and costly for most needs. Contrary to popular belief, the NFPA standard *does not require all suits to have an overcover.* Rather, this configuration has been chosen by some manufacturers to meet the overall requirements of the standard. Other manufacturers have been able to meet all requirements in a single layer.

All NFPA 1991 chemical-protective ensembles are totally encapsulating suits. Although constructing suits that allow for the facepiece of a self-contained breathing apparatus to complete the envelope around the wearer is possible, specific problems occur. First of all, the facepiece would need to be tested to the same requirements as the rest of the suit. Second, several critical parts of the SCBA would be exposed to a contaminated environment. As a consequence, NFPA 1991 suits are designed to completely encapsulate the wearer and the respiratory equipment, providing protection of the equipment and further safeguarding the more sensitive respiratory system of the wearer against chemical hazards.

NFPA 1991 ensemble manufacturers generally configure their suits with multiple glove systems and bootie feet. Unfortunately, no glove manufacturer has found a practical way of creating a single-layer glove that meets the chemical resistance, flame resistance, and puncture resistance requirements imposed on gloves. As a result, two- or three-glove systems are used on chemical-protective suits and have a significant impact on wearer hand function (dexterity and tactility). This problem is compounded by sizing limitations for suits, where gloves sizes are generally proportional to the suit size, which means that gloves generally don't fit well.

All current NFPA 1991 ensemble manufacturers rely on a footwear system that consists of bootie sock extensions of the suit legs for chemical protection and an outer overboot for physical protection. This system works well only when the overboot is properly fitted to the individual's foot and accommodates any excess material in the bootie foot.

Manufacturers have made some improvements in visors, providing options for enlarging the relative size of the visor in the suit. Some visors prevent smaller individuals from seeing near their feet or cut off the upper field of vision for responders who are relatively tall.

Many suits are provided in the following two configurations:

1. The closure across the front of the suit
2. The closure in the back

A front closure makes it easier for the wearer to don and doff the suit without assistance but allows for the possible exposure of a critical suit area (even though a protective flap must cover the closure). Suits with closures in the rear allow easier entry of the responder and better protection of the closure but require assistance in donning and doffing.

Products Compliant with NFPA 1992

Like NFPA 1991, manufacturers of protective suits and garments certified to NFPA 1992, respectively, use different approaches for materials and garment configurations. NFPA 1992 establishes the use of liquid penetration resistance as the qualifying test for material barrier performance. This testing requires that materials prevent the passage of liquid from outside to inside surfaces. Materials prevent *penetration* but can still allow *permeation* of chemicals to occur. This distinction opens new opportunities for different kinds of materials. Materials that are "breathable" yet demonstrate the necessary penetration resistance can be used in NFPA 1992–compliant suits or garments. As a consequence, both coated materials and those incorporating microporous, breathable membrane films are being used in liquid splash–protective clothing. However, the limitation for using NFPA 1992 CPC is that the end user accepts possible exposure to gases or vapors that might be present alongside liquids. Nevertheless, the hazard assessment for many emergency responses often shows little or no vapor hazard.

NFPA 1992 differs from NFPA 1991 by providing for the individual certification of components, such as garments, gloves, and footwear. NFPA 1991 applies to full ensembles where all components must be provided or specified by the manufacturer. This difference recognizes that flexibility is generally provided in using Level B type CPC. Therefore, it is possible to provide garment designs, such as hooded coveralls, or partial body clothing as might be needed in a liquid splash situation. Nevertheless, providing proper interfaces with gloves and footwear to prevent inward leakage of liquids is still important for garments. See Exhibit S5.10.

Garment configurations have been dictated mainly by the liquidtight integrity, or "shower," test used in both standards. Although simple in principle, this test has been the primary stumbling block in many certification efforts. The test involves the following procedures:

EXHIBIT S5.10 *This photograph shows an example of an NFPA 1992-compliant splash suit. (Source: © 2002 W.L. Gore and Associates, Inc.)*

- Place the suit over a mannequin that is already dressed in a water-absorptive inner garment.
- Spray surfactant-treated water at the mannequin from several low-pressure nozzles for a period of 1 hour.
- Examine the inside of the suit and water-absorptive garment for evidence of water penetration.

Any liquid penetration into the garment or watermarks on the inner garment constitute failure of the suit. The test readily detects poorly constructed closure systems, seams, or other design areas capable of allowing liquid penetration. The ramification of this test is that manufacturers have had to pay close attention to integrity issues not generally considered in the past. Practices such as using conventional zippers with cover flaps, two-piece garments, or elasticized hoods no longer provide acceptable performance relative to the test method. Instead, manufacturers use encapsulating designs, "sack" suits, and visors in most liquid splash suit designs. In addition, manufacturers of NFPA 1992–compliant clothing must *specify* other components such as gloves and boots not traditionally included as part of their products.

NFPA 1994 Standard

NFPA 1994 was a new standard, released in August 2001, that sets performance requirements for protective clothing used at chemical and biological terrorism incidents. The standard was updated in 2007 with modification of the requirements to encompass chemicals, biological agents, and radiological particulates (CBRN) hazards and the reclassification of different ensembles. The standard is unique in that it defines three classes of ensembles based on the perceived threat at the emergency scene. Differences between the three classes are based on the following:

1. The ability of the ensemble design to resist the inward leakage of CBRN contaminants
2. The resistance of the materials used in the construction of the ensembles to chemical warfare agents and toxic industrial chemicals
3. The strength and durability of these materials

All NFPA 1994 ensembles are designed for a single exposure use. Ensembles must consist of garments, gloves, and footwear. Table S5.5 summarizes the use of these ensembles.

Changes from 2001 Edition. Significant modifications were made to NFPA 1994 when it was revised in 2007. The largest change was to remove Class 1. In essence, Class 1 was moved to NFPA 1991 so that the vapor-protective suits provide both the highest level of hazardous materials response and CBRN terrorism incident protection. This move was made to recognize the fact that most general first responders do not always have the training and expertise to use totally encapsulating suits. Class 2 and Class 3 protective ensembles were realigned with the type of respirator to be used. A new Class 4 was created to address biological or radiological particulates only (no vapors or liquids).

Class 2 Ensembles. Class 2 ensembles consist of a full-body one- or multipiece suit, gloves, and footwear where the ensemble might be designed with the SCBA worn inside or outside the ensemble. The ensemble is designed to minimize the inward leakage of gases or vapors as demonstrated by a man-in-simulant test where performance levels have been set at levels commensurate with protection in an immediately dangerous to life and health (IDLH) environment. Ensembles are further evaluated for integrity with respect to liquid penetration. Materials are tested for permeation resistance to selected chemical agents and toxic industrial chemicals at concentrations consistent with the same levels used for evaluating CBRN SCBA; materials are also tested for viral penetration resistance and for various physical properties to demonstrate adequate physical hazard resistance and durability for a single use. Ensembles are also tested for functionality.

Class 2 ensembles are intended for circumstances where the agent or threat might be identified, when the actual release has subsided, or in an area where live victims can be rescued. Conditions of exposure include possible contact with residual vapor or gas and highly contaminated surfaces at the emergency scene. Most victims in the response area are alive and show signs of movement but are nonambulatory. For Class 2 ensembles, breathing air from the SCBA can still limit wearing time. However, Class 2 ensembles can also currently be configured with NIOSH-certified CBRN air-purifying or powered air-purifying respirators (APR) that provide longer duration response time.

Class 3 Ensembles. Class 3 ensembles also consist of full body one- or multipiece suit, gloves, and footwear. The ensemble can be designed for use with SCBA or APR, although APR is consistent with the use of this ensemble. The ensemble is designed for protection against lower exposure levels of gases, vapors, liquids, and particulates (as compared to 1994 Class 2 ensembles). The same man-in-simulant test described for Class 2 ensembles is used to measure the inward leakage of gases or vapors, but lower criteria are set for consistency with exposure levels below IDLH levels. Ensembles are evaluated for liquid integrity but at shorter times compared to Class 2 ensembles. Materials are tested for permeation resistance to selected chemical agents and toxic industrial chemicals at concentrations consistent with the same levels used for evaluating CBRN APR; materials are also tested for viral penetration

TABLE S5.5 *Classes of Protective Ensembles Used in NFPA 1994*

Class	*Standard*	*Vapor Threat*	*Liquid Threat*	*Particle Threat*
1	NFPA 1991	Unknown	High	Biological and low energy radiological
2	NFPA 1994	Above IDLH	Likely	Not addressed
3	NFPA 1994	Below IDLH	Primarily contaminated surfaces	Not addressed
4	NFPA 1994	None	Minimal	Biological and low energy radiological

resistance and various physical properties to demonstrate adequate physical hazard resistance and durability for a single use. Ensembles are also tested for functionality.

Class 3 ensembles are intended for use long after the release has occurred, at relatively large distances from the point of release, or in the peripheral zone of the release scene for such functions as decontamination, patient care, crowd control, perimeter control, traffic control, and cleanup. Class 3 ensembles should be used only when there is very little potential for vapor or gas exposure, when exposure to liquids is expected to be incidental through contact with contaminated surfaces, and when dealing with patients or self-evacuating victims. Class 3 ensembles must cover the individual, and it is preferred that this clothing also cover the wearer's respirator to limit its potential for contamination. Because these ensembles are intended for longer wearing periods, the use of NIOSH-approved CBRN air-purifying or powered air-purifying respirators with these suits is likely.

Class 4 Ensembles. Class 4 ensembles consist of full body one- or multipiece garment, gloves, and footwear. The ensemble can be designed for use with SCBA or APR, although APR is consistent with the use of this ensemble. The ensemble is designed to minimize the inward leakage of biological or radiological particulates only by use of a particle-tight integrity test. The suit and component parts do not offer protection from gases, vapors, or aerosols. Limited liquid protection is offered, primarily to enable wet decontamination. Materials are tested for viral penetration resistance and various physical properties to ensure adequate single use durability and resistance to physical hazards. Ensembles are tested for functionality.

Class 4 ensembles are intended for use in situations involving only biological or radiological particulates, where there is no threat of exposure to chemical warfare agents or toxic industrial chemicals. Although Class 4 ensemble materials are evaluated for viral penetration resistance for protection against bloodborne pathogens (OSHA Title 29 CFR 1910.1030) [4], Class 4 ensembles should be used only when there is very little potential for liquid exposure. Class 4 ensembles must cover the individual; however, the respirator certified with the ensemble can cover the face of the wearer. Because these ensembles are intended for longer wearing periods, the use of air-purifying respirators with these suits is likely.

Impact of the Standards in the Marketplace on Selection

Since the adoption of the NFPA standards in 1990, a number of different chemical-protective suits have been developed and certified to each NFPA standard. That each certified product involves either a new material or a novel design or both is significant. In effect, the products that have been certified represent advances in state-of-the-art technology from what existed at the time the standards were first introduced. For that matter, when NFPA 1991 was prepared, no products could meet all of the rigorous requirements for material performance. The fact that now several different products are compliant with the standard is a testament to the ability of the clothing industry to develop "better" products offering improved protection.

The quality of CPC has risen as the result of NFPA standards. The requirements in the NFPA standards are very rigorous, and resulting product designs are significantly more complex than conventional industrial products. For example, because chemical resistance requirements are applied to all the primary materials in the suit (garment, visor, gloves, and boots), all parts of the suit must have the same barrier performance. For NFPA 1991-compliant suit systems, this means that multiple glove systems must be used, which creates dexterity and interface problems. Furthermore, seams for these suits must be taped on both sides to achieve the same level of chemical resistance as the base material.

The NFPA standards have also promoted greater end user confidence. All NFPA standards now have requirements for third-party certification with specific guidelines for independent qualification and follow-up testing, quality assurance audits of manufacturer facilities, and product labeling. These requirements mean that end users can quickly identify compliant products by examining the label inside the suit to see a statement of compliance with the respective NFPA standard along with mark of the certification organization. The product label will further identify all parts of the ensemble to ensure that a complete ensemble is worn that has been certified as a system.

Future Changes to NFPA 1991, NFPA 1992, and NFPA 1994

Each NFPA standard is subject to periodic revision. Once released, all NFPA standards must be reviewed and revised at least every 5 years. This process allows for reconsideration of important requirements in each standard and allows for public proposals and comments on each new edition. This process is now under the responsibility of the Technical Committee on Hazardous Materials Protective Clothing and Equipment.

NFPA 1991 and NFPA 1992 first became effective in 1990; the first edition of NFPA 1994 was available in 2001. At the time they were promulgated, these standards represented the best efforts to develop a set of requirements that would give a minimum level of protection for different emergency response applications. Obviously, a number of improvements in suit design, materials performance, and testing procedures have occurred since the committee finished their early work. These issues have been addressed in subsequent revisions of the two standards both in 1994 and 2000 and more recently in 2005. Revision efforts (as of December 2007) have already begun for the 2010 edition standards.

The most significant changes to the three standards address how permeation testing is conducted. As previously mentioned, there is ongoing research to move from breakthrough times to cumulative permeation. The use of cumulative permeation will allow linking chemical resistance performance criteria with permissible skin exposure levels. This change will have a significant impact on the qualification of materials and will likely open up the industry for a number of new material technologies.

In addition to revising the three standards, the committee is developing a new standard titled *Standard on the Selection, Care, and Maintenance of Hazardous Chemical-Protective Clothing*. This standard is aimed at the response organization and will address the following:

- Clothing selection for purchase
- Clothing selection on scene
- Cleaning and decontamination of clothing after use
- Procedures for inspection, repair, storage, retirement, and disposal of clothing

As with all NFPA standards, the public can participate in the revision process by submitting proposals for revisions of standards, and comments on new drafts created by the responsible committee.

INCORPORATING INFORMATION FOR FINAL SELECTION

End users must judiciously choose appropriate chemical-protective suits based on information supplied by the manufacturer as well as their own evaluations. These suits should be certified as being compliant with the requirements of the applicable NFPA standard. While compliance with NFPA standards should only be a prerequisite, other considerations should include the following:

- Extent and availability of the chemical resistance database to support use of the product
- Expected life cycle cost for the product in terms of cost per use
- Number of sizes in which the product is available and the range of personnel that the suit will fit adequately
- Ease of donning, doffing, and decontamination
- Impact on wearer functioning as determined in trial wearing of the CPC, particularly in the areas of field of vision, hand dexterity, and overall profile in negotiating confined spaces
- Accommodation of other personal protective equipment such as the SCBA, head protection, outer boots, radio systems, cooling devices, and any inner garments
- Supporting technical documentation

Response teams should make their selection decisions only after first seeing and trying out chemical-protective suits, including a review of all technical and support data. Those responsible for the decisions should make choices by conferring with other teams to determine their experience and avoid relying solely on sales or manufacturer's information.

The NFPA standards are by no means all inclusive; they are not a substitute for user education and appropriate training as covered in NFPA 472, *Standard for Professional Competence of Responders to Hazardous Materials/Weapons of Mass Destruction Incidents.* Many response organizations consider these standards to be overly rigorous and as producing expensive products. Nevertheless, the NFPA standards do provide a baseline performance that has spurred the development of chemical-protective suits for improved wearer protection. When used in conjunction with user experience, the process for selecting a chemical-protective suit can become much easier.

REFERENCES

1. NFPA 1991, *Standard on Vapor-Protective Ensembles for Hazardous Materials Emergencies*, National Fire Protection Association, 2005.
2. NFPA 1992, *Standard on Liquid Splash-Protective Ensembles and Clothing for Hazardous Materials Emergencies*, National Fire Protection Association, 2005.
3. NFPA 1994, *Standard on Protective Ensembles for First Responders to CBRN Terrorism Incidents,* National Fire Protection Association, 2007.
4. Title 29, Code of Federal Regulations, OSHA Parts 1910.120 and 1910.1030, U.S. Government Printing Office, Washington, DC.

SUPPLEMENT 6

Response Levels

Editor's Note: *This supplement is taken from the 2002 edition of this handbook. In that handbook, the full text of the 2002 edition of NFPA 471, Recommended Practice for Responding to Hazardous Materials Incidents, was included. Since then, NFPA 471 has been withdrawn. This supplement is a reprint of Chapter 5, including both the standard and commentary text. All text is now considered commentary text, as the standard no longer exists. The material can still be useful to the reader, so it is included as a supplement in this handbook.*

This supplement can be used as a guide for determining the appropriate response levels based on the nature of a hazardous materials/weapons of mass destruction (WMD) incident. Three types of incidents and levels are discussed, along with classes of hazardous materials.

Table S6.1 can be used as a planning guide to assist the user in determining incident levels for response and training. Some response organizations have established a "tiered response plan" to match the response levels listed in Table S6.1. In many cases, the classification of an incident is the responsibility of the emergency call taker. Detailed training is necessary for call takers to make an accurate classification.

As a general guideline, there are three types or levels of incidents: Level 1, Level 2, and Level 3. These levels can be defined as follows:

- **Level 1.** An incident involving hazardous materials that can be contained, extinguished, and/or abated using resources immediately available to the public sector responders having jurisdiction. Level 1 incidents present little risk to the environment and/or to public health with containment and cleanup.
- **Level 2.** An incident involving hazardous materials that is beyond the capabilities of the first responders on the scene and could be beyond the capabilities of the public sector responders having jurisdiction. Level 2 incidents might require the services of a state or regional response team or other state or federal assistance. This level can pose immediate and long-term risk to the environment and public health.
- **Level 3.** An incident involving hazardous materials that is beyond the capabilities of a single state or regional response team and requires additional assistance. Level 3 incidents can require resources from state and federal agencies and private industry. These incidents generally pose extreme, immediate, and/or long-term risk to the environment and public health.

The initial responders to the scene should be alert to signs that hazardous materials are involved. Some signs, such as truck placards, are obvious. Other indications of the presence of hazardous materials can be observed in the type of location of the incident, such as an industrial site, a service station, or a licensed or unlicensed business location. Once the location has been established, then the types of materials normally associated with that type of occupancy, what substances can be present, and in what quantities can be determined.

For example, if the incident is at an unlicensed auto body shop, paints, solvents, and other flammable and combustible liquids are likely to be present but not in the same quantity as would be expected in a larger, licensed facility with the ability for bulk storage. Unlicensed or illegal businesses using hazardous materials are probably not in compliance with any of the local regulations for equipment maintenance, hazardous materials storage, record keeping, or disposal.

First responders to any incident need to be aware of their limitations. For example, if a reported incident is classified as a Level 1 condition, initial responding units should be able to handle it or, after evaluation, call for

TABLE S6.1 Planning Guide for Determining Incident Levels for Response and Training Incident Level

Incident Conditions	*Incident Level One*	*Incident Level Two*	*Incident Level Three*
Product identifications	Placard not required, NFPA 0 or 1 all categories, all Class 9 and ORM-D	DOT placarded, NFPA 2 for any categories, PCBs without fire, EPA regulated waste	Class 2, Division 2.3 — poisonous gases, Class 1, Division 1.1 and 1.2 — explosives, organic peroxide, flammable solid, materials dangerous when wet, chlorine, fluorine, anhydrous ammonia, radioactive materials, NFPA 3 & 4 for any categories including special hazards, PCBs & fire, DOT inhalation hazard, EPA extremely hazardous substances, and cryogenics
Container size	Small [e.g., pail, drums, cylinders except 1-ton (910 kg), packages, bags]	Medium [e.g., 1-ton (910 kg) cylinders, portable containers, nurse tanks, multiple small packages]	Large (e.g., tank cars, tank trucks, stationary tanks, hopper cars/trucks, multiple medium containers)
Fire/explosion potential	Low	Medium	High
Leak severity	No release or small release contained or confined with readily available resources	Release may not be controllable without special resources	Release may not be controllable even with special resources
Life safety	No life-threatening situation from materials involved	Localized area, limited evacuation area	Large area, mass evacuation area
Environmental impact (potential)	Minimal	Moderate	Severe
Container integrity	Not damaged	Damaged but able to contain the contents to allow handling or transfer of product	Damaged to such an extent that catastrophic rupture is possible

additional resources. A Level 2 incident requires greater response capability, and a Level 3 incident needs even more sophisticated equipment and highly trained personnel. Level 2 and Level 3 responses can be served by a regional or nearby mutual aid hazardous materials team. When planning for potential hazardous materials incidents, the planning body should address incidents at all levels to ensure a smooth response coordination.

Some of the terms and conditions in Table S6.1 warrant explanation, and much of the remainder of this supplement provides information about some of the terms in the table. For example, *Placard Not Required* is a term that refers to U.S. Department of Transportation (DOT) regulations. However, the absence of a placard should not be taken as an assurance that the contents are harmless.

The term *NFPA 0 or 1 All Categories* is a reference to NFPA 704, *Standard System for the Identification of the Hazards of Materials for Emergency Response* [1]. NFPA 704 deals with a labeling system, as shown in Exhibit S6.1, that advises on the following three hazard conditions:

- Health
- Flammability
- Reactivity

The five degrees of intensity range from 0 to 4. A 0 or a 1 in all three hazard conditions indicates a relatively low hazard. The following terms also require explanation:

Class 9 (previously identified by the DOT as ORM A, B, and C) and ORM-D: ORM means other regulated materials. Examples of Class 9 materials include adipic acid, hazardous substances such as polychlorinated biphenyls (PCBs), and molten sulfur.

Class 9 Miscellaneous: Examples include miscellaneous hazardous materials, that is, those materials, including the following, that present a hazard during transport but are not included in another hazard class:

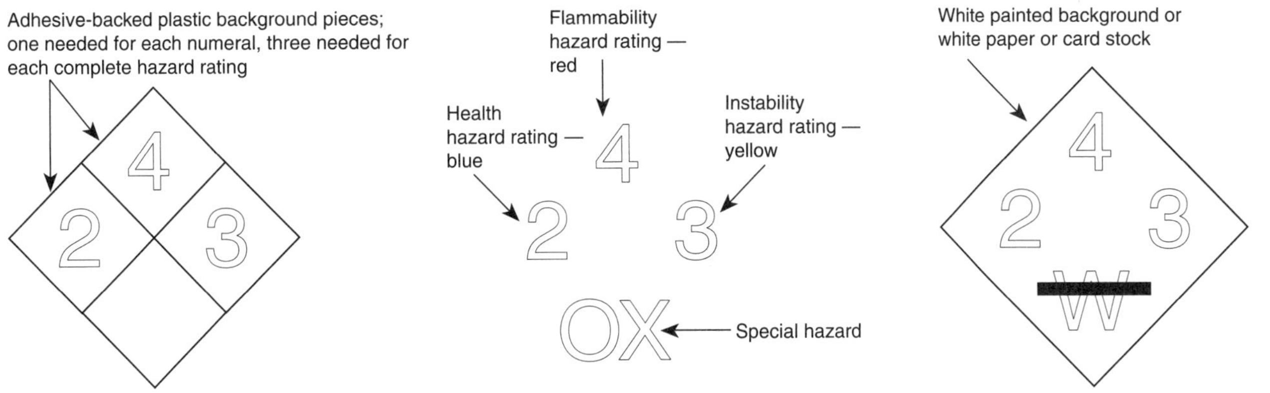

(a) For use where specified color background is used with numerals of contrasting colors

(b) For use where white background is necessary

(c) For use where white background is used with painted numerals or for use when hazard rating is in the form of sign or placard

EXHIBIT S6.1 *The alternate arrangements for display of NFPA 704 Hazard Identification System provide responders with information on the three hazard conditions. [Source: NFPA 704, 2007 ed., Figure 9.1(a)]*

1. Any material that has an anesthetic, noxious, or similar property that could cause a flight crew such annoyance or discomfort as to prevent them from correctly performing their assigned duties
2. Any material, such as a hazardous substance or a hazardous waste, that is not included in any other hazard class but is subject to the DOT requirements

ORM-D Material: A material that presents a limited hazard during transportation due to its form, quantity, and packaging. Examples of ORM-D materials include consumer commodities and small arms ammunition.

PCBs Without Fire: Even without the added hazard of fire, PCBs that are sufficiently harmful to responders to warrant a Level 2 condition. PCBs present serious health threats to skin and the liver.

EPA Regulated Waste: A list of waste regulated by the EPA can be found in Title 40, Code of Federal Regulations, Part 261 [2]. The two classes of EPA regulated wastes are as follows:

1. Class 1, Division 1.1 and 1.2 — Explosives. Examples of Division 1.1 are dynamite and black powder. Examples of Division 1.2 are propellant explosives and rocket motors. Prior to 1991, this class was classified as Explosives A or B.
2. Class 2, Division 2.3 — Poisonous Gases. Examples are arsine, hydrocyanic acid, and phosgene. These are extremely dangerous poisons. Prior to 1991, this class was classified as Poison A.

Organic Peroxide: Organic peroxides that can be highly flammable and most decompose readily when heated. In some cases, the decomposition can be violent.

Flammable Solid: Examples are pyroxylin plastics, magnesium, and aluminum powder.

Materials Dangerous When Wet: A category that includes sodium and potassium metals and calcium.

Chlorine: A greenish yellow gas that is highly toxic and irritating.

Fluorine: An extremely reactive and intensely poisonous yellow gas.

Anhydrous Ammonia: A very toxic and corrosive gas.

Radioactive Materials: Materials that spontaneously emit ionizing radiation having a specific activity greater than 0.002 microcurie per gram.

DOT Inhalation Hazard: Inhalation hazards that are measured in terms of TLV/TWA (threshold limit value/time-weighted average). Shipping papers must indicate the inhalation hazard, and containers must be marked "Inhalation Hazard." Vehicles are placarded "Poison" or "Poison Gas" in addition to the primary hazard listing requirements.

EPA Extremely Hazardous Substances: A list of 366 such substances published by the EPA that can be found at www.epa.gov/swercepp/ehs/ehsalpha.html.

Cryogenics: Extremely cold liquified gases [200°F (129°C)] that can cause severe damage to skin or other body parts.

Container Size: The larger the container, the greater the potential for risk, hence the increase in the level of incident condition. The condition of the container itself can also influence the level of the incident. An example of poor container condition can be seen in Exhibit S6.2. Poorly contained wastes can mix with each other, forming new

EXHIBIT S6.2 Hazardous wastes can be found stored in containers, such as these barrels, that are corroded, lack appropriate identification labels, and are in poor condition.

compounds that are more unstable or toxic than the original components.

Fire/Explosion Potential: The assumption in each case that the incident is not simply a fire but that some hazardous material is involved. Where fire is not present, Level 1 might be appropriate, depending on other prevailing conditions. If a container is involved in fire, Level 3 might be more appropriate. A fire involving a container can be conceivably handled safely by a responding fire department without the assistance of any hazardous materials response personnel. Nonetheless, appropriate authorities have to be notified and alerted to the situation. A vitally important factor to keep in mind is that containers involved in fire can overpressurize and explode, which is especially true of hazardous materials in liquid form stored in tanks and other larger and smaller containers. The explosion can spread hazardous material over a large geographic area as well as spread fire and chemical gases. Thus, every precaution must be taken when approaching containers that are exposed to fire. For the maximum protection of all individuals, the planning committee should address the possibility of fire in each of their hazardous materials response plans.

Leak Severity: The extent of the leak and the likelihood that it can be controlled. The selection of levels obviously depends on this information.

Life Safety: The number of people potentially exposed. This criterion is a major determining factor in selecting the appropriate level.

Environmental Impact: The terms *minimal*, *moderate*, and *severe* are general. Judgment is required, and experts should be consulted. The environmental impacts of an incident might not be known at the start of the incident, and they are frequently more severe than anticipated.

Container Integrity: Extreme care must be taken with damaged containers before allowing them to be transferred. Once with environmental impact, experts in this field should be consulted. If doubt exists, the incident should be considered as Level 3.

Incidents involving damaged containers can be considered as requiring either offensive or defensive operations.

Offensive operations include actions taken by a hazardous materials responder, in appropriate chemical-protective clothing, to handle an incident in such a manner that contact with the released material might result. These actions include patching or plugging to slow or stop a leak; containing a material in its own package or container; and cleanup operations that could require overpacking or transfer of a product to another container.

Defensive operations include actions taken during an incident where there is no intentional contact with the material involved. These actions include elimination of ignition sources, vapor suppression, and diking or diverting to keep a release in a confined area. Defensive operations require notification and possible evacuation, but they do not involve plugging, patching, or cleanup of spilled or leaking materials.

Jurisdictions are responsible for developing standard operating procedures that equate levels of response to levels of training indicated in NFPA 472, *Standard for Competence of Responders to Hazardous Materials/Weapons of Mass Destruction Incidents*. Depending on the capabilities and training of personnel, the first responder operational level can equate to incident level one and the technician level may equate to incident level two.

Response personnel should operate only at the incident level that matches their knowledge, training, and equipment. If conditions indicate a need for a higher response level, additional personnel, appropriate training, and equipment should be summoned.

Potential applications to a jurisdiction's response activities can include development of standard operating procedures, implementation of a training program using the competency levels of NFPA 472, acquisition of necessary equipment, and development of community emergency response plans. When consulting Table S6.1, the user should refer to all of the incident condition criteria to determine the appropriate incident level.

REFERENCES

1. NFPA 704, *Standard System for the Identification of the Hazards of Materials for Emergency Response*, National Fire Protection Association, Quincy, MA. 2007.
2. Title 40, Code of Federal Regulations, Part 261, U.S. Government Printing Office, Washington, DC, July 1, 2001.

SUPPLEMENT 7

Incident Mitigation

Editor's Note: *This supplement is taken from the 2002 edition of this handbook. In that handbook, the full text of the 2002 edition of NFPA 471, Recommended Practice for Responding to Hazardous Materials Incidents, was included. Since then, NFPA 471 has been withdrawn. This supplement is a reprint of Chapter 8, including both the standard and commentary text. All text is now considered commentary text, as the standard no longer exists. The material can still be useful to the reader, so it is included as a supplement in this handbook.*

This supplement addresses those actions necessary to ensure confinement and containment, which is the first line of defense, in a manner that will minimize risk to both life and the environment in the early, critical stages of a spill or leak. Both natural and synthetic methods can be employed to limit the releases of hazardous materials so that effective recovery and treatment can be accomplished with minimum additional risk to the environment or to life.

The well-respected hazardous materials expert Ludwig Benner, Jr. offered a popular definition of hazardous materials: Something that jumps out of its container at you when something goes wrong and hurts or harms the thing it touches [1]. An important element of that definition is that the harmful "something" is normally controlled or contained. Only when the hazardous material is outside its normal controlling element does a hazardous materials incident occur.

It is only reasonable, then, that mitigating an incident must involve controlling the material that is presenting the problem. Control methods are divided into confinement and containment, and the methods of mitigation are either physical or chemical. This approach presents some order to the process and simplifies it for better understanding.

TYPES OF HAZARDOUS MATERIALS

All hazardous materials can be organized into the following three general categories, based on the principal characteristic that makes them harmful or dangerous:

1. Chemical
2. Biological
3. Radioactive

Other safety hazards exist at every emergency site. NFPA 1500, *Standard on Fire Department Occupational Safety and Health Program,* is a complete standard devoted to the elimination or reduction of risks associated with fire fighting and other emergencies [2].

Chemical Materials

Chemical materials are those materials that pose a hazard based upon their chemical and physical properties. Examining the U.S. Department of Transportation (DOT) list of hazard classes reveals that most of the classes would fall under the chemical hazard type of material [3]. The effect of exposure to chemical hazards can be either acute or chronic. An acute exposure occurs in a relatively short period of time. Hazardous materials responders need to consider acute exposure limits when choosing their personal protective equipment, treating the victims, and selecting the methods to mitigate an incident. A worker who is exposed for up to 40 hours per week would run the risk of chronic exposure.

Biological Materials

Biological materials are those organisms that have a pathogenic effect on life and the environment and can exist in normal ambient environments. Examples of biological materials that are hazardous are those whose packaging requires an "Etiologic Agents" label. Biological hazards include toxins or microorganisms that cause diseases, such as anthrax, botulism, cholera, and typhus. Disease-causing organisms are also likely to be found in waste from hospitals, laboratories, and research institutions. Medical waste, such as used needles, hospital dressing gauze, and bandages, should be disposed of in red bags with medical waste and biohazard warnings printed on the plastic bag.

Biological warfare and terrorist devices and the illegal disposal of medical waste are all biological hazards. Such materials are often not labeled and might be packaged in unmarked containers or not packaged at all.

Radioactive Materials

Radioactive materials are those materials that emit ionizing radiation. Ionizing radiation is radiation that has sufficient energy to remove electrons from atoms. One source of radiation is the nuclei of unstable atoms. For these radioactive atoms to become more stable, the nuclei eject or emit subatomic particles and high-energy photons (gamma rays). This process is called radioactive decay. The major types of radiation emitted as a result of spontaneous decay are alpha and beta particles and gamma rays. X-rays, another major type of radiation, arise from processes outside of the nucleus.

Radioactive materials can be generated through nuclear processes. These materials also exist naturally in things such as uranium ore, thorium rock, and some forms of potassium. Some commonly used and transported radioactive materials include radiopharmaceuticals, radiographic sources, radioactive waste, and uranium ores. In transportation, radioactive materials can be identified by markings on the exterior of the package, warning labels attached to the package, or by vehicle placards.

PHYSICAL STATES OF HAZARDOUS MATERIALS

Hazardous materials can be classified into three states: gases, solids, and liquids. They can be stored and contained at a high or low pressure. All three states can be affected by the environment in which the incident occurs. The emergency responder needs to take into account such conditions as heat, cold, rain, or wind, each of which can have a significant effect on the methods used to accomplish a safe operation.

All matter exists as a gas, a solid, or a liquid. Each state has specific properties that have a bearing on how a particular material appears or behaves in the environment. For example, a liquid with a boiling point below 100°F (37.8°C) tends to give off vapor at ambient temperatures. Similarly, a gas with a vapor density that is substantially heavier than the surrounding air can collect in ditches and other low points at the scene and migrate along the ground as it mixes with air. Although in many cases gas cannot be seen, it acts like water that has been poured on the ground. If the gas is a flammable gas such as propane, a fire or explosion occurs if the gas finds an ignition source at ground level.

METHODS FOR MITIGATION

There are two basic methods for mitigation of hazardous materials incidents: physical and chemical. Table S7.1 lists many physical methods for mitigation of hazardous materials incidents, and Table S7.2 lists many chemical methods. Recommended practices should be implemented only by personnel prepared by training, education, or experience.

Many of the methods listed for mitigating an incident require a high degree of specialized training and the use of sophisticated technical equipment, whereas other methods might be carried out by personnel at the first responder operational level. For example, diking or blanketing a liquid spill of diesel fuel can often be accomplished easily by the first responder operations. Plugging a hole in a damaged tank truck to stop a leak, however, would require specialized training and equipment that only a technician would possess. Other operations, such as vent and burn techniques, should be attempted only by technicians with a tank car specialty. In every case, the incident commander should be the primary decision maker on which personnel are assigned to deal with each specific incident.

Physical Methods

Physical methods of control involve any of several processes or procedures to reduce the area of the spill, leak, or other release mechanism. In all cases, methods used should be approved by the incident commander. The selection of personal protective clothing should be based on the hazardous materials and/or conditions present and should be appropriate for the hazards encountered.

The venting of low-vapor-pressure radiological gases is allowed after consultation with a radiation protection technologist and a hazardous materials technician with a radioactive material specialty.

Absorption. Absorption is the process in which materials hold liquids through the process of wetting. Absorption is accompanied by an increase in the volume of the sorbate/sorbent system through the process of swelling. Some of the materials typically used as absorbents are sawdust,

TABLE S.7.1 Physical Methods of Mitigation of Hazardous Materials Incidents

	Chemical				*Biological*				*Radioactive*			
	Gases				*Gases*				*Gases*			
Method	*LVP**	*HVP***	*Liq.*	*Sol.*	*LVP*	*HVP*	*Liq.*	*Sol.*	*LVP*	*HVP*	*Liq.*	*Sol.*
Absorption	yes	yes	yes	no	no	no	yes[4]	no	no	no	yes	no
Covering	no	no	yes	yes	no	no	yes	yes	no	no	yes[3]	yes[3]
Dikes, dams, diversions, and retention	yes	yes[5]	yes	yes	no	no	yes	yes	no	no	yes	yes
Dilution	yes	yes[5]	yes	yes	no	no	no	no	yes	no	yes	yes
Overpack	yes	no	yes	yes	yes	no	yes	yes	yes	no	yes	yes
Plug/patch	yes	yes	yes	yes	yes	yes	yes	yes	yes	yes	yes	yes
Transfer	yes	no	yes	yes	yes	no	yes	yes	yes	no	yes	yes
Vapor suppression (blanketing)	no	no	yes	yes	no	no	yes	yes	no	no	no	no
Vacuuming	no	no	yes	yes	no	no	yes	yes	no	no	yes	yes
Venting[1]	yes	yes	yes	no	yes	no	no	no	yes[2]	no	no	no

Note: For substances involving more than one type, the most restrictive control measure should be used.

*Low Vapor Pressure

**High Vapor Pressure

[1]Venting of low-vapor-pressure gases is recommended only when an understanding of the biological system is known. Venting is allowed when the bacteriological system is known to be nonpathogenic, or if methods can be employed to make the environment hostile to pathogenic bacteria.

[2]Venting of low-vapor-pressure radiological gases is allowed when the gas(es) is/are known to be alpha or beta emitters with short half-lives. Further, this venting is only allowed after careful consultation with a radiation protection technologist, the technician with a radioactive material specialty, or certified health physicist.

[3]Covering should be done only after consultation with experts.

[4]Absorption of liquids containing bacteria is permitted where the absorption medium or environment is hostile to the medium.

[5]Water dispersion on certain vapors and gases only.

TABLE S7.2 Chemical Methods of Mitigation of Hazardous Materials Incidents

	Chemical				*Biological*				*Radioactive*			
	Gases				*Gases*				*Gases*			
Method	*LVP**	*HVP***	*Liq.*	*Sol.*	*LVP*	*HVP*	*Liq.*	*Sol.*	*LVP*	*HVP*	*Liq.*	*Sol.*
Absorption	yes	yes	yes	no	yes[3]	yes	yes[3]	no	no	no	no	no
Burn	yes	yes	yes	yes	yes	yes	yes	yes	no	no	no	no
Dispersion/emulsification	no	no	yes	yes	no	no	yes[3]	no	no	no	no	no
Flare	yes	yes	yes	no	yes	yes	yes	no	no	no	no	no
Gelatin	yes	no	yes	yes	yes[3]	no	yes[3]	yes[3]	no	no	no	no
Neutralization	yes[1]	yes[4]	yes	yes[2]	no	no	no	no	no	no	no	no
Polymerization	yes	no	yes	yes	no	no	no	no	no	no	no	no
Solidification	no	no	yes	no	no	no	yes[3]	no	no	no	yes	no
Vapor suppression	yes	yes	yes	yes	yes	yes	yes	yes	yes	yes	yes	yes
Vent/Burn	yes	yes	yes	no	yes	yes	yes	no	no	no	no	no

*Low Vapor Pressure

**High Vapor Pressure

[1]Technique may be possible as a liquid or solid neutralizing agent, and water can be applied.

[2]When solid neutralizing agents are used, they must be used simultaneously with water.

[3]Technique is permitted only if resulting material is hostile to the bacteria.

[4]The use of this procedure requires special expertise and technique.

clays, charcoal, and polyolefin-type fibers. These materials can be used for confinement, but it should be noted that the sorbed liquid can be desorbed under mechanical or thermal stress. When absorbents become contaminated, they retain the properties of the absorbed hazardous liquids and are, therefore, considered to be hazardous materials and must be treated and disposed of accordingly.

Many commercially available products are suitable for use as absorbents. Different types of absorbents are designed for different types of spilled materials so labels should be carefully checked prior to each use. Absorbents can help reduce vapor generation and can facilitate cleanup procedures.

Absorbents saturated with volatile liquid chemicals can create a more severe vapor hazard than the spill alone because of severely enlarged surface area for vapor release.

Covering. Covering is a temporary form of mitigation for radioactive, biological, and some chemical substances, such as magnesium. It should be done after consultation with a certified health physicist (in the case of radioactive materials) or other experts.

Covering might be used for solids or liquids. The form of covering to be used is influenced by the type of incident. The responder needs to make sure that the hazardous material does not permeate the covering material. At a spill of dust or powder, the cover may be a plastic cover or tarp. Where alpha or beta radioactive materials are involved, a thicker cover might be needed to reduce the radiation emission. Flammable metals could require a covering of an appropriate dry powder agent. Any covering should be accompanied by an effective sealing mechanism to prevent the cover from blowing away or the covered material from escaping freely.

Dikes, Dams, Diversions, and Retention. Dikes, dams, diversions, and retention refer to the use of physical barriers to prevent or reduce the quantity of liquid flowing into the environment. Dikes or dams usually refer to concrete, earth, or other barriers temporarily or permanently constructed to hold back a spill or leak. Vapors from certain materials, such as liquefied petroleum gas (LPG), can be dispersed by means of a water spray.

These techniques are the most commonly employed methods of controlling releases because responders can always improvise, and simple methods of confinement can be devised with a little ingenuity. In the case of substantial liquid spills, hazardous materials can pose significant challenges and, in some rare cases, insurmountable problems for the first hazardous materials response team to arrive on the scene.

In addition to the techniques listed, trenches can be used to collect spilled liquids, and pumps can transfer materials to containers or to a containment system. Earthen dikes or dams can be erected quickly under favorable conditions, and even sandbags can be used in the damming effort. Commercial booms, such as the one shown in Exhibit S7.1, are available and are widely used to control spills, especially spills on waterways.

EXHIBIT S7.1 *Commercial booms can be placed across a waterway to collect hazardous materials spilled into the waterway.*

Dilution. Dilution is the application of water to water-miscible hazardous materials. The goal is to reduce the hazards to safe levels. Responders should not use water indiscriminately or without knowing what effect it has on the hazardous material involved in a hazardous materials incident. Even if a hazardous material is water-soluble, the amount of water necessary to achieve a safe level could render dilution an impractical approach. Adding water to a liquid spill can also increase confinement problems. Unconfined diluted hazardous materials have the potential to cause widespread, long-lasting environmental problems. Many hazardous materials can also react with water, thus increasing the intensity of the incident. Water is nonetheless a viable option for mitigation in many instances and should be considered when appropriate.

Overpacking. The most common form of overpacking is accomplished by the use of an oversized container. Overpack containers should be compatible with the hazards of the materials involved. If the material is to be shipped, DOT specification overpack containers need to be used. The spilled materials should still be treated or disposed of properly.

A leaking drum or container should be temporarily repaired, if possible, to reduce spillage before the container is placed in an overpack container. Reducing a leak can sometimes be accomplished by repositioning the original container. Holes can be covered, and temporary patches can be applied. See Exhibit S7.2 for examples of overpack containers.

EXHIBIT S7.2 *Overpack containers are shown here. (Photo courtesy of WYK Sorbents, LLC, St. Louis, MO)*

A leaking container can be put into an overpack drum or container by placing the overpack on its side and sliding the smaller container into it, by lowering the overpack over the leaking container and then tipping it upright as demonstrated in Exhibit S7.3, or by using mechanical equipment to raise and lower the leaking container into the overpack container. The overpack container must be labeled in accordance with DOT regulations for the particular product carried inside.

EXHIBIT S7.3 *One method of overpacking involves lowering the overpack over the leaking container and then tipping it upright.*

Responders should make sure that other containers without visible leaks or punctures that might be weakened by deterioration or impact do not fail. Responders should also try to avoid physical injury when lifting or moving large or heavy containers. Based on the hazards that are present, personnel should wear the appropriate chemical-protective clothing and respiratory protection.

Plug and Patch. Plugging and patching is the use of compatible plugs and patches to reduce or temporarily stop the flow of materials from small holes, rips, tears, or gashes in containers. The repaired container should not be reused without inspection and certification.

Limiting or restricting a leak is an important condition of the mitigation process, so it is essential that responders master the skill of plugging and patching. At all times, however, the safety of the responder must be paramount.

Plugging involves putting something into a container hole to reduce both the size of the hole and the flow from the hole. Tapered wooden plugs are often used. See Exhibit S7.4 for examples of plugs and wedges. Regardless of the plug's material, however, it must be compatible with both the product and the container. For example, soft pine might

EXHIBIT S7.4 Plugs and wedges are commercially available to fill holes in containers and stop leaks. (Photo courtesy of Michael Callan)

not be appropriate for plugging a strong acid leak. As with overpacking, responders must wear the appropriate chemical-protective clothing and respiratory protection before attempting any plugging activity.

Patching involves placing something over a hole to keep the material inside the container from leaking out. Patches are generally secured with clamps or adhesives. Patches designed to repair leaks in pipes of various sizes are commercially available, as are patching kits as shown in Exhibit S7.5.

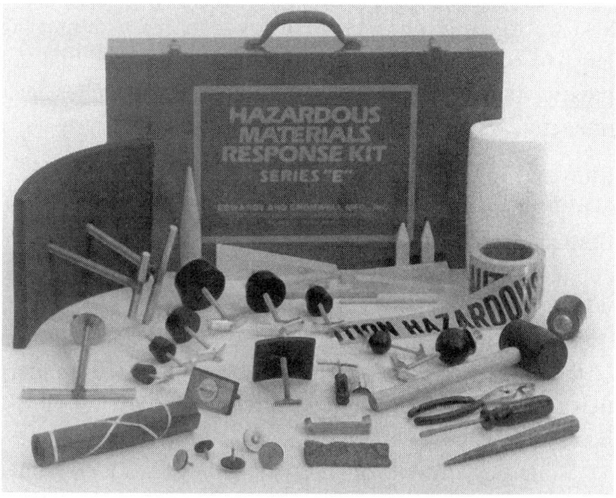

EXHIBIT S7.5 Commercially available patching kits provide a variety of tools and patching materials to stop leaks in pipes and containers. (Photo courtesy of Edwards and Cromwell Spill Control)

Transfer. Transfer is the process of moving a liquid, gas, or some other form of solids, either manually, by pumps, or by pressure, from a leaking or damaged container or tank. Care needs to be taken to ensure that the pump, transfer hoses and fittings, and container selected are compatible with the hazardous material. When a product transfer presents a fire or explosion hazard, concern for electrical continuity, such as bonding and grounding, needs to be observed.

Materials should be transferred from one tank truck to another by personnel who are skilled and practiced in the procedure, such as technicians with a tank car, cargo tank, or intermodal tank specialty. The incident commander is in charge of the transferring operation and is responsible for seeing that proper precautions are taken. Responders must also rely on the experience of industry personnel or specialty technicians who are appropriately trained and equipped to perform the transfer operation.

All electrical equipment used to transfer flammable liquids must be grounded or bonded and all equipment must be approved for such usage.

Vacuuming. Many hazardous materials can be placed in containment simply by vacuuming them up. This method has the advantage of not causing an increase in volume. Care needs to be taken to ensure compatibility of materials. The exhaust air can be filtered, scrubbed, or treated as needed. The method of vacuuming will depend on the nature of the hazardous material.

Vacuuming can reduce the hazard for solid materials, such as fibers and dusts. Specialized equipment offers effective filtering, but ordinary shops vacs are not designed for the filtering required by hazardous materials and could distribute the agent back into the air. Vacuums with HEPA (high-efficiency particulate air) filters provide extra filtering. Some vacuums are made specifically to pick up mercury without releasing mercury vapors into the air.

Vapor Dispersion. Vapors from certain materials can be dispersed or moved by means of a water spray. With other products, such as LPG, the gas concentration can be reduced below the lower flammable limit through rapid mixing of the gas with air, using the turbulence created by a fine water spray. Reducing the concentration of the material through the use of water spray can bring the material into its flammable range. A fine water spray is recommended in some cases to reduce the concentration of a material, as is shown in Exhibit S.7.6.

Vapor Suppression (Blanketing). Vapor suppression is the reduction or elimination of vapors emanating from a spilled or released material through the most efficient method or application of specially designed agents. A rec-

EXHIBIT S7.6 A fine water spray can be used to disperse some hazardous vapors.

ommended vapor suppression agent is an aqueous foam blanket.

While vapor suppression, or blanketing, does not change the nature of a hazardous material, it can greatly reduce the immediate hazard and danger associated with the presence of uncontrolled vapor. In addition, this method buys additional time to use other measures that can control and resolve the incident under safer circumstances. See Exhibit 7.7 for an example of a free-standing foam extinguisher. Hazmat response training should include information on how to identify when to use foam and which type of foam is the most appropriate to use.

EXHIBIT S7.7 Chemical methods of hazardous materials mitigation include the use of foam, available in free-standing foam extinguishers as shown here, or in specialized compartments on other vehicles.

Venting. Venting is the process that is used to deal with liquids or liquefied compressed gases where a danger, such as explosion or mechanical rupture of the container or vessel, is considered likely. The method of venting will depend on the nature of the hazardous material. In general, it involves the controlled release of the material to reduce and contain the pressure and diminish the probability of an explosion.

Chemical Methods

Chemical methods of control involve the application of chemicals to treat spills of hazardous materials. Chemical methods can involve any one of several actions to reduce the involved area affected by the release of a hazardous material. In all cases, methods used should be acceptable to the incident commander. The selection of personal protective clothing should be based on the hazardous materials and/or conditions present and should be appropriate for the hazards encountered.

Adsorption. Adsorption is the process in which a sorbate (hazardous liquid) interacts with a solid sorbent surface. See ASTM F 726, *Standard Test Method for Sorbent Performance of Adsorbents*, for further information [4]. The principal characteristics of this interaction are as follows:

1. The sorbent surface, unlike absorbents, is rigid and no volume increase occurs.
2. The adsorption process is accompanied by heat of adsorption whereas absorption is not.
3. Adsorption occurs only with activated surfaces, for example, activated carbon, alumina, and so forth.

Spontaneous ignition can occur through the heat of adsorption of flammable materials, and caution should be exercised.

Adsorbents saturated with volatile liquid chemicals can create a more severe vapor hazard than the spill alone because of the severely enlarged surface area for vapor release.

The term *sorbents* encompasses both absorbents and adsorbents. Adsorbents act in such a way that the internal structure of the material is not penetrated. They can be natural or synthetic materials and can be used on liquid spills on land and, to some degree, in water. Adsorbents should be nonreactive to the spilled material. Porous clay and sand are two commonly used adsorbents.

Controlled Burning. Controlled combustion is considered a chemical method of control. However, it should be used only by qualified personnel trained specifically in this procedure. In some emergency situations, where extinguishing a fire will result in large, uncontrolled volumes of contaminated water or threaten the safety of responders or the public, controlled burning is used as a technique. It is recommended that consultation be made with the environmental authorities when this method is used. The deliberate burning of a hazardous material should be attempted only by personnel at a technician level with appropriate specialty training or by a specialist employee A.

Some occasions exist when extinguishing a fire is not the proper approach because of the large amount of contaminated runoff that is generated by fire fighting. A Swiss pharmaceutical company fire, for example, was responsible for years of pollution to the Rhine River that was caused by the water runoff from the extinguishment of that fire. However, a similar fire in a paint factory in the United States was allowed to burn out without applying the large quantity of water that would have been necessary to extinguish the fire. The air and area water supply were thus protected from pollution.

Controlled burning can also be used to incinerate the spilled hazardous material. Transportable incinerators are designed to promote combustion of spilled materials, especially oil.

Dispersants, Surface Agents, and Biological Additives. Certain chemical and biological agents can be used to disperse or break up the materials involved in liquid spills. The use of these agents results in a lack of containment and generally results in spreading the liquid over a much larger area. Dispersants are most often applied to spills of liquids on water. The dispersant breaks down a liquid spill into many fine droplets, thereby diluting the material. Use of this method can require prior approval of the environmental authority.

Dispersants are chemical formulations with surfactants, which are chemicals designed to blend oil and water. Surfactants reduce the surface tension of oil and water and result in smaller oil droplets. Smaller droplets result in more oil surface area exposed to the water. These oil droplets can better move into the water, helping to speed the natural biological breakdown and dispersion of the oil.

Dispersants generally result in oil-in-water emulsions because chemicals are used to reduce the surface tension of water so it can mix with oil. Chemical dispersants should not be used in situations where they might produce increased biological damage. Environmental authorities should/must be consulted prior to the use of these agents. Surface-active agents also increase emulsification and dispersion of a spill.

Surface cleaning equipment is available for soil surface cleaning. This equipment agitates the soil's surface with water to form a slurry. The contamination is then removed through a separation process that takes place in a specially designed sand separator.

Biological additives can be used to degrade a hazardous material by biochemical oxidation. Biochemical accelerators are useful for mitigation of certain hazardous materials spilled on land or in water.

Flaring. Flaring is a process that is used with high-vapor-pressure liquids or liquefied compressed gases for the safe disposal of the product. Flaring is the controlled burning of material in order to reduce or control pressure and/or dispose of a product.

Flaring is the controlled destruction and/or consumption by fire of a hazardous material. When this process is chosen, flaring, like any other de-inventory process, has advantages and disadvantages. This technique must be understood, used, and/or controlled to ensure a safe operation. Other factors that play a part in the decision-making process involved in using this type of physical control method include the following:

1. Conditions of the environment
2. Topography
3. Materials to be burned
4. Area demographics
5. Local authorities (i.e., Incident Command System)
6. Applicable regulations
7. Equipment availability

Gelation. Gelation is the process of forming a gel. A gel is a colloidal system consisting of two phases, a solid and a liquid. The resulting gel is considered to be a hazardous material and needs to be disposed of properly.

Gelling agents used on hazardous chemicals produce a gel that is more easily cleaned up by either mechanical or physical methods. Gelling agents can be used on liquids spilled in water and, to a lesser degree, liquid spilled on land.

Neutralization. Neutralization is the process of applying acids or bases to a spill to form a neutral salt. The application of solids for neutralizing can often result in confinement of the spilled material. Special formulations are available that do not result in violent reactions or local heat generation during the neutralization process. In cases where special neutralizing formulations are not available, special consideration should be given to protecting persons applying the neutralizing agent because heat is generated and violent reactions can occur. One of the advantages of neutralization is that a hazardous material can be rendered nonhazardous.

The pH scale is used to categorize compounds as acids or bases. A value of 7 on the scale is neutral, while descending values denote increasing acidity. Levels 8 to 14 denote bases, with the higher values indicating increasingly stronger bases. Neutralizing acidic or base spills is possible by mixing the spilled material with a neutralizing agent.

Polymerization. Polymerization is a process in which a hazardous material is reacted in the presence of a catalyst, in the presence of heat or light, or with itself or another material to form a polymeric system.

Solidification. Solidification is the process whereby a hazardous material is treated chemically so that a solid material results. Adsorbents can be considered an example of a solidification process. There are other materials that can be used to convert hazardous liquids into nonhazardous solids. Examples are applications of special formulations designed to form a neutral salt in the case of spills of acids or caustics. The advantage of the solidification process is that a small-scale spill can be confined relatively quickly and treatment effected immediately.

Commercially available adsorbents include silica, activated carbon, alumina, and zeolite. An online search for "adsorbent" results in many sources.

Adsorbents can be used to solidify oily wastes that are water insoluble. The spilled liquids are adsorbed into granules to form a solid, nonflowing mixture. The resulting product is safer than the spilled material in its liquid form and is more easily transported to an appropriate disposal facility.

Vapor Suppression. Vapor suppression is the use of solid activated materials to treat hazardous materials so as to effect suppression of the vapor off-gassing from the materials. This process results in the formation of a solid that affords easier handling but that can result in a hazardous solid that must be disposed of properly.

Venting and Burning. The process of venting and burning involves the use of shaped charges to vent the high vapor pressure at the top of the container and then the use of additional charges to release and burn the remaining liquid in the container in a controlled fashion.

In a train derailment in the early 1980s, shaped explosive charges were first used successfully to "vent and burn" tank cars that were too badly damaged to attempt transferring the hazardous materials to another container. This procedure occurred eight days into the incident response. Venting and burning, as shown in Exhibit S7.8, is a highly sophisticated technique that should be attempted only under very controlled conditions by trained specialists, such as technicians with a tank car, cargo car, or intermodal tank specialty.

EXHIBIT S7.8 *Venting, shown here, and burning techniques are often used in train accidents involving hazardous chemicals. (Photo courtesy of Tim Campbell/AP Worldwide for the Effingham Daily News)*

REFERENCES CITED

1. Benner, Jr., L., *A Textbook for Use in the Study of Hazardous Materials Emergencies,* 2nd ed., Lufred Industries Inc., Oakton, VA, 1978.
2. NFPA 1500, *Standard on Fire Department Occupational Safety and Health Program,* National Fire Protection Association, Quincy, MA, 2007.
3. Title 49, Code of Federal Regulations, Part 172.101, Subpart B — Table of Hazardous Materials and Special Provisions, U.S. Government Printing Office, Washington, DC.
4. ASTM F 726, *Standard Test Method for Sorbent Performance of Adsorbents*, ASTM International, West Conshocken, PA, 2006.

NFPA 472 Index

P

NFPA 473 Index

Commentary Index

C

N

O

T

U

IMPORTANT NOTICES AND DISCLAIMERS CONCERNING NFPA® DOCUMENTS

NOTICE AND DISCLAIMERS OF LIABILITY CONCERNING THE USE OF NFPA DOCUMENTS

NFPA codes, standards, recommended practices, and guides, including the documents contained herein, are developed through a consensus standards development process approved by the American National Standards Institute. This process brings together volunteers representing varied viewpoints and interests to achieve consensus on fire and other safety issues. While the NFPA administers the process and establishes rules to promote fairness in the development of consensus, it does not independently test, evaluate, or verify the accuracy of any information or the soundness of any judgments contained in its codes and standards.

The NFPA disclaims liability for any personal injury, property or other damages of any nature whatsoever, whether special, indirect, consequential or compensatory, directly or indirectly resulting from the publication, use of, or reliance on these documents. The NFPA also makes no guaranty or warranty as to the accuracy or completeness of any information published herein.

In issuing and making these documents available, the NFPA is not undertaking to render professional or other services for or on behalf of any person or entity. Nor is the NFPA undertaking to perform any duty owed by any person or entity to someone else. Anyone using these documents should rely on his or her own independent judgment or, as appropriate, seek the advice of a competent professional in determining the exercise of reasonable care in any given circumstances.

The NFPA has no power, nor does it undertake, to police or enforce compliance with the contents of these documents. Nor does the NFPA list, certify, test or inspect products, designs, or installations for compliance with these documents. Any certification or other statement of compliance with the requirements of these documents shall not be attributable to the NFPA and is solely the responsibility of the certifier or maker of the statement.

ADDITIONAL NOTICES AND DISCLAIMERS

Updating of NFPA Documents

Users of NFPA codes, standards, recommended practices, and guides should be aware that these documents may be superseded at any time by the issuance of new editions or may be amended from time to time through the issuance of Tentative Interim Amendments. An official NFPA document at any point in time consists of the current edition of the document together with any Tentative Interim Amendments and any Errata then in effect. In order to determine whether a given document is the current edition and whether it has been amended through the issuance of Tentative Interim Amendments or corrected through the issuance of Errata, consult appropriate NFPA publications such as the National Fire Codes® Subscription Service, visit the NFPA website at www.nfpa.org, or contact the NFPA at the address listed below.

Interpretations of NFPA Documents

A statement, written or oral, that is not processed in accordance with Section 6 of the Regulations Governing Committee Projects shall not be considered the official position of NFPA or any of its Committees and shall not be considered to be, nor be relied upon as, a Formal Interpretation.

Patents

The NFPA does not take any position with respect to the validity of any patent rights asserted in connection with any items which are mentioned in or are the subject of NFPA codes, standards, recommended practices, and guides, and the NFPA disclaims liability for the infringement of any patent resulting from the use of or reliance on these documents. Users of these documents are expressly advised that determination of the validity of any such patent rights, and the risk of infringement of such rights, is entirely their own responsibility.

NFPA adheres to applicable policies of the American National Standards Institute with respect to patents. For further information, contact the NFPA at the address listed below.

Law and Regulations

Users of these documents should consult applicable federal, state, and local laws and regulations. NFPA does not, by the publication of its codes, standards, recommended practices, and guides, intend to urge action that is not in compliance with applicable laws, and these documents may not be construed as doing so.

Copyrights

The documents contained in this volume are copyrighted by the NFPA. They are made available for a wide variety of both public and private uses. These include both use, by reference, in laws and regulations, and use in private self-regulation, standardization, and the promotion of safe practices and methods. By making these documents available for use and adoption by public authorities and private users, NFPA does not waive any rights in copyright to these documents.

Use of NFPA documents for regulatory purposes should be accomplished through adoption by reference. The term "adoption by reference" means the citing of title, edition, and publishing information only. Any deletions, additions, and changes desired by the adopting authority should be noted separately in the adopting instrument. In order to assist NFPA in following the uses made of its documents, adopting authorities are requested to notify the NFPA (Attention: Secretary, Standards Council) in writing of such use. For technical assistance and questions concerning adoption of NFPA documents, contact NFPA at the address below.

For Further Information

All questions or other communications relating to NFPA codes, standards, recommended practices, and guides and all requests for information on NFPA procedures governing its codes and standards development process, including information on the procedures for requesting Formal Interpretations, for proposing Tentative Interim Amendments, and for proposing revisions to NFPA documents during regular revision cycles, should be sent to NFPA headquarters, addressed to the attention of the Secretary, Standards Council, NFPA, 1 Batterymarch Park, Quincy, MA 02169-9101.

For more information about NFPA, visit the NFPA website at www.nfpa.org.

A Guide to Using the *Hazardous Materials/Weapons of Mass Destruction Response Handbook*

This fifth edition of the *Hazardous Materials/Weapons of Mass Destruction Response Handbook* contains the complete text of the 2008 editions of NFPA® 472, *Standard for Competence of Responders to Hazardous Materials/Weapons of Mass Destruction Incidents,* and NFPA® 473, *Standard for Competencies for EMS Personnel Responding to Hazardous Materials/Weapons of Mass Destruction Incidents.*

Seven supplements covering a variety of topics appear at the end of the handbook and are not a part of the Standards or commentaries.

PART III

Supplements

The seven supplements included in this part of the *Hazardous Materials/Weapons of Mass Destruction Response Handbook* provide additional information about key areas of concern for hazardous materials responders. The following supplements are not part of the official NFPA documents or the commentary but present additional information for the reader:

1. Excerpts from *Protecting Emergency Responders: Lessons Learned from Terrorist Attacks*
2. Recognizing and Identifying Hazardous Environments
3. Fire Fighter Fatalities, West Helena, Arkansas, May 8, 1997
4. Propane Tank Explosion Results in the Death of Two Volunteer Fire Fighters, Hospitalization of Six Other Volunteer Fire Fighters and a Deputy Sheriff — Iowa
5. Selection of Chemical-Protective Clothing Using NFPA Standards
6. Response Levels
7. Incident Mitigation

541

552 Supplement 2 • Recognizing and Identifying Hazardous Environments

EXHIBIT S2.2 In some cases, physiological monitoring might be required for workers in hot environments.

EXHIBIT S2.3 Radioactive material[s] … many environments.

TABLE 2.4 Characteristics of Ionizing Radiation

Alpha	Two protons/two neutrons charged (+2); heavy, short range in air (4"); 0.01 mm in tissue (stopped by dead skin) (<4 MeV–8 MeV energy)
Beta	Electron emitted upon creation from the nucleus; usually a (−1) charge (positions are +1 change); mid range, (0–20′ in air); .2–0.5 cm in tissue, less energy (1 KeV–1 MeV)
Gamma	No charge/mass (1 KeV–10 MeV energy) large range, small energy deposition
X-Ray	10 eV–120 KeV energy, originates in the electron field of the atom

TABLE S2.5 Effects of Ionizing Radiation

50 R	Decrease in white blood cells
200 R	Nausea, vomiting, dizziness, hair loss, diarrhea, and infection
450 R	$LD_{50/30}$: 50% of those exposed are expected to die within 30 days without medical intervention
600 R	LD_{100}: 100% of those exposed are expected to die without medical intervention
10,000 R	Death results from central nervous system disorder
Note 1:	Biological half life decreases total body burden
Note 2:	Individuals exposed to 5 rem are considered irradiated (individuals evidencing 0.5 mR/Hr [surface] or 0.1 mR/Hr [thyroid] are considered contaminated). The OSHA/EPA Worker Protection Guideline is 5 rem/year. The NRC Public Exposure Limit is 0.5 rem/year. Background levels of ionizing radiation are generally 0.01–0.02 mR/Hr, and as high as 0.05 mR/Hr in some jurisdictions.

takes for the radioactive material to r… its original strength).

Nonionizing Radiation

Nonionizing radiation does not have … break chemical bonds and is therefor… izing. The primary effect of nonion… radio frequency radiation, microwave… heating caused by the increased mov… of molecules.

Various recommended exposure limits are published by the National Council of Radiation Protection and Measurements and have been adopted by the U.S. Federal Communication Commission. The American National Standards Institute, the American Conference of Governmental Industrial Hygienists, the U.S. Department of Defense, and the Institute of Electrical and Electronics Engineers have also published recommended exposure limits.

ASPHYXIANT HAZARDS

Asphyxiant hazards are those hazards associated with the lack of oxygen in the atmosphere, the introduction of particulate matter into air that may obstruct respiration, or the introduction of certain chemicals into breathable air that may prohibit the up-take of oxygen by the blood. See Exhibit S2.4.

The first situation described (low oxygen) is considered immediately dangerous and may be found in many occupational environments. Standard breathing air contains approximately 20.9% oxygen and 79% nitrogen by volume. Nitrogen, the inert component, is not used by the body. The oxygen is circulated and used by the body's car-

2008 Hazardous Materials/Weapons of Mass Destruction Response Handbook

The supplements provide additional, useful information for NFPA 472 and NFPA 473 users and include drawings, photographs, tables, and extended text. The supplements appear in black type inside a brown box.